T0192641

Herbs, Shrubs, and Trees of Potential Medicinal Benefits

Exploring Medicinal Plants

Series Editor

Azamal Husen

Wolaita Sodo University, Ethiopia

Medicinal plants render a rich source of bioactive compounds used in drug formulation and development; they play a key role in traditional or indigenous health systems. As the demand for herbal medicines increases worldwide, supply is declining as most of the harvest is derived from naturally growing vegetation. Considering global interests and covering several important aspects associated with medicinal plants, the Exploring Medicinal Plants series comprises volumes valuable to academia, practitioners, and researchers interested in medicinal plants. Topics provide information on a range of subjects including diversity, conservation, propagation, cultivation, physiology, molecular biology, growth response under extreme environment, handling, storage, bioactive compounds, secondary metabolites, extraction, therapeutics, mode of action, and health care practices.

Led by Azamal Husen, PhD, this series is directed to a broad range of researchers and professionals consisting of topical books exploring information related to medicinal plants. It includes edited volumes, references, and textbooks available for individual print and electronic purchases.

Traditional Herbal Therapy for the Human Immune System, *Azamal Husen*

Environmental Pollution and Medicinal Plants, *Azamal Husen*

Herbs, Shrubs and Trees of Potential Medicinal Benefits, *Azamal Husen*

Herbs, Shrubs, and Trees of Potential Medicinal Benefits

Edited by
Azamal Husen

CRC Press is an imprint of the
Taylor & Francis Group, an **informa** business

First edition published 2022
by CRC Press
6000 Broken Sound Parkway NW, Suite 300, Boca Raton, FL 33487-2742

and by CRC Press
4 Park Square, Milton Park, Abingdon, Oxon, OX14 4RN
CRC Press is an imprint of Taylor & Francis Group, LLC

First edition published by Routledge 2022

Library of Congress Cataloging-in-Publication Data

Names: Husen, Azamal, Editor.
Title: Herbs, shrubs and trees of potential medicinal benefits / edited by Azamal Husen.
Description: First edition. | Boca Raton : CRC Press, 2022. | Series: Exploring medicinal plants | Includes bibliographical references and index. | Summary: "Due to the increasing popularity of herbal-based drugs under a pandemic, the worldwide demand for medicinal plants has increased to aid the immune system. This book provides a thorough and detailed description of many classes of plant-derived natural products for medical use. Chapters describe almost 60 medicinal plant species originating from all over the world and specifically discuss their traditional knowledge, chemical derivatives, and potential health benefits"-- Provided by publisher.
Identifiers: LCCN 2021061219 (print) | LCCN 2021061220 (ebook) | ISBN 9781032068787 (hardback) | ISBN 9781032070360 (paperback) | ISBN 9781003205067 (ebook)
Subjects: LCSH: Materia medica, Vegetable. | Herbs--Therapeutic use. | Trees--Therapeutic use.
Classification: LCC RS164 .H47 2022 (print) | LCC RS164 (ebook) | DDC 615.3/21--dc23/eng/20211220
LC record available at https://lccn.loc.gov/2021061219
LC ebook record available at https://lccn.loc.gov/2021061220

ISBN: 9781032068787 (hbk)
ISBN: 9781032070360 (pbk)
ISBN: 9781003205067 (ebk)

DOI: 10.1201/9781003205067

Typeset in Times
by Deanta Global Publishing Services, Chennai, India

Dedication

To my sister, Akhtar Jahan

Contents

About the Editor xi
Preface xiii
List of Contributors xv

Chapter 1 Exploring Important Herbs, Shrubs, and Trees for Their Traditional Knowledge, Chemical Derivatives, and Potential Benefits 1

Tilahun Belayneh Asfaw, Tarekegn Berhanu Esho, Archana Bachheti, Rakesh Kumar Bachheti, D.P. Pandey, and Azamal Husen

Chapter 2 *Andrographis paniculata* (Creat or Green Chiretta) and *Bacopa monnieri* (Water Hyssop) 27

Pankaj Mundada, Swati Gurme, Suchita Jadhav, Devashree Patil, Nitin Gore, Sumaiya Shaikh, Abhinav Mali, Suraj Umdale, Mahendra Ahire

Chapter 3 *Chlorophytum borivilianum* (Musli) and *Cimicifuga racemosa* (Black Cohosh) 45

Rajib Hossain, Dipta Dey, Partha Biswas, Priyanka Paul, Shahlaa Zernaz Ahmed, Arysha Alif Khan, Tanzila Ismail Ema, and Muhammad Torequl Islam

Chapter 4 *Convolvulus pluricaulis* (Shankhpushpi) and *Erythroxylum coca* (Coca plant) 83

Sashi Sonkar, Akhilesh Kumar Singh, and Azamal Husen

Chapter 5 *Asparagus racemosus* (Shatavari) and *Dioscorea villosa* (Wild yams) 95

Shakeelur Rahman and Azamal Husen

Chapter 6 *Embelia ribes* (False Black Pepper) and *Gymnema sylvestre* (Sugar Destroyer) 113

Chandrabose Selvaraj, Chandrabose Yogeswari, and Sanjeev Kumar Singh

Chapter 7 *Glycyrrhiza glabra* (Licorice) and *Gymnema sylvestre* (Gurmar) 133

Jasbir Kaur, Sana Nafees, Mohd Anwar, Jamal Akhtar, and Nighat Anjum

Chapter 8 *Hydrastis canadensis* (Goldenseal) and *Lawsonia inermis* (Henna) 151

Md. Mizanur Rahaman, William N. Setzer, Javad Sharifi-Rad, and Muhammad Torequl Islam

Chapter 9 *Nardostachys jatamansi* (Spikenard) and *Ocimum tenuiflorum* (Holy Basil) 163

Mani Iyer Prasanth, Premrutai Thitilertdecha, Dicson Sheeja Malar, Tewin Tencomnao, Anchalee Prasansuklab, and James Michael Brimson

Chapter 10 *Panax quinquefolium* (American Ginseng) and *Physostigma venenosum* (Calabar Bean)........ 179

Sushweta Mahalanobish, Noyel Ghosh, and Parames C. Sil

Chapter 11 *Phytolacca dodecandra* (African Soapberry) and *Picrorhiza kurroa* (Kutki)........ 203

K. Meenakshi, Mansi Shah, and Indu Anna George

Chapter 12 *Piper longum* (Long Pepper or Pipli) and *Tinospora cordifolia* (Giloy or Heart-Leaved Moonseed)........ 217

Yashashree Pradhan, Hina Alim, Nimisha Patel, Kamal Fatima Zahra, Belkıs Muca Yiğit, Johra Khan, and Ahmad Ali

Chapter 13 *Plantago ovata* (Isabgol) and *Rauvolfia serpentina* (Indian Snakeroot)........ 235

Ankur Anavkar, Nimisha Patel, Ahmad Ali, and Hina Alim

Chapter 14 *Saussurea costus* (Kust) and *Senna alexandrina* (Senna)........ 261

Amita Dubey, Soni Gupta, Mushfa Khatoon, and Anil Kumar Gupta

Chapter 15 *Swertia chirata* (Chirata) and *Withania somnifera* (Ashwagandha)........ 291

Suchita V. Jadhav, Pankaj S. Mundada, Mahendra L. Ahire, Devashree N. Patil, and Swati T. Gurme

Chapter 16 *Vinca rosea* (Madagascar Periwinkle) and *Adhatoda vesica* (Malabar Nut)........ 301

Rajib Hossain, Md Shahazul Islam, Dipta Dey, and Muhammad Torequl Islam

Chapter 17 *Aegle marmelos* (Bael) and *Annona squamosa* (Sugar Apple)........ 339

Abhidha Kohli, Taufeeq Ahmad, and Sachidanand Singh

Chapter 18 *Azadirachta indica* (Neem) and *Berberis aristata* (Indian Barberry)........ 365

Swati T. Gurme, Devashree N. Patil, Suchita V. Jadhav, Mahendra L. Ahire, and Pankaj S. Mundada

Chapter 19 *Cinchona officinalis* (Cinchona Tree) and *Corylus avellana* (Common Hazel)........ 377

Sawsan A. Oran, Arwa Rasem Althaher, and Mohammad S. Mubarak

Chapter 20 *Crataegus laevigata* (Midland Hawthorn) and *Emblica officinalis* (Indian Gooseberry)........ 395

Bouabida Hayette and Dris Djemaa

Chapter 21 *Eucalyptus spp.* (Eucalypts) and *Ficus religiosa* (Sacred Fig)........ 411

Surendra Pratap Singh, Bhoomika Yadav, and Kumar Anupam

Chapter 22 *Garcinia indica* (Kokum) and *Ilex aquifolium* (European Holly) 427

Dicson Sheeja Malar, Mani Iyer Prasanth, Tewin Tencomnao, James Michael Brimson, and Anchalee Prasansuklab

Chapter 23 *Alnus glutinosa* (Alder) and *Moringa oleifera* (Drumstick Tree) 447

Devashree N. Patil, Swati T. Gurme, Pankaj S. Mundada, and Jyoti. P. Jadhav

Chapter 24 *Madhuca longifolia* (Mahuwa) and *Santalum album* (Indian Sandalwood) 461

Surendra Pratap Singh, Bhoomika Yadav, and Kumar Anupam

Index 471

About the Editor

Professor Azamal Husen (BSc from Shri Murli Manohar Town Post Graduate College, Ballia, UP, MSc from Hamdard University, New Delhi, and PhD from Forest Research Institute, Dehra Dun, India) is a Foreign Delegate at Wolaita Sodo University, Wolaita, Ethiopia. He has served the University of Gondar, Ethiopia, as a Full Professor of Biology and worked there as the Coordinator of the MSc Program and as the Head, Department of Biology. He was a Visiting Faculty of the Forest Research Institute, and the Doon College of Agriculture and Forest at Dehra Dun, India. He has more than 20 years' experience of teaching, research, and administration.

Dr Husen specializes in biogenic nanomaterial fabrication and application, plant responses to nanomaterials, plant adaptation to harsh environments at the physiological, biochemical, and molecular levels, herbal medicine, and clonal propagation for improvement of tree species. He has conducted several research projects sponsored by various funding agencies, including the World Bank, the Indian Council of Agriculture Research (ICAR), the Indian Council of Forest Research Education (ICFRE), and the Japan Bank for International Cooperation (JBIC).

He has published around 200 research papers, review articles, and book chapters, edited books of international repute, presented papers in several conferences, and produced over a dozen manuals and monographs. He received four fellowships from India and a recognition award from the University of Gondar, Ethiopia, for excellence in teaching, research, and community service. An active organizer of seminars/conferences and an efficient evaluator of research projects and book proposals, Dr Husen has been on the Editorial board and the panel of reviewers of several reputed journals published by Elsevier, Frontiers Media SA, Taylor & Francis, Springer Nature, RSC, Oxford University Press, Sciendo, The Royal Society, CSIRO, PLOS, and John Wiley & Sons. He is on the advisory board of Cambridge Scholars Publishing, UK. He is a fellow of the Plantae group of the American Society of Plant Biologists, and a member of the International Society of Root Research, Asian Council of Science Editors, and INPST.

Dr Husen is Editor-in-Chief of the *American Journal of Plant Physiology*. He is also working as series editor of *Exploring Medicinal Plants*, published by Taylor & Francis Group, USA; *Plant Biology, Sustainability, and Climate Change* published by Elsevier, USA; and *Smart Nanomaterials Technology*, published by Springer Nature Singapore Pte Ltd. Singapore.

About the Editor

Preface

Medicinal plants grown in forests or elsewhere are mainly herbs, shrubs, and trees. Various products obtained from these plants have a key impact on human health and the overall ecosystem. In the recent past, considerable progress has been made in plant-based drug development by manipulating a variety of bioactive molecules. Though very many herbs, shrubs, and trees have been identified for developing a number of health care products, a huge number of them are still less exploited or remain unexplored. Forests provide a vast array of medicinal plants used in healing and health care practices. For instance, more than a quarter of modern medicines originate from tropical forest plants. Similarly, some of the selected herbs, shrubs, and trees grown outside the forest, on agricultural land or on domesticated land, often provide the basic material required for the treatment of certain conditions. The use of traditional medicines is no longer confined to rural areas or low-income groups, but prevail in the urban society also.

These plants are used as natural medicines because of their remedial and inherent pharmacological properties and also in several allied industrial products, such as paints, cosmetics, flavoring and fragrance, spices, pesticides, repellents, and herbal beverages. These plants have shown a tremendous potential of bioactive compounds for the development of lifesaving drugs against cancer, hepatitis, asthma, influenza, HIV, and so on. Moreover, they may boost the human immunity, improve mood and brain function, enhance blood and oxygen circulation, boost healing processes, and maintain blood pressure.

The book in hand covers a wide range of frequently used herbs, shrubs, and trees, discussing the relevant traditional knowledge, chemical derivatives, and potential benefits. Some self-explanatory figures and tables are incorporated in chapters in order to strengthen the main text. The selection of herbs, shrubs, and trees is not comprehensive but rather representative. I hope this book would cater to the need of graduate students as a textbook, and for academicians, medical practitioners, foresters as well as researchers working in the area of phytochemistry, pharmacognosy, biotechnology or general botany as a reference book. This must also inspire certain pharmaceutical companies and/or allied industries. With great pleasure, I extend my sincere thanks to all contributors for their timely response along with excellent, up-to-date contributions. I am extremely thankful to Ms. Randy Brehm, Dr Julia Tanner, Mr. Tom Connelly, and all associates at Taylor & Francis Group, LLC/ CRC Press for their sustained cooperation. I shall be happy receiving comments and criticism, if any, from subject experts and general readers of this book.

Azamal Husen
Wolaita, Ethiopia
November, 2021

Preface

List of Contributors

Mahendra L. Ahire
Department of Botany
Yashvantrao Chavan Institute of Science
Satara, India
and
Life Science Laboratory
Rayat Institute of Research and Development
Satara, India

Taufeeq Ahmad
Department of Bioengineering
Faculty of Engineering
Integral University
Lucknow, India

Shahlaa Zernaz Ahmed
Department of Biochemistry and Microbiology
North South University
Dhaka, Bangladesh

Jamal Akhtar
Central Council for Research in Unani Medicine
Ministry of AYUSH
New Delhi, India

Ahmad Ali
Department of Life Sciences
University of Mumbai
Mumbai, India

Hina Alim
Department of Life Sciences
University of Mumbai
Mumbai, India

Arwa Rasem Althaher
Department of Pharmacy
Faculty of Pharmacy
Al-Zaytoonah University of Jordan
Amman, Jordan

Ankur Anavkar
Department of Life Sciences
University of Mumbai
Mumbai, India

Nighat Anjum
Central Council for Research in Unani Medicine
Ministry of AYUSH
New Delhi, India

Kumar Anupam
Chemical Recovery and Biorefinery Division
Central Pulp and Paper Research Institute
Saharanpur, India

Mohd Anwar
Department of Ilaj-Bit-Tadbeer
Aligarh Muslim University
Aligarh, India

Tilahun Belayneh Asfaw
Centre of Excellence in Bioprocess and Biotechnology
Addis Ababa Science and Technology University
Addis Ababa, Ethiopia

Archana Bachheti
Department of Environment Science
Graphic Era University
Dehradun, India

Rakesh Kumar Bachheti
Nanotechnology Centre of Excellence
Addis Ababa Sciences and Technology University
Addis Ababa, Ethiopia

Partha Biswas
Department of Genetic Engineering and Biotechnology
Faculty of Biological Science and Technology
Jashore University of Science and Technology
Jashore, Bangladesh

James Michael Brimson
Department of Clinical Chemistry
Faculty of Allied Health Sciences
Chulalongkorn University
Bangkok, Thailand
and
Natural Products for Neuroprotection and Anti-ageing Research Unit
Faculty of Allied Health Sciences
Chulalongkorn University
Bangkok, Thailand

Dipta Dey
Department of Biochemistry and Molecular Biology
Life Science Faculty
Bangabandhu Sheikh Mujibur Rahman Science and Technology University
Dhaka, Bangladesh

Dris Djemaa
Applied Biology Department
University Larbi Tebessi-Tebessa
Tebessa, Algeria

Amita Dubey
Biosciences Department
Integral University
Lucknow, India

Tanzila Ismail Ema
Department of Biochemistry and Microbiology
North South University
Dhaka, Bangladesh

Tarekegn Berhanu Esho
Centre of Excellence in Bioprocess and Biotechnology
Addis Ababa Science and Technology University
Addis Ababa, Ethiopia

Indu Anna George
University Department of Life Sciences
University of Mumbai
Mumbai, India

Noyel Ghosh
Division of Molecular Medicine
Bose Institute
Kolkata, India

Nitin Gore
Department of Botany
Yashavantrao Chavan Institute of Science
Satara, India

Soni Gupta
Division of Plant Breeding and Genetic Resource Conservation
CSIR-Central Institute of Medicinal and Aromatic Plants
Lucknow, India

Anil Kumar Gupta
Division of Plant Breeding and Genetic Resource Conservation
CSIR-Central Institute of Medicinal and Aromatic Plants
Lucknow, India

Swati T. Gurme
Department of Biotechnology
Yashvantrao Chavan Institute of Science
Satara, India

Bouabida Hayette
Biology of Living Beings Department
University Larbi Tebessi, Tebessa
Tebessa, Algeria

Rajib Hossain
Department of Pharmacy
Life Science Faculty
Bangabandhu Sheikh Mujibur Rahman Science and Technology University
Dhaka, Bangladesh

Azamal Husen
Wolaita Sodo University
Wolaita, Ethiopia

Md Shahazul Islam
Department of Pharmacy
Life Science Faculty
Bangabandhu Sheikh Mujibur Rahman Science and Technology University
Dhaka, Bangladesh

Muhammad Torequl Islam
Department of Pharmacy
Life Science Faculty
Bangabandhu Sheikh Mujibur Rahman Science and Technology University
Dhaka, Bangladesh

Suchita V. Jadhav
Department of Biotechnology
Yashvantrao Chavan Institute of Science
Satara, India

Jyoti. P. Jadhav
Department of Biotechnology
and
Department of Biochemistry
Shivaji University
Kolhapur, India

Meenakshi K
University Department of Life Sciences
University of Mumbai
Mumbai, India

Jasbir Kaur
Department of Ocular Biochemistry
Dr. Rajendra Prasad Centre for Ophthalmic Sciences
All India Institute of Medical Sciences
New Delhi, India

Arysha Alif Khan
Department of Biochemistry and Microbiology
North South University
Dhaka, Bangladesh

Johra Khan
Department of Medical Laboratory Sciences
College of Applied Medical Sciences
Majmaah University
Majmaah, Saudi Arabia

Mushfa Khatoon
Biosciences Department
Integral University
Lucknow, India

Abhidha Kohli
Department of Biotechnology
IMS Engineering College
Ghaziabad, India

Sushweta Mahalanobish
Division of Molecular Medicine
Bose Institute
Kolkata, India

Dicson Sheeja Malar
Department of Clinical Chemistry
Faculty of Allied Health Sciences
Chulalongkorn University
Bangkok, Thailand
and
Natural Products for Neuroprotection and Anti-ageing Research Unit
Faculty of Allied Health Sciences
Chulalongkorn University
Bangkok, Thailand

Abhinav Mali
Department of Botany
Yashavantrao Chavan Institute of Science
Satara, India

Mohammad S. Mubarak
Department of Chemistry
The University of Jordan
Amman, Jordan

Pankaj S. Mundada
Department of Biotechnology
Yashvantrao Chavan Institute of Science
Satara, India

Sana Nafees
Department of Ocular Biochemistry
Dr. Rajendra Prasad Centre for Ophthalmic Sciences
All India Institute of Medical Sciences
New Delhi, India

Sawsan A. Oran
Department of Biolgical Sciences
The University of Jordan
Amman, Jordan

D.P. Pandey
Department of Chemistry
Government P.G. College
Uttarkashi, India

Nimisha Patel
Department of Life Sciences
University of Mumbai
Mumbai, India

Devashree N. Patil
Department of Biotechnology
Shivaji University
Kolhapur, India

Priyanka Paul
Department of Biochemistry and Molecular Biology
Life Science Faculty
Bangabandhu Sheikh Mujibur Rahman Science and Technology University
Dhaka, Bangladesh

Yashashree Pradhan
Department of Life Sciences
University of Mumbai
Mumbai, India

Anchalee Prasansuklab
Natural Products for Neuroprotection and Anti-ageing Research Unit
Faculty of Allied Health Sciences
Chulalongkorn University
Bangkok, Thailand
and
College of Public Health Sciences
Chulalongkorn University
Bangkok, Thailand

Mani Iyer Prasanth
Department of Clinical Chemistry
Faculty of Allied Health Sciences
Chulalongkorn University
Bangkok, Thailand
and
Natural Products for Neuroprotection and Anti-ageing Research Unit
Faculty of Allied Health Sciences
Chulalongkorn University
Bangkok, Thailand

Md. Mizanur Rahaman
Department of Pharmacy
Life Science Faculty
Bangabandhu Sheikh Mujibur Rahman Science and Technology University
Dhaka, Bangladesh

Shakeelur Rahman
Prakriti Bachao Foundation
Ranchi, India

Chandrabose Selvaraj
Computer-Aided Drug Design and Molecular Modeling Lab
Department of Bioinformatics
Alagappa University
Karaikudi, India

William N. Setzer
Department of Chemistry
University of Alabama in Huntsville
Huntsville, Alabama
and
Aromatic Plant Research Center
Lehi, Utah

Mansi Shah
University Department of Life Sciences
University of Mumbai
Mumbai, India

Sumaiya Shaikh
Department of Botany
Yashavantrao Chavan Institute of Science
Satara, India

Javad Sharifi-Rad
Phytochemistry Research Center
Shahid Beheshti University of Medical Sciences
Tehran, Iran

Parames C. Sil
Division of Molecular Medicine
Bose Institute
Kolkata, India

Akhilesh Kumar Singh
Department of Biotechnology
School of Life Sciences
Mahatma Gandhi Central University
Motihari, India

Sachidanand Singh
Department of Biotechnology
Vignan's Foundation for Science, Technology and Research (Deemed to be University)
Guntur, India

Sanjeev Kumar Singh
Computer-Aided Drug Design and Molecular Modeling Lab
Department of Bioinformatics
Alagappa University
Karaikudi, India

Surendra Pratap Singh
Pulp & Paper Research Institute
Rayagada, India

Sashi Sonkar
Department of Botany
Bankim Sardar College
Tangrakhali, India

Tewin Tencomnao
Department of Clinical Chemistry
Faculty of Allied Health Sciences
Chulalongkorn University
Bangkok, Thailand
and
Natural Products for Neuroprotection and Anti-ageing Research Unit
Faculty of Allied Health Sciences
Chulalongkorn University
Bangkok, Thailand

Premrutai Thitilertdecha
Siriraj Research Group in Immunobiology and Therapeutic Sciences
Faculty of Medicine Siriraj Hospital
Mahidol University
Bangkok, Thailand

Suraj Umdale
Department of Botany
Jaysingpur College (Affiliated to Shivaji University)
Kolhapur, India

Bhoomika Yadav
Department of Material Science & Metallurgical Engineering
University Institute of Engineering and Technology
CSJM University
Kanpur, India

Belkıs Muca Yiğit
Vocational School of Technical Sciences
Department of Forestry
Igdir University
Igdir, Turkey

Chandrabose Yogeswari
Ezhilnala Siddha Varma Hospital and Research Centre
Madurai, India

Kamal Fatima Zahra
Hassan First University
Faculty of Sciences and Techniques
Laboratory of Physical Chemistry of Processes and Materials/Agri-food and Health
Settat, Morocco

1 Exploring Important Herbs, Shrubs, and Trees for Their Traditional Knowledge, Chemical Derivatives, and Potential Benefits

Tilahun Belayneh Asfaw, Tarekegn Berhanu Esho, Archana Bachheti, Rakesh Kumar Bachheti, D.P. Pandey, and Azamal Husen

CONTENTS

1.1 Introduction 1
1.1.1 General Importance of Medicinal Plants 2
1.1.2 Significance of Medicinal Plants at the Global Level 2
1.1.3 Important Chemical Constituents and Uses 3
1.2 Descriptions 12
1.2.1 Morphological Descriptions of Some Important Herbs 12
1.2.2 Morphological Descriptions of Some Important Shrubs 13
1.2.3 Morphological Descriptions of Some Important Trees 15
1.3 Traditional Knowledge of Herbs, Shrubs, and Trees 17
1.4 Chemical Derivatives (Bioactive Compounds – Phytochemistry) of Herbs, Shrubs, and Trees 18
1.5 Potential Benefits, Applications, and Uses of Herbs, Shrubs, and Trees 21
1.5.1 General Overview 21
1.5.2 Uses and Bioactive Constituents of Herbs 22
1.5.3 Parts of the Plants, Uses, and Bioactive Constituents of Shrubs 22
1.5.4 Uses and Bioactive Constituents of Trees 22
1.6 Conclusion 22
References 23

1.1 INTRODUCTION

Nature is the original source of elements and molecules that are essential for human health. From time immemorial, medicinal plants (herbs, shrubs, and trees) have played a vital role in the health care of humans and animals and have become integral to life. Methods of application of medicinal plants by local people (traditional healers) vary geographically and socially, depending on time of plant collection, cultural differences, and ecological and biochemical features. Traditional medicinal knowledge within specific geographical locations or tribal groups has been subsequently transferred onto successive generations. According to the World Health Organization (WHO, 2013) definition,

DOI: 10.1201/9781003205067-1

> traditional medicine is the total of the knowledge, skill, and practices based on the theories, beliefs, and experiences indigenous to different cultures, whether explicable or not, used in the maintenance of the health as well as in the prevention, diagnosis, improvement or treatment of physical and mental illness.

Traditional medicinal knowledge can be classified as indigenous and complementary or alternative medicines. Indigenous knowledge refers to a broad set of health care practices based on the culture, religion/belief, and theories or based on the country's tradition or conventional medicines within a dominant health care system. On the other hand, complementary or alternative medicine refers to a broad set of health care practices that are not part of that country's tradition or conventional medicine and are not fully integrated into the dominant health care system. These traditional or complementary medicines are commonly used in developing countries to treat several disorders or physical or spiritual illnesses.

Written evidence from India, China, Ethiopia, and North Africa shows that humans have used medicinal plants throughout history. Developing countries have considerable economic benefits, including both indigenous and exogenous medicines and medicinal plants to treat various diseases. Traditional therapeutic systems are used to prepare the crude extract. They have a synergetic effect of treating several diseases like liver problems, cancer, malaria, obesity, arthritis, stomach, skin problems, pimples, sexually transmitted diseases, tuberculosis, asthma, ulcerations, etc. Medicinal herbs, shrubs, and trees are the best reservoirs of bioactive compounds. Based on the traditional practices, theories, beliefs, and documented or recorded medicinal treatments over time, novel compounds have been discovered and synthesized in modern laboratories and tested against *in vitro* and *in vivo* experiments. Of the main novel drugs isolated from medicinal plants, a few of them have been identified as highly effective natural product compounds in the treatment of various cancers – for example, artemisinin, Taxol, and etoposide. The objective of this chapter is to provide scientific information on some of the frequently used herbs, shrubs, and trees; their traditional knowledge, chemical derivatives/bioactive compounds, biological activities (antibacterial, antifungal, antiviral, and some other disease-causing agents like anxiety, insomnia, pain, muscle tension, oxidative stress, etc.), and other potential benefits.

1.1.1 General Importance of Medicinal Plants

Many people in most parts of the world depend on traditional medicine for their primary health care needs. Traditional medicinal systems are used to prepare crude drugs that have a synergetic effect of treating several diseases like liver problems, cancer, malaria, stomach and skin problems, pimples, ulcerations, etc. There are considerable economic benefits in developing countries, including both indigenous and exogenous medicines and medicinal plants for the treatment of various diseases. This is because medicinal plants are the sources of natural products/ bioactive compounds, which are responsible for treating such conditions. Natural products have been used as beneficial therapeutic agents for treating many diseases for a long time. For drug discovery, bioactive compounds play an essential role in developing the newly formulated drugs that arose from a limited number of basic chemical frameworks.

1.1.2 Significance of Medicinal Plants at the Global Level

The use of plants by man is an ancient practice. Plants are beneficial for humans as sources of medicines, flavors, foods, insect deterrents, ornaments, fumigants, spices, cosmetics, and income. Plants synthesize many secondary metabolites as a part of their regular metabolic activity to prevent themselves from predators, but researchers have demonstrated their use to treat various human illnesses (Gupta et al., 2014). Great emphasis has been given to the secondary metabolites of different natural sources, especially plants (Alassali and Cybulska, 2015). Ethnomedicine refers to natural practices of healing and treating ailments and diseases using various local practices made of wild plant and animal products. According to some studies, about 75–90% of rural populations rely on traditional medicines for their health care system (Assefa et al., 2010). Ethnomedicines/herbal medicines are much in demand as they are affordable and have fewer side effects (Abat et al., 2017). Recently, the WHO has also recognized the importance of traditional medicine in the health care sector and has designed the strategic use of national policies in medicinal plants (WHO, 2013).

1.1.3 Important Chemical Constituents and Uses

Traditional medicine plays a significant role in the health care of the majority of the people in developing countries, including India, China, Ethiopia, and North Africa. In this regard, medicinal plants provide a valuable contribution (Kassaye et al., 2007; Abebe, 2016). There are several classes of medications that traditional healers use throughout the world including anticancer, anti-diabetic, analgesics, antitussives, antihypertensives, cardiotonics, anti-viral, anticoagulant, anti-snake venom, anti-arrhythmic, antifungal, anti-parasitic, antineoplastics, antimalarials, antioxidant, antiseptic, diuretic, nervous system stimulant, sedative, expectorant, and others (Janardhan et al., 2014; Dhananjaya et al., 2016). Herbs, shrubs, and trees have been used as sources of traditional medicine for thousands of years to treat different ailments like stomach disorders, arthritis, syphilis, jaundice, acidity, tumors, piles, boils, inflammations, blood dysentery, hepatitis, malaria, ear ache, sexually transmitted diseases, wounds, tuberculosis, cholera, eczema, diarrhea, and vermifuge. Almost all parts of the plants are used for modern drug sources. For example, the leaves of *Huernia macrocarpa, Rumex nepalensis, Oxalis corniculata, Dodonaea Angustifolia, Justicia schimperiana, Senna septemtrionalis, Croton macrostachyus, Albizia schimperiana, Brucea antidysenterica, Prunus Africana, Syzygium guineense, Artemisia annua*, etc. are some of the common crude drugs of plant origin used for the treatments of skin, breast, colon, cervical, and lung cancer. On the other hand, leaves of *Barleria eranthemoides, Achyranthes Aspera, Allium sativum, Ischnura abyssinica, Kalanchoe laciniata*, etc., as shown in Tables 1.1, 1.2, and 1.3 are some of the common crude drugs from plants that are used for the treatment of wounds, snakebite, abdominal and asthma problems, evil and dust eye, malaria, fever, ear mites, excessive bleeding after birth, brain weakness, sexual diseases, etc. Similarly, the fruits of *Syzygium guineense, Citrullus colocynthis, Foeniculum vulgare, Coriandrum sativum, Physostigma venenosum, Senna* spp.; seeds of *Ricinus communis, Nux vomica, Strophanthus* spp., *Physostigma venenosum*; stem of *Chondrodendron tomentosum*, bark of *Cinchona* spp., *Cinnamomum zeylanicum, Holarrhena antidysenterica, Hyoscyamus niger*; *Atropa belladonna*, wood of *Quassia amara, Santalum album*; roots of *Carapichea ipecacuanha, Rauvolfia serpentine*; rhizome of *Curcuma longa, Zingiber officinale, Valeriana officinalis, Podophyllum peltatum*, etc. are some of the common crude drugs of plant origin. Similarly, flowers, stem barks, areal parts of most herbal drugs, roots, and mixtures of two or more two plants or plant parts are used to treat several diseases (Tables 1.1, 1.2, and 1.3). The morphological descriptions of some of the essential medicinal plants (herbs, shrubs, and trees) are described in section 2.1.

Phytochemicals are naturally occurring plant-derived organic compounds from leaves, vegetables, and roots having defense mechanisms that prevent various diseases. Phytochemicals are primary and secondary compounds which include chlorophyll, proteins, common sugars, alkaloids, coumarins, flavonoids, glycosides, lignans, steroids, sugars, terpenoids, and phenolic compounds, respectively (Joel and Bhimba, 2010; Kumar et al., 2018; Liu, et al., 2018). These secondary metabolites are important classes of natural organic compounds used to treat different viral, bacterial, fungal infections, venom, etc. Alongside modern pharmaceutical medicines, plant-derived medicine containing highly active bioactive compounds has been used for treatment options. These crude cultural medicines are given to the patients by traditional healers with remediation. Natural therapeutic medicines are becoming an attractive option as they are perceived to have fewer incidents of adverse reactions and lower costs associated with remedy preparations when compared to synthetically produced pharmaceuticals. This effectiveness comes from the synergetic effects of bioactive components available in those medicinal plants. For instance, approximately 45% of all anticancer

TABLE 1.1
Uses and Bioactive Constituents of Herbs

Plant Name	Family	English Name	Plant Part Used	Bioactive Compounds	Uses	References
Achyranthes Aspera L.	Amaranthaceae	Devil's Horsewhip	Leaf	Phytosterols, polyphenols, and saponins, glycosides, alkaloids, tannins, saponins, flavonoids, lignin, flavonoids, steroids, and terpenoids	To heal wounds, fever, eye dusts, ear mites, excessive bleeding after birth, cough, bronchitis and rheumatism, malarial fever, dysentery, asthma, hypertension, and diabetes	(Bhosale et al., 2012)
Allium sativum L.	Amaryllidaceae	Garlic	Bulb	Flavonoids such as quercetin and cyanidin, allistatin I and allistatin II, and vitamins C, E, and A	To medicate evil eye, malaria, cardiovascular diseases, regulating blood pressure, lowering blood sugar and cholesterol levels; effective against bacterial, viral, fungal and parasitic infections; enhancing the immune system and having antitumoral and antioxidant features; culinary use and vaginal infections	(Modak et al., 2020)
Andrachne aspera Spreng.	Euphorbiaceae	No associated recorded name	Leaf and root	Piperidine alkaloids, andrachcinine, and andrachcinidine alkaloids: andrachamine, andrachcine, andrachcinine, andrachcinidine, (+)-allosedridine, (-)-8-epi-8-ethylnorlobelol and (-)-8-epihalosaline, aspertin-A, aspertin-B, aspertin-C, and aspertin-D terpenes: lupeol acetate, α-amyrin, β-amyrin, α- taraxerol, stigmasterol, β-stigmasterol, lupeol, oleanolic acid, and germanicol	Snakebite, abdominal pain, and asthma treatments	(Al-snafi, 2021)
Apium graveolens L.	Apiaceae	Celery	Seed	Glycosides, steroids, different types of phenolic including furanocoumarins, flavones, and trace elements (sodium, potassium, and calcium)	To mediate calcium ion suppressing effect through both voltage and receptor-operated calcium channels	(Modak et al., 2020)

(Continued)

TABLE 1.1 (CONTINUED)
Uses and Bioactive Constituents of Herbs

Plant Name	Family	English Name	Plant Part Used	Bioactive Compounds	Uses	References
Artemisia afra Jack. ex Wild	Asteraceae	Wild Wormwood	Leaf	Epoxylinalol and dihydrocostunolide, camphor, davanone, bornyl acetate, 4-terpineol, and chamazulene	To treat evil eye, malaria, analgesic, anti-inflammatory, and antidepressant	(Frimpong et al., 2021; Noronha et al., 2020)
Barleria eranthemoides R. Br. ex C. B. Cl.	Acanthaceae	No associated recorded name	Leaf	--	Wound treatment	(Chekole, 2017)
Centella asiatica L. Urb.	Apiaceae		Whole plant	Saponins (asiatic acid, centelloside, and medecassosides), flavonoid, amino-acids, tannins, and sugar	Herbal extract reduces the resting flux and increases the venoarterial response	(Modak et al., 2020)
Cirsium englerianum O. Hoffm.	Asteraceae		Root and fruit	--	The treatments of intestinal parasite and influenza virus	(Chekole, 2017)
Haplocarpha rueppelii (Sch. Bip.) Beauv.	Asteraceae		Leaf	--	To stop bleeding	(Chekole, 2017)
Huernia macrocarpa (A.Rich) Sprenger	Asclepiadaceae		Latex	phenolic compounds such as kaempferol-3-O-β-D glucoside, gallic acid, norcucurbitacin, fevicordin A and fevicordin A glucosides, terpenes (isoprenoids) compounds, alkaloids compounds, and benzophenone compounds	For the treatments of skin cancer, bone cancer, blood diseases, allergies, and diabetes mellitus	(Tuasha et al., 2018)

(Continued)

TABLE 1.1 (CONTINUED)
Uses and Bioactive Constituents of Herbs

Plant Name	Family	English Name	Plant Part Used	Bioactive Compounds	Uses	References
Kalanchoe laciniata (L.) DC.	Crassulaceae	Christmas tree plant	Leaf and root	Flavonoids (caffeic acid, malic acid, isocitric acid, cyanidin-3-*O*-α-L-glucoside, dihydroquercetin, *Cis-p*-coumaryl glutaric acid, *Trans-p*-coumaryl glutaric acid, isorhamnetin-3-rutinoside, myricitrin, quercetin-3-*O*-α-Larabinopyranosyl-(-1→ 2)-α-Lrhamnopyranoside, kaempferol 3-*O*-α-Larabinopiranosyl-(1→2)-α-Lrhamn opyranoside, quercitrin, I soscoparin-7-*O*-arabinoside), triterpenoids, lignins, phenols, saponins, and glycosides and citric acid	For the treatments of swelling, broken bone, expelled uterus, hip obesity, tonsillitis, toothache, common cough and cold, wounds, inflammation, and diabetes	(Fernandes et al., 2019; Pereira et al., 2018)
Oxalis corniculata L	Oxalidaceae		Leaves and roots	--	Breast cancer treatment and in vivo antitumor activity against Ehrlich ascites carcinoma on mice	(Chekole, 2017)
Phyllanthus amarus Shum. and Thon.	Euphorbiaceae	Nelanelli	Leaf	Lignans, flavonoids, hydrolyzable tannins (ellagitannins), polyphenols, triterpenes, sterols, and alkaloids	Diuresis and anti-atherosclerosis treatments	(Modak et al., 2020)
Rumex nervosus Vahl	Polygonaceae	Nepal Dock	Root and bark, the whole plant	Alkaloids, flavonoids, terpenoids, tannins, glycosides, and volatile oils	Treatments of purgative, antitumor, anti-inflammatory, and treatment for dislocated bones, to treat wounds, pimples, and ringworm,	(Ayele, 2018; Gonfa et al., 2021; Tesfaye et al., 2020)
Rumex nepalensis Spreng.			Root and bark	Anthraquinones, naphthalenes, tannins, and stilbenoids	colon and skin cancer, to treat stomachache, toothache, tonsillitis, ascariasis, rabies, and body swelling, antiproliferative activity on human liver, antioxidant and cytotoxicity on human leukemia cells	

(*Continued*)

TABLE 1.1 (CONTINUED)
Uses and Bioactive Constituents of Herbs

Plant Name	Family	English Name	Plant Part Used	Bioactive Compounds	Uses	References
Ximenia Americana L.	Olacaceae	Wild Plum, Tallowwood, Sour Plum, and Hog Plum	Leaf, fruit and root	Polyphenolics flavonoids ((−) epi-catechin, quercetin), isoprenoids, triterpenes, saponins, alkaloids, simple phenols, glycosides, sterols, and tannins	For the treatments of malaria, leprotic ulcer, skin infections, headaches, cough and fever, venereal diseases, oedema, hepatic effects, hematological effect, skin aches, headaches, leprosy, hemorrhoids and to treat sexually transmitted diseases	(James et al., 2008; Uchôa et al., 2016)

TABLE 1.2
Uses and Bioactive Constituents of Shrubs

Plant Name	Family	English Name	Plant Part Used	Bioactive Compounds	Uses	References
Acokanthera schimperi (A.DC.) Schweinf.	Apocynaceae	Arrow-poison Tree, Common-poison Bush, Round-leaved Poison-bush	Leaf	--	Wound healing, antibacterial, and antiparasitic treatment	(Alemu et al., 2020)
Antirrhinum insipidus E. Mey.	Asclepiadaceae	Dragon Flowers or Snapdragons	Root	---	Breast and skin cancer medication	(Modak et al., 2020)
Calotropis procera (Ait.) Ait.f.	Asclepiadaceae	Date-palm, Swamp Date-palm, Dwarf Date-palm, Senegal Date-palm	Latex	Alkaloids, sterols, fatty acids, starches, sugars, oils, tannins, resins, gums, and enzymatic proteins, such as proteases, chitinases, lipases, peptidases, esterase, peroxidases, papain, hevein, and lectins	Eczema, hemorrhoid, hepatoprotective, antioxidant, and antiapoptotic treatments	(Al Sulaibi et al., 2020; Alhassan et al., 2012; Mainasara et al., 2012)
Carissa spinarum L.	Apocynaceae	Conkerberry or Bush Plum	Leaf and root	Phenolics, Alkaloids, terpenoids, saponins, tannins	To treat snake poison, evil spirit, stomachache, cancer/tumor, and wounds	(Ayele, 2018; Bitew et al., 2019)
Dodonaea angustifolia/ viscosa subsp. (L.f.) J.G.West	Sapindaceae	Mukonachando (Shona) Sand Olive	Leaf and root	Alkaloids, terpenoids, saponins, tannins, sugars, phenolics, and flavonoids, steroids, fixed oil and fat,	Against breast, skin, and cervical cancer and malaria problems	(Al-snafi, 2021; Al-Snafi, 2017; Ayele, 2018)
Crataegus oxycantha	Rosaceae	Hawthorn Berry	Flower, leaves, root, bark, and fruits	Flavonoids (oligomeric proanthocyanidins), citrin bioflavonoid containing vitamin, quercetin, triterpene saponins, vitamin C	Cancer treatment	(Modak et al., 2020; Obakiro et al., 2020)
Justicia schimperiana (Hochst. ex Nees) T. Anders.	Acanthaceae	Water-willow and Shrimp Plant	Root and leaves	alkaloids, lignans, flavonoids, and terpenoids (iridods, diterpenoids, and triterpenoids), essential oils, vitamins, fatty acids (docosanoic acid), and salicylic acid, steroids, campesterol, stigmasterol, sitosterol, and sitosterol-D-glucoside	To treat visual impairment, typhoid and malaria, lung cancer, evil eye, hepatitis B (jaundice), rabies, asthma, common cold, stomachache, diarrhea, tapeworm infestation, anthrax, wound, external parasite, ascariasis, and skin irritation	(Chekole, 2017)

(*Continued*)

TABLE 1.2 (CONTINUED)
Uses and Bioactive Constituents of Shrubs

Plant Name	Family	English Name	Plant Part Used	Bioactive Compounds	Uses	References
Leonotis ocymifolia (Burm.f.)	Lamiaceae	--	Leaf	Phenolics and glucosides	To medicate brain problems, swelling	(Fernandes et al., 2019; Pereira et al., 2018)
Miranda dianthera (Roth ex Roem and Schult.)	Lamiaceae	---	Fruit	Essential oils	To medicate trachoma	(James et al., 2008; Uchôa et al., 2016)
Ocimum lamiifolium Hochst. ex Benth.	Lamiaceae	---	Leaf	Essential oils (more than 80 essential oils)	To medicate headache and febrile illness	(Modak et al., 2020; Runyoro et al., 2010)
Phoenix reclinata Jacq.	Arecaceae	Wild Date Palm or Senegal Date Palm	Root	Flavonoids, saponins, tannins, sterols, and/or triterpenes	To treat sexual diseases	(Hifnawy et al., 2016)
Senna septemtrionalis (Viv.)	Fabaceae	Arsenic Bush	Leaf	Alkaloids, anthraquinones, flavonoids, tannins, glycosides, steroids, terpenoids, saponins, and volatile oils	To medicate cough, lung cancer, and brain problem	(Sinan et al., 2020)

TABLE 1.3
Uses and Bioactive Constituents of Trees

Plant Name	Family	English Name	Plant Part Used	Bioactive Compounds	Uses	References
Abizia schimperiana Oliv.	Fabaceae	Albizia	Leaf	Alkaloids, saponins, tannins, and flavonoids	To treat breast, intestinal, and skin cancers	(Ayele, 2018)
Annona muricata	Annonaceae	Prickly Custard Apple	Fruits and leaves	Fruit contains vitamin C, vitamin B1 and vitamin B2; leaves contain annonamine, an aporphine class alkaloid	Decrease the peripheral vascular resistance	(Modak et al., 2020)
Artemisia annua L.	Asteraceae	Wormwood	Leaf	Artemisinin, artemisinic or arteanuic acid, arteannuin B and dihydroarteannuin, sesquiterpene (Z)-7-acet oxymethyl-11-methyl-3-methylene-dodeca-1,6,10 -triene, quercetagetin 6,7,3′,4′-tetramethyl ether	To treat breast cancer and malaria	(Sinan et al., 2020)
Bombax ceiba L.	Bombacaceae	Cotton Tree	Bark, gum, and root	polyphenols including phenolic acids, tannins, and flavonoids	To treat emetic – a styptic in metrorrhagia	(Hifnawy et al., 2016)
Brucea antidysenterica J.F.Mill.	Simaroubaceae	Vaginosis	Leaf	Alkaloids, saponins, tannins, polyphenols, terpenoids, flavonoids, and anthraquinones	To treat skin cancer	(Zewdie et al., 2020)
Croton macrostachyus Del.	Euphorbiaceae	Broad-leaved Croton	Bark, fruit, and leaf	Alkaloids, amino acids, anthraquinones, carbohydrates, cardiac glycosides, coumarins, essential oil, fatty acids, flavonoids, phenolic compounds, phlobatannins, polyphenols, phytosteroides, saponins, sterols, tannins, terpenoids, unsaturated sterol, and vitamin C	To treat liver problems, stomachache, gonorrhea, malaria, atopic, dermatitis, breast and skin cancer	(Alemu et al., 2020; Mayori, 2017)
Hagenia abyssinica J.F.	Rosaceae	African Redwood, East African Rosewood	Fruit	---	Prevents intestinal parasite	(Modak et al., 2020)
Lobelia gibberoa Hemsl.	Lobeliaceae	Giant Lobelia	Root and seed	Alkaloids, such as lobeline, norlobelanine; flavonoid compounds (apigenin, luteolin, quercetin) and coumarins; and essential oil	To treat eye problems, impotency, malaria, and epilepsy	(Al Sulaibi et al., 2020; Alhassan et al., 2012; Mainasara et al., 2012)

(Continued)

TABLE 1.3 (CONTINUED)
Uses and Bioactive Constituents of Trees

Plant Name	Family	English Name	Plant Part Used	Bioactive Compounds	Uses	References
Myrica salicifolia A. Rich.	Myricaceae	Bayberry, Bay-rum Tree, Candleberry, Sweet Gale, and Wax-myrtle	Bark, Root	Phenolics, alkaloids, terpenoids and glucosides	To treat evil eye and evil spirit, liver problem and joint pain	(Chekole, 2017)
Prunus Africana (Hook.f.) Kalkman	Rosaceae	Prunus Africana, African Cherry	Bark	Ursolic acid, oleanolic acid, atraric acid, ferulic acid, N-butylbenzene-sulfonamide (NBBS), beta-sitosterol lauric acid	To treat breast, skin, and prostate cancer	(James et al., 2008; Komakech et al., 2017; Uchôa et al., 2016)
Rhus retinorrhoea Oliv.	Nacardaceae	--	Leaf and Root	Flavonoids, urushiols, and terpenoid. Flavonoids reported from the plants include kaempferol, 7-O-methyl isokaemferide, quercetin, 7,3′-O dimethylquercetin, quercetin-3-O glucoside, quercitrin, rutin, quercetin-3-O-α-L-(3′′-O-galloyl)-r hamnoside, myricetin, myricetin-3 O-β-glucoside, myricetin 3-rhamnoside, fisetin, apigenin, genkwanin, apigenin dimethyl ether, luteolin, 7-O-methyl, …	Treats tonsillitis and stomachache	(Chekole, 2017; Moremi et al., 2021)
Syzygium guineense (Willd.) DC.	Myrtaceae	Woodland Waterberry, Waterpear	Leaves, roots, fruits, and stem bark	Flavonoid, chromone, terpenoid, steroid, tannin, phenol, acylphloroglucinol, methanol and dichloromethane crude extracts showed *in vitro* cytotoxicity on human leukemia cells and antimicrobial activity	Treats skin cancer and tuberculosis	(Aung et al., 2020; Oladosu et al., 2017; Tesfaye et al., 2020)

drugs have been formulated from plant sources; out of this, 12% are crude natural products, and 32% are synthetic derivatives of plant-based compounds (Gonfa et al., 2021).

1.2 DESCRIPTIONS

1.2.1 Morphological Descriptions of Some Important Herbs

Achyranthes aspera L. (Family: Amaranthaceae) is an erect or ascending multi-branched herb, a commonly used medicinal plant found throughout tropical regions (Figure 1.1). It is mainly native to the Mediterranean region (Morocco, Albania), New Zealand and Macauley Island, Asia, and horns of Africa (Sharma et al., 2015). This herb is approximately 1–2 m in height and has an aromatic odor. The leaves are simple, opposite, oval in shape, with entire margins. Leaves are also simple, opposite, sessile, oval to elliptic in shape or circular. They are 4–9 cm long by 2–4 cm wide. Each leaf has 4–9 arching veins. Stems are square with swollen nodes and the surfaces of the stem have longitudinal grooves (de Lange et al., 2004; Sharma et al., 2015).

Kalanchoe laciniata (L.) DC. (Family: Crassulaceae) (Figure 1.2) is a sub-ligneous and perennial herb usually 30–65 cm high, but up to 100 cm in height in some areas (Fernandes et al., 2019). The leaf morphologies are simple oval or oval to elliptical shape, decussate, succulent – a drought-resistant plant, glabrous, opposite in arrangements, shortly petiolate, crenate, and having a corrugated or sub-crenated border. The flowers are yellow-orange in color, small, abundant, arranged in composite summits of stamps or paniculate, hermaphrodites, gamopetalas with corolla longer than the cup, with the presence of scaly carpels that become polispermous follicles. Their fruit is a follicle 6 cm long that contains brown, oblong seeds (Fernandes et al., 2019).

Oxalis corniculata L. (Family: Oxalidaceae) is a medicinal plant species, and the common name of this herb is creeping woodsorrel (Figure 1.3). The plants are mostly small, semi-erect, and commonly lying on the ground. It is a highly invasive weed species in the horticulture industry worldwide, growing best in spring or fall in warmer climates; but in the other seasons, the plant is a

FIGURE 1.1 *Achyranthes aspera* (L.).

FIGURE 1.2 *Kalanchoe laciniata* (L.).

FIGURE 1.3 *Oxalis corniculata* L.

persistent weed year-round. It is also an annual or perennial much-branched medicinal herb (10–30 cm high). The stem is cylindrical (up to 50 cm in length), solid, pubescent, often tinged with purple. Leaves are composite in their structure, alternately branched, sometimes fiber bundle or fasciculus for young plants. The plants are sometimes supported by a long petiole which may be up to approximately 3 to 8 cm. The leaflet lamina is nearly glabrous and broadly emarginate at the top and wide corner at the base. It is traditionally used as a raw vegetable and, in folk medicine, to treat different human illnesses. The plant species are distributed in Europe and North Africa, Sub-Saharan Africa, Australasia, North America, South America, and Asia (Tibuhwa, 2016; Groom et al., 2019).

Rumex nepalensis Spreng. (Family: Polygonaceae) is commonly known as Nepal dock, a tall, robust plant used for human consumption as a vegetable, a coloring agent, and an herbal medicine used to treat and cure different diseases traditionally, such as purgative, dysentery, venereal diseases, and bacterial and fungal infections (Figure 1.4). It is an erect plant with long taproots and erect stems (50–100 cm tall) that are branched, glabrous, grooved, green or pale brown in color. Basal leaves, petiole 4–10 cm, leaf blade broadly ovate (10–15 cm long and 4–8 cm wide), both the surfaces of leaf are glabrous or abaxially, minutely papillate along veins, base cordate, margin entire, apex acute. This herb has shortly petiolate, membranous, inflorescence paniculate leaves, bisexual pedicellate flowers. *Rumex nepalensis* Spreng. is found worldwide, but especially in East Africa (such as Ethiopia and Nigeria), South Africa, Asia (such as India, China, etc.), and Southwest Asia, (Gonfa et al., 2021; Shaikh et al., 2018)

1.2.2 Morphological Descriptions of Some Important Shrubs

Shrubs are one of the essential sources of traditional medicinal plants. From the Oxford Dictionary definition, a shrub is a woody plant smaller than a tree and has several main stems arising at or near the ground. Shrubs are often survived and flourished or survived under heavy grazing and harsh environmental conditions. The following figures depict a few of these shrubs.

Justicia schimperiana (Hochst. Ex Nees) T. (Family: Acanthaceae) is a common evergreen shrub growing in misty forest regions near streams, rivers on hill slopes, and forest clearings (Figure 1.5). *Adhatoda schimperiana* and *Gendarussa schimperiana* are related names of the plant and are

FIGURE 1.4 *Rumex nepalensis* Spreng.

FIGURE 1.5 *Justicia schimperiana* (Hochst. ex Nees).

usually propagated by cuttings and seedlings. The plant is a leafy shrub growing up to 4 m; the stem is brittle and breaks easily. The leaves of this shrub are simple and opposite; they have a long oval shape up to 13 × 4 cm, the tip is pointed, narrowed to a short stalk. The flower has conspicuous terminal heads on long stems. Ethnobotanical study shows that the plant treats different ailments such as evil eye, hepatitis B (jaundice), rabies, common cold, diarrhea, asthma, stomachache, etc. (Tesfaye et al., 2020; Mekonnen et al., 2018). In modern laboratories, experimental studies indicated that the crude chloroform, methanol, and aqueous extracts of the plant showed antimalarial and antidiarrheal properties and a reduction in blood glucose, respectively (Mekonnen et al., 2018).

Acokanthera schimperi (A.DC.) Schweinf (Family: Apocynaceae) is a small tree (in some areas, like Somalia, it can grow to 9 m), densely branched, and native to Eastern and Central Africa (Figure 1.6). The common names are "arrow-poison" tree "or common poison bush". The plant leaves are glabrous or scabrid up to 2–10 × 1.5–6 cm and petiole up to 1–4 (sometimes up to 9) mm long. The flowers are corolla-tube 8–12.5 mm long or calyx c. 1.5–2 mm long, glabrous or pubescent outside it; lobes ovate-cuspidate 2.5–5 mm in length. The plant has been evaluated against *in vitro* cytotoxicity, antiviral, antimicrobial, and anti-parasitic activities and was effective against all of these (Tesfaye et al., 2020).

Carissa spinarum L. (Family: Apocynaceae) is a small spinous evergreen shrub of tropical deciduous forests that usually grows 1–1.5 m (Figure 1.7). It is native to Africa, Indo-China, Australia to New Caledonia and distributed throughout the world. The plant's leaves are somewhat leathery, blades ovate to elliptic or sub-circular, darker above, 1.7–6.8 × 1–4.6 cm long and acute to obtuse at the apex and glabrous pubescent on both sides. The plant's spines are simple, or rarely forked, and grow 0.4–7 cm long. The flower has shortly pedunculate cymes, is sweetly scented, and can be calyx, when it is 2–4 mm long with lanceolate-subulate lobes, or corolla, when it is 9.5–20 mm long

FIGURE 1.6 *Acokanthera schimperi* (A.DC.) Schweinf.

FIGURE 1.7 *Carissa spinarum* L.

with a white, tinged red, sparsely pubescent outside. Before ripening stages, the fruits have latex, and they are fleshy after ripening. The plants are regenerated or distributed naturally in the forest through seed dispersal and natural propagation (Mishra and Gupta, 2005).

Dodonaea angustifolia L.F. (Family: Sapindaceae) is commonly distributed in tropical and subtropical countries or regions – even if the center of origin is in Australia (Figure 1.8). The English names of the plant are "hop bush", "switch sorrel", and "sand olive". The species is a sticky shrub or a small tree that reaches 4–8 m. The characteristics of the plant leaves are petiole with 1–4 mm long or 1.5–5 tall or in some areas up to 9/12 m long; the leaf blade is 2–5.5 × 0.7–1.7 cm or elliptical up to 1–13 cm × 1.5–4 cm. The leaves are alternate, simple, absent of stipules, margin entire. Flowers are bisexual or unisexual; the colors are whitish to greenish-yellow, pedicel with 8–15 mm long and inconspicuous with no petals. Fruits have 2–3 winged papery capsules (white or straw-colored to brown or purplish) 15–23 mm × 18–25 mm; the pollen is distributed by dispersion, and the plant can be easily propagated from both cuttings and seeds (Al-Snafi, 2017; Beshah et al., 2020; Priya, 2021).

1.2.3 Morphological Descriptions of Some Important Trees

Trees contribute greatly to treatments for several ailments. A tree is defined by the Oxford Dictionary as "a woody perennial plant, typically having a single stem or trunk growing to a considerable height and bearing lateral branches at some distance from the ground". The following offers descriptions of some of the important medicinal tree species used by humans throughout the world.

Croton macrostachyus Del. (Family: Euphorbiaceae) (Figure 1.9) is a common evergreen shrub or tree growing in forests and along rivers or lakes. The plant's common names are "spurge" family, woodland, "forest fever tree", or "broad-leaved Croton". This plant is a deciduous tree that grows to 6–12 m high and is usually propagated by cuttings and seedlings. The plant is a leafy shrub growing up to 4 m, the stem is brittle and breaks easily. The leaves of this shrub are large and green, turning

FIGURE 1.8 *Dodonaea angustifolia/viscosa* subsp. (L.f.) J.G.West.

FIGURE 1.9 *Croton macrostachyus Del.*

to orange before falling, ovate, base subcordate or rounded, apex acuminate, margin crenulated serrulate or sub-entire; with the leaves are 5–19 cm wide and 3.5–15 cm long. The flowers are a creamy yellow–white color, sweetly scented, and up to 3 mm long and these have at least one separate shoot tip. An ethnobotanical study of the fruits showed that the plant has medicinal value in the treatment of malaria, diarrhea, rabies, toothache, cancer, abdominal pain, pneumonia, gastrointestinal disorder, etc. (Abdisa, 2019). Experimental studies indicated that the plant has different medicinal values against foodborne diseases, drug-resistant infectious agents, and abortifacient and uterotonic to expel retained placenta (Abdisa, 2019).

Albizia schimperiana Oliv. (Family: Fabaceae) is a deciduous tree with a flattened, rounded, umbrella-shaped canopy and is commonly known as forest long-pod albizia (Figure 1.10). The species grow up to 30 m (occasionally to 35 m) and are distributed throughout tropical African countries. The tree's canopy is evergreen, leguminous, and relies on nitrogen-fixing Bradyrhizobium bacteria for soil fertility regeneration. *A. schimperiana* is propagated easily via its seeds. The bark is smooth grey or rough brown. The leaves have rachis densely to sparsely pubescent, pinnae which have 2–7 pairs, leaflets of the two distal pairs of pinnae having 6–21/23 pairs that are variable in shape and size, oblong or acute to rounded and mucronate at the apex, which are turned towards the pinna-apex. The leaves are hairy and paler below, shiny dark green above, and each leaflet is less than 2 cm long. The flowers are very many, white, and in round heads. The fruits are also large clusters of dull brown pods about 25 cm long and 3.5 cm across, which have a maximum of 34 × 6 cm, and the seeds are released when the pods break open. The stem bark of *A. schimperiana* is used to treat bacterial and parasitic infections like malaria, pneumonia, fever, and pain (Samoylenko et al., 2009).

Syzygium guineense (Willd.) DC. (Family: Myrtaceae) (Figure 1.11) is commonly known as "water berry", "water pear", or "snake bean tree". The plant is a small to large evergreen, having

FIGURE 1.10 *Albizia schimperiana* Oliv.

FIGURE 1.11 *Syzygium guineense* (Willd.) DC.

15–30 m long, and is mostly found in African countries, such as Ethiopia, Senegal, eastward to Somalia, and southward to Namibia, Botswana, South Africa; it is, however, also found in Saudi Arabia and Yemen. The bark varies in subspecies, greyish and smooth in young trees, turning rough, light grey, black or dark brown in older trees. The leaves are at ends, 5–17.5 cm in length, leaf lamina with 4 × 2–14 × 7 cm, and have a width of 1.3–7.5 cm. While they are very variable in shape, they tend to be broadest at or near the middle. Flowers are 1.5 cm in diameter, receptacle (including pseudopedicle) and calyx of 0.35–0.65 cm long fragrant, creamy white, originating in terminal panicles and forming heads up to 10 × 10 cm or with 4–8 widely spaced flowers in branched heads up to 3 or 4 cm in diameter. The fruits are ovoid or ellipsoid drupes having 1.2–3.5 cm × 1 × 2.5 cm width. The plant is beneficial for treating different ailments, such as skin cancer, tuberculosis, inflammation, wounds, stomach problems, and various female disorders. The plant species are also helpful for flooring, utensils, furniture, construction, etc. (Badou et al., 2020; Nguyen et al., 2016).

1.3 TRADITIONAL KNOWLEDGE OF HERBS, SHRUBS, AND TREES

Traditional medicine is the traditional knowledge of herbs, shrubs, and trees within societies based on cultural and religious beliefs used to prevent, diagnose, improve, or treat physical and mental illnesses to maintain the health of communities. Therefore, traditional knowledge or indigenous knowledge refers to the medicinal uses of plants in traditional ways or systems of medical diagnosis. It includes the knowledge of indigenous or local communities, that is transferred mostly orally only from one generation to the next, particularly to family members, in such a way that the traditional knowledge is kept secret. It is sometimes documented in written materials (for example, Chinese yellow/golden book, Indian Ayurveda, Siddha, Unani, traditional medicinal records) (Salim et al., 2019) that can be transferred secretly from one family to another or from one tribe to another. In some cases, this traditional knowledge is documented in written materials or even electronically, making them open to the public. The ancient Indian medicinal knowledge of Ayurveda medicine, for example, has been considered one of the bases for some current advancements in modern medicines. As a broad description, traditional knowledge includes intellectual (in the religious ways – spiritual beliefs) and intangible cultural heritage practices and knowledge systems of traditional communities that includes botanical – where the medicinal plants are found (hot to cold regions – ecological knowledge), the preparations of medicines (starting from collection to diagnosis – medicinal knowledge), and remedies.

There are different types of medicinal plants traditionally used to treat various illnesses as described in the previous sections. The traditional knowledge of world societies mainly relies on herbs, including climbers, shrubs, trees, and the combinations of two or three of them. These traditional medicines are a vital source of health care and an important source of income for many communities. The use of conventional medicine is even more substantial in the developing world. According to a WHO report, 70% of the population in India and more than 90% of the people in Ethiopia depend on traditional medicine for their primary health care (WHO, 2013).

1.4 CHEMICAL DERIVATIVES (BIOACTIVE COMPOUNDS – PHYTOCHEMISTRY) OF HERBS, SHRUBS, AND TREES

More than 80% of the rural population worldwide, especially in developing countries, depends on traditional medicines for primary health care (WHO, 2013). Medicinal plants can synthesize thousands of diverse phytochemicals or natural products or bioactive constituents, including alkaloids, terpenoids, and phenolics. Phytochemicals are chemical compounds that occur naturally in medicinal and food plants. Plant bioactive compounds or phytochemicals play essential roles in plant survival and provide various valuable natural products (bioactive compounds) to protect themselves against pathogenic attacks and environmental stresses. Phytochemicals are also important ingredients that contribute to plant color, aroma, and flavor (Kumari et al., 2021). The amount of these phytochemicals depends on the types of plants and the climatic growing conditions or geographical variations.

The ancient sources of drugs were plants that provide rich, complex, and highly diversified structures of phytochemicals. Because it is questionable whether some of these natural compounds can be successfully synthesized in chemical laboratories, or even in biosynthetic laboratories, these naturally occurring bioactive compounds could be considered more efficient compared to a modern biosynthetic laboratory.

According to the previously reported studies, there are approximately more than 200,000 secondary metabolites in the plant kingdom. These bioactives are derived from a few primary metabolites within the specific pathways. The plant chemical derivatives can be classified into different types based on their synthetic pathways, containing atoms, elements, or functional groups. Based on the synthetic pathway, they can be classified as phenylpropanoids, terpenoids, and alkaloids. On the other hand, the main chemical groups of bioactive compounds in plants are classified as simple phenolics, polyphenolics, glycosides, tannins, mono- and sequi-terpenoids, phenylpropanoids, diterpenoids, resins, lignans, furocoumarin, glucosides, naphthodianthrones, alkaloids, and terpenoids. Phenol, diphenol, salicylic acid, tannic acid, and phenolic acids, such as benzoic acid derivatives (hydroxybenzoic acids) and cinnamic acid derivatives (hydroxycinnamic acids), are examples of simple phenolic compounds. The glycosides consist of secondary metabolites bound to a mono- or oligosaccharide or uronic acid, such as cardiac glycosides, cyanogenic glycosides, glucosinolates, saponins, anthraquinone glycosides, and flavonoids and proanthocyanidins.

Phenylpropanoids are one of the largest secondary metabolites produced by plants and are classified under polyphenolics. These large classes of bioactive compounds are derived from aromatic amino acids (like phenylalanine) in most plant species (Figure 1.12). Phenylpropanoids (based on their biosynthesis as shown in Figure 1.12) include flavonoids, phenolic acids, stilbenes, coumarins, and monolignols, which can be found in medicinal plants and plant foods (Deng and Lu, 2017). Phenylpropanoids play an important role in the treatments of cancer, stress, having antioxidant, antiviral, and antifungal activities. Resveratrol, flavanonol, coumarin, anthocyanidin, umbelliferone, p-hydroxybenzoic acid, flavonol, flavanone, protocatechuic acid, caffeic acid and sinapic acid, coniferyl alcohol, p-coumaryl alcohol, and sinapyl alcohol are included in this classification.

Based on the distinctive chemical groups, plants' secondary metabolites can be divided into four main classes: terpenes, phenolic compounds, nitrogen-containing compounds, and sulfur-containing compounds (Figure 1.13). The chemical derivatives of phenolic compounds are simple phenylpropanoid, benzoic acid derivatives, anthocyanin, isoflavones, tannins, lignin, flavanones, flavones, and phenyl alanines. The flavonoids are one of the largest classes of plant phenolics that have the basic structure containing 15 carbons arranged in two aromatic rings connected by a three-carbon bridge (C_6-C_3-C_6 arrangements). From the flavonoids, quercetin, catechin, kaempferol, etc.

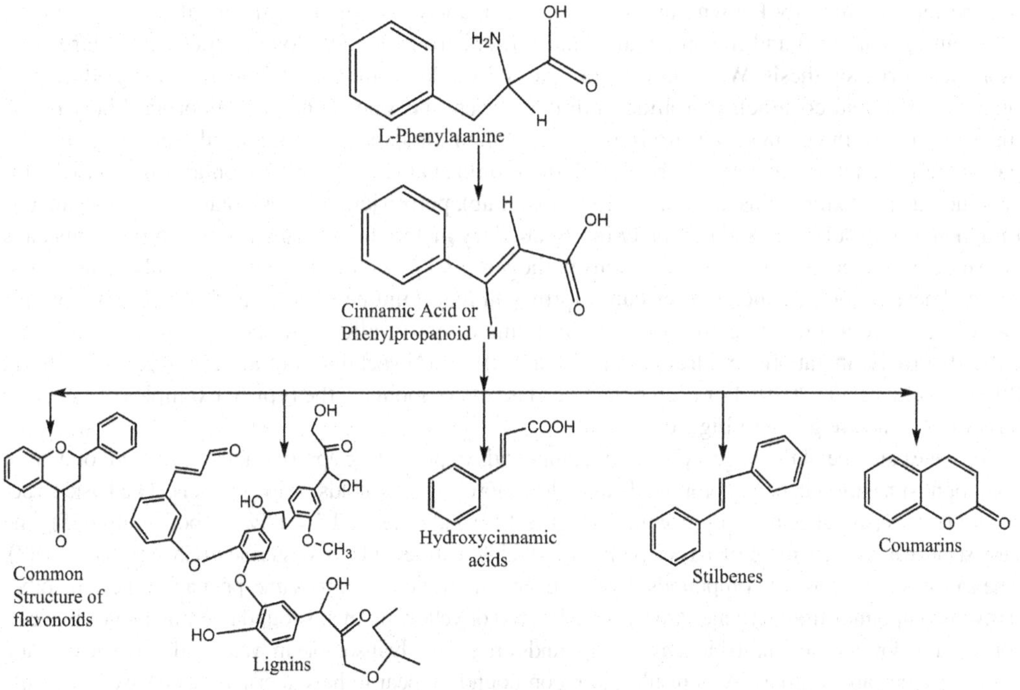

OH
H_2N
O
H
L-Phenylalanine
H
OH
Cinnamic Acid or
Phenylpropanoid H
O
OH
O
O
COOH
OH
HO
OH
$O-CH_3$
O
O
OH
O
HO
O
O
O
Common
Structure of
flavonoids
Lignins
Hydroxycinnamic
acids
Stilbenes
Coumarins

FIGURE 1.12 The systematic classifications of phenylpropanoids and its chemical.

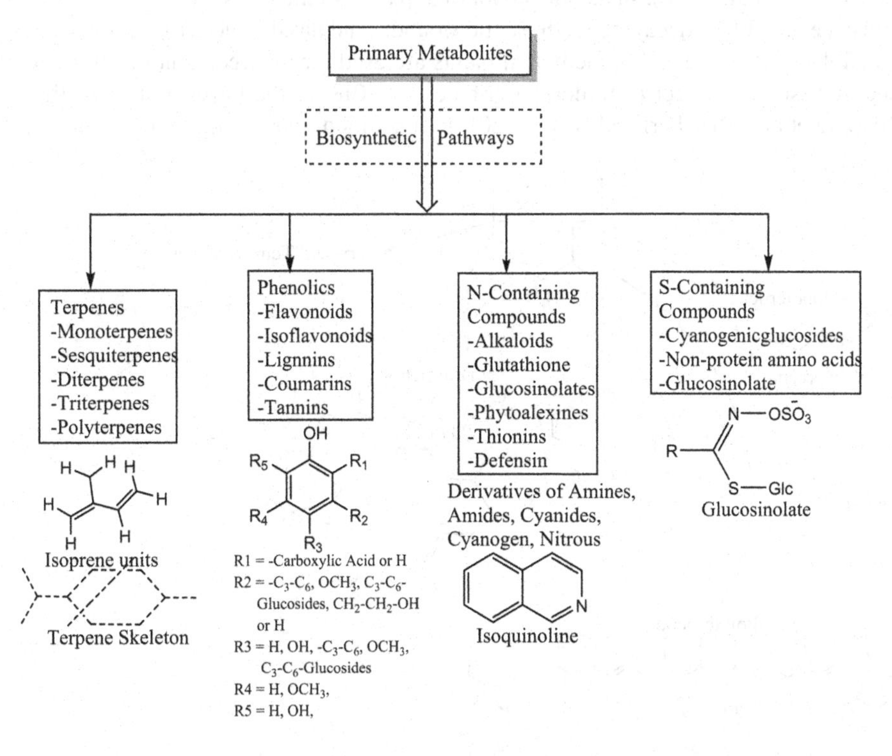

FIGURE 1.13 Basic classifications of secondary metabolites in the plant kingdom.

are the most commonly known compounds in the plant kingdom. The major nitrogen-containing compounds that are found in some plants and flowers are called alkaloids, which play a large role in modern drug synthesis. Well-known examples of alkaloids are pyrrolidine, indole, pyridine, tropine (atropine and cocaine), morphine, quinine, strychnine, ephedrine, and nicotine. Many of the nitrogen-containing compounds are important for their anticancer, antimalarial, antiviral, antifungal, and antibacterial activities (Puri et al., 2018; Gonfa et al., 2021). Sulfur-containing phytochemicals include phytoanticipins, allicin (diallylthiosulfate), petivericin, glucosinolates, isothiocyanates, thiophenes, polysulfanes, and phytoalexins. Secondary metabolites of sulfur-containing plants have active defense mechanisms against various pathogens and pests for plant survival and disease resistance. These secondary metabolites can improve human immune systems and be effective against pathogens, several infectious diseases, autoimmune diseases, allergies, obesity, cancer, cardiovascular disorders, metabolic syndrome, and gastric ulcers (Nwachukwu et al., 2012; Künstler et al., 2020; Mi ekus et al., 2020). For each bioactive class of compounds, the typical examples or the basic structural units are given in Figure 1.13.

Terpenes are one of the largest classes of compounds (representing approximately one-third of natural compounds) next to phenolic compounds and also known as terpenoids or isoprenoids. The basic structure of terpenes is called an "isoprene unit" (Figure 1.14) and is derived from five carbon atoms (C5). The largest derivative structures of terpenes are terpenoids, carotenoids, and steroids (Christianson, 2017). The steroids are a class of compounds derived from tetracyclic triterpenes and phenanthrene. The other derived compounds from terpenes are carotenoids, red or yellow pigments found in many plants, like carrot, tomato, flowers, etc. These bioactive compounds are generally insoluble in water and are toxic to many plants, insects, and animals. As a result, these compounds appear to have a crucial defensive role in the plant kingdom. The primary classifications of terpenes are shown in Figure 1.14.

Because of the content of these secondary metabolites in medicinal herbs, shrubs, and trees, they are used to treat various ailments in the communities. These medicinal plant species are claimed to be the perfect medicine for the treatment of cancer (Rimada et al., 2009; Ayele, 2018; Obakiro et al., 2020; Tesfaye et al., 2020), snakebite, abdominal pain, asthma (Al-snafi, 2021), microbial issues (Kłodzińska et al., 2018), decaying teeth, septic wounds, and nasal issues (Hossain et al., 2014). As shown in Tables 1.1, 1.2, and 1.3, medicinal plants are used for the treatment of different ailments from the highest risk (cancer or tumor) (Tuasha et al., 2018) to the lowest risk (like the common cold) (Pereira et al., 2018; Fernandes et al., 2019). In addition, most of these medicinal plants have

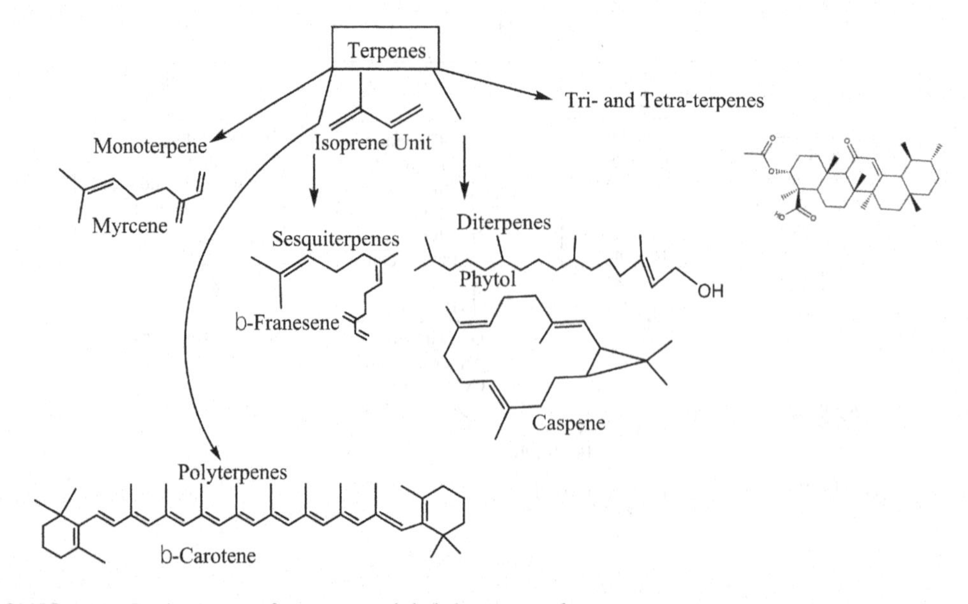

FIGURE 1.14 Basic classes of terpenes and their best examples.

been tested against different disease-causing biological agents (bacterial, fungal, viral, and parasitic) and showed significant effects on reducing and eliminating them. Figure 1.1 shows the types and distributions of secondary metabolites in the plant kingdom.

Synergetic effect (crude extracts – the sum effects of all bioactive components in the plant) and the nature of chemical structures of secondary metabolites affect that drug's ability to eliminate or reduce the target biological agents. Because of those bioactive components available in the parts (leaves, flowers, roots, stems, and aerial parts) of medicinal plants (herbs, shrubs, and trees), the study of these bioactive compounds has a vital role in developing newly formulated drugs for untreated diseases, such as cancer. Some of the derivatives of these bioactive compounds have been demonstrated in clinical trials to be more effective than the other already existing drugs. Traditional herbal medicines have been more effective than modern state-of-the-art clinical drug/pharmaceutical treatments. For example, metabolites that have been cytotoxic to cancer cell lines such as paclitaxel, camptothecin, monocrotaline, and colchicine have been tested and are effective from a few medicinal plants. There are also crude secondary metabolites extracted from African and Indian medicinal plants that are known to be cytotoxic to some cancer cell lines due to bioactive compounds, such as terpenoids (some essential oils), phenolics, and alkaloids. The sources from these plants also include bioactive compounds, such as phenolics/polyphenolics, alkaloids, terpenoids, vitamin A, B, C, D, etc.; and minerals, such as iron, zinc, selenium, calcium, etc.

1.5 POTENTIAL BENEFITS, APPLICATIONS, AND USES OF HERBS, SHRUBS, AND TREES

1.5.1 General Overview

The use of medicinal plants for global communities is therapeutic and income-generating (Sher et al., 2014). The main potential benefits of medicinal plants are human and animal health care, particularly in developing countries. Medicinal plants are the sources of lead compounds – for instance, phenolics, alkaloids, terpenoids, etc. These herbs, shrubs, and trees are the primary sources of bioactive compounds, and therefore, they are rich sources of natural antioxidants.

Bioactive compounds are mainly found in medicinal and dietary plants. They have been reported to possess biological properties, such as antioxidant, antibacterial, antifungal, antiviral, and anti-parasitic activities. For example, numerous common pharmaceuticals, including anticancer, antiviral, and antidiabetic drugs, are derived from traditional plant-derived medicines (Mani et al., 2021). These bioactive compounds constitute a different group of secondary metabolites. These are phenolics, alkaloids, terpenoids, vitamins, and carotenoids. These classes of natural compounds are further divided into various subclasses based on their chemical structure. For instance, phenolics are subdivided into simple phenolics and polyphenols. Polyphenolics are subdivided into phenolic acids, flavonoids, and tannins. Phenolic acids have a simple structure with a single aromatic ring, and they comprise hydroxycinnamic acid and hydroxybenzoic acid derivatives. Flavonoids can be further classified as flavones, flavonols, flavanones, anthocyanidins, catechins (monomeric flavan-3-ols), isoflavonoids, and chalcones (Panche et al., 2016; Singla et al., 2019). Tannins are divided into condensed tannins, i.e., proanthocyanidins and hydrolyzable tannins (Gallo- and ellagitannins) (Singla et al., 2019). Polyphenols are present in almost all plants species.

Nowadays, research into the benefits of medicinal plants is mainly focused on treatments of previously non-cured diseases like diabetes (Bnouham et al., 2006), cancer, asthma, and parasites (helminths) (Athanasiadou et al., 2007) as shown in Tables 1.1, 1.2, and 1.3.

Most of the bioactive compounds are polar and biologically active components such as anthocyanins, caffeic acid, chlorogenic acid, coumaric acid, kaempferol, quercetin, hesperidin, catechin, ellagic acid, condensed tannins, and saponins that are responsible for the therapeutic effects. A few of them are semi- and non-polar compounds, such as monoterpenes (serrulatanes and nerol cinnamates), monoterpenes (1,8-cineole and myodesert-1-ene), sesquiterpenes, diterpenes, and triterpenes that have also shown therapeutic properties related to traditional medicine.

1.5.2 Uses and Bioactive Constituents of Herbs

Herbal medicine is one of the ancient traditional practices and the primary source of traditional medicine relative to the other sources (shrubs and trees). Herbs include crude plant material such as leaves, flowers, fruit, seed, stems, wood, bark, roots, rhizomes, or other plant parts, which may be entire, fragmented, or powdered (WHO, 2013). Table 1.1 shows traditional herbal medicines: their scientific names, families, common names, parts used, bioactive compounds, and uses.

1.5.3 Parts of the Plants, Uses, and Bioactive Constituents of Shrubs

Leaves, flowers, stems/stem barks, and roots of shrubs are used by traditional healers to heal illnesses. Shrubs are also one of the primary sources of conventional medicines. According to ethnobotanical, pharmacological, and chemical studies, several bioactive compounds have been isolated from shrubs, such as phenolics, alkaloids, and essential oils/terpenes. A few types of shrubs and their traditional uses, parts of the plants, and bioactive constituents are shown in Table 1.2.

1.5.4 Uses and Bioactive Constituents of Trees

Trees are also a common source of traditional medicine and food. Several species are given in Table 1.3 for the treatment of ailments, such as cancer, fever, wound, malaria, etc. (Table 1.3). For instance, the plants *S. guineense* (Oladosu et al., 2017; Aung et al., 2020; Tesfaye et al., 2020) and *P. Africana* (Hook.f.) *Kalkman* (James et al., 2008; Komakech et al., 2017; Uchôa et al., 2016) species are traditionally used to treat cancer (skin and breast cancers). Similarly, several other medicinal plant species are used for the treatments of different illnesses and ailments such as *A. annua L.* for the treatment of fever and malaria in which the major secondary metabolites isolated from this species are artemisinin, artemisinic (arteanuic acid), arteannuin B and dihydroarteannuin. Of these chemical substances, artemisinin is a biologically active compound that is used clinically for the treatment of malaria.

1.6 CONCLUSION

This chapter consists of the traditional understanding of medicinal plants and the various applications of these plants, the parts of the plant, the components of bioactive compounds, and which types of diseases can be cured. Based on cultures and geographic regions, several types of medications have been noticed and examined. Thus, herbs, shrubs, and trees are essential parts of culture and geographical environment. Those who practice these traditions have their own unique way of understanding and treating illness. Accordingly, this chapter has described traditional knowledge of some important herb, shrub, and tree species, including chemical derivatives and applications/benefits. The chapter also provides a platform for understanding their phytochemical derivatives and pharmaceutical intervention. Overall, most herbal preparations are considered safe and have been successfully used for thousands of years in the treatment of diseases. Thus, herb, shrub, and tree-based traditional remedies and naturally derived phytochemicals are significant sources for latent and novel drug preparations against several diseases. However, before new herbal-based drug formulations can be commercialized on a large scale, extensive investigations are required on those medicinal plants for higher production of major bioactive compounds and effective treatments.

REFERENCES

Abat, J.K., S. Kumar, and A. Mohanty. 2017. Ethnomedicinal, phytochemical and ethnopharmacological aspects of four medicinal plants of Malvaceae used in Indian traditional medicines: A Review. *Medicines*, 4(4), p.75. https://doi.org/10.3390/medicines4040075

Abdisa, T. 2019. Medicinal value of croton *macrostachyus* and *Solanum incanum* against causative agent of foodborne diseases. *Veterinary Medicine: Open Journal*, 4(2), pp.57–68. https://doi.org/10.17140/vmoj-4-137

Abebe, W. 2016. An overview of Ethiopian traditional medicinal plants used for cancer treatment. *European Journal of Medicinal Plants*, 14(4), pp.1–16. https://doi.org/10.9734/EJMP/2016/25670

Al-Snafi, A.E. 2017. A review on Dodonaea viscosa: A potential medicinal plant. *IOSR Journal of Pharmacy (IOSRPHR)*, 7(2), pp.10–21.

Al-snafi, A.E. 2021. Medicinal plants alkaloids, as a promising therapeutics: A Review (Part 1). *IOSR Journal Of Pharmacy*, 11(2), pp.51–67.

Al Sulaibi, M.A.M., C. Thiemann, and T. Thiemann. 2020. Chemical constituents and uses of *calotropis procera* and *calotropis gigantea*: A Review (Part I: The Plants as Material and Energy Resources). *Open Chemistry Journal*, 7(1), pp.1–15. https://doi.org/10.2174/1874842202007010001

Alassali, A., and I. Cybulska. 2015. Methods for upstream extraction and chemical characterization of secondary metabolites from algae biomass. *Advanced Techniques in Biology and Medicine*, 4(1), pp.1–16. https://doi.org/10.4172/2379-1764.1000163

Alemu, B.K., D. Misganaw and G. Mengistu. 2020. Wound healing effect of *Acokanthera schimperi* schweinf (Apocynaceae) methanol leaf extract ointment in mice and its in-vitro antioxidant activity. *Clinical Pharmacology: Advances and Applications*, 12, pp.213–222. https://doi.org/10.2147/CPAA.S288394

Alhassan, A., M. Sule, M. Atiku, A. Wudil, M. Dangambo, J. Mashi, M. Zaitun, and G. Uba. 2012. Effect of Calitropis Procera aqueous root extract against CCL_4 induced liver toxicity in rabbits. *Bayero Journal of Pure and Applied Sciences*, 5(1), pp.38–43. https://doi.org/10.4314/bajopas.v5i1.8

Assefa, B., G. Glatzel, and C. Buchmann. 2010. Ethnomedicinal uses of *Hagenia abyssinica* (Bruce) J.F. Gmel. among rural communities of Ethiopia. *Journal of Ethnobiology and Ethnomedicine*, 6(20), pp.1–10. https://doi.org/10.1186/1746-4269-6-20

Athanasiadou, S., J. Githiori, and I. Kyriazakis. 2007. Medicinal plants for helminth parasite control: Facts and fiction. *Animal*, 1(9), pp.1392–1400. https://doi.org/10.1017/S1751731107000730

Aung, E.E., A.N. Kristanti, N.S. Aminah, Y. Takaya, and R. Ramadhan. 2020. Plant description, phytochemical constituents and bioactivities of *Syzygium genus*: A review. *Open Chemistry*, 18(1), pp.1256–1281. https://doi.org/10.1515/chem-2020-0175

Ayele, T.T. 2018. A Review on traditionally used medicinal plants/herbs for cancer therapy in Ethiopia: Current status, challenge and future perspectives. *Organic Chemistry: Current Research*, 7(2), pp.1–8. https://doi.org/10.4172/2161-0401.1000192

Badou, R.B., H. Yedomonhan, E.B.K. Ewedje, G.H. Dassou, A. Adomou, M. Tossou, and A. Akoegninou. 2020. Floral morphology and pollination system of *Syzygium guineense* (Willd.) DC. subsp. macrocarpum (Engl.) F. White (Myrtaceae), a subspecies with high nectar production. *South African Journal of Botany*, 131, pp.462–467. https://doi.org/10.1016/j.sajb.2020.04.013

Beshah, F., Y. Hunde, M. Getachew, R.K. Bachheti, A. Husen, and A. Bachheti. 2020. Ethnopharmacological, phytochemistry and other potential applications of *Dodonaea genus*: A comprehensive review. *Current Research in Biotechnology*, 2, pp.103–119. https://doi.org/10.1016/j.crbiot.2020.09.002

Bhosale, U., P. Pophale, R. Somani, and R. Yegnanarayan. 2012. Effect of aqueous extracts of *Achyranthes Aspera* Linn. on experimental animal model for inflammation. *Ancient Science of Life*, 31(4), p.202. https://doi.org/10.4103/0257-7941.107362

Bitew, H., H. Gebregergs, K.B. Tuem, and M.Y. Yeshak. 2019. Ethiopian medicinal plants traditionally used for wound treatment: A systematic review. *Ethiopian Journal of Health Development*, 33(2), pp.102–127.

Bnouham, M., A. Ziyyat, H. Mekhfi, A. Tahri, and A. Legssyer. 2006. Medicinal plants with potential antidiabetic activity: A review of ten years of herbal medicine research (1990–2000). *International Journal of Diabetes and Metabolism*, 14(1), pp.1–25. https://doi.org/10.1159/000497588

Chekole, G. 2017. Ethnobotanical study of medicinal plants used against human ailments in Gubalafto District, Northern Ethiopia. *Journal of Ethnobiology and Ethnomedicine*, 13(1), pp.1–29. https://doi.org/10.1186/s13002-017-0182-7

Christianson, D.W. 2017. Structural and chemical biology of terpenoid cyclases. *Chemical Reviews*, 117(17), pp.11570–11648. https://doi.org/10.1021/acs.chemrev.7b00287

de Lange, P.J., R.P. Scofield, and T. Greene. 2004. *Achyranthes aspera* (Amaranthaceae) a new indigenous addition to the flora of the Kermadec Islands group. *New Zealand Journal of Botany*, 42(2), pp.167–173. https://doi.org/10.1080/0028825X.2004.9512897

Deng, Y., and S. Lu. 2017. Biosynthesis and regulation of phenylpropanoids in plants. *Critical Reviews in Plant Sciences*, 2689(4), pp.257–290. https://doi.org/10.1080/07352689.2017.1402852

Dhananjaya, B.L., S. Sudarshan, Y. Dongol, and S.S. More. 2016. The standard aqueous stem bark extract of *Mangifera indica* L. inhibits toxic PLA2- NN-XIb-PLA2of Indian cobra venom. *Saudi Pharmaceutical Journal*, 24(3), pp.371–378. https://doi.org/10.1016/j.jsps.2016.04.026

Fernandes, J.M., L.M. Cunha, E.P. Azevedo, E.M.G. Lourenço, M.F. Fernandes-Pedrosa, and S.M. Zucolotto. 2019. *Kalanchoe laciniata* and *Bryophyllum pinnatum*: An updated review about ethnopharmacology, phytochemistry, pharmacology and toxicology. *Revista Brasileira de Farmacognosia*, 29(4), pp.529–558. https://doi.org/10.1016/j.bjp.2019.01.012

Frimpong, E.K., J.A. Asong, and A.O. Aremu, (2021). A Review on medicinal plants used in the management of headache in Africa. *Plants*, 10, pp.1–21. https://doi.org/10.3390/plants10102038

Gonfa, Y.H., F. Beshah, M.G. Tadesse, A. Bachheti, and R.K. Bachheti. 2021. Phytochemical investigation and potential pharmacologically active compounds of *Rumex nepalensis*: An appraisal. *Beni-Suef University Journal of Basic and Applied Sciences* 10(1), pp.1–11. https://doi.org/10.1186/s43088-021-00110-1

Groom, Q.J., J. Van Der Straeten, and I. Hoste (2019). The origin of *Oxalis corniculata* L. *Peer J*, 7(2), p.e6384. https://doi.org/10.7717/peerj.6384

Gupta, M., A. Gupta, and S. Gupta. 2014. Characterization of secondary metabolites via LC-MS analysis of DCM extracts of Solanum nigrum. *Biosciences Biotechnology Research Asia*, 11(2), pp.531–535. https://doi.org/10.13005/bbra/1303

Hifnawy, M.S., A.M.K. Mahrous, and R.M.S. Ashour. 2016. Phytochemical investigation of *Phoenix canariensis* Hort. ex Chabaud leaves and pollen grains. *Journal of Applied Pharmaceutical Science*, 6(12), pp.103–109. https://doi.org/10.7324/JAPS.2016.601214

Hossain, M.A., T.H.A. Alabri, A.H.S. Al Musalami, M.S. Akhtar, and S. Said. 2014. Evaluation of in vitro antioxidant potential of different polarities stem crude extracts by different extraction methods of *Adenium obesum*. *Journal of Coastal Life Medicine*, 2(9), pp.699–703. https://doi.org/10.12980/jclm.2.201414d102

James, D.B., A.O. Owolabi, H. Ibiyeye, J. Magaji, and Y.A. Ikugiyi. 2008. Assessment of the hepatic effects, heamatological effect and some phytochemical constituents of *Ximenia americana* (Leaves, stem and root) extracts. *African Journal of Biotechnology*, 7(23), pp.4274–4278. https://doi.org/10.4314/ajb.v7i23.59563

Janardhan, B., V.M. Shrikanth, K.K. Mirajkar, and S.S. More. 2014. In vitro screening and evaluation of antivenom phytochemicals from *Azima tetracantha* Lam. leaves against *Bungarus caeruleus* and *Vipera russelli*. *Journal of Venomous Animals and Toxins Including Tropical Diseases*, 20(1), pp.1–8. https://doi.org/10.1186/1678-9199-20-12

Joel, E.L., and V. Bhimba. 2010. Isolation and characterization of secondary metabolites from the mangrove plant *Rhizophora mucronata*. *Asian Pacific Journal of Tropical Medicine*, 3(8), pp.602–604. https://doi.org/10.1016/S1995-7645(10)60146-0

Kassaye, K., A. Amberbir, B. Getachew, and Y. Museum. 2007. A historical overview of traditional medicine practices and policy in Ethiopia. *Ethiopian Journal of Health Development*, 20(2), pp.127–134. https://doi.org/10.4314/ejhd.v20i2.10023

Kumari, P., and B. B. Ujala. 2021. Phytochemicals from edible flowers: Opening a new arena for healthy lifestyle. *Journal of Functional Foods*, 78, pp.1–18. https://doi.org/10.1016/j.jff.2021.104375

Kłodzińska, S.N., P.A. Priemel, T. Rades, and H.M. Nielsen. 2018. Combining diagnostic methods for antimicrobial susceptibility testing: A comparative approach. *Journal of Microbiological Methods*, 144, pp.177–185. https://doi.org/10.1016/j.mimet.2017.11.010

Komakech, R., Y. Kang, J.H. Lee, and F. Omujal. 2017. A review of the potential of phytochemicals from *Prunus africana* (Hook f.) Kalkman stem bark for chemoprevention and chemotherapy of prostate cancer. *Evidence-Based Complementary and Alternative Medicine*, 2017, pp.1–10. https://doi.org/10.1155/2017/3014019

Kumar, K.S., V. Sabu, G. Sindhu, A.A. Rauf, and A. Helen. 2018. Isolation, identification and characterization of apigenin from *Justicia gendarussa* and its anti-inflammatory activity. *International Immunopharmacology*, 59, pp.157–167. https://doi.org/10.1016/j.intimp.2018.04.004

Künstler, A., G. Gullner, A.L. Ádám, J.K. Nagy, and L. Király. 2020. The versatile roles of sulfur-containing biomolecules in plant defense: A road to disease resistance. *Plants*, 9(12), pp.1–31. https://doi.org/10.3390/plants9121705

Liu, Jun, R., Liu, X. Yunpeng, and C.K. Juan. 2018. Isolation, structural characterization and bioactivities of naturally occurring polysaccharide–polyphenolic conjugates from medicinal plants: A review. *International Journal of Biological Macromolecules*, 107, pp.2242–2250. https://doi.org/10.1016/j.ijbiomac.2017.10.097

Mainasara, M.M., B.L. Aliero, A.A. Aliero, and S.S. Dahiru. 2012. Phytochemical and antibacterial properties of *Calotropis procera* (Ait) R. Br. (Sodom Apple) Fruit and Bark Extracts. *International Journal of Modern Botany*, 1(1), pp.8–11. https://doi.org/10.5923/j.ijmb.20110101.03

Mani, J.S., J.B. Johnson, H. Hosking, N. Ashwath, K.B. Walsh, P.M. Neilsen, D.A. Broszczak, and M. Naiker. 2021. Antioxidative and therapeutic potential of selected Australian plants: A review. *Journal of Ethnopharmacology*, 268, p.113580. https://doi.org/10.1016/j.jep.2020.113580

Mayori, A. 2017. Ethnopharmacological uses, phytochemistry, and pharmacological properties of *Croton macrostachyus* Hochst. Ex Delile: A comprehensive review. *Evidence-Based Complementary and Alternative Medicine*, 2017, pp.1–17. https://doi.org/10.1155/2017/1694671

Mekonnen, B., A.B. Asrie, and Z.B. Wubneh. 2018. Antidiarrheal activity of 80% methanolic leaf extract of *Justicia schimperiana. Evidence-Based Complementary and Alternative Medicine*, 2018, pp.1–10. https://doi.org/10.1155/2018/3037120

Mi ekus, N., K. Marszałek, M. Podlacha, A. Iqbal, C. Puchalski, and A.H. Swiergiel. 2020. Health benefits of plant-derived sulfur compounds, glucosinolates, and organosulfur compounds. *Molecules*, 25(17), pp.1–22. https://doi.org/10.3390/molecules25173804

Mishra, R.M., and P. Gupta. 2005. Frugivory and seed dispersal of *Carissa spinarum* (L.) in a tropical deciduous forest of central India. *Tropical Ecology*, 46(2), pp.151–156.

Modak, P., S. Halder, S.K. Bidduth, A. Das, A.P. Sarkar, and K. Kundu Sukalyan ,. 2020. Traditional antihypertensive medicinal plants: A comprehensive review. *WORLD J. Pharm. Pharm. Sci.*, 9(9), pp.884–912. https://doi.org/10.20959/wjpps20209-17056

Moremi, M.P., F. Makolo, A.M. Viljoen, and G.P. Kamatou. 2021. A review of biological activities and phytochemistry of six ethnomedicinally important South African Croton species. *Journal of Ethnopharmacology*, 280(2), pp.28–36. https://doi.org/10.1016/j.jep.2021.114416

Nguyen, T.L., A. Rusten, M.S. Bugge, K.E. Malterud, D. Diallo, B.S. Paulsen, and H. Wangensteen. 2016. Flavonoids, gallotannins and ellagitannins in *Syzygium guineense* and the traditional use among Malian healers. *Journal of Ethnopharmacology*, 192, pp.450–458. https://doi.org/10.1016/j.jep.2016.09.035

Noronha, M., V. Pawar, A. Prajapati, and R.B. Subramanian. 2020. A literature review on traditional herbal medicines for malaria. *South African Journal of Botany*, 128, pp.292–303. https://doi.org/10.1016/j.sajb.2019.11.017

Nwachukwu, I.D., A.J. Slusarenko, and M.C.H. Gruhlke. 2012. Sulfur and sulfur compounds in plant defence. *Natural Product Communications*, 7(3), pp.395–400. https://doi.org/10.1177/1934578x1200700323

Obakiro, S.B., A. Kiprop, I. Kowino, E. Kigondu, M.P. Odero, T. Omara, and L. Bunalema. 2020. Phytochemistry of traditional medicinal plants used in the management of symptoms of tuberculosis in East Africa: A systematic review. *Tropical medicine and health*, 48(1), pp.1–21.

Oladosu, I.A., L. Lawson, O.O. Aiyelaagbe, N. Emenyonu, and O.E. Afieroho. 2017. Anti-tuberculosis lupine-type isoprenoids from *Syzygium guineense* Wild DC. (Myrtaceae) stem bark. *Future Journal of Pharmaceutical Sciences*, 3(2), pp.148–152. https://doi.org/10.1016/j.fjps.2017.05.002

Panche, A.N., A.D. Diwan, and S.R. Chandra. 2016. Flavonoids: An overview. *Journal of Nutritional Science*, 5, pp.1–15. https://doi.org/10.1017/jns.2016.41

Pereira, K.M.F., S.S. Grecco, C.R. Figueiredo, J.K. Hosomi, M.U. Nakamura, and J.G.L. Henrique. 2018. Chemical composition and cytotoxicity of *Kalanchoe pinnata* leaves extracts prepared using Accelerated System Extraction (ASE). *Natural Product Communications*, 13(2), pp.163–166. https://doi.org/10.1177/1934578x1801300213

Priya, A. 2021. Species distribution models for two subspecies of *Dodonaea viscosa* (Sapindaceae) in Indonesia. *IOP Conference Series: Earth and Environmental Science*, 743(1), pp.1–8. https://doi.org/10.1088/1755-1315/743/1/012027

Puri, S.K., P.V. Habbu, P.V. Kulkarni, and V.H. Kulkarni. 2018. Nitrogen-containing secondary metabolites from endophytes of medicinal plants and their biological/pharmacological activities-a review. *Systematic Reviews in Pharmacy*, 9(1), pp.22–30. https://doi.org/10.5530/srp.2018.1.5

Rimada, R.S., W.O. Gatti, R. Jeandupeux, and L.F.R. Cafferata. 2009. Isolation, characterization and quantification of artemisinin by NMR from Argentinean *Artemisia annua* L. *Boletin Latinoamericano y Del Caribe de Plantas Medicinales y Aromaticas*, 8(4), pp.275–281.

Runyoro, D., O. Ngassapa, K. Vagionas, N. Aligiannis, K. Graikou, and I. Chinou. 2010. Chemical composition and antimicrobial activity of the essential oils of four Ocimum species growing in Tanzania. *Food Chemistry*, 119(1), pp.311–316. https://doi.org/10.1016/j.foodchem.2009.06.028

Salim, M.A., S. Ranjitkar, R. Hart, T. Khan, S. Ali, C. Kiran, A. Parveen, Z. Batool, S. Bano, and J. Xu. 2019. Regional trade of medicinal plants has facilitated the retention of traditional knowledge: Case study in Gilgit-Baltistan Pakistan. *Journal of Ethnobiology and Ethnomedicine*, 15(1), pp.1–33. https://doi.org/10.1186/s13002-018-0281-0

Samoylenko, V., M.R. Jacob, S.I. Khan, J. Zhao, B.L. Tekwani, J.O. Midiwo, L.A. Walker, and I. Muhammad. 2009. Antimicrobial, antiparasitic and cytotoxic spermine alkaloids from Albizia schimperiana. *Natural Product Communications*, 4(6), pp.791–796. https://doi.org/10.1177/1934578x0900400611

Shaikh, S., V. Shriram, A. Srivastav, P. Barve, and V. Kumar. 2018. A critical review on Nepal Dock (*Rumex nepalensis*): A tropical herb with immense medicinal importance. *Asian Pacific Journal of Tropical Medicine*, 11(7), pp.405–414. https://doi.org/10.4103/1995-7645.237184

Sharma, A., S. Kumar, and P. Tripathi. 2015. Impact of *achyranthes aspera* leaf and stem extracts on the survival, morphology and behaviour of an Indian strain of dengue vector, *Aedes aegypti* L. (Diptera: Culicidae). *Journal of Mosquito Research*, 5(7), pp.1–9. https://doi.org/10.5376/jmr.2015.05.0007

Sher, H., A. Aldosari, A. Ali, and H.J. de Boer. 2014. Economic benefits of high value medicinal plants to Pakistani communities: An analysis of current practice and potential. *Journal of Ethnobiology and Ethnomedicine*, 10(1), pp.1–16. https://doi.org/10.1186/1746-4269-10-71

Sinan, K.I., G. Zengin, D. Zheleva-Dimitrova, O.K. Etienne, M.F. Mahomoodally, A. Bouyahya, D. Lobine, A. Chiavaroli, C. Ferrante, L. Menghini, L. Recinella, L. Brunetti, S. Leone, and G. Orlando. 2020. Qualitative phytochemical fingerprint and network pharmacology investigation of *achyranthes aspera* Linn. extracts. *Molecules*, 25(8), pp.1–19. https://doi.org/10.3390/molecules25081973

Singla, R.K.2019. Natural polyphenols: Chemical Classification, definition of classes, subcategories, and structures. *Journal of AOAC International*, 102(5), pp.1397–1400. https://doi.org/10.5740/jaoacint.19-0133

Tesfaye, S., A. Belete, E. Engidawork, T. Gedif, and K. Asres. 2020. Ethnobotanical study of medicinal plants used by traditional healers to treat cancer-like symptoms in eleven districts, Ethiopia. *Evidence-Based Complementary and Alternative Medicine*, 2020, pp.1–23. https://doi.org/10.1155/2020/7683450

Tibuhwa, D.D. 2016. *Oxalis corniculata* L. in Tanzania: Traditional use, cytotoxicity and antimicrobial activities. *Journal of Applied Biosciences*, 105(1), pp.10055–10063. https://doi.org/10.4314/jab.v105i1.2

Tuasha, N., B. Petros, and Z. Asfaw. 2018. Plants used as anticancer agents in the Ethiopian traditional medical practices: A systematic review. *Evidence-Based Complementary and Alternative Medicine*, 2018, pp.1–28. https://doi.org/10.1155/2018/6274021

Uchôa, V.T., C.M.M. Sousa, A.A. Carvalho, A.E.G. Sant'Ana, and M.H. Chaves. 2016. Free radical scavenging ability of *Ximenia americana* L. stem bark and leaf extracts. *Journal of Applied Pharmaceutical Science*, 6(2), pp.91–96. https://doi.org/10.7324/JAPS.2016.60213

World Health Organisation (WHO). 2013. *WHO Traditional Medicine Strategy 2014–2023*. Geneva, Switzerland: World Health Organization (WHO), 1–76. https://doi.org/2013

Zewdie, K.A., D. Bhoumik, D.Z. Wondafrash, and K.B. Tuem. 2020. Evaluation of in-vivo antidiarrhoeal and in-vitro antibacterial activities of the root extract of *Brucea antidysenterica* J. F. Mill (Simaroubaceae). *BMC Complementary Medicine and Therapies*, 20(1), p.201. https://doi.org/10.1186/s12906-020-03001-7

2 *Andrographis paniculata* (Creat or Green Chiretta) and *Bacopa monnieri* (Water Hyssop)

Pankaj Mundada, Swati Gurme, Suchita Jadhav, Devashree Patil, Nitin Gore, Sumaiya Shaikh, Abhinav Mali, Suraj Umdale, Mahendra Ahire

CONTENTS

2.1 Introduction 28
2.2 *Andrographis paniculata* (Burm. f.) Nees 28
 2.2.1 Morphology 28
 2.2.2 Distribution and Common Names 28
 2.2.3 Bioactive Compounds 29
 2.2.4 Pharmacological Activities 30
  2.2.4.1 Antimicrobial Activity 31
  2.2.4.2 Anti-neurodegenerative Activities 31
  2.2.4.3 Antifertility Activities 31
  2.2.4.4 Anti-HIV and Cytotoxic Activity 31
  2.2.4.5 Clinical Trials of Therapeutic Agents 31
  2.2.4.6 Analgesic Activity 32
  2.2.4.7 Antidiabetic Activity 32
  2.2.4.8 Antioxidant Activity 32
2.3 *Bacopa monnieri* (L.) Pennell 32
 2.3.1 Morphology 33
 2.3.2 Distribution and Common Names 34
 2.3.3 Bioactive Compounds 34
 2.3.4 Pharmacological Activities 34
  2.3.4.1 Sedative and Tranquilizing Properties 36
  2.3.4.2 Antidepressant and Antianxiety Effects 36
  2.3.4.3 Antiepileptic Effects 36
  2.3.4.4 Anti-ulcerative Activity 36
  2.3.4.5 Antidiabetic Activity 37
  2.3.4.6 Anticancer Activity 37
  2.3.4.7 Antimicrobial Activity 37
  2.3.4.8 Cardiovascular Activities 37
  2.3.4.9 Analgesic Activities 38
  2.3.4.10 Cognitive Activities 38
2.4 Conclusion 39
Acknowledgments 39
References 39

DOI: 10.1201/9781003205067-2

2.1 INTRODUCTION

Ayurveda is an age-old system of medicine that originated thousands of years ago in India. The ancient Vedic knowledge is considered one of the oldest healing sciences. Ayurveda is also known as the "Mother of All Healing". Plants synthesize an array of secondary metabolites with respect to their defense mechanisms; these secondary metabolites possess certain medicinal properties. It was estimated that there were approximately 100,000 different plant-derived compounds, with many new ones being added to the list every year (Verpoorte et al., 1999).

The World Health Organization (WHO) estimates that 80% of the world's population presently used herbal medicine for some aspects of primary health care. In India, plants of therapeutic potential are widely used by all sections of people as folk medicines and in different indigenous systems of medicine, like Siddha, Ayurveda, Unani as well as processed products of the pharmaceutical industry. Of these, *Andrographis paniculata* (Burm. f.) Nees (Family: Acanthaceae) is an annual herbaceous plant traditionally used to cure bacterial infections and some diseases. *Bacopa monnieri* (L.) Pennell (Family: Scrophulariaceae) is a small, amphibious perennial creeping herb that grows in marshy habitats. The plant is one of the sources of the *medhya rasayan* drugs (that counteract stress and improve intelligence and memory) of Ayurveda. The plant is also used in several other ayurvedic medicinal preparations.

2.2 *ANDROGRAPHIS PANICULATA* (BURM. F.) NEES

Andrographis paniculata (Burm. f.) Nees is an economically important annual. This erect, branched flowering herb belongs to family Acanthaceae and is commonly known as creat or green chiretta. Due to the bitterness of the plant, it is also known as the "King of Bitters". *Andrographis paniculata* has been traditionally used in *Siddha* and *Ayurvedic* medicines in the treatment of bacterial infections (Akbar, 2011). Although the plant is also promoted as a dietary supplement for cancer prevention and cure, there is no evidence to support this claim. Known for its major biological activity as an antidiabetic, it has also been reported to have anti-angiogenic, antibacterial, anticancer, anti-inflammatory, antimalarial, antioxidant, and hepatoprotective activities.

2.2.1 Morphology

Andrographis paniculata is a profusely branched, erect, annual herb (Figure 2.1), which is extremely bitter in taste. It grows to a height of 30–110 cm in moist shady places with glabrous leaves and white flowers that have purple spots on the lower lip. The stem is dark green, 2–6 mm in diameter, quadrangular with longitudinal furrows and wings on the angles of the younger parts, slightly enlarged at the nodes. Leaves glabrous, up to 8.0 cm long and 2.6 cm broad, arranged opposite decussate, lanceolate, and pinnate. Flowers are small and solitary, corolla whitish or light pink in color with hairs, in lax spreading axillary and terminal racemes. Fruits are capsules, linear-oblong, acute at both ends, 1.9–0.3 cm. Seeds are numerous, subquadrate, and yellowish brown in color.

2.2.2 Distribution and Common Names

Andrographis paniculata is a plant native to India and Sri Lanka and is widespread in tropical Asian countries. It is widely distributed in China, Cambodia, Caribbean, Indonesia, Laos, Malaysia, Thailand, Taiwan and Vietnam (Kumar et al., 2021). Its habitat is the plains, hills, coasts, and anthropized and cultivated areas, such as roadsides, farms, and ruined territories, between the plains and hilly areas up to 500 m (Mishra et al., 2007; Niranjan et al., 2010). The plant is commonly known as "creat" or "green chiretta", but its bitterness has earned it the moniker "King of Bitters". The plant has different names in various languages: Kalmegha, Bhunimba, and Yavatikta in Sanskrit; Kalmegh, Kariyat, and Mahatita in Urdu; and Kirayat and Kalpanath in Hindi (Hossain et al., 2014).

FIGURE 2.1 Morphology of *Andrographis paniculata*: A) whole plants; B) flowering branch; and C) flower.

2.2.3 Bioactive Compounds

A. paniculata contains diterpenes, lactones, and flavonoids – which mainly exist in the root but have also been isolated from the leaves. Aerial parts contain alkanes, ketones, and aldehydes, and the bitterness in the leaves is due to the lactone andrographolide kalmegin. Four lactones – Chuaxinlian A (deoxyandrographolide), B (andrographolide), C (neoandrographolide), and D (14-deoxy-11, 12-didehydroandrographolide) – have been isolated from the aerial parts (Figure 2.2) (Hossain et al., 2014).

Andrographis paniculata was screened for various active constituents (andrographolide, flavones, lactones) using routine chemical identification methods. Among them, andrographolide is the main constituent, and it is also an active principle of the plant (Shirisha and Mastan, 2013). Active compounds extracted with ethanol or methanol from the whole plant, leaf, and stem include over 20 diterpenoids and over ten flavonoids. Andrographolide is the major diterpenoid in *A. paniculata*, making up about 4%, 0.8–1.2% and 0.5–6% in dried whole plant, stem, and leaf extracts, respectively. The other main diterpenoids are deoxyandrographolide, neoandrographolide, 14-deoxy-11,12-didehydroandrographide, and isoandrographolide (Chao and Lin, 2010). Some known constituents are: 14-Deoxy-11-dehydroandrographolide, 14-Deoxy-11-oxoandrographolide, 5-Hydroxy-7,8,2',3'-tetramethoxyflavone, 5-Hydroxy-7,8,2'-trimethoxyflavone, andrographan,

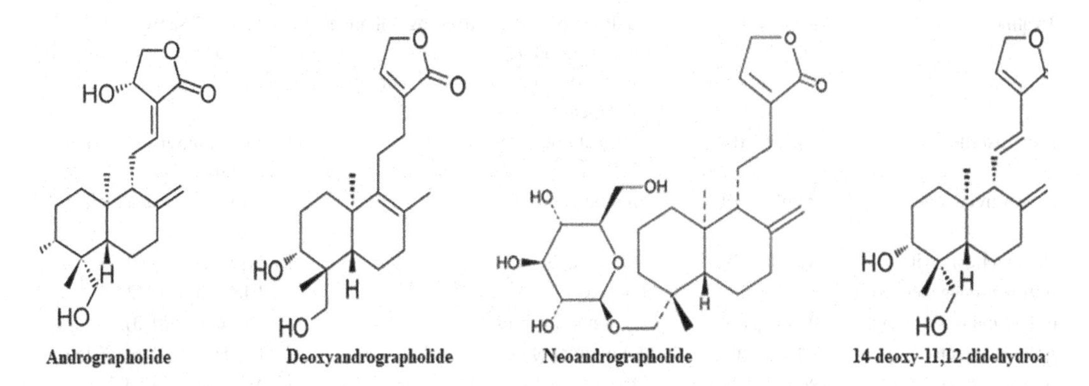

FIGURE 2.2 Major phytoconstituents in *Andrographis paniculata*.

andrographon, andrographine, andrographiside, andrographosterin, deoxyandrographiside, homoandrographolide, panicoline, paniculide-A, paniculide-B, paniculide-C, and stigmasterol (Jalal et al., 1979; Zhang et al., 2006; Xu et al., 2012; Hossain et al., 2014).

2.2.4 Pharmacological Activities

Andrographis paniculata has a wide range of medicinal and pharmacological applications (Table 2.1). It has been used in different traditional systems of medicine and exhibits many activities. The important phytochemicals in *A. paniculata* make the plant useful for treating different ailments and suggest its potential for use in new drugs. The quantitative determination of pharmacognostic parameters will help to set the standards for crude drugs (Joselin and Jeeva, 2014).

The plant is also reported to possess antihepatotoxic, antimalarial, antithrombogenic, anti-inflammatory, antidote, antipyretic, and immunostimulant properties (Perumal Samy et al., 2007; Verma et al., 2013). Current evidence suggests that *A. paniculata* extract alone or in combination with *A senticosus* extract may be more effective than placebo and may be an appropriate alternative treatment of uncomplicated acute upper respiratory tract infection (Poolsup et al., 2004). A primary modern use of *A. paniculata* is for the prevention and treatment of the common cold (Cáceres et al., 1997; Jarukamjorn and Nemoto, 2008). It appears to have antithrombotic actions, suggesting a possible benefit in cardiovascular disease (Zhao and Fang, 1991).

TABLE 2.1
Summary of Some Pharmacological Activities along with Phytochemical Constituents of *Andrographis paniculata*

Pharmacological Activities	Plant Part Used	Bioactive Compounds	References
Liver disorders	Whole plant	Diterpenoid glycosides	Mishra et al. (2007)
Antithrombotic, hyphoglycemic, and antipyretic	Leaf	Andrographolide	Calabrese et al., (2000)
Antibacterial, antifungal	Whole plant	14-deoxyandrographolide	Hossain et al. (2021)
Anti-HIV	Leaves, Root	Andrographolide	Hossain et al. (2021)
Antibacterial	Leaves	Andrographolide	Perumal Samy et al. (2007)
Anti-ulcerogenic activity	Whole plant	Andrographolide, asprin	Perumal Samy et al. (2007)
Treatment for HIV	Aerial parts	Andrographolide, neoandrographolide and 14-deoxy-11,12-didehydroandrographolide, ent-labdene diterpenes	Perumal Samy et al. (2007)
Liver disorder	Aerial parts	Hexachlorocyclohexane	Jarukamjorn and Nemoto, (2008)
Cancer treatment	Whole plant	Andrographolide	Jarukamjorn and Nemoto, (2008)
Antipyretic activity	Aerial parts	Andrographolide	Nyeem et al. (2017)
Anti-neurodegenerative	Aerial parts	Flavonoids	Ishola et al. (2021)
(CTX)-induced toxicity	Whole plant	Cyclophosphamide	Dey et al. (2013)
Anticancer activity	Whole plant	Andrographolide	Pandey and Rao, (2018)
Anticancer activity	Whole plant	Terpenes	Mishra et al. (2007)

2.2.4.1 Antimicrobial Activity

Perumal Samy et al., (2007) reviewed the antimicrobial activity of *A. paniculata* plant extracts, such as antityphoid activity against *Salmonella typhi*; *Micrococcus pyogenes* var. *aureus*, and *Escherichia coli* and antifungal activity against *Helminthosoprium sativum*. The leaf extract of *A. paniculata* showed a broad range of antimicrobial activity against *Staphylococcus aureus* and *Streptococcus pyogenes* (Suparna et al., 2014). Gurupriya and Cathrine (2016) reported antibacterial activity against *Staphylococcus aureus*, *Bacillus subtilis*, *Streptococcus pyogenes*, *Escherichia coli*, and *Pseudomonas aeruginosa* and antifungal activity against *Candida albicans*, *Candida tropicalis*, *Aspergillus flavus*, *Aspergillus niger*, and *Aspergillus fumigatus* of chloroform extract of the *A. paniculata* stem. Leaf methanolic extract of Andrographis paniculata exhibits greater antibacterial activity against *M. tuberculosis*, *E. faecalis*, and *S. aureus* than other extracts (Mishra et al., 2013). Leaf methanolic extracts were also found to be effective against certain fungal species such as *Alternaria solani*, *A. alternata*, *A. brassicae*, *A. tenuissima*, *A. brassicicola*, *Cochliobolus blumae*, *C. lunata*, *C. penniseti*, *Fusarium albizziae*, and *F. udum* (Kumar et al., 2021). Several workers reported antimalarial and filaricidal activity of plant extracts of *A. paniculata* (Dutta and Sukul, 1982; Misra et al., 1992; Najib et al., 1999; Siti Najila et al., 2002; Dua et al., 2004).

2.2.4.2 Anti-neurodegenerative Activities

The Andrographolide present in *A. paniculata* helped to alleviate acute brain injury in a rat model (Tao et al., 2018). Inhibition of cholinesterase and monoamine oxidase activities are effective strategies employed in the treatment of neurodegenerative diseases (Lu et al., 2019). All *Andrographis paniculata*-derived flavonoids used in this study are orally advocated except for rutin. Taxifolin and tangeretin showed a reasonable binding affinity for cholinesterases, while rutin was the most outstanding ligand for cholinesterases and monoamine oxidase. However, only tangeritin possesses the ability to cross the blood–brain barrier. Thus, the study emphasized flavonoids that could be examined using different experimental models to establish their anti-neurodegenerative activities (Ishola et al., 2021).

2.2.4.3 Antifertility Activities

A number of animal studies report an effect of *A. paniculata* on male and female reproduction. Early reports of oral administration of powdered stem indicated an antifertility effect in male Wistar mice, but no impact on fertility in the female mice. It has also been reported that administration of *A. paniculata* resulted in abortion in pregnant rabbits. The anti-fertility effects and hormonal assay in female rats were studied by Sakila et al. (2017). The intraperitoneal injection of the decoction of aerial parts to female albino mice was reported to prevent implantation and caused abortion at different gestation periods (Nyeem et al., 2017).

2.2.4.4 Anti-HIV and Cytotoxic Activity

Reddy et al. (2005) isolated six known compounds from the aerial parts of *A. paniculata*, namely andrographolide, 14-deoxy-11,12-didehydroandrographolide, andrograpanin, 14- deoxyandrographolide, (+/-)-5-hydroxy-7,8-dimethoxyflavanone, and 5-hydroxy- 7,8-dimethoxyflavone, as well as one novel bis-andrographolide and found positive results for anti-HIV and cytotoxic activity. Pharmacological and clinical studies of *A. paniculata* suggest the potential for beneficial effects in diseases, like cancer and HIV infections (Akbar, 2011; Dey et al., 2013). The *A. paniculata* showed anti-HIV activity *in vitro* in a cell-free virus infectivity assay using TZM-bl cells (Uttekar et al., 2012).

2.2.4.5 Clinical Trials of Therapeutic Agents

Clinical trials on andrographolide extract exhibit minor incidences of restenosis, which found that andrographolide extract showed beneficial effects in the prevention of angioplasty (Wang and Zhao,

1993). Standardized *A. paniculata* extract significantly reduced common cold symptoms found in a pilot clinical study (Melchior et al., 1996; Cáceres et al., 1999; Ciampi et al., 2020). The higher dosage of plant Andrographis (6 g) is found to be more effective in decreasing the symptoms of fever and throat pain than acetaminophen. The studies on effect of plant extract for prevention of cardiac and cerebral vascular diseases on animal models suggest that the plant extract significantly prevents constriction of blood vessels, inhibits the clumping of platelets, and induces antihypertensive effects (Talukdar and Banerjee, 1968; Hueng, 1991; Matsuda et al., 1994). The studies on animal models showed that *A. paniculata* extract was more effective than barbital (an anesthetic agent): the animals were more quickly sedated, and the anesthesia lasted longer (Burgos et al., 2001). A study on rabbits showed that *A. paniculata* is found to be safer for the treatment of upper respiratory tract infection. Further, the andrographolide has been used to treat tonsillitis, respiratory infections, and tuberculosis (Guo et al., 1988).

2.2.4.6 Analgesic Activity

The study on mice by Lin et al. (2009) showed that the aqueous whole plant extract of *Andrographis paniculata* exhibits improved analgesic activity over that of the ethanolic extract at a dose of 100 mg/kg. Similarly, Mandal et al. (2001) reported the methanolic extract (100 mg/kg) also showed significant analgesic activity in different experimental animal models using the acetic acid-induced writhing and tail clip method. Suebsasana et al. (2009) report that the analgesic activity of *A. paniculata* was due to the presence of diterpenoid lactones, especially compounds 21 and 28.

2.2.4.7 Antidiabetic Activity

The aqueous and ethanolic extracts of *A. paniculata* showed significant hypoglycemic activity (Hossain et al., 2007). The studies in streptozotocin-diabetic rat model by Zhang and Tan (2000) reported that the ethanolic extract of the aerial plant parts significantly decreases the fasting serum glucose level compared with the control. Similarly, decoction of aerial parts of *A. paniculata* and the aqueous extract has shown significant antidiabetic effect on diabetic rats (Reyes et al., 2005; Akhtar et al., 2016) and proved the antidiabetic potential of *A. paniculata*. This antidiabetic potential was found to be due to the presence of diterpenoid lactones or may be a synergistic effect of flavonoids (Zhang et al., 2009).

2.2.4.8 Antioxidant Activity

To date several researchers have attempted to test the antioxidant activity of extracts obtained from different plant parts of *Andrographis paniculata* using different solvent systems and assays, including 2, 2-diphenyl-1-picryl-hydrazyl-hydrate (DPPH), lipid peroxidation inhibition, free radical activities, nitric oxide, ferric reducing/antioxidant power (FRAP), cupric ion reducing antioxidant capacity (CUPRAC), and 2, 20-azino-bis(3-ethylbenzothiazoline-6-sulfonic acid) (ABTS), and the plant exhibits significant antioxidant activity (Akowuah et al., 2008; Lin et al., 2009; Tewari et al., 2010; Vasu et al., 2010; Sharma and Joshi, 2011; Sangeetha et al., 2014; Zhao et al., 2014; Kurzawa et al., 2015; Rao and Rathod, 2015).

2.3 *BACOPA MONNIERI* (L.) PENNELL

Bacopa is a traditional medicinal plant in Ayurveda. It is called Brahmi, after Lord Brahma, the mythological creator of the world and originator of the science of Ayurveda. *Bacopa* is frequently mentioned in the religious, social, and medical treatises of India since the time of Vedic civilization. Brahmi is a useful herbal drug for neurological disorders. It is used as a nerve tonic and to improve memory in ayurvedic medicines. Brahmi is now called the "brain booster" of the new millennium, as ayurvedic medicine has known for many centuries. Brahmi has more value in treating conditions affecting the nervous system and brain. This review summarizes our current knowledge of the major bioactivities of *B. monnieri*. It is prescribed for a variety of curative indications, including epilepsy, madness, and memory enhancement (Satyavati et al., 1976). The drug is included in several Ayurvedic formulations such as Brahmi Ghritam and Saraswatharishtam (Anonymous, 1978; Sivarajan and Balachandran, 1994), and it is also effective in cases of anxiety neurosis (Singh et al., 1979).

2.3.1 Morphology

The genus *Bacopa* (Family: Scrophulariaceae) consists of 20 species living in warm parts of the world; out of these, three are found in India (Sharma, 2003). The whole plant is used as medicine. *Bacopa monnieri* is a small herb, known as "Brahmi" in Hindi, and sometimes Indian pennywort. It is a perennial, creeping herb with small leaves and white or purple flowers. Brahmi is a small, smooth, creeping fleshy plant with numerous branches. It grows to a height of 60–90 cm, and its branches are 5–35 cm long. Roots are thin, wiry, small, branched, and creamy yellow. The bloom of this plant occurs from July to December. Its stem is thin, glabrous, green or purplish green, 1–2 mm thick, soft, with prominent nodes and internodes, and the taste is slightly bitter. Leaves are sessile, simple, opposite, decussate, green, 8–15 mm long and 4 mm wide, ovate-oblong, and taste slightly bitter. Flowers are small, axillary and solitary, five-petaled white, purple, pink, or pale violet in color; pedicels 6–30 mm long, bracteoles shorter than pedicels. Fruits are capsules up to 5 mm long, ovoid, glabrous, and sharp at apex (Trivedi et al., 2011) (Figure 2.3). Native to India and Australia and also found cultivated in the United States (US) and East Asia (Barrett and Strother, 1978), it has been used as a medicinal herb in Ayurveda since time immemorial. Brahmi grows in humid climates, mainly distributed in damp and marshy places in the subtropical region of the Indian subcontinent. It is an amphibious plant of the tropics and normally found growing on the banks of rivers and lakes. It requires a well-drained, moist, sandy loam soil, rich in organic matter, and grows well at a temperature from 30°C to 40°C.

FIGURE 2.3 Morphology of *Bacopa monnieri*: A) whole plants; B) flowering branch; and C) flower.

2.3.2 Distribution and Common Names

Bacopa monnieri is native to India, Bangladesh, and Southern Asia, but it also grows in Australia, Europe, and Africa. It grows sporadically in moist, wet, and marshy areas across India. The several natural populations of Brahmi were recorded from West Bengal, Punjab, Haryana, and Himachal Pradesh (Chauhan, 1999). It grows naturally up to a height of 1000 m in marshy areas in subtropical regions (Sharma, 2003). *Bacopa monnieri* is found growing in Nepal, Sri Lanka, China, Taiwan and Vietnam. In Florida and other southern states of the US, the herbs are recognized as weeds in rice fields and found growing abundantly in marshes and wetlands of warmer regions. Warm (30–40°C) and humid (65–80%) climatic conditions with plenty of sunshine and abundant rainfall are ideal for growing Brahmi (Barrett and Strother, 1978). This herb is also commonly known by various names such as, Brahmi-Sak, Kiru-Brahmi, Neer-Brahmi, Samrani, and Safed Chamani. Its English names are "thyme-left gratiola" and "water-hyssop". The vernacular names of *B. monnieri* are Indian pennywort, water hyssop (English); farfakh (Arabic); brahmi (Assamese); aaghabini (Bengali); jia ma chi xian (Chinese); petite *Bacopa* (French); kleine fettblatt (German); baam (Gujrati); psheta srua (Hebrew); adha birni (Hindi); bakopa (Japanese); jala *brahmi* (Kannada); barna (Malayalam); ghola (Marathi); medha giree (Nepalese); bakopa drobnolistna (Polish); brahmibuti (Punjabi); adha birni (Sanskrit); ahaznda poozndu (Tamil); neeri sambraani mokka (Telugu); phrommi (Thai); and rau dang bien (Vietnamese).

2.3.3 Bioactive Compounds

Brahmi contains alkaloid brahmine, nicotinine, herpestine, bacosides A and B, saponins A, B, and C, triterpenoid saponins, stigmastanol, sitosterol, betulinic acid, D-mannitol, stigmasterol, alanine, aspartic acid, glutamic acid, and serine and pseudojujubogenin glycoside (Chatterji et al., 1963; Devishree et al., 2017). Bacoside A yields bacogenins A1, A2, A3, and A4 upon hydrolysis (Chatterji et al., 1965). The other chemical components include bacoside A1, hersaponin, betulinic acid, stigmasterol, b-sitosterol, and stigmastenol (Chatterji et al., 1963; Jain and Kulshreshtha, 1993). *Bacopa monnieri* contains alkaloid brahmine, nicotinine, and herpestine. Bacosides A [3-(α-L-arabinopyranosyl)-*O*-β-D-glucopyranoside-10, 20-dihydroxy-16-keto-dammar-24-ene]. Triterpenoid saponins, saponins A, B, and C, and pseudojujubogenin glycoside were also isolated from *Bacopa monniera*. They identified as 3-*O*-α-L-arabinopyranosyl-20-*O*-α-L-arabinopyrasonyl-jujubogenin, 3-*O*-[α-L-arabinofuranosyl-(1→2)-α-L arabinopyranosyl] pseudojujubogenin, 3-*O*-β-D-glucopyranosyl (1→3)-{α-L-arabinofuranosyl-(1→2)}-α-Larabinopyrasonyl] pseudojujubogeninand, and 3-*O*-[α-L-arabinofuranosyl-(1→2)-β-D-glucopiranosyl] pseudojujubogenin. Bacopasides I, II, III, IV, and V were also isolated from Brahmi, which identified as 3-*O*-α-L- arabi nofuranosyl-(1→2)-β-D-glucopyranosyljujubogenin, 3-*O*-β-D-glucopyranosyl-(1→3)-α-L-arabi nopyranosyl jujubogenin, and 3-*O*- β-D-glucopyranosyl-(1→3)-α-L-arabinofuranosyl pseudojujubogenin (Chatterji et al., 1965). *Bacopa monnieri* also contains betulinic acid, D-mannitol, stigmastanol, β-sitosterol, and stigmasterol (Avanigadda and Vangalapati, 2011) (Figure 2.4).

2.3.4 Pharmacological Activities

Bacopa monnieri consists of a variety of active components and shows a number of pharmaceutical and pharmacological activities (Table 2.2), including memory enhancement, antioxidant, tranquilizing, sedative, antidepressant, cognitive, anticancer, antianxiety, adaptogenic, antiepileptic, gastrointestinal effects, endocrine, smooth muscle relaxant, cardiovascular, analgesic, antipyretic, antidiabetic, antiarthritic, anticancer, antihypertensive, antimicrobial, antilipidemia, anti-inflammatory, neuroprotective, and hepatoprotective activities (Russo and Borrelli, 2005; Sinha and Saxena, 2006; Ramasamy et al., 2015). The Brahmi showed diverse mechanisms of action for cognitive effects, including acetylcholinesterase (AChE) inhibition,

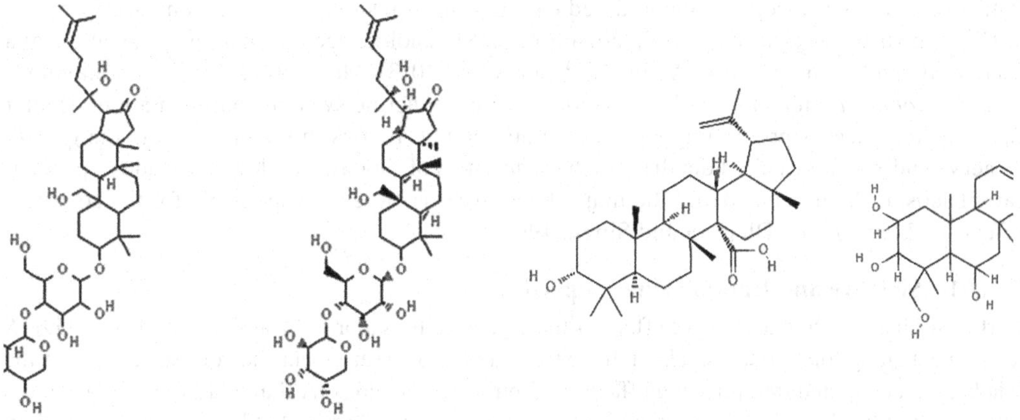

Bacoside A Bacoside B Bacosine Brahmic

FIGURE 2.4 Major phytoconstituents in *Bacopa monnieri*.

TABLE 2.2
Summary of Some Pharmacological Activities along with Phytochemical Constituents of *Bacopa monnieri*

Pharmacological activities	Plant Part Used	Bioactive Compounds	Reference
Sedative effect	Whole plant	Glycosides – Hersaponin	Malhotra and Das (1959)
Anxiolytic activity comparable to Lorazepam drug	Whole plant	Benzodiazepine Anxiolytic	Bhattacharya and Ghoshal (1998); Shankar and Singh (2000)
Antidepressant activity	Whole plant	Serotonin and GABA (Gamma Amino Butyric Acid)	Sairam et al. (2002)
Epilepsy	Whole plant	Hersaponin	Martis et al. (1992)
Antiulcer activity	Whole plant	Ethanol, Aspirin	Rao et al. (2000)
Antidiabetic activity	Whole plant	Bacosine, a triterpene	Sabina et al. (2014); Ghosh et al. (2011a)
Antitumor activity	Aerial parts	Stigmasterol	Ghosh et al. (2011b)
Anticancerous activity	Aerial parts	Bacoside-A	Elangovan et al. (1995)
Antibacterial activity	Aerial parts	Diethyl ether, ethyl acetate extract	Sampathkumar et al. (2008)
Antifungal activity	Aerial parts	Betulinic acid, Oroxindin	Chaudhuri et al. (2004)
Analgesic effect	Aerial parts	Bacosine	Vohora et al. (1997)
Cognition facilitating activity	Aerial parts	Saponins, Bacoside A, and Bacoside B	Sairam et al. (2001); Roodenrys et al. (2002)
Cardiovascular	Whole plant	inhibition of calcium ions	Channa et al. (2003)
Antioxidant properties	Whole plant	Saponins, Bacoside A, and Bacoside B	Negi et al. (2000); Raghav et al. (2006)
Memory enhancer	Whole plant	Saponins, such as bacosides A, B, C, and D	Singh and Dhawan, (1982)

antioxidant neuroprotection, amyloid reduction, neurotransmitter modulation (acetylcholine [ACh], 5-hydroxytryptamine [5-HT], dopamine [DA]), choline acetyltransferase activation, and increased cerebral blood flow (Aguiar and Borowski, 2013). Metabolites or active compounds from *B. monnieri* interact with the dopamine and serotonergic systems, but its main molecular mechanism concerns promoting neuron communication. It does this by increasing the growth of nerve endings, also called dendrites. Characteristics of saponins called "bacosides", particularly bacoside A, are considered the major bioactive constituents responsible for the cognitive effects of *B. monnieri* (Dhawan and Singh, 1996).

2.3.4.1 Sedative and Tranquilizing Properties

Earlier studies reported a sedative effect of the glycoside hersaponin (Malhotra and Das, 1959). A subsequent study found that the alcoholic extract, and to a lesser extent the aqueous extract of the whole plant, exhibited tranquilizing effects on albino rats and dogs (Aithal and Sirsi, 1961). On the other hand, it has been found that the alcoholic extract of the plant and chlorpromazine improved the performance of rats in motor learning (Prakash and Sirsi, 1962). A previous study reported that a single dose of hersaponin is better than pentobarbitone in facilitating acquisition and retention of brightness discrimination reaction. Furthermore, Brahmi showed its potential in positive enhancement during cognitive disorders (Jeyasri et al., 2020).

2.3.4.2 Antidepressant and Antianxiety Effects

In one study, *Bacopa monnieri* extract in a dose range of 20–40 mg/kg was given once daily for 5 days, and it was found comparable to the standard antidepressant drug imipramine in rodents. The same study suggested the role of serotonin and gamma amino butyric acid (GABA) as the mechanism of action attributed for its antidepressant action along with its anxiolytic potential, based on the compelling evidence that the symptoms of anxiety and depression overlap each other (Shader and Greenblatt, 1995). Research using a rat model of clinical anxiety demonstrated that a *B. monnieri* extract containing 25% bacoside A exerted anxiolytic activity comparable to lorazepam, a common benzodiazepine anxiolytic drug, and it was noted that the *B. monnieri* extract did not induce amnesia (a side effect associated with lorazepam) but instead had a memory *enhancing* effect (Bhattacharya and Ghoshal, 1998). The antidepressant potential of *B. monnieri* has been evaluated in an earlier study, wherein it showed a significant antidepressant activity in the most commonly used behavior paradigms in animal models of depression, namely, forced swim test and learned helplessness tests (Sairam et al., 2002).

2.3.4.3 Antiepileptic Effects

Brahmi is used as a remedy for epilepsy in Ayurvedic medicine, although research in animals has showed anticonvulsant activity only at high doses over extended periods of time. The hersaponin present in Brahmi exhibited protection against seizures in mice, and the possibility of its use as an adjuvant in the treatment of epilepsy has been mentioned (Martis et al., 1992). The study examined the anticonvulsant properties of *Bacopa monnieri* extracts in mice and rats. The intraperitoneal injections of high doses of *B. monnieri* extract (close to 50% of LD50) given for 15 days demonstrated anticonvulsant activity. When administered acutely at lower doses (approaching 25% of LD50), anticonvulsant activity was not observed (Ganguly and Malhotra, 1967). It has been suggested that the anticonvulsive effects may be mediated through GABA which is involved in neural impulse transmission, because substances which stimulate GABA are known to possess anticonvulsant, pain relieving, and sedative activities. Mathew et al. (2012) reported the antiepileptic effect of *B. monnieri* in the cerebral cortex of epileptic rats.

2.3.4.4 Anti-ulcerative Activity

The fresh extract of Brahmi was given to treat gastric ulcer models induced by ethanol and aspirin (Rao et al., 2000). Brahmi juice was given orally twice daily for 5 days at a dose of 100 and 300 mg/

Kg and sucralfate at 250 mg/Kg. The juice showed antiulcer activity in the aspirin-induced gastric ulcer models but not for the ethanol-induced ulcers. The effect is attributed to mucosal defensive factors like enhanced mucin secretion, mucosal glycoprotein, and decreased mucosal cell exfoliation. Dorababu et al. (2004) performed an experiment using methanolic extract of *B. monnieri* on the susceptibility of NIDDM in normal rats and found a 50 mg/kg body weight dose to be effective in healing penetrating ulcers induced by acetic acid and HCl after 5–10 days of treatment. Goel et al. (2003) showed anti-helicobacter pylori activity at 1000 μg/ml of extract, while at the dose of 10 μg/ml, it showed an increase in prostanoids. Also, Brahmi is advised in irritable bowel syndrome (IBS) (Dar and Channa, 1999).

2.3.4.5 Antidiabetic Activity

Diabetes mellitus is a metabolic disorder affecting carbohydrate, fat, and protein metabolism. Approximately 1% of the population suffers from this disease. It was reported that bacosine, a triterpene found in the ethanolic extract of Brahmi, is responsible for an increase in glycogen content in diabetic rats (Ghosh et al., 2011a; Sabina et al., 2014). The saponins present in Brahmi exhibited insulin-like activity in alloxan-induced diabetic rats (Ghosh et al., 2008). Taznin et al. (2015) demonstrated that the methanolic extract of Brahmi exhibited antihyperglycemic activity in Swiss albino mice.

2.3.4.6 Anticancer Activity

Ghosh et al. (2011b) examined the anti-tumor activity of stigmasterol isolated from the aerial parts of Brahmi against Ehrlich ascites carcinoma in Swiss albino mice. Kumar et al. (1998) showed the anti-tumor activity of ethanolic extract of Brahmi: oral administration of extract delayed the development of solid tumors. The study was carried out on *in vitro* short-term chemosensitivity and *in vivo* tumor model test systems. D'Souza et al. (2002) assessed ethanolic extracts, which are saponin-rich fractions that showed potential antitumor activity. D'Souza et al. (2002) reported Bacoside A as an active component accountable for anticancerous activity. Elangovan et al. (1995) reported alcoholic (ethanol) extract of *Bacopa monnieri* as an anticancerous drug tested for sarcoma 180 cell culture, where cell growth was inhibited with increasing concentration of extracts. Mallick et al. (2015) demonstrated that the ethanolic extract of Brahmi showed anticancer activity against MCF-7 and MDA-MB 231 cell lines.

2.3.4.7 Antimicrobial Activity

The potential antibacterial and antifungal activity of ethyl acetate and n-butanol fractions of ethanol extract of aerial parts of Brahmi were reported by Ghosh et al. (2007). Sampathkumar et al. (2008) evaluated the antimicrobial activity of ethanolic, diethyl acetate, ethyl acetate, and aqueous extracts of the aerial parts of *B. monnieri*. Diethyl ether extract showed antibacterial activity against gram positive while ethyl acetate extract showed activity against gram negative organisms. The test was carried out on bacteria such as *Staphylococcus aureus*, *Proteus vulgaris*, etc. Ethanolic extract exhibited antifungal activity against *Aspergillus niger* and *Candida albicans*. Antifungal activity was shown against *Alternaria alternata* and *Fusarium fusiformis* by phytochemicals, like betulinic acid and oroxindin, isolated from the aerial parts of *B. monnieri* has also been reported (Chaudhuri et al., 2004). Khan and Ahmed, (2010) reported the antibacterial activity of ethyl acetate and methanol extracts of *Bacopa monnieri* against 7-gram negative and 11-gram positive bacteria by the disk diffusion method. Fazlul et al. (2019) reported a potential antibacterial and antifungal activity of the ethanolic, diethyl ether, and ethyl acetate extracts of *B. monnieri*.

2.3.4.8 Cardiovascular Activities

The cardiovascular activity of crude ethanolic extract of Brahmi on rabbit showed cardiac depressive activity (Rashid et al., 1990). The effect of hydro-alcoholic extract of Brahmi in isoproterenol (ISP)-induced myocardial necrosis exhibited cardioprotective effects (Nandave et al., 2007). Kamkaew et al. (2011) studied the effect of intravenous *B. monnieri* extract administered in a dose of 20–60 mg/Kg on arterial blood pressure and heart rate of anesthetized rats. The extract decreased systolic and diastolic pressures without disturbing heart rate. It was concluded that blood pressure reduced partly via releasing nitric oxide from the endothelium and partly by actions on vascular smooth muscle Ca^{2+} homeostasis. Channa et al. (2003) reported the broncho vasodilatory activity of *B. monnieri* extract in anesthetized rats. The work was done on various fractions derived from *Bacopa monnieri*, and the activity was observed because of inhibition of calcium ions. The significant *in vitro* cytotoxic and cardiac depressant activity of hydro-ethanolic extract of Brahmi was reported by Sai Krishna et al. (2015). The ethanolic extract of Brahmi improves the myocardial function in the rat heart and is reported to be a novel cardioprotectant (Srimachai et al., 2017).

2.3.4.9 Analgesic Activities

Vohora et al. (1997) isolated bacosine from aerial parts of *Bacopa monnieri* and found that it possesses analgesic effects. The effect was opioidergic in nature. It was also reported that bacosine didn't show any effect on barbiturate narcosis, haloperidol-induced catalepsy, or conditioned avoidance response. Siraj et al. (2012) investigated the analgesic effect of ethanol extract of *B. monnieri*. The extract yielded noteworthy writhing inhibition in acetic acid-induced writhing in mice at the oral dose (250 and 500 mg/kg) compared to standard drug diclofenac sodium (25 mg/kg).

2.3.4.10 Cognitive Activities

Cognition is the mental processing that includes the attention of working memory, comprehending and producing language, calculating, reasoning, problem-solving, and decision-making. It is the process by which the sensory input is transformed, reduced, elaborated, stored, recovered, and used. In general, cognition refers to an information processing view of an individual (Bhattacharya et al., 1998). *Bacopa monnieri* demonstrated enhanced attention and cognitive processing capability together with enhanced working memory (Singh and Dhawan, 1982). According to one study, standardized bacoside-rich extract of *B. monnieri* reversed the cognitive deficits induced by intracerebroventricularly administered colchicines and injection of ibotenic acid into the nucleus *Basalis magnocellularis* (Singh and Dhawan, 1982). Also, *Bacopa monnieri* has been shown to reverse the depletion of acetylcholine, the reduction in choline acetylase activity, and the decrease in muscarinic cholinergic receptor-binding in the frontal cortex and hippocampus (Singh et al., 1988). The cognition facilitating activity of the *B. monnieri* extract is attributed to the saponins, Bacoside A and Bacoside B, which are effective in much lower doses (Sairam et al., 2001; Roodenrys et al., 2002). In a study conducted on 36 children diagnosed with attention deficit or hyperactivity, a significant benefit was observed in *B. monnieri*-treated subjects, as evidenced by improvement in sentence repetition, logical memory, and paired associated learning tasks hyperactivity (Singh and Dhawan, 1997). The efficacy of standardized *Bacopa monnieri* in subjects with age-associated memory impairment (AAMI), without any evidence of dementia or psychiatric disorder, was evaluated in one of the studies. *Bacopa monnieri* produced significant improvement in mental control, logical memory, and paired associated learning during the 12-week drug therapy. Brahmi was found to be efficacious in subjects with AAMI (Martis et al., 1992). Numerous clinical studies have been carried out, to date, to establish the efficacy of *B. monnieri* in memory and attention disorders and to study its acute and chronic effects clinically on cognitive function. *B. monnieri* decreases the rate of forgetting in adults when chronic administration was shown to enhance cognitive effects (Stough et al., 2001). In another study on 38 healthy volunteers (ages 18–60), subjects were given a single dose of 300 mg *B. monnieri* (Stough et al., 2001). These results were attributed to Brahmi antioxidant properties and/or its effect on the cholinergic system (Negi et al., 2000; Raghav et al., 2006).

2.4 CONCLUSION

A. paniculata and *B. monnieri* have demonstrated a broad spectrum of pharmacological activities. The commercial value of *A. paniculata* is increasing due to its broad range of therapeutic values. The available data on pharmacological activities proves its extensive benefits. Several clinical studies have proven the safety of its use in treatment of upper respiratory tract infection under different conditions. However, the validation of several therapeutic activities such as antidiabetic, anticancer, anti-inflammatory, and hepatoprotective activities via clinical study is still lagging. *B. monnieri* exhibits enormous potential in the amelioration of cognitive disorders, along with prophylactic reduction of oxidative damage, modulation of neurotransmitters, and cognitive enhancement among healthy people. However, there is much less information available about biomedical research on *Bacopa monnieri*. We assume that both the medicinal plants will be highly useful in treatment of several health aliments in the near future. Along with pharmacological studies, researchers might provide attention to *in vitro* multiplication of these plants via tissue culture techniques for their commercial exploitation.

ACKNOWLEDGMENTS

The authors are grateful to Yashavantrao Chavan Institute of Science, Satara for financial assistance and providing necessary laboratory facilities. Financial assistance to the faculty under self-funded project is gratefully acknowledged.

REFERENCES

Aguiar, A., and T. Borowski. 2013. Neuropharmacological review of the nootropic herb *Bacopa monnieri*. *Rejuvenation Research*, 16, pp.313–326.

Aithal, H.N., and M. Sirsi. 1961. Pharmacological investigations on *Herpestis monniera* H.B. & K. *Indian Journal of Pharmacology*, 23, pp.2–5.

Akbar, S. 2011. *Andrographis paniculata*: A review of pharmacological activities and clinical effects. *Alternative Medicine Review*, 16, pp.66–77.

Akhtar, M.T., M.S.Bin Mohd Sarib, I.S. Ismail, F. Abas, A. Ismail, N.H. Lajis, and K. Shaari. 2016. Anti-diabetic activity and metabolic changes induced by *Andrographis paniculata* plant extract in obese diabetic rats. *Molecules*, 21, pp.1026.

Akowuah, G.A., I. Zhari, and A. Mariam. 2008. Analysis of urinary andrographolides and antioxidant status after oral administration of *Andrographis paniculata* leaf extract in rats. *Food and Chemical Toxicology*, 46, pp.3616–3620.

Anonymous. 1978. *The Ayurvedic Formulary of India, Part I*. 1st edn. Faridabad, India: Government of India Press.

Avanigadda, S., and M. Vangalapati. 2011. A Review on pharmacological studies of *Bacopa monniera*. *Journal of Chemical, Biological and Physical Sciences*, 1, pp.250–259.

Barrett, S.C.H. and J.L. Strother. 1978. Taxonomy and natural history of *Bacopa* (Scrophulariaceae) in California. *Systematic Botany*, 3, pp.408–419.

Bhattacharya, S.K., and S. Ghoshal. 1998. Anxiolytic activity of a standardized extract of *Bacopa monniera*: An experimental study. *Phytomedicine*, 5, pp.77–82.

Bhattacharya, S.K., A. Kumar, and S. Ghosal. 1999. Effect of *Bacopa monniera* on animal models of Alzheimer's disease and perturbed central cholinergic markers of cognition in rats. *Research Communications in Pharmacology and Toxicology*, 4, pp.1–12.

Burgos, R.A., M.J. Aguila, E.T. Santiesteban, N.S. Sanchez, and J.L. Hancke. 2001. *Andrographis paniculata* Nees induces relaxation of uterus by blocking voltage operated calcium channels and inhibits $Ca^{(+2)}$ influx. *Phytotherapy Research*, 15, pp.235–239.

Cáceres, D.D., J.L. Hancke, and R.A. Burgos. 1999. Use of visual analogue scale measurements (VAS) to assess the effectiveness of standardized *Andrographis paniculata* extract SHA-10 in reducing the symptoms of common cold. A randomized double blind-placebo study. *Phytomedicine*, 6, pp.217–223.

Cáceres, D.D., J.L. Hancke, R.A. Burgos, and G.K. Wikman. 1997. Prevention of common colds with *Andrographis paniculata* dried extract. A pilot double blind trial. *Phytomedicine*, 4, pp.101–104.

Calabrese, C., S.H. Berman, J.G. Babish, X. Ma, L. Shinto, M. Dorr, K. Wells, C.A. Wenner, and L.J. Standish. 2000. A phase I trial of andrographolide in HIV positive patients and normal volunteers. *Phytotherapy Research*, 14, pp.333–338.

Channa, S., A. Dar, and M. Yaqoob. 2003. Bronchovasodilatory activity of fractions and pure constituents isolated from *Bacopa monniera*. *Journal of Ethnopharmacology*, 86, pp.27–35.

Chao, W.W., and B.F. Lin. 2010. Isolation and identification of bioactive compounds in *Andrographis paniculata* (Chuanxinlian). *Chinese Medicine*, 5, pp.1–15.

Chatterji, N., R.P. Rastogi, and M.L. Dhar. 1963. Chemical examination of *Bacopa monniera* Wettst. Part I. Isolation of chemical constituents. *Indian Journal of Chemistry*, 1, pp. 212–215.

Chatterji, N., R.P. Rastogi, and M.L. Dhar. 1965. Chemical examination of *Bacopa monniera* Wettst: Parti-isolation of chemical constituents. *Indian Journal of Chemistry* 3, pp.24–29.

Chaudhuri, P.K., R. Srivastava, S. Kumar, and S. Kumar. 2004. Phytotoxic and antimicrobial constituents of *Bacopa monnieri* and *Holmskioldia sanguinea*. *Phytotherapy Research*, 18, pp.114–117.

Chauhan, N.S. 1999. *Medicinal and Aromatic Plants of Himachal Pradesh*. New Delhi: Indus Publishing Company, pp.234–237.

Ciampi, E., R. Uribe-San-Martin, C. Carcamo et al. 2020. Efficacy of andrographolide in not active progressive multiple sclerosis: A prospective exploratory double-blind, parallel-group, randomized, placebo-controlled trial. *BMC Neurology*, 20, pp.173.

D'Souza, P., M. Deepak, P. Rani, S. Kadamboor, A. Mathew, A.P. Chandrashekar, and A. Agarwal. 2002. Brine shrimp lethality assay of *Bacopa monnieri*. *Phytotherapy Research*, 16, pp.197–198.

Dar, A., and S. Channa. 1999. Calcium antagonistic activity of *Bacopa monnieri* on vascular and intestinal smooth muscles of rabbit and guinea-pig. *Journal of Ethnopharmacology*, 66, pp.167–174.

Devishree, R.A., S. Kumar, and A.R. Jain. 2017. Short-term effect of *Bacopa monnieri* on memory: A brief review. *Journal of Pharmacy Research*, 11,, pp.1447–1450.

Dey, Y.N., S. Kumari, S. Ota, and N. Srikanth. 2013. Phytopharmacological review of *Andrographis paniculata* (Burm.f) Wall. ex Nees. *International Jouranl of Nutrition, Pharmacology and Neurological Diseases*, 3, pp.3–10.

Dhawan, B.N., and H.K. Singh. 1996. Pharmacology of Ayurvedic nootropic *Bacopa monniera*. *International Convention of Biological Psychiatry, Bombay*, Abst. NR 59, pp.21

Dorababu, M., T. Prabha, S. Priyambada, V.K. Agrawal, N.C. Aryya, and R.K. Goel. 2004. Effect of *Bacopa monnieri* and *Azadirachta indica* on gastric ulceration and healing in experimental NIDDM rats. *Indian Journal of Experimental Biology*, 42, pp. 389–397.

Dua, V.K., V.P. Ojha, R. Roy, B.C. Joshi, N. Valecha, C.U. Devi, M.C. Bhatnagar, V.P. Sharma, and S.K. Subbarao. 2004. Anti-malarial activity of some xanthones isolated from the roots of *Andrographis paniculata*. *Journal of Ethnopharmacology*, 95, pp.247–251.

Dutta, A., and N.C. Sukul. 1982. Filaricidal properties of a wild herb, *Andrographis paniculata*. *Journal of Helminthology*, 56, pp.81–84.

Elangovan, V., S. Govindaswamy, N. Ramamoorthy, and K. Balusubramaniam. 1995. *In vitro* studies on anti-cancer activity of *Bacopa monnieri*. *Fitoterapia*, 3, pp.211–215.

Fazlul, M.K.K., S.P. Deepthi, I. Mohammed, Y. Farzana, B. Munira, and Nazmulmhm. 2019. Antibacterial and antifungal activity of various extracts of *Bacopa monnieri*. *International Journal of Pharmaceutical Research*, 11, pp.1698–1702.

Ganguly, D.K., and C.L. Malhotra. 1967. Some behavioural effects of an active fraction from *Herpestis monniera*, Linn. (Brahmi). *Indian Journal of Medical Research*, 55, pp.473–482.

Ghosh, T., T.K. Maity, A. Bose, G.K. Dash, and M. Das. 2007. Antimicrobial activity of various fractions of ethanol extract of *Bacopa monnieri* Linn. aerial parts. *Indian Journal of Pharmaceutical Sciences*, 69, pp.312–314.

Ghosh, T., T.K. Maity, and J. Singh. 2011a. Antihyperglycemic activity of bacosine, a triterpene from *Bacopa monnieri*, in alloxan-induced diabetic rats. *Planta Medica*, 77, pp, 804–808.

Ghosh, T., T.K. Maity, and J. Singh. 2011b. Evaluation of antitumor activity of stigmasterol, a constituent isolated from *Bacopa monnieri* Linn aerial parts against Ehrlich Ascites Carcinoma in mice. *Oriental Pharmacy and Experimental Medicine*, 11, pp.41–49.

Ghosh, T., T.K. Maity, P. Sengupta, D.K. Dash, and A. Bose. 2008. Antidiabetic and *in vivo* antioxidant activity of ethanolic extract of *Bacopa monnieri* Linn. aerial parts: A possible mechanism of action. *The Iranian Journal of Pharmaceutical Research*, 7, pp.61–68.

Goel, R.K., K. Sairam, M.Dora Babu, I.A. Tavares and A. Raman. 2003. In vitro evaluation of *Bacopa monniera* on anti-Helicobacter pylori activity and accumulation of prostaglandins. *Phytomedicine*, 10, pp. 523–527.

Guo, S.Y., D.Z. Li, W.S. Li, A.H. Fu, and L.F. Zhang. 1988. Study of the toxicity of andrographolide in rabbits. *Journal of Beijing Medical University*, 5, pp.422–428.

Gurupriya, S., and L. Cathrine. 2016. Antimicrobial activity of *Andrographis paniculata* stem extracts. *International Journal of Scientific and Engineering Research*, 7, pp. 105.

Hossain, M.A., B.K. Roy, K. Ahmed, A.S. Chowdhury, and M.A. Rashid. 2007. Antidiabetic activity of *Andrographis paniculata. Dhaka University Journal of Pharmaceutical Sciences*, 6, pp.5–20.

Hossain, M.D., Z. Urbi, A. Sule, and K.M. Rahman. 2014. *Andrographis paniculata* (Burm. f.) Wall. ex Nees: A review of ethnobotany, phytochemistry, and pharmacology. *The Scientific World Journal*, 1, pp.28.

Hossain, S., Z. Urbi, H. Karuniawati, R.B. Mohiuddin, A. Moh Qrimida, A.M.M. Allzrag, L.C. Ming, E. Pagano, and R. Capasso. 2021. *Andrographis paniculata* (Burm. f.) Wall. ex Nees: An updated review of phytochemistry, antimicrobial pharmacology, and clinical safety and efficacy. *Life*, 11, pp.348.

Hueng, S.J. 1991. Treating anal tumor by washing using *Andrographis paniculata* extractions plus vinegar. *Chinese Journal of Analytical Chemistry*, 11, pp.40.

Ishola, A.A., B.E. Oyinloye, B.O. Ajiboye, and A.P. Kappo. 2021. Molecular docking studies of flavonoids from *Andrographis paniculata* as potential acetylcholinesterase, butyrylcholinesterase and monoamine oxidase inhibitors towards the treatment of neurodegenerative diseases. *Biointerface Research in Applied Chemistry*, 11, pp.9871–9879.

Jain, P., and D.K. Kulshreshtha. 1993. Bacoside A1, A minor saponin from *Bacopa monnieri. Phytochemistry*, 33, pp.449–451.

Jalal, M.A.F., K.H. Overton, and D.S. Rycroft. 1979. Formation of three new flavones by differentiating callus cultures of *Andrographis paniculata. Phytochemistry*, 18, pp.149–151.

Jarukamjorn, K., and N. Nemoto. 2008. Pharmacological aspects of *Andrographis paniculata* on health and its major diterpenoid constituent andrographolide. *Journal of Health Sciences*, 54, pp.370–381.

Jeyasri, R., M. Pandiyan, V. Suba, M. Ramesh, and J.T. Chen. 2020. *Bacopa monnieri* and their bioactive compounds inferred multi-target treatment strategy for neurological diseases: A cheminformatics and system pharmacology approach. *Biomolecules*, 10, pp.536.

Joselin, J., and S. Jeeva. 2014. *Andrographis paniculata*: A review of its traditional uses, phytochemistry and pharmacology. *Medicinal and Aromatic Plants*, 3, pp.4.

Kamkaew, N., C.N. Scholfield, K. Ingkaninan, P. Maneesai, H.C. Parkington, M. Tare, and K. Chootip. 2011. *Bacopa monnieri* and its constituents is hypotensive in anaesthetized rats and vasodilator in various artery types. *Journal of Ethnopharmacology*, 137, pp.790–795.

Khan, A.V., and Q.U. Ahmed. 2010. Antibacterial efficacy of *Bacopa monnieri* leaf extracts against pathogenic bacteria. *Asian Biomedicine*, 4, pp.651–655.

Kumar, E.P., A.A. Elshurafa, K. Elango, T. Subburaju, and B. Suresh. 1998. Cytotoxic and anti-tumour activities of Ethanolic extract of *Bacopa monnieri* (L) Penn. *Ancient Science of Life*, 17, pp.228–234.

Kumar, S., B. Singh, and V. Bajpai. 2021. *Andrographis paniculata* (Burm.f.) Nees: Traditional uses, phytochemistry, pharmacological properties and quality control/quality assurance. *Journal of Ethanopharmacology*, 275, pp.114054.

Kurzawa, M., A. Filipiak-Szok, E. Kłodzińska, and E. Szłyk. 2015. Determination of phytochemicals, antioxidant activity and total phenolic content in *Andrographis paniculata* using chromatographic methods. *Journal of Chromatography B*, 995, pp.101–106.

Lin, F.L., S.J. Wu, S.C. Lee, and L.T. Ng. 2009. Antioxidant, antioedema and analgesic activities of *Andrographis paniculata* extracts and their active constituent andrographolide. *Phytotherary Research*, 23, pp.958–964.

Lu, J., Y. Ma, J. Wu, H. Huang, X. Wang, Z. Chen, J. Chen, H. He, and C. Huang. 2019. A review for the neuroprotective effects of andrographolide in the central nervous system. *Biomedicine and Pharmacotherapy*, 117, pp.109078.

Malhotra, C.L., and P.K. Das. 1959. Pharmacological studies of *Herpestis monniera* Linn (Brahmi). *Indian Journal of Medical Research*, 4, pp.294–305.

Mallick, M.N., M.S. Akhtar, M.Z. Najm, E.T. Tamboli, S. Ahmad, and S.A. Husain. 2015. Evaluation of anticancer potential of Bacopa monnieri L. against MCF-7 and MDA-MB 231 cell line. *Journal of Pharmacy and Bioallied Sciences*, 7, pp.325–328.

Mandal, S.C., A.K. Dhara, and B.C. Maiti. 2001. Studies on psychopharmacological activity of *Andrographis paniculata* extract. *Phytotherapy Research*, 15, pp.253–256.

Martis, G., A. Rao and K.S. Karanth. 1992. Neuropharmacological activity of *Herpestis monniera. Fitoterapia*, 62, pp.399–406.

Mathew, J., S. Balakrishnan, S. Antony, P.M. Abraham, and C.S. Paulose. 2012. Decreased GABA receptor in the cerebral cortex of epileptic rats: Effect of *Bacopa monnieri* and Bacoside-A. *Journal of Biomedical Science*, 19, pp.25.

Matsuda, T., M. Kuroyanagi, S. Sugiyama, K. Umehara, A. Ueno, and K. Nishi. 1994. Cell differentiation-inducing diterpenes from *Andrographis paniculata* Nees. *Chemical and Pharmaceutical Bulletin*, 42, pp.1216–1225.

Melchior, J., S. Palm, and G. Wikman. 1996. Controlled clinical study of standardized *Andrographis paniculata* extract in common cold a pilot trial. *Phytomedicine*, 34, pp.315–318.

Mishra, P.K., R.K. Singh, A. Gupta, A. Chaturvedi, R. Pandey, S.P. Tiwari, and T.M. Mohapatra. 2013. Antibacterial activity of Andrographis paniculata (Burm. f.) Wall ex Nees leaves against clinical pathogens. *Journal of Pharmacy Research*, 7, pp. 459–462.

Mishra, S.K., N.S. Sangwan, and R.S. Sangwan. 2007. *Andrographis paniculata* (Kalmegh): A review. *Pharmacognosy Reviews*, 1, pp.283–298.

Misra, P., N.L. Pal, P.Y. Guru, J.C. Katiyar, V. Srivastava, and J.S. Tandon. 1992. Antimalarial activity of *Andrographis paniculata* (Kalmegh) against *Plasmodium berghei* NK 65 in *Mastomys natalensis*. *International Journal of Pharmacology*, 30, pp.263–274.

Najib, N.N., A. Rahman, T. Furuta, K. Kojima, T.K. Kikuchi, and M.A. Mohd. 1999. Antimalarial activity of extracts of Malaysian medicinal plants. *Journal of Ethnopharmacology*, 64, pp.249–254.

Nandave, M, S.K. Ojha, S. Joshi, S. Kumari, and D.S. Arya. 2007. Cardioprotective effect of B. *monneira* against Isoproterenol-Induced Myocardial Necrosis in rats. *International Journal of Pharmacology*, 3, pp.385–392.

Negi K., Y. Singh, K. Kushwaha, C. Rastogi, A. Rathi, J. Srivastava, O. Asthana, R. Gupta, J.C. Rathi, and R. Gupta. 2000. Clinical evaluation of memory enhancing properties of memory plus in children with attention deficit hyperactivity disorder. *Indian Journal of Psychiatry*, 2002, pp.42.

Niranjan, A., S.K. Tewari, and A. Lehri. 2010. Biological activities of Kalmegh (*Andrographis paniculata* Nees) and its active principles-A review. *Indian Journal of Natural Products and Resources*, 1, pp.125–135.

Nyeem, M.A.B., M.A. Mannan, M. Nuruzzaman, K.M. Kamrujjaman, and S.K. Das. 2017. Indigenous king of bitter (*Andrographis paniculata*): A review. *Journal of Medicinal Plants Studies*, 5, pp.318–324.

Pandy, G., and C.H. Rao. 2018. Andrographolide: Its pharmacology, natural bioavailability and current approaches to increase its content in *Andrographis paniculata*. *Journal of Integrative and Complementary Medicine*, 11, pp.355–360.

Perumal Samy, R., M.M. Thwin, and P. Gopalakrishnakone. 2007. Phytochemistry, pharmacology and clinical use of Andrographis paniculata. *Natural Product Communications*, 2, pp.607–618.

Poolsup, N., C. Suthisisang, S. Prathanturarug, A. Asawamekin, and U. Chanchareon. 2004. Andrographis paniculata in the symptomatic treatment of uncomplicated upper respiratory tract infection: Systematic review of randomized controlled trials. *Journal of Clinical Pharmacy and Therapeutics*, 29, pp.37–45.

Prakash, J.C., and M. Sirsi. 1962. Comparative study of the effects of Brahmi (*Bacopa monniera*) & Chlorpromazine on motor learning in rats. *Journal of Scientific and Industrial Research*, 21C, pp.93–96.

Raghav, S., H. Singh, P.K. Dalal, J.S. Srivastava, and O.P. Asthana. 2006. Randomized controlled trial of standardized *Bacopa monniera* extract in age-associated memory impairment. *Indian Journal of Psychiatry*, 48, pp.238–242.

Ramasamy, S., S.P. Chin, S.D. Sukumaran, M.J.C. Buckle, L.V. Kiew and L.Y. Chung. 2015. *In Silico* and *in vitro* analysis of Bacoside A aglycones and its derivatives as the constituents responsible for the cognitive effects of *Bacopa monnieri*. *PLoS One*, 10, pp.e0126565.

Rao, C.V., K. Sairam, and R. Goel. 2000. Experimental evaluation of *Bacopa monniera* on rat gastric ulceration and secretion. *Indian Journal of Physiology and Pharmacology*, 44, pp.435–441.

Rao, P.R., and V.K. Rathod. 2015. Rapid extraction of andrographolide from *Andrographis paniculata* Nees by three-phase partitioning and determination of its antioxidant activity. *Biocatalysis and Agricultral Biotechnology*, 4, pp.586–593.

Rashid, S., F. Lodhi, M. Ahmad, and K. Usmanghani. 1990. Cardiovascular effects of *Bacopa monnieri* (L.) pennel extract in rabbits. *Pakistan Journal of Pharmaceutical Sciences*, 3, pp.57–62.

Reddy, N.V.L., S. Malla Reddy, V. Ravikanth, P. Krishnaiah, T. Venkateshwar Goud, T.P. Rao, T. Siva Ram, R.G. Gonnade, M. Bhadbhade, and Y. Venkateswarlu. 2005. A new bis-andrographolide ether from *Andrographis paniculata* Nees and evaluation of anti-HIV activity. *Natural Product Research*, 19, pp.223–230.

Reyes, B.A., N.D. Bautista, N.C. Tanquilut, R.V. Anunciado, A.B. Leung, G.C. Sanchez, R.L. Magtoto, P. Castronuevo, H. Tsukamura, and K.I. Maeda. 2005. Anti-diabetic potentials of *Momordica charantia* and *Andrographis paniculata* and their effects on estrous cyclicity of alloxan-induced diabetic rats. *Journal of Ethnopharmacology*, 105, pp.196–200.

Roodenrys, S., D. Booth, S. Bulzomi, A. Phipps, C. Micallef, and J. Smoker. 2002. Chronic effects of Brahmi (*Bacopa monniera*) on human memory. *Neuropsychopharmacology*, 27, pp.279–281.

Russo, A., and F. Borrelli. 2005. *Bacopa monniera*, a reputed nootropic plant: An overview. *Phytomedicine*, 12, pp.305–317.

Sabina, E.P., U.L. Baskaran, S.J. Martin, M. Swaminathan, Y. Bhattacharya, and S. Tandon. 2014. Assessment of antidiabetic activity of the traditional Indian ayurvedic formulation Brahmigritham in streptozotocin-induced diabetic rats. *International Journal of Pharmacy and Pharmaceutical Sciences*, 6, pp.347–351.

Sai Krishna K., G. Srividya, and K.R. Kumar. 2015. Cardiac depressant and cytotoxic activities of Hydro-methanolic extract of *Bacopa monnieri* L. *International Journal of Pharmacy and Pharmaceutical Research*, 4, pp.140–149.

Sairam, K., C.V. Rao, M.D. Babu, and R.K. Goel. 2001. Prophylactic and curative effects of *Bacopa monniera* in gastric ulcer models. *Phytomedicine*, 8, pp.423–430.

Sairam, K., M.D. Babu, R.K. Goel, and S.K. Bhattacharya. 2002. Antidepressant activity of Standardized extract of *Bacopa monniera* in experimental models of depression in rats. *Phytomedicine*, 9, pp.207–211.

Sakila, S., N. Begum, S. Kawsar, Z.A. Begum, and M.S. Zoha. 2017. Relationship of anti-fertility effects of *Andrographis paniculata* and hormonal assay in female rats. *Bangladesh Journal of Medical Sciences*, 8, pp.1–2.

Sampathkumar, P., B. Dheeba, Z.V. Vidhyasagar, T. Arulprakash, and R. Vinothkannan. 2008. Potential antimicrobial activity of various extracts of *Bacopa monnieri* (Linn.). *International Journal of Pharmacology Research*, 4, pp.230–232.

Sangeetha, S., R. Archit, and A. SathiaVelu. 2014. Phytochemical testing, antioxidant activity, HPTLC and FTIR analysis of antidiabetic plants *Nigella sativa*, *Eugenia jambolana*, *Andrographis paniculata* and *Gymnema sylvestre*. *Journal of Biotechnology*, 9, pp.1–9.

Satyavati, G.V., M.K. Raina, and M. Sharma. 1976. *Indian Medicinal Plants*, vol 1. New Delhi: Indian Council of Medical Research.

Shader, R.I., and D.J. Greenblatt. 1995. Pharmacotherapy of acute anxiety. In: Bloom, F.E., and D.J. Kupfer (eds.) *Psychopharmacology: Fourth Generation of Progress*. New York: Raven Press, pp.1341–1348.

Shankar, G., and H.K. Singh. 2000. Anxiolytic profile of standardized Brahmi extract. *Indian Journal of Pharmacology*, 32, pp.152.

Sharma, R. 2003. *Medicinal Plants of India*. New Delhi: Daya Publishing House, pp 30–31.

Sharma, M. and S. Joshi. 2011. Comparison of antioxidant activity of *Andrographis paniculata* and *Tinospora cordifolia* leaves. *Journal of Current Chemical and Pharmaceutical Sciences*, 1, pp.1–8.

Shirisha, K., and M. Mastan. 2013. *Andrographis paniculata* and its bioactive phytochemical constituents for oxidative damage: A systemic review. *Pharmacophore*, 4, pp.212–229.

Singh, H.K., and B.N. Dhawan. 1982. Effect of *Bacopa monniera* (Brahmi) on avoidance response in rat. *Journal of Ethnopharmacology*, 5, pp.205–214.

Singh, H.K., and B.N. Dhawan. 1997. Neuropsychopharmacological effects of the ayurvedic nootropic *Bacopa monniera* Linn. (Brahmi). *Indian Journal of Pharmacology*, 29, pp.S359–S365.

Singh, H.K., R.P. Rastogi, R.C. Srimal, and B.N. Dhawan. 1988. Effect of Bacoside A and B on avoidance response in rats. *Phytotherapy Research*, 2, pp.70–75.

Singh, R.H., R.L. Singh, and P.O. Seni. 1979. Studies on the anti-anxiety effect of the medha rasayana drug brahmi (*Bacopa monniera*). Part II – Experimental studies. *Journal of Research in Indian Medicine, Yoga, and Homeopathy*, 14, pp.1–6.

Sinha, S., and R. Saxena. 2006. Effect of iron on lipid peroxidation and enzymatic and non-enzymatic anti-oxidants and bacoside content in medicinal plant *Bacopa monniera* L. *Chemosphere*, 62, pp.1340–1350.

Siraj, M.A., N. Chakma, M. Rahman, S. Malik, and S.K. Sadhu. 2012. Assessment of analgesic, antidiarrhoeal and cytotoxic activity of ethanolic extract of the whole plant of *Bacopa monnieri* Linn. *International Research Journal of Pharmacy*, 3, pp.98–101.

Siti Najila, N.J., A. Noor Rain, A.G.Mohamad Kamel, S.I. Syed Zahir, S. Khozirah, S.Lokman Hakim, I. Zakiah, and A.K. Azizol. 2002. The screening of extracts from *Goniothalamus scortechinii*, *Aralidium pinnatifidum* and *Andrographis paniculata* for anti-malarial activity using the lactate dehydrogenase assay. *Journal of Ethnopharmacology*, 82, pp.239–242.

Sivrajan, V.V., and I. Balachandran. 1994. *Ayurvedic Drugs and Their Plant Sources*. New Delhi: Oxford and IBH Publ.

Srimachai, S., S. Devaux, C. Demougeot, S. Kumphune, N.D. Ullrich, E. Niggli, K. Ingkaninan, N. Kamkaew, C.N. Scholfield, S. Tapechum, and K. Chootip. 2017. *Bacopa monnieri* extract increases rat coronary flow and protects against myocardial ischemia/reperfusion injury. *BMC Complementary Medicine and Therapies*, 17, pp.117.

Stough, C., J. Lloyd, J. Clarke, L.A. Downey, C.W. Hutchison, T. Rodgers, and P.J. Nathan. 2001. The chronic effects of an extract of *Bacopa monniera* (Brahmi) on cognitive function in healthy human subjects. *Psychopharmacology*, 156, pp. 481–484.

Suebsasana, S., P. Pongnaratorn, J. Sattayasai, T. Arkaravichien, S. Tiamkao, and C. Aromdee. 2009. Analgesic, antipyretic, anti-inflammatory, and toxic effects of andrographolide derivatives in experimental animals. *Archives of Pharmacal Research*, 32, pp.1191–1200.

Suparna, D., A. Pawar, and P. Shinde. 2014. Study of antioxidant and antimicrobial activities of *Andrographis paniculata*. *Asian Journal of Plant Science Research*, 4, pp. 31–41.

Talukdar, P.B., and S. Banerjee. 1968. Studies on the stability of andrographolide. *Indian Journal of Chemistry*, 6, pp.252–254.

Tao, L., L. Zhang, R. Gao, F. Jiang, J. Cao, and H. Liu. 2018. Andrographolide alleviates acute Brain injury in a Rat model of traumatic brain injury: Possible involvement of inflammatory signalling. *Frontiers in Neuroscience*, 12, pp. 657.

Taznin, I., M. Mukti, and M. Rahmatullah. 2015. *Bacopa monnieri*: An evaluation of antihyperglycemic and antinociceptive potential of methanolic extract of whole plants. *Pakistan Journal of Pharmaceutical Sciences*, 28, pp.2135–2139.

Tewari, S.K., A. Niranjan, and A. Lehri. 2010. Variations in yield, quality, and antioxidant potential of kalmegh (*Andrographis paniculata* Nees) with soil alkalinity and season. *Journal of Herbs, Spices & Medicinal Plants*, 16, pp.41–50.

Trivedi, M.N., A. Khemani, U.D. Vachhni, C.P. Shah, and D.D. Santani. 2011. Comparative pharmacognostic and phytochemical investigation of two plant species valued as Medhya Rasayanas. *International Journal of Applied Biology and Pharmaceutical Technology*, 2, pp.28–36.

Uttekar, M.M., T. Das, R.S. Pawar, B. Bhandari, V. Menon Nutan, S.K. Gupta, S.V. Bhat. 2012. Anti-HIV activity of semisynthetic derivatives of andrographolide and computational study of HIV1 gp120 protein binding. *European Journal of Medicinal Chemistry*, 56, pp.368–374.

Vasu, S., V. Palaniyappan, and S. Badami. 2010. A novel microwave-assisted extraction for the isolation of andrographolide from *Andrographis paniculata* and its in vitro antioxidant activity. *Natural Product Research*, 24, pp.1560–1567.

Verma, V.K., K.K. Sarwa, A. Kumar, and M. Zaman. 2013. Comparison of hepatoprotective activity of *Swertia chirayita* and *Andrographis paniculata* plant of North-East India against CCl4 induced hepatotoxic rats. *Journal of Pharmacy Research*, 7, pp.647–653.

Verpoorte, R., R. van der Heijden, H. ten Hoopen, and J. Memelink. 1999. Metabolic engineering of plant secondary metabolite pathways for the production of fine chemicals. *Biotechnology Letters*, 21, pp.467–479.

Vohora, S.B., T. Khanna, M. Athar, and B. Ahmad. 1997. Analgesic activity of bacosine, a new triterpene isolated from *Bacopa monnieri*. *Fitoterapia*, 68, pp. 361–365.

Wang, D., and H. Zhao. 1993. Experimental studies on prevention of atherosclerotic arterial stenosis and restenosis after angioplasty with *Andrographis paniculata* Nees and Fish Oil. *Journal of Tongji Medical University*, 13, pp.193–198.

Xu, C., G.X. Chou, C.H. Wang, and Z.T. Wang. 2012. Rare noriridoids from the roots of *Andrographis paniculata*. *Phytochemistry*, 77, pp. 275–279.

Zhang, X., B.K.H. Tan. 2000. Anti-diabetic property of ethanolic extract of *Andrographis paniculata* in streptozotocin-diabetic rats. *Acta Pharmacologica Sinica*, 21, pp.1157–1164.

Zhang, X.Q., G.C. Wang, W.C. Ye, Q. Li, G.X. Zhou, and X.S. Yao. 2006. New diterpenoids from *Andrographis paniculata* (Burm. F.) Nees. *Journal of Integrative Plant Biology*, 48, pp. 1122–1125.

Zhang, Z., J. Jiang, P. Yu, X. Zeng, J.W. Larrick, and Y. Wang. 2009. Hypoglycemic and beta cell protective effects of andrographolide analogue for diabetes treatment. *Journal of Translational Medicine*, 7, pp.1–13.

Zhao, H.Y., and W.Y. Fang. 1991. Antithrombotic effects of *Andrographis paniculata* Nees in preventing myocardial infarction. *Chinese Medical Journal*, 104, pp.770–775.

Zhao, Y., C.P. Kao, K.C. Wu, C.R. Liao, Y.L. Ho, and Y.S. Chang. 2014. Chemical compositions, chromatographic fingerprints and antioxidant activities of *Andrographis* Herbs. *Molecules*, 19, pp.18332–18350.

3 *Chlorophytum borivilianum* (Musli) and *Cimicifuga racemosa* (Black Cohosh)

Rajib Hossain, Dipta Dey, Partha Biswas, Priyanka Paul, Shahlaa Zernaz Ahmed, Arysha Alif Khan, Tanzila Ismail Ema, and Muhammad Torequl Islam

CONTENTS

3.1 Introduction 46
3.2 Plant Description 47
3.2.1 Chlorophytum borivilianum 47
3.2.2 *Cimicifuga racemosa* 49
3.3 TraditionalKnowledge 49
3.3.1 *Chlorophytum borivilianum* 49
3.3.2 *Cimicifuga racemosa* 50
3.4 Chemical Derivatives (Bioactive Compounds – Phytochemistry) 51
3.4.1 Chlorophytum borivilianum 51
3.4.2 Cimicifuga racemosa 52
3.5 Potential Benefits, Applications, and Uses 54
3.5.1 *Chlorophytum borivilianum* 54
3.5.1.1 Antioxidant Effect 54
3.5.1.2 Immunomodulatory Effect 58
3.5.1.3 Antidiabetic Effect 60
3.5.1.4 Anticancer Effect 60
3.5.1.5 Anti-UlcerativeEffect 60
3.5.1.6 Analgesic Effect 61
3.5.1.7 Antimicrobial Effect 61
3.5.1.8 Anti-Stress Effect 61
3.5.1.9 Anthelmintic Effect 61
3.5.1.10 AntidyslipidemicEffect 62
3.5.1.11 Larvicidal Effect 62
3.5.1.12 Aphrodisiac Effect 62
3.5.1.13 Anxiolytic Effect 63
3.5.1.14 HepatoprotectiveEffect 63
3.5.1.15 Toxicity 63
3.5.2 *Cimicifuga racemosa* 64
3.5.2.1 Anti-Allergic Effect 64
3.5.2.2 Antiestrogenic and Estrogenic Effect 64
3.5.2.3 Antihyperglycemic Effect 65
3.5.2.4 Antimicrobial Effect 66
3.5.2.5 Anti-Osteoporosis Effect 66
3.5.2.6 Antioxidant Effect 67

DOI: 10.1201/9781003205067-3

3.5.2.7 Anticancer Effect 67
3.5.2.8 Antidiabetic Effect 68
3.5.2.9 Anti-Inflammatory Effect 69
3.5.2.10 Anxiolytic Effect 69
3.5.2.11 GABA Receptor Modulating Effect 69
3.5.2.12 Menopause Effect 70
3.5.2.13 Neuroprotective Effect 71
3.5.2.14 Serotonin Receptor Effect 71
3.6 Conclusion 72
Acknowledgment 72
References 72

3.1 INTRODUCTION

Chlorophytum borivilianum (also known as Safed Musli; Family: Liliaceae) is a potential medicinal plant (Purohit et al., 1994). The name *Chlorophytum* is derived from the Greek word, *Chloros* meaning green, and *phyton* meaning plant (Vijaya and Chavan, 2009). Medicinally, the tuberous root of *C. borivilianum* is the most important part). It exerts a plethora of biological effects with its root extract and isolated bioactive compounds in an *in-vivo* and *in vitro*experimental model (Acharya et al., 2008b; 2009).Traditionally, *C. borivilianum* is used for its rejuvenating, aphrodisiac, and natural sex tonic properties (as it is effective in alleviating sexual disorders) in India and the Indian subcontinent (Khanam et al., 2013).It has also demonstrated a variety of pharmacological activities, most importantly antimicrobial(Dabur et al., 2007), antistress (Kenjale et al., 2007), aphrodisiac (Kenjale et al., 2008), immunomodulatory (Thakur et al., 2007), anticancer (Kumar et al., 2010), antioxidant (Kenjale et al., 2007), antiulcer (Arif, 2005), antidiabetic (Giribabu et al., 2014), anthelmintic effects (Deore and Khadabadi, 2010), antidyslipidemic (Giribabu et al., (2014),analgesic (Kumari et al., 2015), anxiolytic (Swami et al., 2014), and hepatoprotective activities due to its chemical compounds(Deore and Khadabadi, 2010).*C. borivilianum* tubers have traditionally been used to treat illnesses such as diabetes, dysuria, diarrhea, and dysentery (Dabur et al., 2007; Negi et al., 1993; Oudhia, 2001,). The information of traditional usage of *C. borivilianum* has largely been passed down the generations in local speaking vernacular and has already been left unrecorded by diverse populations. *C. borivilianum* has typically been enough to treat a wide variety of male sexual issues and is regarded as an overall health booster (Thakur et al., 2009a). Due to its well-known aphrodisiac and immunomodulatory qualities, *C. borivilianum* is presently the most economically utilized species (Figure 3.1).

Cimicifuga racemosa synonyms name *Actaea racemosa* is a major perineal flowering plant under the division of Tracheophyta (Family: Ranunculaceae).Popularly, it is known as black cohosh, black snakeroot, black bugbane (Li et al., 2003;Kennelly et al., 2002). Although this herbal medicine is native to the US,(mainlycentral Georgia, Arkansas, and others) and Ontario (Canada), some countries, like China, haveintroduced its medicinal properties and currently commercially cultivate this plant (Kennelly et al., 2002; Gardner et al., 2012). Native American women used this herbal medicine to address menopausal symptoms, but some research findings demonstrateits capacity to minimize joint pain, several inflammations (mainly inflammation within the lung), myalgias,and neuralgias (Rhyu et al., 2006; Knoess, 2010). Pengelly et al.2014 reported that the Native American population uses this medicinal plant in the case of snakebite, smallpox, measles, old ulcers, and scarlet fever. *C. racemosa* is a shiny dark green plant, which grows up to 3–8feet in height and-2feet in width; the leaves have outstanding features,alternating with sharp notched and separated into 2–5 leaflets inthreegroups (Predny, 2006). *C. racemosa*flowers look buttery greenish-white with a sweet, fetid smell, and the flowers arise from June to September along with buds and seedpods. But most interestingly,beneficial insects, namely bees, pollinate these flowers (Lonner, 2007). Additionally, the fruit of this herbal plant is 5–10mm long, containing multiple seeds, and only one

FIGURE 3.1 (A) Leaves (B) flower, and (C) root of *Chlorophytum borivilianum*(musli).

carpel. Consequently, branching rootstocks are generated with the black rhizome. The rootstocks contain leafstalk and the underside is overcast with several rootlets (Applequist, 2003) (Figure 3.2).

3.2 PLANT DESCRIPTION

3.2.1 CHLOROPHYTUM BORIVILIANUM

The *Chlorophytum* genus is said to have originated in tropical and subtropical Africa and was transported to India through South Africa. The species *C. borivilianum* became popular in the late 1980s. The genus *Chlorophytum* has approximately 300 species of rhizomatous plants that are mainly found in tropical rainforests up to 1500 meters high (Oudhia, 2001). Chlorophytum is found in Asia, tropical Africa, America, and Australia, according to the Genera Plantarum. Although there are 13 species of Chlorophytum identified in India, only around 6–7 of them are used in folk medicine (Table 3.1). Some *Chlorophytum* species are cultivated for their aesthetic value, while others are cultivated for their therapeutic use.

C. borivilianum (Family: Asparagaceae; Sub family: Anthericoideae) has 6–16, radical, 13–23cm × 1–2.5cm in size, spirally imbricate at the base, sessile, linear ovate leaves; small, white, bracteates, pedicellate with joints, zygomorphic flowers arranged in alternate clusters; brown to black-skinned, white after peeling, characteristic odor, tasteless, 3–20in number, fleshy, 8–25cm long roots; greenish-yellow colored, loculicidal capsule, triquetrous, bear 3–12 seeds in each fruit and small, black, angular in shape, endospermic seeds (). For further detail botanical classification of *C. borivilianum* has been explained by Vijaya and Chavan (2009).

C. borivilianum has been cultivated in Australia, Africa, Bangladesh, India, andseveral countries. For its tuberous roots, it is generally grown in an area of more than 400 hectares. Plants stay in an aggressive growth phase or achieve maturation 2.5–3months after germination, with new fleshy roots developing after one month and tendays. Harvesting generally takes place in March and April. During October–November, the produced crop's long and heavy tuberousroots are separated (50–70%) from the disc, and the remaining tiny, thin tubers (fingers) and disc are saved for sowing in the next growing season. To achieve a good yield, *C. borivilianum* requires a variety of

FIGURE 3.2 (A) Leaves, (B) flower and (C) root of *Cimicifuga racemosa*(black cohosh).

TABLE 3.1
Distribution of Various Species of *Chlorophytum*

Species	Distribution
*C. arundinaceum*Baker	Districts of Chhota Nagpur, parts of centralIndia, foothills of Northeast Himalaya inAssam, West Bengal, and Bihar.
C. attentuatum Baker	Western Ghats from Karnataka southwardto Coimbatore.
C. breviscapum Dalz.	Sikkim Himalaya, Belgaum, and WestPenninsula.
C. borivilianum Santand Fern.	Madhya Pradesh, North Gujarat,the subtropical Himalayas from Kumaon,Khasia hills, Bengal, Assam, Kokan,Kanara, West Peninsula,and Chennaiextending to Kanyakumari.
C. glaucum Dalz., *C.orcbidastrum* Lindley	Hilly ranges of Sahyadris in western India.
C. kbasianum Hooker,*C. undulatum* Wall. syn. *C. nepalense*(Lindley) Baker	Eastern parts of India.
C. laxum R.Br.	Kakti Hills, Belgaum, Dharwar, DeccanPeninsula
C. malabaricum Baker	Nilgiris and western Ghats
C. tuberosum Baker	Parts of Konkan to Travancore in Kerla,Eastern Himalaya, Bihar, and West Bengal

circumstances for proper growth and development. Weather and soil types, as well as propagating, plant techniques, fertilizer treatment, weed, and insect control, and intercropping, all have a part in achieving higher yields.*C. borivilianum* has many common names depending upon the languages used in a particular region. The names used in different languages of India and other countries are given in Table 3.2.

TABLE 3.2
Vernacular Names of *C. borivilianum* (Adopted from Singh et al., 2012)

Country	Vernacular names
India	Swetha musli; Safed musli, Hazarmuli, Satmuli, Ujlimusli, Dholi musali; Shedeveli, Shedheveli; Safed musli, Sufed musli, Kuli; Tannirvittang, Tannirvittan-Kizhangu, Vipurutti, Taniravi thang; Tsallogadda, Swetha musli; Hirtha-wariya, Mushali; Jhirna; Khairuwa
Bangladesh	Shatamuli
Saudia Arabia	Shaqaqule-hindi, Shaqaqule
England	Indian spider plant, spider plant, white musale
France	Chlorophytum medicinal

3.2.2 *Cimicifuga racemosa*

In the eastern US as well as Canada, *C. racemosa*is a tropical plant, and these two countries are known as commercial storehouses of *C. racemosa* (Huntley, 2004; Barceloux, 2008). Not only in these two but also many otherareas of the world propagate this plant. Since thelast decade, the *C. racemosa* plant has spread worldwideincluding the distribution range of this herb from Massachusetts to Ontario as well as Missouri and Georgia, west to Illinois and Arkansas (Gardner et. al., 2012; Gafner, 2016). Additionally, China is nowcommercially growing some species of *C. racemose* (Jiang B et. al., 2006). Thousands of populations of several species of this plant are found worldwide. Around 100 species are found in Indiana, 750 to 1000 arefound inNorth Carolina national forest land. There are a thousand species in New York, a hundred in Tennessee and Maryland, and 20–30 in South Carolina (Predny, 2006).*C. racemosa*taxonomical details are illustrated by McKenna et al. (2001) and Guo et al. (2017).

*C. racemosa*is a perennial vertical herb,a plant that contains a bland stalk (Lonner, 2007). This plant grows 3 to 8feet in height and2feet in width (Predny, 2006),leaves are alternated with sharp notchesand separated into 2–5 leaflets which are placed into three groups. The terminal of the leaves consists of three lobes with threebasal veins. The leaves areshiny dark green (Lonner, 2007). *C. racemosa*bearsspike-like buttery greenish-white flowers (Papps, 2000). Petal-less flowers arise from June to September along with buds and seedpods; it has an outstanding sweet, fetid smell that attracts flies and beetles. Importantly, beneficial insects, namely beespollinate these flowers (Barceloux DG., 2008). The fruits of this plant are oval-shaped, and the ovary of this fruit is incisive. The bald-pated seeds of these fruits are placed in two rows. Branching rootstocks are generated with a black rhizome that is an important part of it. The rootstocks contain leafstalk and the underside is overcast with several rootlets (Pengelly et al., 2014).

3.3 TRADITIONALKNOWLEDGE

3.3.1 *Chlorophytum borivilianum*

Since the 11th century, the root tubers of *C. borivilianum* have been utilized as a traditional medicine in India for a variety of illnesses and health issues (Vijaya and Chavan, 2009).Ayurveda, an ancient Indian traditional system of medicine, maintains a particular place for *C. borivilianum*, which is connected with the divine plant "Caitha". In Ayurvedic, it also is known as "Divya aushad" (holy medication). *C. borivilianum* is sweet, bitter (Rasa), moist, unctuous (Guna), cold (heavy Virya), with apleasant post-digestive impact (Vipaka) according to Ayurveda (Paques and Boxus, 1985). *C. borivilianum* has been used for around 4000 years, based on the current Ancient epic Srimad Bhagawat. It belongs to the "Vajikaran Rasayana" family of medicinal plants, which are utilized for renewing and revitalizing qualities to improve sexual dynamics (Puri, 2003) and alleviate erectile dysfunction (Triveni, 1977).

This is also the foundation for the Kamasutra's medication recommendations. According to Indian literature, Ashwini Kumars, a heavenly physician, made the "chyawanprash" from safed musli for Chyavanrishi, who married at the age of 80, to improve sexuality. Other traditional Indian works, such as the Bhavaprakash Nighantu, Rajendra Sarsangrah, and Raja Ballabh Nighantu, identified *C. borivilianum* as "Vajikaran".There are 30–35 herbal medicines with "Rasayana"characteristics, especially *C. borivilianum*, listed in several Ayurvedic and Siddha treatises. Sushruta described Rasayana as a remedy that slows down the aging process (Vayasthapam), extends life (Ayushkaram), improves knowledge (Medha), and strengthens the body (Ayushkaram) (Bala) (Pushpangadan et al., 2012).*C. borivilianum* roots have long been used to treat impotency and oligospermia in males, and they're also a sort of immunomodulator (Kothari and Singh, 2004; Puri, 2003; Sharma et al., 1999; Tandon and Shukla, 1995; Triveni, 1977).

Having astringent, carminative, anti-pyretic, diuretic, and carminative properties, the roots are also utilized as a galactagogue and an aphrodisiac (Bhandary et al., 1995; Chetty and Rao, 1989). Rhizomes have been known to cure premature ejaculation in some parts of Mewar, India (Deshwal and Trivedi, 2011). *C. borivilianum* roots have an identical effect on the central nervous system as ginseng and are commonly referred to as "Indian ginseng" due to their miraculous medicinal power. *C. borivilianum*not only rejuvenates the reproductive organs, but also inhibits erectile dysfunction and enhances the body's overall immune response. It is well-known for increasing muscle strength and endurance, combating overall body fragility, increasing muscular mass, and assisting in the recovery from muscular tiredness and fatigue.

The tubers are also used to improve strength in asthma patients. The rhizome powder has been used in the production of numerous,ancient Indian health tonics, together with other botanical ingredients (Haque et al., 2011). Drinking a spoonful of safed musli twice a day with milk is a component of the daily health care program in western India, particularly in Gujrat. The indigenous inhabitants of western Asiaconsume the leaves as an expectorant. (Sebastian and Bhandari, 1988). During pregnancy, the roots are used as a nutritional stimulant for both the pregnant woman and the fetus, and they are also used to restore bodily fluids in the postpartum phase.

The fresh roots are one of the components in the production of a particular form of dessert-that women eat after giving birth as an invigorating meal. The dried root powder promotes the production of milkin breastfeeding women andcows and is used to treat a variety of gynecological problems, including gonorrhea, leucorrhea, and pre- and post-natal symptoms(Deshwal and Trivedi, 2011). Besides, *C. borivilianum* roots are often used to treat arthritis, rheumatism, and joint discomfort. If consumed regularly with milk, it is known to relieve knee discomfort within a week (Elizabeth, 2001; Singh and Chauhan, 2003). *C. borivilianum* roots have indeed been utilized to treat bone fractures, leucorrhea, and as a male stimulant (Meena and Rao, 2010). Furthermore, *C. borivilianum* has been discovered to be a primary source of aphrodisiac and metabolism-boosting compounds. It also provides strong immunity and wound-healing as a native therapy for health care. Some other studies have demonstrated that the tuberous root of *C. borivilianum* exertsethno-pharmacological effect against diabetes mellitus, dysuria, diarrhea, and dysentery (Dabur et al., 2007; Negi et al., 1993; Oudhia, 2001,), and is also administered orally for purifying the blood and delaying the aging process as a supplementary therapy. In obesity and its related side effects, it is considered an effective medication to treat and prevent.

Additionally, *C. borivilianum* has been reported as a cosmetic agent for men and women. To lighten the skin tone, apply a slurry of the root mixed with goat milk or honey to the face. Evidence suggests that *C. borivilianum* exhibits no adverse response but sometimes over-dosing may cause GI disorders. Besides, leaves are used as vegetables in several preparations (Patil, 2003; Vartak, 1981).

3.3.2 *CIMICIFUGA RACEMOSA*

Native Americans were the first to notice that this herb was helpful forpain relief and several inflammations. Women used it for pain duringthe menstrual cycle (Pengelly and Bennett, 2014). In 1828, *C.*

*racemosa*was first used by Americans for medical purposes; and since1860, Europe has broadly used this plant for various disorders (Papps, 2000). Consistently, native North American used this herb in severaldisorders including joint pain, myalgias and neuralgias, general gynecological symptoms, and so on (Huntley and Ernst, 2003). The World Health Organization (WHO) and European scientific cooperative on phytotherapy reported that this green plant is beneficial for the treatment of menopausal symptoms (Huntley, 2004). Additionally, this herb is also used to treat rheumatoid arthritis, sciatica, whooping cough (Guo etal., 2017). *C. racemosa*treatsseveral other ailmentsincluding malaria, kidney function diseases, uterine disorders, musculoskeletal problems, sore throat (used as a gargle), and complications in childbirth (Papps, 2000; Dugoua et al., 2006; Predny, 2006). It acts to stimulate milk production of new mothers (Firenzuoli et al., 2011). American people used it to treat snakebites and smallpox, old ulcers, inflammation of nerves, measles, and scarlet fever (Pengelly and Bennett, 2014).

3.4 CHEMICAL DERIVATIVES (BIOACTIVE COMPOUNDS – PHYTOCHEMISTRY)

3.4.1 CHLOROPHYTUM BORIVILIANUM

C. borivilianum is a valued medicinal plant, having copious bioactive components, such as phenols, saponins, flavonoids, alkaloids, tannins, steroids, triterpenoids, and vitamins (Visavadiya et al., 2010). Another phytochemical evaluation of several extracts of *C. borivilianum*suggested thatit contains carbohydrates, protein, fibers, phytosterols, saponins, polyphenols, flavonoids, and ascorbic acid (Huang et al., 2019). Primarily saponins and alkaloids impart medicinal value. It is a rich source of over 25 alkaloids, vitamins, proteins, calcium, magnesium, phenol, resins, mucilage, and polysaccharides and also contains a high quantity of simple sugars, mainly sucrose, glucose, fructose, galactose, mannose, and xylose. Recently,stigmasterol and saponin were namfurostanol and Chlorophytoside-I(3b,5a,22R,25R)-26-(β-Dglucopyranosyloxy)-22-hydroxy-furostan-12-one-3ylO -β-D-galactopyranosyl (1-4) glucopyranoside has been isolated.

Some studies have shown that the roots of *C. borivilianum* contain42% carbohydrate, 8–9% protein, 3–4% fiber, and 2–17% saponin. Other researchersreported that the root constitutes about 30% alkaloids, 10–20%saponins, 40–45% polysaccharide (mucilage), and 5–7% protein (Deore and Khadabadi,2009b). In addition, *C.borivilianum* contains steroids, triterpenoids, gallo-tannins, vitamins, potassium, calcium,magnesium, rare elements (such as Zn, Cu, and P), resins, and a high quantity of simple sugars(mainly sucrose, glucose, fructose, galactose, mannose, and xylose).

Kothari and Singh (2004) were the first to report the inulin-type 2->1 linked fructans by comparative reverse-phase high-pressure anion exchange (RP-HPAE) chromatography. Later Narasimhan et al. (2006) successfully isolated for the first time, the same fructooligosaccharides from *C. borivilianum* and identified them as *O*-β-*D*fructofuranosyl-(2->1)-(β-*D*-fructofuranosyl)n-(2->1)-α-*D*-glucopyranoside (n = 5–30) using HPAE chromatography.

From the roots and rhizomes of *C. borivilianum* several bioactivechemicals have been found, including phytosterol (stigmasterol and hecogenin) (Bathoju and Giri, 2012a; 2012b), steroidal saponins (Borivillianoside A, B, C, D, E, F, G and H) (Acharya et al., 2008a; Acharya et al., 2009), saponin (25R) -3β,5α,22α-22-methoxyfurostan-3,22,26-triol3-*O*-β-D-xylopyranosyl-(1->3)-[β-D-glucopyranosyl-(1->2)]-β-D-glucopyranosyl-(1->4)-β-D galactopyranosyl 26-*O*-β-D-glucopyranoside;(25R)-3β,5α,furost-en-20(22)-3,26-diol 3-O-β-D-xylopyranosyl-(1->3)-[β-D-glucopyranosyl-(1->2)]-β–D-glucopyranosyl -(1->4)-β-D-galactopyranosyl-26-O-β-D-glucopyranoside (5);(25R)-3β-5α-spirostan-3-ol 3-*O*- α-L-rhamnopyranosyl(1->4)-[β-D-glucopyranosyl-(1->3)]-β-D-glucopyranosyl-(1->3)-α-L-arabinopyranosyl-(1->3)-[β-D-xylopyranosyl-(1->2)]-β–D-glucopyranosyl-(1->2)-[α–L-rhamnopyranosyl-(1->6)-β–D -galactopyranoside;3β-5α- spirost-ene-25(27)-3-ol 3-*O*-α-L-rhamnopyranosyl(1->4)-[β-D-glucopyranosyl-(1->3)]-β-D-glucopyranosyl-(1->3)-α-L-arabinopyranosyl-(1->3)-[β-D-xylopyranosyl-(1->2)]-β–D-glucopyranosyl-(1->2)-[α-L-rhamnopyranosyl-(1->6)]-β–D-galactopyranoside; 3β-5α- spirost-ene-25(27)-2,3-diol3-O-β-D-glucopyranosyl-(1->3)-[β-D-glucopyranosyl-(1->2)]-β-D-glucopyranosyl-(1->4) -β-D- galactopyranoside; (25S)-3β-5α-spirostane-3-ol3-*O*-β-D-xylopyranosyl-(1->3)-β-D-glucopyran

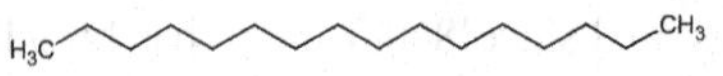

Hexadecane

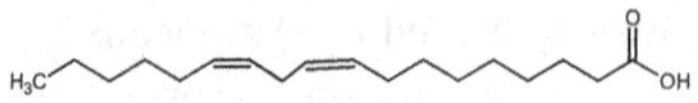

Linoleic Acid

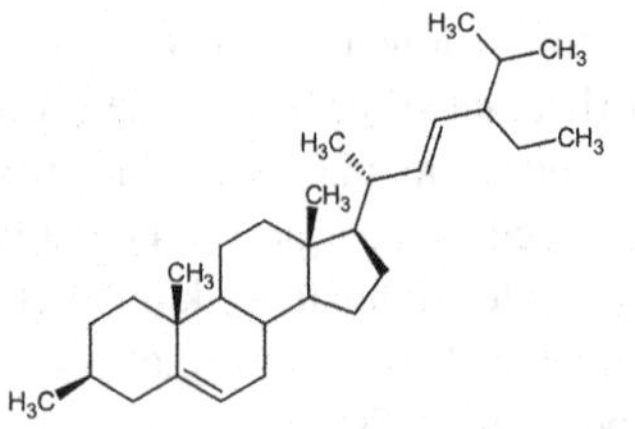

Stigmasterol

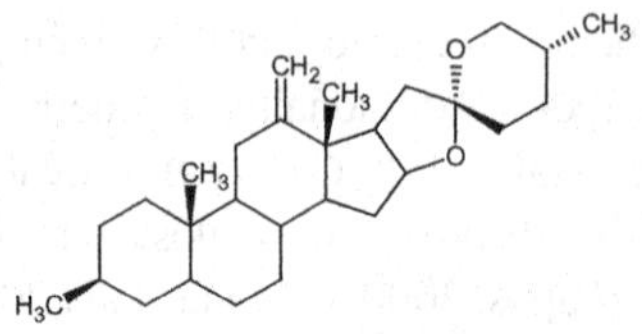

Hecogenin

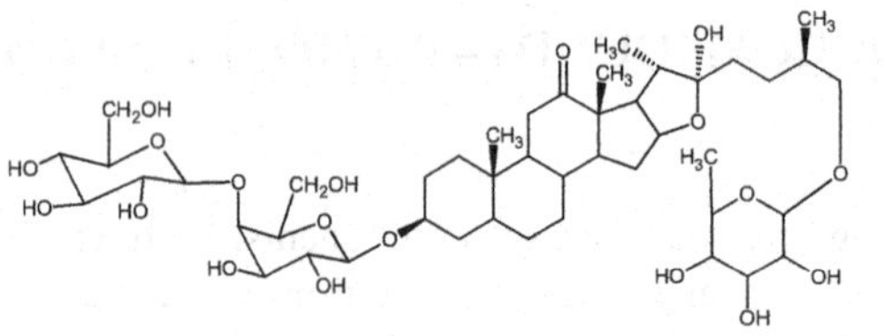

Chorophytoside I

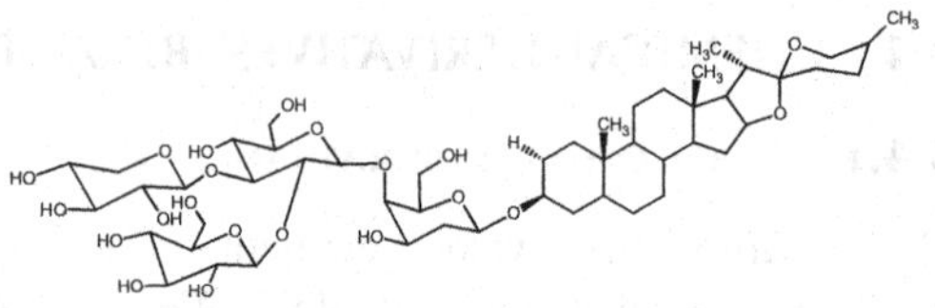

25(R,S)-5α–spirostan-3-ol-*O*-β-D-xylopyranosyl-(1->3)-[β-D-glucopyranosyl-(1->2)]-β-D-glucopyranosyl-(1->4)-β-D- galactopyranoside

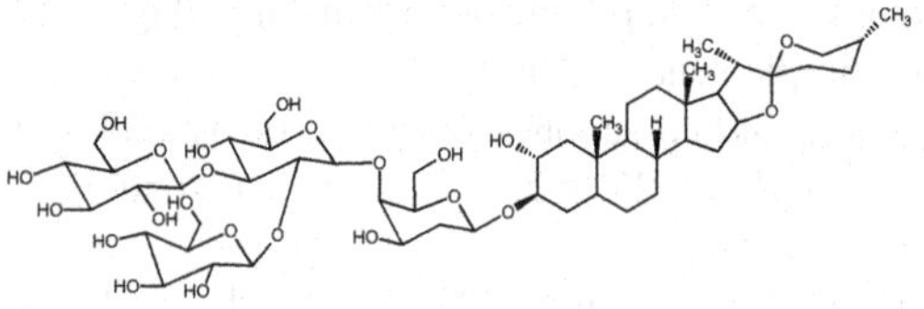

Gitogenin 3-*O*-β-D-glucopyranosyl-(1->3)-[β-D-glucopyranosyl-(1->2)]-β-D-glucopyranosyl-(1->4)-β-D- galactopyranoside

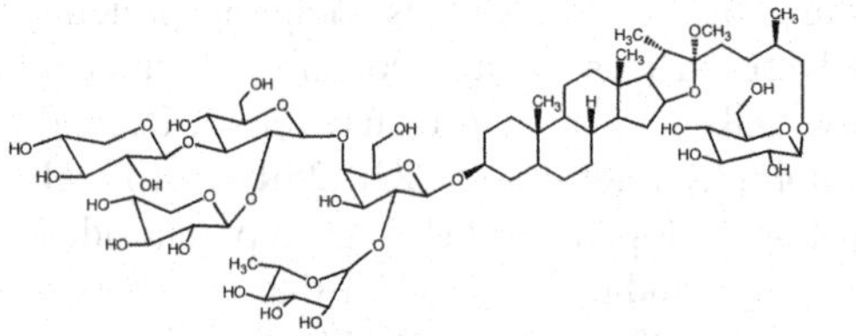

(3β,5α,22R,25R)-26-(β-D-glucopyronosyloxy)-22-methoxyfurostan-3-yl-*O*-β-D-xylopyranosyl-(1->2)-O-[β-D-xylopyranosyl-(1->3)]-O-β-D-glusopyranosyl-(1->4)-O-[α-L-rhamnopyranosyl-(1->2)-β-D- galactopyranoside

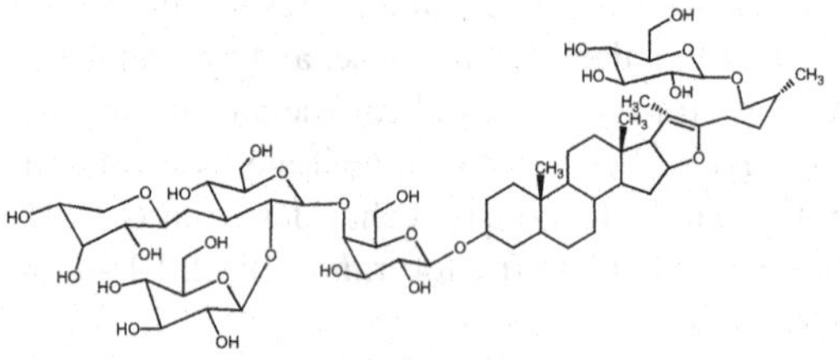

(25R)-3β,5α,furost-en-20(22)-3,26-diol3-*O*-β-D-xylopyanosyl-(1->3)-[β-D-glucopyranosyl-(1->2)]-β-D-glucopyranosyl-(1->4)-β-D-galactopyra nosyl 26-O-β-D-glucopyranoside

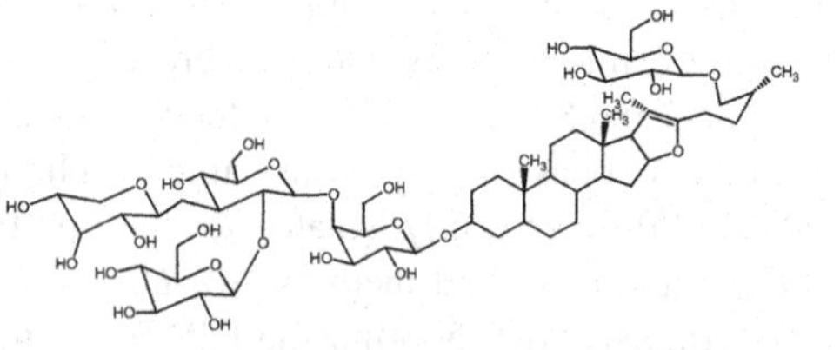

(25R)-3β,5α,furost-en-20(22)-3,26-diol3-*O*-β-D-xylopyanosyl-(1->3)-[β-D-glucopyranosyl-(1->2)]-β-D-glucopyranosyl-(1->4)-β-D-galactopyra nosyl 26-O-β-D-glucopyranoside

FIGURE 3.3 Chemical compounds isolated from *Chlorophytum borivilianum*.

osyl-(1->2)-β-D-glucopyranosyl-(1->4)-β-D-galactopyranoside) (Acharya et al., 2008b); fatty acids (11, 14 eicosadienoic acidand linoleic acid)(Deore andKhadabadi, 2010; hydrocarbon (Hexadecane, 1'-acetoxychavicol acetate, Di -92-ethylhexyl); and phthalate (Chua et al., 2015; Chua et al., 2017) (Figure 3.3).

3.4.2 Cimicifuga racemosa

A large number of potential phytoconstituents have been isolated from various parts of the *C. racemosa* plant, which is a tropical tree native to North America. In the total plant parts,

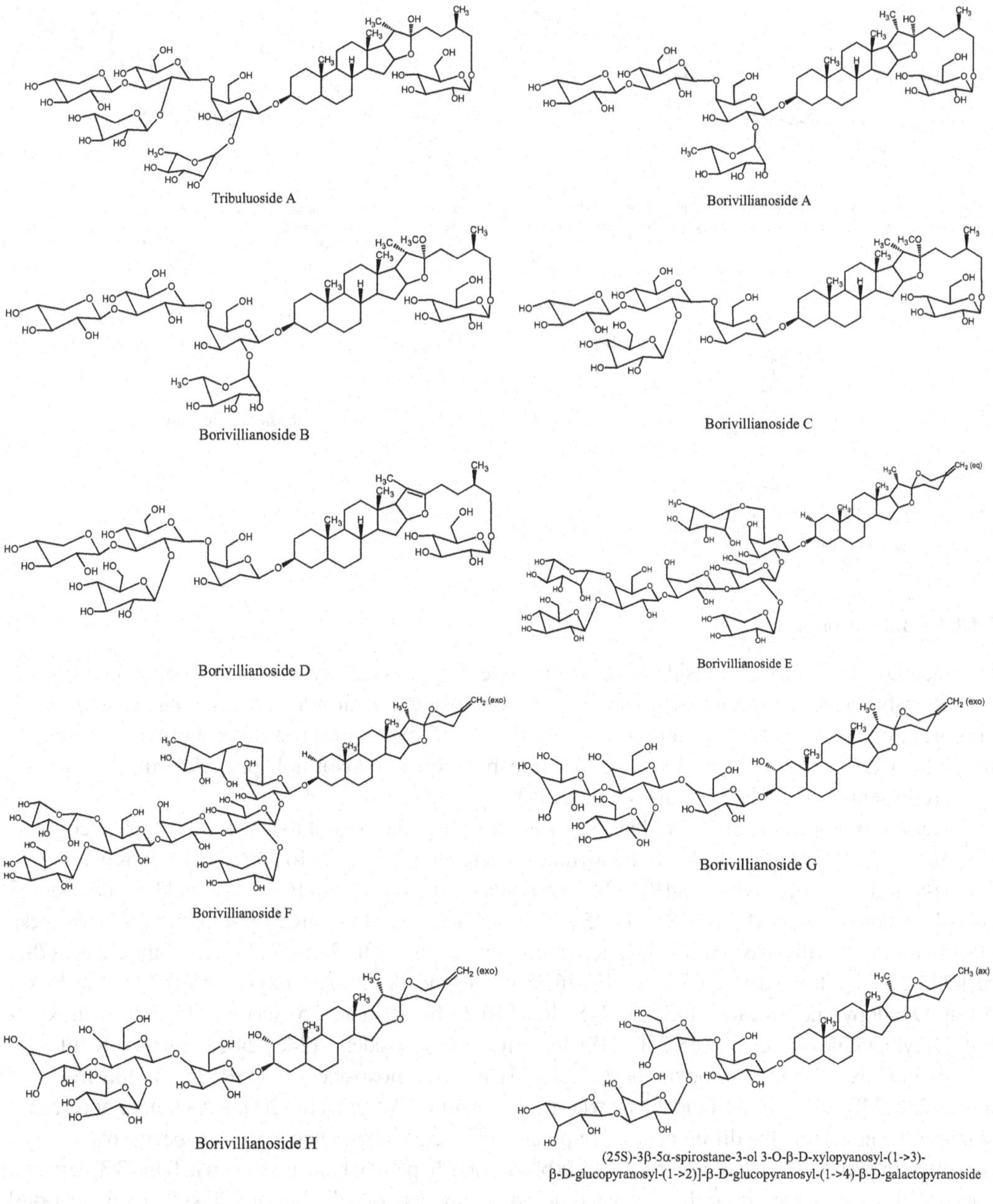

Tribuluoside A

Borivillianoside A

Borivillianoside B

Borivillianoside C

Borivillianoside D

Borivillianoside E

Borivillianoside F

Borivillianoside G

Borivillianoside H

(25S)-3β-5α-spirostane-3-ol 3-O-β-D-xylopyanosyl-(1->3)-β-D-glucopyranosyl-(1->2)]-β-D-glucopyranosyl-(1->4)-β-D-galactopyranoside

FIGURE 3.3 (Continued)

the rhizome and root are the major sources of the potential bioactive phytochemicals noted as cimipronidine, fukinolic acid, cimicifugic acids A, B, and F, fukinolic acid, ferulic acid, isoferulic acid are alkaloid in nature (Fabricant et al., 2005),as well as, cimiracemosides A–H, cimicifugic acid, rosmarinic acid, sinapic acid, cimiciphenol, cimicipgycolate, P-coumaric acid are the cimicifugic acid derivatives (Godecke et al., 2009). On the contrary, the rhizome and root also contain a large number of polyphenolic compounds, including the cimicifugic acid A, B,E,F andfukinolic acid, cimiracemoside N, cimiracemoside O, cimiracemoside L, cimicifugoside H-2, cimifugoside H-1, 23-O-acetylshengmanol-3-D-xyloside, 23-epi-26-deoxyactein, 23-O-acetylshengmanol-3-D-arabinoside, 25-anhydrocimigenol-3-O-xyloside,cimiracemoside P,

3β-5α-spirost-ene-25(27)-2,3-diol 3-O-β-D-glyucopyanosyl-(1->3)-[β-D-glucopyranosyl-(1->2)]-β-D-glucopyranosyl-(1->4)-β-D-galactopyranoside

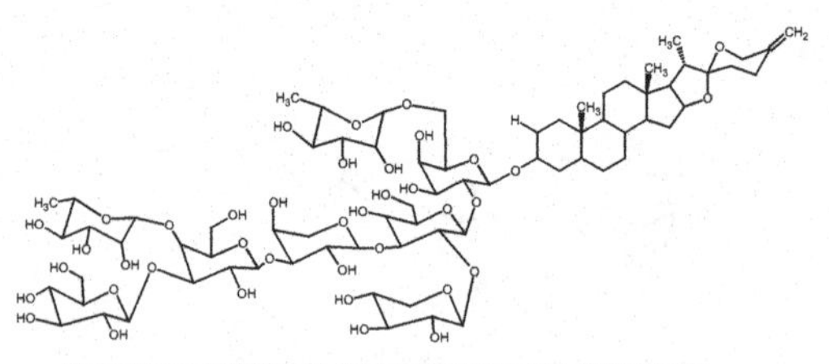

3β-5α-spirost-ene-25(27)-3-ol-3-O-α-L-rhamnopyanosyl-(1->4)-[β-D-glucopyranosyl-(1->3)]-β-D-glucopyranosyl-(1->3)-α-L-arabinopyranosyl-(1->3)-[β-D-glucopyranosyl-(1->2)-[α-L-rhamnopyranosyl-(1->6)]-β-D-galactopyranoside

(25R)-3β-5α-spirostan-3ol-3-O-a-L-rhamnopyranosyl(1->4)-[β-D-glucopyranosyl-(1->3)]-β-D-glucopyranosyl-(1->3)-α-L-arabinopyranosyl-(1->3)-[β-D-xylopyranosyl-(1->2)]-β-D-glucopyranosyl-(1->2)-[α-L-rhamnopyranosyl-(1->6)-β-D-galactopyranoside

11,14 eicosadienoic Acid

FIGURE 3.3 (Continued)

cimiracemoside J,cimiracemoside K, 25-O-acetylcimigenol-3-O-xyloside, 25-O-acetylcimigenol-3-O-R-arabinoside, 25-anhydrocimigenol-3-O-arabinoside, pyridoxine, adenine, panthotenic acid, cimipronidine methyl ester, N-methyl cyclocimipronidine, cyclocimipronidine, choline, γ-guanidino butyraldehyde, salsolinol, cytidine, γ-guanidino butanol, norsalsolinol, phenylalanine and many more represented in Table 3.3 (Imai et al., 2016).

Diverse types of research studies have reported that the aerial part of theblack cohosh plant contains (1*S*,15*R*)-1,15,25-trihydroxy-3-*O*-β-D-xylopyranosyl-acta-(16*S*,23*R*,24*R*)-16,23;16,24-binoxoside; and 3-*O*-α-L-arabinopyranosyl-(1*S*,24*R*)-1,24,25-trihydroxy-15-oxo-acta-(16*R*,23*R*)-16,23-monoxoside;3-O-α-L-Arabinopyranosyl-(1S,15R)-1,15,25-trihydroxy-acta- (16S,23R,24S)-16,23;16,24-binoxoside; (1S,15R)-1,15,25-Trihydroxy-3-O-β-D-xylopyranosyl-acta- (16S,23R,24S)-16,23;16,24-binoxoside; (7S,15R)-7,15,25-Trihydroxy-acta-(16S,23R,24S)-16,23;16,24-binoxol; 25-Acetoxy-(7S,15R)-7,15-dihydroxy-3-O-β-D-xylopyranosyl-acta- (16S,23R,24S)-16,23;16,24-binoxoside: 25-Acetoxy-(15R)-15-hydroxy-3-O-β-D-xylopyranosylacta- (16S,23R,24S)-16,23;16,24-binoxoside: (24S)-24-Acetoxy-(15R,16R)-15, 16,25-trihydroxy-3-O-β-D-xylopyranosyl-acta-(23S)-16,23-monoxoside: (15R)-15,25-Dihydroxy-acta-(16S,23R,24S)-16,23;16,24-binoxol are triterpene in nature (Powell et al., 2008). A number of research studies have noted that the different types of potential bioactive phytochemicals have been isolated from the different parts of the *Cimicifuga racemosa*(black cohosh) plant which are shown inTable 3.3, depicted with the chemical structure of these isolated bioactive phytochemicals. In 2012, Nikolić et al., reported that the triterpenoid compounds 27-deoxyactein, 27-desoxyacetylacteol, and carboxylic acid (likeacetic acid) were isolated from the black cohosh fruits. Recently, some major compounds, such as cimigoside, formic acid, isoferulic-acid, palmitic-acid, and racemosin were isolated from the fruits (Imai et al., 2020) Data is tabulated in Table 3.3 (Figure 3.4).

3.5 POTENTIAL BENEFITS, APPLICATIONS, AND USES

3.5.1 *CHLOROPHYTUM BORIVILIANUM*

3.5.1.1 Antioxidant Effect

C. borivilianum plant contains several phytochemical constituents such as phytosterols, saponins, polyphenols, flavonoids, and ascorbic acid (Giribabu et al., 2014). Through these bioactive compounds

TABLE 3.3
The Tabular Representation of Isolated Phytochemicals from the Total Plant of *Cimicifuga racemosa*

Plant Part	Class	Compounds	References
Rhizome and root	Alkaloid	Cimipronidine, Fukinolic acid, Cimicifugic acid A, B, and F, Ferulic acid, Isoferulic acid, Cyclo-cimipronidine	Fabricant et al., 2005; Godecke et al., 2009
	Triterpene glycosides	Cimiracemoside A-H, Rosmarinic acid, Sinapic acid, Cimiciphenol, Cimicipgycolate, P-coumaric acid	Shao et al., 2000; Godecke et al., 2009;
	polyphenolic compound	Cimicifugic acidA, B, D-F, fukinolic acid; Cimiracemoside J-L, N-P; Cimifugoside H-1; Cimicifugoside H-2; 23-O-acetylshengmanol-3-D-xyloside; 23-epi-26-deoxyactein; 23-O-acetylshengmanol-3--D-arabinoside; 25-anhydrocimigenol-3-O-xyloside;25-O-acetylcimigenol-3-O-xyloside; 25-O acetylcimigenol-3-O-R-arabinoside; 25-anhydrocimigenol-3-O-arabinoside; 17β-estradiol; Pyridoxine; Adenine; Panthotenic acid; Cimipronidine methyl ester; N-methyl cyclocimipronidine; Cyclocimipronidine; Choline; γ-Guanidino butyraldehyde;Salsolinol; Cytidine; γ-Guanidino butanol; Norsalsolinol; Phenylalanine; Guanosine; 5′-O-(β-D-glucopyranosyl) Pyridoxine; Glycine betaine; Proline betaine; Trigonelline; δ-guanidinovaleric acid; Cimipronidine; γ-Guanidino butyric acid; Pipecolic acid; L-carnitine; α-N-acetyl arginine; N-formyl arginine; Pyroglutamic acid; Histidine betaine; Choline hexoside; Arginine; Adenosine; γ-Guanidino butyric acid methylester; Caffeoyl arginine; N-isoferuloyl histidine; N-feruloyl arginine; N-isoferuloyl arginine; Magnoflorine; Cyclanoline; γ-Guanidino butyric acid ethylester; δ-Guanidinovaleric acid methyl ester; 2′-O-methyladenosine; N-methyladenosine; Feruloyl putrescine; Isoferuloyl putrescine; Benzoyl choline; Magnocurarine; 1,2,3,4,-Tetrahydro-β-carboline-3- carboxylic acid; Feruloyl choline; Isoferuloyl choline; N-feruloyl dopamine-4′-O-hexoside; N-isoferuloyl dopamine-4′-Ohexoside; 1,2-Dehydrosalsolinol; N-phenylacetyl acetamide; N(2)-methyl-6-hydroxy-3,4- dihydro-β-carboline; N-isoferuloyl glutamic acid, Laurifoline; N-feruloyl phenylalanine-4′-O hexoside; N-cyclohexyl-4-hydroxy benzylamine; N(2)-methyl-6-hydroxy-1,2,3,4- tetrahydro-β-carboline; Cimitrypazepine; Norcoclaurine; Laurolitsine; Reticuline; oblongine; N-isoferuloyl arginine methylester; Menisperine; N-feruloyl tyramine-4‴-O-hexoside; N-isoferuloyl arginine ethylester; Laurotetanine; N-feruloyl-3‴-methoxytyramine-4‴- O-hexoside; N-methyl tetrahydrocolumbamine or isomer; Xanthoplanine; Protopine; Allocryptopine; (3s)-2-(3”,4”-dihydroxyphenylm thylene -3-hydroxy l-3-(3’,4’-dihydroxybenzonyl-gama-butyrolactone; (2R,3S)-2-o-(3’,4’-dimethoxy-E-cinnamoyl)-3-hydroxy-3-[(3”,4”-dihydroxyphenyl)-methyl]-butanedioic acid;rotocatechuic acid; Protocatechualdehyde; P-coumaric acid; Caffeic acid; Methyl caffeate; Ferulic acid; Ferulate-1-methyl ester; Isoferulic acid; 1-isoferuloyl-â-D-glucopyranoside; Fukinolic acid	Li et al., 2003; Nuntanakorn et al., 2006; Nikolić et al., 2012; Imai et al., 2016
	Phenol carboxylic compounds	Dehydrocimicifugic acid A, B	Li et al., 2003

(*Continued*)

TABLE 3.3 (CONTINUED)
The Tabular Representation of Isolated Phytochemicals from the Total Plant of *Cimicifuga racemosa*

Plant Part	Class	Compounds	References
	Triterpeneglycosides	23-epi-26-deoxyactein, Cimigenol-3-*O*-xyloside	Schmid et al., 2009; Shao et al., 2000
	Cimipronidine	Cimipronidine methyl ester; Cimipronidine; Salsolinol; Dopargine; γ-guanidinobutyraldehyde; 3-hydroxytyrosol 3-*O*-glucoside	Godecke et al., 2009
	Cycloartane glycosides	Cimigenol 3-O-a-L-arabinopyranoside; 25-Omethoxycimigenol 3-O-a-L-arabinopyranoside 12β-hydroxycimigenol 3-O-a-L-arabinopyranoside; 27-deoxyactein; Actein; Cimiracemoside F-H 12b,21-dihydroxycimigenol 3-O-a-L-arabinopyranoside;23-O-acetylshengmanol 3-O-a-L-arabinopyranoside; 25-O-acetyl-12-b hydroxycimigenol 3-O-a-L-arabinopyranoside (22R,23R,24R)-12b acetyloxy-16b ,23:22,25-diepoxy-23,24-dihydroxy-9,19- cyclolanostan-3b-yl a-L-arabinopyranoside; *N*-methylserotonin	Watanabe et al., 2002
	Triterpene glycosides	23-epi-26-deoxyactein,actein, 26-deoxyactein	Powell et al., 2008
Aerial part	Triterpene	(1*S*,15*R*)-1,15,25-trihydroxy-3-*O*-β-D-xylopyranosyl-acta-(16*S*,23*R*,24*R*)-16,23;16,24-binoxoside 3-*O*-α-L-arabinopyranosyl-(1*S*,24*R*)-1,24,25-trihydroxy-15-oxo-acta-(16*R*,23*R*)-16,23-monoxoside 3-O-α-L-Arabinopyranosyl-(1S,15R)-1,15,25-trihydroxy-acta- (16S,23R,24S)-16,23;16,24-binoxoside; (1S,15R)-1,15,25-Trihydroxy-3-O-β-D-xylopyranosyl-acta- (16S,23R,24S)-16,23;16,24-binoxoside; (7S,15R)-7,15,25-Trihydroxy-acta-(16S,23R,24S)-16,23;16,24-binoxol 25-Acetoxy-(7S,15R)-7,15-dihydroxy-3-O-β-D-xylopyranosyl-acta- (16S,23R,24S)-16,23;16,24-binoxoside; 25-Acetoxy-(15R)-15-hydroxy-3-O-β-D-xylopyranosylacta- (16S,23R,24S)-16,23;16,24-binoxoside; (24S)-24-Acetoxy-(15R,16R)-15,16,25-trihydroxy-3-O-β-D-xylopyranosyl-acta-(23S)-16,23-monoxoside; (15R)-15,25-Dihydroxy-acta-(16S,23R,24S)-16,23;16,24-binoxol	Imai et al., 2016
	isoflavone	Formononetin	Li et al., 2002; McCoy et al., 2019
	Aromatic acids	Caffeic acid; Isoferulic acid; Ferulic acid	
	Cycloartane triterpene glycosides	Cimiracemoside A, C;Acetyl shengmanol xyloside; 26- deoxycimicifugoside; Caffeic acid; Ferulic acid, Isoferulic acid, Cimicifugoside H-1, Cimicifugoside H-2, (26*R*)-actein, (26*S*)-actein, 26-deoxycimicifugoside, 23-*epi*-26-deoxyactein, 23-OAc-shengmanol-3-*O*-β-D-xyloside, 26-deoxyactein, 25-OAc-cimigenol-3-*O*-α-L-arabinoside, 25-OAc-cimigenol-3-*O*-β-D-xyloside, Cimigenol-3-*O*-α-L-arabinoside, Cimigenol-3-*O*-β-D-xyloside	
Fruit	Terpenoid	27-deoxyactein, 27- desoxyacetylacteol, cimigoside	Nikolić et al., 2012
	carboxylic acid	acetic-acid Formic acid, isoferulic-acid, palmitic-acid, starch, racemosin	Imai et al., 2020

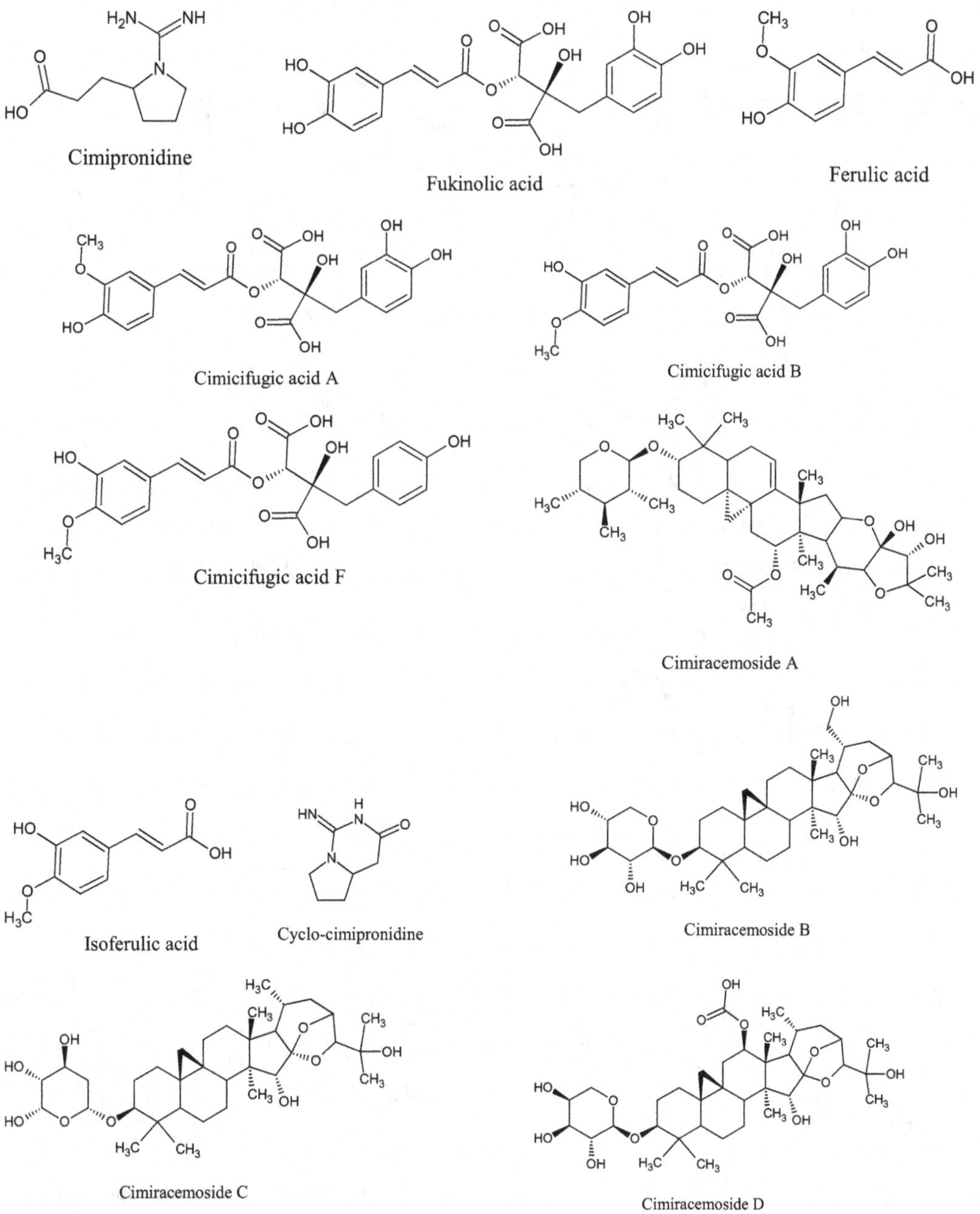

Cimipronidine

Fukinolic acid

Ferulic acid

Cimicifugic acid A

Cimicifugic acid B

Cimicifugic acid F

Cimiracemoside A

Isoferulic acid

Cyclo-cimipronidine

Cimiracemoside B

Cimiracemoside C

Cimiracemoside D

FIGURE 3.4 Chemical compounds isolated from *Cimicifuga racemosa*.

(polyphenols, flavonoids, and ascorbic acid) *C. borivilianum* has strong antioxidant potential. In 1, 1-diphenyl-2-picrylhydrazyl (DPPH) free radical scavenging assay and lipid peroxidation assay, *C. borivilianum* aqueous extract scavenges free radicals (ROS) and thiobarbituric acid reactive substances, at 250mg/kg concentration in a dose-dependent manner (Kenjale et al., 2007). Further, *C. borivilianum* aqueous root extract (25 to 1000 μg/ml) scavenges nitric oxide, superoxide, hydroxyl, DPPH, and ABTS [2, 20-azinobis (3-ethylbenzothiazoline 6-sulfonic acid)] radicals as long as attenuating lipid peroxidation in mitochondria. In addition, aqueous extract of *C. borivilianum* drastically ameliorates LDL oxidation which was mediated by copper in serum and kidney (Visavadiya et al., 2010). Glutathione (GH),

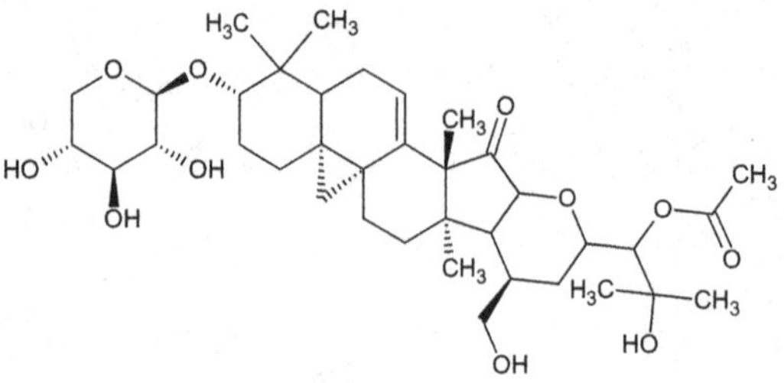

Cimiracemoside E

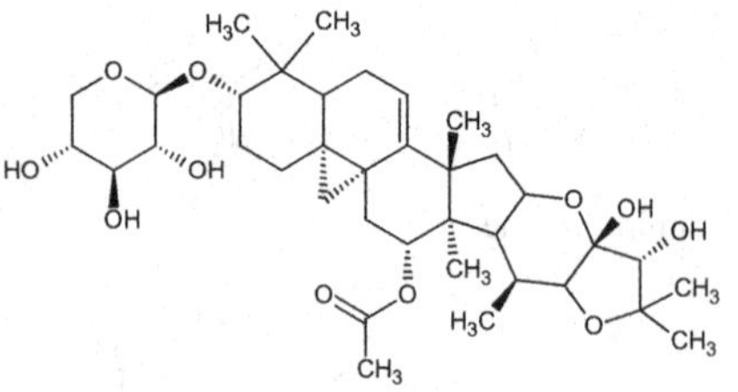

Cimiracemoside F

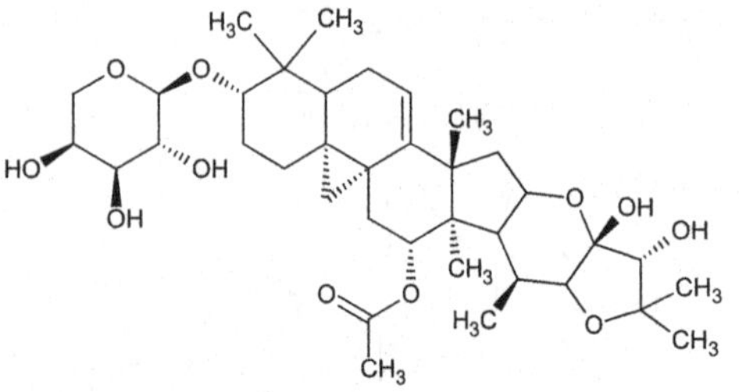

Cimiracemoside G

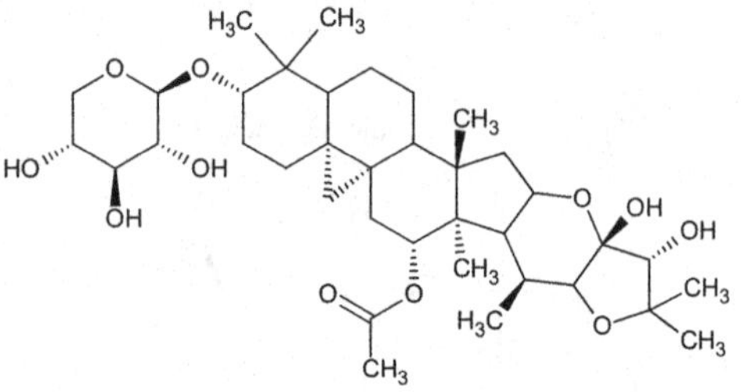

Cimiracemoside H

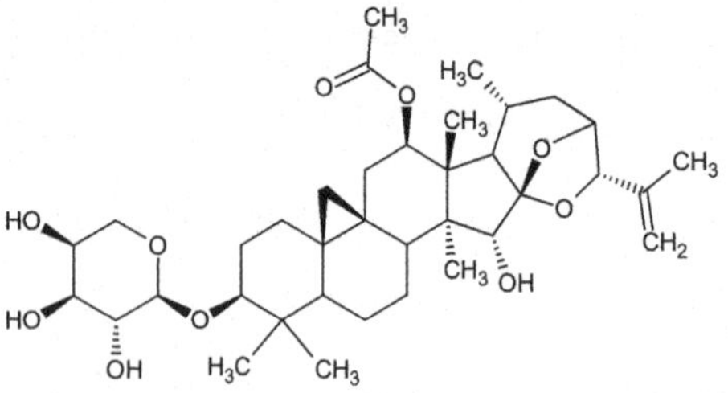

Cimiracemoside J

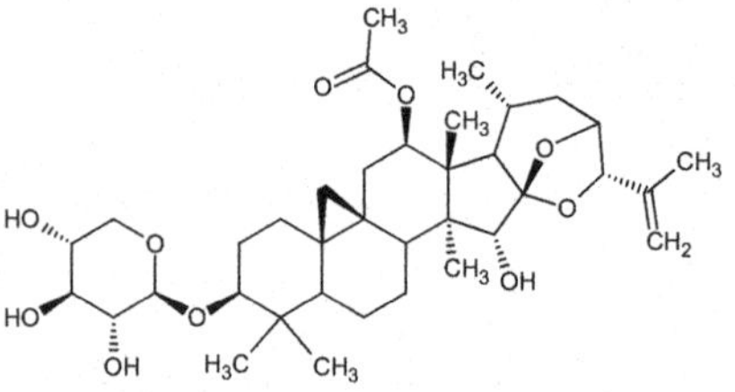

Cimiracemoside K

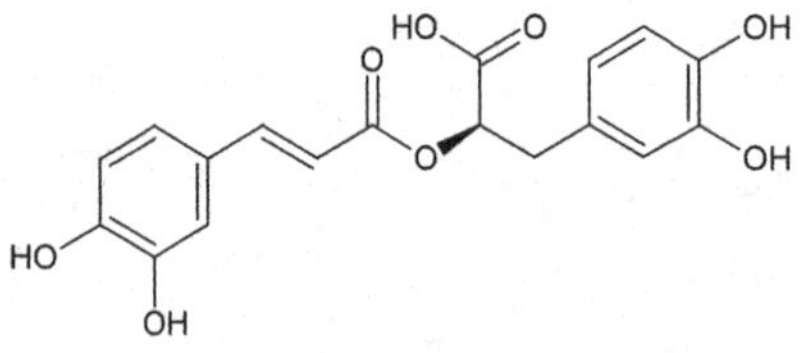

Rosmarinic acid

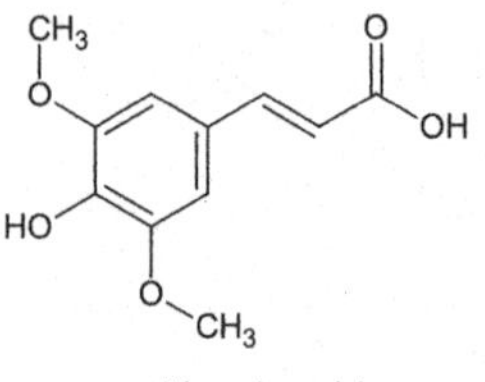

Sinapic acid

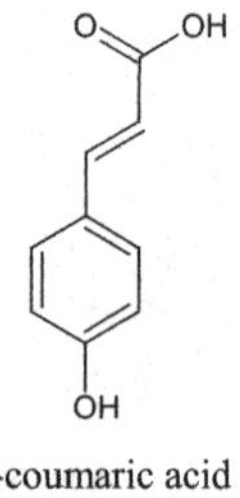

p-coumaric acid

FIGURE 3.4 (Continued)

catalase (CAT), and superoxide dismutase (SOD) are antioxidant enzymes that play an essential role to suppress oxidative free radicals. These increased antioxidant enzymes reduce malondialdehyde (MDA) and protect cellular molecules from oxidative stress. Hence, at 100, 400, and 800mg/kg dose of root extract of *C. borivilianum* elevates GH, CAT, SOD activity (Kumar et al., 2010).

3.5.1.2 Immunomodulatory Effect

Immunomodulators are natural or manufactured substances that can stimulate, inhibit, or control any aspect of the immune system, including adaptive and innate responses (Kumar et al., 2012).

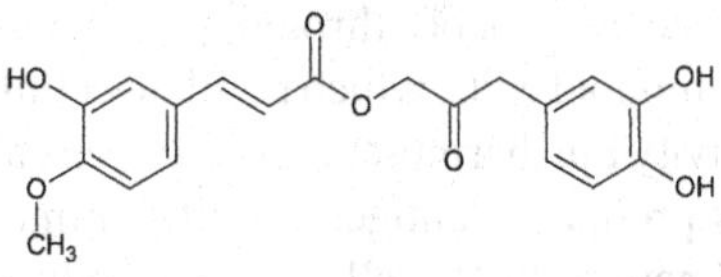

Cimiciphenol

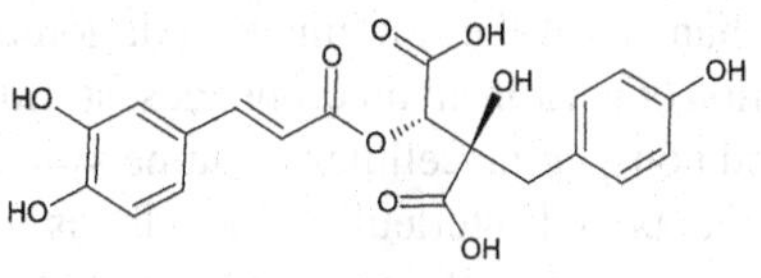

Cimicifugic acid D

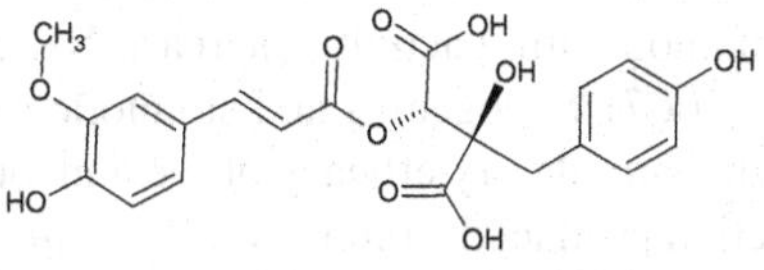

Cimicifugic acid E

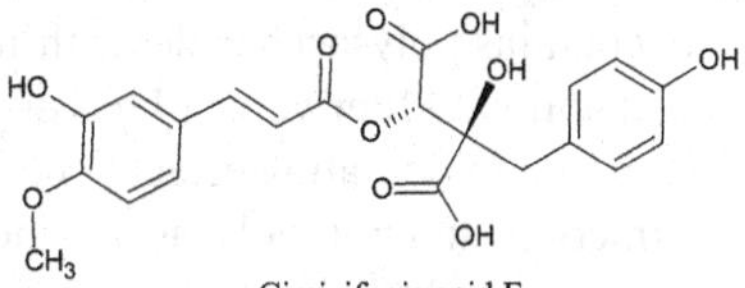

Cimicifugic acid F

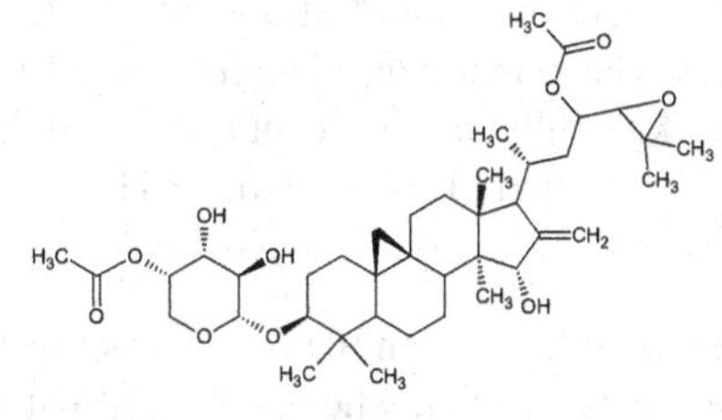

Cimiracemoside L

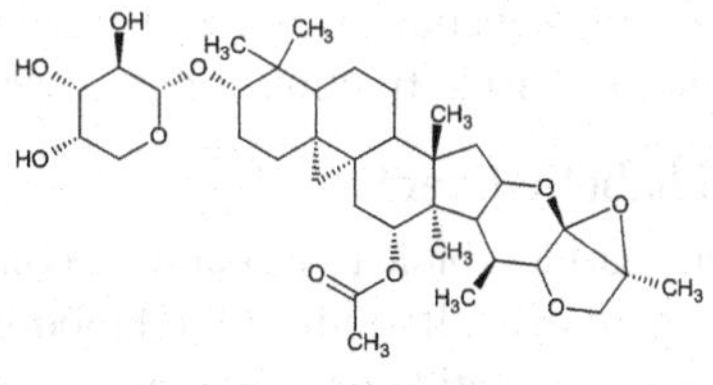

Cimiracemoside N

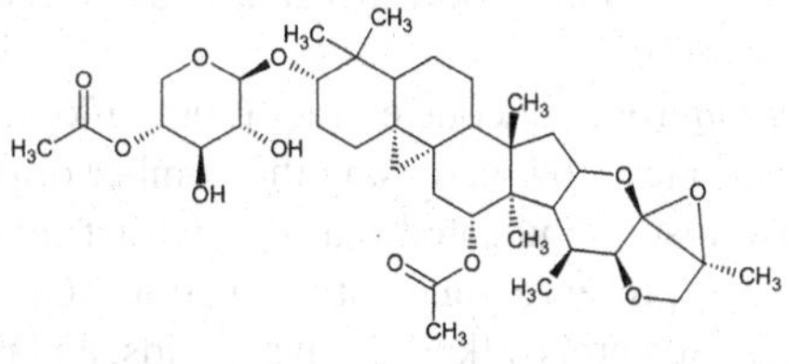

Cimiracemoside O

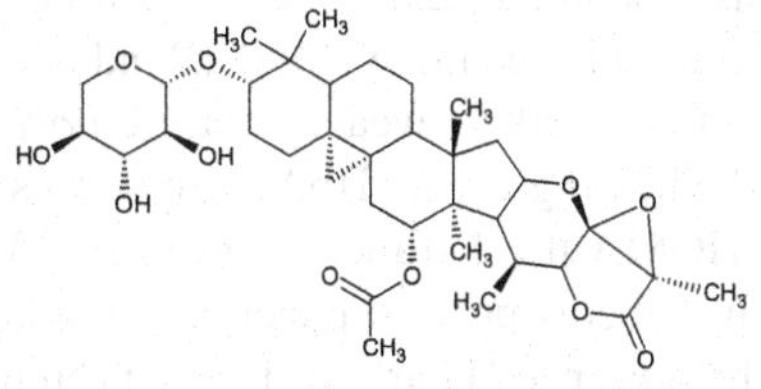

Cimiracemoside P

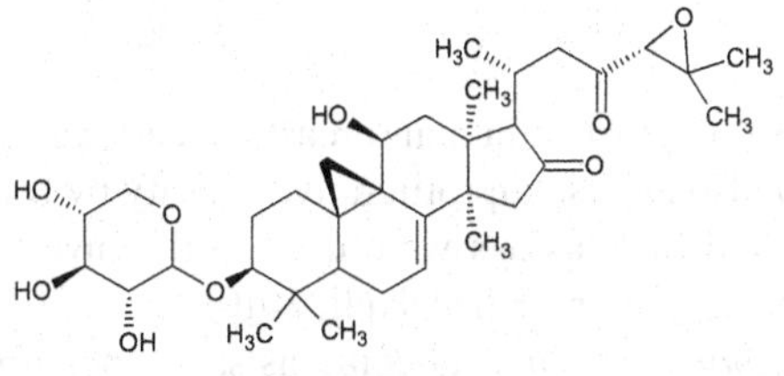

Cimifugoside H1

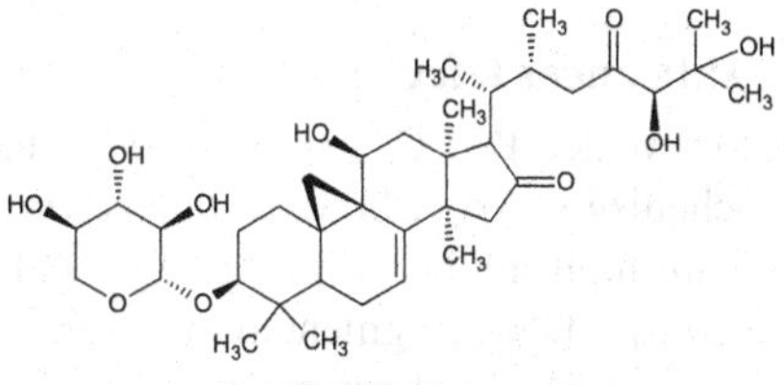

Cimicifugoside H2

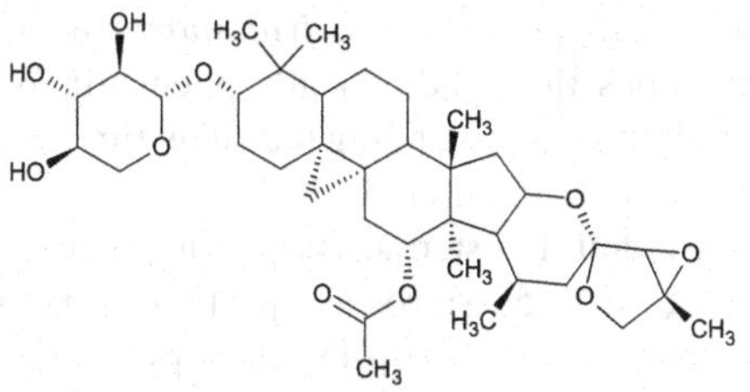

Cimicifugoside H2

FIGURE 3.4 (Continued)

The mechanism of the immunomodulation effect occursprobably through phagocytosis regulation, activation of peritoneal macrophages, stimulate lymphoid cells, which regulate cellular immune response and nonspecific cellular immune system activity.Furthermore, increasing circulating total white cell counts and interleukin-2 levels, as well as producing antigen-specific immunoglobulin (IgG), inducing non-specific immunity mediators and natural killer cell numbers, reducing chemotherapy-induced leukopenia, and reducing chemotherapy-induced leukopenia (Kumar et al., 2012;). *C. borivilianum* contains polysaccharideswithimmunomodulating activity, increase NK cell effect, phagocytosis, and antibody titer valve (Thakur et al., 2007). Sapogenins and ethanolic root extract also provide an immunomodulating effect. The immunostimulatory efficacy of the aqueous extract, polysaccharide fraction, and non-polysaccharide fraction produced from *C. borivilianum* hot water extraction was investigated (Thakur et al., 2011a). The phytochemical examination of *C. borivilianum*' shot water-soluble root extracts resulted in the quantification of almost 60% w/w polysaccharides, namely 31% inulin-type fructans and 25% acetylated mannans (Thakur et al., 2009b). The administration of the aqueous extract at 50 and 100mg/kg exhibit a significant increase in IgG level, followed by polysaccharide fraction and polysaccharide fraction (Thakur et al., 2011a).

3.5.1.3 Antidiabetic Effect

Fructo-oligosaccharides, bioactive chemical constituents isolated from the root extract of *C. borivilianum*which can drastically reduce the blood glucose (sugar) level, when compared with positive control groupexhibited antidiabetic activity with prolongedantioxidant activity in a diabetic animal model (Narasimhan et al., 2006). Furthermore, *C. borivilianum* root extract attenuated blood glucose concentration in the mice model (Panda et al., 2007).

Giribabu et al., (2014) revealed that *C. borivilianum*root aqueous extract (500mg/kg/day, p.o.) amelioratesthe blood glucose, HbA1c, increases insulin level by increasing the number of pancreas β-cells and HOMA-β cell functioning index. These results indicated that the extract might help patients with diabetes prevent pancreatic damage and preserve pancreatic function. (Giribabu et al., 2014). The presence of carbohydrates, proteins, phytosterols, alkaloids, flavonoids, and phenolic acids was discovered in the aqueous extract of *C. borivilianum* root. (Giribabu et al., 2014)

3.5.1.4 Anticancer Effect

Arif (2005) revealed that *C. borivilianum* induces apoptosisin human breast cancer cell lines via several mechanisms (Arif, 2005). *C. borivilianum* also exertsantiproliferative activity by fragmenting DNA with high inhibition percentage. MTT and SRB assays were used in the investigation, accompanied by a DNA fragmentation experiment (Deore and Khadabadi, 2010).

The phytochemical evaluation revealed that *C. borivilianum* root contains saponinchloromaloside-A (steroidal glycoside) and β-D-apiofuranose (spirostanolpentaglycosides), which produce cytotoxicity, are responsible constituents for anticancer property (Mimaki et al., 1996; Qiu et al., 2000).In skin papillomagenesis investigations, *C. borivilianum* aqueous root extract showed antitumor and antimutagenic capabilities throughdrastically attenuatingpapilloma, tumor incidence, tumor burden, tumor size, and weight, and prolonging the latent time in an animal model at 800mg/kg body weight/day p.o. concentration (Kumar et al., 2010).

Acharya et al., (2009) demonstrated that steroidal saponins, such as borivilianosides F, G, and H, and three known compounds (25(*R,S*)-5α-spirostan-3β-ol 3-O-β-D-xylopyranosyl-(1->3)-[β-D-glucopyranosyl-(1->2)]-β-D-glucopyranosyl-(1->4)-β-D galactopyranoside, borivilianosides A and C isolated from C. borivilianum root extract exert cytotoxicity in human colon cancer cell lines (HT-29 and HCT 116). Among them, borivilianoside H displayed the best cytotoxicity (IC50 = 0.38 μM in HCT 116 cells and IC50 = 2.6 μM in HT-29 cells) (Acharya et al., 2009).

3.5.1.5 Anti-UlcerativeEffect

The alcoholic extract of *C. borivilianum* possesses ulcer-healing effects. A cold stress-generated gastric ulcer model was utilized to investigate the anti-ulcer activity. A single oral administration of alcoholic extracts at a dose of 200mg/kg decreases the ulcer index significantly as compared to the control group (Deore and Khadabadi, 2009a).

3.5.1.6 Analgesic Effect

An analgesic or painkiller is a class of a diverse group of medicationsthat have the ability to attenuate pain sensation (Bi et al., 2021). The peripheral (PNS) and central nervus systems (CNS) are affected differently by analgesic medications through many mechanistic pathways, such as NF-kB (Wang et al., 2021), anti-inflammatory, GR, GABA, opioid receptors pathways (Bae et al., 2020), dopaminergic (Liu et al., 2019). Paracetamol, non-steroidal anti-inflammatory drugs (NSAIDs), narcotic drugs, synthetic medicines, and a variety of other medications have analgesic effects.

Panda et al. found that a methanolic root extract of *C. borivilianum*helped alleviate pain at (100mg and200mg/kg, p.o.) concentration in a mice experimental model by utilizing tail flick and tail immersion methods.This research was developed based on ethno-pharmacologicaluse of the *C. borivilianum* plant to treat rheumatoid arthritis. The analgesic activity might have contributed to the plant's steroidal bioactive constituents (Panda et al., 2007). Furthermore, *C. borivilianum* provides an analgesic effect when it is co-administered with *Lawsonia inermis* Linn plant extract via the hot plate analgesic method (Kumari et al., 2015).

3.5.1.7 Antimicrobial Effect

Using a micro broth dilution experiment, the antibacterial ability of *C. borivilianum* was tested against eight bacteria and four harmful fungi. The extract's minimum inhibitory concentration (MIC) was determined to be the lowest concentration that inhibited any visible microbial growth after treatment with p-iodo-nitrotetrazolium violet. *C. borivilianum* extracts had antimicrobial activity at 75–1200 µg/ml concentration (Dabur et al., 2007).

Another study reports that *C. borivilianum* extract exerted a potent antiviral effect against the BHV-1 virus (Goel et al., 2011). Moreover, C. borivilianum extract also showed antibacterial effect against 4 bacteria species *Staphylococcus aureus, E. coli, Pseudomonasaeruginosa*, and *Bacillus substalis*when thecup diffusion method is used (Sundaran et al., 2011). In *Pseudomonas aeruginosa, Bacillus subtilis, Methicillin-resistant Escherichia coli, Staphylococcus aureus*, and *Candida albicans* pathogens, *C. borivilianum* also showed antibacterial effect at 100, 200, and 300 µg/mL.*C. borivilianum* inhibits the growth of the pathogen (Huang et al., 2019).

3.5.1.8 Anti-Stress Effect

The chronic cold restraint stress rat model was used for the anti-stress effect. In comparison to the control group, *C. borivilianum* displayed substantial increases in plasma glucose, plasma cholesterol, triglycerides, serum corticosterone, and adrenal gland weight at ranges from 125 to250mg/kg p.o. (Kenjale et al., 2007). This anti-stress activity may be demonstrated due to the presence of alkaloids and steroidal saponins (stigmasterol) active compounds in *C. borivilianum* plant root (Acharya et al., 2009).

3.5.1.9 Anthelmintic Effect

Several preliminary phytochemical studiesrevealed that *C. borivilianum* contains a plethora of photochemical such as saponin, steroids, carbohydrates, alkaloids, and proteins. (Visavadiya and Narasimhacharya, 2007). The methanolic root extract of *C. borivilianum* exertsan anthelmintic effectagainst Indian earthworms*Pheritima pasthumaa* and *Ascardia Galli.* At several extract concentrations (25, 50, 100mg/ml), it exhibited an anthelmintic effectby paralyzing the earthworms, and death was observed. All extracts showed significant anthelmintic activity on selected worms. This anthelmintic effect was observed due to the saponin, alkaloids, and steroidal compounds (Deore and Khadabadi, 2010).

3.5.1.10 AntidyslipidemicEffect

C. borivilianum tuber has potent efficacy to modulate blood cholesterol levels. Aqueous and alcoholic root extract (100 and 300mg/kg for both) lowered serum cholesterol and triglycerides and increased serum HDL levels as well as lipid profiles in a dose-dependent manner. Thus, the hypocholesterolemic and increased HDL-cholesterol levels have a protective role in cardiovascular disease possibly due tothe presence of phytosterols and saponins in *C. borivilianum* (Deore and Khadabadi, 2009c). Furthermore, *C. borivilianum* elevated hepatic 3- hydroxy-3-methylglutaryl coenzyme reductase enzyme activity, which is responsible for increasing both cholesterol consumption and bile acid production in the liver. Those are related to lowering the lipid and cholesterol in the blood. Saponin from medicinal plants has cholesterol and bile acid-lowering potentials (Oakenfull and Sidhu, 1990; Harwood et al., 1993;Visavadiya and Narasimhacharya, 2007). In addition, the levels of superoxide dismutase and ascorbic acid were elevated by *C. borivilianum*. However, the existence of fructans might potentially be a key contributing element in improved hypercholesteremia treatment. The bioactive components of *C. borivilianum*, such as phytosterols, saponins, polyphenols, flavonoids, and ascorbic acid, may be responsible for this positive impact (Visavadiya and Narasimhacharya, 2007). Additionally, Giribabu et al. claimed that *C. borivilianum*lowered TC, TG, VLDL, and LDL at 250 and 500mg/kg/day in the rat model (Giribabu et al., (2014).

3.5.1.11 Larvicidal Effect

C. borivilianum tuber extract has a larvicidal effect which was examined for the mosquito species *Anopheles stephensi*, *Culex quinquefasciatus*, and *Aedes aegypti*. Methanol extract, crude saponin extract, and pure saponin fraction were used to evaluate the larvicidal effect on mosquito species. All extracts had larvicidal efficacy, with pure saponin fraction being the most efficient. (Deore and Khadabadi, 2009a).

3.5.1.12 Aphrodisiac Effect

The word aphrodisiac comes from the Greek word Aphrodite, which honors the goddess of love and passion. This word is now used to describe drugs that improve sexual behavior and aid in the treatment of sexual malfunction. *C. borivilianum* is commonly found in woodlands and belongs to the Vajikaran Rasayana group of herbal ingredients, which can be used to improve efficacy and treat erectile problems. The usage of various herbs is discussed in the Kamasutra of Vatsyayan, a book on the art and science of lovemaking. Spermatogenesis ability of *C. borivilianum* was also found in animals (Kenjale et al., 2008). *C. borivilianum*increases sperm count at 125mg/kg and 250mg/kg,can be useful in the treatment of certain forms of sexual inadequacies, such as premature ejaculation and oligospermia(Kenjale et al., 2008).Kenjale et al. (2008) compared *C. borivilianum* root extract's putative aphrodisiac and spermatogenesis activities to that of the conventional aphrodisiac medication, sildenafil citrate (Viagra).The effect of ethanolic and sapogenin extract from *C. borivilianum* roots on sexual activity and sperm production in male albino rats was investigated at two different doses (100mg/kg and 200mg/kg). This investigation proved to be one of the first pieces of empirical data in favor of *C. borivilianum* traditional usage as an aphrodisiac (Thakur and Dixit, 2006).

The improved reproductive organs, as well as histological activity, suggest that the animals treated experienceda strong spermatogenic impact. There was a significant decrease in mounting, ejaculation, post-ejaculatory, and intromission delay, as well as an improvement in mount rate and attraction to females, indicating improved sexuality. *C. borivilianum* (200mg/kg b.w.) pedunculated activities and sperm count increased (Thakur and Dixit, 2007). At 14 days after starting the medication, there was an increase in sperm count and pedunculated response (Thakur and Dixit, 2007). A further study was run by Thakur and his coworkers and explained that *C. borivilianum* changes the sexual behavior in an animal model through testosterone-like effects (Thakur et al., 2009). It can attenuate testicular damage of the male genital organ by reducing physical stress and inducingspermatogenesis (Thakur et al., 2008). At 100mg/kg in male albino rats, *C. borivilianum* root extract increases seminal fructose levels and sperm count and reducesnitric oxide (NO) release. In this investigation, *C. borivilianum* exhibited drastically positive responsiveness for all variables examined, which would be strongly associated with erectile function (Thakur et al., 2011b). *C. borivilianum* has been found to help with diabetes-related sexual dysfunction. Another study reported that *C. borivilianum* elevated semen quality and motility when administered with other medicinal plants, *Mucuna pruriens* (Linn), C. and *Eulophiacampestris* (Wall), in oligozoospermic patients (Thakur et al., 2009b). The Rho-kinase 2 (ROCK-II) enzyme can promote smooth muscle contraction in the corpora cavernosa, and inhibiting ROCK-II enzyme activities could be a potential target in erectile dysfunction treatment. Along with 15 more medicinal plants, *C. borivilianum* suppressed the ROCK-II enzyme and androgen synthesis which inhibited the enzyme effect (Thakur et al., 2009b; Goswami et al., 2012).

3.5.1.13 Anxiolytic Effect

The main objective of this research was to evaluate the anxiolytic action of *C. borivilianum* root extract and its impact on brain GABAat 20, 60, and 200mg/kg concentration in the mice model. The root extract was given by intraperitoneal injection (short-term therapy) or orally (long-term administration). Several parameters were used, such as elevated plus maze test, open field test, light and dark test, actophotometer, and rotarod, which have been used to assess exploratory, sedation, and muscle relaxant actions in mice experimental models.

Swami et al. (2014) demonstrated that *C. borivilianum* root aqueous extract administered acutely (60 and 200mg/kg, i.p.) in the open arm of the EPM, raised the frequency of squares traveled in the center in the OFT, and extended the timespent and entries in the light region in the L/D model. Doses of 60 and 200mg/kg p.o. demonstrated anxiolytic effect in a chronic trial, with 60mg/kg p.o. being more persistent and noticeable. After acute and chronic administration, the aqueous extract exhibited substantial anxiolytic efficacy at 60mg/kg and 200mg/kg, with no sedative impact (Swami et al., 2014).

3.5.1.14 HepatoprotectiveEffect

Arsenic induces ATP depletion by disrupting mitochondrial membrane potential and increasing ROS production (Larochette et al., 1999; Csanaky et al., 2003). It inhibits energy-linked NAD+ reduction, mitochondrial biogenesis, and ATP production via uncoupling oxidative phosphorylation. As a result, after arsenic poisoning, ATPase activity is considerably decreased. Thus, it impaired hepatic normal functions (Klassen and Watkins, 2003). *C. borivilianum* possesses hepatoprotective activityagainst arsenic-induced toxicity. In comparison to the arsenictoxicgroup (800mg/kg p.o.), it improved overall bodyand hepatic mass, as well as ATPase enzymelevels and restored hepatic functions. *C. borivilianum* tuber extract can reduce arsenic toxicity.

3.5.1.15 Toxicity

Animals demonstrated good resistance to a single dose of *C. borivilianum* root aqueous extract up to levels of 2000mg/kg, which were determined to be non-lethal. Until the end of the trial period, the maximum dose of the extract did not show any symptoms of toxicity or death (Giribabu et al., 2014). No negative impacts were detected in rats given *C. borivilianum*root extract (orally) for 7 days at doses of 100, 400, and 800mg/kg body weight/day. It appears that mice may withstand preparations of *C. borivilianum*root up to 800mg/kg body weight each day (Kumar et al., 2010). Up to a dosage of 2000mg/kg b.w., the test medication (alcoholic/aqueous extract of *C. borivilianum* roots) was shown to be safe. Changes in skin, fur, eyes, mucous membranes, respiration, heart rate, blood pressure, salivation, lacrimation, perspiration, piloerection, urine incontinence, defecation, ptosis, sleepiness, gait, tremors, and convulsions were seen over 14 days (Thakur et al., 2009b).

3.5.2 *Cimicifuga racemosa*

3.5.2.1 Anti-Allergic Effect

Substances with anti-allergic activity can relieve the symptoms caused by allergies and prevent allergic response, i.e., fever, inflammation, hives, etc. In traditional medicine, black cohosh has been used to treat pain and inflammation. However, the effects of black cohosh extract on mast cell-dependent allergic reactions are yet to be fully understood. It is also known that *C. racemosa* root extract limits nitric oxide generation by lowering iNOS expression without altering the enzyme's activity. This might have a role in *C. racemosa*'s anti-inflammatory properties. Mast cells play a significant rolein allergic responses by releasing a variety of inflammatory and immunomodulatory mediators in response to antigen-induced IgE receptor cross-linking, which is caused by antigen (Yi et al., 2001). Histamines, proteases, leukotrienes, prostaglandins, and cytokines are among them. Clinical symptoms of acute allergic reactions such as vasodilation, increased vascular permeability, broncho-contraction, and edema are generated as a result. Several inflammatory cells, including eosinophils and lymphocytes, are also stimulated and activated by allergic reactions (Kim et al., 2004). Black cohosh has been shown to have the ability to scavenge reactive oxygen species (ROS), and its polyphenolic components, including caffeic acid, ferulic acid, and cimiracemate A, contributing to its antioxidant potential (Burdette et al., 2002). Mast cells have been shown to produce intracellular ROS in response to antigen exposure, which might play a role in histamine release (Matsui et al., 2000). Evidence showed that naturally present polyphenolic antioxidants decreased ROS levels and prevented histamine release in antigen-IgE-activated mast cells (Chen et al., 2000). As a result, black cohosh's anti-allergic properties may be attributable in part to the polyphenolic component's antioxidant capabilities. The specific action mechanism of black cohosh extract and its constituentshas yet to be determined. In conclusion, it has beenshown that black cohosh extract exhibits powerful anti-allergic properties in mast cell-dependent test models, indicating that black cohosh extract might be a promising choice for the treatment of allergic inflammation (Kim et al., 2004). On the other hand, another study conducted later suggests that the suppression of iNOS expression by *C. racemosa* extracts appears to be the cause of the inhibitory impact on inducible NO generation. 23-epi-26-deoxyactein, a triterpene glycoside, was shown to suppress inducible NO generation in LPS-stimulated macrophages. The inhibition of iNOS by aqueous extracts of*C. racemosa*, as well as our discovery of the active components in *C. racemosa* extracts, contribute to our knowledge of black cohosh's anti-inflammatory properties (Schmid et al., 2009). This could also play a role in using *C. racemosa* extract as an anti-allergic compound as inflammation is a symptom of allergic infections.

3.5.2.2 Antiestrogenic and Estrogenic Effect

Antiestrogenic activity refers to the action of a compound that prevents estrogens such as estradiol from mediating their biological effects in the body. Antiestrogens, also called estrogen blockers or estrogen antagonists, are compounds that work by blocking estrogen receptors (ER) and decreasing estrogen synthesis; they have a positive impact on the body. For the treatment of menopausal symptoms, particularly hot flashes, black cohosh is becoming a more popular alternative to estrogen replacement therapy. In any of these test methods, black cohosh extracts showed no estrogenic action. This is a positive step forward in the evaluation of black cohosh's safety as a treatment for menopausal hot flashes (Lupu et al., 2003).A study was conducted using two independent cell-based estrogen inducible assays – transactivation assay and proliferation essay – and integrating data from an *in vitro* cell proliferation,providing evidence for potential antiestrogenic activities of extracts from the rhizome of *C. racemosa*(Zierau et al., 2002). Thus, it is believed that black cohosh may have antiestrogenic properties, but the researchers are trying to find out the exact mechanism.

Chemicals with estrogenic activity (EA) or antiestrogenic activity are those that imitate or antagonize the *in vitro* and/or *in vivo* activities of naturally occurring estrogens, like 17β-estradiol (E2), and impacts on estrogen signaling are the most frequent and well-studied endocrine disruptor activity. Early-onset of puberty in females, reduced sperm counts, altered reproductive organ functions, obesity, altered sex-specific behaviors, and higher rates of some breast, ovarian, testicular, and prostatic cancers are just a few of the biological and adverse health effects that xenobiotic chemicals with EA can cause in mammals (Bittner et al., 2014). *In vitro* and *in vivo* investigations of black cohosh's impact on estrogen receptor-positive breast cancer cell lines have yielded conflicting findings. Treatment of human mammary cancer cells with <2.5μg/mL of SBC-R did not improve cell growth *in vitro*, while higher doses of the extract (≥2.5 μg/mL) inhibited cell proliferation significantly (Mahady et al., 2002). However, recent research has shown no evidence that black cohosh has an estrogen receptor-mediated mechanism of action. Preclinical results show that compounds with dopaminergic or serotoninergic activity may be involved in the effectiveness of black cohosh extracts in alleviating menopausal symptoms.

It's been suggested that it should be categorized as a selective estrogen receptor modulator (SERM) since it has ER-binding and potential estrogen agonistic activity on bone tissue but no stimulatory effects on the breast or endometrial tissue (Viereck et al., 2005). The ideal SERM works as an estrogen in the bone and brain but not the breast or uterus. Black cohosh extracts may include compounds that meet the SERM requirements (Ruhlen et al., 2008). The lack of estrogenic substances in *C. racemosa* is crucial from a therapeutic standpoint. Antiestrogens like tamoxifen or aromatase inhibitors are being used as adjuvant treatments for breast cancer patients all over the world. These individuals typically experience significant climacteric symptoms, especially hot flashes. For these people, traditional HRT is contraindicated. As a result, gynecologic oncologists are frequently faced with the circumstance when a patient seeks assistance to alleviate these distressing psycho-vegetative symptoms. The extracts of the rhizome *C. racemosa* are the only non-estrogenic option (Wuttke et al.,2014).

3.5.2.3 Antihyperglycemic Effect

Both type I and type II diabetes, when uncontrolled, result in hyperglycemia, and if this persists for too long, accumulation of methylglyoxal (MG) occurs. MG is a highly robust dicarbonyl metabolite that is generated during the metabolism of glucose and is considered a major precursor to advanced glycated end products (AGEs). AGEs are proved to promote the infestation of diabetes and inflammation (Wu et al., 2011). AGEs and MG activates the receptor for AGE (RAGE) which generates inflammatory molecules via oxidative stress. AGEs also cause a decrease in insulin synthesis by the pancreatic beta cells by causing repression of the gene expression of pancreatic duodenal homeobox-1 (PDX-1). This PDX-1 plays a major role in both the development of the pancreas and maintenance of the beta cells (Shu et al., 2011). Oxidative stressinduced by glucotoxicity causes damage to the function of pancreatic islets as well as reducing insulin secretion via overproduction of ROS. There is low expression ofantioxidant enzymes in the pancreatic betacells; antioxidant agents like catalase (CD), superoxide dismutase (SOD), and glutathione peroxidase (GPX) are quite low in number under physiological conditions (Tanaka et al., 2002). Hence, when the pancreatic beta cells are exposed to an increasing number of ROS, their function declines, whichleads to the accumulation of glucose in the bloodstream, thereby kick-starting the onset of type 2 diabetes (Bonora, 2008). Preventing the rise of ROS and maintaining the pancreatic betacell function seems to be the most viable way to manage hyperglycemia. Insulin and oral therapeutic agents have been prescribed to patients with diabetes, but they exhibited adverse side effects and alleviated certain symptoms, so a vast amount of research has now gone into finding antihyperglycemic phytochemicals, expecting that it will activate regeneration of pancreatic beta-cells while preventing the apoptosis resulting in the regain of the control of the glucose homeostasis (Lee et al., 2012). There was an experiment in streptozotocin (STZ)-induced diabetic rats on the activity of isoferulic acid. The plasma levels in these rats decreased in a dose-dependent manner because of the raised expression of the mRNA level of glucose transporter subtype 4 form (GLUT4) located in the soleus muscle; after one day of repeated treatment, the increased level of mRNA phosphoenolpyruvate carboxykinase in the liver was reversed to the normal level (Liu et al., 2000). In several other studies, it was indicated that isoferulic acid extracted from *Cimicifuga* decreased the plasma level of glucose with isoferulic acid of 5.0mg/kg, intravenously; this caused the plasma level of glucose to drop from 14.5±0.5 to 12.9±0.4 mmol/L (N=8) markedly ($p<0.05$) in non-insulin-dependent diabetic rats (Liu et al., 1999).

The *C. racemosa* extracts (CRE) are presumed to have significant antimicrobial characteristics as well (Li et al., 1993). One experiment was conducted where 15 triterpene glycosides were isolated from the black cohosh rhizomes, and an agar diffusion assay was done to assess their bacterial activity. 12β-acetoxycimigenol-3-O-b-D-xylopyranoside (45mg), 25-acetylcimigenol xyloside (350mg), actein (33mg), 27-deoxyactein (124mg), foetidinol-3-O-b-xyloside (26mg), cimidahurine (49mg), and cimifugin (820mg) (with the amount used for the paper disk diffusion assay on agar plates) were the few compounds that showed weak activity against *Shigella flexneri*, *Shigella dysenteriae*, *Shigella sonnei*, *Mycobacterium tuberculosis*, *a-hemolytic Streptococcus*, and *Streptococcus pneumoniae* strains of bacteria (Lai et al., 2005). *C. racemosa* extracts were used to witness its inhibitory activity against the *Neisseria gonorrhoeae* which is a causative agent of gonorrhea, a common sexually transmitted disease (STD) worldwide (WHO, 2009). Another experiment was conducted on ferulic acid and isoferulic acid for their inhibitory effect on Interleukin-8 production in the course of influenza virus infection taking place both *in vitro* and *in vivo*. This investigation gave positive results on the antiviral effects of *C. racemosa* extracts (Hirabayashi et al., 1995).

3.5.2.4 Antimicrobial Effect

An investigation was performed to assess the preventative properties of *C. racemosa* extracts in lipopolysaccharide-stimulated (LPS) murine macrophages. The inducible isoform of NOS (iNOS) promotes the production of a great amount of NO over a long period via a calcium-independent pathway (Xie et al.,1992). This expression of iNOS is seen in murine macrophages. All the while, the presence of human iNOS is observed in macrophages of patients who are suffering from infectious or inflammatory diseases. The prolonged production of NO gives the macrophages an upper hand, for they achieve cytotoxic activity against microbes like viruses, bacteria, fungi, and protozoa (MacMicking et al., 1997). Hence, NO plays the role of mediator and an inflammatory response regulator (Korhonen et al., 2005). The experiment was performed using a cell line of mouse macrophage called RAW 264.7. The results described non-significant inhibition of the iNOS enzyme by the *C. racemosa* extracts, and a concentration-dependent reduction was observed in IFN-β and IRF-1 mRNA expression when the macrophages were incubated in increasing amounts of *C. racemosa* extracts. When 5–20mg/ ml of 23-epi-26-deoxyactein was used, NO production by LPS-activated RAW 264.7 cells experienced a reduction; the documented percentage inhibition was 23.5 ± 0.1% at 5mg/ml. Reduced intracellular iNOS protein concentration was observed using Western blot and semiquantitative competitive RT-PCR analysis (Schmida et al., 2009).

3.5.2.5 Anti-Osteoporosis Effect

Osteoporosis is a significant health disorder manifesting a decrease in bone density and increased risk of fractures. It is prevalent in women after menopause and correlates with estrogen deficiency (Qiu et al., 2007). The basal cellular machinery of lower bone mass in osteoporosis is due to an imbalance of osteoclast-mediated bone resorption concerning the osteoblast-mediated bone formation, which leads to bone loss (Qi et al., 2010). Drugs that inhibit osteoclast resorption are bisphosphonates, such as parathyroid hormone (PTH), which acts as a possible therapeutic agent (Conte and Guarneri, 2004). Bisphosphonates, however, have a poor absorption rate from the gastrointestinal (GI) tract and have been associated with GI-related diseases whereas, PTH, a therapeutic peptide, is given through subcutaneous injection, not orally (Seldova-Wuttke et al., 2003). Black cohosh extract consists of 25-acetylcimigenol xylopyranoside (ACCX), which can downregulate osteoclast genesis. ACCX inhibits it through receptor activators of nuclear factor-kappa-B (RANKL) and tumor necrosis factor-alpha (TNF-α) (Roggia et al., 2001). Inhibition of osteoclast genesis through induction of ACCX results from the negation of the NF-kB and ERK pathways by either RANKL or

TNF-α, respectively. Significantly, this compound attenuates TNF-α-mediated bone loss (Redlich et al.,2002). So, ACCX is a prospective lead for the development of a new class of antiosteoporosis agents.TNF-α is the crucial inflammatory cytokine that stimulates inflammation-regulated osteoclast genesis and bone loss. This cytokine is upregulated in estrogen-deficient mice and causes bone loss through estrogen deficiency (Zheng et al.,2006). RANKL induces osteoclast genesis under physiological conditions. Although the exact mechanism operated by RANKL stimulates osteoclast genesis is not entirely known. RANKL arbitrates activation of the NF-kB and MAPK pathways required for osteoclast genesis (Ikeda et al.,2004). ACCX eradicates RANKL-induced stimulation of the NF-kB pathway, as demonstrated through lack of phosphorylation and NF-kB DNA-binding activity and absence of RANKL-induced expression of the NF-kB-dependent genes (Singh et al., 2020). Although the exact molecular markers of ACCX on the NF-kB pathway need to be known, our data suggest the upstream of IkBa phosphorylation (Ruocco et al.,2005). The significance of the NF-kB pathway in osteoclast genesis demonstrates that mice lacking both NF-kB p50 and p52 are osteoporotic (Lam et al.,2001). Therefore, it is rational to deduce that inhibition of the NF-kB pathway is responsible for ACCX blockage of osteoclast genesis (Jimi et al.,2004).

3.5.2.6 Antioxidant Effect

There are two groups of secondary metabolites in black cohosh, triterpene glycosides and phenolic acids (Jiang et al.,2006).These medicinal phyto-compounds produce free radicals and other reactive oxygen species (ROS), are constantly produced *in vivo*, and cause oxidative stress to biomolecules. Various antioxidants, DNA repair systems, and the replacement of damaged lipids and proteins modulate the process (Slamenova et al., 2002). ROS scavenging can effectively protect against chronic diseases such as cardiovascular disease, cancer, and aging. Elevation in oxidative stress can substantially affect repair mechanisms, which can diminish cellular processes (Silva et al., 2000). DNA is a potential target of oxidative damage because continuous oxidative stress leads to colon, breast, rectum, and prostate cancers (Villares et al., 2011). Black cohosh and its isolated compounds could protect DNA from single-strand breaks and oxidation of bases or not. In that case, S30 breast cancer cells are prepared with the pro-oxidant menadione, a quinone known tocause oxidative damage in cells (He et al., 2000). Single-strand breaks regulated by menadione are found through single-cell gel electrophoresis assay (Borrelli et al., 2003; Nikam and Chavan, 2009). It is observed that black cohosh contains antioxidants, mainly derived from hydroxycinnamic acids, which serve as a protective barrier from menadione-induced DNA damage (Son and Lewis, 2002). Studies suggest that black cohosh extracts contain efficacious ROS scavengers that shield cellular DNA against oxidative stress. Methyl caffeate, caffeic acid, ferulic acid, fukinolic acid, cimicifugic acid A, and cimiracemate B are all cinnamic acid derivatives that possess a phenol group at the conjugated ethylene side chain (Kikuzaki et al., 2002). Through analysis, it is said that the hydrogen on the phenol can be retrieved by a radical, such as hydroxyl radical, resulting in phenoxy radical. Confirmation is stabilized through resonance as the unpaired electron can delocalize across the entire molecule (Yumrutas et al., 2012). ROS from these phenolic compounds is a significant factor in fortifying biomolecules andalso shields the body's cells fromthe development of specific cancers (Burdette et al., 2002).

3.5.2.7 Anticancer Effect

Cancer is the uncontrolled proliferation of the body's normal cells and may exhibit metastasis rapidly. Cancer is the major leading cause of death in the modern world (Anand et al., 2008). Prostate cancer (PCa) is the most recurrent cancer found in males and one of the prime causes of death in western countries (Aus et al. 2002). 90% of patients respond to androgen deprivation therapy strategies. However, novel therapeutic strategies are incorporated for PCa improvement (Rice et al.,2007). Several *in vitro* studies depict promising proliferation inhibition effects of *C. racemosa* extract on androgen-dependent and independent prostate cancer cells (Aus et al. 2002). BNO-1055 is a dry ethanolic rhizome extract obtained from *C. racemosa* to impair proliferation in PCa cells. The mechanism remains unknown, but it deduces that cimicifugoside, a triterpenoid xyloside, inhibits cellular uptake of nucleosides through an unfamiliar process (Dueregger et al., 2013). Nucleosides are vital components of cellular processes such as ATP, GTP, adenosine, inosine, cAMP, cGMP. Although most cells can synthesize through *de novo* biosynthesis, the nucleotides pool is preserved through an energy-efficient "salvage pathway" that involves recycling and uptake of extracellular hydrophilic nucleosides by membrane proteins (Molina-Arcas et al., 2009). Herein, the effects of the *C. racemosa* extract BNO-1055 on PCa cells and assesses its therapeutic prospects in treating PCa cell lines and, most importantly, BNO-1055 rapidly influences the weakening of extracellular thymidine,hampering their integration into DNA in benign and malignant prostate cell lines (Betz et al., 2009). However, the extension of the treatment has a noteworthy decrease in cell numbers. The effects of BNO-1055 is limitedto prostate carcinoma cells since thymidine addition was attenuated in benign prostate cells and other carcinoma cells (Nuntanakorn, 2005; Sampson et al., 2012). Incidentally, at concentrations that lessened nucleoside uptake and incorporation, BNO-1055 did not diminish cell apoptosis. However, research findings demonstrate proliferative and anti-tumor effects of BNO-1055 on prostate and BCa cells both *in vitro* and *in vivo* (Jiang et al., 2011). Studies observe that extracts of black cohosh suppress the growth of human breast cancer cells. However, the accurate mechanism is not definite but researchers are trying to identify the accurate mechanism (Betz et al., 2009). The growthinhibition by actein and a MeOH extract of black cohosh on human breast cancer cells corresponds with activation of stress response pathways (Benjamin, 2006). These agents stimulate the integrated stress response phases, either the survival or the apoptotic stage. It depends on the duration of treatment and the dose of the medicine actein (Wu et al. 2005, 2016). Breast cell lines were prepared, and the extract of black cohosh was added, which hinders the higher levels of Her 2in MDA-MB-453 breast cancer cells at an IC50 value. It stimulates 50% inhibition of cell proliferation (Spangler et al., 2007). These studies indicate that black cohosh has the potential to treat breast cancer. In animal studies, *A. racemosa* did not demonstrate any mutagenic, carcinogenic, or toxic effects (Davis et al., 2008).

3.5.2.8 Antidiabetic Effect

Diabetes has been deemed one of the leading causes of fatality worldwide. The rate of death is predicted to peak at 438 million including misdiagnosed and undiagnosed cases. It is defined as a disorder of the metabolic system identified via hyperglycemia, resulting from an insufficient amount of insulin from defectivepancreatic β-cells (Brownlee, 2005). AMP-activated protein kinase (AMPK) is a key sensor of cellular energy and also a regulator of maintaining energy. *C. racemosa*extracts Ze 450 have exhibited AMPK-activating properties *in vitro* to the same extent as Metformin (Moser et al., 2014). Increased activity of the AMPK is directly related to improved glucose uptake by cells and better insulin sensitivity, hence making the *C. racemosa* extract Ze 450 a potential agent for managing type 2 diabetes (Hardie et al., 2012). These antidiabetic effects of the *C. racemosa* extract Ze 450 were assessed in an*in vivo* mouse model suffering from type 2 diabetes along with insulin resistance. The experimental results showed significant AMPK activation via the Ze 450 and its components like 23-epi-26-deoxyactein, hydrochloride, and protopine. The data described that the Ze 450 has antidiabetic effects equal to or even greater than metformin (Moser et al., 2014). While metformin showed a negligible effect on cumulative weight gain, daily water, and food intake, *C. racemosa* extracts Ze 450 oral or intraperitoneal administration showed a notable decrease in all these aforementioned factors. In other words, experimental data exhibited that Ze 450 markedly elevated plasma glucose, enhanced glucose metabolism, and better insulin sensitivity (Thorup et al., 2015). The food and water intake were reduced at 30mg/kg of Ze 450, indicating it helped with polyuria and polyphagia.This meansthe target tissues are getting sufficient glucose, so the plasma level glucose has gone down. Dietary *C. racemosa* BNO 1055 extract was tested on 27 female Sprague–Dawley rats to assess the effects of *C. racemosa* BNO 1055 on glucose tolerance and fasting plasma insulin (FPI) levels. The experimental results represented lowered FPI levels, better glucose tolerance, and improved insulin sensitivity. The mechanism applied by the *C. racemosa* extract is complex and is assumed to be involved with the RANK-RANKL (Receptor activator of nuclear factor NFκB–RANK ligand) system (Rachoń et al., 2008). The RANKL system plays its role in lipogenesis, also in lipocyte metabolism regulation (Roche et al., 2004; Rohit Singh and Ezhilarasan, 2020). *C. racemosa* is predicted to possess antioxidant properties which influence fat and glucose metabolism positively. Regardless, like the *C. racemose* extract Ze 450, CR BNO 105 also needs further investigation to assess its biological and chemical properties to define the unclear mechanism usedby the *C. racemose* extract and to further determine what components are participating to bring about these effects (Kirtikar, 1975; Enbom et al., 2014).

3.5.2.9 Anti-Inflammatory Effect

Black cohosh is used to treat a variety of conditions, including malaria, rheumatism, menstrual irregularities, and menopause. Additionally, *C. racemosa* and its constituents act as anti-inflammatory agents (Cheng et al., 2007). For this study, a bioactivity-guided fractionation analysis involves partitioning extraction and high-performance liquid chromatography (HPLC) to purify the specific compound that can inhibit TNF-α production (Montesano et al., 2005). TNF-α exemplifies the development of chronic diseases, such as tumorigenesis and rheumatoid arthritis. The uncontrollable levels of TNF-α production participated in different stages of tumorigenesis, including the instigation of tumor growth, cell proliferation, and invasion (Komatsu et al., 2001). For cell expansion, TNF-α activates specific growth factors to regulate the development of malignant tumors. The cytokine promotes angiogenesis, thus playing a vital role in tumor metastasis (Lee et al., 2005). Triterpene glycosides are the major components of black cohosh extract. Experiments concluded that phenolic acids, such as ferulic and isoferulic acids, have an anti-inflammatory property where the production of macrophage inflammatory protein-2 is dysregulated (Li et al., 2005).

3.5.2.10 Anxiolytic Effect

Anxiolytic effects demonstrate a compound or drug's ability to prevent or treat anxiety disorders. Despite the prevalence of benzodiazepine (BZ)-induced dependency and antidepressant-induced sexual side effects, weight gain, and withdrawal, BZ anxiolytics and many types of antidepressants have become routine therapy for anxiety symptoms. It is known that black cohosh is used to treat menopausal symptoms. These symptoms may include anxiety disorders as well. In recent studies, it is suggested that black cohosh may contain anxiolytic effects as well (Amsterdam et al., 2009). Actein is a bioactive triterpene glycoside derived from *Cimicifuga* species (such as *C. racemosa*) that has been shown to inhibit breast cancer cell growth as well as osteoblastic cell proliferation (Yue et al., 2016). Along with 5-HT7, recent research from molecular docking and molecular dynamics simulations suggests that actein binds to 5-HT2C, another serotonin receptor subtype. Serotonin blocks dopamine release in particular regions of the brain when this receptor is activated, resulting in the depressed and anxious symptoms that are common in women throughout menopause (Poschner et al., 2020). As a result, black cohosh extract and its major component actein may inhibit serotonin absorption via 5-HT2C, improving mood in postmenopausal women (Liao et al., 2020). Peroral administration of black cohosh extract at a very high dosage (1,000mg/kg) to BALB/c mice was shown to have stress-relieving effects, with plasma corticosterone and aspartate aminotransferase levels considerably reduced. Furthermore, bioassay-guided fractionation of black cohosh extract revealed that actein, 23-epi-26-deoxyactein, and cimiracemoside were responsible for the majority of the anti-stress benefits. With a low risk of significant side effects, black cohosh looks to be worth considering for the treatment of depression and anxiety. For patients who want a natural approach, these appear to be viable choices. The mechanisms of action, as well as the safety and efficacy of these plants, should be the focus of future studies (Yeung et al., 2018).

3.5.2.11 GABA Receptor Modulating Effect

GABA is a neurotransmitter that regulates fast inhibitory neurotransmission in the mammalian brain (Hossain et al., 2021). GABA receptors are composed of five subunits that form a chloride-conducting ion pore. Nineteen different GABA receptor subunits have been identified in the human genome (Simon et al.,2004). Black cohosh consists of high amounts of triterpenoids and these triterpenoids are designated by a 9,19-cycloartane group which is the plant's active site. Because of the hormone-like activity of black cohosh preparations, an estrogen mediates the alleviating effects resulting from GABA(A)- receptor modulation (Cicek et al.,2010). *A. racemosa* comprises triterpenoids and divergent triterpene types, which differ in conformation. Acteol and cimigenole triterpenoids have a six-ring structure, whereas shengmanol-type triterpenoids have a four-ring structure resembling the steroid (Choi, 2013). Concerning polarity, acteol- and shengmanol-type triterpenoids exhibit higher oxidation and are usually more polar than cimigenol (Guo et al.,2017). The variants in chemical structure and polarity obviously can lead to diverging activities, as observed in GABA receptors, where the effect of 23-O-acetylshengmanol-3-O-β-d-xylop yranoside suppresses that of actein and 25-O-acetylcimigenol-3-O-α-l-arabinopyranoside (3) by five times (Strommer et al.,2014). The active components in black cohosh behave as neurotransmitters, such as5-hydroxytryptamine (5-HT) and γ -aminobutyric acid (GABA) (Lee et al.,2013). GABAergic activities in the brain, including the hypothalamus, are crucial in maintaining the sleep cycle, but ovarian disorder disrupts it in menopausal women (Monti,2011). To investigate the effect of black cohosh on GABA activity-induced chloride currents,the GABA receptors are expressed in Xenopus oocytes, andthe active principles are detected (Garcia et al.,2006). Allosteric regulation of GABA receptors by the identified compounds may develop anxiolytic, sedative, and hypnotic effects. These potentially contribute to the beneficial impacts described for black cohosh extracts in the treatment of postmenopausal symptoms (Granger et al., 2005).

3.5.2.12 Menopause Effect

C. racemosa is utilized for gynecologic and pain disorders. Black cohosh has been used for the cure of various ailments, such as amenorrhea, menopause, and childbirth (Borrelli and Ernst 2008). Menopause happens aroundthe age range of 45 to54 years, where the function of the ovaries begins to diminish. By 2030, the population of menopausal women will be 1.2 billion globally. Estrogen deficiency which results from menopause is coupled with vasomotor, vaginal, and psychological symptoms, including hot flashes, vaginal dryness, and bone loss. Hormone replacement therapy (HRT) is a constructive intervention for such difficulties (MacLennan et al., 2004). However, clinical trials implied that the risk associated with HRT might outweigh the pros for women on continuous estrogen and progestin treatment. So, clinicians are uncertain about HRT (Schwartz and Woloshin,2004). Black cohosh is a promising alternative in relieving menopausal disorders. Extracts of *C. racemosa* have reduced the pain experienced during childbirth and uterine colic and dysmenorrhea (Viereck et al.,2005). The chemical components present in black cohosh are triterpene glycosides (actein, 23-epi26-deoxyactein cimicifugoside), phenolic acids (isoferulic acid and fukinolic acid), flavonoids, volatile oils, tannins, and other pharmacological activators (Nuntanakorn et al., 2006).These constituents play a pivotal role in the modulation of SER, and therefore, regulate the central nervous system along with minimizing menopausal problems. Since there is a correlation between CNS regulation in menopause and sex hormones and CNS receptors, it is feasible that black cohosh may interact with the central nervous system in treating menopausal symptoms (Weiss et al., 2004). Furthermore, reports depict that black cohosh acts as an agonist at specific receptors, such as serotonin receptors, and demonstrates a high binding affinity for the dopamine D2 receptor (Burdette et al., 2003). The opioid system, especially the μ opiate receptor, is vital for regulating temperature and hormone homeostasis. It concludes as one of the mechanisms for the action of black cohosh in menopausal issues (Borelli et al., 2003). The current study shows that the components of black cohosh have a considerable affinity for the human μ opiate receptor. The opiate receptor system serves various features of female reproductive neuroendocrinology, such as controlling sex hormones (Acosta-Martinez and Etgen,2002). On the other hand, the endogenous opiate system modulates the CNS temperature control center through catecholamines. Opiates can adjust core temperature directly or indirectly through the release of catecholamines. Remarkable resemblances exist between opiate withdrawal and menopausal hot flashes (Manson et al., 2013). Surprisingly, opiate dependence withdrawal is seen as an animal model of menopausal hot flashes. Therefore, botanical dietary supplements – mainlyblack cohosh extracts – containing opiates are advantageous in relieving menopausal symptoms, including eliminating hot flashes (Cybulska et al. 2011; Kreek et al., 2002).

3.5.2.13 Neuroprotective Effect

Traumatic brain injury (TBI) is the primary cause of mortality and morbidity, seen at 14–45 years of age. TBI is responsible for 10 million deaths annually (Langlois et al., 2006). Biomechanical and neurochemical damage results from TBI leads to irreparable behavioral, neuropsychiatric, cognitive, and physical changes (Masel and DeWitt,2010).The chemical constituents of *A. racemosa* are the followingcaffeic acid, ferulic acid, phenylpropanoids, triterpenoids, cimigenol, and formononetin.Black cohosh has exhibited several therapeutic uses, such as neuroprotection (Rathore et al., 2012). A study with *C. racemosa* depicts how formononetin, a component of black cohosh, significantly increased the neurological severity score (NSS) and cortical neuronal numbers in rats in 7 days(Baez-Jurado et al.,2018). Additionally, formononetin lowered the levels of interleukin 6 (IL-6) and TNF-α and induced the synthesis of IL-10in the cerebral cortex (Li et al., 2017). Menopause-associated estrogen deficiency aggravates the aging process of the female brain and accompanies a progressive decline in cognition and memory, respectively (Hui et al., 2012). Additionally, lower estrogen levels lead to a defect in temperature-sensitive neurons in the hypothalamus. This disorder reduces the ability to adjust to changes in ambient temperature,resultingin hot flashes, the most dominant menopausal symptom (Stute et al., 2021). The hippocampus plays a crucial role in learning and memory. Hippocampal outputs are primarily inhibitory on neuroendocrine activity (Nicken et al., 2017). For menopause, black cohosh extract is given to treat hot flashes. Estrogens and black cohosh extract activate the release of c-fos protein (a marker of neuronal activity) within the hypothalamic nuclei (Rance et al., 2013). Furthermore, black cohosh also interacts with the hypothalamic-pituitary-adrenal axis under stress leading to elevation of acute stress responses in rats. The neuroprotectivity in Sprague Dawley rats compares the impact of estradiol and black cohosh on the hypothalamus with ovariectomized (OVX) and intact rats (Sturdee et al.,2017). Both estradiol and black cohosh avert dysregulation of the hypothalamic neurons through different mechanisms. Black cohosh applies changes to brain activity in the same way asdrugs used to treat Parkinson's disease, dementia, or depression and enhances mood and cognitive behavior (Skorupskaite et al., 2018).

3.5.2.14 Serotonin Receptor Effect

Serotonin receptors are found in the peripheral and central nervous systems as a group of ligand-gated ion channels and G-protein coupled receptors. Natural ligand and serotonin neurotransmitters are responsible for regulating such receptors via inhibitory and excitatory neurotransmission (Stiedl et al., 2015). The release of neurotransmitters, such as glutamate, dopamine, norepinephrine, epinephrine, acetylcholine, and GABA, are modulated by serotonin receptors. Thus, these groups of receptors have a significant impact on various neurological and biological processes such as mood, anxiety, cognition, learning, etc., and are targeted for drugs addressing the aforementioned issues (Meltzer and Massey, 2011).Previous studies have suggested that black cohosh might act through the serotonin pathway in an attempt to relieve hot flashes.The estrogenic and serotonergic systems are known to be interconnected as well (Burdette et al., 2003). Serotonin production is reduced when estrogen levels decline because the level of tryptophan hydroxylase in the body, which is responsible for converting tryptophan into serotonin, is also decreased. In addition, the amount of serotonin in menopausal women's circulation decreases when estrogen levels drop. It is possible to increase serotonergic activity with HRT (Loprinzi et al., 2002). Both healthy women and cancer patients using selective estrogen receptor modulators have had successin decreasing menopausal hot flashes with selective serotonin reuptake inhibitors (SSRIs) (Stearns et al., 2000). It has been seen in experiments that black cohosh contains serotonin receptor ligands and has a high affinity for the 5-HT1A, 5-HT1D, and 5-HT7 receptors at a high dose (250mg/mL) (Burdette et al., 2003). Black cohosh extracts in animal experiments and an estrogen receptor bioassay demonstrated estrogenic-like activity. Researchers found that black cohosh contains three chemicals that worked together to lower serum luteinizing hormone (LH) levels and bind to estrogen receptors, according to their findings. Because of the decrease in LH and the binding of formononetin, triterpenoid glycosides were thought to be involved. It has been suggested that black cohosh works through estrogen receptors (Ruhlen et al., 2008). For two reasons, however, this idea has not been accepted by other researchers. In animal investigations, cell binding tests, and cell proliferation assays, more recent research has revealed a lack of estrogenic action. Second, extracts of black cohosh did not contain formononetin, a recognized estrogenic component. Due to the selective ER modulator SERM-like activity of black cohosh extracts, some researchers believe that black cohosh may bind to an undiscovered ER receptor in humans. According to current research, it is doubtful that black cohosh has direct estrogenic action, although it might have a central activity (Powell et al., 2008).

3.6 CONCLUSION

Nowadays, natural food products are very popular in the fight against diseases, such as cardiovascular disease, immune dysfunction, and cancer. Moreover, due to the lower risk of side effects and lower cost of these biomolecules, consumers increasingly turn to compounds derived from medicinal plants to treat a wide range of conditions, including malignant tumors. It was recently discovered that scientists in the pharmaceutical and medical fields are looking for natural compounds to use as medicinal agents, as chemical constituents have significant adverse effects on patients' bodies. As a result, theaim of this chapter was to investigate natural phytochemicals that have therapeutic potential. *C. borivilianum* and *C. racemosa* were examined. Both plants were found to contain a significant number of diverse chemical compounds. As a primary source of flavonoid and phenolic compounds, these plants exhibited various biological properties, most notably strong anticancer and antidiabetic properties. They may also be useful as potential sources of phyto-therapeutic lead molecules in the future. However, prior published research indicates that the pharmacokinetic evidence for this promising, highly nutritional medicinal plant and its derivatives is insufficient in this instance. To better understand these natural compounds, particularly their toxicogenetical profiles, more research studies need to be conducted to elucidate their specific disease-controlling and toxicological mechanisms, as well as their pharmacokinetic properties.

ACKNOWLEDGMENT

We would like to express our sincere gratitude to the Department of Pharmacy, Life Science Faculty, Bangabandhu Sheikh Mujibur Rahman Science and Technology University, Gopalganj (Dhaka)-8100, and International Centre for Empirical Research and Development, Bangladesh.

REFERENCES

Acharya, D., Mitaine-Offer, A.C., Kaushik, N., Miyamoto, T., Paululat, T. and Lacaille-Dubois, M.A., 2008a. Furostane-type steroidal saponins from the roots of *Chlorophytum borivilianum*. *Helvetica Chimica Acta*, 91(12), pp.2262–2269.

Acharya, D., Mitaine-Offer, A.C., Kaushik, N., Miyamoto, T., Paululat, T. and Lacaille-Dubois, M.A., 2008b. Steroidal saponins from the roots of *Chlorophytum borivilianum*. *Planta Medica*, 74(9), pp.PB23.

Acharya, D., Mitaine-Offer, A.C., Kaushik, N., Miyamoto, T., Paululat, T., Mirjolet, J.F., Duchamp, O. and Lacaille-Dubois, M.A., 2009. Cytotoxic spirostane-type saponins from the roots of *Chlorophytum borivilianum*. *Journal of Natural Products*, 72(1), pp.177–181.

Acosta-Martinez, M. and Etgen, A.M., 2002. Activation of μ-opioid receptors inhibits lordosis behavior in estrogen and progesterone-primed female rats. *Hormones and Behavior*, 41(1), pp.88–100.

Amsterdam, J.D., Yao, Y., Mao, J.J., Soeller, I., Rockwell, K. and Shults, J., 2009. Randomized, double-blind, placebo-controlled trial of *Cimicifuga racemosa* (black cohosh) in women with anxiety disorder due to menopause. *Journal of Clinical Psychopharmacology*, 29(5), pp.478.

Anand, P., Kunnumakara, A.B., Sundaram, C., Harikumar, K.B., Tharakan, S.T., Lai, O.S., Sung, B. and Aggarwal, B.B., 2008. Cancer is a preventable disease that requires major lifestyle changes. *Pharmaceutical Research*, 25(9), pp.2097–2116.

Applequist, W.L., 2003. Rhizome and root anatomy of potential contaminants of *Actaea racemosa* L.(black cohosh). *Flora-Morphology, Distribution, Functional Ecology of Plants*, 198(5), pp.358–365.

Arif, J.M., 2005. Effects of Safed musli on cell kinetics and apoptosis in human breast cancer cell lines. In International Conference on Promotion and Development of Botanicals with International Co-ordination Exploring Quality, Safety, Efficacy and Regulation Organized by School of Natural Product Study, Jadavpur University, Kolkata, India Feb (pp. 25–26).

Aus, G., Adolfsson, J., Selli, C. and Widmark, A., 2002. Treatment of patients with clinical T3 prostate cancer. *Scandinavian Journal of Urology and Nephrology*, 36(1), pp.28–33.

Bae, A.H., Kim, G., Seol, G.H., Lee, S.B., Lee, J.M., Chang, W. and Min, S.S., 2020. Delta-and mu-opioid pathways are involved in the analgesic effect of Ocimum basilicum L in mice. *Journal of Ethnopharmacology*, 250, pp.112471.

Baez-Jurado, E., Vega, G.G., Aliev, G., Tarasov, V.V., Esquinas, P., Echeverria, V. and Barreto, G.E., 2018. Blockade of neuroglobin reduces protection of conditioned medium from human mesenchymal stem cells in human astrocyte model (T98G) under a scratch assay. *Molecular Neurobiology*, 55(3), pp.2285–2300.

Barceloux, D.G., 2008. *Medical Toxicology of Natural Substances: Foods, Fungi, Medicinal Herbs, Plants, and Venomous Animals*. New York: John Wiley & Sons..

Bathoju, G. and Giri, A., 2012a. Production of medicinally important secondary metabolites (stigmasterol and hecogenin) from root cultures of Chlorophytum borivilianum (Safed musli). *Recent Research in Science and Technology*, 4(5).

Bathoju, G. and Giri, A, 2012b. Production of stigmasterol and hecogenin from *in vitro* cultures of Chlorophytum borivilianum. *Journal of Pharmacognosy*, 3(2), pp.101–103.

Benjamin, I.J., 2006. Viewing a stressful episode of ER: is ATF6 the triage nurse?. *Circulation Research*, 98(9), pp.1120–1122.

Betz, J.M., Anderson, L., Avigan, M.I., Barnes, J., Farnsworth, N.R., Gerdén, B., Henderson, L., Kennelly, E.J., Koetter, U., Lessard, S. and Dog, T.L., 2009. Black cohosh: considerations of safety and benefit. *Nutrition Today*, 44(4), pp.155–162.

Bhandary, M.J., Chandrashekar, K.R. and Kaveriappa, K.M., 1995. Medical ethnobotany of the siddis of Uttara Kannada district, Karnataka, India. *Journal of Ethnopharmacology*, 47(3), pp.149–158.

Bi, Y., Wei, Z., Kong, Y. and Hu, L., 2021. Supraspinal neural mechanisms of the analgesic effect produced by transcutaneous electrical nerve stimulation. *Brain Structure and Function*, 226(1), pp.151–162.

Bittner, G.D., Denison, M.S., Yang, C.Z., Stoner, M.A. and He, G., 2014. Chemicals having estrogenic activity can be released from some bisphenol a-free, hard and clear, thermoplastic resins. *Environmental Health*, 13(1), pp.1–18.

Bonora, E., 2008. Protection of pancreatic beta-cells: is it feasible?. *Nutrition, Metabolism and Cardiovascular Diseases*, 18(1), pp.74–83.

Borrelli, F. and Ernst, E., 2008. Black cohosh (Cimicifuga racemosa) for menopausal symptoms: a systematic review of its efficacy. *Pharmacological Research*, 58(1), pp.8–14.

Borrelli, F., Izzo, A.A. and Ernst, E., 2003. Pharmacological effects of Cimicifuga racemosa. *Life Sciences*, 73(10), pp.1215–1229.

Brownlee, M., 2005. The pathobiology of diabetic complications: a unifying mechanism. *Diabetes*, 54(6), pp.1615–1625.

Burdette, J.E., Chen, S.N., Lu, Z.Z., Xu, H., White, B.E., Fabricant, D.S., Liu, J., Fong, H.H., Farnsworth, N.R., Constantinou, A.I. and Van Breemen, R.B., 2002. Black cohosh (Cimicifuga racemosa L.) protects against menadione-induced DNA damage through scavenging of reactive oxygen species: bioassay-directed isolation and characterization of active principles. *Journal of Agricultural and Food Chemistry*, 50(24), pp.7022–7028.

Burdette, J.E., Liu, J., Chen, S.N., Fabricant, D.S., Piersen, C.E., Barker, E.L., Pezzuto, J.M., Mesecar, A., Van Breemen, R.B., Farnsworth, N.R. and Bolton, J.L., 2003. Black cohosh acts as a mixed competitive ligand and partial agonist of the serotonin receptor. *Journal of Agricultural and Food Chemistry*, 51(19), pp.5661–5670.

Chen, S.S., Gong, J., Liu, F.T. and Mohammed, U., 2000. Naturally occurring polyphenolic antioxidants modulate IgE-mediated mast cell activation. *Immunology*, 100(4), pp.471–480.

Cheng, S.M., Xing, B., Li, J.C., Cheung, B.K. and Lau, A.S., 2007. Interferon-γ regulation of TNFα-induced matrix metalloproteinase 3 expression and migration of human glioma T98G cells. *International Journal of Cancer*, 121(6), pp.1190–1196.

Chetty, K.M. and Rao, K.N., 1989. Ethnobotany of Sarakallu and adjacent areas of Chittoor district, Andhra Pradesh. *Vegetos*, 2(51), p.58.

Choi, E.M., 2013. Deoxyactein Isolated from *Cimicifuga racemosa* protects osteoblastic MC3T3-E1 cells against antimycin A-induced cytotoxicity. *Journal of Applied Toxicology*, 33(6), pp.488–494.

Chua, B., Abdullah, Z., Pin, K.Y., Abdullah, L.C., Choong, T.S.Y. and Yusof, U.K., 2017. Isolation, structure elucidation, identification and quantitative analysis of 1′-acetoxychavicol (ACA) from the roots of chlorophytum boriviliuanum (SAFED MUSLI). *Journal of Engineering Science and Technology*, 12(1), pp.198–213.

Chua, B., Zunoliza, A., Pin, K., Luqman, C.A., Choong, S. and Umi, K.Y., 2015. Isolation, structure elucidation, identification and quantitative analysis of di (2-ethylhexyl) phthalate (DEHP) from the roots of *Chlorophytum boriviliuanum* (safed musli). *Research Journal of Pharmaceutical, Biological and Chemical Sciences*, 6(5), pp.1090–1095.

Cicek, S.S., Khom, S., Taferner, B., Hering, S. and Stuppner, H., 2010. Bioactivity-guided isolation of GABAA receptor modulating constituents from the rhizomes of *Actaea racemosa*. *Journal of Natural Products*, 73(12), pp.2024–2028.

Conte, P. and Guarneri, V., 2004. Safety of intravenous and oral bisphosphonates and compliance with dosing regimens. *The Oncologist*, 9(S 4), pp.28–37.

Csanaky, I., Németi, B. and Gregus, Z., 2003. Dose-dependent biotransformation of arsenite in rats: not S-adenosylmethionine depletion impairs arsenic methylation at high dose. *Toxicology*, 183(1–3), pp.77–91.

Cybulska, P., Thakur, S.D., Foster, B.C., Scott, I.M., Leduc, R.I., Arnason, J.T. and Dillon, J.A.R., 2011. Extracts of Canadian first nations medicinal plants, used as natural products, inhibit Neisseria gonorrhoeae isolates with different antibiotic resistance profiles. *Sexually Transmitted Diseases*, 38(7), pp.667–671.

Dabur, R., Gupta, A., Mandal, T.K., Singh, D.D., Bajpai, V., Gurav, A.M. and Lavekar, G.S., 2007. Antimicrobial activity of some Indian medicinal plants. *African Journal of Traditional, Complementary and Alternative Medicines*, 4(3), pp.313–318.

Davis, V.L., Jayo, M.J., Ho, A., Kotlarczyk, M.P., Hardy, M.L., Foster, W.G. and Hughes, C.L., 2008. Black cohosh increases metastatic mammary cancer in transgenic mice expressing c-erbB2. *Cancer Research*, 68(20), pp.8377–8383.

Deore, S.L. and Khadabadi, S.S., 2009a. Larvicidal activity of the saponin fractions of Chlorophytum borivilianum santapau and Fernandes. *Journal of Entomology and Nematology*, 1(5), pp.064–066.

Deore, S.L. and Khadabadi, S.S., 2009b. Screening of antistress properties of Chlorophytum borivilianum tuber. *Pharmacology Online*, 1, pp.320–328.

Deore, S.L. and Khadabadi, S.S., 2009c. A study of hypocholesteremic effects of Chlorophytum borivilianum tubers in rats. *Pharmacology Online*, 1, pp.398–403.

Deore, S.L. and Khadabadi, S.S., 2010. *In vitro* anthelmintic studies of Chlorophytum borivilianum Sant. & Fernandez tubers. *Indian Journal of Natural Products and Resources*, 1(1), pp.53–56.

Deshwal, R.K. and Trivedi, P., 2011. Effect of kinetin on enhancement of tuberous root production of Chlorophytum borivilianum. *International Journal of Innovations in Biological and Chemical Sciences*, 1, pp.28–31.

Dueregger, A., Guggenberger, F., Barthelmes, J., Stecher, G., Schuh, M., Intelmann, D., Abel, G., Haunschild, J., Klocker, H., Ramoner, R. and Sampson, N., 2013. Attenuation of nucleoside and anti-cancer nucleoside analog drug uptake in prostate cancer cells by Cimicifuga racemosa extract BNO-1055. *Phytomedicine*, 20(14), pp.1306–1314.

Dugoua, J.J., Seely, D., Perri, D., Koren, G.K. and Mills, E., 2006. Safety and efficacy of black cohosh (Cimicifuga racemosa) during pregnancy and lactation. *Journal of Population Therapeutics and Clinical Pharmacology*, 13(3), pp. e257–261.

Elizabeth, K.G., 2001. Safed musli: a promising medicinal plant. *Ind. J. Arecanut, Spices Med. Plants*, 5(2), pp.65–69.

Enbom, E.T., Le, M.D., Oesterich, L., Rutgers, J. and French, S.W., 2014. Mechanism of hepatotoxicity due to black cohosh (Cimicifuga racemosa): histological, immunohistochemical and electron microscopy analysis of two liver biopsies with clinical correlation. *Experimental and Molecular Pathology*, 96(3), pp.279–283.

Fabricant, D.S., Nikolic, D., Lankin, D.C., Chen, S.N., Jaki, B.U., Krunic, A., van Breemen, R.B., Fong, H.H., Farnsworth, N.R. and Pauli, G.F., 2005. Cimipronidine, a Cyclic Guanidine Alkaloid from Cimicifuga racemosa. *Journal of Natural Products*, 68(8), pp.1266–1270.

Firenzuoli, F., Gori, L. and di Sarsina, P.R., 2011. Black cohosh hepatic safety: follow-up of 107 patients consuming a special Cimicifuga racemosa rhizome herbal extract and review of literature. *Evidence-Based Complementary and Alternative Medicine*, 2011.

Gafner, S., 2016. Adulteration of Actaea racemosa. *Botanical Adulterants Bulletin.*

García, D.A., Bujons, J., Vale, C. and Suñol, C., 2006. Allosteric positive interaction of thymol with the GABAA receptor in primary cultures of mouse cortical neurons. *Neuropharmacology*, 50(1), pp.25–35.

Gardner, Z.E., Lueck, L., Erhardt, E.B. and Craker, L.E., 2012. A morphometric analysis of *Actaea racemosa* L.(Ranunculaceae). *Journal of Medicinally Active Plants*, 1(2), pp.47–59.

Giribabu, N., Kumar, K.E., Rekha, S.S., Muniandy, S. and Salleh, N., 2014. *Chlorophytum borivilianum* root extract maintains near normal blood glucose, insulin and lipid profile levels and prevents oxidative stress in the pancreas of streptozotocin-induced adult male diabetic rats. *International Journal of Medical Sciences*, 11(11), p.1172.

Godecke, T., Lankin, D.C., Nikolic, D., Chen, S.N., van Breemen, R.B., Farnsworth, N.R. and Pauli, G.F., 2009. Guanidine Alkaloids and Pictet– Spengler Adducts from Black Cohosh (Cimicifuga racemosa). *Journal of Natural Products*, 72(3), pp.433–437.

Goel, A., Singh, R., Dash, S., Gupta, D., Pillai, A., Yadav, S.K. and Bhatia, A.K., 2011. Antiviral activity of few selected indigenous plants against Bovine Herpes Virus-1. *Journal of Immunology and Immunopathology*, 13(1), pp.30–37.

Goswami, S.K., Kumar, P.M., Jamwal, R., Dethe, S., Agarwal, A. and Naseeruddin, I.M., 2012. Screening for Rho-kinase 2 inhibitory potential of Indian medicinal plants used in management of erectile dysfunction. *Journal of Ethnopharmacology*, 144(3), pp.483–489.

Granger, R.E., Campbell, E.L. and Johnston, G.A., 2005. (+)-And (–)-borneol: efficacious positive modulators of GABA action at human recombinant $\alpha 1\beta 2\gamma 2L$ GABAA receptors. *Biochemical Pharmacology*, 69(7), pp.1101–1111.

Guo, Y., Yin, T., Wang, X., Zhang, F., Pan, G., Lv, H., Wang, X., Orgah, J.O., Zhu, Y. and Wu, H., 2017. Traditional uses, phytochemistry, pharmacology and toxicology of the genus Cimicifuga: a review. *Journal of Ethnopharmacology*, 209, pp.264–282.

Haque, R., Saha, S. and Bera, T., 2011. A peer reviewed of general literature on Chlorophytum borivilianum commercial medicinal plant. *International Journal of Drug Development and Research*, 3(1), pp.165–177.

Hardie, D.G., Ross, F.A. and Hawley, S.A., 2012. AMPK: a nutrient and energy sensor that maintains energy homeostasis. *Nature Reviews Molecular Cell Biology*, 13(4), pp.251–262.

Harwood, H.J., Chandler, C.E., Pellarin, L.D., Bangerter, F.W., Wilkins, R.W., Long, C.A., Cosgrove, P.G., Malinow, M.R., Marzetta, C.A. and Pettini, J.L., 1993. Pharmacologic consequences of cholesterol absorption inhibition: alteration in cholesterol metabolism and reduction in plasma cholesterol concentration induced by the synthetic saponin beta-tigogenin cellobioside (CP-88818; tiqueside). *Journal of Lipid Research*, 34(3), pp.377–395.

He, K., Zheng, B., Kim, C.H., Rogers, L. and Zheng, Q., 2000. Direct analysis and identification of triterpene glycosides by LC/MS in black cohosh, Cimicifuga racemosa, and in several commercially available black cohosh products. *Planta Medica*, 66(07), pp.635–640.

Hirabayashi, T., Ochiai, H., Sakai, S., Nakajima, K. and Terasawa, K., 1995. Inhibitory effect of ferulic acid and isoferulic acid on murine interleukin-8 production in response to influenza virus infections in vitro and in vivo. *Planta Medica*, 61(03), pp.221–226.

Hossain, R., Al-Khafaji, K., Khan, R.A., Sarkar, C., Islam, M., Dey, D., Jain, D., Faria, F., Akbor, R., Atolani, O. and Oliveira, S.M., 2021. Quercetin and/or ascorbic acid modulatory effect on phenobarbital-induced sleeping mice possibly through GABAA and GABAB receptor interaction pathway. *Pharmaceuticals*, 14(8), p.721.

Huang, F., Long, Y., Liang, Q., Purushotham, B., Swamy, M.K. and Duan, Y., 2019. Safed Musli (Chlorophytum borivilianum L.) callus-mediated biosynthesis of silver nanoparticles and evaluation of their antimicrobial activity and cytotoxicity against human colon cancer cells. *Journal of Nanomaterials*, 2019.

Hui, Z., Xiaoyan, M., Mukun, Y., Ke, W., Liyuan, Y., Sainan, Z., Jing, J., Lihua, Q. and Wenpei, B., 2012. Effects of black cohosh and estrogen on the hypothalamic nuclei of ovariectomized rats at different temperatures. *Journal of Ethnopharmacology*, 142(3), pp.769–775.

Huntley, A., 2004. The safety of black cohosh (Actaea racemosa, Cimicifuga racemosa). *Expert Opinion on Drug Safety*, 3(6), pp.615–623.

Huntley, A. and Ernst, E., 2003. A systematic review of the safety of black cohosh. *Menopause*, 10(1), pp.58–64.

Ikeda, F., Nishimura, R., Matsubara, T., Tanaka, S., Inoue, J.I., Reddy, S.V., Hata, K., Yamashita, K., Hiraga, T., Watanabe, T. and Kukita, T., 2004. Critical roles of c-Jun signaling in regulation of NFAT family and RANKL-regulated osteoclast differentiation. *Journal of Clinical Investigation*, 114(4), pp.475–484.

Imai, A., Lankin, D.C., Gödecke, T., Chen, S.N. and Pauli, G.F., 2020. NMR based quantitation of cycloartane triterpenes in black cohosh extracts. *Fitoterapia*, 141, p.104467.

Imai, A., Lankin, D.C., Nikolić, D., Ahn, S., van Breemen, R.B., Farnsworth, N.R., McAlpine, J.B., Chen, S.N. and Pauli, G.F., 2016. Cycloartane triterpenes from the aerial parts of *Actaea racemosa. Journal of Natural Products*, 79(3), pp.541–554.

Jiang, B., Kronenberg, F., Balick, M.J. and Kennelly, E.J., 2006. Analysis of formononetin from black cohosh (*Actaea racemosa*). *Phytomedicine*, 13(7), pp.477–486.

Jiang, B., Kronenberg, F., Nuntanakorn, P., Qiu, M.H. and Kennelly, E.J., 2006. Evaluation of the botanical authenticity and phytochemical profile of black cohosh products by high-performance liquid chromatography with selected ion monitoring liquid Chromatography–Mass spectrometry. *Journal of Agricultural and Food Chemistry*, 54(9), pp.3242–3253.

Jiang, B., Ma, C., Motley, T., Kronenberg, F. and Kennelly, E.J., 2011. Phytochemical fingerprinting to thwart black cohosh adulteration: a 15 Actaea species analysis. *Phytochemical Analysis*, 22(4), pp.339–351.

Jimi, E., Aoki, K., Saito, H., D'Acquisto, F., May, M.J., Nakamura, I., Sudo, T., Kojima, T., Okamoto, F., Fukushima, H. and Okabe, K., 2004. Selective inhibition of NF-κB blocks osteoclastogenesis and prevents inflammatory bone destruction in vivo. *Nature Medicine*, 10(6), pp.617–624.

Kenjale, R., Shah, R. and Sathaye, S., 2008. Effects of Chlorophytum borivilianum on sexual behaviour and sperm count in male rats. *Phytotherapy Research: An International Journal Devoted to Pharmacological and Toxicological Evaluation of Natural Product Derivatives*, 22(6), pp.796–801.

Kenjale, R.D., Shah, R.K. and Sathaye, S.S., 2007. Anti-stress and anti-oxidant effects of roots of *Chlorophytum borivilianum* (Santa Pau & Fernandes). *Indian Journal of Experimental Biology*, 45(11), pp. 974–979.

KennellyEJ, BaggettS, NuntanakornP, OsoskiAL, MoriSA, DukeJ, ColetonM, KronenbergF., 2002 Jan 1. Analysis of thirteen populations of black cohosh for formononetin. *Phytomedicine*, 9(5), pp. 461–467.

Khanam, Z., Singh, O., Singh, R. and Bhat, I.U.H., 2013. Safed musli (Chlorophytum borivilianum): a review of its botany, ethnopharmacology and phytochemistry. *Journal of Ethnopharmacology*, 150(2), pp.421–441.

Kikuzaki, H., Hisamoto, M., Hirose, K., Akiyama, K. and Taniguchi, H., 2002. Antioxidant properties of ferulic acid and its related compounds. *Journal of Agricultural and Food Chemistry*, 50(7), pp.2161–2168.

Kim, C.D., Lee, W.K., Lee, M.H., Cho, H.S., Lee, Y.K. and Roh, S.S., 2004. Inhibition of mast cell-dependent allergy reaction by extract of black cohosh (Cimicifuga racemosa). *Immunopharmacology and Immunotoxicology*, 26(2), pp.299–308.

Kirtikar, K.R., 1975. Liliaceae: chlorophytum. *Indian Medicinal Plants*.

Klassen, C.D. and Watkins, J.B., 2003. *Casarett and Doull's Essentials of Toxicology*. New York: McGraw Hill Professional.

Knoess, W., 2010. *Assessment Report on Cimicifuga racemosa (L.) Nutt., rhizoma*. Amsterdam: European Medicines Agency, Committee on Herbal Medicinal Products (HPMC), p.39.

Komatsu, M., Kobayashi, D., Saito, K., Furuya, D., Yagihashi, A., Araake, H., Tsuji, N., Sakamaki, S., Niitsu, Y. and Watanabe, N., 2001. Tumor necrosis factor-α in serum of patients with inflammatory bowel disease as measured by a highly sensitive immuno-PCR. *Clinical Chemistry*, 47(7), pp.1297–1301.

Korhonen, R., Lahti, A., Kankaanranta, H. and Moilanen, E., 2005. Nitric oxide production and signaling in inflammation. *Current Drug Targets-Inflammation & Allergy*, 4(4), pp.471–479.

Kothari, S. and Singh, K., 2003. Production techniques for the cultivation of safed musli (Chlorophytum borivilianum). *The Journal of Horticultural Science and Biotechnology*, 78(2), pp.261–264.

Kumar, D., Arya, V., Kaur, R., Bhat, Z.A., Gupta, V.K. and Kumar, V., 2012. A review of immunomodulators in the Indian traditional health care system. *Journal of Microbiology, Immunology and Infection*, 45(3), pp.165–184.

Kumar, M., Meena, P., Verma, S., Kumar, M. and Kumar, A., 2010. Anti-tumour, anti-mutagenic and chemomodulatory potential of *Chlorophytum borivilianum. Asian Pacific Journal of Cancer Prevention*, 11(2), pp.327–334.

Kumari, P., Singh, N. and Kumar, D., 2015. Synergistic analgesic activity of chloroform extract of *Lawsonia inermis* Linn. and *Chlorophytum borivilianum* Sant. *Journal of Pharmacognosy and Phytochemistry*, 3(6).

Kreek, M.J., LaForge, K.S. and Butelman, E., 2002. Pharmacotherapy of addictions. *Nature Reviews Drug Discovery*, 1(9), pp.710–726.

Lai, G.F., Wang, Y.F., Fan, L.M., Cao, J.X. and Luo, S.D., 2005. Triterpenoid glycoside from Cimicifuga racemosa. *Journal of Asian Natural Products Research*, 7(5), pp.695–699.

Lam, J., Nelson, C.A., Ross, F.P., Teitelbaum, S.L. and Fremont, D.H., 2001. Crystal structure of the TRANCE/RANKL cytokine reveals determinants of receptor-ligand specificity. *Journal of Clinical Investigation*, 108(7), pp.971–979.

Langlois, J.A., Rutland-Brown, W. and Wald, M.M., 2006. The epidemiology and impact of traumatic brain injury: a brief overview. *The Journal of Head Trauma Rehabilitation*, 21(5), pp.375–378.

Larochette, N., Decaudin, D., Jacotot, E., Brenner, C., Marzo, I., Susin, S.A., Zamzami, N., Xie, Z., Reed, J. and Kroemer, G., 1999. Arsenite induces apoptosis via a direct effect on the mitochondrial permeability transition pore. *Experimental Cell Research*, 249(2), pp.413–421.

Lee, D.C., Cheung, C.Y., Law, A.H., Mok, C.K., Peiris, M. and Lau, A.S., 2005. p38 mitogen-activated protein kinase-dependent hyperinduction of tumor necrosis factor alpha expression in response to avian influenza virus H5n1. *Journal of Virology*, 79(16), pp.10147–10154.

Lee, D.Y., Roh, C.R., Kang, Y.H., Choi, D., Lee, Y., Rhyu, M.R. and Yoon, B.K., 2013. Effects of black cohosh on the plasminogen activator system in vascular smooth muscle cells. *Maturitas*, 76(1), pp.75–80.

Lee, S.H., Park, M.H., Park, S.J., Kim, J., Kim, Y.T., Oh, M.C., Jeong, Y., Kim, M., Han, J.S. and Jeon, Y.J., 2012. Bioactive compounds extracted from Ecklonia cava by using enzymatic hydrolysis protects high glucose-induced damage in INS-1 pancreatic β-cells. *Applied Biochemistry and Biotechnology*, 167(7), pp.1973–1985.

Li, J.C., Lee, D.C., Cheung, B.K. and Lau, A.S., 2005. Mechanisms for HIV Tat upregulation of IL-10 and other cytokine expression: kinase signaling and PKR-mediated immune response. *FEBS Letters*, 579(14), pp.3055–3062.

Li, J.X., Kadota, S., Hattori, M., Yoshimachi, S., Shiro, M., Oogami, N., Mizuno, H. and Namba, T., 1993. Constituents of Cimicifugae rhizoma. I. Isolation and characterization of ten new cycloartenol triterpenes from *Cimicifuga heracleifolia* Komarov.*Chemical and Pharmaceutical Bulletin*, 41(5), pp.832–841.

Li, W., Chen, S., Fabricant, D., Angerhofer, C.K., Fong, H.H., Farnsworth, N.R. and Fitzloff, J.F., 2002. High-performance liquid chromatographic analysis of Black Cohosh (Cimicifuga racemosa) constituents with in-line evaporative light scattering and photodiode array detection. *Analytica Chimica Acta*, 471(1), pp.61–75.

Li, W., Sun, Y., Liang, W., Fitzloff, J.F. and van Breemen, R.B., 2003. Identification of caffeic acid derivatives in *Actea racemosa* (*Cimicifuga racemosa*, black cohosh) by liquid chromatography/tandem mass spectrometry. *Rapid Communications in Mass Spectrometry*, 17(9), pp.978–982.

Li, Z., Wang, Y., Zeng, G., Zheng, X., Wang, W., Ling, Y., Tang, H. and Zhang, J., 2017. Increased miR-155 and heme oxygenase-1 expression is involved in the protective effects of formononetin in traumatic brain injury in rats. *American Journal of Translational Research*, 9(12), p.5653.

Liao, X., Zhang, Q.Y., Xu, L. and Zhang, H.Y., 2020. Potential Targets of Actein Identified by Systems Chemical Biology Methods. *ChemMedChem*, 15(6), pp.473–480.

Liu, I.M., Chi, T.C., Hsu, F.L., Chen, C.F. and Cheng, J.T., 1999. Isoferulic acid as active principle from the rhizoma of Cimicifuga dahurica to lower plasma glucose in diabetic rats. *Planta Medica*, 65(08), pp.712–714.

Liu, I.M., Hsu, F.L., Chen, C.F. and Cheng, J.T., 2000. Antihyperglycemic action of isoferulic acid in streptozotocin-induced diabetic rats. *British Journal of Pharmacology*, 129(4), pp.631–636.

Liu, S., Tang, Y., Shu, H., Tatum, D., Bai, Q., Crawford, J., Xing, Y., Lobo, M.K., Bellinger, L., Kramer, P. and Tao, F., 2019. Dopamine receptor D2, but not D1, mediates descending dopaminergic pathway-produced analgesic effect in a trigeminal neuropathic pain mouse model. *Pain*, 160(2), p.334.

Lonner, J., 2007. *Medicinal Plant Fact Sheet: Cimicifuga racemosa/Black Cohosh. A Collaboration of the IUCN Medicinal Plant Specialist Group, PCA-Medicinal Plant Working Group, and North American Pollinator Protection Campaign.* Arlington, VA: PCA-Medicinal Plant Working Group.

Loprinzi, C.L., Sloan, J.A., Perez, E.A., Quella, S.K., Stella, P.J., Mailliard, J.A., Halyard, M.Y., Pruthi, S., Novotny, P.J. and Rummans, T.A., 2002. Phase III evaluation of fluoxetine for treatment of hot flashes. *Journal of Clinical Oncology*, 20(6), pp.1578–1583.

Lupu, R., Mehmi, I., Atlas, E., Tsai, M.S., Pisha, E., Oketch-Rabah, H.A., Nuntanakorn, P., Kennelly, E.J. and Kronenberg, F., 2003. Black cohosh, a menopausal remedy, does not have estrogenic activity and does not promote breast cancer cell growth. *International Journal of Oncology*, 23(5), pp.1407–1412.

MacLennan, A.H., Taylor, A.W. and Wilson, D.H., 2004. Hormone therapy use after the Women's Health Initiative. *Climacteric*, 7(2), pp.138–142.

MacMicking, J., Xie, Q.W. and Nathan, C., 1997. Nitric oxide and macrophage function. *Annual Review of Immunology*, 15(1), pp.323–350.

Mahady, G.B., Fabricant, D., Chadwick, L.R. and Dietz, B., 2002. Black cohosh: an alternative therapy for menopause?. *Nutrition in Clinical Care*, 5(6), pp.283–289.

Manson, J.E., Chlebowski, R.T., Stefanick, M.L., Aragaki, A.K., Rossouw, J.E., Prentice, R.L., Anderson, G., Howard, B.V., Thomson, C.A., LaCroix, A.Z. and Wactawski-Wende, J., 2013. Menopausal hormone therapy and health outcomes during the intervention and extended poststopping phases of the Women's Health Initiative randomized trials. *JAMA*, 310(13), pp.1353–1368.

Masel, B.E. and DeWitt, D.S., 2010. Traumatic brain injury: a disease process, not an event. *Journal of Neurotrauma*, 27(8), pp.1529–1540.

Matsui, T., Suzuki, Y., Yamashita, K., Yoshimaru, T., Suzuki-Karasaki, M., Hayakawa, S., Yamaki, M. and Shimizu, K., 2000. Diphenyleneiodonium prevents reactive oxygen species generation, tyrosine phosphorylation, and histamine release in RBL-2H3 mast cells. *Biochemical and Biophysical Research Communications*, 276(2), pp.742–748.

McCoy, J.A., Young, J.H., Nifong, J.M., Hummer, K., DeNoma, J., Avendaño-Arrazate, C.H., Greene, S.L. and Kantar, M.B., 2019. Species for medicinal and social use with an emphasis on Theobroma cacao L.(Cacao), Nicotiana tabacum L.(Tobacco), Actaea racemosa L.(Black Cohosh), and Humulus lupulus L.(Hops). In *North American Crop Wild Relatives*, Vol. 2. Cham: Springer, pp.645–692.

McKenna, D.J., Jones, K., Humphrey, S. and Hughes, K., 2001. Black cohosh: efficacy, safety, and use in clinical and preclinical applications. *Alternative Therapies in Health and Medicine*, 7(3), p.93.

Meena, A.K. and Rao, M.M., 2010. Folk herbal medicines used by the Meena community in Rajasthan. *Asian Journal of Traditional Medicines*, 5(1), pp.19–31.

Meltzer, H.Y. and Massey, B.W., 2011. The role of serotonin receptors in the action of atypical antipsychotic drugs. *Current Opinion in Pharmacology*, 11(1), pp.59–67.

Mimaki, Y., Kanmoto, T., Sashida, Y., Nishino, A., Satomi, Y. and Nishino, H., 1996. Steroidal saponins from the underground parts of Chlorophytum comosum and their inhibitory activity on tumour promoter-induced phospholipids metabolism of hela cells. *Phytochemistry*, 41(5), pp.1405–1410.

Molina-Arcas, M., Casado, F.J. and Pastor-Anglada, M., 2009. Nucleoside transporter proteins. *Current Vascular Pharmacology*, 7(4), pp.426–434.

Montesano, R., Soulié, P., Eble, J.A. and Carrozzino, F., 2005. Tumour necrosis factor α confers an invasive, transformed phenotype on mammary epithelial cells. *Journal of Cell Science*, 118(15), pp.3487–3500.

Monti, J.M., 2011. Serotonin control of sleep-wake behavior. *Sleep Medicine Reviews*, 15(4), pp.269–281.

Moser, C., Vickers, S.P., Brammer, R., Cheetham, S.C. and Drewe, J., 2014. Antidiabetic effects of the *Cimicifuga racemosa* extract Ze 450 in vitro and in vivo in ob/ob mice. *Phytomedicine*, 21(11), pp.1382–1389.

Narasimhan, S., Govindarajan, R., Madhavan, V., Thakur, M., Dixit, V.K., Mehrotra, S. and Madhusudanan, K.P., 2006. Action of (2→ 1) fructo-oligopolysaccharide fraction of Chlorophytum borivilianum against streptozotocin-induced oxidative stress. *Planta Medica*, 72(15), pp.1421–1424.

Negi, K.S., Tiwari, J.K., Gaur, R.D. and Pant, K.C., 1993. Notes on ethnobotany of five districts of Garhwal Himalaya, Uttar Pradesh, India. *Ethnobotany*, 5, pp.73–81.

Nicken, P., Kuchernig, J.C., Pickartz, S., Henneicke-von Zepelin, H.H. and Nolte, K.U., 2017. Cimicifuga racemosa for treatment of vasomotor symptoms: mode of action. *Maturitas*, 100, p.152.

Nikam, V. and Chavan, P., 2009. Influence of water deficit and waterlogging on the mineral status of a medicinal plant Chlorophytum borivilianum. *Acta Botanica Hungarica*, 51(1–2), pp.105–113.

Nikolić, D., Gödecke, T., Chen, S.N., White, J., Lankin, D.C., Pauli, G.F. and van Breemen, R.B., 2012. Mass spectrometric dereplication of nitrogen-containing constituents of black cohosh (Cimicifuga racemosa L.). *Fitoterapia*, 83(3), pp.441–460.

Nuntanakorn, P., 2005. *Polyphenolic Constituents from Black Cohosh (Actaea racemosa) and Related Species*. New York: City University of New York.

Nuntanakorn, P., Jiang, B., Einbond, L.S., Yang, H., Kronenberg, F., Weinstein, I.B. and Kennelly, E.J., 2006. Polyphenolic constituents of Actaea racemosa. *Journal of Natural Products*, 69(3), pp.314–318.

Oakenfull, D. and Sidhu, G.S., 1990. Could saponins be a useful treatment for hypercholesterolaemia?. *European Journal of Clinical Nutrition*, 44(1), pp.79–88.

Oudhia, P., 2001, June. My experiences with wonder crop Safed Musli. In *Sovenier*. Chennai, India: International Seminar on Medicinal Plants & Quality Standardization VHERDS, pp.9–10.

Panda, S.K., Si, S.C. and Bhatnagar, S.P., 2007. Studies on hypoglycaemic and analgesic activities of Chlorophytum borivilianum Sant and Ferz. *Journal of Natural Remedies*, 7(1), pp.31–36.

Papps, F.A., 2000. Therapeutic use and associated biochemistry of Cimicifuga racemosa in the treatment of menopausal symptoms. *Australian Journal of Medical Herbalism*, 12(1), pp. 22–26.

Paques, M. and Boxus, P., 1985, September. A model to learn" vitrification", the rootstock apple M. 26 present results. In Symposium on In Vitro Problems Related to Mass Propagation of Horticultural Plants 212, pp.193–210.

Patil, D.A., 2003. *Flora of Dhule and Nandurbar Districts (Maharashtra).* M/S Bishen Singh Mahendra Pal Singh.

Pengelly, A. and Bennett, K., 2014. *Appalachian plant monographs. Black cohosh Actaea racemosa L.*

Poschner, S., Wackerlig, J., Dobusch, D., Pachmann, B., Banh, S.J., Thalhammer, T. and Jäger, W., 2020. Actaea racemosa L. extract inhibits steroid sulfation in human breast cancer cells: effects on androgen formation. *Phytomedicine*, 79, p.153357.

Powell, S.L., Gödecke, T., Nikolic, D., Chen, S.N., Ahn, S., Dietz, B., Farnsworth, N.R., Van Breemen, R.B., Lankin, D.C., Pauli, G.F. and Bolton, J.L., 2008. In vitro serotonergic activity of black cohosh and identification of N ω-methylserotonin as a potential active constituent. *Journal of Agricultural and Food Chemistry*, 56(24), pp.11718–11726.

Predny, M.L., 2006. *Black Cohosh (Actaea racemosa): An Annotated Bibliography*, Vol. 97. Southern Research Station.

Puri, H.S., 2003. *RASAYANA: Ayurvedic Herbs of Rejuvenation and Longevity.* London: Taylor &Francis.

Pushpangadan, P., Dan, V.M., Ijinu, T.P. and George, V., 2012. Food, nutrition and beverage. *Indian Journal of Traditional Knowledge*, 11(1), pp. 26–34..

Qi, F., Li, A., Inagaki, Y., Gao, J., Li, J., Kokudo, N., Li, X.K. and Tang, W., 2010. Chinese herbal medicines as adjuvant treatment during chemoor radio-therapy for cancer. *Bioscience Trends*, 4(6),pp. 297–307.

Qiu, S.X., Dan, C., Ding, L.S., Peng, S., Chen, S.N., Farnsworth, N.R., Nolta, J., Gross, M.L. and Zhou, P., 2007. A triterpene glycoside from black cohosh that inhibits osteoclastogenesis by modulating RANKL and TNFα signaling pathways. *Chemistry & Biology*, 14(7), pp.860–869.

Qiu, S.X., Li, X.C., Xiong, Y., Dong, Y., Chai, H., Farnsworth, N.R., Pezzuto, J.M. and Fong, H.H., 2000. Isolation and characterization of cytotoxic saponin chloromaloside A from Chlorophytum malayense. *Planta Medica*, 66(06), pp.587–590.

Rachoń, D., Vortherms, T., Seidlová-Wuttke, D. and Wuttke, W., 2008. Effects of black cohosh extract on body weight gain, intra-abdominal fat accumulation, plasma lipids and glucose tolerance in ovariectomized Sprague–Dawley rats. *Maturitas*, 60(3–4), pp.209–215.

Rance, N.E., Dacks, P.A., Mittelman-Smith, M.A., Romanovsky, A.A. and Krajewski-Hall, S.J., 2013. Modulation of body temperature and LH secretion by hypothalamic KNDy (kisspeptin, neurokinin B and dynorphin) neurons: a novel hypothesis on the mechanism of hot flushes. *Frontiers in Neuroendocrinology*, 34(3), pp.211–227.

Rathore, R., Rahal, A. and Mandil, R., 2012. Cimicifuga racemosa potentiates antimuscarinic, anti-adrenergic and antihistaminic mediated tocolysis of buffalo myometrium. *Asian Journal of Animal and Veterinary Advances*, 6, pp.300–308.

Redlich, K.R., Hayer, S., Ricci, R., David, J.P., Tohidast-Akrad, M., Zwerina, J., Kollias, G., Steiner, G., Smolen, J.S., Wagner, E. and Schett, G., 2002, February. Osteoclasts are essential for TNF-mediated joint destruction. In *Arthritis Research & Therapy*, 4(1), pp.1–38.

Rhyu, M.R., Lu, J., Webster, D.E., Fabricant, D.S., Farnsworth, N.R. and Wang, Z.J., 2006. Black cohosh (Actaea racemosa, Cimicifuga racemosa) behaves as a mixed competitive ligand and partial agonist at the human μ opiate receptor. *Journal of Agricultural and Food Chemistry*, 54(26), pp.9852–9857.

Rice, S., Amon, A. and Whitehead, S.A., 2007. Ethanolic extracts of black cohosh (Actaea racemosa) inhibit growth and oestradiol synthesis from oestrone sulphate in breast cancer cells. *Maturitas*, 56(4), pp.359–367.

Roche, H.M., 2004. Dietary lipids and gene expression. *Biochemical Society Transactions*, 32(6), pp.999–1002.

Roggia, C., Gao, Y., Cenci, S., Weitzmann, M.N., Toraldo, G., Isaia, G. and Pacifici, R., 2001. Up-regulation of TNF-producing T cells in the bone marrow: a key mechanism by which estrogen deficiency induces bone loss in vivo. *Proceedings of the National Academy of Sciences*, 98(24), pp.13960–13965.

Rohit Singh, T. and Ezhilarasan, D., 2020. Ethanolic extract of Lagerstroemia Speciosa (L.) Pers., induces apoptosis and cell cycle arrest in HepG2 cells. *Nutrition and Cancer*, 72(1), pp.146–156.

Ruhlen, R.L., Haubner, J., Tracy, J.K., Zhu, W., Ehya, H., Lamberson, W.R., Rottinghaus, G.E. and Sauter, E.R., 2007. Black cohosh does not exert an estrogenic effect on the breast. *Nutrition and Cancer*, 59(2), pp.269–277.

Ruocco, M.G., Maeda, S., Park, J.M., Lawrence, T., Hsu, L.C., Cao, Y., Schett, G., Wagner, E.F. and Karin, M., 2005. IκB kinase (IKK) β, but not IKKα, is a critical mediator of osteoclast survival and is required for inflammation-induced bone loss. *Journal of Experimental Medicine*, 201(10), pp.1677–1687.

Sampson, N., Berger, P. and Zenzmaier, C., 2012. Therapeutic targeting of redox signaling in myofibroblast differentiation and age-related fibrotic disease. *Oxidative Medicine and Cellular Longevity*, 2012.

Schmid, D., Gruber, M., Woehs, F., Prinz, S., Etzlstorfer, B., Prucker, C., Fuzzati, N., Kopp, B. and Moeslinger, T., 2009. Inhibition of inducible nitric oxide synthesis by Cimicifuga racemosa (Actaea racemosa, black cohosh) extracts in LPS-stimulated RAW 264.7 macrophages. *Journal of Pharmacy and Pharmacology*, 61(8), pp.1089–1096.

Schmid, D., Woehs, F., Svoboda, M., Thalhammer, T., Chiba, P. and Moeslinger, T., 2009. Aqueous extracts of Cimicifuga racemosa and phenolcarboxylic constituents inhibit production of proinflammatory cytokines in LPS-stimulated human whole blood. *Canadian Journal of Physiology and Pharmacology*, 87(11), pp.963–972.

Schmida, D., Grubera, M., Woehsa, F., Prinzb, S., Etzlstorfera, B., Pruckera, C., Fuzzatic, N., Koppb, B. and Moeslingera, T., 2009. Inhibition of inducible nitric oxide synthesis by Cimicifuga racemosa (Actaea racemosa, black cohosh) extracts in LPS-stimulated RAW 264.7 macrophages. *JPP*, 61, pp.1089–1096.

Schwartz, L.M. and Woloshin, S., 2004. The media matter: a call for straightforward medical reporting. *Annals of Internal Medicine*, 140(3), pp.226–228.

Sebastian, M.K., Bhandari, M.M., 1988. Medicinal plant flore of Udaipur district, Rajasthan. *Bull. Med. Ethanobot. Res.*5, 133–134.

Seidlova-Wuttke, D., Jarry, H., Becker, T., Christoffel, V. and Wuttke, W., 2003. Pharmacology of Cimicifuga racemosa extract BNO 1055 in rats: bone, fat and uterus. *Maturitas*, 44(S1), pp.S39–S50.

Shao, Y., Harris, A., Wang, M., Zhang, H., Cordell, G.A., Bowman, M. and Lemmo, E., 2000. Triterpene glycosides from Cimicifuga racemosa. *Journal of Natural Products*, 63(7), pp.905–910.

Sharma, S.K., Chunekar, K.C. and Paudal, K., 1999. Plants of Sharangdhar Samhita. *National Academy of Ayurveda*, 1, p.289.

Sharma, S.K. and Kumar, M., n.d. Hepatoprotective effect of Chlorophytum Borivilia Um root extract Agai St Arse Ic I Toxicatio. *Pharmacology Online* 3, pp. 1021–1032

Shu, T., Zhu, Y., Wang, H., Lin, Y., Ma, Z. and Han, X., 2011. AGEs decrease insulin synthesis in pancreatic β-cell by repressing Pdx-1 protein expression at the post-translational level. *PLoS One*, 6(4), p.e18782.

Silva, F.A., Borges, F., Guimarães, C., Lima, J.L., Matos, C. and Reis, S., 2000. Phenolic acids and derivatives: studies on the relationship among structure, radical scavenging activity, and physicochemical parameters. *Journal of Agricultural and Food Chemistry*, 48(6), pp.2122–2126.

Simon, J., Wakimoto, H., Fujita, N., Lalande, M. and Barnard, E.A., 2004. Analysis of the set of GABAA receptor genes in the human genome. *Journal of Biological Chemistry*, 279(40), pp.41422–41435.

Singh, A. and Chauhan, H.S., 2003. Safed musli (Chlorophytum borivilianum): distribution, biodiversity and cultivation. *Journal of Medicinal and Aromatic Plant Sciences*, 25, pp.712–719.

Skorupskaite, K., George, J.T., Veldhuis, J.D., Millar, R.P. and Anderson, R.A., 2018. Neurokinin 3 receptor antagonism reveals roles for neurokinin B in the regulation of gonadotropin secretion and hot flashes in postmenopausal women. *Neuroendocrinology*, 106(2), pp.148–157.

Slameňová, D., Kuboškova, K., Horváthová, E. and Robichová, S., 2002. Rosemary-stimulated reduction of DNA strand breaks and FPG-sensitive sites in mammalian cells treated with H2O2 or visible light-excited Methylene Blue. *Cancer Letters*, 177(2), pp.145–153.

Son, S. and Lewis, B.A., 2002. Free radical scavenging and antioxidative activity of caffeic acid amide and ester analogues: structure–activity relationship. *Journal of Agricultural and Food Chemistry*, 50(3), pp.468–472.

Spangler, L., Newton, K.M., Grothaus, L.C., Reed, S.D., Ehrlich, K. and LaCroix, A.Z., 2007. The effects of black cohosh therapies on lipids, fibrinogen, glucose and insulin. *Maturitas*, 57(2), pp.195–204.

Stearns, V., Isaacs, C., Rowland, J., Crawford, J., Ellis, M.J., Kramer, R., Lawrence, W., Hanfelt, J.J. and Hayes, D.F., 2000. A pilot trial assessing the efficacy of paroxetine hydrochloride (Paxil®) in controlling hot flashes in breast cancer survivors. *Annals of Oncology*, 11(1), pp.17–22.

Stiedl, O., Pappa, E., Konradsson-Geuken, Å. and Ögren, S.O., 2015. The role of the serotonin receptor subtypes 5-HT1A and 5-HT7 and its interaction in emotional learning and memory. *Frontiers in Pharmacology*, 6, p.162.

Strommer, B., Khom, S., Kastenberger, I., Cicek, S.S., Stuppner, H., Schwarzer, C. and Hering, S., 2014. A cycloartane glycoside derived from Actaea racemosa L. modulates GABAA receptors and induces pronounced sedation in mice. *Journal of Pharmacology and Experimental Therapeutics*, 351(2), pp.234–242.

Sturdee, D.W., Hunter, M.S., Maki, P.M., Gupta, P., Sassarini, J., Stevenson, J.C. and Lumsden, M.A., 2017. The menopausal hot flush: a review. *Climacteric*, 20(4), pp.296–305.

Stute, P., Ehrentraut, S., Henneicke-von Zepelin, H.H. and Nicken, P., 2021. Gene expression analyses on multi-target mode of action of black cohosh in menopausal complaints–a pilot study in rodents. *Archives of Gynecology and Obstetrics*, pp.1–11.

Sundaram, S., Dwivedi, P. and Purwar, S., 2011. Antibacterial activities of crude extracts of Chlorophytum borivilianum to bacterial pathogens. *Research Journal of Medicinal Plant*, 5(3), pp.343–347.

Swami, U.S., Lande, A.A., Ghadge, P.M., Adkar, P.P. and Ambavade, S.D., 2014. Pharmacological evaluation of Chlorophytum borivilianum Sant. & Fern. for anxiolytic activity and effect on brain GABA level. *Oriental Pharmacy and Experimental Medicine*, 14(2), pp.169–180.

Tanaka, Y., Tran, P.O.T., Harmon, J. and Robertson, R.P., 2002. A role for glutathione peroxidase in protecting pancreatic β cells against oxidative stress in a model of glucose toxicity. *Proceedings of the National Academy of Sciences*, 99(19), pp.12363–12368.

Tandon, M. and Shukla, Y.N., 1995. Phytoconstituents of Asparagus adscendens, Chlorophytum arundinaceum and Curculigo orchioides: a review. *Curr Res Med Aromat Plants*, 17, pp.42–50.

Thakur, M. and Dixit, V.K., 2006. Effect of Chlorophytum borivilianum on androgenic & sexual behavior of male rats. *Indian Drugs-Bombay*, 43(4), p.300.

Thakur, M. and Dixit, V.K., 2007. Effect of some vajikaran herbs on pendiculation activities and in vitro sperm count in male. *Sexuality and Disability*, 25(4), pp.203–207.

Thakur, M., Bhargava, S. and Dixit, V.K., 2007. Immunomodulatory activity of Chlorophytum borivilianum Sant. F. *Evidence-Based Complementary and Alternative Medicine*, 4(4), pp.419–423.

Thakur, M., Bhargava, S., Praznik, W., Loeppert, R. and Dixit, V.K., 2009. Effect of Chlorophytum borivilianum Santapau and Fernandes on sexual dysfunction in hyperglycemic male rats. *Chinese Journal of Integrative Medicine*, 15(6), pp.448–453.

Thakur, M., Chauhan, N.S., Bhargava, S. and Dixit, V.K., 2009. A comparative study on aphrodisiac activity of some ayurvedic herbs in male albino rats. *Archives of Sexual Behavior*, 38(6), pp.1009–1015.

Thakur, M., Connellan, P., Deseo, M.A., Morris, C. and Dixit, V.K., 2011. Immunomodulatory polysaccharide from Chlorophytum borivilianum roots. *Evidence-based Complementary and Alternative Medicine*, 2011.

Thakur, M., Loeppert, R., Praznik, W. and Dixit, V.K., 2008. Effect of some ayurvedic vajikaran rasayana herbs on heat induced testicular damage in male albino rats. *Journal of Complementary and Integrative Medicine*, 5(1).

Thakur, M., Thompson, D., Connellan, P., Deseo, M.A., Morris, C. and Dixit, V.K., 2011. Improvement of penile erection, sperm count and seminal fructose levels in vivo and nitric oxide release in vitro by ayurvedic herbs. *Andrologia*, 43(4), pp.273–277.

Thorup, A.C., Lambert, M.N., Kahr, H.S., Bjerre, M. and Jeppesen, P.B., 2015. Intake of novel red clover supplementation for 12 weeks improves bone status in healthy menopausal women. *Evidence-based Complementary and Alternative Medicine*, 2015.

Triveni, A., 1977. *Rasendrasarasangrah: Vajikaranadhikar.* Rajkot, India: Nutan Press.

Vartak, V.D., 1981. Observations on wild edible plants from the hilly regions of Maharastra and Goa: resume and future prospects. In *Glimpses of Indian Ethnobotany.* New Delhi: Oxford & IBH, pp.261–271.

Viereck, V., Emons, G. and Wuttke, W., 2005. Black cohosh: just another phytoestrogen?*Trends in Endocrinology & Metabolism*, 16(5), pp.214–221.

Vijaya, K. and Chavan, P., 2009. Chlorophytum borivilianum (Safed musli): a review. *Pharmacognosy Reviews*, 3(5), p.154.

Villares, A., Rostagno, M.A., García-Lafuente, A., Guillamón, E. and Martínez, J.A., 2011. Content and profile of isoflavones in soy-based foods as a function of the production process. *Food and Bioprocess Technology*, 4(1), pp.27–38.

Visavadiya, N.P. and Narasimhacharya, A.V., 2007. Ameliorative effect of Chlorophytum borivilianum root on lipid metabolism in hyperlipaemic rats. *Clinical and experimental pharmacology & physiology*, 34(3), pp.244–249.

Visavadiya, N.P., Soni, B., Dalwadi, N. and Madamwar, D., 2010. Chlorophytum borivilianum as potential terminator of free radicals in various in vitro oxidation systems. *Drug and Chemical Toxicology*, 33(2), pp.173–182.

Wang, Y.H., Tang, Y.R., Gao, X., Liu, J., Zhang, N.N., Liang, Z.J., Li, Y. and Pan, L.X., 2021. The anti-inflammatory and analgesic effects of intraperitoneal melatonin after spinal nerve ligation are mediated by inhibition of the NF-κB/NLRP3 inflammasome signaling pathway. *Brain Research Bulletin*, 169, pp.156–166.

Watanabe, K., Mimaki, Y., Sakagami, H. and Sashida, Y., 2002. Cycloartane glycosides from the rhizomes of Cimicifuga racemosa and their cytotoxic activities. *Chemical and Pharmaceutical Bulletin*, 50(1), pp.121–125.

Weiss, G., Skurnick, J.H., Goldsmith, L.T., Santoro, N.F. and Park, S.J., 2004. Menopause and hypothalamic-pituitary sensitivity to estrogen. *JAMA*, 292(24), pp.2991–2996.

Wu, C.H., Huang, S.M., Lin, J.A. and Yen, G.C., 2011. Inhibition of advanced glycation endproduct formation by foodstuffs. *Food & Function*, 2(5), pp.224–234.
Wu, Y., Zhang, H., Dong, Y., Park, Y.M. and Ip, C., 2005. Endoplasmic reticulum stress signal mediators are targets of selenium action. *Cancer Research*, 65(19), pp.9073–9079.
Wuttke, W., Jarry, H., Haunschild, J., Stecher, G., Schuh, M. and Seidlova-Wuttke, D., 2014. The non-estrogenic alternative for the treatment of climacteric complaints: black cohosh (Cimicifuga or Actaea racemosa). *Journal of Steroid Biochemistry and Molecular Biology*, 139, pp.302–310.
Xie, Q.W., Cho, H.J., Calaycay, J., Mumford, R.A., Swiderek, K.M., Lee, T.D., Ding, A., Troso, T. and Nathan, C., 1992. Cloning and characterization of inducible nitric oxide synthase from mouse macrophages. *Science*, 256(5054), pp.225–228.
Yeung, K.S., Hernandez, M., Mao, J.J., Haviland, I. and Gubili, J., 2018. Herbal medicine for depression and anxiety: a systematic review with assessment of potential psycho-oncologic relevance. *Phytotherapy Research*, 32(5), pp.865–891.
Yi, J.M., Kim, M.S., Seo, S.W., Lee, K.N., Yook, C.S. and Kim, H.M., 2001. Acanthopanax senticosus root inhibits mast cell-dependent anaphylaxis. *Clinica Chimica Acta*, 312(1–2), pp.163–168.
Yue, G.G.L., Xie, S., Lee, J.K.M., Kwok, H.F., Gao, S., Nian, Y., Wu, X.X., Wong, C.K., Qiu, M.H. and Bik-San Lau, C., 2016. New potential beneficial effects of actein, a triterpene glycoside isolated from Cimicifuga species, in breast cancer treatment. *Scientific Reports*, 6(1), pp.1–11.
Yumrutas, O., Sokmen, A., Akpulat, H.A., Ozturk, N., Daferera, D., Sokmen, M. and Tepe, B., 2012. Phenolic acid contents, essential oil compositions and antioxidant activities of two varieties of Salvia euphratica from Turkey. *Natural Product Research*, 26(19), pp.1848–1851.
Zheng, H., Yu, X., Collin-Osdoby, P. and Osdoby, P., 2006. RANKL stimulates inducible nitric-oxide synthase expression and nitric oxide production in developing osteoclasts: an autocrine negative feedback mechanism triggered by RANKL-induced interferon-β via NF-κB that restrains osteoclastogenesis and bone resorption. *Journal of Biological Chemistry*, 281(23), pp.15809–15820.
Zierau, O., Bodinet, C., Kolba, S., Wulf, M. and Vollmer, G., 2002. Antiestrogenic activities of Cimicifuga racemosa extracts. *Journal of Steroid Biochemistry and Molecular Biology*, 80(1), pp.125–130.

4 *Convolvulus pluricaulis* (Shankhpushpi) and *Erythroxylum coca* (Coca plant)

Sashi Sonkar, Akhilesh Kumar Singh, and Azamal Husen

CONTENTS

4.1 Introduction 83
4.2 Description of the Plant 84
4.2.1 *Convolvulus pluricaulis* Choisy 84
4.2.2 *Erythroxylum coca* Lam. 85
4.3 Traditional Knowledge 85
4.3.1 *Convolvulus pluricaulis* 85
4.3.2 *Erythroxylum* coca 86
4.4 Phytochemistry 86
4.4.1 *Convolvulus pluricaulis* 86
4.4.2 *Erythroxylum* coca 87
4.5 Potential Benefits, Applications, and Uses 87
4.5.1 *Convolvulus pluricaulis* 87
4.5.1.1 Enhances Memory 87
4.5.1.2 Reduces Hypertension 87
4.5.1.3 Reduces Body Cholesterol and Acts as an Antidiabetic 88
4.5.1.4 Improves the Reproductive System 88
4.5.1.5 Antiulcer and Anti-catatonic Properties 88
4.5.1.6 Effects on the Thyroid Gland 88
4.5.1.7 Enhances Beauty 88
4.5.2 *Erythroxylum.* coca 88
4.5.2.1 Treats Gastrointestinal Disorders, Oral Sores, and Toothaches 88
4.5.2.2 Relieves Environmental Stress 89
4.5.2.3 Alleviates Hunger 89
4.5.2.4 Relieves Altitude Illness 89
4.5.2.5 Fast-Acting Antidepressant 89
4.5.2.6 Other Uses 90
4.6 Conclusion 90
References 90

4.1 INTRODUCTION

Plants are critical to the survival of life on Earth and are essential to people's livelihoods. All humans rely on plants to satisfy a variety of survival needs (Phillips and Meilleur, 1998). Plants with therapeutic and medicinal characteristics have been recognized and used by humans in some way or another since ancient times (Jain and Saklani, 1991; Silori and Badola, 2000). In underdeveloped nations, traditional folk remedies derived from natural resources are still used by 80% of the population (Farnsworth et al., 1985; Silori and Badola, 2000). India is widely recognized for its great

DOI: 10.1201/9781003205067-4

geographical variety, which has aided in the development of many ecosystems and varied vegetation. India is a country rich in indigenous herbal resources that thrive on its varied topography and agro-climatic conditions, allowing the development of about 20,000 plant species, approximately 2,500 of which are medicinal in nature (Choudhari, 1980; Chitravadivu et al., 2009). Traditional healing techniques are critical to the physical and psychological well-being of India's overwhelming majority of tribal people (Reddy et al., 2009). Approximately 400 plants are utilized in the creation of Ayurvedic, Unani, Siddha, and tribal medicine daily. Around 75% come from tropical woods, whereas 25% are from temperate forests. Furthermore, 30% of preparations are produced from roots, 14% from bark, 16% from entire plants, 5% from flowers, 10% from fruits, 6% from leaves, 7% from seeds, 3% from wood, 4% from rhizomes, 6% from stems, and fewer than 20% from cultivated species (Verma et al., 2007). Herbal medicines derived from plants are thought to be safer; this has been demonstrated in the treatment of a variety of illnesses (Mitalaya et al., 2003). Today, India's population is rising significantly, resulting in deforestation and ecological destruction in several regions of the country, resulting in a loss of medicinal plants and accompanying knowledge. Furthermore, the use of contemporary synthetic medications by the locals has resulted in the loss of traditional wisdom. Therefore, it is essential to chronicling traditional knowledge of plant uses for the treatment of various diseases before it is lost to future generations. In addition, there is need to collect information on the plant species particularly *Convolvulus pluricaulis* and *Erythroxylum coca*, their traditional knowledge, chemical derivatives, and potential benefits to mankind.

4.2 DESCRIPTION OF THE PLANT

4.2.1 *Convolvulus pluricaulis* Choisy

The plant *Convolvulus pluricaulis* Choisy (*C. microphyllus* Sieb ex. Spreng.) (Figure 4.1), commonly known as "Shankhapushpi" in India and "Aloe" weed in English, belongs to the family convolvulaceae (Agaewa et al., 2014). The stem is herbaceous, weak, prostrate, diffuse, cylindrical, branched, solid, hairy, and green. The leaf is cauline and ramal, alternate, exstipulate, simple, sessile, lanceolate, margin entire and hairy, acute, surface hairy, unicostate, and reticulate. The inflorescence is axillary dichasial cyme. Flower is bracteate, bracteolate, pedicellate, complete, actinomorphic, hermaphrodite, pentamerous, hypogynous, and cyclic. The calyx consists of five sepals, polysepalous, quincuncial, hairy, and persistent. The corolla is composed of five petals, gamopetalous, induplicate valvate, infundibuliform, and light purple. The stamens are five, polyandrous, epipetalous,

FIGURE 4.1 *Convolvulus pluricaulis.*

FIGURE 4.2 Pictorial view of the *Erythroxylum coca.*

filaments unequal consists of three short and two long, broader at the base, dithecous, dorsifixed, and introrse. Gynoecium is bicarpellary, syncarpous, ovary superior, bilocular with two ovules in each locule, placentation axile, an annular nectary is present below the ovary, style short, stigma bifid, and spreading. The fruit is capsule (Bendre and Kumar, 2009).

4.2.2 *Erythroxylum coca* Lam.

The plant *Erythroxylum coca* Lam. (Figure 4.2), commonly known as "Sivadari" in Tamil/Siddha and "coca" or "cocaine plant" in English, belongs to the family Erythroxylaceae. The plant is native to Peru and Bolivia and was introduced in India and cultivated in Kerala, Tamil Nadu, Uttar Pradesh, Bihar, West Bengal, and Assam experimentally (Khare, 2007). The plants are shrubs or small trees of about 5 m (Bhattacharya et al., 2017), which resembles a blackthorn bush (AP, 2014). The branches are straight, and the green-tinged leaves are thin, opaque, oval, and taper towards the ends. An areolated region of the leaf is distinguished by two longitudinal curving lines, one on each side of the midrib and more visible on the underside of the leaf. Inflorescence is axillary clusters. The flower is small, creamy white, hypogynous, petals five, polypetalous, and sepals five. Stamens are two sets of which one set of stamens faces the petals, while the other, generally shorter set, faces the sepals. The filaments are connate, producing a small staminal tube basally. Each petal has a ventral three-lobed, ligule-like nectariferous appendage placed on the claw's tip. The ligules of the five petals are upright and, while not united, overlap to form an erect tube. The ovary is superior, three celled, with one erect ovule in each cell, only one of which is typically viable. Pistil with carpel three and united, ovary three-chambered (Ganders, 1979; AP, 2014). The flowers mature into red berries (AP, 2014).

4.3 TRADITIONAL KNOWLEDGE

4.3.1 *Convolvulus pluricaulis*

C. pluricaulis has traditionally been used to treat nervous debility, sleeplessness, tiredness, and a lack of vitality. In fever, nervous debility, and memory loss, the entire plant is used medicinally in the form of a decoction with cumin and milk. *C. pluricaulis* is utilized as a tonic for the brain. It's a tonic, an alterative, and a febrifuge. It is a powerful treatment for gastrointestinal problems, particularly dysentery. The herb is said to be a popular memory enhancer. It's a psychostimulant and a tranquilizer. It is said to relieve mental strain. The plant's ethanolic extract lowers total serum cholesterol, triglycerides, phospholipids, and nonesterified fatty acid (Bhowmik et al., 2012). It boosts strength, digestion, complexion, and voice, as well as treating intestinal worms, dysuria, animal poisoning, dyspnea, cough, diabetes, and uterine disorders. It is beneficial in the treatment of epilepsy, sleeplessness, heart disease, and hematemesis. The leaves and blossoms have hypotensive qualities that are used to treat anxiety neurosis. The tribals of Chhindwara, Madhya Pradesh, India, describe it as an anthelmintic, a treatment for diarrhea, and a single herb that treats skin diseases and lowers high blood pressure. The leaves are advised for mental instability and depression in Gonda, Uttar Pradesh, India. The plant is nontoxic, and its use has no negative side effects. On the other side, there is a stimulating impact on health and weight increase. Rasayana treatment, according to Ayurvedic theory, affects both the body and the mind, resulting in psychological and physical improvement. This therapy slows the aging process, improves intellect, and strengthens the body's resilience to disease. It is one of Ayurveda's most significant Medhya Rasayana medicines. Its usage promotes Kaphavata-pitta dosha balance and vitiation, and the plant is astringent and bitter. Shankhpushpi, according to herbalists, soothes the nerves by controlling the body's production of the stress chemicals cortisol and adrenaline (Jalwal et al., 2016).

4.3.2 *ERYTHROXYLUM COCA*

The archaeological research in South and Central America has proved the history of coca usage (Biondich and Joslin, 2016). The usage of coca in Northern Chilean mummified human remains dates back to 1000 BC (Rivera et al., 2005). These documents show that the indigenous populations of the Andes have been using coca for almost 3,000 years (Biondich and Joslin, 2016). Coca was utilized for a variety of functions under the Incan reign, including physiological, social, and ritual uses (Cobo, 1979). Coca has been utilized for medical purposes over history and continues to do so now. It acts as a general anesthetic, in local applications, by chewing against pain in the throat and mouth or toothache, or as a tea for gastrointestinal problems (Naranjo, 1981; Bauer, 2019); it relieves childbirth pain and hastens labor (Stolberg, 2011; Bauer, 2019). Even today, Andean immigrants in the United Kingdom employ coca, a culturally significant plant, in legal goods such as flour, chocolates, and tea bags (Ceuterick et al., 2011; Bauer, 2019). Historically, coca was eaten three times each day to assist work: before beginning, midway through the day, and soon before completing (Unanue, 1794; Bauer, 2019). Coca, both then and now, acts as a stimulant, suppresses hunger and tiredness, and reduces the effects of altitude (Rivera et al., 2005; Bauer, 2019). Coca is a prominent cultural identification marker for many Andean tribes (Allen, 1981; Stolberg, 2011; Bauer, 2019). It is critical for recognizing and sustaining social relationships (Rivera et al., 2005; Bauer, 2019). Chewing together shows friendship and affection; refusing is considered antisocial (Stolberg, 2011; Bauer, 2019). Religious features of coca can be seen in shamans' rituals, during animal sacrifices and burials, but also in daily life, such as a gift to a possible bride's father, at carnivals and festivals, and even as motivation for weavers (Stolberg, 2011; Bauer, 2019). The chewing of a cocada is a time measurement representing 45 minutes of comfortable walking for 2 km of hilly terrain or 3 km on level land (Stolberg, 2011; Bauer, 2019). The highly ceremonial handling, sharing, and usage of the leaves is governed by strict protocol (Allen, 1981; Bauer, 2019). Coca is used in key economic activities both locally, such as rural women dealing in traditional remedies (Bauer, 2019; Sikkink, 2000), and nationally (Morales, 1986; Bauer, 2019).

4.4 PHYTOCHEMISTRY

4.4.1 *CONVOLVULUS PLURICAULIS*

The chemical constituents of *Convolvulus pluricaulis* Choisy show the presence of amino acids, protein, and carbohydrates, such as starch, sucrose, rhamnose, maltose, D-glucose, and other carbohydrates (Deshpande and Srivastava, 1969a; Shah and Quadry, 1990). Although only convolamine has been identified, additional alkaloids (convosine, confoline, convolvine, convolidine, convoline, and so on) have been discovered in other species of this family. The plant includes the alkaloid shankhapushpine ($C_{17}H_{25}NO_2$), which has a melting point ranging from 162°C to 164°C (Basu and Dandiya, 1948;

Prasad et al., 1974; Bisht and Singh, 1978; Lounasmaa, 1988; Mirzaev and Aripova, 1998; Singh and Bhandari, 2000; Razzakov and Aripova, 2004). The plant also contains fatty alcohols, fatty acids, and volatile oil as well as hydrocarbons, such as linoleic acid (2.3%), palmitic acid (66.8%), and myristic acid (30.9%), in addition to straight chain hydrocarbon and hextriacontane (Gapparov et al., 2007; Gapparov et al., 2008). The analysis of phytoconstituents also shows the occurrence of ceryl alcohol β-sitosterol and scopoletin (Deshpande and Srivastava, 1969b). The chloroform fraction of this contains β-sitosterol (phytosterols), kampferol (flavonoid), oxodotriacontanol, tetratriacontanoic acid, and oxodotriacontanol (Srivastava and Deshpande, 1975). The estimation of scopoletin and scopoletin were conducted by HPTL and spectrofluorimetry (Kapadia et al., 2006), respectively, in *C. pluricaulis* (Singh and Bhandari, 2000; Patil and Dixit, 2005). The phytochemical marker (CP-1) has also been isolated and characterized by HPTLC technique (Deshpande and Srivastava, 1969a). Ethanol derived from CP aids in the reduction of total serum cholesterol, phospholipids, and certain kinds of toxic fatty acids in the body (Dandekar, 1992). Convolvine's unique pharmacological activity has been discovered to inhibit M2 and M4 cholinergic muscarinic receptors. Convolvine was also shown to increase the effects of arecoline, a muscarinic memory enhancer that improves cognitive impairments in Alzheimer's disease (Zafar et al., 2005; Kapadia et al., 2006; Amin et al., 2014).

4.4.2 *Erythroxylum coca*

The *E. coca* leaf contains 0.7–1.5% of total alkaloids, with (-)-cocaine (pharmacologically active ingredient), a diester of (-)-ecgonine. Ecgonine has four chiral centers and is hence optically active. Other minor components of coca leaves include tropacocaine, methylecgonine, β-truxilline, α-truxilline, and cinnamonylcocaine (Christen, 2000). Cocaine was the most abundant alkaloid, accounting for 0.56% of the total dry weight, and other alkaloids include anhydroecgonine methyl ester (0.2% dry weight), ecgonine methyl ester (0.18% dry weight), *trans*-cinnamoylcocaine (0.4% dry weight), and *cis*-cinnamoylcocaine (0.7% dry weight) (Penny et al., 2009). Additional alkaloids also found in coca leaf in coca tea bags include hygrine, nicotine, dihydrocuscohygrine, cuscohygrine, ecgonine, hydroxytropacocaine, benzoylecgonine, and methylecgonine cinnamate (Jenkins et al., 1996). The leaves also had magnesium, zinc, iron, calcium, vitamin D, vitamin E, β-carotene, and protein, with lysine being the limiting amino acid (Penny et al., 2009). So far, 18 alkaloids from pyridines, pyrrolidines, and tropanes have been identified in a cultivated variety of *E. coca* (Novak et al., 1984). Several alkaloids were discovered and identified in the seeds of *E. coca*, including *trans*-cinnamoylcocaine, *cis*-cinnamoylcocaine, cocaine, hexanoylecgonine methyl ester, benzoyltropine, N-norbenzoyltropine, cuscohygrine, ecgonine methyl ester, 3α-acetoxytropane, tropine, and methylecgonidine (Casale et al., 2005).

4.5 POTENTIAL BENEFITS, APPLICATIONS, AND USES

4.5.1 *Convolvulus pluricaulis*

4.5.1.1 Enhances Memory

The chemical composition of *C. pluricaulis* includes phytonutrients such as confoline, convoline, phyllabine, convolvine, subhirsine, convolvidine, β-sitosterol, making it one of the best and most well-known natural remedies for enhancing memory. *C. pluricaulis* is most commonly used as a brain tonic and stimulant. Memory loss can be avoided by consuming *C. pluricaulis* on a daily basis. *C. pluricaulis* plant is also beneficial in the treatment of neurodegenerative illnesses, such as Alzheimer's disease, due to its memory-enhancing properties (Bhowmik et al., 2012).

4.5.1.2 Reduces Hypertension

The *C. pluricaulis* plant is also one of the most significant elements in the treatment of diseases such as neurosis, anxiety, hypertension, hypotension, and stress. *C. pluricaulis* assists to reduce stress and anxiety by controlling the synthesis of stress hormones, such as cortisol and adrenaline, in our bodies. It functions as a rejuvenation treatment, as well as a psychostimulant and a tranquilizer. The plant promotes healthy sleep and relieves mental tiredness and stress by inducing a sense of serenity and tranquility. It significantly reduces neuroticism and anxiety caused by varying stress levels (Bhowmik et al., 2012).

4.5.1.3 Reduces Body Cholesterol and Acts as an Antidiabetic

The extraction of *C. pluricaulis* plant aids in the reduction of blood cholesterol levels, including phospholipids and triglycerides, as well as the removal of specific types of fatty acids that are detrimental to the body (Bhowmik et al., 2012). It also acts as an excellent therapy for diabetes (Sharma et al., 1965; Amin et al., 2014).

4.5.1.4 Improves the Reproductive System

A fine paste prepared by crushing the plant is useful for the treatment of abscesses, and the juice of the whole plant prevents heavy menstruation (Rakhit and Basu, 1958; Amin et al., 2014). It is advised in cases of seminal and sexual debility (Sharma et al., 1965; Amin et al., 2014).

4.5.1.5 Antiulcer and Anti-catatonic Properties

The antiulcerogenic impact was found due to the enhancement of mucosal defensive factors such as mucin production, mucosal cell lifespan, and glycoprotein rather than offensive ones, like acid pepsin (Rakhit and Basu, 1958; Amin et al., 2014). It also enhances the quality of bone marrow and nerve tissues. It has also been discovered that *C. pluricaulis* is the most effective treatment for hypothyroidism (Bhowmik et al., 2012).

4.5.1.6 Effects on the Thyroid Gland

The suppression of thyroid function caused by the plant's root extract is largely mediated via T4-T3 conversion (Nahata et al., 2009; Amin et al., 2014). Furthermore, the powerful impact was seen in the treatment of thyrotoxicosis (Bhakuni et al., 1996; Amin et al., 2014).

4.5.1.7 Enhances Beauty

The herb *C. pluricaulis* is often used for replenishing all layers of skin and improving attractiveness. *C. pluricaulis* is prepared as a decoction with cumin and milk (Bhowmik et al., 2012).

4.5.2 *Erythroxylum.* coca

4.5.2.1 Treats Gastrointestinal Disorders, Oral Sores, and Toothaches

Coca is still used as a medical treatment by South American Indians, in addition to its general stimulant and social uses (Martin, 1970; Biondich and Joslin, 2016). One of the most traditional functions of coca in Andean living is the treatment of gastrointestinal distress. Coca leaf tea is used to treat stomach discomfort, intestinal spasms, nausea, indigestion, as well as constipation and diarrhea (Weil, 1981; Biondich and Joslin, 2016). Coca seems to restore normal tone to the smooth muscle of the gastrointestinal system. It may be preferable to belladonna, atropine, and other parasympathetic blocking medications in that it has no unpleasant side effects. In reality, most people see coca leaf's "side effects" as positive. It is used as a replacement stimulant for coffee by those who consume a lot of coffee but have exacerbations of gastrointestinal problems from it. Coffee is highly inflammatory to the stomach mucosa while also being extremely stimulating to the intestines. Coca can give beneficial CNS stimulation while also being a traditional treatment for gastrointestinal problems. Furthermore, coffee can cause significant physiological dependency. However, coca addiction is significantly less prevalent (Weil, 1981). It is primarily regarded as a complete treatment that restores digestive system equilibrium. Coca is chewed or kept in the mouth to relieve pain from oral sores and to help in the healing of oral lesions (Weil, 1981; Biondich and Joslin, 2016). Similarly, this herb is used to treat toothaches (Biondich and Joslin, 2016).

4.5.2.2 Relieves Environmental Stress

One of the applications of coca that continues to draw the attention of the medical community is as a treatment for the strains of living at high altitudes (Torchetti, 2011; Biondich and Joslin, 2016). Given the widespread belief in coca's intrinsic stimulating properties, this might explain the advantage of consuming this leaf in the high-stress environment characterized by high altitude. Coca is also thought to have characteristics that assist users in coping with hypoxia, cold, and hunger. In the 1970s, a series of tests were conducted to determine if coca chewing was related with the sensation of warmth. Hanna and Little discovered that coca users had lower hand and foot temperatures (Hanna, 1971; Little, 1970; Biondich and Joslin, 2016). Although the temperature difference between control and experimental participants was modest, this little variation might be beneficial in reducing heat loss in severe conditions (Biondich and Joslin, 2016).

4.5.2.3 Alleviates Hunger

Coca chewing is used to alleviate hunger among Andean peoples. Further research into this phenomenon has revealed that coca has an influence on glucose homeostasis. Chewing coca leaves has been shown to raise blood glucose levels above fasting levels (Bolton, 1976; Biondich and Joslin, 2016). This discovery led Bolton to think that coca has a basic metabolic purpose for Indians who struggle with glucose homeostasis. The discovery of increased glucose levels after coca chewing adds scientific credence to the indigenous notion that coca alleviates hunger (Bolton, 1979; Biondich and Joslin, 2016).

4.5.2.4 Relieves Altitude Illness

Coca chewers report reduced head discomfort and dizziness associated with working at high altitudes (Monge, 1943; Mortimer, 1901; Biondich and Joslin, 2016), which is specific to the field of high-altitude medicine. Fuchs et al. proposed a working theory of how coca chewing aided persons who work and live in high-altitude settings using existing anthropological and physiologic evidence (Fuchs et al., 1978; Biondich and Joslin, 2016). Data collection revealed that the percentage of local peoples consuming coca increased with altitude. This generalization applies to other cultural groups and the feminine gender, which do not generally accept and utilize coca in the same way as Quechua or other high altitude cultural groups do (Biondich and Joslin, 2016). Although earlier research has indicated that coca usage may improve physiological responses to cold (Little, 1970; Hanna, 1971; Biondich and Joslin, 2016), Fuchs et al. (1978) highlighted that the frequency of coca use does not always correspond with climate. Mineworkers use coca frequently, yet the setting is marked by more than just cold temperatures and high elevation (Galeano, 1973; Biondich and Joslin, 2016). An intriguing new idea proposed that coca's therapeutic qualities may be found in the reduction of the consequences of hypoxia. Polycythemia is a common reaction to the prolonged hypoxia of high-altitude settings (Hurtado, 1945; Biondich and Joslin, 2016). However, there is now debate about whether this positive physiologic alteration may become a maladaptation when carried to its logical conclusion. Polycythemia has a negative influence on blood viscosity, which is considered to be the mechanism behind persistent mountain sickness. Finally, Fuchs et al. (1978) suggested that the alkaloids contained in whole leaf coca pharmacologically prevent hypoxia-induced stimulation of excessive red blood cell formation. Reduced polycythemic stress alleviates symptoms of mountain sickness and modifies the mechanisms by which the body responds to the high-altitude environment (Fuchs et al., 1978; Biondich and Joslin, 2016).

4.5.2.5 Fast-Acting Antidepressant

Coca has definite mood-lifting qualities, particularly when the set and environment are helpful. The impact is felt within minutes of ingestion and does not cause patients to feel drugged or inebriated. Current antidepressant medicines (tricyclics), on the other hand, may not show an effect for weeks, have considerable toxicity, and may cause patients to feel intoxicated (Weil, 1981).

4.5.2.6 Other Uses

The other uses of coca include as an adjuvant therapy in weight loss and physical fitness regimens. Coca is anorexic and stimulating; unlike amphetamines, it offers some nutrients and does not cause toxicity or dependency. Coca is utilized as a laryngeal tonic in people who have to use their voices more frequently than usual, such as public speakers and professional singers. Coca might also be used to alleviate acute motion sickness and as a tonic and normalizer of bodily functioning. It is used as an energizer for those who do a lot of physical labor, such as athletes. Nevertheless, coca is utilized as replacement stimulant for amphetamines and cocaine, which are more hazardous and have a far higher potential for misuse. Further, it can also be used as a carbohydrate metabolism normalizer and therapy for hypoglycemia and diabetes mellitus (Weil, 1981).

4.6 CONCLUSION

The present study discusses the general description, traditional knowledge, chemical derivatives, and potential benefits to mankind of the plants *Convolvulus pluricaulis* and *Erythroxylum coca*. The chemical constituents of *C. pluricaulis* show the presence of amino acids, protein, carbohydrates, convolamine, shankhapushpine, fatty acids, volatile oil, hydrocarbons, ceryl alcohol, β-sitosterol, scopoletin, kampferol, oxodotriacontanol, tetratriacontanoic acid, and oxodotriacontanol. However, the *E. coca* leaf contains 0.7–1.5% of total alkaloids, with cocaine as well as tropacocaine, methylecgonine, β-truxilline, α-truxilline, cinnamonylcocaine, hygrine, nicotine, dihydrocuscohygrine, cuscohygrine, ecgonine, hydroxytropacocaine, benzoylecgonine, methylecgonine cinnamate, magnesium, zinc, iron, calcium, vitamin D, vitamin E, β-carotene, and protein. Moreover, the seeds of *E. coca* contain trans-cinnamoylcocaine, cis-cinnamoylcocaine, cocaine, hexanoylecgonine methyl ester, benzoyltropine, N-norbenzoyltropine, cuscohygrine, ecgonine methyl ester, 3α-acetoxytropane, tropine, and methylecgonidine. This study looked at the advantages of *C. pluricaulis* and *E. coca* as alternative medicines for a variety of illnesses. The potential benefits and uses of these plants have been thoroughly discussed, providing references to studies that have been performed utilizing these plants.

REFERENCES

Agaewa, P., Sharma, B., Fatima, A., and Jain, S.K. 2014. An update on Ayurvedic herb *Convolvulus pluricaulis* Choisy. *Asian Pacific Journal of Tropical Biomedicine*, 4, pp.245–252.

Allen, C. 1981. To be Quechua: The symbolism of coca chewing in highland Peru. *American Ethnologist*, 8, pp.157–171.

Amin, H., Sharma, R., Vyas, M., Prajapati, P., and Dhiman, K. 2014. Shankhapushpi (*Convolvulus pluricaulis* Choisy): Validation of the Ayurvedic therapeutic claims through contemporary studies. *International Journal of Green Pharmacy*, 8, pp.193–200.

AP. 2014. *African Pharmacopoeia. General methods for Analysis. African Pharmacopoeia*. 2nd edn. Abuja, Nigeria: AU/STRC Publications Division PMB 5368, pp.220–222.

Basu, N.K. and Dandiya, P.C. 1948. Chemical investigation of *Convulvulus pluricaulis*. *Journal of the American Pharmacists Association*, 37, p.27.

Bauer, I. 2019. Travel medicine, coca and cocaine: Demystifying and rehabilitating *Erythroxylum*: A comprehensive review. *Tropical Diseases, Travel Medicine and Vaccines*, 5, p.20.

Bendre, A.M. and Kumar, A. 2009. *A Text Book of Practical Botany*, vol. 2. Meerut, india: Rastogi Publication, pp.158–159.

Bhakuni, R.S., Tripathi, A.K., Shukla, Y.N., and Singh, S.C. 1996. Insect antifeedant constituent from *Convolvulus microphyllus* (L) Sieb. *Phytotherapy Research*, 10, pp.170–171.

Bhattacharya, K., Ghosh, A.K., and Hait, G. 2017. *A Textbook of Botany*, vol 4. Kolkata, India: New Central Book Agency (P) Ltd, p.527.

Bhowmik, D., Kumar, K.P.S., Paswan, S., Srivatava, S., Yadav, A.P., and Dutta, A. 2012. Traditional indian herbs *Convolvulus pluricaulis* and its medicinal importance. *Journal of Pharmacognosy and Phytochemistry*, 1, pp.44–51.

Biondich, A.S. and Joslin, J.D. 2016. Coca: The history and medical significance of an ancient Andean tradition. *Emergency Medicine International*, 2016, pp.1–5.

Bisht, N.P. and Singh, R. 1978. Chemical studies of *Convulvulus microphyllus* Sieb. *Planta Medica*, 34, pp.222–223.

Bolton, R. 1976. Andean coca chewing: A metabolic perspective. *American Anthropologist*, 78, pp.630–634.

Bolton, R. 1979. On coca chewing and high-altitude stress. *Current Anthropology*, 20, pp.418–420.

Casale, J.F., Toske, S.G., and Colley, V.L. 2005. Alkaloid content of the seeds from *Erythroxylum Coca* var. *Coca*. *Journal of Forensic Science*, 50, pp.1402–1406.

Ceuterick, M., Vandebroek, I., and Pieroni, A. 2011. Resilience of Andean urban ethnobotanies: A comparison of medicinal plant use among Bolivian and Peruvian migrants in the United Kingdom and in their countries of origin. *Journal of Ethnopharmacology*, 136, pp.27–54.

Chitravadivu, C., Manian, S., and Kalaichelvi, K. 2009. Qualitative analysis of selected medicinal plants, Tamilnadu, India. *Middle-East Journal of Scientific Research*, 4, pp.144–146.

Choudhari, M.M. 1980. Tribes of Assam Plains Guwahati Assam. New vistas in ethnobotany. In J.K. Maheswari (ed.), *Ethnobotany in South Asia*. India, Jodhpur: Scientific Publishers, pp.1–11.

Christen, P. 2000. Tropane alkaloids: Old drugs used in modern medicine. *Studies in Natural Products Chemistry*, 22, pp.717–749.

Cobo, B. 1979. *History of the Inca Empire: An Account of the Indian's Customs and their Origin, Together with a Treatise on Inca Legend*. Austin, TX: University of Texas Press.

Dandekar, U.P. 1992. Analysis of a clinically important interaction between phenytoin and Shankhapushpi, an ayurvedic preparation. *Journal of Ethnopharmacology*, 35, pp.285–288.

Deshpande, S.M. and Srivastava, D.N. 1969a. Chemical examination of the fatty acids of *Convulvulus pluricaulis*. *Indian Oil & Soap Journal*, 34, pp.217–218.

Deshpande, S.M. and Srivastava, D.N. 1969b. Chemical studies of *Convulvulus pluricaulis* Choisy. *Journal of Indian Chemical Society*, 46, pp.759–760.

Farnsworth, N.R., Akereele, O., and Bingel, A.S. 1985. Medicinal plants in therapy. *Bulletin of the World Health Organization*, 63, pp.965–981.

Fuchs, A., Burchard, R.E., Curtain, C.C., Azeredo P.R.D., Frisancho, A.R., Gagliano, J.A., Katz, S.H., Little, M.A., Mazess, R.B., Picón-Reátegui, E., Sever, L.E., Tyagi, D., and Wood, C.S. 1978. Coca chewing and high-altitude stress: Possible effects of coca alkaloids on erythropoiesis [and comments and reply]. *Current Anthropology*, 19, pp.277–291.

Galeano, E. 1973. *Open Veins of Latin America: Five Centuries of the Pillage of a Continent*. New York: Monthly Review Press.

Ganders, F.R. 1979. Heterostyly in *Erythroxylum coca* (Erythroxylaceae). *Bolanical Journal of the Linnean Society*, 78, pp.11–20.

Gapparov, A.M., Aripova, S.F., Razzakov, N.A., and Khuzhaev, V.U. 2008. Conpropine, a new alkaloid from the aerial part of *Convolvulus subhirsutus* from Uzbekistan. *Chemistry of Natural Compounds*, 44, pp.743–744.

Gapparov, A.M., Razzakov, N.A., and Aripova, S.F. 2007. Alkaloids of *Convolvulus subhirsutus* from Uzbekistan. *Chemistry of Natural Compounds*, 43, pp.291–292.

Hanna, J.M. 1971. Responses of Quechua Indians to coca ingestion during cold exposure. *American Journal of Physical Anthropology*, 34, pp.273–277.

Hurtado, A. 1945. Influence of anoxemia on the hemopoietic activity. *Archives of Internal Medicine*, 75, pp.284–323.

Jain, S.K. and Saklani, A. 1991. Observations on the ethnobotany of the tons valley region in the Uttarkashi District of the Northwest Himalaya, India. *Mountain Research and Development*, 11, pp.157–161.

Jalwal, P., Singh, B., Dahiya, J., and Khokhara, S. 2016. A comprehensive review on shankhpushpi a morning glory. *Pharma Innovation Journal*, 5, pp.14–18.

Jenkins, A.J., Llosa, T., Montoya, I., and Cone, E.J. 1996. Identification and quantitation of alkaloids in coca tea. *Forensic Science International*, 77, pp.179–189.

Kapadia, N.S., Acharya, N.S., Acharya, S.A., and Shah, M.B. 2006. Use of HPTLC to establish a distinct chemical profile for Shankhpushpi and for quantification of scopoletin in *Convulvulus pluricaulis* choisy and in commercial formulations of Shankhpushpi. *Journal of Planar Chromatography*, 19, pp.195–199.

Khare, C.P., 2007. *Indian Medicinal Plants: An Illustrated Dictionary*. Berlin/Heidelberg: Springer, p.246.

Little, M.A. 1970. Effects of alcohol and coca on foot temperature responses of highland Peruvians during a localized cold exposure. *American Journal of Physical Anthropology*, 32, pp.233–242.

Lounasmaa, M. 1988. The tropane alkaloids. In A. Brossi (ed.), *The Alkaloids: Chemistry and Pharmacology*. New York: Academic Press, pp.72–74.

Martin, R.T. 1970. The role of coca in the history, religion, and medicine of South American Indians. *Economic Botany*, 24, pp.422–438.

Mirzaev, Y.R. and Aripova, S.F. 1998. Neuro and psychopharmacological investigation of the alkaloids convolvine and atropine. *Chemistry of Natural Compounds*, 34, pp.56–58.

Mitalaya, K.D., Bhatt, D.C., Patel, N.K., and Didia, S.K. 2003. Herbal remedies used for hair disorders by tribals and rural folk in Gujarat. *Indian Journal of Traditional Knowledge*, 2, pp.389–392.

Monge, C. 1943. Chronic mountain sickness. *Physiological Reviews*, 23(Supplement), 166–184.

Morales, E. 1986. Coca and cocaine economy and social change in the Andes of Peru. *Economic Development and Cultural Change*, 35, pp.143–161.

Mortimer, W.G. 1901. *Peru: History of Coca, "the Divine Plant" of the Incas*, J. H. Vail: New York.

Nahata, A., Patil, U.K., and Dixit, V.K. 2009. Anxiolytic activity of *Evolvulus alsinoides* and *Convulvulus pluricaulis* in rodents. *Pharmaceutical Biology*, 47, pp.444–451.

Naranjo, P. 1981. Social function of coca in pre-Columbian America. *Journal of Ethnopharmacology*, 3, pp.161–172.

Novák, M., Salemink, C.A., and Khan, I. 1984. Biological activity of the alkaloids of *Erythroxylum coca* and *Erythroxylum novogranatense*. *Journal of Ethnopharmacology*, 10, pp.261–274.

Patil, U.K. and Dixit, V.K. 2005. Densitometric standardization of herbal medical products containing *Evolvulus alsinoides* by quantification of a marker compound. *Journal of Planar Chromatography*, 18, pp.234–239.

Penny, M.E., Zavaleta, A., Lemay, M., Liria, M.R., Huaylinas, M.L., Alminger, M., McChesney, J., Alcaraz, F., and Reddy, M.B. 2009. Can coca leaves contribute to improving the nutritional status of the Andean population? *Food and Nutrition Bulletin*, 30, pp.205–216.

Phillips, O.L. and Meilleur, B.A. 1998. Usefulness and economic potential of the rare plants of the United States: A statistical survey. *Economic Botany*, 52, pp.57–67.

Prasad, G.C., Gupta, R.C., Srivastava, D.N., Tandon, A.K., Wahi, R.S., and Udupa, K.N. 1974. Effect of Shankhpushpi on experimental stress. *Journal of Research Indian Medicine*, 9, pp.19–27.

Rakhit, S. and Basu N.K. 1958. *Convolvulus pluricaulis*. *Indian Y Pharm*, 20, pp.357–359.

Razzakov, N.A. and Aripova, S.F. 2004. Confolidine, a new alkaloid from the aerial part of *Convolvulus subhirsutus*. *Chemistry of Natural Compounds*, 40, pp.54–55.

Reddy, C.S., Reddy, K.N., Murthy, E.N., and Raju, V.S. 2009. Traditional medicinal plants in Seshachalam hills, Andhra Pradesh, India. *Journal of Medicinal Plants Research*, 3, pp.408–412.

Rivera, M., Aufderhaide, A., Cartmell, L., Torres, C., and Langsjoen, O. 2005. Antiquity of coca-leaf chewing in the south central Andes: A 3,000 year archaeological record of coca-leaf chewing from northern Chile. *Journal of Psychoactive Drugs*, 37, pp.455–458.

Shah, S.C. and Quadry, S.J. 1990. *A Textbook of Pharmacognosy*. 7th edn. New Delhi: CBS Publishers, pp.388–389.

Sharma, V.N., Barar, F.S., Khanna, N.K., and Mahawar, M.M. 1965. Some pharmacological actions of *Convolvulus pluricaulis*: An Indian indigenous herb. *Indian Journal of Medical Research*, 53, pp.871–876.

Sikkink, L. 2000. Ethnobotany and exchange of traditional medicines on the Southern Bolivian Altiplano. *High Altitude Medicine & Biology*, 1, pp.115–123.

Silori, C.S. and Badola, R. 2000. Medicinal plant cultivation and sustainable development: A case study in the buffer zone of the Nanda Devi biosphere reserve, Western Himalaya, India. *Mountain Research and Development*, 20, pp.272–279.

Singh, G.K. and Bhandari, A. 2000. *Text Book of Pharmacognosy*. New Delhi: CBS Publishers, pp.193–194.

Srivastava, D.N. and Deshpande, S.M. 1975. Gas chromatographic identification of fatty acids, fatty alcohols, and hydrocarbons of *Convulvulus pluricaulis* (Choisy). *Journal of the American Oil Chemists' Society*, 52, pp.318–319.

Stolberg, V. 2011. The use of coca: Prehistory, history, and ethnography. *Journal of Ethnicity in Substance Abuse*, 10, pp.126–146.

Torchetti, T. 2011. Coca chewing and high altitude adaptation. *Totem: The University of Western Ontario Journal of Anthropology*, 1, p.16.

Unanue, H. 1794. Disertación sobre el aspecto cultivo, comercio y virtudes de la famosa planta del Perú nombrada COCA. *Mercurio Peruano*, 11, pp.205–257.

Verma, A.K., Kumar, M., and Bussmann, R.W. 2007. Medicinal plants in an urban environment: The medicinal flora of Banares Hindu University, Varanasi, Uttar Pradesh. *Journal of Ethnobiology and Ethnomedicine*, 3, p.35.

Weil, A.T. 1981. The therapeutic value of coca in contemporary medicine. *Journal of Ethnopharmacology*, 3, pp.367–376.

Zafar, R., Ahmad, S., and Mujeeb, M. 2005. Estimation of scopoletin in leaf and leaf callus of *Convolvulus microphyllus* Sieb. *Indian Journal of Pharmaceutical Sciences*, 67, pp.600–603.

5 *Asparagus racemosus* (Shatavari) and *Dioscorea villosa* (Wild yams)

Shakeelur Rahman and Azamal Husen

CONTENTS

5.1 Introduction 95
5.1.1 *Asparagus racemosus* 95
5.1.2 *Dioscorea villosa* 96
5.2 Description 96
5.2.1 *Asparagus racemosus* 96
5.2.2 *Dioscorea villosa* 97
5.3 Traditional Knowledge 98
5.3.1 *Asparagus racemosus* 98
5.3.2 *Dioscorea villosa* 99
5.4 Phytochemicals 100
5.4.1 *Asparagus racemosus* 100
5.4.2 *Dioscorea villosa* 101
5.5 Potential Benefits, Applications, and Pharmacological Activities 102
5.5.1 *Asparagus racemosus* 102
5.5.2 *Dioscorea villosa* 106
5.6 Conclusion 106
References 107

5.1 INTRODUCTION

5.1.1 *ASPARAGUS RACEMOSUS*

There are about 300 species in genus *Asparagus* widespread all over the world, of which 22 species are documented from India. These are herbs, shrubs, and vines which are very important plant species having dietetic and curative value. *A. racemosus* (Family: Asparagaceae) is also known as shatamul, shatuli, vrishya, kurilo, and shatavari. Shatavari is an inhabitant of the Indian subcontinent. The plant grows wild in tropical and subtropical regions of India including Andaman and Nicobar Islands. The plant species are found from mean sea level up to 1500 m in the Himalayas from Kashmir eastward to Sri Lanka, northern Australia, and Africa (Shasnay et al., 2003). *A. racemosus* can be planted in upland rocky, sandy soils at 4300–4500 ft altitude. However, black fertile soil is also very suitable for the cultivation of *A. racemosus*. The crop mostly requires hot tropical climatic conditions and low irrigation. Nursery saplings are raised May–June and transplanted July–August. Usually, the standing crop is not infested with diseases and pest. The gestation period of the crop is about 1.5–2 years. *A. racemosus* is a climber, thin branched with soft leaves (nmpb). Because of overharvesting, destruction of natural habitat, and fast deforestation, the plant species is endangered (Warrier et al., 2001). The name Shatawari means (shat "hundred"; awari "curer") or "curer of a hundred diseases" (Shashi et al., 2013). *A. racemosus* is an aphrodisiac herb that helps

DOI: 10.1201/9781003205067-5

human beings to cope with emotional and physical stress. The tuberous roots and leaves of the species are medicinally significant in some diseases. The wonderful herb is also known as the "Queen of Herbs" in the *Ayurvedic* system of medicine with ubiquitous use as a cooling agent, antenatal care, and diuretic. The herb is also used to increase vitality and female fertility (Sharma and Dash, 2003). *A. racemosus* has a broad range of secondary metabolites that can be effective in urinary tract infections, hypertension, hyperlipidemia, angina, dysmenorrhea, anxiety disorders, benign prostatic hyperplasia, and leucorrhea. The dried herb is generally used in male genital dysfunctions, spermatogenic irregularities, oligospermia, and painful micturition. It is also used in traditional formulations for indigestion, debility, amoebiasis, and piles (Hussain et al., 2011). *Asparagus racemosus* contains minerals (0.71%), moisture (90.75%), crude fiber (0.82%), crude fat (0.78%), carbohydrate (5.21%), and crude protein (2.35%) (Khan, 2002).

5.1.2 *Dioscorea villosa*

A large number of wild plants remain unknown in this world, and most of them have tremendous nutritional and medicinal values. Wild yam of the genus Dioscorea has recorded about 600 species distributed all over the world (Sautour et al., 2007), of which only about ten species have been domesticated across Asia, Africa, and Latin America for herbal use and food purposes (Scarcelli et al., 2017). Most of the species of *Dioscorea* with its antioxidant and nutritive properties are not only supplementing the food and nutritional demand of the poor people but are also important as traditional medicine (Son et al., 2014). Different tribal and rural communities of various geographic regions use species of Dioscorea as an herbal medicines, food, and nutritional sources (Dutta, 2015; Trimanto and Hapsari, 2015). *D. villosa* is native to eastern North America and commonly known as wild yam, four leaf yam, colic root, rheumatism root and devil's bones. The species is listed on the United Plant Savers "at risk" plant list. The plant species grows well in moist sunny-shade conditions. The shape of *D. villosa* is moderately variable from a small vine to winged stems, compact groups of flowers, hairy leaves and large vines with stems without wing, leaves are large and smooth, and broad spread of flowers. The rhizome of the species is simple or branched that grows underground. The rhizomes vary in shape from linear to highly contort and brown in color. *D. villosa* contains diosgenin, is not a phytoestrogen, and does not work together with estrogen receptors. Other steroidal saponins are also found in the plant species. There is little information on floral-faunal relationships of this species. *Thrips crawfordi*, an insect species, is occasionally found on the plant parts. Mammalian herbivores stay away from the plant to use it as a food source. *D. villosa* is a monocot plant species. (https://www.illinoiswildflowers.info/savanna/plants/wild_yam.html).

5.2 DESCRIPTION

5.2.1 *Asparagus racemosus*

Asparagus racemosus was first described botanically in 1799. The plant species belongs to the genus Asparagus of the recently created family Asparagaceae from Liliaceae. The plant species is a woody climber having stems up to 4 m in height, much branched, under-shrub, green, spinosus, and terete. *A. racemosus* has uniform shiny green phylloclades like a pine needle. Cladodes from the axils of scale leaves in clusters of 2–6, 0.8–1.5 × 0.1–0.3 cm, slightly triquetrous, linear-falcate, apex acute, base narrow. Racemes are 2.5–5 cm long, slender, solitary, and axillary. Flowers are bisexual, 5–6 mm across; bracts triangular; pedicel 1 mm long. Perianth lobes are 6, white, 3 × 0.5 mm, oblong, acute. Stamens are 6, adnate to the perianth lobes; filaments subulate. Ovary 2–3 mm long, globose to slightly 3-gonous, 3-celled; ovules 2 per cell; stigma 3, recurved. Berry 4–6 mm diam., globose, purple on ripening. Seeds 2–5, 2 mm across and globose. Minute, white flowers rise up from the short, spiky stems in July–August. Fruits are produced in September as blackish-purple, globular berries. Rootstock is tuberous with fascicled fibrous roots. The adventitious root

FIGURE 5.1 Morphology of *Asparagus racemosus*.

system is measured about one meter in length, tapering at both ends, with almost a hundred on each plant. The roots are ash-color or silvery white on the surface and white on the inside. Fresh roots are more or less smooth, which develop longitudinal folds upon drying. Microscopically, the inner parenchymatous zone of the cortex is made of 18–24 layers in the upper part and 42–47 layers in the middle tuberous part of the roots. Cells are thin-walled and made of cellulosic fibers with circular to oval outlines and visible intercellular spaces. In some roots, 3–4 layers of cortex adjoining to the endodermis are modified into a sheath of stone cells around the endodermis. The number of vascular bundles ranges from 30 to 35 in the upper levels and 35 to 45 in the middle tuberous portions of the roots. The roots upon grinding are light brown in color with a coarse texture (NPGS/GRIN. 2009) (Figure 5.1).

Microscopically, the transverse section of the root is circular or elliptical; periderm is made of 5–6 layers of compact cells, tangentially elongated thin-walled phloem. Single layer of phelloderm comes after the 2–3 peripheral layers of cork cells. Similarly, the phelloderm is followed by 6–7 layers of cortical cells. A circular ring is formed by the vascular bundles as it is arranged in the center. Protoxylems are arranged toward the center, while the metaxylem is situated in the outer side. There is a wide zone of secondary phloem composed of sieve tubes, companion cells, and phloem parenchyma. A wide zone of secondary xylem, which is composed of vessels, tracheids, and xylem parenchyma, follows secondary phloem. Several epidermal hairs are present on the epidermal layers (Ahmad et al., 1991).

5.2.2 *Dioscorea villosa*

Dioscorea villosa (Family: Dioscoreaceae) is a twining herbaceous vine with pale-brown branches, 5–30 feet in height, woody, knotty, and cylindrical tubers. Leaves of the plants are heart-shaped, acuminate point, symmetrical with gradual tapering to a sharp point. *D. villosa* is found in different geographical areas with morphological variability. The plant exists in various soil types either in full shade or full sun. The cylindrical stems can climb upward and outward to twine counterclockwise around adjoining plants. The stems are light green, reddish green, dark red, or pale yellow and glabrous. The stems are narrowly ridged, bluntly angular, and terete. The lower stem portion

FIGURE 5.2 Morphology of *Dioscorea villosa.*

is somewhat hard, woody, and ridged but the upper portions are flexible and smooth (Kaimal and Kemper, 1999). Leaves found on the lower parts of the stem can be alternate to sub-opposite or in whorls of 3–7 leaves. Stem leaves of upper parts are always alternate. The long leaf stalk is generally ridged and frequently longer than the leaf blade which is ovate to cordate in shape, 5–10cm long, abruptly acuminate apex and well differentiated into a petiole and blade, margins entire, 9–11 prominent veins, leaf venation is palmate, primary veins converge at the apex of the leaf tip, and the base is heart-shaped. The young leaves have a golden-green cast, the upper surface is generally smooth and bright green, but the inner side can have dense, short, thin hair (Figure 5.2).

D. villosa is a dioecious plant that flowers from late June to July. There are six stamens in male flowers adnate to the tepals. The female flowers have three connate carpels with inferior ovary, flowers are radially symmetrical with six glandular tepals. The staminate flowers occur in a solitary inflorescence from the leaf axils, in the form of a drooping panicle that is widely branched, each cyme on the panicle having 1–3 flowers. Pistillate flowers occur in a drooping spike-like raceme, 2–4 inches long, with 4–18 flowers. Greenish-white perianth is found on staminate flowers, bell to funnel shaped, very small about 2 mm in diameter. Petals and sepals are combined as 6 tepals, ovate to elliptic in shape with the tips rounded to a slight point, united only at their bases. Stamens are erect but in two whorls of unequal length with the anthers about 1/2 the length of the filaments. The pistillate flowers are similar in design to the staminate and have six sterile staminodes with the anthers less than half the length of the filaments. The ovary is 2 to 3 times the length of the perianth and has a 3-branched style. It has been proposed that the pollinating genus is insect or fly because of the inconspicuous flowers. Fruits are three-winged, broad, and capsulate, 7–11mm long. Greenish-gold papery capsule or seed is developed from the fertile flowers, ovoid to broader at the apex in shape, varying much in size, from 1 to 3 cm wide by 1 to 3.5 cm long. The capsule is 3-sectioned with generally two thin winged seeds per section. The capsule changes into dark brown when it falls and opens along the seams of the section to release the seeds. The remains of the perianth and style are generally still attached. Seeds take 60 days of cold stratification to germinate (www.friendsofthewildflowergarden.org).

5.3 TRADITIONAL KNOWLEDGE

5.3.1 *Asparagus racemosus*

Asparagus was described as a "cleansing and healing" herb by a second-century physician, Galen. The plant species is commonly used in about 64 Ayurvedic formulations which include traditional formulations such as Phalaghrita, Vishnu taila and Shatavari Kalpa, Shatavarighrut, Bala Tail, Narayan tail, and Chyawanprash (Bopana and Saxena, 2007; Sharma and Dash, 2003; Shastri and Ratnavali, 2014). *A. racemosus* is an important medicinal plant in the Indian system of medicine.

Its root juice or root paste are used in a range of diseases and as an herbal tonic (Krtikar and Basu, 1975). The plant species is generally used as a "rasayana" in indigenous medicine to promote physical fitness by increasing cellular vitality and immunity (Goyal et al., 2003). It prevents aging, increases longevity, imparts immunity, improves mental function, vigor, and vitality to the body, and it is also used in nervous disorders, dyspepsia, tumors, inflammation, neuropathy, and hepatopathy. Traditional use of *A. racemosus* is also mentioned in the ancient literature "Charaka samhita" (Chawla et al., 2011). The plant root is used as a brain tonic and for treatment of epilepsy, hypertension, and cardiac disorders by traditional practitioners (Gomase and Sherkhane, 2010). This wonderful herbal drug is known as the "Queen of herbs" in Ayurveda because it creates soothing effects such as love and devotion. *A. racemosus* is an important Ayurvedic tonic especially for the female. Traditionally, it is also used in treating internal heat and chronic fever (Frawley, 1997). The plant is listed in Charak Samhita as a part of the formulas to treat women's health disorders (Garde and Vagbhat, 1970). Shathavari is a popular traditional herb used in increasing lactation and helping menopausal symptoms (Mayo and Facog, 1998; Mitra et al., 1999). In Ayurveda, the plant is used as a strong rejuvenating, nurturing, and stabilizing effect on excessive air, gas, dryness, and agitation in body and mind (Karmakar et al., 2012). The root of the plant is suggested in ancient Ayurvedic literature to be used in cases of threatened abortion and as a galactogogue, nervous disorders, dyspepsia, diarrhea, dysentery, tumors, inflammations, hyper dipsia, neuropathy, hyperacidity, hepatopathy, cough, bronchitis, etc (Sharma and Chakra, 2001; Sairam et al., 2003; Goyal et al., 2003). In the Indian traditional medicine system *A. racemosus* along with *Chlorophytum arundinaceum* and *Asparagus adscendens* are taken to improve physical health and stress-related immune disorders (Kanwar and Bhutani, 2010). Conventionally the roots have been utilized during internal pain, tumors, fever, and as a tonic (Kala, 2009). The people of Thailand use the decorticated roots of the species in the prevention of miscarriage, liver, and spleen diseases (Wiboonpun et al., 2004).

Ayurvedic products based on traditional formulations are available in markets, including Abana® (containing 10 mg Satavari root extract per tablet), Diabecon® (containing 20 mg Satavari root extract per tablet), EveCare® (containing 32 mg Satavari root extract per 5ml syrup), Geriforte® (containing 20 mg Satavari root powder per 50 mg Satavari root extract per tablet) and Menosan® (containing 110 mg Satavari root extract per tablet). Medicine developed *A. racemosus* by Himalaya Herbal Healthcare, India is Ricalex-Lactation (Joglekar et al., 1967) which is effective in increasing milk secretion (Sharma et al., 1996). Geriforte is a combination of a number of plants including *A. racemosus* given as a soothing tonic in old age (Singh et al., 1978). It has been observed in Indian markets that apart from *A. racemosus*, the roots of *A. sarmentosus* Linn, *A. filicinus* Ham, *A. curillus* Ham, and *A. sprengeri* Regel are most likely sold in the name of Shatavari (Sharma et al., 2000).

5.3.2 *Dioscorea villosa*

Since prehistoric times, indigenous peoples have used the rhizomes and roots of Dioscorea spp. as a traditional medicine and food (Singh, 1960). *D. villosa* is used as an herbal medicine for centuries for a wide range of diseases (Fisher and Painter, 1996). In the traditional system of medicines, the plant is used in gallbladder inflammation, rheumatoid arthritis, and irritable bowel syndrome (Sautour et al., 2004; Olayemi and Ajaiyeoba, 2007). The roots of this species are not directly edible because of its acrid properties, but medicinal uses have been reported from folk medicine. Fresh roots of the plant are more effective than the dried one as the root chemicals are ineffective after a year. It is used to treat spasms, colic pain, and nausea during pregnancy. Regular doses in small amounts are given to the mother during the gestation period. The root of *D. villosa* was used by the Meskwaki tribe to reduce pain during childbirth. Tribal people of America used to boil the root to treat menstrual pains, menopause, and morning sickness (Duke, 2000). Aboriginal Hawaiians used the tuber as a drink for the sick and as a food (Nelson, 2003). The plant is also used in the treatment of bilious gastrointestinal complaints. The action of the root of plant is noted as effective and quick in relieving bilious colic (Hutchen, 1969).

5.4 PHYTOCHEMICALS

5.4.1 *Asparagus racemosus*

A. racemosus possesses a large range of phytochemicals including alkaloids, steroids, flavonoids, essential oils, dihydrophenanthrene, and furan derivatives. The plant species has 29 steroidal saponins, 55 essential oil constituents extracted from aerial parts with a diverse range of chemical classes, such as hydrocarbon, acids, alcohol, aldehyde, *N*-containing compounds, ester, and ketone. The concentrations of ester, S, and *N*-containing compounds were found in low amounts (Gyawali and Kim, 2011). There was variation in the content of trace elements (Negi et al., 2010) (Figure 5.3).

Steroidal saponins (Shatavarins I-IV) are the major active constituents present in the root of *Asperagus racemosus.* The main glycoside is Shatvarin I with 3-glucose and rhamnose moieties attached to sarsapogenin (Gaitonde and Jetmalani, 1969; Joshi, 1988; Nair and Subramanian, 1969; Patricia et al.,2006). Oligospirostanoside is referred to as immunoside (Handa et al., 2003). Polycyclic alkaloid-aspargamine A is a cage-type pyrrolizidine alkaloid (Sekine, 1994). Isoflavones-8-methoxy-5, 6, 4-trihydroxy isoflavone-7 0-beta-D-glucopyranoside (Saxena and Chourasia 2001), cyclic hydrocarbon-racemosol, dihydrophenantherene (Sekine and Fukasawa, 1995), furan compound (Wiboonpun et al., 2004), carbohydrates-polysacharides and mucilage (Acharya et al., 2012) have also been reported in this plant. Flavanoids-glycosides of quercitin, rutin, and hyperoside are present in the

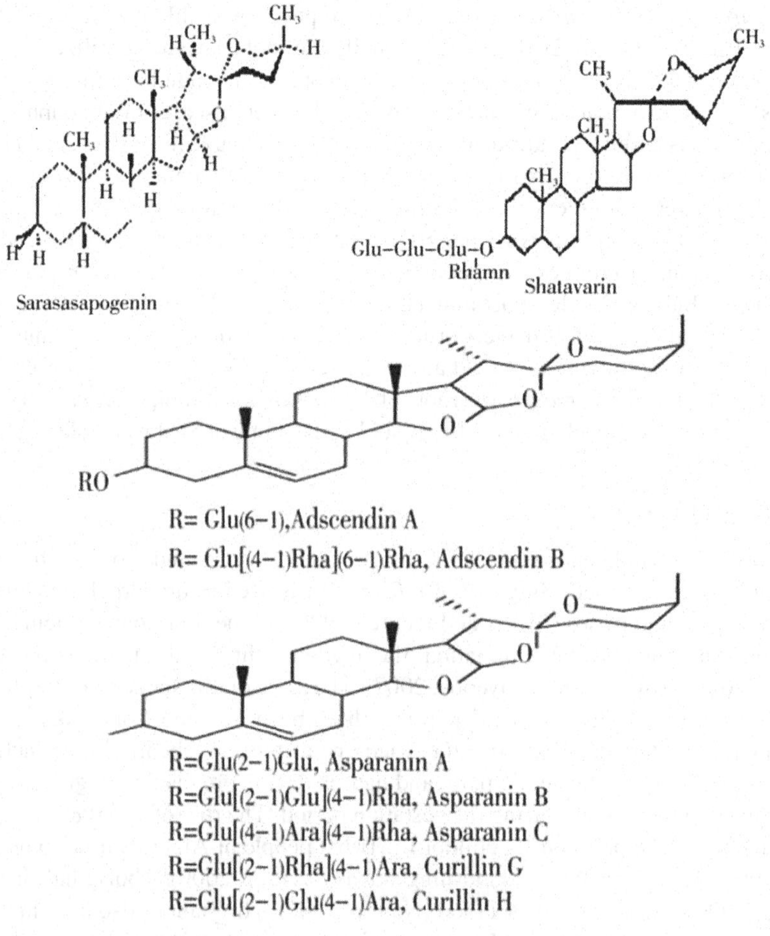

R= Glu(6−1),Adscendin A
R= Glu[(4−1)Rha](6−1)Rha, Adscendin B

R=Glu(2−1)Glu, Asparanin A
R=Glu[(2−1)Glu](4−1)Rha, Asparanin B
R=Glu[(4−1)Ara](4−1)Rha, Asparanin C
R=Glu[(2−1)Rha](4−1)Ara, Curillin G
R=Glu[(2−1)Glu(4−1)Ara, Curillin H

FIGURE 5.3 Phytochemical constituents' structure of *Asparagus racemosus.*

TABLE 5.1
Phytochemicals Present in the Different Parts of *Asparagus racemosus*

Roots	Asparagamine A, rutin, asparagan, 9,10- dihydro 1, 5 methoxy- Quercetin3 glucouronides, 8-methyl-2, 7- phenenthrenediol, racemofuron, ncoumertans, shatavarin V, shatavarin I, II, III,IV (steroid glycosides), immunoside, sitosterol, undecanyl cellanoate, shatavari, 4,6- dihydroxy-2-0 (2- hydroxyl isobutyl) benzaldehyde, secoisolariciresinol, diosgenin, racemosol, 4- trihydro isoflavine 7-0-beta-D-glucopyranoside, sterols, alkaloid, tannins, carbohydrates, flavonoids, isoflavones, coumestans, prenylated. lactones, amino acids and rutin.
Shoot	Sarsasapogenin and kaempferol thiophenes, thiazole, aldehyde, ketone, Gamma-linolenic acids, undecanyl cetamoate,
Leaves	vanillin, asparagusic acid, and methyl/ethyl esters
Flower	Diosgenin, quercetin-3-glucuronide
Fruits	Quercetin, rutin, hyperoside, Racemoside A, B, and C [9] sarsasapogenin, the aglycone of racemosides A-

flower and fruits (Sharma, 1981). Sterols-roots also contain sitosterol, 4, 6-dihydryxy2-O (-2-hydroxy isobutyl) benzaldehyde, and undecanyl cetanoate (Singh and Tiwari, 1991). Trace minerals are found in roots-zinc (53.15), manganese (19.98 mg/g), copper (5.29 mg/g), cobalt (22.00 mg/g) along with calcium, magnesium, potassium, zinc, and selenium (Choudhary and Kar, 1992; Mohanta et al., 2003). Kaepfrol-kaepfrol along with sarsapogenin from the woody portions of the tuberous roots could be isolated (Ahmad et al., 1991). Miscellaneous essential fatty acids, gamma-linolenic acids, vitamin A, diosgenin, quercetin 3-glucourbnides are also noticed in this plant (Subramanian and Nair, 1969). Shatavarin IV has been reported to exhibit significant activity as an inhibitor of core Golgi enzymes transferase in cell free assays and recently exhibited immuno-modulation activity against specific T-dependent antigens in immuno-compromised animals (Kamat et al., 2000). Due to the availability of isoflavones in the roots, *A. racemosus* is known for its phytoestrogenic characteristics (Ashajyothi et al., 2009; Saxena and Chourasia, 2001) (Table 5.1).

5.4.2 *Dioscorea villosa*

The phytochemical screening of *D. villosa* confirmed the presence of flavonoids, steroids, alkaloids glycosides, saponins, phytosterols, and tannins, which may be responsible for its pharmacological activities (Satija et al., 2018). Saponin-rich extract of *D. villosa* has been reported to contain prosapogenin A, dioscin, deltonin, and diosgenin 3-O- [alpha-L-rhamnopyranosyl (1 –> 2)]- [beta-D-glucopyranosyl (1 –> 3)-beta-D-glucopyranosyl (1 –> 4)]-beta-D-glucopyranoside, and spirostanol glycoside (Yoon and Kim, 2008). Methanolic extract of *D. villosa* roots has been reported to contain methyl parvifloside, methyl protodeltonin, deltonin, and zingiberensis saponin I (Hayes et al., 2007). Diosgenin is the primary active ingredient in *Dioscorea villosa* (Dweck, 2002). More than 50 steroid saponins of furostan-, spirostan-, and pregnane-type skeletons have been extracted and characterized from various Dioscorea species, which are major physiologically active constituents (Cayen et al., 1979). Most of the yam species restrain steroid saponins and also sapogenins, such as Diosgenin, the starting material of industrial production of steroids which are used as contraceptive drugs, and are anti-inflammatory, androgenic, and estrogenic (Sautour et al., 2007; Mbiantcha et al., 2010). The root contains the diosogenin type of saponin to make progesterone and other steroids in the lab (Figure 5.4).

It has been demonstrated in various research that diosgenin and chemicals of *D. villosa* have tremendous anti-inflammatory qualities; these can be used as an aid to intestinal, kidney, and spleen functions (http://www.rain-tree.com/dioscorea.htm). The production of hormonal drugs and cortisone, such as progestational and sex hormone as well as other steroids with diosgenin extracted from *D. villosa* have been reported (Cheng et al., 2015). The efficacy of diosgenin extracted from *D.*

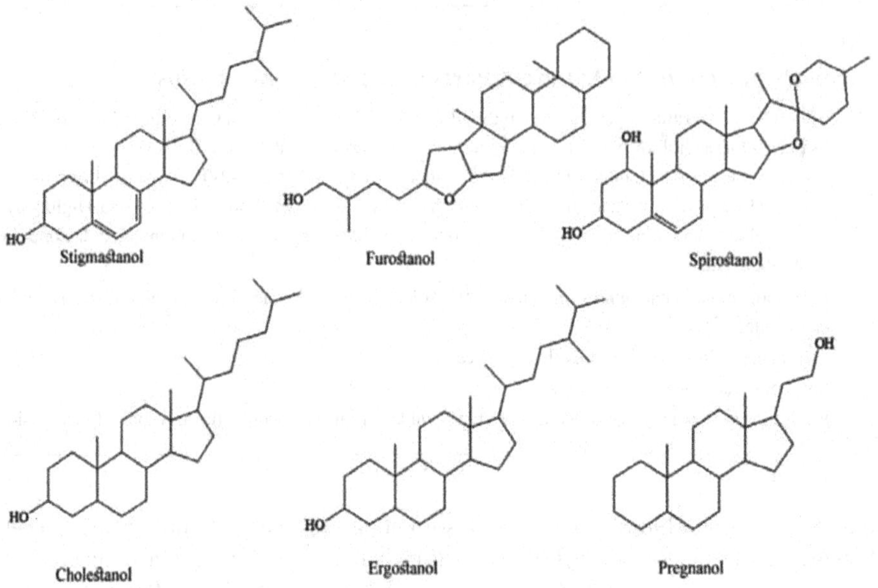

Stigmastanol Furostanol Spirostanol

Cholestanol Ergostanol Pregnanol

FIGURE 5.4 Phytochemical constituents' structure of *Dioscorea villosa.*

villosa against skin aging revealed its potential to enhance DNA synthesis of skin using a human 3D skin equivalent model and a restoration of keratinocyte proliferation in aged skin (Tada et al., 2009).

5.5 POTENTIAL BENEFITS, APPLICATIONS, AND PHARMACOLOGICAL ACTIVITIES

5.5.1 *Asparagus racemosus*

Asparagus racemosus has been included as one of the 32 highly prioritized medicinal plants of India by the National Medicinal Plants Board; government of India has included (NMPB, 2002). The root of *A. racemosus* is one of the major parts used in pharmacological studies. The root tuber of the plant in its powdered form is light yellow, slightly sweet, and odorless with the presence of a good quantity of starch grains. The root is used as a rejuvenating herb in female infertility. It is equally effective in inflammation of sexual organs, increasing libido, moistening of dry tissues of the sexual organs, improves folliculogenesis and ovulation to prepare the womb for conception, prevents miscarriage, works as post-partum tonic by increasing lactation, normalizing the uterus, changing hormones, and is prescribed for menorrhagia and leukorrhea (Sharma and Bhatnagar, 2011). *A. racemosus* is helpful in menstrual disorders, such as premenstrual syndrome, dysmenorrhea, irregular bleeding during peri-menopausal period, and in post menopause. Saponins content of the plant hinders oxytocic activity on uterine musculature to maintain spontaneous uterine motility. A good source of phytoestrogen of the plant is effective in reducing difficult menopausal situations (Aarti, 2015).

It was observed that the administration of 'U-3107' in normal rats increased wet and dry uterine weights and also resulted in an increase in estrogen levels without a change in progesterone levels as compared to the control. The initial changes in uterine tissues are controlled by progesterone and estrogen. The rats from both the control and treated groups resulted in normal estrous cycles (Mitra et al., 1999). It was proved in a study by Nevrekar et al. (2002) that EveCare capsules are effective in the treatment of dysfunctional uterine bleeding. This was due to healing of the endometrium stimulated by endometrial microvascular thrombosis as a result of high doses of phytoestrogens. A drug (Menosan) patented by Dhaliwal based on the formulation of *A. racemosus* has been observed to be effective in PMS in human females (Dhaliwal, 2003). Menosan is effective for those women in menopause who frequently face mood swings, lack of concentration, and sleep deprivation (Singh and Kulkarni, 2002). *A. racemosus* also has antibacterial properties responsible for relief from symptoms like night sweats and hot flashes. EuMil-Neurochemical modulator, a polyherbal formulation containing the standardized extracts of *A. racemosus, Withania somnifera, Emblica officinalis, and Ocimum sanctum*, was found effective in antistress activity in rats (Bhattacharya et al., 2002). Similarly, Mentat is an herbal psychotropic drug having *A. racemosus* that is effective in Mentat-Neurological disorder (Kulkarni and Verma, 1993). Satavari mandur was found effective in the treatment of cold restraint stress-induced gastric ulcer in rats (Datta et al., 2002). Abana is a mineral-based herbal formulation containing 10 mg *A. racemosus* constituent per tablet and was found useful for its hypocholesterolemic effect in rats and therefore exhibited a potential for use as a cardioprotective drug (Khanna et al., 1991). The chemical extracts of *A. racemosus* have been studied to be safe at therapeutic doses but toxicity at much higher doses has been observed (Goel et al., 2006). More details of pharmacological activities of *A. racemosus* are mentioned below.

Antisecretory and antiulcer activity: *A. racemosus* may have cytoprotective action in binding of bile salts as well as healing duodenal ulcers without inhibiting acid secretion (Sairam et al., 2003).

Galactogogue effect: The root extract of *A. racemosus* is advised in Ayurveda during lactation for increasing milk secretion (Nadkarni, 1954). A significant increase in milk yield and growth of alveolar tissue, mammary glands, and acini in guinea pigs was observed (Narendranath et al., 1986). Galactogogue effect of *A. racemosus* in buffaloes was observed by Patel and Kanitkar (1969). The alcoholic extract of the roots of the plant at 250 mg/kg given intramuscularly causes increase in lobuloaveolar tissue of mammary gland and increase in the milk yield of estrogen-primed rats (Joglekar et al., 1967).

Antitussive effect: Methanolic extract of roots, at a dose of 200 and 400 mg/kg p.o., observed considerable antitussive activity on sulphur dioxide-induced cough in mice (Mandal et al., 2000).

Antibacterial activity: Methanolic extract of *A. racemosus* roots at 50, 100, and 150 mg/ml showed significant *in vitro* antibacterial efficacy against *Escherichia coli, Shigella dysenteriae, Shegella sonnei, Shigella flexneri, Vibriocholerae, Salmonella typhi, Salmonella typhimurium, Pseudomonas pectida, Bacillus subtilis*, and *Staphylococcus aureus* (Mandal et al., 2000).

Antiprotozoal activity: An aqueous solution of the crude alcoholic extract of the *A. racemosus* roots exhibited an inhibitory effect on the growth of *Eintamoeba histolytica in vitro* (Roy et al., 1971).

Anti-plasmodial activity: The ethyl acetate extract of the roots of *A. racemosus* has been reported for anti-plasmodial activity. The extract with yield value of 7.9% per 100g has shown dose-dependent inhibition of chloroquine resistant strain of *Plasmodium falciparum* (3D7) with an IC50 value of 29μg/mL (Kaushik et al., 2013).

Antileshmanial activity: Racemoside A-treated *L. donovani* promastigotes showed signs of programmed cell death, i.e., the flagellated promastigotes shrank and became aflagellated, oval, or round with increased vacuoles. Phosphatidylserine was translocated from the inner side of the outer layer of the plasma membrane which is observed during cell death (Koonin and Aravind, 2002).

Antidiarrheal activity: The efficacy of *A. racemosus* was evaluated for antidiarrheal activity in castor oil-induced diarrheal rats (Venkatesan et al., 2005).

Gastrointestinal effects: Powdered dried root of *A. racemosus* is given orally to promote gastric emptying. Its action is comparable with the synthetic dopamine antagonist metoclopromide (Dalvi et al., 1990). The fresh root juice of *A. racemosus* has shown to have a definite curative effect in patients with duodenal ulcers (Kishore et al., 1980). Mixed powder of

A. racemosus and *Terminalia chebula* protected gastric mucosa against pentagastrin and carbachol-induced ulcer (Dahanukar et al., 1986). It is also effective in the proper function of rats' fundal strip, rabbit's duodenum, and guinea pig ileum without affecting peristaltic movement (Jetmalani et al., 1967).

Effect on uterus: Ethyl acetate and acetone extracts of the root of *A. recemosus* blocked spontaneous motility of the virgin rat's uterus. Shatavari receptor can be applied as a uterine sedative (Jetmalani et al., 1967).

Antihepatotoxic activity: Alcoholic extract of root of *A. racemosus* has been observed to significantly reduce increased levels of aspartate transaminase, alanine transaminase, and alkaline phosphate in CCl_4-induced liver damage in rats (Muruganadan et al., 2000). It indicates the antihepatotoxic potential of *A. racemosus*. The aqueous root extract of the plant has potential as an effective formulation to prevent hepatocarcinogenesis (Agrawal et al., 2008). Paracetamol-induced liver damage in rats increases levels of SGOT, SGPT, serum bilirubin, and serum alkaline phosphatase. The ethanolic roots extract of *A. racemosus* was observed hepatoprotective (Rahiman et al., 2011).

Cardiovascular effects: It has been reported that the alcoholic extract of the root of *A. racemosus* produces positive ionotropic and chronotropic effect on frog heart with lower doses and cardiac arrest with higher doses (Mandal et al., 2000). It has been demonstrated that the root extract of *A. racemosus* affects lipid-lowering in hypercholesteremic rats (Visavadiya and Narasimhacharya, 2009).

Neurodegenerative disorders: The potential of methanolic root extract against kainic acid-induced hippocampal and striatal neuronal damage in mice has been evaluated. It was found effective in neurodegenerative disorders like Parkinson's disease and Alzheimer's (Parihar and Hemnani, 2004).

Immunomodulatory activity: *A. racemosus* has immunomodulating effect in rats and mice against experimentally-induced abdominal sepsis. The decoction of powdered root of *A. racemosus* used orally was reported to produce leucocytosis and predominant neutrophilia along with enhanced phagocytic activity of the macrophages and polymorphs. Mortality rate of *A. racemosus*-treated animals was significantly reduced (Dahanukar et al., 1986). Polyhydroxylated steroidal sapogenin acids were studied on the immune system of normal and cyclosporine-A induced immune-suppressed animals and it was observed that the compound is a potent immune system stimulator (Sharma et al., 2011). Steroidal saponins, shatavaroside A, and shatavaroside B, isolated from the methanolic extract of *A. racemosus* and their immunomodulatory activity, have been evaluated using poly-morphonuclear leukocyte function test and some more sensitive assays such as nitroblue tetrazolium, nitrous oxide, and chemiluminescence assays were used as a confirmatory test for the activity. The steroidal saponins isolated were found to be active at nano concentrations (5ng/mL) and can act as a potent immune-stimulant (Sharma et al., 2009).

Antioxidant effects: The aqueous extract of *A. racemosus* against membrane damage induced by the free radicals generated during gamma radiation was effective in rat liver mitochondria. (Kamat et al., 2000; Wiboonpun et al., 2004; Visavadiya et al., 2009). Crude extract and purified aqueous fraction of *A. racemosus* have been demonstrated for their antioxidant effect. The crude and purified extracts indicated protection against radiation-induced loss of protein thiols and inactivation of superoxide dismutase (Kamat et al., 2000).

Antilithiatic effects: The ethanolic extract of *A. racemosus* has been reported for its inhibitory potential of lithasis or stone formation. It was observed that the ethanolic extract significantly reduced the elevated level of urinary concentration of magnesium, as it is one of the inhibitors of crystallization (Christina et al., 2005).

Antidepressant activity: *A. racemosus*'s use in psychological disorders like depression has not been scientifically evaluated. Methanolic extract of roots of *A. racemosus* was given to rats in doses of 100, 200, and 400 mg/kg daily for seven days and then subjected to a forced swim test (FST) and learned helplessness (LH) test. It was observed that MAR decreased immobility in FST and increased avoidance response in LH indicating antidepressant activity (Singh et al., 2009). Methanolic extract of *A. racemosus* showed antidepressant-like activity almost certainly by inhibiting MAO-A and MAO-B; and through interaction with adrenergic, dopaminergic, serotonergic, and GABAergic systems (Dhingra and Kumar, 2007).

Enhances memory and protects against amnesia: MAR also significantly reversed scopolamine and sodium nitrite-induced increase in transfer latency on elevated plus maze indicating anti-amnesic activity. Further, MAR dose-dependently inhibited acetylcholinesterase enzyme in specific brain regions (prefrontal cortex, hippocampus, and hypothalamus). Thus, MAR showed nootropic and anti-amnesic activities in the models tested, and these effects may probably be mediated through augmentation of the cholinergic system due to its anticholinesterase activity. Post-trial administration of *Convolvulus pluricaulis* and *A. racemosus* extract demonstrated significant decreases in latency time during retention trials. Hippocampal regions associated with the learning and memory functions showed a dose-dependent increase in AChE activity in carbonic anhydrase 1 with *A. racemosus* and Carbonic anhydrase 3 area with *C. pluricaulis* treatment. The underlying mechanism of these actions of *A. racemosus* and *C. pluricaulis* may be attributed to their antioxidant, neuroprotective, and cholinergic properties (Sharma et al., 2010).

Aphrodisiac activity: Lyophilized aqueous extracts of some of the plant species like *A. racemosus*, *Chlorophytum borivilianum*, and rhizomes of *Curculigo orchioides* were acquired from their roots for sexual behavior effects in male albino rats and compared with untreated control group animals. The effect of treatments on anabolic consequences of rats was evaluated along with seven other measures of sexual behaviors. Application of 200 mg/kg body weight of the aqueous extracts had resulted anabolic effect in treated animals as evidenced by weight gains in reproductive organs and body. The significant variation in the sexual behavior of animals was observed as reflected by reduction of ejaculation latency, mount latency, increase of mount frequency, post ejaculatory latency and intromission latency. Penile erection of the animals was also significantly enhanced. The result appeared to be attributable to the testosterone-like effects of the root extracts. The claim of traditional importance of the herbal extracts used as aphrodisiac is supported by the present study (Thakur et al., 2009).

Antenatal tonic: The clinical trial of a drug called Sujat containing *A. racemosus* extract was reported to decrease the rate of prenatal deaths and increase the fetal weight as well as pregnancy-induced hypertension (PIH) was also reduced (Bhasale et al., 1994).

Antiulcer: Ayurvedic formulation with *A. racemosus* is useful in peptic ulcers and healing of the peptic ulcers was verified by the endoscopic method (Sairam et al., 2003). The positive impact was indicated by the increase in mucin secretion and marked decrease in cell shedding (Goel and Sairam, 2002; Mangal et al., 2006). The anti-ulcerogenic effect of *A. racemosus* was observed by inhibitory effect on release of gastric hydrochloric acid and protects gastric mucosal damage (Bhatnagar and Sisodia, 2006). *Asparagus racemosus* Willd. (Shatavari) was tried for its ulcer healing properties. Root powder of *Asparagus racemosus* is effective in chronic peptic ulcers (Mangal et al., 2006).

Anti-aflatoxigenic activity: From the biodeteriorated *A. racemosus*,14 essential oil constituents have been obtained with their anti-aflatoxigenic activity. Out of the 14 constituents, thymol and eugenol have potent fungicidal activity because both are responsible for blocking the growth of spores, and the remaining essential oil constituents showed moderate antifungal activity (Mishra et al., 2013).

Anticancer property: The root extract of *A. racemosus* was shown to have a protective effect in mammary cell carcinoma (Rao1981). Steroidal constituents of the *A. racemosus* were studied for the apoptotic activity or effect tumor cell death (Bhutani et al., 2010).

Anticancer activity of Shatavarin IV which was extracted from the plant roots has potent anticancer activity (Mitra et al., 2012).

5.5.2 *Dioscorea villosa*

Dioscorea villosa has been used as a raw material to produce contraceptive. Sometimes it is used as a substitute for hormone replacement therapy. The chemical constituents of *D. villosa* are produced in the hormones estrogen or progesterone as the human body is unable to use it in the same way. The plant extract is effective in menopausal symptoms and premenstrual syndrome. It is also recommended in labor pain, prevention of early miscarriage, and nausea during pregnancy (Krochmal and Connie, 1973). *D. villosa* is applied in rheumatic pain and arthritis as well as intestinal cramps and bilious. As an herbal treatment, the root is used for indigestion, dysmenorrhea, dispel uterine, ovarian pain, and asthma (Howell, 2006; Foster and Duke, 1990; Crellin and Philpott, 1990; Chevallier, 1996). It was found in a study that the root extract of *D. villosa* is anti-inflammatory, antispasmodic, cholagogue, diaphoretic, and vasodilator. It is used orally in the treatment of arthritis, irritable bowel syndrome, gastritis, gall bladder complaints, and painful menstruation (Bown, 1995). The plant is also a visceral relaxant (Mills, 1988). Storing the roots of *D. villosa* for more than one year results in the loss of its medicinal properties (Grieve, 1984). The plant root is used to make childbirth easier and for the treatment of infant colic (Moerman, 1998; Castro, 1990).

The tubers of *D. villosa* are effective in antibacterial activity. It works against *S. dysenteriae*, *E. coli*, *V. cholerae*, *K. pneumoniae*, *P. aeruginosa*, and *S. aureus* (Roy et al., 2012). The root extract of *D. villosa* contains the active ingredient diosgenin, which acts as mild anti-estrogenic (Rosenberg et al., 2001). The root extract of *D. villosa* has demonstrated mild phytoestrogenic effects on human breast cancer cells (Park et al., 2009). It is also used in cough, hiccups, muscular spasms, and gas problems. It is advised for loosening phlegm, increasing urine flow, and inducing vomiting (Moalica et al., 2001; Foster and Duke, 1990).

Dioscorine is the key toxic content (Lu et al., 2012) and possesses secondary metabolites and several antinutritionals that make it bitter in taste and decrease palatability. Fatal paralysis of the nervous system is triggered by dioscorine. Similarly, histamine is the main allergen and creates itching and mild inflammation (Reddy, 2015; Shim and Oh, 2008).

5.6 CONCLUSION

A. racemosus is a valuable medicinal plant, as it is used in the traditional or indigenous system of medicines like Ayurveda, Unani, and Sidha. Traditional use of the plant is proven by various experimental and scientific reports. This presents the plant with wonderful potential in both therapeutic use and trade. *A. racemosus* is an important plant having numerous therapeutic properties such as antibacterial, anti-ulcerative, neurodegenerative, antioxidant, antidepressant, antiepileptic, antitussive, diuretic, anti-HIV, immunostimulant, hepatoprotective and cardioprotective. The phytochemical constituents of the plant are found effective against the abovementioned ailments. The safety profile analysis of *A. racemosus* has proved that it is safe in therapeutic doses and can even be used during pregnancy. The optimum production of phytochemicals of the plant can be enhanced by using biotechnological approaches like micropropagation and callus culture as well as optimization of environmental conditions and adoption of appropriate agricultural techniques to assure quality and quantity of phytochemicals. The huge demand for *A. racemosus* makes it vulnerable and endangered in its natural habitat. Therefore, the information on genetic variability is very important for its conservation, long-term survival, and genetic improvement. Evaluation of genetic variation in a species gives information about the level of genetic divergence and is a requirement for initiating a competent breeding program and molding desirable genotypes. Review of literatures disclosed inadequate work on explaining the degree of genetic diversity in *A. racemosus*. The active principle is the main phytochemicals of the medicinal plant that can be improved by cultivating superior varieties. Such varieties can be identified through molecular marker techniques and chemo-profiling. Those varieties can be developed through micropropagation methods, which are proficient *in vitro* procedures. The cell suspension culture system can be applied for the mass production of secondary metabolites from the plant cells. *D. villosa* is a wild tuber used by the aboriginal people during drought and famine. The plant is useful in the treatment of various kinds of ailments and disorders due to the presence of a number of bioactive compounds. Diosgenin is the most valuable chemical compound of *D. villosa* used in the synthesis of steroidal drugs. The drugs have anti-inflammatory, estrogenic, and anticancer potential. Plant scientists have conducted substantial work to investigate the biological activity and medicinal applications of the *A. racemosus* and *D. villosa*. But still there are huge potentials of pharmacological applications which need to be explored. The ethno-medicinal potential of both of the plants need to be authenticated, and a detailed study on the composition and pharmacological significance of the medicinal plants along with the standardization of the formulations used must be undertaken extensively. Similarly, validation of all the secondary metabolites, such as saponin, alkaloids, flavonoids, phenols, and tannins of this plant should be studied carefully by highly advanced analytical techniques. More studies are also required to address various issues regarding the composition of the extracts used, explicability of the preclinical experiments and lack of conversion of the preclinical results to clinical effectiveness. Effort should also be made to conduct careful human trials and to determine the mechanism of action, bioavailability, pharmacokinetics, and the physiological pathways for various types of bioactive compounds for their potential applications in drug discovery and for curing various life-threatening diseases. Studies should also be carried out to utilize the bioactive compounds present in these tubers for formulation of new drugs to fight against pathogenic multidrug-resistant microorganisms and antimicrobial resistance. Research on these plants will open up new views in the study of biodiversity management for germplasm conservation, sustainable development, pharmacology, and other maiden fields of research in plant science and pharmaceutics.

REFERENCES

Aarti, K. 2015. *Asparagus racemosus* (shatavari): a multipurpose plant. *European Journal of Pharmaceutical and Medical Research*:599–613.

Acharya S.R., N.S. Acharya, J.O. Bhangale, S.K. Shah, and S.S. Pandya. 2012. Antioxidant and hepatoprotective action of *Asparagus racemosus* Willd. Root extracts. *Indian Journal of Experimental Biology* 50(11):795–801.

Agrawal, A., M. Sharma, S.K. Rai, B. Singh, M. Tiwari, and R. Chandra. 2008. The effect of the aqueous extract of the roots of *Asparagus racemosus* on hepatocarcinogenesis initiated by diethyl nitrosamine. *Phytotherapy Research*22(9):1175–1182.

Ahmad, S., S. Ahmed, and P.C. Jain. 1991. Chemical examination of Shatavari (*Asparagus racemosus*). *Bulletin of Medico-Ethano Botanical Research* 12(3–4):157–160.

Ashajyothi, V., R.S. Pippalla, and D. Satyavati. 2009. *Asparagus racemosus*: a Phytoestrogen. *International Journal of Pharmacy and Technology* 1(1):36–47.

Bhasale, L., D. Padia, H. Malhotra, D. Thakkar, H. Palep, and K. Algotar. 1994. Capsule 'Surat' for comprehensive antenatal care and prevention of pregnancy induced hypertension. *Lancet* 343:619–629.

Bhatnagar, M., and S. Sisodia. 2006. Antisecretory and antiulcer activity of *Asparagus racemosus* Wild. Against indomethacin plus pyloric ligation-induced gastric ulcer in rats. *Journal of Herbal Pharmacotherapy* 6:13–20.

Bhattacharya, A., A.V. Muruganandam, V. Kumar, and S.K. Bhattacharya. 2002. Effect of poly herbal formulation, EuMil, on neurochemical perturbations induced by chronic stress. *Indian Journal of Experimental Biology* 40(10):1161–1163.

Bhutani, K.K., A.T. Paul, W. Fayad, and S. Linder. 2010. Apoptosis inducing activity of steroidal constituents from *Solanum xanthocarpum* and *Asparagus racemosus*. *Phytomedicine* 17:789–793.

Bopana, N., and S. Saxena. 2007. *Asparagus racemosus*-Ethnopharmacological evaluation and conservation needs. *Journal of Ethnopharmacology* 110:1–15.

Bown, D. 1995. *Encyclopaedia of Herbs and their Uses*. London: Dorling Kindersley. ISBN 0-7513-020-31

Castro, M. 1990. *The Complete Homeopathy Handbook*. London: Macmillan. ISBN 0-333-55581-3

Cayen, M.N., E.S. Ferdinandi, E. Greselin, and D. Dvornik. 1979. Studies on the disposition of diosgenin in rats, dogs, monkeys and man. *Artherosclerosis* 33:71–87.

Chawla, A., P. Chawla, and R. Mangalesh. 2011. *Asparagus racemosus* (Wild): biological activities and its active principles. *Indo-Global Journal of Pharmacological Science* 2:113–120.

Cheng, Y., C. Dong, C. Huang, and Y. Zhu. 2015. Enhanced production of diosgenin from *Dioscorea zingiberensis* in mixed culture solid state fermentation with *Trichoderma reesei* and Aspergillus fumigatus. *Biotechnology and Biotechnological Equipment* 29:773–778.

Chevallier, A. 1996. *The Encyclopedia of Medicinal Plants.* London: Dorling Kindersley. ISBN 9-780751-303148.

Choudhary, B.K., and A. Kar. 1992. Mineral contents of *Asparagus racemosus. Indian Drugs* 29(13):623.

Christina, A.J., K. Ashok, and M. Packialashmi. 2005. Antilithiatic effect of *Asparagus racemosus* Willd on ethylene glycol-induced lithiasis in male albino Wistar rats. *Experimental and Clinical Pharmacology* 27(9):633–638.

Crellin, J.K., and J. Philpott. 1990. *A Reference Guide to Medicinal Plants.* Durham, NC: Duke University Press.

Dahanukar, S., U. Thatte, N. Pai, P.B. Mose, and S.M. Karandikar. 1986. Protective effect of *Asparagus racemosus* against induced abdominal sepsis. *Indian Drugs* 24:125–128.

Dalvi, S.S., P.M. Nadkarni, and K.C. Gupta. 1990. Effect of *Asparagus racemosus* (Shatavari) on gastric emptying time in normal healthy volunteers. *Journal of Postgraduate Medicine* 36:91–94.

Datta, G.K., K. Sairam, S. Priyambada, P.K. Debnath and R.K.Goel. 2002. Antiulcerogenic activity of Satavari mandur: an Ayurvedic herbo-mineral preparation. *Indian Journal of Experimental Biology* 40(10):1173–1177.

Dhaliwal, K.S. 2003. Method and composition for treatment of premenstrual syndrome in women. US Patent number 698662.

Dhingra, D., V. Kumar. 2007. Pharmacological evaluation for antidepressant like activity of *Asparagus racemosus* wild in mice. *Pharmacology Online* 3:133–152.

Duke, J.A. 2000. *The Green Pharmacy Herbal Handbook: Your Comprehensive Reference to the Best Herbs for Healing.* Emmaus, PA: Rodale Inc, 223–224.

Dutta, B. 2015. Food and medicinal values of certain species of Dioscorea with special reference to Assam. *Journal of Pharmacognocy and Phytochemistry* 3:15–18.

Dweck, A.C. 2002. The wild yam: a review. *Personal Care Magazine* 3:7–9.

Fisher, C. and G. Painter. 1996. *Materia Medica of Western Herbs for the Southern Hemisphere.* New Zealand: Aukland, 91.

Foster, S. and J.A. Duke. 1990. *A Field Guide to Medicinal Plants.* Eastern and Central N. America: Houghton Mifflin Co.

Frawley, D. 1997. *Ayurvedic Healing: A Comprehensive Guide.* New Delhi, India: Motilal Banarsidass Publishers Private Limited.

Gaitonde, B.B. and M.H. Jetmalani 1969. Antioxycytocic action of saponin isolated from *Asparagus racemosus* Willd. (Shatavari) on uterine muscle. *Archives Internationales de Pharmacodynamie et de Therapie* 179:121–129.

Garde, G.K., and S. Vagbhat. 1970. Marathia translation of vagbhat's astangahridya. *Uttarstana: Aryabhushana Mudranalaya*:40–48.

Goel, R., and K. Sairam. 2002. Anti-ulcer drugs from indigenous sources with emphasis on *Musa sapientum*, tamrahbasma, *Asparagus racemosus* and *Zingiber officinale. Indian Journal of Pharmacology* 34:100–110.

Goel, R., T. Prabha, M.M. Kumar, M. Dorababu and G. Singh. 2006. Teratogenicity of *Asparagus racemosus* Wild. Root, a herbal medicine. *Indian Journal of Experimental Biology* 44:570–573.

Gomase V. and A. Sherkhane. 2010. Isolation, structure elucidation and biotransformation studies on secondary metabolites from *Asparagus racemosus. International Journal of Microbiology Research* 2:7–9.

Goyal, R.K., J. Singh, and H. Lal. 2003. *Asparagus racemosus*: an update. *Indian Journal of Medical Sciences* 57:408–414.

Grieve. 1984. *A Modern Herbal.* London: Penguin. ISBN 0-14-046-440-9

Gyawali, R., and K.S. Kim. 2011. Bioactive volatile compounds of three medicinal plants from Nepal, Kathmandu. University. *Journal of Science, Engineering and Technology* 8:51–62.

Handa, S.S., O.P. Suri, V.N. Gupta, K.A. Suri, N.K. Satti, and V. Bhardwaj. 2003. Oligospirostanoside from *Asparagus racemosus* as immunomodulator. US Patent number 6649745.

Hayes, P.Y., L.K. Lambert, R. Lehmann, K. Penman, W. Kitching, and J.J. De Voss. 2007. Assignments of the four major saponins from *Dioscorea villosa* (wild yam). *Magnetic Resonance in Chemistry* 45(11):1001–1005.

Howell, P. 2006. Medicinal plants of the southern Appalachians. Mountain City, GA: Botano Logos Books.
Hussain, A., M.P. Ahmad, S. Wahab, M.S. Hussain, and M. Ali. 2011. A review on pharmacological and phytochemical profile of *Asparagus racemosus* Wild. *Pharmacologyonline* 3:1353–1364.
Hutchen, A. 1969. *Indian Herbalogy of North America*. Boston, MA: Shambala, 301–302.
Jetmalani, M.H., P.B. Sabins, and B.B. Gaitonde. 1967. A study on the pharmacology of various extracts of Shatavari-*Asparagus racemosus* (Willd). *Journal of Research in Indian Medicine* 2:1–10.
Joglekar, G.V., R.H. Ahuja, and J.H. Balwani. 1967. Galactogogue effect of *Asparagus racemosus*. Preliminary communication. *Indian Medical Journal* 7(61):165.
Joshi, J.D.S. 1988. Chemistry of ayurvedic crude drugs: part VIII: Shatavari 2 structure elucidation of bioactive shatavarin I and other glycosides. *Indian Journal of Chemistry Section B Organic Chemistry* 27(1):12–16.
Kaimal, A., and K.J. Kemper. 1999. *Wild Yam (Dioscoreaceae). The Longwood Herbal Taskforce and the Center for Holistic Pediatric Education and Research*. New York: William Morrow and Company, 1–11.
Kala, C.P. 2009. Aboriginal uses and management of ethnobotanical species in deciduous forests of Chhattisgarh state in India. *Journal of Ethnobiology and Ethnomedicine* 5:1–9.
Kamat, J.P., K.K. Boloor, T.P. Devasagayam, and S.R. Venkatachalam. 2000. Antioxidant properties of *Asparagus racemosus* against damaged induced by gamma radiation on rat liver mitochondria. *Journal of Ethanopharmacology* 71:425–435.
Kanwar, A.S., and K.K. Bhutani. 2010. Effects of *Chlorophytum arundinaceum*, *Asparagus adscendens* and *Asparagus racemosus* on pro inflammatory cytokine and corticosterone levels produced by stress. *Phytotherapy Research* 24(10):1562–1566.
Karmakar, U.K., S.K. Sadhu, S.K. Biswas, A. Chowdhury, M.C. Shill, and J. Das. 2012. Cytotoxicity, analgesic and antidiarrhoeal activities of *Asparagus racemosus*. *Journal of Applied Science* 12:581–586.
Kaushik, N.K., A. Bagavan, A.A. Rahuman, D. Mohanakrishnan, C. Kamaraj, G. Elango, A.A. Zahir, and D. Sahal. 2013. Antiplasmodial potential of selected medicinal plants from Eastern Ghats of South India. *Experimental Parasitology* 134:26–32.
Khan, I.A. 2002. Estimation of Nutritive Constituents of Exotic Vegetables. *Progressive Horticulture* 34(1):95–98.
Khanna, A.K., R. Chander, and N.K. Kapoor. 1991. Hypolipidaemic activity of Abana in rats. *Fitoterapia* 62:271–275.
Kishore, P., P.N. Pandey, S.N. Pandey, and S. Dash. 1980. Treatment of duodenal ulcer with *Asparagus racemosus* Linn. *Journal of Research in Indian Medicine, Yoga and Homeopathy* 15:409–415.
Koonin, E.V., and L. Aravind. 2002. Origin and evolution of eukaryotic apoptosis: the bacterial connection. *Cell Death and Differentiation* 9:394–404.
Krochmal, A., and A. Connie. 1973. *A Guide to the Medicinal Plants of Appalachia*. USA: Quadrangle, the Times Book Co.
Krtikar, K.R., and B.D. Basu. 1975. *Indian Materia Medica*, Vol. 3. Dehra Dun: Bishen Singh, Mahendra Pal Singh, 2499–2501.
Kulkarni, S.K., and A. Verma. 1993. Protective effect of Mentat (BR-16A) A herbal preparation, on alcohol abstinence-induced anxiety and convulsions. *Indian Journal of Experimental Biology*. 31:435–442.
Lu, Y.L., Chia, C.Y., Liu, Y.W., and Hou, W. 2012. Biological activities and applications of dioscorins, the major tuber storage proteins of yam. *Journal of Traditional and Complementary Medicine* 2:41–46.
Mandal, S.C., A. Nandy, M. Pal, and B.P. Saha. 2000. Evaluation of antibacterial activity of *Asparagus racemosus* Willd root. *Phytotherapy Research* 14(2):118–119.
Mandal, S.C., C.K.A. Kumar, L.S. Mohana, S. Sinha, T. Murugesan, and B.P. Saha. 2000. Antitussive effect of *Asparagus racemosus* root against sulfur dioxide-induced cough in mice. *Fitoterapia* 71(6):686.
Mangal, A., D. Panda, and M.C. Sharma. 2006. Peptic ulcer healing properties of Shatavari (*Asparagus racemosus* Wild.). *Indian Journal of Traditional Knowledge* 5:229–236.
Mayo, J.L., and M.D. Facog. 1998. Black cohosh and chasteberry: herbs valued by women for centuries. *Clinical Nutrition Insights* 6(15):1–4.
Mbiantcha, M., A. Kamanyi, R.B. Teponno, A.L. Tapondjou, P. Watcho, and T.B. Nguelefack. 2010. Analgesic and anti-inflammatory properties of extracts from the bulbils of *Dioscorea bulbifera* L. var sativa (Dioscoreaceae) in mice and rats. *Evidence Based Complementary and Alternative Medicine* 2010:1–9.
Mills, S.Y. 1988. *The Dictionary of Modern Herbalism: A Comprehensive Guide to Practical Herbal Therapy*. Rochester, VT: Healing Arts Press.
Mishra, P.K., P. Singh, B. Prakash, A. Kedia, N.K. Dubey, and C.S. Chanotiya. 2013. Assessing essential oil components as plant-based preservatives against fungi that deteriorate herbal raw materials. *International Biodeterioration and Biodegradation* 80:16–21.

Mitra, S.K., N.S. Prakash, and R. Sundaram. 2012. Shatavarins (containing Shatavarin IV) with anticancer activity from the roots of *Asparagus racemosus*. *Indian Journal of Pharmacology* 44:732–736.

Mitra, S.K., S. Gopumadhavan, M.V. Venkataranganna, D.N.K. Sarma, and S.D. Anturlikar. 1999. Uterine tonic activity of U-3107, a herbal preparation, in rats. *International Journal of Pharmacology* 31(3):200–203.

Moalica, S., B. Liagrea, A. Bianchib, and M. Dauc. 2001. A plant steroid, diosgenin, induces apoptosis, cell cycle arrest and COX activity in osteosarcoma cells. *FEBS Letters* 506:225–230.

Moerman, D. 1998. *Native American Ethnobotany*. Portland, OR: Timber Press.

Mohanta, B., A. Chakraborty, M. Sudarshan, R.K. Dutt, and K.M. Baruah. 2003. Elemental profile in some common medicinal plants of India. Its correlation with traditional therapeutic usage. *Journal of Radio Analytical and Nuclear Chemistry* 258(1):175–179.

Muruganadan, S., H. Garg, J. Lal, S. Chandra, and D. Kumar. 2000. Studies on the immunostimulant and antihepatotoxic activities of *Asparagus racemosus* root extract. *Journal of Medicinal and Aromatic Plant Science* 22:49–52.

Nadkarni, A.K. 1954. *Indian Materia Medica*. Bombay: Popular Book Depot, 153–155.

Nair, A.G.R., and S.S. Subramanian. 1969. Occurrence of diosgenin in *Asparagus racemosus*. *Current Science* 17:414.

Narendranath, K.A., S. Mahalingam, V. Anuradha, and I.S. Rao. 1986. Effect of herbal galactogogue (Lactare) a pharmacological and clinical observation. *Medical Surgical Nursing* 26:19–22.

National Medicinal Plants Board. 2002. http://www.nmpb.nic.in

Negi, J.S., P. Singh, G.J.N. Pant, M.S.M. Rawat, and H. Pandey. 2010. Variation of trace elements contents in *Asparagus racemosus* (Wild). *Biological Trace Element Research* 135:275–282.

Nelson, W. 2003. *Native American Garden Plants*. Herb Society of America: New England Unit.

Nevrekar, P., N. Bai, and S. Khanna. 2002. Eve Care capsules in DUB. *Obstetrics and Gynaecology Communications* 3:51–53.

NPGS/GRIN. 2009. *Asparagus Racemosus Information from Germplasm Resources Information Network*. Aberdeen, ID: Department of Agriculture.

Olayemi, J.O., and E.O. Ajaiyeoba. 2007. Anti-inflammatory studies of yam (*Dioscorea esculenta*) extract on wistar rats. *African Journal of Biotechnology* 16:1913–1915.

Parihar, M., and T. Hemnani. 2004. Experimental excitotoxicity provokes oxidative damage in mice brain and attenuation by extract of *Asparagus racemosus*. *Journal of Neural Transmission* 111:1–12.

Park, M.K., H.Y. Kwon, W.S. Ahn, S. Bae, M.R. Rhyu, and Y.J. Lee. 2009. Estrogen activities and the cellular effects of natural progesterone from wild yam extract in MCF-7 human breast cancer cells. *American Journal of Chinese Medicine* 37(1):159–167.

Patel, A.B., and U.K. Kanitkar. 1969. *Asparagus racemosus* Willd form Bordi, as a galactogogue, in buffaloes. *Indian Veterinary Journal* 46:718–721.

Patricia, Y.H., A.H. Jahidin, R. Lehmann, K. Penman, W. Kitchinga, and J.J. De Vossa. 2006. Asparinins, asparosides, curillins, curillosides and shavatarins. Structural clarification with the isolation of shatavarin V, a new steroidal saponin from the root of *Asparagus racemosus*. *Tetrahedron Letters* 47:8683–8687.

Rahiman, O., M.R. Kumar, T.T. Mani, K.M. Niyas, B.S. Kumar, P. Phaneendra, and B. Surendra. 2011. Hepatoprotective activity of *Asparagus Racemosus* root on liver damage caused by paracetamol in rats. *Indian Journal of Novel Drug Delivery* 3:112–117.

Reddy, D.S. 2015. Ethnobotanical studies of *Dioscorea hispida* Dennst. In Nallamala forest area AP, India. *Review Research* 4:1–4.

Rosenberg, Z.R.S, D.J.A. Jenkins, and E.P. Diamandis. 2001. Effects of natural products and nutraceuticals on steroid hormone-regulated gene expression. *Clinica Chimica Acta* 312:213–219.

Roy, A., R.V. Geetha, and T. Lakshmi. 2012. Valuation of the antibacterial activity of ethanolic extract of *Dioscorea villosa* tubers-an in vitro study. *International Journal of Pharmacy and Pharmaceutical Sciences* 4, 314–316.

Roy, R.N., S. Bhagwager, S.R. Chavan, and N.K. Dutta. 1971. Preliminary pharmacological studies on extracts of root of *Asparagus racemosus* (Satavari) Willd, Lilliaceae. *Journal of Research in Indian Medicine* 6:132–138.

Sairam, K.S., N.C. Priyambada, and R.K, Goel. 2003. Gastroduodenal ulcer protective activity of *Asparagus racemosus*. An experimental, biochemical and histological study. *Journal of Ethnopharmacology* 86(1):1–10.

Sautour, M., A. Mitaineoffer, and M. Lacailledubois. 2007. The Dioscorea genus: a review of bioactive steroid saponins. *Journal of Natural Medicines* 61:91–101.

Sautour, M., M. Miyamoto, A. Dongmo, and M.A. Lacaille. 2004. Antifungal steroidal saponins from *Dioscorea cayenensis*. *Planta Medica* 70:90–1.

Saxena, V.K., and S. Chourasia. 2001. A new isoflavone from the roots of *Asparagus racemosus*. *Fitoterapia* 72:307–309.
Scarcelli, N., H. Chair, S. Causse, R. Vesta, T.L. Couvreur, and Y. Vigouroux. 2017. Crop wild relative conservation: wild yams are not that wild. *Biological Conservation* 210:325–333.
Sekine, T., F.N. Ikegami, Y. Fukasawa, T. Kashiwagi, Y. Aizawa, Y. Fujii, N. Ruangrungsi, and I. Murakoshi. 1995. Structure and relative stereochemistry of a new polycyclic alkaloid, asparagamine A, showing anti-oxytocin activity, isolated from *Asparagus racemosus*. *Journal of Chemical Society, Perkin Transaction*:391–393.
Sekine, T.N. 1994. Fukasawa Structure of asparagamine A, a novel polycyclic alkaloid from *Asparagus racemosus*. *Chemical and Pharmaceutical Bulletin* 42(6):1360–1362.
Sharma, K., and M. Bhatnagar. 2011. *Asparagus racemosus* (Shatavari): a versatile female tonic. *International Journal of Pharmaceutical and Biological Archive* 2(3):855–863.
Sharma, K., M. Bhatnagar, and S.K. Kulkarni. 2010. Effect of *Convolvulus pluricaulis* Choisy and *Asparagus racemosus* Willd on learning and memory in young and old mice, a comparative evaluation. *Indian Journal of Experimental Biology* 48(5):479–485.
Sharma, P., P.S. Chauhan, P. Dutt, M. Amina, K.A. Suri, B.D. Gupta, O.P. Suri, K.L. Dhar, D. Sharma and V. Gupta. 2011. A unique immuno-stimulant steroidal sapogenin acid from the roots of *Asparagus racemosus*. *Steroids* 76:358–364.
Sharma P.V., and S. Charaka. 2001. Chaukhambha Orientalis. Varanasi, India: Chowkhamba, 7–14.
Sharma, R.K., and B. Dash. 2003. *Charaka Samhita-text with English Translation and Critical Exposition based on Chakrapani Datta's Ayurveda Dipika*. Varanasi, India: Chowkhamba.
Sharma, S., S. Ramji, S. Kumari, and J.S. Bopana. 1996. Randomized controlled trial of *A. racemosus* (Shatavari) as a lactogogue in lactational inadequacy. *Indian Journal of Pediatrics* 33(8) 675–677.
Sharma, S.C. 1981. Constituents of the fruits of *Asparagus racemosus* Willd. *Pharmazie* 36(10):709.
Sharma, U., N. Kumar, B. Singh, R.K. Munshi, and S. Bhalerao. 2009. Immunomodulatory active steroidal saponins from *Asparagus racemosus*. *Medical Chemistry Research* 121:1–7.
Shashi, A., S.K. Jain, A. Verma, M. Kumar, A. Mahor, and M. Sabharwal. 2013. Plant profile, phytochemistry and pharmacology of *Asparagus racemosus* (shatavari): a review. *Asian Pacific Journal of Tropical Disease* 3(3):242–251.
Shasnay, A.K., P. Darokar, D. Saikia, S. Rajkumar, V. Sundaresan, and S.P.S. Khanuja. 2003. Genetic diversity and species relationship in *Asparagus spp*. Using RAPD analysis. *Journal of Medicinal and Aromatic Plant Science* 25:698–704.
Shastri, A.A., and B. Ratnavali. 2014. *Chaukhamba Prakashan, Varanasi*, 2014:559.
Shim, W.S., and U. Oh. 2008. Histamine-induced itch & its relationship with pain. *Molecular Pain* 4:29–32.
Singh, G.K., D. Garabadu, A.V. Muruganandam, V.K. Joshi, and S. Krishnamurthy. 2009. Antidepressant activity of *Asparagus racemosus* in rodent models. *Pharmacology Biochemistry and Behaviour* 91(3):283–290.
Singh, J., and H.P. Tiwari. 1991. Chemical examination of roots of *Asparagus racemosus*. *Journal of Indian Chemical Society* 68(7):427–428.
Singh, N., R. Nath, and R.P. Kohil. 1978. An Experimental Evaluation of Anti-stress effects of Geriforte.*Quarterly Journal of Crude Drug Research* 31:125.
Singh, S.C. 1960. Some wild plants of food value in Nepal. *Journal of Tribhuban University, Kathmandu* 4:50–56.
Singh, S.K. and K.S. Kulkarni. 2002. Evaluation of the efficacy and safety of Menosan in post-menopausal symptoms: a short-term pilot study. *Obstetrics Gynaecology Today* 12:727–730.
Son, I.S., J.S. Lee, J.Y. Lee, and C.S. Kwon. 2014. Antioxidant and anti-inflammatory effects of yam (*Dioscorea batatas* Decne.) on azoxymethane induced colonic aberrant crypt foci in F344 rats. *Preventive Nutrition and Food Science* 19:82–88.
Subramanian, S.S., and A.G.R. Nair. 1969. Occurrence of Diosegenin in *Asparagus racemosus* leaves. Current Science 38(17):414.
Tada, Y., N. Kanda, A. Haratake, M. Tobiishi, H. Uchiwa, and S. Watanabe. 2009. Novel effects of diosgenin on skin aging. *Steroids* 74:504–511.
Thakur, M., N.S. Chauhan, S. Bhargava, and V.K. Dixit. 2009. A comparative study on aphrodisiac activity of some ayurvedic herbs in male albino rats. *Archives of Sexual Behaviour* 38(6):1009–1015.
Trimanto, T., and L. Hapsari. 2015. Diversity and utilization of *Dioscorea spp*. Tuber as alternative food source in Nganjuk Regency, East Java. *Agrivita* 37:97–107.
Venkatesan, N., V. Thiyagarajan, S. Narayanan, A. Arul, S. Raja, and S. Gurusamy. 2005. Anti-diarrhoeal potential of *Asparagus racemosus* wild root extracts in laboratory animals. *Journal of Pharmacy and Pharmaceutical Sciences* 8:39–46.

Visavadiya, N.P., and A. Narasimhacharya. 2009. *Asparagus* root regulates cholesterol metabolism and improves antioxidant status in hypercholesteremic rats. *Evidence Based Complementary and Alternative Medicine* 6:219–226.

Visavadiya, N.P., B. Soni, and D. Madamwar. 2009. Suppression of reactive oxygen species and nitric oxide by *Asparagus racemosus* root extract using in vitro studies. *Cellular and Molecular Biology* 55(Supplement):1083–1095.

Warrier, P.K., V.P.K. Nambiar, and P.M. Ganapathy. 2001. *Some Important Medicinal Plants of the Western Ghats, India: A Profile.* New Delhi: International Development Research Centre, Artstock 398.

Wiboonpun, N., P. Phuwapraisirisan, and S. Tip-pyang. 2004. Identification of antioxidant compound from *Asparagus racemosus. Phytotherapy Research* 18:771–773.

Yoon, K.D., and J. Kim. 2008. Preparative separation of dioscin derivatives from *Dioscorea villosa* by centrifugal partition chromatography coupled with evaporative light scattering detection. *Journal of Separation Science* 31(12):2486–2491.

https://www.nmpb.nic.in/Write Read Data /links/3733877856 Shatavari.pdf

https://www.illinoiswildflowers.info/savanna/plants/wild_yam.html

http://www.rain-tree.com/dioscorea.htm

6 *Embelia ribes* (False Black Pepper) and *Gymnema sylvestre* (Sugar Destroyer)

Chandrabose Selvaraj, Chandrabose Yogeswari, and Sanjeev Kumar Singh

CONTENTS

6.1 Introduction ... 114
6.2 Description of *Embelia ribes* ... 114
6.3 Description of *Gymnema sylvestre* ... 115
6.4 Phytochemical Constituents of *Embelia ribes* ... 116
6.5 Formulations of *Embelia ribes* ... 116
6.6 Application in Ayurveda ... 117
6.7 Application in Traditional Uses ... 117
6.8 Pharmacological Uses ... 118
6.8.1 Antibacterial ... 118
6.8.2 Anthelmintic ... 118
6.8.3 Antidiabetic ... 119
6.8.4 Hepatoprotective ... 119
6.8.5 Antifertility ... 119
6.8.6 Antitumor ... 120
6.9 Toxicological Effect of *E. ribes* ... 120
6.10 Chemical Constituents of *Gymnema sylvestre* ... 120
6.11 Traditional Use ... 122
6.12 Pharmacological Actions ... 122
6.12.1 Application in Diabetes Mellitus ... 122
6.12.2 Anticancer Activity ... 123
6.12.3 Lipid-Lowering Activity ... 123
6.12.4 Antimicrobial Activity ... 123
6.12.5 Antioxidant Activity ... 124
6.12.6 Antiarthritic Activity ... 124
6.12.7 Immunomodulating Effect of *G. sylvestre* ... 124
6.12.8 Anti-Inflammatory Activity ... 125
6.12.9 Hepatoprotective Activity ... 125
6.13 Toxicological Effect of *Gymnema sylvestre* ... 125
6.14 Conclusion ... 126
Acknowledgments ... 126
References ... 126

DOI: 10.1201/9781003205067-6

6.1 INTRODUCTION

Herbal plants have long been used to treat and prevent many illnesses due to their high medicinal value. Herbal products and remedies made from various herbal plants are well described in ancient texts, like the Vedas and Scriptures (Hoareau and DaSilva, 1999). Recently, developed countries have also widely accepted and used herbal plants as a traditional medicinal system to treat various diseases. Historically, rural and local communities have used crude extracts of herbal plants for medicinal purposes (Rai and Nath, 2005; Street and Prinsloo, 2013). More than 15,000 different plants are used in India's traditional medicinal system due to their high concentration of bioactive molecules, such as phenols, flavonoids, tannins, and alkaloids (Singh et al., 2003). As a significant source of molecules with great therapeutic value, medicinal plants have vast potential of interest and commercialization, and thus, they take a keen interest in modern drug discovery methods. In recent years, herbal and natural products have played an increasingly important role in modern medicine. About 25% of the highly active drug molecules used in modern medicine, as well as over 15% of the plant-based bioactive molecules, are currently being studied in clinical trials for their potential pharmaceutical use (Gurnani et al., 2014). For decades, researchers have relied on herbal medicines and natural products for their work because of their wide range of pharmacological properties (Goto et al., 2010; Sarker and Gohda, 2013; Rouhi et al., 2017; Chen et al., 2018; Shiekh et al., 2017; Imam et al., 2013; Yasmin et al., 2009). *Gymnema sylvestre* (*G. sylvestre*) and *Embelia ribes* burm (*E. ribes*) are two of the most widely used botanicals in Ayurvedic and folk medicine, respectively (Singh et al., 2008). The Apocynaceae family includes the medicinal plant *G. sylvestre*, which is found in India, China, and Australia, and this *G. sylvestre* is widely familiar as gurmur owing to its sugar-lowering property (Christopoulos et al., 2010; Tiwari et al., 2014). It also shows significant inhibitory properties against significant diseases, such as asthma, cancer, cardiovascular, obesity, and diabetes. Various formulations, including tea bags, food supplements, and tablets are available for *Gymnema sylvestre*, which has been reported to be effective against anemia, hypercholesterolemia, cardiopathy, and bacterial infection (Tiwari et al., 2014). Another medicinal plant, *Embelia ribes* (Family: Myrsinaceae), has potential medicinal value against various diseases, like diabetes mellitus, due to active components, such as embelin and alkyl-substituted hydroxyl benzoquinone (Warrier et al., 2001). Embelin is a powerful antidiabetic and hypoglycemic agent derived from the fruits of the *E. ribes* plants (Ved et al., 2006). *Embelia ribes* and *Gymnema sylvestre* are discussed in this chapter for their traditional uses, pharmacological effects, and the crucial phytoconstituents and chemical structures of these medicinal plants.

6.2 DESCRIPTION OF *EMBELIA RIBES*

E. ribes is a large woody climbing shrub commonly known as Baobarang that is widely distributed in the Western Ghats of India, as well as Malaysia, Sri Lanka, and South China (Guhabakshi et al., 2001). An essential ingredient in Ayurvedic medicines, it is known as Vaibidang in Sanskrit. The woody, slender, and flexible stem is highly branched with gland-dotted leaves that are lanceolate, coriaceous with a length of 6–14 cm and breadth of 2–4 cm. Meghalaya, Arunachal Pradesh, and Mizoram are just a few of the North-East Indian states where it's found in abundance, it can also be found in India's Western Ghats regions in Tamil Nadu and Karnataka. (Figure 6.1). The petiole is cylindrical, and the flowers are small and numerous, growing at the ends of the upper axils. There is a 1–2.5 mm long calyx with connate triangular-ovate sepals that are ciliate. Each petal is free-elliptical, measuring 4 mm long and pubescent on both sides, and the petals are a greenish-yellow color. The *E. ribes* fruit looks globular and wrinkled, with a color range from dull red to black, and the ripe fruits turn brown once they have been ripped from their smooth, globose skins (Kirtikar and Basu, 1975). The fruit has a small pedicel and pericarp, and the seed is enclosed in a membrane-sealed capsule. The plant's root system reaches deep into the soil, which aids in their inclination to climb (Latha, 2007).

FIGURE 6.1 Represents the fresh and dry fruits of *Embelia ribes.*

6.3 DESCRIPTION OF *GYMNEMA SYLVESTRE*

G. sylvestre, a slow-growing plant found in the dry forests of central and southern India and other parts of Asia, resembles a woody climbing shrub with many branches (Wu et al., 2012; Kapoor, 2017). The plant's stem is cylindrical and highly branched, measuring 0.7–17.2 cm in length and 2–10 mm in diameter, and the woody parts of the plant have taproots and are pubescent in nature (Najafi and Deokule, 2011). The leaf structure is simple, elliptical, shortly acuminate, and has petioles 1–2 cm long. Phyllotactic leaves show opposite arrangement patterns with 2.5–6 cm long leaves. Velvety pubescence covers the entire upper surface, the margins are ciliate, and it shows transverse venation and a reticulate type with the marginal vein (Pramanick, 2016; Kanetkar et al., 2007). Seeds are flat, 1.3 cm long, and have thin peripheral wings (Chopra et al., 1992). Flowers are small, axillary, and yellow in color and lateral umbel in cymes (Figure 6.2). Follicles are lanceolate, terete; the calyx is five-lobed, ovate, and ciliated, where corolla is campanulate (Pramanick, 2016).

FIGURE 6.2 Represents a morphological image of *Gymnema sylvestre.*

6.4 PHYTOCHEMICAL CONSTITUENTS OF *EMBELIA RIBES*

Bioactive constituents found in *E. ribes* include vilangin, which is often extracted from the dried ripened fruits of the plant (Latha, 2007). Berries of *E. ribes* were also found to contain a variety of phytoconstituents, including tannin, resin, alkaloid, phenolic acid, vanillic acid, and o-coumaric acid (Haq et al., 2005). *E. ribes* seeds yielded the bioactive compounds embelinol, embeliaribyl ester, embeliol, and the fruits were found to contains 4.33% of bioactive compound embelin (Indrayan et al., 2005). Others such as 2-hydroxy-4-undecyl-3-6-benzoquinone, quercetin, embalate, and fatty acids were found in *E. ribas* in addition to those mentioned previously (Lin et al., 2006). Seeds are rich in inorganic metals, like Ca, Zn, Mn, Cr, K, and C, as well as steroids, cardiac glycosides, and anthraquinone compounds (Tambekar et al., 2009; McErlean and Moody, 2007). In the root region, it has several nitrogen-containing compounds 3-alkyl1, 4-benzoquinone derivative, 4-benzoquinone, gomphilactone derivatives, 5, 6-dihydroxy-7-tridecyl-3- [4-tridecyl-3-hydroxy-5-oxo-2(5H)-furylidene]-2-oxo 3(2H)-benzofuran, daucosterol, and sitosterol (Raja et al., 2005; Dang et al., 2014). Aerial parts of the plant have novel embeliphenol A and embelanide, which are bioactive compounds. The most active bioactive compounds of the *E. ribes* are represented in Figure 6.3.

6.5 FORMULATIONS OF *EMBELIA RIBES*

E. ribes is frequently used to prepare various ayurvedic formulations, such as Lauha, Vidangadi Lauha, Agnimandyahara, Vahnikara, Sara, Krumikushta Krumihara, Vidanghrista, Brahmarasayana, Vidangadi churna, Shulahara, Vahnikara, Shirorogahara, Shleshma Krumihara, and Guduchi. The formulation of *E. ribes* Jantugna is widely used against microbial infection and is also effectively used as an anthelmintic agent that inhibits worm populations; another formulation, Kushthgna, is commonly used to treat skin disorders; and Shirovirochan is used to treat the doshas accumulated above the chest level. Deepana is a formulation of *E. ribes*, one of the most essential medicines in the Ayurvedic system, which is used as an appetizer (to stimulate digestion); digestive disorders are treated with anuloman. Pachan, a different type of formulation, is used to treat digestive problems, while Balys is prescribed to improve brain function. Kamla is used to treat jaundice, while Mootravaah raises the pH and turns the urine acidic. Apsmaar is used to treat epilepsy, and Schirog is widely used to treat central nervous complaints and head-related ailments, like migraines and headaches. Another formulation of *E. ribes*, Raktashodhal, is used as an effective blood purifier. Both Pakshaghaat and Udrashool are used to treat paralysis and abdominal pain, respectively (Asadulla and Ramandang, 2011).

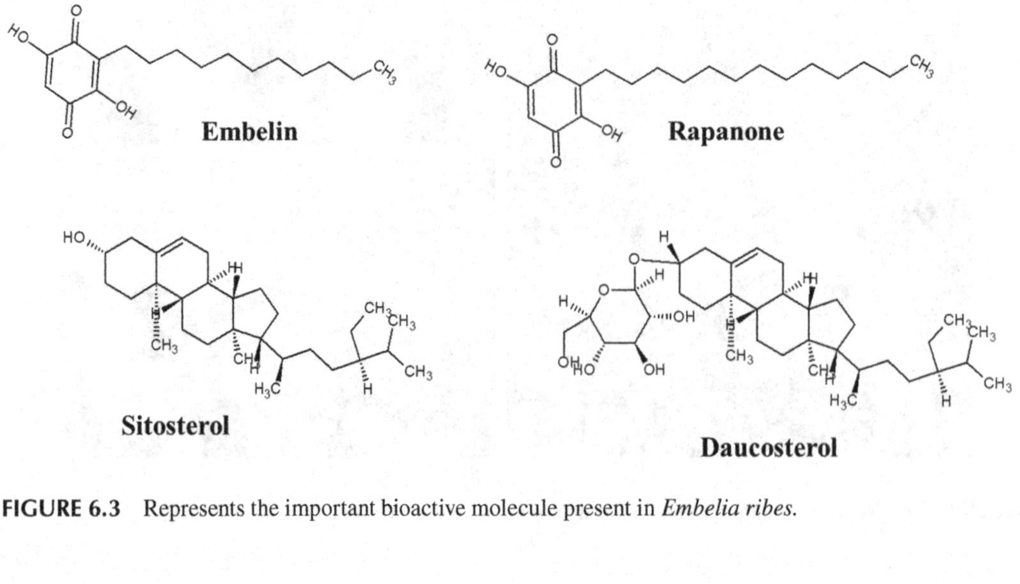

FIGURE 6.3 Represents the important bioactive molecule present in *Embelia ribes*.

6.6 APPLICATION IN AYURVEDA

Since ancient times *E. ribes* has been used in Ayurveda in the name of Charaka Samhita, Ashtanga hirdayam, and Sushruta Samhita and also used as Vidanga or Baibidanga in ayurvedic medicine. It is considered a key ingredient in ayurvedic formulations (Bhandari et al., 2002). Doctors of Ayurveda claim that the fruit extract of *E. ribes* has strong anthelmintic properties, so it is used as a tonic along with the root of licorice, which is known to have antiaging properties as well as being a powerful body strengthener. In general, it is used to treat various ailments, including gastrointestinal pains, skin diseases, and parasitic infections (Thippeswamy et al., 2011). In addition, it is widely used in the treatment of cancer, cardiovascular diseases, mental disorders, bronchitis, and jaundice (Shirole et al., 2015). Among its many uses, the fruit extract of *E. ribes* is effective against parasites, such as pinworms, tapeworms, and rounds, as well as a laxative and diuretic. Gargling with *E. ribes* juice gives effective relief of tonsillitis. Table 6.1 discusses the various properties of *E. ribes* (Souravi and Rajasekharan, 2014).

6.7 APPLICATION IN TRADITIONAL USES

E. ribes is used as an effective laxative, anthelmintic, carminative, and appetizer in traditional medicine, in the same way that ayurvedic medicine is used to treat tumors and mental disorders, dyspnea, and heart disease along with hemicrania, urinary discharges, and worm infections. It is common to use the *E. ribes* fruit extract in various formulations, including those for treating ringworm and other skin conditions. This plant's seed extract is commonly used to treat tapeworms, and the *E. ribes* fresh juice is used as an effective diuretic and cooling agent. The pulp from *E. ribes* fruit is an effective laxative, and the seeds' mucilage is used to treat mouth ulcers, aphthae, and sore throat. *E. ribes* root extract is used to treat influenza viral infections effectively, and dried root bark

TABLE 6.1
Pharmacological Activity of *Embelia ribes*

Plant Extract/Compound	Pharmacological Activity	References
Embelin	Analgesic activity	Atal et al., 1984
Aqueous and alcoholic extracts	Anthelmintic activity	Jalalpure et al., 2007; Khare, 2004
Embelin	Antianxiety activity	Sajith Mohandas et al., 2013
Aqueous and ethanolic extracts	Antibacterial activity	Mohammad Alam Khan et al., 2010; Radhakrishnan et al., 2011
Seed oil	Antinematodal activity	Brahmeshwari and Kumaraswamy, 2012
Seed oil	Acaricidal properties	Javed and Akhtar, 1990
Embelin	Anticancer activity	Chitra et al., 1995
Embelin	Anticonvulsant activity	Reuter et al., 2010
Embelin	Antidepressant activity	Mahendran et al., 2011
Embelin	Antifertility activity	Chauhan et al., 1979
Seed extract	Antifungal activity	Sabitha Rani et al., 2011
Embelin	Antigenotoxicity activity	Pankaj Tripathi and Rina Tripathi, Patel, 2010
Embelin/ ethanol extract	Antioxidant and Neuroprotective activity	Joshi et al., 2007
Aqueous and alcoholic extracts	Cardioprotective activity	Uma Bhandari et al., 2008; Nazam Ansari and Bhandari, 2008
Ethanolic extract, Embelin	Nephroprotective activity	Ashish et al., 2013
Ethanolic extract, Embelin	Antidiabetic activity	Uma Bhandari et al., 2007; Ashok Purohit et al., 2008

of this plant is used to treat toothache. A prepared paste made from the root of *E. ribes* is used to treat lung ailments, and fruit juice mixed with butter is used to treat headaches (Javed and Akhtar, 1990). The essential oil extracted from *E. ribes*, along with Croton tiglilium and sodium carbonate, is used to treat headaches and hemicrania (Nadkarni, 1982). The fruit extract has purgative and analgesic properties as well as significant effects on parasitic worms and bronchitis. It also effectively removes bad humor from the body and reddens urine. It's used to treat children's flatulence, as well as diarrhea, cough, and fever (Saikia et al., 2006). The berries of *E. ribes* and those of *Piper longum, Terminalia bellerica*, and *Emblica officinalis* are used in equal parts to treat carbuncle infection (Chopda and Mahajan, 2009). *E. ribes* is used to treat stomach problems in Arunachal Pradesh and heal wounds in Maharashtra using fruit extracts. This plant is used to treat age-related cognitive disorders in the eastern Shimoga district, and 1–3 tablespoons of *E. ribes* root and lemon juice are taken twice daily for two days to treat cold and cough (Rajkumar and Shivanna, 2009). Oral administration of a paste made from *E. ribes* leaves with the roots of *Asparagus racemosus* and *Withania somnifera*, along with hot water, twice daily, reduces the risk of paralysis.

6.8 PHARMACOLOGICAL USES

Several studies on the pharmacological effect of *E. ribes* have been published, demonstrating effective therapeutic properties such as anticancer, anti-indigestion, antimicrobial, antioxidant, antihyperlipidemia, anti-inflammatory, and antiprotozoal activity. Table 6.1 discusses the various pharmacological effects of *E. ribes*.

6.8.1 ANTIBACTERIAL

Researchers have found that the antimicrobial properties of *E. ribes* have only moderate activity against bacteria that are resistant to multiple drugs, such as *Salmonella typhi* (Ansari and Bhandari, 2008). Researchers found that the main bioactive component of *E. ribes* had significant antibacterial effects on the populations of *Shigella flexneri*, *Streptococcus pyogenes*, and *Pseudomonas aeruginosa* (Chitra et al., 2003). There is effective growth inhibition on *Enterobacter aerogenes*, *Staphylococcus aureus*, and *Klebsiella pneumonia* caused by the ethanol extract of *E. ribes* seeds (Gajjar et al., 2009). According to Feresin et al., (2003), embelin of *E. ribes* effectively inhibits the growth of methicillin-sensitive and methicillin-resistant *Staphylococcus aureus* with MICs of 250 and 62 μg/ml, respectively and also shows 50 μg/ml of MIC for *E. coli*. Schrader (2010) described the antimicrobial activity of *E. ribes* extract against *Edwardsiella ictaluri*, an effective causative organism of enteric septicemia, with a MIC value of 294.4 μg/ml. The ethanolic extract of *E. ribes* has significant inhibitory activity against *Staphylococcus aureus*, *E. coli*, *Streptococcus faecalis*, and *B. subtilis* at MBC concentrations ranging from 16 to 18.5 mg/ml (Khan et al., 2010). In 2011, a study by Rathakrishnan et al. found that embelin had a significant antibacterial effect against gram-positive bacteria and bacteriostatic activity against gram-negative strains of bacteria.

6.8.2 ANTHELMINTIC

E. ribes has been used as an effective anthelmintic in both traditional medicine and the Ayurvedic medicine system. The essential oil extracted from the seeds of *E. ribes* has been shown to be effective against *Pheretima posthuma*. The seed oil of *E. ribes,* provided orally thrice a day with an increasing dosage of 10, 50, and 100 mg/ml, effectively kills the worms compared to other plants, such as *Impatiens balsamina* and *Mucuna pruries*, *Gynandropsis gynandra*, and *Celastrus paniclulata* (Jalalpure et al., 2007). The ethanolic extract of *E. ribes* fruits exhibited significant anthelmintic activity against *Haemonchus contortus* and gastrointestinal nematodes, with up to 93% activity (Tambekar et al., 2009). The combination of *E. ribes* fruit extract and *Vernonia anthelmintica* seed extract has antinematodal activity in treating goats. The treatment with methanol extract (2 g/kg) of *E. ribes* on goats revealed a significant reduction in fecal EPG counts in goats infected with gastrointestinal nematodes before and after treatment (Hördegen et al., 2006; Javed and Akhtar, 1990).

6.8.3 Antidiabetic

Diabetes is one of the most severe complications in India and a major, troubling human disease in a number of other countries (Modak et al., 2007). Even though a variety of methods have been used to reduce the illness of diabetes, herbal formulations are highly preferred due to their secondary metabolites and fewer side effects (Shukla et al., 2011). Since medicinal plants and their extracts are a rich source of bioactive compounds, they are widely used in diabetic research around the world. A variety of medicinal plants, including *E. ribes*, *E. officianalis*, *Gymnema sylvetre*, and *Curcoma longa*, have hypoglycemic properties (Bhandari et al., 2007; 2008). The aqueous extract of the fruit of *E. ribes* shows significant antidiabetic activity in diabetic rats, and is orally administered at 100–200 mg/kg for 40 days. It effectively reduces blood glycosylated hemoglobin, blood glucose, creatine kinase, systolic blood pressure, heart rate, and serum lactate while significantly increasing blood glutathione levels in streptozotocin-induced diabetic rats. The orally administered ethanolic extract of *E. ribes* fruits for six weeks at 100 and 200 mg/kg concentration effectively decreases blood glucose level, systolic blood pressure, and heart rate in streptozotocin-induced rats. Also, it decreases pancreatic thiobarbituric acid in diabetic rat pancreatic tissue (Nazam Ansari et al., 2008).

6.8.4 Hepatoprotective

Liver damage and associated diseases are extremely dangerous and necessitate the use of an effective therapeutic agent as well as medical treatment. Herbal formulations, medicinal plant-based agents, and traditional approaches are used to treat liver diseases. Several studies have been conducted to investigate the hepatoprotective effects of various medicinal plants. In mouse models, an ethanolic extract of *E. ribes* shows significant hepatoprotective effects against paracetamol-induced liver damage. It effectively reduces serum glutamate pyruvate transaminase in a concentration-dependent manner. *E. ribes* extract has hypolipidemic activity and reduces serum lipids. Histopathological studies revealed significant reductions in fat deposits and average liver weight. Treatment with *E. ribes* extracts at doses ranging from 25 to 200 mg/kg per day results in a significant reduction in SGPT (serum glutamic pyruvic transaminase) levels in paracetamol-induced rats (Swamy et al., 2007).

6.8.5 Antifertility

Since ancient times, a variety of medicinal plants and their extracts have been widely used as effective antifertility agents in folk medicine systems all over the world. A variety of herbal formulations are commonly used in contraception to suppress ovulation and fertilization in women effectively. In India, *E. ribes* is used in ayurvedic contraceptive formulations. The extract of *E. ribes* has effective antispermatogenic activity, and bioactive compounds such as embelin and 2, 5 dihydroxy-3-undecyl-p-benzoquinone have antifertility effects (Khan et al., 2010). The powder made from the root of *E. ribes* has a strong anti-pregnancy effect in female albino rats. At a 4 g/day concentration, the powder of *E. ribes* fruits has an antifertility effect of more than 50%. The addition of embelin at a concentration of 50–100 mg/kg for 7 days results in an anti-implantation activity of 85.71% (Kumar et al., 2012). In another study, embelin supplementation significantly increases acids and alkaline phosphate levels in the testis and prostate glands. A mixture of *E. ribes*, *Piper longum*, and borax is effective as a contraceptive (Prakash and Mathur, 1979). Ethanol and benzene extracts of *E. ribes* significantly increased glycogen and non-protein content, methanol, petroleum ether; chloroform extract of *E. ribes* fruits exhibits effective antifertility activity. The fruits of *E. ribes* have potent antifertility properties, influencing the motility, quality, and quantity of sperm in male bonnet monkeys while also lowering hormone levels. The bioactive compound isolated from fruits of *E. ribes* effectively alters the testicular histopathology (Chitra et al., 1994).

6.8.6 Antitumor

Natural products have played an essential role in advancing anticancer drugs in recent years due to their powerful pharmacological effects. Several plant-derived anticancer agents, such as paclitaxel, camptothecin, indole alkaloids, vinca alkaloid, and their derivatives, are currently being tested in clinical trials as effective anticancer agents (Paul et al., 2010; Kuno et al., 2012). Traditional chemotherapy, radiotherapy, and surgery are commonly used in cancer treatment, despite having adverse side effects. As a result, there is a need to develop potential bioactive molecules that have enhanced efficiency against cancer from medicinal plants, which is already widely practiced but causes fewer side effects. Embelin, a major bioactive molecule found in the fruits of *E. ribes*, has been shown to have potent anticancer properties against cancer cell lines, effectively decreasing tumor size and inhibiting the function of serum enzymes such as t-glutamyl transferase, acid phosphatase, lactate dehydrogenase, and aldose in rats with fibrosarcoma. Embelin strongly influences carbohydrate and amino acid metabolism in tumor-bearing rats (Joya and Lakshmi, 2010). Hexane extract of E. ribes fruits has strong anticancer activity against leukemic and lymphoma ascites cells. The experimental studies revealed that embelin has an anticancer effect on these two cells. Several studies have found that Embelia species have potent anticancer properties due to phytoconstituents like embelin. Chitra et al. (1994) discovered that embelin significantly inhibits fibrosarcoma cell proliferation in albino rats, while also increasing their survival. Dai et al. (2011) reported that embelin has anticancer properties against chemical carcinogen-induced colon cancer and that an *in vitro* study with methanolic extract of *E. ribes* inhibits 50% of the proliferation of MCF-7 cells (Suvarna, 2014). According to various studies, the major bioactive compound embelin has significant anticancer activity against fibrosarcoma in albino mice (Seshadri et al., 1980).

6.9 TOXICOLOGICAL EFFECT OF *E. RIBES*

Though *E. ribes* has a variety of pharmacological activities, it has no toxic effect on various cell types. The alcoholic and aqueous extracts of *E. ribes* fruits have no toxic effect on the reproductive organs of rats (Low et al., 1985). Whereas the treatment of embelin in chicks shows some defects in visual behavior, a high dosage of anthelmintics influences the visual ability in chicks. Embelin is responsible for the visual defects, as indicated by the difficulty in distinguishing between feed grains and pebbles. In animal models, treatment with a low dose of embelin resulted in no lesions. The study with embelin caused some pathological changes in perinuclear vacuolation after six weeks of short-term toxicity. The use of seed and fruit extracts in ayurvedic contraception results in low birth weight as well as defects in the development of soft tissues and the skeleton (Chaudhury et al., 2001).

6.10 CHEMICAL CONSTITUENTS OF *GYMNEMA SYLVESTRE*

The phytoconstituents of *Gymnema sylvestre* include soluble resin, tartaric acid, stigmasterol, saponins, calcium oxalate, gurmarin, quercitol, and amino acid derivatives, which can reduce hyperglycemia in both animals and humans (Liu et al., 2004; Ye et al., 2001). Gymnemic acid is abundant in the aerial parts of *G. sylvestre*; structurally, Gymnemic acid contains tri-terpinoids as shown in Figure 6.4, combined with a number of fatty acids and glucuronic acid. The HPLC analytical method was used to analyze the extract of refined *G. sylvestre*, and the results show that Gymnemic acid is a major bioactive compound found in the plant (Kanetkar et al., 2004). The primary chemical constituents of *G. sylvestre* extracts were identified as gurmarin, gymnemic acid, calcium oxalate, stigmasterol, choline, betaine, and tartaric acid. The presence of a water-soluble acidic fraction of

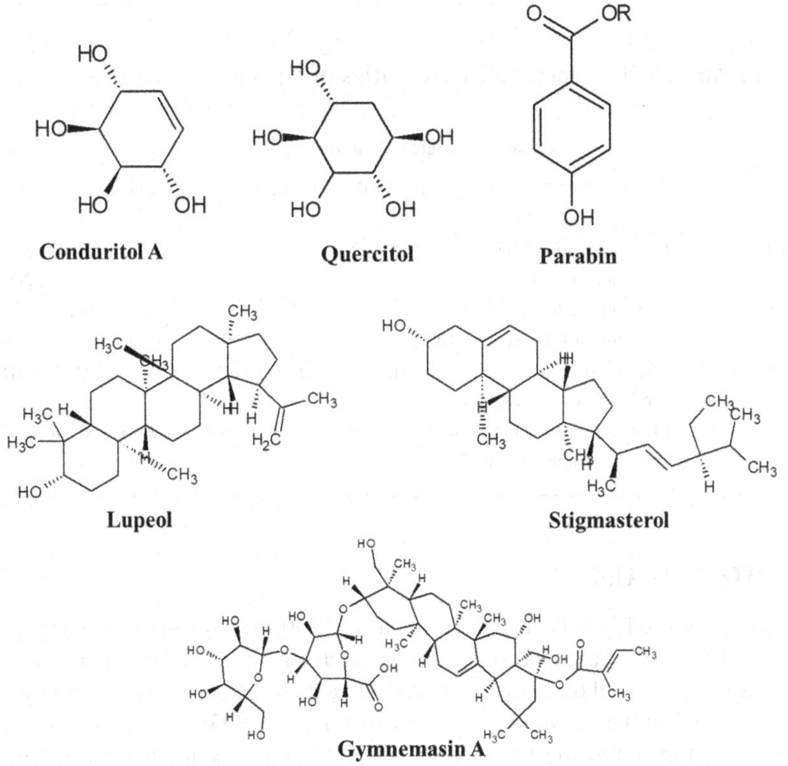

FIGURE 6.4 Represents the important bioactive components present in *Gymnema sylvestre*.

E. ribes has been shown to have effective hypoglycemic action. Several studies have found that gymnemic acids are the most abundant bioactive compound in *G. sylvestre*; however, more research is needed to back this up (Kanetkar et al., 2004). Gumarin, which is found in the leaves of *G. sylvestre*, is another important compound that inhibits the sweet taste in humans (Persaud et al., 1999). The hypoglycemic effect of *G. sylvestre* leaf extract was first reported in 1920; treatment with this leaf extract significantly increased insulin levels in healthy volunteers. This evidence is supported by animal studies, which show that the action of this extract may influence the regeneration capacity of pancreatic cells to secrete insulin. The *in vivo* studies also revealed that the supplementation of gymnemic extract improves glucose uptake into cells and effectively prevents the production of adrenaline from the liver, lowering blood glucose levels (Preuss et al., 2004). *G. sylvestre* extract has also been found to contain quercitol, lupeol, stigmasterol, and amyrin, in addition to the major bioactive compound. A series of gymnemic acids (GAI, GA II, III, IV) have also been reported in leaf extract, which inhibits the sweet taste. All of these compounds have glucuronic acid moiety at C-21 and C-28. Another series of gymnemic acid, such as V-VII, contain a 3-O-glucuronide of gymnemagenin and 3-O-glycuronyl-22, 21-bis-O-tigloyl substitution pattern. The third series of gymnemic acid VIII-IX are saponin derivatives that have oxoglycoside moiety on the glucuronic acid residue. Gurmarin is the major component present in leaves, which is also responsible for blocking sweet taste in humans (Kerry, 2007). Several acylated derivatives of gymnemagenin, such as tigloyl, methyllbutyroyl, and deceylgymnemic acid, have also been reported in *G. sylvestre*. The leaf extract also contains triterpenes and saponins from the oleanane (gymnemic acid, gymnemasaponins) and dammarene (gymnemasides) classes, as well as other plant components such as flavones, hentriacontane, anthraquinones, and chlorophylls phytin, resins, formic acids, and butyric acid (Senthilkumar, 2015). Table 6.2 discusses the various pharmacological activites of *Gymnema sylvestre*.

TABLE 6.2
Represents the Various Pharmacological Activities of *Gymnema sylvestre*

Plant Extract/ Compound	Pharmacological Activity	References
Water extract	Antidiabetic effect, antioxidant activity, anti-allergic	(Sathya et al., 2008; Ahmed et al., 2010
Alcoholic extract	Hypolipidemic activity	(Shigematsu et al., 2001)
Acetone extract	Larvicidal activity	Elumalai et al., 2013)
Methanol extract	Larvicidal activity	Elumalai et al., 2013)
Gymnemic acids	Decrease blood glucose level	(Shimizu et al., 1997)
Dihydroxy gymnemic	Effectively inhibits the production of sucrase, maltase, and pancreatic α-amylase	(Shenoy et al., 2018)
Triterpene glycoside	Effectively inhibits the expression of glucose-stimulated gastric-inhibitory peptide	(Fushiki et al., 1992)

6.11 TRADITIONAL USE

G. sylvestre has been used in folk medicine for nearly 2000 years to treat diabetes. *G. sylvestre* leaves were first shown to lower blood sugar levels effectively in 1920. This plant's use in the treatment of diabetes has increased dramatically in recent years. *G. sylvestre* leaf extract has been used in both ayurvedic and Indian traditional systems (Nadkarni, 1993). *G. sylvestre* leaves are commonly used in Indian medicine to treat stomach ailments, water retention, liver diseases, and constipation. This plant is known as a "destroyer of sugar" in traditional medicine because it effectively eliminates the sweet taste in humans (Vaidyaratnam, 1995). For over 1000 years, *G. sylvestre* has been used as a non-toxic remedy for various ailments, including digestion, urinary tract infection, allergies, cholesterol, hyperactivity, diabetes, and anemia (Chopra et al., 1992; Joy and Thomas, 1998). In India, *G. sylvestre* is frequently used to prepare formulations for modulating taste, sweet taste sensation, and as an anti-allergic, antiviral, and lipid-lowering agent, as well as for the treatment of diabetes mellitus (Ekka and Dixit, 2007).

6.12 PHARMACOLOGICAL ACTIONS

6.12.1 Application in Diabetes Mellitus

G. sylvestre has been used for a variety of pharmacological purposes. It has, however, been widely used to lower blood sugar levels. The extract of *G. sylvestre* root and leaf and their bioactive molecules can bind with sugar molecular acceptors 20 times better than dextrose. The sugar acceptor molecule of the small intestine cannot absorb sugar molecules, whereas *G. sylvestre* treatment can highly occupy sugar molecules, lowering blood sugar levels. In the diabetic condition, where β–cells have been damaged, the supplement of *G. sylvestre* in this condition helps β–cell generation, reduces the symptoms of diabetic mellitus, and effectively reduces blood sugar in both infant and adult diabetics (Sugihara et al., 2000). Tea prepared with *G. sylvestre* leaves effectively suppresses glucose absorption and may provide significant health assistance to persons who need to diminish their blood sugar levels. *G. sylvestre* is commonly grown in tropical regions and widely used in various herbal preparations (Spasov et al., 2008). Intake of leaf extract of *G. sylvestre* significantly increases the insulin level via enhancing the regeneration capacity of β–cells in the pancreas and also stimulates the glucose uptake into cells by raising the expression of glucose-utilizing enzymes and effectively suppresses the secretion of adrenaline from the liver (Aralelimath and Bhise, 2012). The bioactive compound gymnemic acid in *G. sylvestre* leaf extract suppresses and neutralizes sweet cravings, inhibits hyperglycemia, and acts as a critical stimulant in cardiovascular diseases (Baskaran et al., 1990). *G. sylvestre* has been used in primary clinical applications to treat both type I and type II diabetes since 1930 due to its ability to control blood sugar levels without causing harm. It can be used safely in conjunction with the regimen to promote proper pancreatic function.

6.12.2 Anticancer Activity

Several studies have reported *G. sylvestre* anticancer activity due to its main constituent gymnemagenol, which has a potential anticancer effect on HeLa cells (Khanna, 2010). The ethanolic and chloroform extracts have antiproliferative properties against the A549 and MCF7 cell lines. This study revealed that both extracts show similar impact at the same IC_{50} value against MCF and in lung cancer cells; chloroform extract shows more activity than ethanolic extract (Srikanth et al., 2010). *G. sylvestre* ethanolic extract has potent activity against human skin melanoma and skin papilloma models but has no toxic effect on normal liver cell lines (Chakraborty et al., 2013); it also has a significant inhibitory effect on breast cancer resistant protein (BCRP). The administration of *G. sylvestre* flavonoids effectively suppresses BCRP expression, which improves the activity of BCRP substrates methotrexate, topotecan, epirubicin, and daunorubicin, resulting in increased systemic absorption (Tamaki et al., 2010; Imai et al., 2004; Mao, 2005). Another study, with ethanolic extract, shows significant antiproliferative activity on mice with IC_{50} value of 50–55 nmol/ear. The identification and isolation of polysaccharides from *G. sylvestre*, such as GSP11, GSP22, GSP33, GSP44, and GSP55 also had potential anticancer activity via modulating the immunological function by raising the process of phagocytosis, improving serum hemolysin levels, and thymus and spleen indexes. Among the five types of polysaccharides, GDP11 and GP33 inhibit the proliferation of U937 cells with potent inhibitory activity, whereas GSP33 showed effective inhibitory activity against SGC cells with 78% inhibitory rate (Wu et al., 2012). In another study, methanolic extract of *G. sylvestre* shows significant anticancer activity in Swiss albino mice where papilloma genesis was prompted by 7, 12-dimethylbenz (a) anthracene (DMBA). The administration of methanolic extract of *G. sylvestre* effectively decreases tumor incidence and cumulative number of papillomas (Agarwal et al., 2016).

6.12.3 Lipid-Lowering Activity

G. sylvestre ethanolic extract has potent activity against human skin melanoma and skin papilloma models, but has no toxic effect on normal liver cell lines (Chakraborty et al., 2013). It also has a significant inhibitory effect on breast cancer resistant protein (BCRP). The administration of flavonoids from *G. sylvestre* effectively suppresses BCRP expression, which improves the activity of BCRP substrates methotrexate, topotecan, epirubicin, and daunorubicin, resulting in increased systemic absorption. Several studies have revealed *G. sylvestre* hypolipidemic activity. The administration of a *G. sylvestre* leaf extract to Wister female rats significantly reduced cholesterol, low-density lipoprotein, and triglyceride levels while increasing HDL levels (Singh et al., 2017). *G. sylvestre* hydroalcoholic leaf extract also significantly lowers LDL. The study was conducted with supplements of a higher level of cholesterol, LDL, triglyceride, and effectively lowering the level of HDL for seven days, after which the mice were treated with *G. sylvestre* leaf extract, which results in lowering the level of triglycerides, LDL, and cholesterol and also increasing the level of HDL due to the presence of chemical constituents such as saponins, tannins, and flavonoids (Rachh et al., 2010; Dholi and Raparla, 2014). Similar results were reported in several studies with diabetic rats (Bishayee and Chatterjee, 1994; Kumar et al., 2013).

6.12.4 Antimicrobial Activity

Antimicrobial properties of *G. sylvestre* were reported in several studies with different extracts and bioactive compounds against both gram-negative and gram-positive. Methanolic extract of leaf of *G. sylvestre* shows significant inhibition on *E. coli, C. albicans, B. cereus, and C. kefyr.* The aqueous extract exhibits moderate inhibitory activity against *C. perfringens, S. aureus, and C. Kefyr.* And the hexane extract shows significant activity against *H. paragallinarum, S. entrica S. aureus, S. enterica*, and *C. perfringens* type-A (David and Sudarsanam, 2013; Tahir et al., 2017). A study done with aqueous and ethanol extract reported an effective antibacterial effect against pathogenic microorganisms such as *Salmonella typhi, S. typhimurium*, and *S. paratyphi.* The ethanolic extract, ethyl acetate, and chloroform extracts also reported potential inhibitory activity against *E. coli, P. aeruginosa, K. pneumoniae, P. vulgaris*, and *S. aureus* (Pasha et al., 2009; Paul and Jayapriya, 2009. The study carried out by Swami and Prabakaran (2012) reported that *G. sylvestre* has good inhibitory effect against both gram-negative and gram-positive bacteria including *K. pneumoniae, E. coli, S. aureus*, and *P. aeruginosa.* The bioactive compound isolated from *G. sylvestre*, gymnemic acid, shows significant activity against *V. cholera, S. mutans, A. niger, C. albicans*, and *S. aureus* with a zone of inhibition ranging from 6.00mm to 9.25mm (Gupta and Singh, 2014). Thanwar et al. (2016) reported the antimicrobial effect of *G. sylvestre* on both gram-positive and gram-negative bacteria and the compound gymnemic acid and a triterpene show very active against microorganisms such as *E. coli and B. aureus* (Shivanna and Raveesha, 2009). Ramalingam et al. (2019) revealed that the leaf extract of *G. sylvestre* enhances the microbial activity with poly-ε-caprolactone nanofibers. In another study, ZnO nanoparticles were synthesized with *G. sylvestre* extract and found effective against a wide range of bacterial strains (Karthikeyan et al., 2019). The nanoformulation with poly-ε-caprolactone nanofibers and *G. sylvestre* effectively inhibit the growth of methicillin-resistant *S. aureus, S. epidermidis, E. coli*, and *P. aeruginosa* (Ramalingam et al., 2019).

6.12.5 Antioxidant Activity

Antioxidants are the essential molecules or substances that highly influence the reduction of oxidative stress in a cell by which they help treat several health issues, including cardiovascular disease, cancer, inflammatory diseases, and diabetes. Antioxidants potentially inhibit the process of oxidation. The chemical constituents of medicinal plants act as radical scavengers including butylated hydroxytoluene (BHT) and butylated hydroxyanisole (BHA) (Mandal et al., 2009). The ethanol extract of *G. Sylvestre* shows substantial antioxidant activity, effectively scavenging 1,1-diphenyl-2-picrylhydrazyl (DPPH) than other free radicals BHT and significantly reducing LDL oxidation (Rahman et al., 2014; Rupanar et al., 2012; Ohmori et al., 2005). Another study reported that *G. Sylvestre* shows potential antioxidant activity on hydroxyl free radicals at 59.8% and DPPH at 87.3% (Gunasekaran et al., 2019). It also significantly scavenges the ferric radical, superoxide radical, and hydrogen peroxide (Rachh et al., 2009).

6.12.6 Antiarthritic Activity

A numbers of studies have stated that the antiarthritic activity of aqueous and petroleum ether extracts of *G. sylvestre* shows significant antiarthritic activity, and it reduces the discharge of inflammatory mediators that are crucial for bone destruction in antiarthritic complaint (Malik et al., 2010). Shankar and Rao (2008) reported that the ethanolic extract of the root of *G. sylvestre* effectively reduced paw edema and inhibited histamine-induced paw edema up to 39–75 %.

6.12.7 Immunomodulating Effect of *G. sylvestre*

In addition to antioxidant, anticancer, and antiarthritic activity, *G. sylvestre* plant extract shows a significant immunomodulatory effect. Methanolic leaf extract revealed effective immunosuppressive activity in albino mice after treatment. Various parameters, like hemagglutination, antibody titer, and delayed-type hypersensitivity were evaluated, and the other immune modulators such as

B-lymphocytes, interleukins 2 and 4, IFN-γ, and Th2 cytokines are estimated via flow cytometer. After treatment with plant extract, the expression of primary and secondary antibodies is significantly reduced with increasing concentrations of plant extract. At higher dose (200 mg/kg), it was observed that maximal reduction was noticed in the production of CD3, CD19, IL-2, and 4, IFN-γ (Ahirwal et al., 2015). It also promotes the level of myeloid and lymphoid components. In another study, it was described that the methanolic extract of *G. sylvestre* effectively increased the induction of free radicals, such as nitric oxide and reactive oxygen species (ROS) via stimulating the function of macrophages (Singh et al., 2015). The aqueous extract also significantly enhances the activity of phagocytosis in human neutrophils, by which it modulates the immunostimulatory effect (Jitender et al., 2009). Ethanol extract of *G. sylvestre* also improves the immunosuppressed condition via induction of cyclophosphamide in albino mice: after treatment with plant extract, it was observed that the activity of hemagglutination titer and phagocytosis significantly increased the paw edema (Kar et al., 2019).

6.12.8 Anti-Inflammatory Activity

Since ancient times, medicinal plants have been extensively used to treat various diseases; in India, several thousand plant species have various pharmacological activities. *G. sylvestre* also reported potential anti-inflammatory activity owing to the number of bioactive components present in it. The methanolic extract shows a significant anti-inflammatory effect in Wistar rats that have carrageenan-induced inflammation, and the results revealed that the methanolic extract effectively reduces rat paw edema in peritoneal ascites in mice (Diwan et al., 1995; Kumar et al., 2012). The ethanolic extract also shows substantial anti-inflammatory activity against TPA-induced inflammation with an IC_{50} value of 50–55 nmol/ear (Yasukawa et al., 2014).

6.12.9 Hepatoprotective Activity

The hepatoprotective activity of *G. sylvestre* for both cell line and animal model research has been extensively reported. Srividya et al. (2010) stated that the hydroalcoholic extract of *G. sylvestre* shows substantial hepatoprotective effect, where hepatocytes of rats were administered with different concentrations of the hydroalcoholic extract. The cells showed a substantial restoration and had changes in biochemical parameters in a dose-dependent manner. Dholi and Raparla (2014) reported the low level of urea and creatinine after treatment of methanolic extract of *G. sylvestre* in Wister rats that were experiencing acute and chronic inflammation. The presence of a polyherbal preparation with *G. sylvestre* shows reverse hepatotoxicity in albino rats (Yogi and Mishra, 2016).

6.13 TOXICOLOGICAL EFFECT OF *GYMNEMA SYLVESTRE*

The toxicological effect of *G. sylvestre* was investigated to show its safety and recommended doses. The high dose of *G. sylvestre* shows moderate toxic effects and also causes side effects such as excessive sweating, weakness, shakiness, and muscular dystrophy. In another study, the supplement of *G. sylvestre* – powder in the diet of Wistar rats – for 52 weeks shows no toxic effects or death (Ogawa et al., 2004). In the case of diabetic patients, administration of *G. sylvestre* shows toxic hepatitis and liver injury (Shiyovich et al., 2010). The *in vivo* model with albino mice treated with *G. sylvestre* showed LD50 level at 3900 mg/kg with no behavioral, autonomic, neurologic, or harmful effects. Another study reported the LD50 value at 375mg/kg in mice treated with *G. sylvestre* via the intraperitoneal route. A single case of drug-induced liver injury was observed in diabetic patients treated with *G. sylvestre* (Shiyovich et al., 2010). Ogawa et al. (2004) reported that the treatment of both male and female Wistar rats with *G. sylvestre* extract for 52 weeks has shown no toxic effects.

6.14 CONCLUSION

Medicinal plants and their bioactive components have several pharmacological properties and have been widely used to treat cancer, cardiovascular diseases, neurological disorders, inflammation, diabetes, and microbial infection. The major bioactive components like flavonoids, alkaloids, tannins, and phenolic compounds play a crucial role in treating the diseases mentioned above. It has been reported that about 80% of people extensively use natural medicine worldwide to treat primary health complaints (Hamilton, 2004). However, only 10% of the plants have great interest in treatment and are investigated for their therapeutic potential. They act as a good source of novel bioactive molecules but have limits due to their habitat destruction and unsustainable use (Brower, 2008). The Indian traditional plants *G. sylvestre* and *E. ribes* are therapeutically important plant species that contain several biologically active molecules, including saponin, tanins, alkaloids, gymnemic acid, embelin, gurmarin, gymnemanol, etc., that can provide a wide range of pharmacological activities. However, owing to their unsustainable use, over-exploitation, and deforestation, both plants are disappearing very fast, and a number of unauthorized preparations of these plants are available in the local market of many countries. People are utilizing these plants as a substitution for antidiabetic medicine, which then leads to the destruction of the plants. Hence, the preservation and sustainable use of these plants are monitored very strictly.

ACKNOWLEDGMENTS

The author, CS, and SKS thankfully acknowledge the Tamil Nadu State Council for Higher Education (TANSCHE) for the research grant (Au/S.o. (P&D): TANSCHE Projects: 117/2021).

Declaration of Interest: The authors declare there is nothing to declare with this chapter.

REFERENCES

Agrawal, R.C., Soni, S., Jain, N., Rajpoot, J., and S.K. Maheshwari. 2016. Chemopreventive effect of *Gymnema sylvestre* in Swiss albino mice. *International Journal of Scientific and Research Publications*. 6:78–83.

Ahirwal, L., Singh, S., Kumar, M.D., Bharti, V., Mehta, A., and S. Shukla. 2015. *In vivo* immunomodulatory effects of the methanolic leaf extract of Gymnema sylvestre in Swiss albino mice. *Archives of Biological Sciences*. 67:561–570.

Ahmed, A.B., Rao, A., and M. Rao. 2010. *In vitro* callus and in vivo leaf extract of Gymnema sylvestre stimulate β-cells regeneration and anti-diabetic activity in Wistar rats. *Phytomedicine*. 17:1033–1039.

Ansari, M.N., and U. Bhandari. 2008. Antihyperhomocysteinemic activity of an ethanol extract from *Embelia ribes* in albino rats. *Pharmaceutical Biology*. 46:283–7.

Aralelimath, V.R., and S.B. Bhise. 2012. Anti-diabetic effects of *Gymnema sylvestre* extract on streptozotocin induced diabetic rats and possible b-cell protective and regenerative evaluations. *Digest Journal of Nanomaterials and Biostructures*. 7:135–142.

Asadulla, S., and R. Ramandang. 2011. Pharmacognosy of *Embelia ribes* Burm f. *International Journal of Research in Pharmacy and Chemistry*. 1:1236–51.

Ashish, K.S., Gautam, M.K., Pradeep, T.D., Lokendra, S.K., Silawat, N., Akbar, Z., and M.S. Muthu. 2013. Effect of embelin on lithium-induced nephrogenic diabetes insipidus in albino rats. *Asian Pacific Journal of Tropical Disease*. 2013:S729–S733

Ashok, P., Keshav, B.V., and S.K. Vyas. 2008. Hypoglycaemic activity of *Embelia ribes* berries (50% etoh) extract in alloxan induced diabetic rats. *Ancient Science of Life*.27:41–4.

Atal, C.K., Siddiqui, M.A., Zutshi, U., Amla, V., Johri, R.K., Rao, P.G., and S. Kour. 1984. Non-narcotic orally effective, centrally acting analgesic from an Ayurvedic drug. *Journal of Ethnopharmacology*. 11:309–17.

Baskaran, K., Ahamath, B.K., Shanmugasundaram, K.R., and E.R.B. Shanmugasundaram. 1990. Antidiabetic effect of a leaf extract from *Gymnema sylvestre* in non-insulin-dependent diabetes mellitus patients. *Journal of Ethnopharmacology*. 30:295–305.

Bhandari, U., Kanojia, R., and K.K. Pillai. 2002. Effect of ethanolic extract of *Embelia ribes* on dyslipidemia in diabetic rats. *International Journal of Experimental Diabetes Research.* 3:159–62.

Bhandari, U., Nazam Ansari, M., and F. Islam. 2008. Cardioprotective effect of aqueous extract of *Embelia ribes* Burm fruits against isoproterenol- induced myocardial infarction in albino rats. *Indian Journal of Experimental Biology.* 46:35–40.

Bhandari, U., Neeti J., and K.K. Pillai. 2007. Further studies on antioxidant potential and protection of pancreatic beta-cells by *Embelia ribes* in experimental diabetes. *Experimental Diabetes Research.* 2007:1–6.

Bishayee, A., and M. Chatterjee. 1994. Hypolipidaemic and antiatherosclerotic effects of oral *Gymnema sylvestre* R. Br. Leaf extract in albino rats fed on a high fat diet. *Phytotherapy Research.* 8:118–120.

Brahmeshwari, G., and G. Kumaraswamy. 2012. Anti-bacterial activity of benzoxadiazines derived from Embelin. *IJPBS.* 2:284–287.

Brower, V. 2008. Back to nature: Extinction of medicinal plants threatens drug discovery. *Journal of the National Cancer Institute.* 100: 838–839.

Chakraborty, D., Ghosh, S., Bishayee, K., Mukherjee, A., Sikdar, S., and A.R. Khuda- Bukhsh. 2013. Antihyperglycemic drug *Gymnema sylvestre* also shows anticancer potentials in human melanoma A375 Cells *via* reactive oxygen species generation and mitochondria-dependent caspase pathway. *Integrative Cancer Therapies.* 12:433–441.

Chaudhury, M.R., Chandrasekaran, R., and S. Mishra. 2001. Embryotoxicity and teratogenicity studies of an ayurvedic contraceptive—pippaliyadi vati. *Journal of Ethnopharmacology.* 74:189–93.

Chauhan, S., Agrawal, S., Mathur, R., and R.K. Gupta. 1979. Phosphatase activity in testis and prostate of rats treated with embelin and *Vinca rosea* extract. *Experientia.* 35:1183–1185.

Chen, Y., Liu, Y., Sarker, M.M.R., Yan, X., Yang, C., and L. Zhao. 2018. Structural characterization and anti-diabetic potential of a novel heteropolysaccharide from Grifola frondosa *via* IRS1/PI3K-JNK signaling pathways. *Carbohydrate Polymer.* 198:452–461.

Chitra, M., Shyamala Devi, C.S., and E. Sukumar. 2003. Anti-bacterial activity of embelin. *Fitoterapia.* 74:401–403.

Chitra M, Sukumar E, and C.S. Shyamala Devi. 1995. [3H]-Thymidine uptake and lipid peroxidation by tumor cells on embelin treatment: An in vitro study. *Oncology.* 52:66–68.

Chitra, M., Sukumar, E., Suja, V., and S. Devi. 1994. Anti-tumor, anti-inflammatory and analgesic property of embelin, a plant product. *Chemotherapy.* 40:109–113.

Chopda, M.Z., and R.T. Mahajan. Wound healing plants of Jalgaon district of Maharashtra state, India. *Ethnobotanical Leaflets.* 2009:1.

Chopra, R.N., Nayar, S.L., and I.C. Chopra. 1992. *Glossary of Indian Medicinal Plants*, 3rd ed. New Delhi: Council of Scientific and Industrial Research, 319–322.

Christopoulos, M.V., Rouskas, D., Tsantili, E., and P.J. Bebeli. 2010. Germplasm diversity and genetic relationships among walnut (*Juglans regia* L.) cultivars and Greek local selections revealed by Inter-Simple Sequence Repeat (ISSR) markers. *Scientia Horticulturae.* 125:584–592.

Dai, Y., De Sano, J., and Y. Qu. 2011. Natural IAP inhibitor embelin enhances therapeutic efficacy of ionizing radiation in prostate cancer. *American Journal of Cancer Research.* 1:128–143.

Dang, P.H., Nguyen, H.X., Nguyen, N.T., Le, H.N., and M.T. Nguyen. 2014. α-Glucosidase Inhibitors from the Stems of *Embelia ribes. Phytotherapy Research.* 28:1632–1636.

David, B.C., and G. Sudarsanam 2013. Antimicrobial activity of *Gymnema sylvestre* (*Asclepiadaceae*). *Journal of Acute Disease.* 3:222–225.

Dholi, S.K., and R.K. Raparla. 2014. *In vivo* anti-diabetic evaluation of gymnemic acid in streptozotocin induced rats. *Journal of Pharmaceutical Innovation.* 3:82–86.

Diwan, P.V., Margaret, I., and S. Ramakrishna. 1995. Influence of *Gymnema sylvestre* on inflammation. *Inflammopharmacology.* 3:271–277.

Ekka, N.R., and V.K. Dixit. 2007. Ethno-pharmacognostical studies of medicinal plants of Jashpur district (Chhattisgarh). *International Journal of Green Pharmacy.* 1:1–4.

Elumalai, K., Dhanasekaran, S., and K. Krishnappa. 2013. Larvicidal activity of Saponin isolated from *Gymnema sylvestre* R. Br. (Asclepiadaceae) against Japanese Encephalitis vector, *Culex tritaeniorhynchus Giles* (Diptera: Culicidae). *European Review for Medical and Pharmacological Sciences.* 17:1404–1410.

Feresin, G.E., Tapia, A., Sortino, M., Zacchino, S., de Arias, A.R., and A. Inchausti. 2003. Bioactive alkyl phenols and embelin from Oxalis erythrorhiza. *Journal of Ethnopharmacology.* 88:241–247.

Fushiki, T., Kojima, A., Imoto, T., Inoue, K., and E. Sugimoto. 1992. An extract of *Gymnema sylvestre* leaves and purified gymnemic acid inhibits glucose- stimulated gastric inhibitory peptide secretion in rats. *Journal of Nutrition.* 122:2367–2373.

Gajjar, U.H., Khambholja, K.M., and R.K. Patel. 2009. Comparison of anti microbial activity of Bhallataka Rasayana and its ingredient. *International Journal of PharmTech Research.* 1:1594–1597.

Goto, T., Sarker, M.M.R., Zhong, M., Tanaka, S., and E. Gohda. 2010. Enhancement of immunoglobulin M production in B cells by the extract of red bell pepper. *Journal of Health Science.* 56:304–309.

Guhabakshi, D.N., Sensarma, P., and D.C. Pal. 2001. *A Lexicon Medicinal Plants of India.* Calcutta, India: Naya Prakashan, 135–136.

Gunasekaran, V., Srinivasan, S., and S.S. Rani. 2019. Potential antioxidant and antimicrobial activity of *Gymnema sylvestre* related to diabetes. *Journal of Medicinal Plants.* 7:05–11.

Gupta, P., and P. Singh. 2014. Antimicrobial activity of Gymnemic acid on pathogens- *Gymnema sylvestre* R.Br. *International Journal of Current Microbiology and Applied Sciences.* 3:40–45.

Gurnani, N., Mehta, D., Gupta, M., and B.K. Mehta. 2014. Natural products: Source of potential drugs. *African Journal of Basic & Applied Sciences.* 6:171–186.

Hamilton, A.C. 2004. Medicinal plants, conservation and livelihoods. *Biodiversity and Conservation.* 8:1477–1517.

Haq, K., Ali, M., and A.W. Siddiqui. 2005. New compounds from the seeds of *Embelia ribes* Burm. *Die Pharmazie-An International Journal of Pharmaceutical Sciences.* 60:69–71.

Hoareau, L., and E.J. DaSilva. 1999. Medicinal plants: A re-emerging health aid. *Electronic Journal of Biotechnology.* 2:3–4.

Hördegen, P., Cabaret, J., Hertzberg, H., Langhans, W., and V. Maurer. 2006. In vitro screening of six anthelmintic plant products against larval Haemonchus contortus with a modified methyl-thiazolyl-tetrazolium reduction assay. *Journal of Ethnopharmacology.* 108:85–89.

Imai, Y., Tsukahara, S., Asada, S., and Y. Sugimoto. 2004. Phytoestrogens/ flavonoids reverse breast cancer resistance protein/ABCG2-mediated multidrug resistance. *Cancer Research.* 64:4346–4352.

Imam, H., Mahbub, N.U., Khan, M.F., Hana, H.K., and M.M.R. Sarker. 2013. Alpha amylase enzyme inhibitory and anti-inflammatory effect of Lawsonia inermis. *Pakistan Journal of Biological Science.* 16:1796–1800.

Indrayan, A.K., Sharma, S., Durgapal, D., Kumar, N., and M. Kumar. Determination of nutritive value and analysis of mineral elements for some medicinally valued plants from Uttaranchal. *Current Science.* 10:1252–5.

Jalalpure, S.S., Alagawadi, K.R., Mahajanashetti, C.S., Shah, B.N., Salahuddin, S.V., and J.K. Patil. *In vitro* anthelmintic property of various seed oils against *Pheritima posthuma. Indian Journal of Pharmaceutical Sciences.* 2007. 69:158–160.

Javed, I., and M.S. Akhtar. 1990. Screening of Veronica anthelmintica seed and *Embella ribes* fruit mixed in equal parts against gastrointestinal nematodes. *Pakistan Journal of Biological Science.* 3:69–74.

Jitender, K.M., Manvi, F.V., Nanjwade, B.K., Alagawadi, K.R., and S. Sanjiv. 2009. Immuno-modulatory activity of *Gymnema sylvestre* leaves extract on In vitro human neutrophils. *Journal of Pharmacy Research.* 2:1284–1286.

Joshi, R., Kamat, J.P., and T. Mukherjee. 2007. Free radical scavenging reactions and antioxidant activity of embelin: Biochemical and pulse radiolytic studies. *Chemico-Biological Interactions.* 167:125–134.

Joy, P.P., and J. Thomas. 1998. *Medicinal Plants.* Kerala Agriculture University, Aromatic and Medicinal Plants Research Station, 16.

Joya, B., and S. Lakshmi. 2010. Anti-proliferative properties of *Embelia ribes. Open Process Chemistry Journal.* 3:17–22.

Kanetkar, P., Singhal, R., and M. Kamat. 2007. *Gymnema sylvestre*: A memoir. *Journal of Clinical Biochemistry and Nutrition.* 41:77–81.

Kanetkar, P.V., Laddha, K.S., and M.Y. Kamat. 2004. Gymnemic acids: A molecular perspective of its action on carbohydrate metabolism. Poster presented at the 16th ICFOST meet organized byCFTRI and DFRL, Mysore, India.

Kapoor, L.D. 2017. *CRC Handbook of Ayurvedic Medicinal Plants.* Boca Raton, FL: CRC Press.

Kar, P.P., Rath, B., Ramani, Y.R., and C.S. Maharana. 2019. Amelioration of Cyclophosphamide induced immunosupression by the hydroalcoholic extract of *Gymnema sylvestre* leaves in albino rats. *Biomedical and Pharmacology Journal.* 11:251–258.

Karthikeyan, M., Ahamed, A.J., Karthikeyan, C., and P.V. Kumar. 2019. Enhancement of anti-bacterial and anti-cancer properties of pure and REM doped ZnO nanoparticles synthesized using *Gymnema sylvestre* leaves extract. *SN Applied Sciences.* 1:355.

Kerry, B. 2007. *Gymnema: A Key Herb in the Management of Diabetes. (Phytotherapy Review & Commentary).* National Institute of Herbalists, National Herbalists Association of Australia. www.mediherb.com.au

Khan, M.A., Naidu, M.A., and Z. Akbar. 2010. In-vitro antimicrobial activity of fruits extract of *Embelia ribes* Burm. *International Journal of Pharmaceutical and Biological Archives*. 1:267–270.

Khanna, G. 2010. Non-proliferative activity of saponins isolated from the leaves of *Gymnema sylvestre* and *Eclipta prostrata* on HepG2 cells-In vitro study. *International Journal of Pharmaceutical Sciences and Research*. 1:38–42.

Khare, C.P. 2004. Indian herbal remedies. In *Rational Western Therapy, Ayurvedic and Other Traditional Usage*. New York: Springer, 200–201.

Kirtikar, K.R., and B.D. Basu. 1975. Indian medicinal plants. In *Periodical Experts D-42*, vol. III. New Delhi, India: Vivek Vihar.

Kumar, A.R., Rathinam, K.S., and C.A. Kumar. 2012. Evaluation of anti-inflammatory activity of some selected species of Asclepiadaceae family. *International Journal of Chemical Sciences*. 10:548–556.

Kumar, D., Kumar, A., and O. Prakash. 2012. Potential antifertility agents from plants: A comprehensive review. *Journal of Ethnopharmacology*. 140:1–32.

Kumar, V., Bhandari, U., Tripathi, C., and G. Khanna. 2013. Anti-obesity effect of *Gymnema sylvestre* extract on high fat diet-induced obesity in wistar rats. *Drug Research*. 63:625–632.

Kuno, T., Testuya, T., and H. Akira. 2012. Cancer chemoprevention through the induction of apoptosis by natural product. *Journal of Biophysical Chemistry*. 3:156–173.

Latha, C. 2007. Microwave-assisted extraction of embelin from *Embelia ribes*. *Biotechnology Letters*. 29:319–322.

Lin, P., Li, S., Wang, S., Yang, Y., and J. Shi. 2006. A nitrogen-containing 3-alkyl-1, 4-benzoquinone and a gomphilactone derivative from *Embelia ribes*. *Journal of Natural Products*. 69:1629–1632.

Liu X, Ye W, Yu B, Zhao S, Wu H, and C. Che. 2004. Two new flavonol glycosides from *Gymnema sylvestre* and Euphorbia ebracteolata. *Carbohydrate Research*. 339(4):891–895.

Low, G., Rogers, L.J., Brumley, S.P., and D. Ehrlich. 1985. Visual deficits and retinotoxicity caused by the naturally occurring anthelmintics, *Embelia ribes* and Hagenia abyssinica. *Toxicology and Applied Pharmacology*. 81:220–230.

Mahendran, S., Thippeswamy, B.S., Veerapur, V.P., and S. Badami. 2011. Anticonvulsant activity of embelin isolated from *Embelia ribes*. *Phytomedicine*. 18:186–188.

Malik, J.K., Manvi., F.V., Nanjware, B.R., Dwivedi, D.K., Purohit, P., and S. Chouhan. 2010. Anti-arthritic activity of leaves of *Gymnema sylvestre* R.Br. leaves in rats. *Pharmacist's Letter*. 2:336–341.

Mandal, S., Yadav, S., Yadav, S., and R.K. Nema. 2009. Antioxidants: A Review. *Journal of Chemical and Pharmaceutical Research*. 1:102–104.

Mao, Q. 2005. Role of the breast cancer resistance protein (ABCG2) in drug transport. *AAPS J*. 7:E118–E133.

McErlean, C.S., and C.J. Moody. 2007. First synthesis of N-(3-carboxylpropyl)-5-amino-2-hydroxy-3-tridecyl-1, 4-benzoquinone, an unusual quinone isolated from *Embelia ribes*. *Journal of Organic Chemistry*. 72:10298–10301.

Modak, M., Dixit, P., Londhe, J., Ghaskadbi, S., and T.P. Devasagayam. 2007. Indian herbs and herbal drugs used for the treatment of diabetes. *Journal of Clinical Biochemistry and Nutrition*. 40:163–173.

Mohandas, S., Sreekumar, T.R., and V. Prakash. 2013. Anthelmintic activity of vidangadi churna. *Asian Journal of Pharmaceutical and Clinical Research*. 6:94–95.

Nadkarni, A.K. 1982. Indian material medica, popular prakashan pvt ltd. Bombay, India.1:1199.

Nadkarni, K.M. 1993. *Indian Materia Medica*. Bombay: Popular Prakashan, 596–599.

Najafi, S., and S.S. Deokule. 2011. Studies on *Gymnema sylvestre*-a medicinally important plant of the family Asclepiadaceae. *Trakia Journal of Sciences*. 9:26–32.

Nazam Ansari, M., and U. Bhandari. 2008. Effect of an ethanol extract of *Embelia ribes* fruits on isoproterenol-induced myocardial infarction in albino rats. *Pharmacutical Biology*. 46:928–932.

Nazam Ansari, M., Bhandari, U., Islam, F., and C.D. Tripathi. 2008. Evaluation of antioxidant and neuroprotective effect of ethanolic extract of *Embelia ribes* Burm in focal cerebral ischemia/reperfusion-induced oxidative stress in rats. *Fundamental & Clinical Pharmacology*. 22:305–14.

Ogawa, Y., Sekita, K., and T. Umemura. 2004. *Gymnema sylvestre* leaf extract: A 52-week dietary toxicity study inwistar rats. *Shokuhin Eiseigaku Zasshi*. 45:8–18.

Ohmori, R., Iwamoto, T., Tago, M., Takeo, T., Unno, T., and H. Itakura. 2005. Antioxidant activity of various teas against free radicals and LDL oxidation. *Lipids*. 40:849–853.

Pasha, C., Sayeed, S., Ali, M.S., and M.Z. Khan. 2009. Antisalmonella activity of selected medicinal plants. *Trakia Journal of Sciences*. 33:59–64.

Paul, G.G., Cragg, M.G, and N.J. David. 2010. Plant natural products in anticancer drug discovery. *Current Organic Chemistry*. 14:1781–1791.

Paul, J., and K.P. Jayapriya. 2009. Screening of anti-bacterial effects of *Gymnema sylvestre* (L.) R.Br.: A medicinal plant. *Pharmacology Online* 3:832–836.

Persaud, S.J., Majed, H.A., Raman, A., and P.M. Jones. 1999. *Gymnema sylvestre* stimulates insulin release in vitro by increased membrane permeability *Journal of Endocrinology.* 163:207–212.

Prakash, A.O., and R. Mathur. 1979. Biochemical changes in the rat uterine tissue following *Embelia ribes* burm. extracts. *Indian Journal of Pharmacology.* 11:127.

Pramanick, D.D. 2016. Anatomical studies on the leaf of Gymnema sylvestre (Retz.) R. Br. ex Schult. (Apocynaceae), in A magical herbal medicine for diabetes. *International Journal of Herbal Medicine.* 4:70–72.

Preuss, H.G., Bagchi, D., Bagchi, M., Rao, C.V., Dey, D.K., and S. Satyanarayana. 2004. Potential additive effects of garcinia cambogia on atorvastatin treated hyperlipidemic patients: Randomized crossover clinical study. *Diabetes, Obesity and Metabolism.* 6:171–180.

Rachh, P., Rachh, M., Ghadiya, N., Modi, D., Modi, K., and N. Patel. 2010. Antihyperlipidemic activity of Gymenma sylvestre R. Br. leaf extract on rats fed with high cholesterol diet. *International Journal of Pharmacology.* 6:138–141.

Rachh, P.R., Patel, S.R., Hirpara, H.V., Rupareliya, M.T., Rachh, M.R., and A.S. Bhargava. 2009. In-vitro evaluation of antioxidant activity of *Gymnema sylvestre* R.Br. leaf extract. *Romanian Journal of Biology.* 54:141–148.

Radhakrishnan, N., Gnanamani, A., and A.B. Mandal. 2011. A potential anti-bacterial agent Embelin a natural benzoquinone extracted from *Embelia ribes. Biological Medicine.* 3:1–7.

Rahman, M.M., Habib, M.R., Hasan, M.A., Saha, A., and A. Mannan. 2014. Comparative assessment on In vitro antioxidant activities of ethanol extracts of *Averrhoa bilimbi, Gymnema sylvestre* and *Capsicum frutescens. Pharmacogency Research.* 6:36–41.

Rai, R., and V. Nath. 2005. Use of medicinal plants by traditional herbal healers in Central India. *Indian Forester.* 131:463–8.

Raja, S.S., Unnikrishnan, K.P., Ravindran, P.N., and I. Balach. 2005. Determination of embelin in *Embelia ribes* and Embelia tsjeriam-conttam by HPLC. *Indian Journal of Pharmaceutical Sciences.* 67:734.

Rajakumar, N., and M.B. Shivanna. 2009. Ethno-medicinal application of plants in the eastern region of Shimoga district, Karnataka, India. *Journal of Ethnopharmacology.* 126:64–73.

Ramalingam, R., Dhand, C., Leung, C.M., Ong, S.T., Annamalai, S.K., and M. Kamruddin. 2019. Antimicrobial properties and biocompatibility of electrospun poly-ε-caprolactone fibrous mats containing *Gymnema sylvestre* leaf extract. *Materials Science and Engineering.* 98:503–514.

Reuter, S., Prasad, S., Phromnoi, K., Kannappan, R., Yadav, V.R., and B.B. Aggarwal. 2010. Embelin suppresses osteoclastogenesis induced by receptor activator of NF-κB ligand and tumor cells in vitro through inhibition of the NF-κB cell signaling pathway. *Molecular Cancer Research.* 8:1425–1436.

Rouhi, S.Z.T., Sarker, M.M.R., Rahmat, A., Alkahtani, S.A., and F. Othman. 2017. The effect of pomegranate fresh juice versus pomegranate seed powder on metabolic indices, lipid profile, inflammatory biomarkers, and the histopathology of pancreatic islets of Langerhans in streptozotocin-nicotinamide induced type 2 diabetic Sprague–Dawley rats. *BMC Complementary and Alternative Medicine.* 17:156.

Rupanar, S.V., Pingale, S.S., Dandge, C.N., and D. Kshirsagar. 2012. Phytochemical screening and in vitro evaluation of antioxidant antimicrobial activity of *Gymnema sylvestre. International Journal of Current Research.* 8:43480–43486.

Sabitha Rani, A., Saritha, K., Nagamani, V., and G. Sulakshana. 2011. *In vitro* evaluation of antifungal activity of the seed extract of *Embelia ribes. Indian Journal of Pharmaceutical Sciences.* 73:247–249.

Saikia, A.P., Ryakala, V.K., Sharma, P., Goswami, P., and U. Bora. 2006. Ethnobotany of medicinal plants used by Assamese people for various skin ailments and cosmetics. *Journal of Ethnopharmacology.* 106:149–57.

Sarker, M.M.R., and E. Gohda. 2013. Promotion of anti-keyhole limpet hemocyanin IgM and IgG antibody productions in vitro by red bell pepper extract. *Journal of Functional Foods.* 5:1918–1926.

Sathya, S., Kokilavani, R., and K. Gurusamy. 2008. Hypoglycemic effect of *Gymnema sylvestre* (retz.), R. Br leaf in normal and alloxan induced diabetic rats. *Ancient Science of Life.* 28:12–14.

Schrader, K.K. 2010. Plant natural compounds with anti-bacterial activity towards common pathogens of pond-cultured channel catfish (Ictalurus punctatus). *Toxins.* 2:1676–1689.

Senthilkumar, M. 2015. Phytochemical screening and anti-bacterial activity of *Gymnema sylvestre* R.Br. Ex Schult. *International Journal of Pharmaceutical Sciences and Research.* 6:2496–2503.

Seshadri, C., Sitaram, R., Pillai, S.R. and S. Venkataraghavan. 1980. Effect of aqueous and alcoholic extract of the berries of Embelia ribes on male reprocuctive organs in adult rats. A preliminary study. *Bulletin of Medico-Ethno-Botanical Research.* 1:272–280.

Shankar, K.R., and B.G. Rao. 2008. Anti-arthritic activity of *Gymnema sylvestre* root extract. *Biosciences Biotechnology Research Asia*. 5:469–471.

Sheikh, B.Y., Sarker, M.M.R., Kamarudin, M.N.A. and A. Ismail. 2017. Prophetic medicine as potential functional food elements in the intervention of cancer: A review. *Biomedicine & Pharmacotherapy*. 95:614–648.

Shenoy, R.S., Prashanth, K.V., and H.K. Manonmani. 2018. *In vitro* anti-diabetic effects of isolated triterpene glycoside fraction from *Gymnema sylvestre*. *Evidence-Based Complementary and Alternative Medicine*. 2018:1–12.

Shigematsu, N., Asano, R., Shimosaka, M., and M. Okazaki. 2001. Effect of administration with the extract of *Gymnema sylvestre* R. Br leaves on lipid metabolism in rats. *Biological and Pharmaceutical Bulletin*. 24:713–717.

Shimizu, K., Iino, A., Nakajima, J., Tanaka, K., Nakajyo, S., and N. Urakawa. 1997. Suppression of glucose absorption by some fractions extracted from *Gymnema sylvestre* leaves. *Journal of Veterinary Medical Science*. 59:245–251.

Shirole, R.L., Shirole, N.L., and M.N. Saraf. 2015. *Embelia ribes* ameliorates lipopolysaccharide-induced acute respiratory distress syndrome. *Journal of Ethnopharmacology*. 168:356–363.

Shivanna, Y., and K.A. Raveesha. 2009. In-vitro anti-bacterial effect of selected medicinal plant extracts. *Journal of Natural Products*. 2:64–69.

Shiyovich, A., Sztarkier, I., and L. Nesher. 2010. Toxic hepatitis induced by *Gymnema sylvestre*, a natural remedy for type 2 diabetes mellitus. *American Journal of the Medical Sciences*. 340:514–517.

Shukla A, Bukhariya V, Mehta J, Bajaj J, Charde R, Charde M., and B. Gandhare. 2011. Herbal remedies for diabetes: An overview. *International Journal of Biomedical & Advance Research*. 2:57–58.

Singh, D.K., Kumar, N., Sachan, A., Lakhani, P., Tutu, S., and R. Nath. 2017. Hypolipidaemic effects of *Gymnema sylvestre* on high fat diet induced dyslipidaemia in wistar rats. *Journal of Clinical and Diagnostic Research*. 11:FF01–FF05.

Singh, J., Singh, A.K., and R. Pravesh. 2003. Production and trade potential of some important medicinal plants: An overview. *Proceedings of the 1st National Interactive Meet on Medicinal and Aromatic Plants*. 50–58.

Singh, V.K., Dwivedi, P., Chaudhary, B.R., and R. Singh. 2015. Immunomodulatory effect of *Gymnema sylvestre* (R.Br.) Leaf Extract: An *in vitro* study in rat model. *PLoS One*. 10:1–15.

Singh, V.K., Umar, S., Ansari, S.A., and M. Iqbal. 2008. *Gymnema sylvestre* for diabetics. *Journal of Herbs, Spices & Medicinal Plants*. 14:88–106.

Souravi, K., and P.E. Rajasekharan. 2014. Ethnopharmacological uses of *Embelia ribes* Burm. F. A review. *IOSR Journal of Pharmacy and Biological Sciences*. 9:23–30.

Spasov, A.A. Samokhina, M.P., and A.E. Bulanov. 2008. Anti-diabetic properties of *Gymnema sylvestre* (a review). *Pharmaceutical Chemistry Journal*. 42:22–26.

Srikanth, A.V., Sayeeda, M., Lakshmi, N., Ravi, M., Kumar, P., and R.B. Madhava. 2010. Anticancer activity of *Gymnema sylvestre* R.Br. *International Journal of Pharmaceutical Sciences and Nanotechnology*. 3:897–899.

Srividya, A.R., Varma, S.K., Dhanapal, S.P., Vadivelan, R. and P. Vijayan. 2010. *In vitro* and *in vivo* evaluation of hepatoprotective activity of *Gymnema sylvestre*. *International Journal of Pharmaceutical Sciences and Nanotechnology*. 2:768–773.

Street, R.A., and G. Prinsloo. 2013. Commercially important medicinal plants of South Africa: A review. *Journal of Chemistry*. 1:1–16.

Sugihara, Y., Nojima, H., Matsuda, H., Murakami, T., Yoshikawa, M., and I. Kimura. 2000. Antihyperglycemic effects of gymnemic acid IV, a compound derived from *Gymnema sylvestre* leaves in streptozotocin-diabetic mice. *Journal of Asian Natural Products Research*. 2:321–327.

Suvarna, V. 2014. *In-Vitro*cytotoxic activity of methanolic extract of *embelia tsjeriam* cottam against human breast and colon cancer cell lines. *RJPBCS*. 5:131–135.

Swami, J., and Prabakaran, G. 2012. Studies on anti-bacterial activity of *Gymnema sylvestre* against respiratory infection causing bacteria. *International Journal of Current Advanced Research*. 1:1–4.

Swamy, H.K., Krishna, V., Shankarmurthy, K., Rahiman, B.A., Mankani, K.L., Mahadevan, K.M., Harish, B.G., and H.R. Naika. 2007. Wound healing activity of embelin isolated from the ethanol extract of leaves of *Embelia ribes* Burm. *Journal of Ethnopharmacology*. 109:529–534.

Tahir, M., Rasheed, M.A., Niaz, Q., Ashraf, M., Anjum, A.A., and M.U. Ahmed. 2017. Evaluation of anti-bacterial effect of *Gymnema sylvestre* R.Br. species cultivated in Pakistan. *Pakistan Veterinary Journal*. 37:245–250.

Tamaki, H., Satoh, H., Hori, S., Ohtani, H., and Y. Sawada. 2010. Inhibitory effects of herbal extracts on breast cancer resistance protein (BCRP) and structure: Inhibitory potency relationship of isoflavonoids. *Drug Metabolism and Pharmacokinetics*. 25:170–179.

Tambekar, D.H., Khante, B.S., and B.R. Chandak. 2009. Screening of anti-bacterial potentials of some medicinal plants from Melghat forest in India. *African Journal of Traditional, Complementary and Alternative Medicines*. 6:228–232.

Thanwar, M., Dwivedi, D., Gharia, D., and S. Chouhan. 2016. Antibacterial study of *Gymnema sylvestre* plant. *International Journal of Chemistry* 4:80–83.

Thippeswamy, B.S., Mahendran, S., Biradar, M.I., Raj, P., Srivastava, K., Badami, S., V.P. Veerapur. 2011. Protective effect of embelin against acetic acid induced ulcerative colitis in rats. *European Journal of Pharmacology*. 654:100–5.

Tiwari, P., Mishra, B.N., and N.S. Sangwan. 2014. Phytochemical and pharmacological properties of *Gymnema sylvestre*: An important medicinal plant. *BioMed Research International*. 1–18.

Tripathi, P., Tripathi, R., and R.K. Patel. 2010. Investigation of antimutagenic potential of *Embelia ribes* fruit extract against genotoxicity and oxidative stress induced by cyclophosphamide. *Pharmacologyonline*. 3:867–885.

Vaidyaratnam, P. 1995. *Indian Medicinal Plants*. Madras: Orient Longman Publisher, 107–9.

Ved, D.K., and A. Singh. 2006. Identity of vidanga: A plant drug in trade. *Newsletter- Medicinal Plants of Conservation Concern*, April-June 2006.

Warrier, P.K., Nambiar, V.P.K., and P.M. Ganapathy. 2001. *Some Important Medicinal Plants of Western Ghats, India: A Profile*. Canada: International Development Research Centre (IDRC), 139–156.

Wu, X., Mao, G., Fan, Q., Zhao, T., Zhao, J., and F. Li. 2012. Isolation, purification, immunological and anti-tumor activities of polysaccharides from *Gymnema sylvestre*. *Food Research International*. 48:935–939.

Yasmin, H., Kaiser, M.A., Sarker, M.M.R., Rahman, M.S., and M.A. Rashid. 2009. Preliminary anti-bacterial activity of some indigenous plants of Bangladesh. *Dhaka University Journal of Pharmaceutical Sciences*. 8:61–65.

Yasukawa, K., Okuda, S., and Y. Nobushi. 2014. Inhibitory effects of Gymnema (*Gymnema sylvestre*) leaves on tumour promotion in two-stage mouse skin carcinogenesis. *Evidence-Based Complementary and Alternative Medicine*. 2014:1–5.

Ye, W., Liu, X., Zhang, Q., Che, C.T., and S. Zhao. 2001. Antisweet saponins from *Gymnema sylvestre*. *Journal of Natural Products*. 64:232–235.

Yogi, B., and A. Mishra. 2016. Hepatoprotective effects of polyherbal formulation against carbon tetrachloride-induced hepatic injury in albino rats: A toxicity screening approach. *Asian Journal of Pharmaceutical and Clinical Research*. 10:192–198.

7 *Glycyrrhiza glabra* (Licorice) and *Gymnema sylvestre* (Gurmar)

Jasbir Kaur, Sana Nafees, Mohd Anwar, Jamal Akhtar, and Nighat Anjum

CONTENTS

7.1 Introduction 133
7.2 *Glycyrrhiza glabra* L. 133
7.2.1 Ethnobotanical Description 134
7.2.1.1 Macroscopic 134
7.2.1.2 Microscopic 134
7.2.1.3 Ethnobotanical Uses of *G. glabra* 135
7.2.1.4 Chemical Constituents 135
7.2.1.5 Scientific Studies 136
7.2.2 Clinical Studies 138
7.3 *Gymnema sylvestre* R.Br. 138
7.3.1 Ethnobotanical Description 139
7.3.1.1 Macroscopic 139
7.3.1.2 Microscopic 139
7.3.1.3 Ethnobotanical Uses 140
7.3.1.4 Chemical Constituents 140
7.3.1.5 Pharmacological Activities 141
7.3.2 Clinical Studies 144
7.4 Conclusion 144
References 145

7.1 INTRODUCTION

Traditional herbal medicines have gained popularity as a potent alternative treatment for diverse ailments across numerous countries. Nowadays, traditional medicinal plants are much more popular and have wider acceptance due to the perception that they have lesser side effects and better efficacy in comparison to allopathic medicines (Beshbishy et al., 2019). Pharmacological activities of these plant species are due to the presence of numerous active phytoconstituents, like alkaloids, flavonoids, glycosides, saponins, terpenes, and tannins (Batiha et al., 2019). Widely available documented data are evidence that these plants are important sources for drug discovery as new pharmaceutical moieties are developed to treat various severe diseases.

7.2 *GLYCYRRHIZA GLABRA* L.

Glycyrrhiza glabra is a perennial herb, growing to 2 m high, that belongs to the class: Magnoliopsida; order: fabales; family: Fabaceae; genus: *Glycyrrhiza*; species: *glabra* L. Common names include

DOI: 10.1201/9781003205067-7

licorice, sweet wood or mulaithi; in Unani, it is known as *Aslus-Soos* Rubb-us-Soos, Mulethi; English: Liquorice, Licorice, Liquorice root, Liqourica; Sanskrit: Yashti-madhu, Madhuka, Madhuyasti; Kannada: Yashti madhuka, Madhuka; Arabic: Asl-us-soos, Aslussus; Urdu: Mulethi, Asl-us-sus; Ayurvedic: Yashtyaahva, Madhuli, Madhurasaa, Yashti Madhuka; Hindi: Jethi-madh, Mulhatti, Mulethi, Mulathi, Muleti; Siddha: Athimathuram (Anonymous 2007a). This herb is found in Iran, Iraq, Afghanistan, Europe, and in a few parts of India (Shah et al., 2018). This drug is known as the "grandfather of herbs", and it is one of the most commonly used traditional medicines in the world. For more than 4000 years, *Glycyrrhiza glabra* has been used medicinally in both western and eastern countries (Hosseinzadeh and Asl, 2008). This genus is extensively distributed throughout the world and comprises about 30 species. Its nomenclature was derived from the word "glykys", the Greek word for sweet, and "rhiza", meaning root; whereas, the word glabra is derived from the Latin word "glaber", which means slick or bare (Chopra et al., 2002).

7.2.1 Ethnobotanical Description

7.2.1.1 Macroscopic

G. glabra is a perennial plant, or subshrub, that is usually found in temperate and subtropical regions, attaining a maximum height of about 2 m. Its underground stem grows horizontally with a highly branched short tap root system, which has a large number of rhizomes. The shape of the root is cylindrical, color is grayish brown on the outside and yellowish-green on the inside. It runs to a varied length and depth, and the diameter ranges from 0.75 to 2.5 cm. The stem is erect, about 2 feet in length, smooth, and the color is dull glaucous gray. Leaves are pinnate, leaflets are oval in shape, entire, obtuse, viscid, and underneath enclosed with soft hairs (Bhattacharjee 2004). Flowers are pale lilac in axillary, erect, stalked racemes (Lindley 2000) as shown in Figure 7.1.

7.2.1.2 Microscopic

The transverse section of stolon of *G. glabra* shows cork made up of tabular cells that are arranged in 1–20 layers, reddish-brown with amorphous substances in the outer layer, while inner layer is thick with 3 or 4 rows of colorless walls. 1–3 layers of parenchymatous cells in secondary cortex cells that are radially arranged containing calcium oxalate crystals, secondary phloem, inner cells are of cellulose and outer are lignified, 10–50 radially arranged fibers enclosed by a layer of parenchymatous cells, each containing a calcium oxalate prism. The root transverse section shows a similar structure as that of stolon, but it does not have medulla, tetrarch xylem; medullary rays are

FIGURE 7.1 *Glycyrrhiza glabra.*

perpendicular to one another, and sometimes it is without secondary phloem (Anonymous 2001; Anonymous 2003b; Anonymous 2007a).

7.2.1.3 Ethnobotanical Uses of *G. glabra*

The drug licorice has been used in traditional systems of medicine for various diseases like bronchial asthma, sore throat, tonsillitis, cough, fever, flatulence, hyperdipsia, epilepsy, paralysis, sexual debility, gastric ulcers, colic, rheumatism, skin diseases, jaundice, leucorrhoea, bleeding, and hemorrhagic conditions (Khare, 2004; Damle, 2014; Kaur and Dhinds, 2013). Moreover, it is also used as an anti-arthritic, anti-inflammatory, laxative, antibacterial, antiviral, insecticidal, and memory enhancer because it inhibits the monoamine oxidase (MOA). It also has several other actions like anti-cholinergic, antioxidant, anticancer, antimycotic, antidiuretic, hypolipidemic, prevents dental caries, and acts as an estrogenic agent (Zadeh et al., 2013). Furthermore, this is used in industries like tobacco manufacturing and sweets, such as soft drinks and alcohol. This drug is widely used in traditional systems of medicines, including Unani and Ayurvedic systems of medicines. Its traditional actions and uses are described in Table 7.1.

7.2.1.4 Chemical Constituents

G. glabra roots contain a variety of active chemical components including flavonoids, such as liquirtin, liquirtigenin, glucoliquirtin, rhamnoliquirtin, shinpterocarpin, prenyllicoflavone A, 1-methoxy-xyphaseolin, shinflavanone, apioside, licoarylcoumarin, glisoflavone, licopyranocoumarin, coumarin-G-12, and saponins, viz. glycyrrhizin, which is 60 times sweeter than sugarcane. Phenolic compounds present are isoangustone A, semilicoisoflavone B, licoriphenone, kanzonol R, and 1-methoxyficilinnol. Volatile compounds include terpineol, pentanol, geraniol, hexanol, tetramethyl pyrazine, terpinen-4-ol, and linalool oxides A and B. The components that are isolated from the essential oil are propionic acid, benzoic acid, ethyl linoleate, maltol, methyl ethylketone, furfuryl formate, furfuraldehyde, 1-methyl-2-formylpyrrole, 2,3-butanediol, and trimethylpyrazie. The chief compound of *G. glabra* is a saponin compound "glycyrrhizin", whereas its aglycone part is "glycyrrhetinic acid". Glycyrrhizin is composed of glycyrrhetic acid and triterpenoid aglycone, which bind with glucuronic acid disaccharide that has been found naturally as potassium and calcium salts (Biondi et al., 2003; Washington, 2003). Because glycyrrhizin is metabolized and transformed to glycyrrhetinic acid in humans, its pharmacological activity is comparable to that of glycyrrhetinic acid (Shah et al., 2018). Protein, carbohydrates, fat, moisture, fiber, and minerals, such as sodium, calcium, potassium, phosphorus, copper, and zinc, are all included in raw and tea licorice infusions. Glycine, glutamic, aspartic, serine, valine, threonine, proline, leucine, alanine, isoleucine, lysine, tyrosine, phenylalanine, and histidine are among the amino acids found in them.

TABLE 7.1
Traditional Actions and Uses of *Glycyrrhiza glabra*

Actions and Uses	References
Anti-inflammatory	Kabeeruddin (2010)
Anti-spasmodic	Kabeeruddin (2010)
Asthma, Bronchitis	Kritikar and Basu (2008)
Burning micturition	Anonymous (2007a)
Chronic fever	Saeed (2007), Sina (1998), Baitar (1985)
Expectorant	Kabeeruddin (2010), Anonymous (2000)
Laxative	Anonymous (2000)
Nervine tonic	Kabeeruddin (2010), Hakim (2009), Hakeem (2002)
Pterygium	Sina (1998)
Ulcers	Kabeeruddin (2010), Saeed (2007), Sina (1998)

Several organic acids were discovered in the HPLC study of licorice methanolic extract, including citric, propanoic, fumaric, butyric, tartaric, and malic acids (Isbrucker and Burdock, 2006; Badr et al., 2013).

7.2.1.5 Scientific Studies

Antitussive and expectorant activity: *G. glabra* extract showed antitussive and expectorant effects by relaxing and helping to get rid of the congestion in the upper respiratory tract (URT) and accelerating mucus secretion of trachea. This softening effect is due to glycyrrhizin. The licensed and listed powder has been shown to help treat throat ulcers, cough, and bronchial parotitis (Hikino et al., 1985). Apioside liquid was recently discovered to be a highly active ingredient in the extraction of methanol from alcohol. The capsule induces capsaicin-induced asthma (Kamei et al., 2003).

Immunomodulatory activity: Roots of *G. glabra* aqueous extract showed immunomodulatory activity at a dose of 1.5g/kg b.w. of mice. Viruses like swine flu are extremely contagious diseases with 1–4% low mortality. This is a very common disease in temperate regions, especially in the autumn and winter; usually the outbreak of this disease occurs once a year. H1N1 virus is one of these viruses and can cross the species barrier, migrating from pigs to humans, where it is widely dispersed. It is evident that stimulation of macrophage occurs by the polysaccharide fractions of *G. glabra* and thus promote and maintain immunity (Wagner and Jurcic, 2002). The N-acetylmuroamoyl peptide is an analog of glycyrrhizin with potent *in vitro* immunostimulatory properties (Blatina, 2003). Animal studies have shown efficacy against the influenza virus, which is mediated by viral replication. Inhibition of viral growth and activation of viral particles by the glycyrrhizinic acid and may act as potent immunomodulators (Arora et al., 2011).

Antioxidant and anti-inflammatory: Hydromethanol extract of *G. glabra* root showed significant antioxidant activity in the tube system (Sharma et al., 2013). Licorice root contains many polyphenolic compounds that are the potent antioxidants, like inhibition of lipid peroxidation in microorganisms by licochalcones B and D, whereas retrochalcones exhibit lipid peroxidation in mitochondria, thus preventing oxidative hemolysis of red blood cells (Haraguchi et al., 1998). Isoflavones of *G. glabra* such as glabridin, histapalabridin A and 30-hydroxy-4-O-methylglabridin have also shown efficient antioxidant activity (Biondi et al., 2003). Research shows that when glycyrrhizin is broken down in the gut, it has anti-inflammatory effects like hydrocortisone and other corticosteroid hormones (Masoomeh and Kiarash, 2007; Adel et al., 2005).

Antibacterial activity: *In vitro* studies have shown significant bactericidal activity against two gram-positive pathogens example *Bacillus subtilis* and *Staphylococcus aureus* with two gram-negative, *Escherichia coli* and *Pseudomonas aeruginosa* (Nitalikar et al., 2010).

Antifungal activity: *Glycyrrhiza glabra* showed significant biological activity against Candida compared to Candida isolated from vaginal vulvar candidiasis. Several components isolated from licorice root have been shown to have antimicrobial activity *in vitro*, including glabridin, gabrine, glabrol, glabren, istanabridin A, isphalabridin B, 40-methylglabridine, and 3-hydroxyglabryl. Itching, atopic dermatitis, and cysts caused by skin parasites have been treated by glycyrrhizic acids (Mahmoudabadi et al., 2009).

Antiviral: Glycyrrhizin works by preventing the viral cell from binding to the receptor. Recently, the antiviral effects of glycyrrhizin, ribavirin, mycophenolic acid, 6-azauridine, and pyrazofurin were evaluated against two clinical isolates of the severe acute respiratory syndrome virus (SARS), namely FFM-1 and FFM-2 (Badam 1997). Glycyrrhizin is considered to be the most effective way to control viral replication and can be used as a preventative measure. Previously, glycyrrhizin has been reported in hepatitis C virus and HIV-1 virus (De Clercq, 2000).

Antiulcer activity: The glycyrrhizin produced by deglycyrrhizinated licorice root is commonly used to treat wounds effectively. Carbenoxolone, derived from the root system, has an antiulcerogenic impact inhibiting gastric secretion (Masoomeh and Kiarash, 2007). Prostaglandin and mucus secretion in the digestive tract is increased by the licorice. It increases the life of cells on the surface of the stomach and has an antiseptic effect (Adel et al., 2005).

Anticoagulant activity: The glycyrrhizin was the first plant thrombin; it increases the typing time of fibrinogen and thrombin and increases the duration of plasma recalculation. Glycyrrhizin has been shown to inhibit trebine-induced platelet aggregation, but glycyrrhizin-induced platelet aggregation factor (PAF) and collagen-induced agglomeration are not affected (Mauricio et al., 1997; Mendes-Silva et al., 2003).

Anticarcinogenic and antimutagenic activity: Rathi et al., (2009) investigated the anticancer activity of *G. glabra*, and results show that chloroform extract has good cytotoxicity against cancerous MCF7 cells (human breast cancer), as it consists of a high quantity of β-glycyrrhetic acid. In another study, Ehrlich ascites tumor cells were inhibited by the aqueous extract as shown in *in vivo* and *in vitro* proliferation, preventing angiogenesis and *in vivo* assay choreo-allantoic and peritoneal membrane assay (Sheela et al., 2006). Furthermore, there is considerable evidence of anticancer effects of its derivatives both in *in vivo* and *in vitro* studies. Studies show that the proapoptotic pathway could be triggered by the glycyrrhetic acid by inducing mitochondrial membrane permeability transition, which is helpful in inducing apoptosis of tumor cells (Salvi et al., 2003; Fiore et al., 2004). The compound licocoumarone induces G2/M cell cycle arrest and Bcl_2 phosphorylation and apoptosis in human monoblastic leukemia U937 cells. This compound is also known to have antioxidant and antimicrobial activity (Watanabe et al., 2002). Sharma et al. (2014) evaluated the antimutagenic effect of hydromethanolic extract of *G. glabra* roots, and results showed the potent antimutagenic potential through inhibition of micronucleus development and chromosomal aberration in bone marrow cells of albino mice. It has been found that glycyrrhizin encourages the activator protein-1 (AP-1) activity in untreated cells; however, it suppressed 12-O-tetradecanoylphorbal-13-acetate (TPA) induced AP-1 activity in TPA cells. This mechanism could serve as a model for the development of new chemoprotective agents (Hsiang et al., 2002).

Hepatoprotective activity: Viral or non-viral chronic hepatitis is a gradually developing liver diseases that may convert into cirrhosis (or liver failure) or hepatocellular carcinoma (HCC). For more than 60 years in Japan, glycyrrhizin has been used as a remedy for chronic hepatitis under the name of SNMC (stronger neo-minophagen-C) (Acharaya et al., 1993). A significant decrease in serum aminotransferases and liver histology was improved on the administration of glycyrrhizin as compared to a placebo. HCC from chronic hepatitis C has been prevented by the long-term use of glycyrrhizin, and *in vitro* data suggests that the intracellular transport system is modified by the glycyrrhizin (Sato et al., 1996, Van Rossum et al., 1998). The aglycone part of glycyrrhizin, i.e., 18β-glycyrrhetinic acid (GA), suppresses the expression of P450 E1, thus shielding the liver (Jeong et al., 2002). GA also inhibits oxidative and liver damage caused by aflatoxins by increasing the activity of CYP1A1 and GST (glutathione-S-transferase) and can promote anticancer activity through the metabolic inactivation of epitotoxin (Chan et al., 2003). One study showed that *G. glabra* root hydromethanol extract showed significant protection against CCl4 hepatotoxicity in the liver of experimental mice (Sharma and Agrawal, 2014). Glycyrrhizin and its analogues via epidermal growth factor receptors have also been studied experimentally, exciting the mitogen-activated protein (MAP) kinase pathway followed by stimulation of hepatocyte DNA synthesis and proliferation (Kimura et al., 2001).

Antidiabetic: Insulin dependent type 2 diabetes is an insulin-resistant disorder, an emerging health problem in modern-day society. PPARs (peroxisomal receptor antagonists) are ligand-dependent changes that regulate the expression of a group of genes that play an essential role in glucose metabolism. PPAR receptors are of 3 types: PPAR-α, PPAR-γ, and PPAR-δ. PPAR-α is present in muscles, liver, and kidney; PPAR-γ is found in adrenal glands, fat cells, and small intestine; and PPAR-δ is ubiquitous. PPAR-γ are the primary targets for insulin sensitization like pioglitazone and rosiglitazone. Ethyl acetate extract showed significant binding of PPAR-γ as seen in GAL-4-PPAR-γ-Flue test. This activity is due to the phenolic compounds like glycycoumarin, glycyrin, isolglycyrol, dehydroglyasperin, glyasperin B, and glyasperin D. Pioglitazone and glycine have been shown to prevent elevated blood sugar levels in rats after sucrose exposure in the oral sucrose tolerance test. The potent PPAR-γ agonist, i.e., pioglitazone, improves resistance of insulin, hence improves type 2 diabetes, and glycyrrhizin showed intense activity bound to the PPAR-γ ligand, thus lowering blood sugar in KK-Ay (Knockout diabetes mice). This finding is important because traditionally licorice has been used as an artificial sweetening agent and can help in treating insulin-resistant disorders common in modern society (Takii et al., 2000).

7.2.2 Clinical Studies

Numerous clinical trials on the effectiveness of *G. glabra* L. on allergic rhinitis, diabetes mellitus, dyspepsia, etc. are listed in Table 7.2.

7.3 *GYMNEMA SYLVESTRE* R.BR.

G. sylvestre is a native plant belongs to the class: Magnoliopsida; order: gentianales and family: Apocynaceae. The plant is having abundant amount of biologically active constituents (Manohar et al., 2009). It is found in the sub-tropical and tropical regions and it is widely distributed in Southern and central areas of India, also found in Sri Lanka, tropical Africa, southern China and Malaysia (Singh et al., 2008). *G. sylvestre* R.Br. is a slow growing plant that occurs preferably in humid sub-tropical and tropical climates and is common in green mountain forests. This plant is also known by many vernacular names, such as Waldschlinge in German, Chigengteng or Australian Cowplant, Gur-mar, merasingi in Hindi, Periploca Forest in English, Barkista in Arabic, Meshashringi, madhunashini in Sanskrit, Gurmar, Gurmar buti, Kakrasingi in Urdu, Kavali, kalikardori in Marathi, Dhuleti, mardashingi in Gujrathi, Podapatri in Telugu, Adigam, cherukurinja in Tamil, Sannagerasehambu in Kanada (Kanetkar et al., 2007). It has been used to treat diabetes in India

TABLE 7.2
Clinical Studies of *Glycyrrhiza glabra*

Preparation of the Plant Given	Number of Subjects	Duration	Therapeutic Potential	Reference
Licorice nasal irrigation	20 allergic rhinitis patients	1 month	• Superior to steroid and saline nasal irrigation. • It remains on nasal mucosa after irrigation and continues the anti-inflammatory effects. • It tastes sweet and has fragrant smell, which is highly accepted by patients.	Chang et al. (2021)
Licorice root powder in the hard gelatin capsule	14 type 2 diabetic patients with HSD11B1 gene polymorphism	3 weeks	• Significant reduction in serum insulin levels ($p = 0.03$).	Devang et al. (2021)
Licorice root extract	90 menopausal	8 weeks	• Frequency of hot flash decreased significantly. • Severity of hot flash decreased significantly.	Nahidi et al. (2012)
GutGard- Extract of G. glabra root	54 patients of functional dyspepsia	30 days	• Significant decrease ($P \leq .05$) in total symptom scores. • Significant decrease ($P \leq .05$) in the Nepean dyspepsia index. • Improvement in the global assessment of efficacy.	Raveendra et al. (2012)

for over a thousand years. It was named "*Gurmar*" because it removes the taste of sugar, and it is believed that excess sugar can be neutralized in people with diabetes.

7.3.1 Ethnobotanical Description

7.3.1.1 Macroscopic

Gymnema sylvestre R.Br. is fragile in nature, and it is a slow-growing species that usually need support to grow. It looks very branched, tree-like, and can climb to the top of trees growing in dry forests. The plant is a shrub and hairy, having branches and young stems (Kanetkar et al., 2007) as shown in Figure 7.2. It has a tap root system. The internodes are branched, hard, cylindrical, convoluted, having dimensions of 2–10 mm in diameter and 0.7–17.2 cm long. The leaves are oppositely arranged, generally elliptical or oval (1.25–2.0 x 0.5–1.25). They are pointed, with petioles 1–2 cm long, smooth on the upper part, base is rounded, lower surface is densely pubescent and ciliated along the margin, especially on the veins. The reticular vein is transpedicular to a peripheral vein (Kirtikar and Basu, 1975). The flowers are small, yellow in color, beautifully arranged laterally, and axillary on the tips; follicles terrestrial and lanceolate, up to 3 cm in height (Kanetkar et al., 2007). The crown is pale yellow, folded, with a single crown with five fleshy scales. The stems of the calyx are long, oval, bald, hairy. They can be Carpels-2 front connectors, one-eyed, egg meal, webbed tip (Gurav et al., 2007; Potawale et al., 2008). The seeds are 1.3 cm long, flattened, having thin margin, ovate elongated (Chopra et al., 2002; Kirtikar and Basu, 1975). Flowers bloom from August to March.

While *Gymnema sylvestre* can be propagated by seed, germination is difficult due to low seed viability. Hence the alternative to propagation is by root seedlings, which are usually planted from June and July. For vegetative propagation, 3 or 4 nodes are sufficient and planted in February and March. Seeds are sown in November and December, and harvesting is done from September to February (Reddy et al., 2004).

7.3.1.2 Microscopic

The hairs are not glandular and are present on the whole surface. The leaves have five vascular bundles and are surrounded by two small bundles on each side that resemble a fan. Midvein has a ventral bulge The lamina contains rosette calcium crystals. Idoblasts are present in the parenchyma... They also indicate the presence of a spongy parenchyma (Gurav et al., 2007).

FIGURE 7.2 *Gymnema sylvestre.*

TABLE 7.3
Traditional Actions and Uses of *Gymnema sylvestre*

Actions and Uses	References
Antidiabetic	Anonymous (2003a)
Anti-inflammatory	Ghani (NM), Anonymous (2007b)
Amenorrhoea	Srivastava et al. (2009),
Asthma, bronchitis	Kirtikar and Basu (2003), Chopra et al. (1992)
Cardiotonic	Ghani NM, Anonymous (2007b)
Conjunctivitis, eye sore	Anonymous (2003b), Kirtikar and Basu (2003)
Diabetes Mellitus	Kabeeruddin (2010)
Laxative	Nadkarni (1982)
Lecucoderma	Nadkarni (1982), Kirtikar and Basu (2003), Chopra et al. (1992)
Opacities of cornea and vitreous	Kirtikar and Basu (2003), Srivastava et al. (2009),
Renal calculi	Chopra et al. (1992)

7.3.1.3 Ethnobotanical Uses

Traditionally, the leaves of *G. sylvestre* R.Br. have been used to treat diabetes and other ailments, and the flowers and bark have been used for phlegm-related illnesses. (Kirtikar and Basu, 1975). Ancient Indian medical literature, Sushruta, describes the gurmar as madhumeha gum (*glucosuria*) and other urinary ailments (Nadkarni, 1986). In addition, various other parts of the plant, like roots, stems, and leaves, have been used in traditional systems, such as tonic, uterine tonic, digestive, diuretic, laxative, stomach, and stimulant (Mathew, 2004). The plant also has medicinal value in treating constipation, heart diseases, jaundice, asthma, amenorrhea, conjunctivitis, kidney and bladder stones, indigestion, leukoderma, and Parkinsonism (Chopra et al., 1956). The traditional actions and uses in Unani's medical system are listed in Table 7.3.

7.3.1.4 Chemical Constituents

The leaves of *G. sylvestre* contain triterpenic saponins linked to olean and dammaran classes. The main components, like gymnemic acids and gymnemasaponins, belong to oleanane saponins and gymnemasides belongs to dammarane saponins (Foster, 2002; Khramov et al., 2008). Other phytocomponents include anthraquinone, flavone, hentriacontane, calcium oxalate, phytin, resins, tartaric acid, pentatriacontane, formic acid, lupeol- and amirine-related glycosides, butyric acid, and stigmasterol (Sinsheimer et al., 1970). The leaves of *G. sylvestre* contain anthraquinones and their derivatives, also having acidic glycosides (Dateo and Long, 1973). Its primary and secondary metabolites include nine groups closely related to acid glycosides, mainly A–D gymnemics and the whole plant.

The highest percentage of gymnemic acid was found in shoot tips andlowest in seeds. The anti- saccharin properties of gymnemic acid A1 were decreased significantly after converting to A2, and no activity was observed in A3. This suggests that the ester group in the genetic part of the gymnemic acid gives an anti-sweat property to aqueous "triterpene saponins". The acids of the A2 and A3 genes contain galactose and glucuronic acid in their molecular structures and glucuronic acid is the only component of A1 glyceric acid (Chakravarti and Debnath, 1981). In addition, several gymnemic acids (gymnemic acids I, II, III, IV, V, VI, and VII) have been characterized and isolated by hot aqueous extract of dried leaves of *G. sylvestre* (Yoshikawa et al., 1989). It consists of several members called gymnemic acids, gymnemic acids I–VII, gymnemosids A–F, and gymnemasaponin. Gymnemic acid derivatives are a few members of gymnemagen which are a 3-O-β-glucuronide (3β, 16β, 21β, 22α, 23, 28-hexahydroxy-olean) with tigloyl group, methylbutyryl, deacylgimemic acid (DAGA). Gymnemic acid A contains the gymnemic acids, namely, A1, A2, A3, and A4 and is called "gymnemagenin". This ingredient is an acid-esterified D-glucuronide hexahydroxytriterpene. Then five other gymnemic acids, VIII, IX, X, XI, and XII, were isolated and characterized (Yoshikawa et al., 1992). Gymnemasaponin III, is another anti-sweetener compound derived from the *G. sylvestre*, consisting of 23-hydroxyspispinogenin as a glycosylated aglycone molecule with 1 or 2 molecules of glucose in 23 and 28 hydroxyl groups (Murakami et al., 1996). These compounds had a lower resistance than dimer acids (Yoshikawa et al., 1991). Gourmarin, a powerful 35 amino-acid peptide having a molecular weight of 4209, was isolated from *G. sylvestre* (Imoto et al., 1991). The sugar suppression activity of this constituent was determined by the electrophysiological property in the taste response of rats (Gent et al., 1999). The anti-sweet effect of this compound is particular for the tongue's sweet taste, which is affected by variations in pH. Polypeptide is said to have the highest anti-sweetener properties near its isoelectric point (Chattopadhyay, 1999). In addition to ionic interactions, hydrophobic interactions play an essential role in the appropriate binding of gourmarin to target molecules (Imoto et al., 1991; Arai et al., 1995). Other essential components which were isolated from the leaves are lyceum and alkaloids A, B, C, and D (Suttisri et al., 1995). Various saponins, such as gymnemic acid, deacylamic acid, gymnemagenine (Rao and Sinsheimer, 1971), 23-hydroxyl-ispinogenin, and gynestrogenin (Yoshikawa et al., 1997), have been purified by *G. sylvestre*. Phytoconstituents were isolated from the leaves, analyzed by several methods like gas chromatography and spectrometry, and identified as fatty acids, terpenoids, glycosides, and alkaloids in 3 different solvents namely methanol, chloroform, and ether (Sathya et al., 2010).

7.3.1.5 Pharmacological Activities

Antidiabetic property: The plant explains the moderately inactive properties of triterpene saponins called gymnemic, gymnemic, and gourmarinic acids. Experimental tests confirmed the hypoglycemic effect of *G. sylvestre* in rats treated with beryllium nitrate and streptozotocin. A slight increase in the body weight and protein and significant decrease in fasting blood sugar were observed in diabetic rats having treatment with *G. sylvestre*, *C. auriculata*, *S. reticulata*, and *E. jambolanum*, and the effects were very close to those found in insulin treatment and glibenclamide. A study was conducted by Kang et al., using ethanolic extracts of gymnema leaf, to determine the antioxidant activity and also the role of antioxidants in diabetic rats (Kang et al., 2012). Many antioxidant models, such as modified thiobarbituric acid (TBA) assay with superoxide dismutase (SOD), 2-deoxyribose (associated with lipid peroxidation) as well as sample reactions with 2.2-Azinobis (3 ethylbenzothiazoline-6-sulfonic acid (ABTS) test described the importance of antioxidant activity of ethanolic extract. The activity of transaminases in Ketogenesis and Gluconeogenesis in diabetes, such as glutamate pyruvate transaminase (GPT) and cytotonic hepatic glutathione peroxidase, returned to their normal levels after the administration of an ethanol extract of the leaves to experimental animals (Patil et al., 2012). The anti-hyperglycemic effect of a crude fraction of saponin and 5 triterpene glycosides (gymnemic acids I-IV and gymnemasaponin V) isolated from the methanol extract of the leaves has been documented (Sugihara et al., 2000). Intravenous dimnemic acid in the doses of 3.4–13.4 mg/kg reduced glucose levels in blood within 6 hrs of ingestion by 14.0–60.0%, compared with the glibenclamide. Intravenous gymnemic acid increases plasma insulin levels in STZ-induced diabetic mice by 13.4 mg/kg, with no inhibitory effect on β-glucosidase activity in small intestinal brush vesicles in rats. A pilot study also investigated the hypoglycemic and lipid-lowering activity of dried leaves of *G. sylvestre*. The leaf extract of the plant was given to both diabetic and non-diabetic rats with alloxan. It has been found that the leaf extract of the plant had no effect on attenuated blood glucose levels due to a balanced meal or amylose or glucose, but it did increase serum lipid levels after SOC treatment. Although, in non-diabetic and alloxan rats, the subacute therapy and the chronic therapy with plant extract did not affect water and food intake, elevated body weight, and levels of glucose and lipids in blood. But scientific validation and clinical approval of the herbal before it can be used to treat diabetes and hyperlipidemia is needed (Galletto et al., 2004). Finally, research has shown that this plant has antidiabetic effects and activates sugar.

Anti-arthritic activity: *G. sylvestre* leaf extract was tested for anti-arthritic activity in rats. The water-soluble extract and petroleum ether extract (40–60°C) have been found highly efficacious against arthritis. Results suggest that anti-arthritic activity of the plant leaves may be related to the nature of the steroids, glycosides, triterpenoids, and saponins (Malik et al., 2010). Several extracts were suspended in 1% Tween 80; the drug diclofenac sodium was given orally once a day, and the effects were observed for 21 days. Following adjuvant administration, rats developed edema in several joints and cellular inflammation, destruction, and bone remodeling were observed. The petroleum ether extract group exhibited inflammation in the legs, possibly be due to inhibition of the inflammatory response of the cells or the release of mediators, such as GM-CSF, interferon, cytokines (IL-Iβ and TNF-a), PGDF, and the significant reduction causes of pain and disability due to bone and cartilage destruction (Eric and Lawrence, 1996). Another possible mechanism of action suggested the preservation of articular cartilage release and bone recovery in the chronic arthritis model (Malik et al., 2010). Numerous studies by researchers using solvents which are polar in nature in the preparation of extracts have illustrated the anti-arthritis activity of the extract of the leaf.

Antibiotic and antimicrobial activity: The antibiotic and antimicrobial activities of various extracts of *G. sylvestre* have been determined against multiple pathogens such as *E. coli, B. subtilis and S. aureus* but no action has been taken against gram-negative bacteria. The leaf extracts of *G. sylvestre* have a promising role as effective herbal supplements in the treatment of microbial infections (Saumendu et al., 2010). In one study, the antibacterial activity of *G. sylvestre* and gymnemic acid against *B. cereus* and *E. coli* was determined, and results showed significant antimicrobial effect (Yogisha and Raveesha, 2009). Bhuvaneswari et al. (2011) reported that methanol extracts from *G. sylvestre* were tested separately for antimicrobial activity of shoots and roots. The results showed that methanol extracts in the acidic range are highly effective against all pathogens with a broad spectrum of activity. In another study, the antimicrobial activity of the ethanolic extract of *G. sylvestre* was determined against *B. subtilis, S. aureus, P. aeruginosa*, and *Bacillus pumilus*, with results showing significant antimicrobial activity (Satdive et al., 2003). From the research, it can be concluded that the methanol and ethanolic extract of the leaves of *Gymnema sylvestre* have effective antibiotic and antimicrobial activity.

Anti-inflammatory activity: *G. sylvestre* leaf is widely used in Ayurvedic medicine, where it is considered bitter, acidic, thermal, tonic, digestive, harmless, and anti-inflammatory (Kokate, 1999). The active components of *G. sylvestre* are tannins and saponins, and these constituents are mainly responsible for its anti-inflammatory activity (Diwan et al., 1995). In this study, paw edema caused by carrageenan and granulomas caused by cotton balls the anti-inflammatory effect of the aqueous extract of leaves of *G. sylvestre* was seen at different doses viz. 200, 300. and 500 mg/kg. At 300 mg/kg, the extract reduced the amount of leg edema by 48.5%, and phenylbutazone reduced the volume of leg edema by 57.6% fours hours after ingestion. In addition, the aqueous extract of the plant at concentrations of 200 and 300 mg/kg showed a decrease in granuloma when compared with the control group (Malik et al., 2007).

Antioxidant activity: The ethanolic extract of the plant showed significant antioxidant activity ($p<0.05$) to radically remove '1,1-diphenyl-2-picrylhydrazyl (DPPH)' and exhibits higher antioxidant potential than *C. frutescens* and *A. bilimbi* (Rahman et al., 2014). The antioxidant activities from the *G. sylvestre* against DPPH were observed in an investigation by Rupanar et al., (2012). This plant has a higher affinity for DPPH than butadiene hydroxy ethylene (BHT), and another study has found that it reduces LDL oxide. (Ohmori et al., 2005; Rupanar et al., 2012). Another study reported hydroxyl free radical-scavenging activity and significant antioxidant potential of this plant against DPPH when DPPH inhibition was 87.3% and hydroxyl free radical inhibition at 59.8% (Gunasekaran et al., 2019). The significant radical-scavenging activity was also found for iron ($p<0.05$), superoxide ($p<0.05$), and hydrogen peroxide ($p<0.05$) (Rachh et al., 2009). The plant showed antioxidant activity in rats under various conditions, including fats rich in fat, nitric oxide, hydrogen peroxide, and superoxide, induced by radical oxidative stress. (Arun et al., 2014; Kishore and Singh, 2015; Chakrapani and Periandavan, 2018).

Hepatoprotective activity: Srividya and others have evaluated the hepatoprotective effects of *G. sylvestre* hydro-alcohol extract (Srividya et al., 2010). Acute (new) hepatocytes have been treated to remove alcohol from water containing various doses found in hot water. Manufacturers targeting different concentrations such as 200, 400, 600 μg/ml exhibit inhibition of the acetoxicity of the "D-galactosamine" coating, and at the 800 μg/ml harvest was considered "cytotoxic". The cells showed a significant reversal of the standard chemical threshold ($p<0.001$) compared to different groups of galactosamine receptors when *G. sylvestre* was excluded from alcohol.

Anticancer and cytotoxic activity: The anticancer effect of this plant is due to phytoconstituents, like saponins and ginsenosides. Jain et al. (2007) investigated the gymnemagenol anticancer activity in HeLa cancer cell lines. This cytotoxic activity of saponin has been confirmed by MTT cell proliferation assay. Varied concentrations of gymnemagenol (5, 15, 25, and 50 μg/ml) have been incubated at different time points. At 96-hour IC50 value was 50μg/ml with good cytotoxic effect of 73% in HeLa cells. Another photoactive component "gymnemagenol" effectively inhibited the growth of HeLa cancer cell lines. Furthermore, these saponins do not show toxicity to average cell growth *in vitro* culture (Khanna and Kannabiran, 2009). Herbal medicines have become potential treatment candidates as the rate of incidence of cancer increases, and drug-resistance of conventional drugs has been documented in malignancies.

Antihyperlipidemic activity: Higher mortality rates have been seen in coronary heart diseases (CHD) than any other cause (Hardman et al., 2001). The major risk factor in atherosclerosis and coronary heart disease is hyperlipidemia (Kaushik et al., 2011). Lowering of serum cholesterol can be a better choice in reducing the risk of CHD (Hardman et al., 2001). Herbal medicines have good prospects in treating cardiovascular ailments (Yoshikawa et al., 1997). Rach et al. (2010) investigated the antihyperlipidemic activity of gimnema hydrochloride leaf extract in female rats. Results showed a decrease in serum cholesterol, serum triglycerides, low-density lipoproteins and very low-density lipoproteins, and higher level of high-density lipoprotein, an effect that is comparable to the standard drug atorvastatin. Another study demonstrated the anti-obese effect of hexane extract of *G. sylvestre* leaves. Results showed significant decrease in body weight and improvement in HDL levels after 45 days of drug administration. This effect is similar to that of atorvastatin (Shivaprasad et al., 2006). This research demonstrates that the plant leaf of *G. sylvestre* has significant action in reducing blood cholesterol levels and can provide herbal remedy for obesity.

Immunostimulatory activity: Gupta et al. (2010) considered *G. sylvestre* an immune plant, as its leaves showed immunostimulatory effects. The aqueous leaf extract at different doses (10, 25, 50, 100, and 1000μg/ml) was tested by different assays like neutrophil movement, chemotaxis tests, phagocytosis of dead *C. albicans*, and nitro tetrazolium blue tests; the results showed significant immunostimulatory activity (Malik et al., 2009).

Wound-healing activity: The alcoholic extract of the leaves of *G. sylvestre* exhibits significant wound healing activity in rats as well (Alam et al., 2011). According to Kiranmai et al. (2011), the hydrophilic extract of the plant showed significant wound-healing properties when compared to the control group. The TLC analysis of wound reduction and the quality tests confirmed the synergistic wound-healing activity of the plant. The greater curative activity of hydro-alcoholic extracts might be associated with the eliminating effect of the free radicals and phytotoxins (flavonoids) that are present, which are effective alone or have additional effects. The flavonoids in the alcohol extract were also determined with TLC and phytochemical analysis.

Treatment of dental caries: Dental cary is defined as an infection of the teeth caused by various gram-positive cariogenic bacteria (Marsh and Martin, 1992), such as *S. aureus*, *S. mitis*, and *S. mutans*, as well as the fungal microorganisms *albicans*. Fungus adheres to the surface of the tooth, releasing extra-cellular polysaccharides from sucrose, converting sugars into organic acids (especially lactic acid), causing tooth enamel to decentralize (Akhtar and Bhakuni, 2004). Chloroform, petroleum ether, and *G. sylvestre* leaves methanolic extract at concentrations 25, 50, and 100 mg/ml were tested for microbial dental diseases and proved to be very effective against the cariogenic bacteria, in particular the production of methanol, which they demonstrated maximum yield at

TABLE 7.4
Clinical Studies of *Gymnema sylvestre*

Preparation of the Plant Given	Number of Subjects	Duration	Therapeutic Potential	Reference
Leaf extract in capsule form	58 human subjects with type-2 diabetes	90 days	• Reduction in hyperglycemia and hypertriglyceridemia ($p < 0.05$) • Reduced fasting and post-prandial blood glucose levels significantly ($p < 0.005$ and $p < 0.001$, respectively) • Significantly increased HbA1c level ($p < 0.001$) • Significantly reduced insulin resistance ($p < 0.05$)	Kumar et al. (2010)
Leaf powder in gelatin capsule	32 human subjects with type-2 diabetes	30 days	• Significantly reduced glucose level ($p < 0.05$) • Reduced triglyceride, cholesterol and LDL level	Li et al. (2015)
Leaf extract	64 individuals with type 1 diabetes	6 to 30 months	• HbA1c level was reduced significantly ($p < 0.001$) • Reduced glucose level • Reduced requirement of insulin	Shanmugasundaram et al. (1990)
GS leaves, Swanson Health Products, USA	24 diabetic patients	12 weeks	• Reduction in body weight, body mass index (BMI) significantly ($p < 0.05$ and $p < 0.05$, respectively) • Decreased level of very low-density lipoprotein (VLDL) significantly ($p < 0.05$)	Zuñiga et al. (2017)

least observation. The excellent potential of hydrophobic plant extracts led to the development and production of Gurmar toothpaste, marketed under the name "Gurmar herbal toothpaste". These herbal formulations open up new perspectives in treating tooth decay when clinically approved by the scientific community (Devi and Ramasubramaniaraja, 2010).

7.3.2 Clinical Studies

In addition to various experiments conducted in animal models, multiple products of this plant have been tested in humans to confirm its therapeutic potential in humans. Clinical studies of *Gymnema sylvestre* R.Br. have shown their ability in the reduction of body weight and levels of glucose, LDL, cholesterol, total cholesterol and triglycerides in blood, increased levels of C-peptide and insulin in the blood (Kumar et al., 2010; Li et al., 2015), reduced doses of external insulin, and significant reduction of HbA1c having value $p<0.001$ (Shanmugasundaram et al., 1990). The clinical studies conducted to date are summarized in Table 7.4.

7.4 CONCLUSION

This chapter examined the ethnobotanical uses with brief phyto-chemistry, pharmacological activities, and clinical trials of two traditional medicinal plants *G. glabra* and *G. sylvestre*. The plant *G. glabra* has a strong ethnobotanical history. In Europe and in eastern countries, both root and rhizome are used as folklore medicine; whereas, in the Unani system, only the rhizome is used as medicine.

It possesses a wide range of pharmacological actions like anticancer, antimutagenic, antiulcer, antifungal, antioxidant, due to the presence of active chemical constituents like glycyrrhizin, glabridin, glycyrrhetic acid, etc. This plant has been used since ancient times and has been mentioned in traditional pharmacopoeias. The other important medicinal plant is *G. sylvestre*, which has antidiabetic, anti-inflammatory, antioxidant, hepatoprotective, antibacterial, lipid-lowering activities; the chief chemical constituents are gymnemanol, gurmarin, saponins, flavonol, and glycosides. Because this plant is an inexpensive substitute for antidiabetic drugs, people use it without any prior knowledge of which part of this plant is to be used. This leads to the redundant ruining of the whole plant. Due to overexploitation and over usage to meet demand, this plant is disappearing quickly. Hence, legal production and thorough monitoring of medicinal preparation should be done. Further comprehensive scientific studies regarding the mechanism of action of these medicinal plants, their phytochemical compounds, drug interactions, effective dose determination, and side effects are required.

REFERENCES

Acharaya, S.K., Dasarathy, S., Tandon, A., Joshi, Y.K., and B.N. Tandon. 1993. Preliminary open trial on interferon stimulator (SNMC) derived from *Glycyrrhiza glabra* in the treatment of subacute hepatic failure. *Indian Journal of Medical Research* 98:69–74.

Adel, M., Alousi, L.A., and H.A. Salem. 2005. Licorice: a possible anti-inflammatory and anti-ulcer drug. *American Association of Pharmaceutical Scientists PharmSciTech* 6:74–82.

Akhtar, M.S., and V. Bhakuni. 2004. Streptococcus pneumoniae hyaluronate lyase: an overview. *Current Science* 86(2):285–295.

Alam, G., Singh M.P., and A. Singh. 2011. Wound healing potential of some medicinal plants. *International Journal of Pharmaceutical Sciences Review and Research* 9:136–145.

Anonymous. 2000. *The Useful Plant of India*. New Delhi, India: National Institute of Science Communication and Information Resources. pp.109.

Anonymous. 2001. *The Ayurvedic Pharmacopoeia of India*, Vol. 1. New Delhi, India: Ministry of Health and Family Welfare, Department Of AYUSH, Government of India, pp.108.

Anonymous. 2003a. *Quality Standard of Indian Medicinal Plants*, Vol. 1. New Delhi, India: ICMR, pp.102–108.

Anonymous. 2003b. *The Wealth of India (Raw Materials)*, Vol. 4. New Delhi, India: National Institute of Science Communication and Information Resources, pp.276–277.

Anonymous. 2007a. *The Unani Pharmacopoeia of India*, Vol. 2. New Delhi, India: Ministry of Health and Family welfare, Department of AYUSH, Government of India, pp.45–46.

Anonymous. 2007b. *The Unani Pharmacopoeia of India*, Vol. 3. New Delhi, India: Ministry of Health and Family Welfare, Department of AYUSH, pp.86–89.

Arai, K., Ishima, R.S., Morikawa, S., Miyasaka, A., Imoto, T., Yoshimura, S., Aimoto, S., and K. Akasaka. 1995. Three-dimensional structure of gurmarin, a sweet taste-suppressing polypeptide. *Journal of Biomolecular NMR* 5(3):297–305.

Arora, R., Chawla, R., Marwah, R., Arora, P., Sharma, R.K., Kaushik, V., Goel, R., Kaur, A., Silambarasan, M., Tripathi, R.P., and J.R. Bhardwaj. 2011. Potential of complementary and alternative medicine in preventive management of novel H1N1 Flu (Swine Flu) pandemic: thwarting potential disasters in the bud. *Evidence Based Complementary and Alternative Medicine* 2011:1–16.

Arun, L.B., Arunachalam, A.M., Arunachalam, K.D., Annamalai, S.K., and K.A. Kumar. 2014. *In vivo* anti-ulcer, anti-stress, anti-allergic and functional properties of Gymnemic Acid Isolated from *Gymnema sylvestre* R.Br. *BMC Complementary and Alternative Medicine* 14:70.

Badam, L. 1997. *In vitro* antiviral activity of indigenous glycyrrhizin, licorice and glycyrrhizic acid (Sigma) on Japanese encephalitis virus. *Journal of Communicable Diseases* 29:91–99.

Badr, S.E.A., Sakr, D.M., Mahfouz, S.A., and M.S. Abdelfattah. 2013. Licorice (*Glycyrrhiza glabra* L.) Chemical composition and biological impacts. *Research Journal of Pharmaceutical, Biological and Chemical Sciences* 4:606–621.

Baitar, I.Al. 1985. *Jame ul Mufradat ul Advia wal Aghzia (Urdu translation)*, Vol. 1. New Delhi, India: Central Council for Research in Unani Medicine, Ministry of Health and Family Welfare, Govt. of India, pp.185.

Batiha, G.E.S., Beshbishy, A.A., Tayebwa, D.S., Shaheen, M.H., Yokoyama, N., and I. Igarashi. 2019. Inhibitory effects of *Syzygium aromaticum* and *Camellia sinensis* methanolic extracts on the growth of Babesia and Theileria parasites. *Ticks and Tick-borne Diseases* 10:949–958.

Beshbishy, A.M., Batiha, G.E.S., Adeyemi, O.S., Yokoyama, N., and I. Igarashi. 2019. Inhibitory effects of methanolic *Olea europaea* and acetonic *Acacia laeta* on the growth of Babesia and Theileria. *Asian Pacific Journal of Tropical Medicine* 12:425–434.

Bhattacharjee, P.S.K. 2004. *Handbook of Medicinal Plants*, 4th ed. Jaipur: CRC Press, Pointer Publishers, pp.206.

Bhuvaneswari, C.H., Rao, K., and A. Giri. 2011. Evaluation of *Gymnema sylvestre* antimicrobial activity in methanol. *Recent Research in Science and Technology* 3:73–75.

Biondi, D.M., Rocco, C., and G. Ruberto. 2003. New dihydrostilbene derivatives from the leaves of *Glycycrrhiza glabra* and evaluation of their antioxidant activity. *Journal of Natural Products* 66:477–480.

Blatina, L.A. 2003. Chemical modification of glycyrrhizic acid as a route to bioactive compounds for medicine. *Current Medicinal Chemistry* 10:155–171.

Chakrapani, L.N., and K. Periandavan. 2018. Protective role of gymnemic acid in curbing high fat diet and high fructose induced pancreatic oxidative stress mediated type-2 diabetes in wistar rats. *International Journal of Pharmaceutical Sciences and Research* 9(5):2130–2139.

Chakravarti, D., and N.B. Debnath. 1981. Isolation of gymnemagenin the sapogenin from *Gymnema sylvestre* R.Br. (Asclepiadaceae). *Journal of the Institution of Chemists* 53:155–158.

Chan, H.T., Chan, C., and J.W. Ho. 2003. Inhibition of glycyrrhizic acid on alfatoxin B1-induced cytotoxicity of hepatoma cells. *Toxicology* 188:211–217.

Chang, G.H., Lin, Y.S., Hsu, K.H., Cheng, Y.C., Yang, P.R., and M.S. Tsai. 2021. Nasal irrigation with *Glycyrrhiza glabra* extract for treatment of allergic rhinitis – A study of *in vitro, in vivo* and clinical trial. *Journal of Ethnopharmacology* 15(275):114–116.

Chattopadhyay, R.R. 1999. Comparative evaluation of some blood sugar lowering agents of plant origin. *Journal of Ethnopharmacology* 67(3):367–372.

Chopra, R.N., Nayar, S.L., and I.C. Chopra. 1956. *Glossary of Indian Medicinal Plants.* New Delhi, India: Council of Scientific and Industrial Research, pp.1624–1627.

Chopra, R.N., Nayar, S.L., and I.C. Chopra. 1992. *Glossary of Indian Medicinal Plants.* New Delhi, India: Council of Scientific and Industrial Research, pp.319–322.

Chopra, R.N., Nayar, S.L., and I.C. Chopra. 2002. *Glossary of Indian Medicinal Plants.* New Delhi, India: Council of Scientific and Industrial Research, pp.1956–1992.

Damle, M. 2014. *Glycyrrhiza glabra* (Liquorice)-A potent medicinal herb. *International Journal of Herbal Medicine* 2:132–136.

Dateo, G.P. Jr., and L. Long. 1973. Gymnemic acid, the antisaccharine principle of *Gymnema sylvestre.* Studies on the isolation and heterogeneity of gymnemic acid A1. *Journal of Agricultural and Food Chemistry* 21(5):899–903.

De Clercq, E. 2000. Current lead natural products for the chemotherapy of human immunodeficiency virus (HIV) infection. *Medicinal Research Reviews* 20:323–349.

Devi, B.P., and R. Ramasubramaniaraja. 2010. Pharmacognostical and antimicrobial screening of *Gymnema sylvestre* R.Br and evaluation of Gurmar herbal tooth paste and powder, composed of *Gymnema sylvestre* R. Br, extracts in dental caries. *International Journal of Pharma and Bio Sciences* 1(3):1–16.

Devang, N., Adhikari, P., Nandini, M., Satyamoorthy, K., & Rai, P.S. 2021. Effect of licorice on patients with HSD11B1 gene polymorphisms- a pilot study. *Journal of Ayurveda and Integrative Medicine* 12(1):131–135.

Diwan, P.V., Margaret, I., and S. Ramakrishna. 1995. Influence of *Gymnema sylvestre* on inflammation. *Inflammopharmacology* 3:271–277.

Eric, G.B., and J.L. Lawrence. 1996. *Rheumatoid Arthritis and Its Therapy: The Text Book of Therapeutics: Drug and Disease Management*, 16th ed. Baltimore, MD: Williams and Wilkins Company, p.16.

Fiore, C., Salvi, M., Palermo, M., Sinigagliab, G., Armaninia, D., and A. Toninello. 2004. On the mechanism of mitochondrial permeability transition induction by glycyrrhetinic acid. *Biochimica et Biophysica Acta* 1658:195–201.

Foster, S. 2002. *Gymnema sylvestre in Alternative Medicine Reviews Monographs.* USA: Thorne Research Inc., pp.205–207.

Galletto, R., Siqueira, V.L.D., Ferreira, E.B., Oliveira, A., and R. Bazotte. 2004. Absence of antidiabetic and hypolipidemic effect of *Gymnema sylvestre* in non-diabetic and alloxan-diabetic rats. *Brazilian Archives of Biology and Technology* 47(4):545–551.

Gent, J.F., Hettinger, T.P., Frank, M.E., and L.E. Marks. 1999. Taste confusions following gymnemic acid rinse. *Chemical Senses* 24(4):393–403.

Ghani, N.M. 2010. *Khazainul Advia.* Vol. 2. New Delhi, India: Idara Kitab ul Shifa Publication, pp.554–569.

Gunasekaran, V., Srinivasan, S., and S.S. Rani. 2019. Potential antioxidant and antimicrobial activity of *Gymnema sylvestre* related to diabetes. *Journal of Medicinal Plants* 7(2):5–11.
Gupta, S.N., Pramanik, S., Tiwari, O.P., Thacker, N., Pande, M., and N. Upmanyu. 2010. Immunomodulatory activity of *Gymnema sylvestre* leaves. *International Journal of Pharmacology* 8(2):1531–2976.
Gurav S., Gulkari V., Duragkar N., and A. Patil. 2007. Systemic review: pharmacognosy, phytochemistry, pharmacology and clinical applications of *Gymnema sylvestre* R.Br. *Pharmacognosy Reviews* 1(2):338–343.
Hakeem, A.H. 2002. *Bustanul Mufradat*. New Delhi, India: Idara Kitabus Shifa, pp.43–44.
Hakim, M.A.H. 2009. *Mufradat Azeezi*. New Delhi, India: Central Council for Research in Unani Medicine, Ministry of AYUSH, Government of India, pp.104.
Haraguchi, H., Tanimato, K., Tamura, Y., and T. Kinoshita. 1998. Antioxidative and superoxide scavenging activities of retrochalcones in Glycyrrhiza inflate. *Phytochemistry* 48:125–129.
Hardman, J.G., Limbird, L.E., Goodman, L.S., and A.G Gilman. 2001. *Goodman and Gilman's the Pharmacological Basis of Therapeutics*, 10th ed. New York: McGraw Hill, pp.543.
Hikino, H., Wagner, H., and N.R Farnsworth. (eds). 1985. Recent research on oriental medicinal plants. In *Economic and Medicinal Plant Research*, Vol. 1. London: Academic, pp.53–85.
Hosseinzadeh, H., and M.N Asl. 2008. Review of Pharmacological Effects of Glycyrrhiza sp. and its Bioactive Compounds. *Phytotherapy Research* 22:709–724.
Hsiang, C.Y., Lai, I.L., Chao, D.C., and T.Y Ho. 2002. Differential regulation of activator protein-1 activity by glycyrrhizin. *Life Sciences* 70:1643–1656.
Imoto, T., Miyasaka, A., Ishima, R., and K. Akasaka. 1991. A novel peptide isolated from the leaves of *Gymnema sylvestre*-characterization and its suppressive effect on the neural responses to sweet taste stimuli in the rat. *Comparative Biochemistry and Physiology* 100(2):309–314.
Isbrucker, R.A., and G.A. Burdock. 2006. Risk and safety assessment on the consumption of Licorice root (Glycyrrhiza sp.), its extract and powder as a food ingredient, with emphasis on the pharmacology and toxicology of glycyrrhizin. *Regulatory Toxicology and Pharmacology* 4:167–192.
Jain, K.S., Kathiravan, M.K., Somani, R.S., and C.J. Shishoo. 2007. The biology and chemistry of hyperlipidemia. *Bioorganic and Medicinal Chemistry* 15(14):4674–4699.
Jeong, H.G., You, H.J., Park, S.J., Moon, A.R., Chung, Y.C., Kang, S.K., and H.K Chun. 2002. Hepatoprotective effects of 18β-glycyrrhetinic acids on carbon tetrachloride-induced liver injury, inhibition of cytochrome P450 2E1 expression. *Pharmacology Research* 46(3):221–227.
Kabeeruddin, A.H.M. 2010. *Makhzan-Ul-Mufradat*. New Delhi, India: Idara Kitab-Us-Shifa, pp.65.
Kamei, J., Nakamura, R., Ichiki, H., and M. Kubo. 2003. Anti-tussive principles of *Glycyrrhiza radix*, a main component of Kampo preparations Bakumondo-to. *European Journal of Pharmacology* 69:159–163.
Kanetkar, P., Singhal, R., and M. Kamat. 2007. *Gymnema sylvestre*: a memoir. *Journal of Clinical Biochemistry and Nutrition* 41(2):77–81.
Kang, M.H., Lee, M.S., Choi, M.K., Min, K.S., and T. Shibamoto. 2012. Hypoglycemic activity of *Gymnema sylvestre* extracts on oxidative stress and antioxidant status in diabetic rats. *Journal of Agricultural and Food Chemistry* 60(10):2517–2524.
Kaur, R., and A.S. Dhinds. 2013 *Glycyrrhiza glabra*: a phytopharmacological review. *International Journal of Pharmaceutical Sciences and Research* 4:2470–2477.
Kaushik, M., Kaushik, A., Arya, R., Singh, G., and P. Malik. 2011. Anti-obesity property of hexane extract from the leaves of *Gymnema sylvestre* in high fed cafeteria diet induced obesity rats. *International Research Journal of Pharmacy* 2:112–116.
Khanna, V., and K. Kannabiran. 2009. Anticancer-cytotoxic activity of saponins isolated from the leaves of *Gymnema sylvestre* and *Eclipta prostrata* on HeLa cells. *International Journal of Green Pharmacy* 3(3):227–229.
Khare, C.P. 2004. *Encyclopaedia of Indian Medicinal Plants*. New York: Springer, pp.233–235.
Khramov, V.A., Spasov, A.A., and M.P. Samokhina. 2008. Chemical composition of dry extracts of *Gymnema sylvestre* leaves. *Pharmaceutical Chemistry Journal* 42(1):30–32.
Kimura, M., Inoue, H., Hirabayahi, K., Natsume, H., and M. Ogihara. 2001. Glycyrrhizin and some analogues induce growth of primary cultured adult rat hepatocytes via epidermal growth factor receptors. *European Journal of Pharmacology* 431:151–161.
Kiranmai, M., Kazim, S.M., and M. Ibrahim. 2011. Combined wound healing activity of *Gymnema sylvestre* and *Tagetes erecta* linn. *International Journal of Pharmaceutical Applications* 2:135–140.
Kirtikar, K.R., and B.D. Basu. 1975. *Indian Medicinal Plants*, Vol. 3. New Delhi, India: Periodicals Experts, pp.1625.
Kirtikar, K.R., and B.D. Basu. 2003. *Indian Medicinal Plants*.7 ed. Vol. 3. Dehradun: Oriental Press, pp.2240–2244.

Kishore, L., and R. Singh. 2015. Protective effect of *Gymnema sylvestre* L. against advanced glycation end-product, sorbitol accumulation and aldose reductase activity in homoeopathic formulation. *Indian Journal of Research in Homoeopathy* 9(4):240–248.

Kokate, C.K. 1999. *Pharmacognosy.* New Delhi, India: Nirali Prakashan, p.12.

Kritikar, K.R., and B.D. Basu. 2008. *Indian Medicinal Plants.* Dehradun: International Book Distributors, pp.1632–1635.

Kumar, S.N., Mani, U.V., and I. Mani. 2010. An open label study on the supplementation of *Gymnema sylvestre* in type 2 diabetics. *Journal of Dietary Supplements* 7(3):273–282.

Li, Y., Zheng, M., Zhai, X., Huang, Y., Khalid, A., and A. Malik. 2015. Effect of *Gymnema sylvestre*, *Citrullus colocynthis* and *Artemisia absinthium* on blood glucose and lipid profile in diabetic human. *Acta Poloniae Pharmaceutica* 72:981–985.

Lindley, J. 2000. *Flora Medica.* New Delhi, India: Ajay Book Service, pp.46.

Mahmoudabadi, A.Z., Iravani, M., and A. Khazrei. 2009. Anti-fungal activity of *Glycyrrhiza glabra* (Licorice) against vaginal isolates of Candida. *BioTechnology: An Indian Journal* 3(2):75–77.

Malik, J.K., Manvi, F.V., and K.R. Alagawadi. 2007. Evaluation of anti-inflammatory activity of *Gymnema sylvestre* leaves extract in rats. *International Journal of Green Pharmacy* 2:114–115.

Malik, J.K., Manvi, F.V., and B.R. Nanjware. 2009. Wound healing properties of alcoholic extract of *Gymnema sylvestre* R.Br. leaves in rats. *Journal of Pharmacy Research* 2:1029–1030.

Malik, J.K., Manvi, F.V., Nanjware, B.R., Dwivedi, D.K., Purohit, P., and S. Chouhan. 2010. Anti-arthritic activity of leaves of *Gymnema sylvestre* R.Br. leaves in rats. *Der Pharmacia Lettre* 2:336–341.

Manohar, S.H., Naik, P.M., Praveen, N., and H.N Murthy. 2009. Distribution of gymnemic acid in various organs of Gymnema sylvestre. *Journal of Forestry Research* 20(3):268–270.

Marsh, P., and M. Martin. 1992. *Oral Microbiology*, Vol. 3. London: Chapman and Hall.

Masoomeh, M.J., and G. Kiarash. 2007. *In vitro* susceptibility of Helicobacter pylori to licorice extract. *Iranian Journal of Pharmaceutical Research* 6:69–72.

Mathew, M. 2004. Aromatic and medicinal plants research station, Odakkali: a centre for promoting medicinal and aromatic plants. *Indian Coconut Journal* 34(10):10–15.

Mauricio, I., Francischett, B., Monterio, R.Q., and J.A. Guimaraeas. 1997. Identification of glycyrrhizin as thrombin inhibitor. *Biochemical and Biophysical Research Communications* 235:259–263.

Mendes-Silva, W., Assafim, M., Ruta, B., Monteiro, R.Q., Guimaraes, J.A., and R.B. Zingali. 2003. Antithrombotic effect of glycyrrhizin, a plant-derived thrombin inhibitor. *Thrombosis Research* 112:93–98.

Murakami, N., Murakami, T, Kadoya M., Matsuda, H., Yamahara, J., and M. Yoshikawa. 1996. New hypoglycemic constituents in gymnemic acid from *Gymnema sylvestre. Chemical and Pharmaceutical Bulletin* 44(2):469–471.

Nadkarni, K.M. 1982. *Indian Materia Medica*, Vol. 1, 2nd ed. (3rd revised). Mumbai: Popular Prakashan, 596–599.

Nadkarni, K.M. 1986. *Gymnema sylvestre.* In *Indian Materia Medica with Ayurvedic Unani*, Vol. 1. Bombay, India: Popular Prakashan, pp.596–599.

Nahidi, F., Zare, E., Mojab, F., & Alavi-Majd, H. 2012. Effects of licorice on relief and recurrence of menopausal hot flashes. *Iranian Journal of Pharmaceutical Tresearch* 11(2):541–548.

Nitalikar, M.M., Munde, K.C., Dhore, B.V., and S.N. Shikalgar. 2010. Studies of antibacterial activities of *Glycyrrhiza glabra* root extract. *International Journal of Pharmacy and Technology* 2:899–901.

Ohmori, R., Iwamoto, T., Tago, M., Takeo, T., Unno, T., and H. Itakura. 2005. Antioxidant activity of various teas against free radicals and LDL oxidation. *Lipids* 40(8):849–853.

Patil, P.M., Chaudhari, P.D., Duragkar, N.J., and P.P. Katolkar. 2012. Formulation of anti-diabetic liquid preparation of *Gymnema sylvestre* and qualitative estimated by TLC. *Asian Journal of Pharmaceutical and Clinical Research* 5(1):16–19.

Potawale, S.E., Shinde, V.M., Anandi, L., Borade, S., Dhalawat, H., and R.S. Deshmukh. 2008. *Gymnema sylvestre*: a comprehensive review. *Pharmacology Online* 2:144–157.

Rachh, P.R., Patel, S.R., Hirpara, H.V., Rupareliya, M.T., Rachh, M.R., and A.S. Bhargava. 2009. In-vitro evaluation of antioxidant activity of *Gymnema sylvestre* R.Br. leaf extract. *Romanian Journal of Biology-Plant Biology* 54(2):141–148.

Rachh, P.R., Rachh, M.R., Ghadiya, N.R., Modi, D.C., Modi, K.P., Patel, N.M., and M.T. Rupareliya. 2010. Antihyperlipidemic activity of *Gymenma sylvestre* R.Br. leaf extract on rats fed with high cholesterol diet. *International Journal of Pharmacology* 6(2):138–141.

Rahman, M.M., Habib, M.R., Hasan, M.A., Saha, A., and A. Mannan. 2014. Comparative assessment on In vitro antioxidant activities of ethanol extracts of *Averrhoa bilimbi. Gymnema sylvestre* and *Capsicum frutescens. Pharmacognosy Research* 6:36–41.

Rao, G.S., and J.E. Sinsheimer. 1971. Constituents from *Gymnema sylvestre* leaves. Isolation, chemistry, and derivatives of gymnemagenin and gymnestrogenin. *Journal of Pharmaceutical Sciences* 60(2):190–193.

Rathi, S.G., Suthar, M., Patel, P., Bhaskar, V.H., and N.B. Rajgor. 2009. *In-vitro* cytotoxic screening of *Glycyrrhiza glabra* L. (Fabaceae): a natural anticancer drug. *Journal of Young Pharmacist* 1(3):239–243.

Raveendra, K.R., Jayachandra, S.V., Sushma, K.R., Allan, J.J., Goudar, K.S., Shivaprasad, H.N., and K. Venkateshwarlu. 2012. An extract of *Glycyrrhiza glabra* (gutguard) alleviates symptoms of functional dyspepsia: a randomized, double-blind, placebo-controlled study. *Evidence-Based Complementary and Alternative Medicine* 2012:1–9.

Reddy, P.S., Gopal, G.R., and G.L. Sita. 2004. *In vitro* multiplication of *Gymnema sylvestre* R.Br. an important medicinal plant. *Current Science* 10:1–4.

Rupanar, S.V., Pingale, S.S., Dandge, C.N., and D. Kshirsagar. 2012. Phytochemical screening and *in vitro* evaluation of antioxidant antimicrobial activity of *Gymnema sylvestre. International Journal of Current Research* 8(12):43480–43486.

Saeed, A. 2007. *Kitab Al Fatah Fi Al Tadawi (Urdu Translation).* New Delhi, India: NCPC Printers, pp.89.

Salvi, M., Fiore, C., Armanini, D., and A. Toninello. 2003. Glycyrrhetinic acid induced permeability transition in rat liver mitochondria. *Biochemical Pharmacology* 66:2375–2379.

Satdive, R.K., Abhilash, P., and D.P. Fulzele. 2003. Antimicrobial activity of *Gymnema sylvestre* leaf extract. *Fitoterapia.* 74(7–8):699–701.

Sathya, A., Ramasubramaniaraja, R., and P. Brindha. 2010. Pharmacognostical, phytochemical and GC-MS investigation of successive extract of *Gymnema sylvestre* R.Br. *Journal of Pharmacy Research* 3:984–987.

Sato, H., Goto, W., Yamamura, J., Kurokawa, M., Kageyama, S., Takahara, T., Watanabe, A., and K. Shiraki. 1996. Therapeutic basis of glycyrrhizin on chronic hepatitis B. *Antiviral Research* 30:171–177.

Saumendu, D.R., Sarkar, K., Dipankar, S., Singh, T., and B. Prabha. 2010. *In vitro* antibiotic activity of various extracts of *Gymnema sylvestre. International Journal of Pharmaceutical Research and Development* 2:1–3.

Shah, S.L., Wahid, F., Khan, N., Farooq, U., Shah, A.J., Tareen, S., Ahmad, F., and T. Khan. 2018. Inhibitory effects of *Glycyrrhiza glabra* and its major constituent Glycyrrhizin on inflammation-associated corneal neovascularization. *Evidence Based Complementary and Alternative Medicine* 23:843–850.

Shanmugasundaram, E., Rajeswari, G., Baskaran, K., Kumar, B., Shanmugasundaram, K., and B. Ahmath. 1990. Use of *Gymnema sylvestre* leaf extract in the control of blood glucose in insulin-dependent diabetes mellitus. *Journal of Ethnopharmacology* 30(3):281–294.

Sharma, V., and R.C. Agrawal 2014. *In vivo* antioxidant and hepatoprotective potential of *Glycyrrhiza glabra* extract on carbon tetra chloride (CCl4) induced oxidative stress mediated hepatotoxicity. *International Journal of Research in Medical Sciences* 2:314–320.

Sharma, V., Agrawal, R.C., and S. Pandey. 2013. Phytochemical screening and determination of antibacterial and antioxidant potential of *Glycyrrhiza glabra* root extracts. *Journal of Environmental Research and Development* 7(4A):1552–1558.

Sharma, V., Agrawal, R.C., and V.K. Shrivastava. 2014. Assessment of median lethal dose and antimutagenic effects of *Glycyrrhiza glabra* root extract against chemically induced micronucleus formation in swiss albino mice. *International Journal of Basic Clinical Pharmacology* 3:292–29.

Sheela, M.L., Ramakrishna, M.K., and B.P. Salimath. 2006. Angiogenic and proliferative effects of the cytokine VEGF in Ehrlich ascites tumor cells is inhibited by *Glycyrrhiza glabra. International Immunopharmacology* 6:494–498.

Shivaprasad, H.N., Kharya, M.D., Rana, A.C., and S. Mohan. 2006. Preliminary immunomodulatory activities of the aqueous extract of *Terminalia chebula. Pharmaceutical Biology* 44(1):32–34.

Sina, I. 1998. *Al-Qanoon, Fit Tib. (English translation By Dept. of Islamic Studies Jamia Hamdard).* New Delhi, India: S. Waris Nawab, Senior Press Superintendent.

Singh, V.K., Umar, S., Ansari, S.A., and M. Iqbal. 2008. *Gymnema sylvestre* for diabetics. *Journal of Herbs, Spices and Medicinal Plants* 14(1–2):88–106.

Sinsheimer, J.E., Rao, G.S., and H.M. McIlhenny. 1970. Constuents from *Gymnema sylvestre* leaves. V: isolation and preliminary characterization of the gymnemic acids. *Journal of Pharmaceutical Sciences* 59(5):622–628.

Srivastava, R., Solanki, S.S., Tomar, V., Garud, A., Garud, P., Kannojia, P., and N. Jain. 2009. Comparative evaluation of polyherbal combination for hypolipidaemic activity. *International Journal of Pharmaceutical Sciences and Drug Research* 1(1):9–12.

Srividya, A.R., Varma, S.K., Dhanapal, S.P., Vadivelan, R., and P. Vijayan. 2010. *In vitro* and *in vivo* evaluation of hepatoprotective activity of *Gymnema sylvestre. International Journal of Pharmaceutical Sciences and Nanotechnology* 2:768–773.

Sugihara, Y., Nojima, H., Matsuda, H., Murakami, T, Yoshikawa, M., and I. Kimura. 2000. Antihyperglycemic effects of gymnemic acid IV, a compound derived from Gymnema sylvestre leaves in streptozotocin-diabetic mice. *Journal of Asian Natural Products Research* 2(4):321–327.

Suttisri, R., Lee, I-S., and A.K. Douglas. 1995. Plant-derived triterpenoid sweetness inhibitors. *Journal of Ethnopharmacology* 47(1):9–26.

Takii, H., Kometani, T., Nishimura, T., Nakae, T., Okada, S., and T. Fushiki. 2000. Anti-diabetic effect of glycyrrhizin in genetically diabetic KK-Ay mice. *Biological and Pharmaceutical Bulletin* 24:484–487.

Van Rossum, T.G., Vulto, A.G., De Man, R.A., Brouwer, J.T., and S.W. Schalm. 1998. Glycyrrhizin as a potential treatment of chronic hepatitis C. *Alimentary Pharmacology and Therapeutics* 12:199–205.

Wagner, H., and K. Jurcic. 2002. Immunological studies of Revitonil: a phytopharmaceutical containing *Echinacea purpurea* and *Glycyrrhiza glabra* root extract. *Phytomedicine* 9(5):390–397.

Washington, D.C. 2003. *Food Chemicals Codex*, 5th ed. Washington, DC: National Academy Press, pp.25.

Watanabe, M., Hayakawa, S., Isemura, M., Kumazawa, S., Nakayama, T., Mori, C., and T. Kawakami. 2002. Identification of licocoumarone as an apoptosis inducing component in licorice. *Biological and Pharmaceutical Bulletin* 25:1388–1390.

Yogisha, S., and K.A. Raveesha. 2009. *In vitro* antibacterial effect of selected medicinal plant extracts. *Journal of Natural Products* 2:64–69.

Yoshikawa, K., Amimoto, K., Arihara, S., and K. Matsuura. 1989. Structure studies of new antisweet constituents from *Gymnema sylvestre. Tetrahedron Letters* 30(9):1103–1106.

Yoshikawa, K., Arihara, S., and K. Matsuura. 1991. A new type of antisweat principles occurring in *Gymnema sylvestre. Tetrahedron Letters* 32(6):789–792.

Yoshikawa, K., Murakami, T., and H. Matsuda. 1997. Medicinal food stuffs. IX. The inhibitors of glucose absorption from the leaves of *Gymnema sylvestre* R.Br. (Asclepiadaceae): structures of gymnemosides A and B. *Chemical and Pharmaceutical Bulletin.* 45(10):1671–1676.

Yoshikawa, K., Nakagawa, M., Yamamoto, R., Arihara, S., and K. Matsuura. 1992. Antisweet natural products. V. Structures of gymnemic acids VIII-XII from *Gymnema sylvestre* R.Br. *Chemical and Pharmaceutical Bulletin* 40(7):1779–1782.

Zadeh, J.B., Kor, Z.M., and M.K. Goftar. 2013. Licorice (*Glycyrrhiza glabra* Linn) as a valuable medicinal plant. *International Journal of Advanced Biological and Biomedical Research* 1(10):1281–1288.

Zuñiga, L., González-Ortiz, Y., and E. Martínez-Abundis. 2017. Effect of *Gymnema sylvestre* administration on metabolic syndrome, insulin sensitivity, and insulin secretion. *Journal of Medicinal Food* 20(8):750–754.

8 *Hydrastis canadensis* (Goldenseal) and *Lawsonia inermis* (Henna)

Md. Mizanur Rahaman, William N. Setzer, Javad Sharifi-Rad, and Muhammad Torequl Islam

CONTENTS

8.1 Introduction 152
8.2 Description 152
8.2.1 *Hydrastis canadensis* (Goldenseal) 152
8.2.2 *Lawsonia inermis* (Henna) 152
8.3 Traditional Knowledge 152
8.3.1 *Hydrastis canadensis* (Goldenseal) 152
8.3.2 *Lawsonia inermis* (Henna) 153
8.4 Chemical Derivatives of *H. canadensis* (Goldenseal) and *L. Inermis* (Henna) 153
8.5 Potential Benefits of *H. canadensis* (Goldenseal) 154
8.5.1 Antibacterial Effect 154
8.5.2 Upper Respiratory Tract Infection and Colds 157
8.5.3 Diabetes 157
8.5.4 Chlamydia and Herpes 157
8.5.5 Acne and Psoriasis 157
8.5.6 Oral Health 158
8.5.7 Cardiovascular Effects 158
8.5.8 Immune Modulation 158
8.6 Potential Benefits of *L. inermis* (Henna) 158
8.6.1 Anti-Aging Properties 158
8.6.2 Wound Healing 158
8.6.3 Memory Enhancement 158
8.6.4 Antibacterial Activity 158
8.6.5 HypoglycemicActivity 159
8.6.6 Immunomodulatory Effect 159
8.6.7 Abortifacient Activity 159
8.6.8 Hepatoprotective Activity 159
8.6.9 Antioxidant Activity 159
8.6.10 Antifungal Activity 159
8.7 Conclusion 159
References 160

DOI: 10.1201/9781003205067-8

8.1 INTRODUCTION

Medicinal plants have been used worldwide for thousands of years as alternative or complementary medicines. They contain natural substances that are valuable due to their biological activities. These plants are promising sources of chemical components that can used for drug development and synthesis. However, a few medicinal plants are considered due to their nutrition, and as a result of that, these plants (e.g., ginger, green tea, and walnuts) are recommended for their therapeutic values (Hassan, 2012). Most people who turn each year to medicinal plants do so because they believe these plant remedies will cause fewer or no undesirable side effects (Philomena, 2011). However, medicinal herbs are important as a primary health care mode for approximately 85% of the world's total population (Pesic, 2015) and also an important resource for drug discovery; approximately 80% of all synthetic drugs are derivedfrom them (Bauer and Bronstrup, 2014). Both *Hydrastis canadensis* (goldenseal) and *Lawsonia inermis* (henna) are well known medicinal herbs due to their high physiological activities. *Hydrastis canadensis*,commonly known as "goldenseal", is broadly used for its important therapeutic activitiesin humans, such as antimicrobial, anti-inflammatory, anti-spasmodic, anti-parasitic, anticancer, and anti-secretory activities. Additionally, it is sometimes used for the treatment of digestive disorders, nasal congestion, snakebite, chronic candidiasis, canker sores, and vaginitis (Asmi and Lakshmi, 2013). On the other hand, pharmacological studies have shown that henna shows different activities, such as antibacterial, anti-parasitic, antifungal, antioxidant, molluscicidal, antipyretic, immunomodulatory, hepatoprotective, antidiabetic, analgesic, anti-inflammatory, central nervous, wound and burn healing, hypolipidemic, anti-urolithiatic, anti-diarrheal, antiulcer, diuretic, anticancer, and also some other pharmacological effects (Al-Snafi et al.,2019).

8.2 DESCRIPTION

8.2.1 *Hydrastis canadensis* (Goldenseal)

Hydrastis canadensis (Figure 8.1) is an herbaceous plant that is annually growing from a horizontal, yellowish rhizome, Plantae kingdom and Ranunculaceae family. Fine hairy, erect, unbranched, stalks are 15–50 cm long (www.efloras.org. Retrieved January 1, 2021). Goldenseal is a small, long-lived herb with two large leaves (Gentry et al., 1998). It has one small white flower and a single fruit that is similar in shape, and its color is raspberry (Lloyd and Lloyd, 1884).

8.2.2 *Lawsonia inermis* (Henna)

Lawsonia inermis (Kingdom: Plantae, Family: Lythraceae) (Figure 8.2) is a branched, deciduous, glabrous plant that sometimes occurs as a spinescent shrub or small tree with greyish brown type bark, and its height is 2.4–5 m (Borade et al., 2011). It is cultivated all over India as a hedge plant and as a commercial crop in several states of India for its dye properties (Dev, 2006). The fruits of henna are small, brownish capsules, 4–8 mm (0.16–0.31 in) in diameter, containing 32–49 seeds per fruit, opening irregularly into four fragments (Kumar et al., 2005).

8.3 TRADITIONAL KNOWLEDGE

8.3.1 *Hydrastis canadensis* (Goldenseal)

Hydrastis canadensis is a medicinal plant that is also known as "orange root" or "yellow puccoon" (United States Department of Agriculture (USDA). Retrieved December 12, 2017). It is one of the most commonly used food supplements in the USA (Weber et al., 2003). This plant is nativeto southeastern Canada and the eastern United States. Goldenseal has been usedby Native Americans to treat different types of disease and several health conditions, like gastric and digestive disorders, wound healing, skin andeye ailments, peptic ulcers and colitis, cancers, liver conditions, and

FIGURE 8.1 Different parts of *Hydrastis canadensis*.

diarrhea (Weber et al., 2003; Predny, 2005; Hwang et al., 2003). However, preparations of goldenseal rhizome extracts have been used for centuries as traditional medicine for the treatment of different conditions including ulcers, gastritis, intestinal catarrh, urinary disorders, muscular debility, constipation, nervous prostration, dysmenorrhea, hepatic congestion, blood stasis, skin, mouth, and eye infections;upper respiratory disorders;cancer;and diarrhea (Weber et al., 2003; Predny, 2005; Hwang et al., 2003; Inbaraj et al., 2006; Leyte-Lugo et al., 2017).

8.3.2 *Lawsonia inermis* (Henna)

Lawsonia inermis is a commonly known medicinal plant all over the world. This plant is also known as "mehndi" or "henna". *L. inermis* has been widely used to adorn hair, nails, beards, fingernails, leather, silk, and wool for many centuries ((Hema et al., 2010; Gupta, 2003; Ali, 1996; Anand et al., 1992; Auboyer, 2002). For the treatment of poliomyelitis and measles, the leaves of henna are used among the Yoruba tribe of southwestern Nigeria. Henna seeds are also used for their deodorant action and sometimes are used in cases of gynecological disorders especially for vaginal discharge, menorrhagia, and leukorrhea (Nawagish et al., 2007). In India, henna is widely used in the cosmetic industry as a dyeing agent (Chengaiah et al., 2010).

8.4 CHEMICAL DERIVATIVES OF *H. CANADENSIS* (GOLDENSEAL) AND *L. INERMIS* (HENNA)

The major chemical compounds of goldenseal are found in theroots and rhizomes (Pengelly et al., 2012). The plant contains the isoquinoline alkaloids, such as hydrastine, berberine, berberastine, hydrastinine, tetrahydroberberastine, canadine, and canalidine (Weber et al., 2003).The three major alkaloids are berberine, hydrastine, and canadine (Pengelly et al., 2012). *Hydrastis canadensis* has

FIGURE 8.2 Different parts of *Lawsonia inermis*.

been indicated to contain isoquinoline alkaloids inthe ranges of 1.5–4% hydrastine, 0.5–6% berberine, and 2–3% berberastine (Hamon, 1990). One study also found the compound 8-oxotetrahydrothalifendine (Gentry et al., 1998).

Lawsonia inermis contains 2-hydroxy-1,4-napthoquinone (lawsone) which is the main natural dye (Kirkland and Marzin, 2003). Different types of compounds also present in henna include1,4-dihydroxynaphthalene, 1,4-naphthoquinone, β-sitosterol, betulinicacid, xanthones, quercetin, lupeol, fraxetin, stigmasterol, and flavonoids (e.g., luteolins, apigenin, and their glycosides) (Borade et al., 2011; Sing et al., 2015). Henna leaves also contain substances which are soluble in water, alcohol, and other solvents, such as resin, gallic acid, tannin, fat, glucose, mannitol, and mucilage (Nayak et al., 2007; Dilworth et al., 2017). Coumarins, lacoumarin (5-allyoxy-7-hydroxycoumarin), two new isocoumarin carbonates (inermis carbonates A and B) are also found inthehenna plant (Bhardwaj et al., 1976, Yang et al., 2016). Henna bark contains some valuable chemical derivatives, such as naphthoquinone, isoplumbagin, triterpenoids-hennadiol, and aliphatics (3-methylnonacosan-1-ol) (Dev, 2006). A number of important chemical constituents of *H. canadensis* and *L. inermis* are shown in Figure 8.3.

8.5 POTENTIAL BENEFITS OF *H. CANADENSIS* (GOLDENSEAL)

Goldenseal is valued mainly for its anti-inflammatory and antibacterial properties. It has been used primarily to treat upper respiratory tract infections and the common cold (Ettefagh et al., 2011). The health benefits of goldenseal are outlined below.

8.5.1 Antibacterial Effect

Goldenseal is beneficial for its antibacterial activity; most of the research shows evidence that goldenseal has antibacterial effects due to the presence of the active compounds berberine and hydrastine

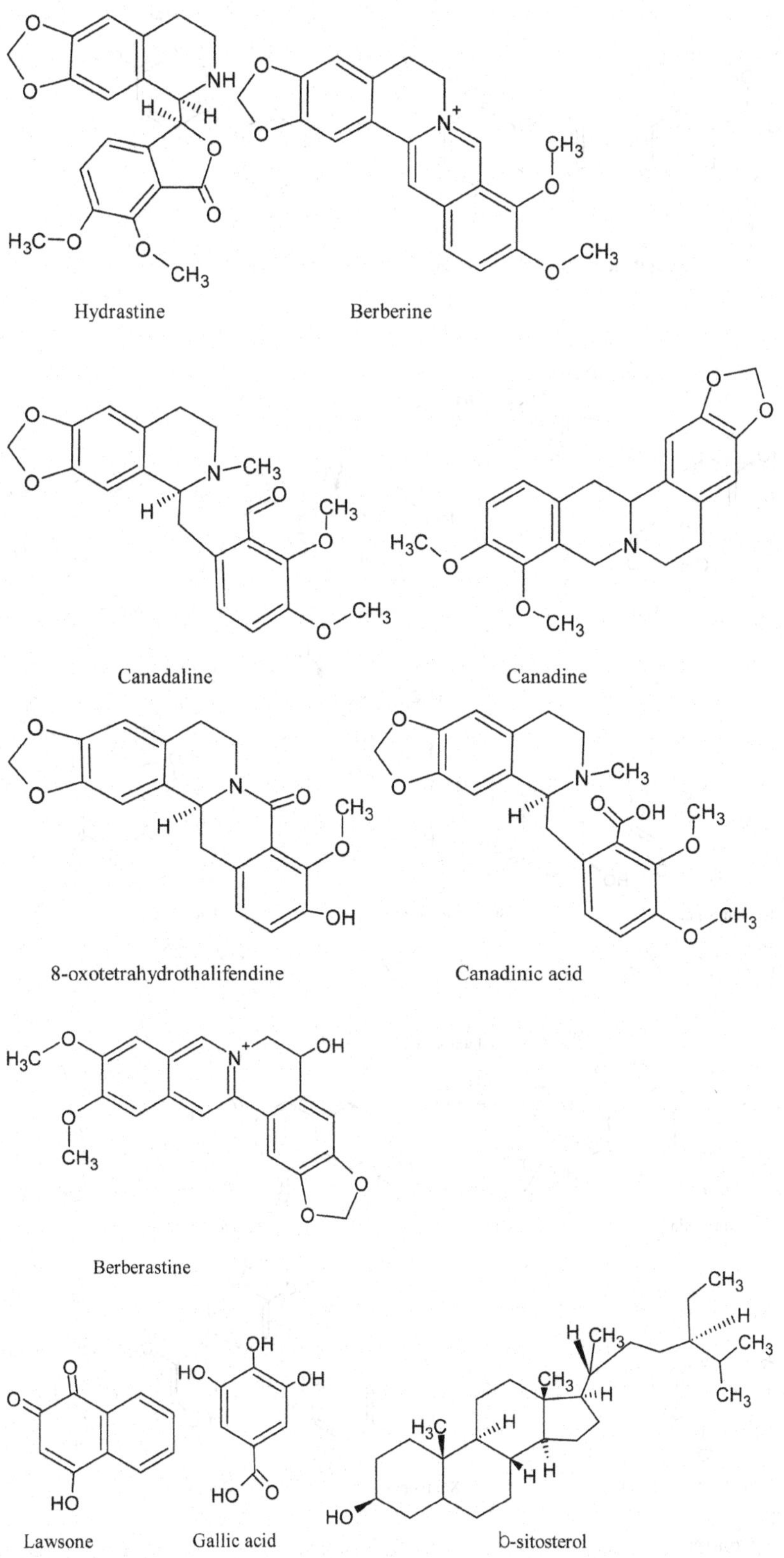

Hydrastine

Berberine

Canadaline

Canadine

8-oxotetrahydrothalifendine

Canadinic acid

Berberastine

Lawsone

Gallic acid

b-sitosterol

FIGURE 8.3 Structure of some important compounds found in *H. canadensis* and *L. inermis*.

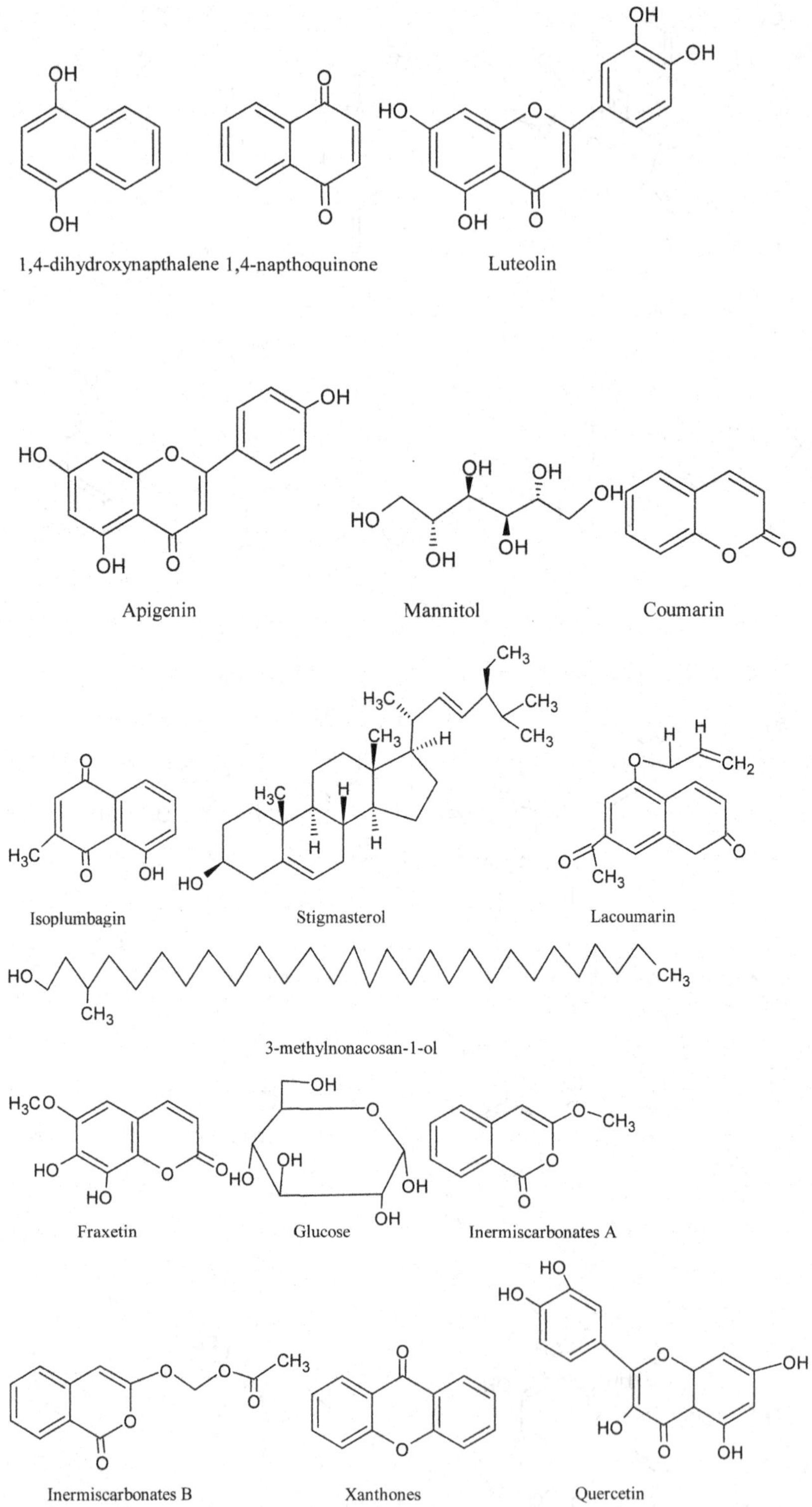

1,4-dihydroxynapthalene 1,4-napthoquinone Luteolin

Apigenin Mannitol Coumarin

Isoplumbagin Stigmasterol Lacoumarin

3-methylnonacosan-1-ol

Fraxetin Glucose Inermiscarbonates A

Inermiscarbonates B Xanthones Quercetin

FIGURE 8.3 (Continued)

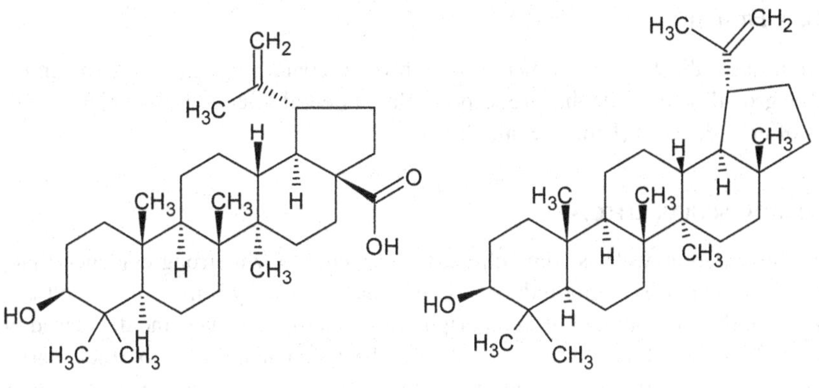

Betulinic acid

Lupeol

FIGURE 8.3 (Continued)

(Asmi and Lakshmi, 2013). Berberine has shown antibacterial activity against a few types of bacteria, such as *Vibrio cholerae, Escherichia coli,* which in turn significantly reduces smooth muscle contractions (Sack and Froelich, 1982; Swabb et al., 1981; Amin et al., 1969; Akhteret al., 1979). This antibacterial activity is presumably attributed to the inhibitory effect of berberine on the fimbrial structure formation on the surface of the bacteria (Sun et al., 1988).

8.5.2 Upper Respiratory Tract Infection and Colds

Goldenseal is one of the most useful natural treatments for upper respiratory tract infections including the common cold (National Center for Biotechnology Information). Cell-based and animal studies have shown that berberine is one of the major active compounds in goldenseal, which are used to fight against bacterial and viral infections, including the common cold virus (Enkhtaivan et al., 2017).

8.5.3 Diabetes

Berberine is the main compound in goldenseal that helps to reduce sugar absorption from the gut, lower insulin resistance, and promote insulin secretion (Pang et al., 2015). A few studies have suggested that the bloodsugar-lowering effects of berberine show activity comparable to metformin, which is used as a common antidiabetic medication (Pang et al., 2015).

8.5.4 Chlamydia and Herpes

Research suggests that berberine is one of the main active compounds in goldenseal that may help to treat herpes and chlamydia infections (https://www.healthline.com/health/goldenseal-cure-for-everything#benefits-uses). Vaginal chlamydia infections may be medicated with berberine-containing douches, vaginal suppositories, and also various types of oral goldenseal supplements (Vermani and Garg, 2002). One study demonstrated that goldenseal mixed with myrrh and thyme helped to treat oral herpes (Chin et al., 2010).

8.5.5 Acne and Psoriasis

Berberine which is the well-known active compound in goldenseal that may help to fight against *Cutibacterium acnes*, a bacterium responsible for acne (Sinha et al., 2014). In addition, some animal research suggests that the anti-inflammatory effect of berberine may help to treat inflammatory skin conditions like psoriasis (Nimisha et al., 2017).

8.5.6 Oral Health

Some research has indicated that an herbal mouth rinse containing various herbs and goldenseal attenuated the growth of bacteria that are responsible for dental infections, like plaque and gingivitis (mild forms of gum disease) (Kumar et al., 2013).

8.5.7 Cardiovascular Effects

Much animal research as well as some clinical trials have shown strong evidence that berberine administration prevented ischemia-induced ventricular tachyarrhythmia, stimulated cardiac contractility, and that this compound lowered peripheral vascular resistance and decreased blood pressure (Chun et al., 1978; Marin-Neto et al., 1988). The mechanism of berberine's antiarrhythmic effect is not yet clear, but an animal study indicated that it may be attributed to the suppression of delayed after-depolarization in the ventricular muscle (Wang et al., 1994).

8.5.8 Immune Modulation

The natural effects of *H. canadensis* on pro-inflammatory cytokines, which are produced by cultured macrophages, were examined. The results indicated that this plant was found to exhibit abilities that modulate macrophage responses during stimulation;it was also found to govern macrophage immune responses and activation events (Kruzel et al., 2008).

8.6 POTENTIAL BENEFITS OF *L. INERMIS* (HENNA)

Despite the fact that henna has been used primarily as a dye plant, it is also known for its medicinal properties. The medicinal properties with the most scientific investigations are outlined below.

8.6.1 Anti-Aging Properties

Though the antioxidant capacity of henna has not been widely researched, the oil has been known to be an astringent, and many people use its juice and oil on the skin to reduce the signs of aging and wrinkles(https://www.healthbenefitstimes.com/henna-plant/). It is supplemented by antiviral and antibacterial activity that can defend the largest organ ofthe body, the skin (Malekzadeh, 1968).

8.6.2 Wound Healing

Henna is a well-known medicinal plant, which is considered most beneficial for protecting the skin against infections and eliminating inflammation. Arecent study showed that plant leaf extracts were able to inhibit the growth of microorganisms that are responsible for causing burn wound infections (Muhammad and Muhammad, 2005).

8.6.3 Memory Enhancement

A recent study suggests that *L. inermis* leaves may have a memory-enhancing effect due to an increase of cholinergic neurotransmission through inhibition of acetylcholinesterase (AChE) activity and by stabilizing the antioxidant system (Rajesh et al., 2017).

8.6.4 Antibacterial Activity

Alcoholic extract of henna leaves has shown antibacterial effects on four bacterial strains, *Staphylococcus aureus*, *Staphylococcus epidermidis* (coagulase–negative staphylococci (cons), ß-hemolytic streptococci and *Pseudomonas aeruginosa*) (Al-Rubiay et al., 2008). It has shown *in vitro* antibacterial action against tested bacterial strains (Hemem, 2002).

8.6.5 HypoglycemicActivity

Syamsudin and Winarno (2008) ran a study to determine the effect of *L. inermis* leaf ethanol extract on glucose levels of artificiallyinduced diabetes in rats. The extract decreased the glucose level significantly, and it showed notable hypoglycemic activity. Henna ethanol extract also demonstrated hypolipidemic activity. Arayne et al. (2007) have provided evidence that *L. inermis* methanolic leaf extract also showed significant *in vitro* antihyperglycemic activity.

8.6.6 Immunomodulatory Effect

Mikhaeil et al. (2004), have shown that the methanolic extract of henna leaves at 1 mg/mL concentration produced immunomodulatory activity, which was determined by stimulation of T-lymphocyte proliferative responses. According to Dikshit et al. (2000), naphthoquinone is also an important component of *L. inermis* that shows significant immunomodulatory effects.

8.6.7 Abortifacient Activity

Aguwa (1987) has suggested that the methanolic extract of *L. inermis* root shows abortifacient activity; the methanolic extract demonstrated dose-dependent activity in the induction of abortion in rats, mice, and guinea pigs.

8.6.8 Hepatoprotective Activity

Hossain et al. (2011) demonstrated that a warm aqueous extract of *L. inermis* leaves showed protection against carbon tetrachloride (CCl_4)-induced liver damage. The significant hepatoprotective effects of henna were obtained as evidenced by decreased levels of serum enzymes, glutamate pyruvate transaminase (SGPT), glutamate oxaloacetate transaminase (SGOT), serum alkaline phosphatase (SAKP), and serum bilirubin (SB).

8.6.9 Antioxidant Activity

Al-Damegh (2014) provided evidence that *L. inermis* leaf extract has high antioxidant activity that may inhibit oxidative toxicity. High doses of henna may enhance the antioxidant defense system against reactive oxygen species and can be shown as a potential treatment to elevate the toxic effects associated with liver and kidney disease.

8.6.10 Antifungal Activity

Rahmoun et al. (2013) found that ethanol extract of *L. inermis* leaf shows strong antifungal activity against *Fusarium oxysporum* (MIC 230 μg/mL), which can be pathogenic for plants. This antifungal activity may be related to the presence of the active component lawsone.

8.7 CONCLUSION

Most of the studies suggest that both goldenseal and henna are biologically very active, which has attracted much attention. Goldenseal has shown very strong antimicrobial, hypoglycemic, and anti-mycotic activity due to the presence of highly active alkaloids. On the other hand, henna has demonstrated different types of medicinal properties including antibacterial, antiviral, abortifacient, memory enhancing, antimycotic, and antimicrobial. More research is required to determine the additional activities of *H. canadensis* and *L. inermis* for possible medicinal applications.

REFERENCES

Akhter, M.H., M. Sabir, and N.K. Bhide. 1979. Possible mechanism of antidiarrhoel effect of berberine. *Indian Journal of Medical Research*, 70, pp. 233–241.

Aguwa, C.N.1987. Toxic effects of the methanolic extract of *Lawsonia inermis* roots. *Pharmaceutical Biology*, 25, pp. 241–245.

Al-Damegh, M.A. 2014. Evaluation of the antioxidant activity effect of Henna (*Lawsonia inermis* Linn.) leaves and or vitamin C in rats. *Life Science Journal*, 3, pp. 234-241.

Ali, M.1996. Potential of *Lawsonia inermis* L. as a medicinal plant. *Hamdard Medicus* 39, pp. 43–48.

Al-Rubiay, K.K., N.N. Jaber, B.H. Al-Mhaawe, and L.K. Alrubaiy. 2008. Antimicrobial efficacy of henna extracts. *Oman Medical Journal*, 23(4), pp. 253–256.

Al-Snafi, A.E.2019. A Review on *Lawsonia inermis*: A potential medicinal plant. *International Journal of Current Pharmaceutical Research*, 11(5), pp. 1–13.

Amin, A.H., T.V. Subbaiah, and K.M. Abbasi. 1969. Berberinesulfate: antimicrobial activity, bioassay, and mode of action. *Canadian Journal of Microbiology*, 15, pp. 1067–1076.

Anand, K.K., B. Singh, D. Chand, and B.K. Chandan. 1992. An evaluation of *Lawsonia alba* extract as hepatoprotective agent. *Planta Medica*, 58, pp. 22–25.

Arayne, M.S., N. Sultana, A.Z. Mirza, M.H. Zuberi, and F.A. Siddiqui. 2007. In vitro hypoglycemic activity of methanolic extract of some indigenous plants. *Pakistan Journal of Pharmaceutical Sciences*, 20(4), pp. 268–273.

Asmi, S., and Lakshmi. 2013. Therapeutic aspects of goldenseal. *International Research Journal of Pharmacy*, 4(9), pp. 41-3.

Auboyer, J.2002. *Daily Life in Ancient India: from 200 BC to 700AD*. London: Phoenix.

Bauer, A., and M. Brönstrup. 2014. Industrial natural product chemistry for drug discovery and development. *Natural Product Reports*, 31(1), pp. 35–60.

Bhardwaj D.K., R. Murari, T.R. Seshadri, and R. Singh. 1976. Lacoumarin from *Lawsonia inermis. Phytochemistry*, 15, p. 1789.

Borade, A.S., B.N. Kale, and R.V. Shete. 2011. A phytopharmacological review on *Lawsonia inermis* (Linn.). *International Journal of Pharmaceuticaland Life Sciences*, 2(1), pp. 536–541.

Chengaiah, B., K.M. Rao, K.M. Kumar, M. Alagusundaram, and C.M. Chetty, 2010. Medicinal importance of natural dyes: A review. *International Journal of PharmTech Research*, 2, pp. 144–154.

Chin, L.W., Y.W. Cheng, S.S. Lin, Y.Y. Lai, L.Y. Lin, M.Y. Chou, M.C. Chou, and C.C. Yang. 2010. Anti-herpes simplex virus effects of berberine from Coptidisrhizoma, a major component of a Chinese herbal medicine, Ching-Wei-San. *Archives of Virology*, 155(12), pp. 1933–41.

Chun, Y.T., T.T. Yip, K.L. Lau, and Y.C. Kong. 1978. A biochemical study on the hypotensive effect of berberine in rats. *General Pharmacology*, 10, pp. 177–182.

Dev, S. 2006. *A Selection of Prime Ayurvedic Plant Drugs, Ancient- Modern Concordance*. New Delhi: Anamaya Publishers, pp. 276–279.

Dikshit, V., J. Dikshit, M. Saraf, V. Thakur, and K. Sainis. 2000. Immunomodulatory activity of naphthoquinone fraction of *Lawsonia inermis* Linn. *Phytomedicine*, 7, pp. 102–103.

Dilworth L.L, C.K. Riley, and D.K. Stennett. 2017. Chapter 5 - Plant Constituents: Carbohydrates, Oils, Resins, Balsams, and Plant Hormones. *Pharmacognosy*. Academic Press, pp. 61–80.

Enkhtaivan, G., P. Muthuraman, D.H. Kim, and B. Mistry. 2017. Discovery of Berberine Based Derivatives as Anti-influenza Agent through Blocking of Neuraminidase. *Bioorganic & Medicinal Chemistry*, 25(20), pp. 5185-5193..

Ettefagh, K., J. Burns, H. Junio, G. Kaatz, and N. Cech. 2011. Goldenseal (*Hydrastis Canadensis* L.) extracts synergistically enhance the antibacterial activity of berberine via efflux pump inhibition. *Planta Medica*, 77, pp. 835–840.

Gentry, E.J., H.B.Jampani, A. Keshavarz-Shokri, et al. 1998. Antitubercular natural products: berberine from the roots of commercial *Hydrastis canadensis* powder. Isolation of inactive 8- oxotetrahydrothalifendine, canadine, beta-hydrastine, and two new quinic acid esters, hycandinic acid esters-1 and –2. *Journal of Natural Products*, 61(10), pp. 1187–1193.

Gupta, A.K.2003. Quality standards of Indian medicinal plants. *Indian Council of Medical Research*, 1, pp. 123–129.

Hamon, N.W. 1990. Goldenseal. *Canadian Pharmaceutical Journal*, 123(11), pp. 508–510.

Hassan, B.A.R.2012. Medicinal Plants (Importance and Uses). *Pharmaceutica Analytica Acta*, 3(10), pp. 1000e139.

Hema, R., S. Kumaravel, S. Gomathi, and C. Sivasubramaniam. 2010. Gas chromatography: mass spectroscopic analysis of *Lawsonia inermis* leaves. *New York Science Journal*, 3, pp. 142–143.

Hemem, S.S.2002. *Activity of Some Plant Extracts Against Common Pathogens In Bacterial Skin Infection.* thesis MSc, College of Education, Basra University, Iraq.

Hossain, C.M., S. Himangshu, Maji, and P. Chakraborty. 2011. Hepatoprotective activity of *Lawsonia inermis* Linn, warm aqueous extract in Carbon tetrachloride-induced hepatic injury in Wister rats. *Asian Journal of Pharmaceutical and Clinical Research*, *4*(3), pp. 106-109.

https://www.healthline.com/health/goldenseal-cure-for-everything#benefits-uses

Hwang, B.Y., S.K.Roberts, L.R. Chadwick, C.D. Wu, and A.D. Kinghorn. 2003. Antimicrobial constituents from goldenseal (the Rhizomes of *Hydrastis canadensis*) against selected oral pathogens. *Planta Medica*, 69(7), pp. 623–627.

Hydrastis canadensis. n.d. Germplasm Resources Information Network (GRIN). Agricultural Research Service (ARS), United States Department of Agriculture (USDA). Retrieved 2017-12-12.

Hydrastis canadensis in Flora of North America @ efloras.org. www.efloras.org. Retrieved 2021-01-16. https:/ /sustainabledevelopment.un.org/content/documents/6544118_Pesic_Development%20of%20natural %20product%20drugs%20in%20a%20%20sustainable%20manner.pdf. (Accessed August 15, 2018).

Inbaraj, J., B. Kukielczak, P. Bilski, Y.-Y. He, R. Sik, and C. Chignell. 2006. Photochemistry and photocytotoxicity of alkaloids from Goldenseal (*Hydrastis canadensis* L.). Palmatine, hydrastine, canadine, and hydrastinine. *Chemical Research in Toxicology*, 19(6), pp. 739–744.

Kumar S., Y.V. Singh, and M. Singh. 2005. Agro-history, uses, ecology and distribution of henna (*Lawsonia inermis* L. syn. Alba Lam). In *Henna: Cultivation, Improvement, and Trade.* Jodhpur: Central Arid Zone Research Institute, pp. 11–12.

Kirkland D., and B. Marzin. 2003. An assessment of the genotoxicity of 2-hydroxy-1, 4-naphthoquinone, the natural dye ingredient of Henna. *Mutation Research*, 537, pp. 183–199.

Kruzel, S.C., S.A. Hwang, M.C. Kruzel, A. Dasgupta, and J.K. Actor. 2008. Immune modulation of macrophage pro-inflammatory response by goldenseal and astragalus extracts. *Journal MedicinalFood*, 11(3), pp. 493–498.

Kumar, G, M. Jalaluddin, P. Rout, R. Mohanty, and C.L. Dileep. 2013. Emerging trends of herbal care in dentistry. *Journal of Clinical and Diagnostic Research*, 7(8), pp. 1827–1829.

Leyte-Lugo, M., E.R. Britton, D.H. Foil, A.R. Brown, D.A. Todd, J. Rivera-Chávez, N.H. Oberlies, and N.B. Cech. 2017. Secondary metabolites from the leaves of the medicinal plant goldenseal (*Hydrastis canadensis*). *Phytochemistry Letters*, 20, pp. 54–60.

LloydJ.U., and C.G. Lloyd. 1884. *Drugs and Medicines from North America.* Cincinnati, OH: Robert Clarke & Company Press, pp. 76–184.

Malekzadeh, F.1968. Antimicrobial activity of *Lawsoniainermis* L. *Applied Microbiology*, 16(4), pp. 663–664.

Marin-Neto, J.A., B.C. Maciel, A.L. Secches, and L. Gallo. 1988. Cardiovascular effects of berberine in patients with severe congestive heart failure. *Clinical Cardiology*, 11, pp. 253–260.

Mikhaeil, B.R., F.A. Badria, G.T. Maatooq, and M.M.A. Amer. 2004. Antioxidant and immunomodulatory constituents of henna leaves. *Z Naturforsch C Journal of Bioscience*, Jul-Aug, 59(7–8), pp. 468–76.

Muhammad, H.S., and S. Muhammad. 2005. The use of *Lawsonia inermis* Linn. (Henna) in the management of burn wound infections. *African Journal of Biotechnology*, 4(9), pp. 934–937.

National Center for Biotechnology Information, U.S. National Library of Medicine 8600. n.d.Rockville Pike, Bethesda MD, 20894 USA.

Nawagish, M., S.H. Ansari, and S. Ahmad. 2007. Preliminary pharmacognostical standardisation of *Lawsonia inermis* Linn. seeds. *Research Journal of Botany*, 2, pp. 161–164.

Nayak B.S., G. Isitor, E.M. Davis, and G.K. Pillai. 2007. The evidence based wound healing activity of *Lawsonia inermis* Linn. *Phytotherapy Research*, 21, pp. 827–831.

Nimisha, R.D.A, F.Z. Neema, and C.D. Kaur. 2017. Antipsoriatic and Anti-inflammatory Studies of Berberisaristata Extract Loaded Nanovesicular Gels. *Pharmacognosy Magazine*, 13(Suppl 3), pp. S587–S594.

Pang, B., L.H. Zhou, Q. Zhao, T.Y. Zhao, H. Wang, C.J. Gu, and X.L. Tong. 2015. Application of Berberine on Treating Type 2 Diabetes Mellitus. *International Journal of Endocrinology*, 2015(905749), pp. 1–12.

Pengelly, A., K. Bennett, K. Spelman, and M. Tims. 2012. *Appalachian plant monographs: Hydrastis canadensis L., goldenseal.* Appalachian Center for Ethnobotanical Studies. USA.

Pešić, M.2015. *Development of Natural Product Drugs in a Sustainable Manner. Brief for United Nations Global Sustainable Development Report 2015.*

Philomena, G.2011. Concerns regarding the safety and toxicity of medicinal plants: An overview. *Journal of Applied Pharmaceutical Science*, 1(6), pp. 40–44.

Predny, M.L.2005. *Goldenseal (Hydratis canadensis): An Annotated Bibliography.* US Department of Agriculture, Forest Service, Southern Research Station.

Rajesh, V., T. Riju, S. Venkatesh, et al. 2017. Memory enhancing activity of *Lawsonia inermis* Linn. Leaves against scopolamine induced memory impairment in *Swiss albino* mice. *Oriental Pharmacy and Experimental Medicine*, 17, pp. 127–142.

Rahmoun, N., Z.Boucherit-Otmani, K. Boucherit, M. Benabdallah, and N. Choukchou-Braham. 2013. Antifungal activity of the Algerian *Lawsonia inermis* (henna). *Pharmaceutical Biology*, 51(1), pp. 131–135.

Sack, R.B., and J.L.Froelich. 1982. Berberine inhibits intestinal secretory response of *Vibrio cholerae* and *Escherichia coli* enteroxins. *Infection and Immunity*, 35, pp. 471–475.

Singh, D.K., S. Luqman, and A.K. Mathur. 2015. *Lawsonia inermis* L. A commercially important primaeval dying and medicinal plant with diverse pharmacological activity: A review. *Industrial Crops and Products*, 65, pp. 269–286.

Sinha, P., S. Srivastava, N. Mishra, and N.P. Yadav. 2014. New perspectives on antiacne plant drugs: contribution to modern therapeutics. *Biomed Research International*, 301304, 19 pages.

Sun, D., S.N. Abraham, and E.H. Beachey. 1988. Influence of berberine sulfate on synthesis and expression of Pap fimbrialadhesin in uropathogenic *Escherichia coli*. *Antimicrobial Agents Chemotherapy*, 32, pp. 1274–1277.

Swabb, E.A., Y.H. Tai, and L. Jordan. 1981. Reversal of cholera toxin-induced secretion in rat ileum by luminal berberine. *American Journal of Physiology*, 241, pp. G248–G252.

Syamsudin, I., and H. Winarno. 2008. The effects of Inai (*Lawsonia inermis*) leave extract on blood sugar level: An experimental study. *Research Journal of Pharmacology*, 2(2), pp. 20–23.

Vermani, K., and S.Garg. 2002. Herbal medicines for sexually transmitted diseases and AIDS. *Journal of Ethnopharmacology*, 80(1), pp.49–66.

Wang, Y.X., X.J. Yao, and Y.H. Tan. 1994. Effects of berberine on delayed after depolarizations in ventricular muscles in vitro and in vivo. *Journal of Cardiovascular Pharmacology*, 23, pp. 716–722.

Weber, H.A., M.K. Zart, A.E. Hodges, H.M. Molloy, B.M. O'Brien, L.A. Moody, A.P. Clark, R.K. Harris, J.D. Overstreet, and C.S. Smith. 2003. Chemical comparison of goldenseal (*Hydrastis canadensis* L.) root powder from three commercial suppliers. *Journal of Agricultural and Food Chemistry*, 51(25), pp. 7352–7358.

Yang, C.S., H.C. Huang, S.Y. Wang, P.J. Sung, G.J. Huang, J.I. Chen, et al. 2016. New diphenol and isocoumarins from the aerial part of *Lawsonia inermis* and their inhibitory activities against NO production. *Molecules*, 21, p. E1299.

9 *Nardostachys jatamansi* (Spikenard) and *Ocimum tenuiflorum* (Holy Basil)

Mani Iyer Prasanth, Premrutai Thitilertdecha, Dicson Sheeja Malar, Tewin Tencomnao, Anchalee Prasansuklab, and James Michael Brimson

CONTENTS

9.1 Introduction 163
9.2 Description.......... 164
 9.2.1 *Nardostachys jatamansi* 164
 9.2.2 *Ocimum tenuiflorum* 164
9.3 Traditional Knowledge 164
9.4 Chemical Derivatives (Bioactive Compounds – Phytochemistry) 166
9.5 Potential Benefits, Applications, and Uses 168
 9.5.1 Hepatoprotective Activity and Cardioprotective Activity 169
  9.5.1.1 *Nardostachys jatamansi* 170
  9.5.1.2 *Ocimum tenuiflorum* 170
 9.5.2 Neuroprotective Activities 170
  9.5.2.1 *Nardostachys jatamansi* in Cerebral Ischemia 171
  9.5.2.2 *Ocimum tenuiflorum* in Cerebral Ischemia 171
  9.5.2.3 *Nardostachys jatamansi* in Parkinson's Disease 171
  9.5.2.4 *Nardostachys jatamansi* in Alzheimer's Disease 171
  9.5.2.5 *Ocimum tenuiflorum* in Alzheimer's Disease 172
 9.5.3 Nootropic Activity 172
  9.5.3.1 *Nardostachys jatamansi* 173
  9.5.3.2 *Ocimum tenuiflorum* 173
9.6 Conclusion 173
References 173

9.1 INTRODUCTION

Herbs and edible plants have been used as medicine for millennia, whether directly by eating the herb or plant or by extracting the active compounds using various methods, including making infusions or teas, extracting essential oils, or using solvents to extract the active compounds. One of the most well-known and successful drugs derived from plants is aspirin, with pharmacopeias from ancient Suma, Egypt, and Greece describing willow extract and other salicylate-rich plants to treat pain (Goldberg, 2009; Mehta, 2005; Nunn, 1996). By the 19th century, salicylic acid had been isolated. Bayer marketed it as aspirin; it is now used to treat multiple diseases, from fever and pain to reducing the risk of blood clots and heart attack (National Clinical Guideline, 2013), and cancer prevention (Cuzick et al., 2015). Another immensely successful group of compounds that are very important in modern medicine, originally isolated from plants, are opiates. Opioids are one of the world's oldest known drugs (Chevalier et al., 2014;

DOI: 10.1201/9781003205067-9

Manglik et al., 2012), with *Papaver somniferum* (poppy seeds) being found at archaeological sites dating back to Neolithic times. Opium was known to the ancient Egyptians, Greeks, and Sumerians as a sedative and pain relief medicine (Brownstein, 1993; Kritikos and Papadaki, 1967). Morphine was isolated from raw opium extract in the 19th century and was the first alkaloid isolated from any medicinal plant and thus marked the start of the modern era of drug discovery (Atanasov et al., 2015). Without the discovery of opium and the isolation of morphine, modern surgeries would be excruciating and recovery complicated. The addictive properties of morphine and subsequent findings of semi-synthetic opioids, such as diamorphine (heroin), and synthetic opioids (fentanyl), have led to significant addiction problems worldwide.

The synthesis of an entirely new chemical compound is an expensive and time-consuming process that leading chemists estimate to require seven or more individual chemical synthesis steps (Baxendale et al., 2007). All plants produce chemical compounds that give them an evolutionary advantage, such as salicylic acid, which prevents the consumption of the plant by animals or insects (Hayat and Ahmad, 2007). The major classes of pharmacologically active phytochemicals are alkaloids, glycosides, polyphenols, and terpenes. Using traditional knowledge of medicinal plants, a shortcut is provided, giving rise to active compounds that may be built upon to produce highly functional and effective new drugs for a wide range of diseases. One example is the treatment of Alzheimer's disease (AD): the group of herbs in the *Narcissus* (daffodil) genus contains nine groups of alkaloids, including galantamine, which have been licensed for the treatment of AD (Birks, 2006). There are estimated to be 50,000 medicinal plants in use worldwide (Schippmann et al., 2002), and approximately a quarter of all modern medicines are derived in some form from a medicinal plant (Farnsworth et al., 1985). There are currently 374,000 known plant species (Christenhusz and Byng, 2016), and a large number are yet to be discovered (Joppa et al., 2011), which could provide novel drugs or starting points for new drug development. This chapter discusses the potential pharmacological significance of two medicinal herbs, *Nardostachys jatamansi* and *Ocimum tenuiflorum*, focusing on their cytoprotective activities in hepatoxicity, cardiotoxicity, and neurotoxicity related diseases.

9.2 DESCRIPTION

9.2.1 *Nardostachys jatamansi*

N. jatamansi (Figure 9.1a) is a flowering plant of the *Caprifoliaceae* (honeysuckle) family that grows in the Himalayas, typically found at an altitude of 3,000–5,000 m in countries including Kumaon, Nepal, Sikkim, and Bhutan (Bakhru, 1992). The plant grows 10–50 cm (4–20 in) in height and has pink, bell-shaped flowers and a rhizomatous root (Figure 9.1b). The leaves are elongated and spatulated, while few leaves are sessile, oblong, or subovate.

9.2.2 *Ocimum tenuiflorum*

O. tenuiflorum (Figure 9.1c), commonly known as "Tulasi", "holy basil", or "Thai holy basil" (a distinct variety from Thai basil, Italian basil, and sweet basil (*O. basilicum*)), is an aromatic perennial plant in the family Lamiaceae ("mint", "deadnettle" or "sage family"). The herb is native to the Indian subcontinent and widespread as a cultivated plant throughout Southeast Asia. Holy basil stands erect between 30–60 cm tall; the stems contain tiny hairs and multiple branches containing the leaves. The leaves are approximately five centimeters long and maybe green or purple. The leaves are an oval shape with a slightly toothed margin, and the flowers are small and tubular with a light shade of purple or pink. The fruits are nutlets and produce numerous seeds.

9.3 TRADITIONAL KNOWLEDGE

The two herbs discussed in this chapter have multiple synonyms by which they may be referred to and many common names; these are listed in Table 9.1. *N. jatamansi* has a long history of

FIGURE 9.1 An *N. jatamansi* (Spikenard) flower and leaves B Dried roots of *N. jatamansi* C *O. tenuiflorum* (Holy Basil), leaves and flower.

health promotion and treatment of various ailments in different traditional medicines, including Chinese, Tibetan, Nepalese, Bhutanese, Indian, Japanese, and Thai. It has been well documented in Ayurvedic classics, the Old Testament, Ben-Cao-Shi-Yi, Homer's Iliad, and Phra-Osod-Phra-Narai textbooks (Dhiman and Bhattacharya, 2020). The plant parts used for medicinal purposes are roots and rhizomes, which can be effectively formulated in either single herb or multi-herbal recipes. The traditional preparations are decoctions and powders. In India, *N. jatamansi* has been generally used as a bitter tonic, stimulant, and antispasmodic. It is also employed to treat epilepsy, seizures, hysteria, syncope, convulsion, heart palpitation, and mental weakness (Bagchi et al., 1991; Pandey et al., 2013). In Thai folklore, the dried roots and rhizomes have been recorded to treat abscess, sepsis, infections with parasitic worms, and poisoning. They also act as carminative, digestive, and hemagogic (Kot Chadamangsi, 2000; Kot Chadamangsi, 2001). Besides, *N. jatamansi* is well-known as one of the active ingredients in a multi-herbal formulation called "Nine Kots" (Kot-Tung-Khao or Na-Wa-Kot), which is used to relieve fever, asthma, and flatulence as well as to nourish the blood (Kot Chadamangsi, 2001).

O. tenuiflorum (or *O. sanctum*) is one of the most holistic herbs used over the years in Ayurvedic (e.g., India, Bangladesh, Bhutan, Malaysia, Nepal, and Sri Lanka), Unani (e.g., Bangladesh, India, Malaysia, Pakistan, and Sri Lanka) and Thai systems. In Ayurveda, this plant's best oral administration form is believed to be a hot-water infusion using fresh and whole parts (Engels and Brinckmann, 2013). Decoctions of leaves and juices of fresh leaves are also traditionally prepared. The indications are unsurprisingly broad, covering fever, cold, cough, constipation, headache, abdominal pain, asthma, diabetes, tuberculosis, rheumatism, syphilis and fertility, and memory enhancement (Engels and Brinckmann, 2013). It is even considered an elixir of life due to its promotion of longevity (Pattanayak et al., 2010). Unlike in the Ayurvedic system, only leaves and roots, either fresh or dried, are recognized for therapeutic uses in Thai traditional medicine. Twenty-five grams of fresh leaves or four grams of dried leaves are recorded to be effective doses that are prepared as decoctions with water (Kaprao, 1999). *O. tenuiflorum* has been widely consumed daily as a part of Thai cuisine and prescribed to alleviate chronic fever, diarrhea, stomachache, flatulence,

TABLE 9.1
Herb Synonyms and Common Names Plus Their Chemical Constituent Compounds Found in *N. jatamansi* and *O. tenuiflorum*

Plant Name	Common names	Phytochemical Constituent Compounds
Nardostachys jatamansi	Spikenard	Beta sitosterol
Synonyms:	Nard	ursolic acid
• *Fedia grandiflora* Wall.	Nardin Muskroot	Lupeol
• *Fedia jatamansi* Wall. ex DC.	Jatamansi	kanshone B
• *Nardostachys chinensis* Batalin		7methoxydesox-narchiol
• *Nardostachys gracilis* Kitam.		nardosinone
• *Nardostachys grandiflora* DC.		Narchinol
• *Nardostachys jatamansi* C.B.Clarke		isonardosinone
• *Patrinia jatamansi* D.Don		beta-sitosterol
• *Valeriana jatamansi* D.Don		Kanshone E
• *Valeriana jatamansi* (D.Don) Wall.		Kanshone N
Ocimum tenuiflorum	Holy Basil	Rosmarinic acid
Synonyms:	Tulsi	Ursolic acid
• *Geniosporum tenuiflorum* (L.) Merr.	Tulasi	Apigenin
• *Lumnitzera tenuiflora* (L.) Spreng.		Eugenol
• *Moschosma tenuiflorum* (L.) Heynh.		Chlorrogenic acid
• *Ocimum anisodorum* F.Muell.		Cirsimaritin
• *Ocimum caryophyllinum* F.Muell.		Cirsilineol
• *Ocimum hirsutum* Benth.		Isothymusin
• *Ocimum inodorum* Burm.f.		Isothymonin
• *Ocimum monachorum* L.		Orientin
• *Ocimum sanctum* L.		Vicenin-2
• *Ocimum scutellarioides* Willd. ex Benth.		
• *Ocimum subserratum* B. Heyne ex Hook.f.		
• *Ocimum tomentosum* Lam.		
• *Ocimum villosum* Roxb. nom. illeg.		
• *Plectranthus monachorum* (L.) Spreng.		

heartburn, nausea, and vomiting in households (Kaprao, 1999). It is also an essential ingredient in pregnant women's meals to improve breast milk supply during breastfeeding (Karaöz et al., 2017). Regarding topical applications, fresh leaves can be ground, squeezed for juices and mixed with asafetida for applying on the abdomen in children to relieve stomachache and flatulence. The juices alone are used to cure skin diseases involving ringworm and *Pityriasis Versicolor* (Karaöz et al., 2017). Again, *O. tenuiflorum* can be formulated with other medicinal plants for multi-herbal formulations called "Phrasa-Mawang", "Phrasa-Kaprao", and "Lued-Ngam" that are mainly prescribed to reduce cough, mucous, and period cramps and are registered in the National List of Essential Medicines by Thai Food and Drug Administration (Karaöz et al., 2017; National List of Essential Medicines (NLEM), 2013).

9.4 CHEMICAL DERIVATIVES (BIOACTIVE COMPOUNDS – PHYTOCHEMISTRY)

Phytochemical constituents in roots and rhizomes of *N. jatamansi* are found to be a variety of sesquiterpenes, coumarins, lignans, and others including valeranol (Sahu et al., 2016), valeranone (Hoerster and Ruechker, 1997), spirojatamol (Anjana Bagchi et al., 1990), jatamansic acid (Anjana Bagchi et al., 1990), jatamols A and B (Bagchi et al., 1991), jatamansinol (Sahu et al., 2016), jatamansin (Shanbhag et al., 1964), jatamansone (Arora et al., 1962), calarenol (Sastry et al., 1967), pinoresinol (Bagchi et al., 1991), virolin (Bagchi et al., 1991), patchouli alcohol (Rucker G et al., 1993), angelicin (Sahu et al., 2016), actinidine (Sahu et al., 2016), nardal (Sahu et al., 2016), β-sitosterol (S. Chaudhary et al., 2015), lupeol (S. Chaudhary et al., 2015), kanshones A, B, D, E, L, M, and N (Ko et al., 2018; Yoon et al., 2018), narchinol A (Yoon et al., 2018), 7-methoxydesoxo-narchiol (Yoon et al., 2018), nardosinanone G (Yoon et al., 2018), nardoaristolone B (Yoon et al., 2018), nardostachysin (Chatterjee et al., 2000), nardostachone (Sahu et al., 2016), nardosinone (Ko et al., 2018), isonardosinone (Ko et al., 2018), and ursolic acid (Bose et al., 2019). Because *N. jatamansi* is rich in sesquiterpenes, it is a promising medicinal plant candidate in many therapeutic fields, such as anti-oxidation, anti-inflammation, neuroprotection, and cancer drug discovery. Some of its naturally occurring compounds have been investigated for these biological potentials (Figure 9.2). Kanshone N, narchinol A, and 7-methoxydesoxo-narchiol exhibited anti-neuroinflammatory effects in LPS-stimulated BV2 microglial cells through inhibiting of NF-κB signaling pathway (Yoon et al., 2018). Likewise, nardosinone, isonardosinone, and kanshones B and E inhibited NF-κB and MAPK-mediated signaling pathways in the same settings using LPS-induced BV2 microglial cells (Ko et al., 2018). Lupeol and β-sitosterol were reported to possess anti-oxidative properties via free radical scavenging activity (Chaudhary et al., 2015). Ursolic acid was also proved to be a potent antioxidant through tyrosinase inhibition (Bose et al., 2019), as well as inhibiting the production of IL-2 and TNF-α production in response to inflammatory stimulation by

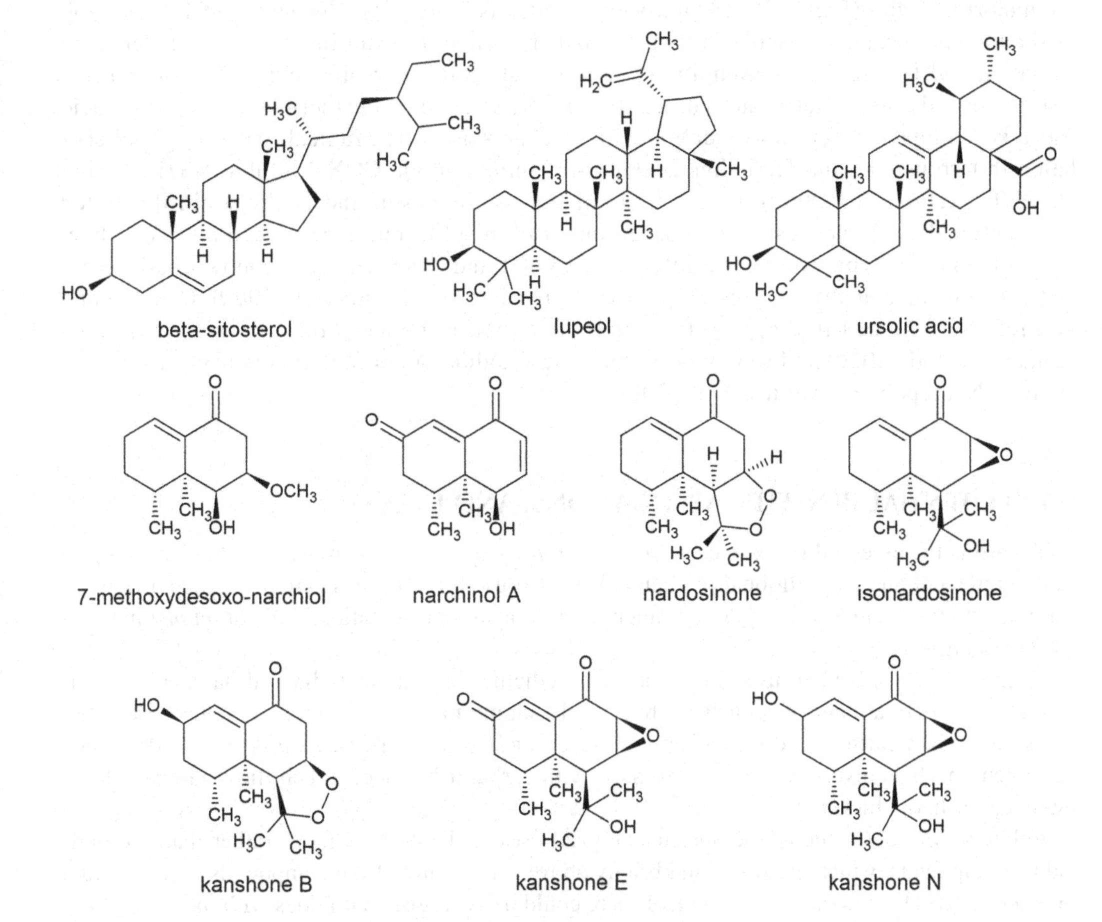

FIGURE 9.2 Structures of compounds found in *N. jatamansi*.

Phorbol 12-Myristate 13-Acetate and lectin from *Phaseolus vulgaris*/Leucoagglutinin (PMA/PHA) (Kaewthawee and Brimson, 2013). Furthermore, compounds such as β-sitosterol and lupeol have been shown to reduce amyloid beta-induced cell death in a cultured neuron cell line (HT-22) (Brimson et al., 2012).

Concerning *O. tenuiflorum*, the stems and leaves contain several naturally occurring compounds comprising volatile oils, saponins, flavonoids, triterpenoids, sesquiterpenes, and tannins. Rosmarinic acid (Kelm et al., 2000; Pattanayak et al., 2010), propanoic acid (Pattanayak et al., 2010), ursolic acid (Anandjiwala et al., 2006; Batra and Sastry, 2014; Vetal et al., 2012), oleanolic acid (Anandjiwala et al., 2006), gallic acid (Hussain et al., 2017), chlorogenic acid (Hussain et al., 2017), *p*-hydroxy benzoic acid (Hussain et al., 2017), caffeic acid (Hussain et al., 2017), vanillic acid (Hussain et al., 2017), *p*-coumaric acid (Hussain et al., 2017), sinapic acid (Hussain et al., 2017), ferulic acid (Hussain et al., 2017), quercetin (Hussain et al., 2017), myricetin (Hussain et al., 2017), apigenin (Kelm et al., 2000; Pattanayak et al., 2010), luetolin (Anandjiwala et al., 2006), cirsimaritin (Kelm et al., 2000; Pattanayak et al., 2010), isothymusin (Kelm et al., 2000; Pattanayak et al., 2010), isothymonin (Kelm et al., 2000; Pattanayak et al., 2010), orientin (Pattanayak et al., 2010), vicenin (Pattanayak et al., 2010), eugenol (Anandjiwala et al., 2006; Hussain et al., 2017; Kelm et al., 2000), carvacrol (Pattanayak et al., 2010), linalool (Ijaz et al., 2017), β-elemene (Hussain et al., 2017), β-caryophyllene (Hussain et al., 2017), germacrene D (Hussain et al., 2017), and cirsilineol (Kelm et al., 2000) are characterized. Some studies have explored the anti-oxidative and anti-inflammatory properties of *O. tenuiflorum* through its active components, especially polyphenolic compounds, which are the majority within (Figure 9.3). Chlorogenic acid was a prominent phenolic acid in ethanol, methanol, and hexane extracts of this plant and showed significant inhibition of linoleic acid oxidation and free radical scavenging (Hussain et al., 2017). Ursolic acid and its derivatives also possessed anti-oxidative activity (Batra and Sastry, 2014). Furthermore, rosmarinic acid was a key component for anti-oxidation with mechanisms in free radical scavenging and chelation in ferrous ions and ferric ions and anti-inflammation via COX-2 inhibition (Hakkim et al., 2007; Kelm et al., 2000). Eugenol, cirsilineol, isothymusin, and isothymonin illustrated good antioxidant properties at the concentration of 10 μM, and eugenol exhibited excellent anti-inflammation with over 97% inhibition of COX-1 and -2 activities. In contrast, irsimaritin and apigenin specifically suppressed the COX-2 mechanism (Kelm et al., 2000). Orientin and vicenin also inhibited free radical formation through scavenging at the equivalent efficiency (Uma Devi et al., 2000) and were reported to act as modulators of radiation injury and enhancers for DNA repair (Satyamitra et al., 2014).

9.5 POTENTIAL BENEFITS, APPLICATIONS, AND USES

Spikenard is the essential oil extracted from *N. jatamansi* and has been used worldwide as a perfume ingredient and as traditional medicine. It is also used in religious ceremonies from India to Europe with the coat of arms of Pope Francis containing a representation of *N. jatamansi* in reference to Saint Joseph.

O. tenuiflorum has been used in traditional medicine throughout India and has been recommended for the treatment of bronchitis, bronchial asthma, malaria, diarrhea, dysentery, skin diseases, arthritis, painful eye diseases, chronic fever, and insect bites. *O. tenuiflorum* is also a key ingredient in Thai cuisine, used as an aromatic with garlic, fish sauce, and chillies to impart flavor into meat, fish, or chicken.

Below we discuss some of the pharmacological uses of the herb extracts, rather than the individual compounds, which are discussed briefly above since many of the compounds may be found in other herbs. There is the possibility that there could be synergistic activities with the select compound found in these herb extracts (Madan Mohan Pandey et al., 2013).

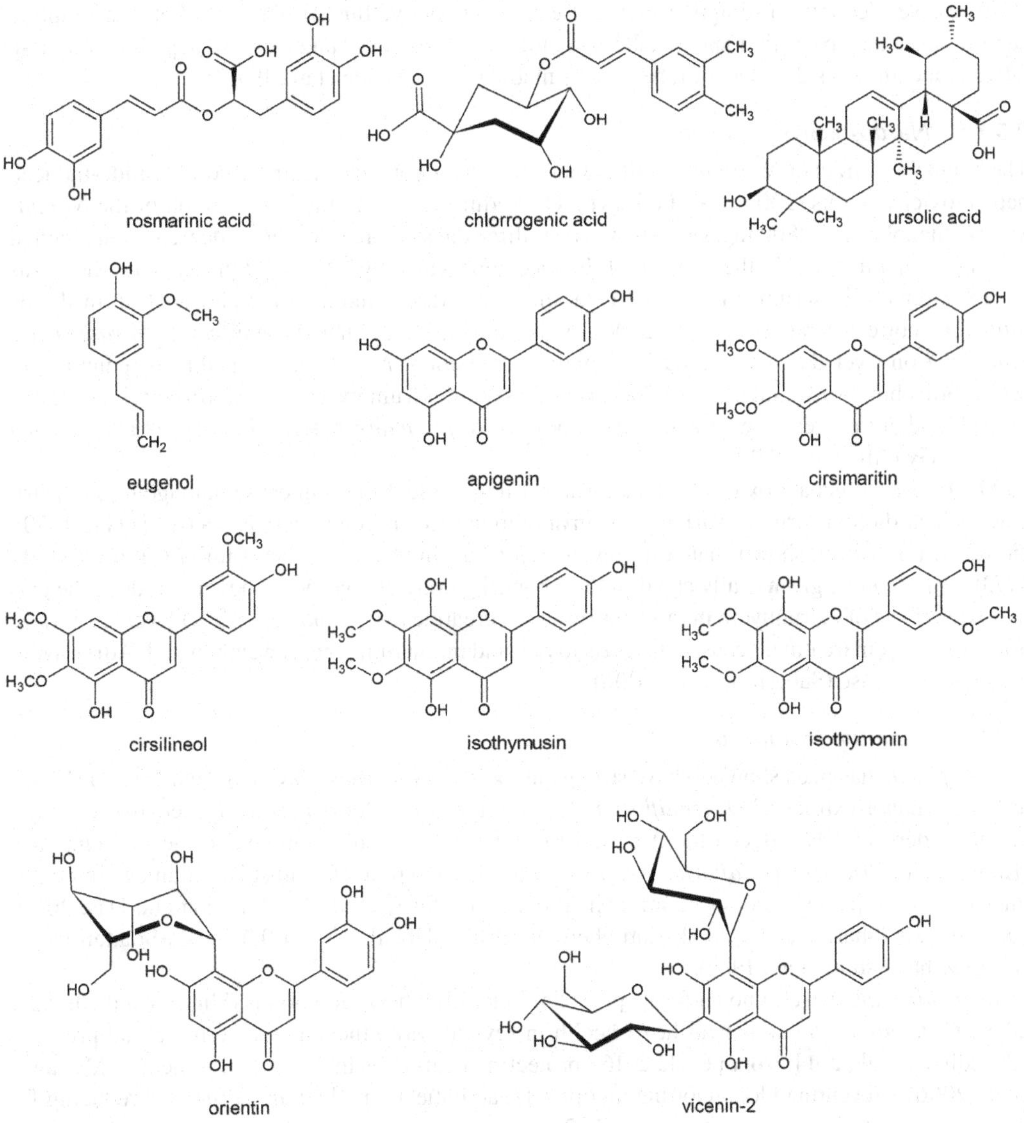

rosmarinic acid chlorrogenic acid ursolic acid eugenol apigenin cirsimaritin cirsilineol isothymusin isothymonin orientin vicenin-2

FIGURE 9.3 Structures of compounds found in *O. tenuiflorum*.

9.5.1 Hepatoprotective Activity and Cardioprotective Activity

An estimated 17.9 million people died from cardiovascular disease in 2019, representing 32% of all global deaths. Of these deaths, 85% were due to heart attack and stroke, making cardiovascular disease the deadliest worldwide (World-Health-Organization, 2020). Therefore, it is essential to improve strategies for the prevention of heart disease and develop cardioprotective medicines.

Many drugs can induce hepatotoxicity (liver damage) when taken in excess over a long time. Therefore, it is sometimes necessary to counteract the hepatoxicity of a drug to take a higher dose or be treated for a longer time. There are more than 900 drugs known to induce hepatoxicity; some examples for hepatotoxic medicines include analgesics such as paracetamol aspirin and phenylbutazone (Manov et al., 2006), the monoaminoxidase inhibitor antidepressant iproniazid (Schläppi, 1985), and even some antibiotics (Westphal et al., 1994).

Excessive alcohol consumption can cause hepatotoxicity; furthermore, alcohol consumption can influence the absorption and metabolism of tuberculosis drugs and increase hepatotoxicity risk (already a common side effect of tuberculosis medications) (Pande et al., 1996).

9.5.1.1 *Nardostachys jatamansi*

The ethanolic extract of *N. jatamansi* rhizomes are hepatoprotective against thioacetamide-induced hepatotoxicity in rats (200 mg/kg body wt.). Oral administration of a 50% solution of the *N. jatamansi* ethanol extract (800 mg/kg body wt.) over three days before thioacetamide treatment resulted in a significant increase in the survival of the rats (approximately 20–75%). Furthermore, there was an improvement in serum transaminases (aminotransferases) and alkaline phosphatase in the *N. jatamansi* cotreated rats (Ali et al., 2000). A separate study evaluated the effect of *Nardostachys jatamansi* on liver damage caused by ionizing irradiation. *N. jatamansi* mediated reduction in serum bilirubin levels, glutamic oxallotransaminase, and serum pyruvic oxallotransaminase levels in the blood of rats treated with 200 mg/kg body wt. *N. jatamansi* before whole-body radiation at a dose of 3Gy (Ali et al., 2000).

Much like the hepatotoxic effects of certain drugs, some drugs can cause damage to the heart. One such cardiotoxic drug is Adriamycin, an antibiotic with antitumor activity (Suzuki et al., 1990). However, it has been shown that Adriamycin can result in congestive heart failure (Lefrak et al., 1973). *N. jatamansi* given orally at a dose of 500mg/kg body wt. has been shown to reduce the cardiotoxic effects of Adriamycin in rats, returning the creatine phosphokinase (CPK) levels to normal after Adriamycin treatment caused its levels to rise and maintaining heart weight which Adriamycin caused to decrease (Subashini et al., 2006).

9.5.1.2 *Ocimum tenuiflorum*

O. tenuiflorum has been shown to have strong antioxidant properties (Shyamala and Devaki, 1996), and the ethanol extract of *O. tenuiflorum* protects rats from Atorvastatin induced liver toxicity and dose-dependently reduced the Atorvastatin induced elevated serum level of hepatic enzymes (Kumar et al., 2013). *O. tenuiflorum* has also been shown to protect against liver damage in rats or mice caused by paracetamol (acetaminophen) (Chattopadhyay et al., 1992; Lahon and Das, 2011), heavy metals (Sharma et al., 2002), antitubercular drugs (Ubaid et al., 2003) and carbon tetrachloride (Seethalakshmi et al., 1982).

O. tenuiflorum is well known for its potential in treating heart conditions (Sharma and Chanda, 2018). *O. tenuiflorum* can benefit heart health in several ways, including lowering blood pressure (Chaudhary et al., 2014; Irondi et al., 2016), protecting heart cells in the case of ischemia (Mohanty et al., 2006), preventing blood clotting (antiplatelet activities) (Singh et al., 2001), and reducing fat deposition in arteries (Suanarunsawat et al., 2010).

9.5.2 Neuroprotective Activities

Neurodegenerative diseases are a modern problem; while there are advances in modern medicine and improvements in nutrition, people live longer, and the world's population ages. Diseases such as AD and Parkinson's disease PD (where age is a key risk factor) are on the rise, with 44 million people currently suffering from AD, and this is expected to rise to 135 million by 2050 (Guerchet et al., 2013). With few current treatment options and none that can reverse or halt the progression of neurodegenerative disease, there will be a future drain on health care resources in caring for these patients who will need specialized care. Therefore, it is essential to continue searching for new drugs and therapies that can reduce the burden of neurodegenerative disease in the world's health care systems.

Multiple studies have investigated the neuroprotective properties of *N. jatamansi* regarding many neurological diseases, including stroke/cerebral ischemia, Parkinson's disease, Alzheimer's disease, and depression. Many of the neuroprotective activities seen with *N. jatamansi* are predicted to result from improved cellular glutathione reduced lipid peroxidation and action on the Na+/K+, ATPase, and catalase enzyme systems (Salim et al., 2003).

9.5.2.1 *Nardostachys jatamansi* in Cerebral Ischemia

N. jatamansi treatment of rats at 250 mg/kg for 15 days before a two-hour occlusion of the middle cerebral artery followed by 22 hours reperfusion attenuated the decrease in Na^+K^+ ATPase and catalase activities. It reversed the reduction of spontaneous motor activity and motor coordination (Salim et al., 2003). Two other studies focusing on the antioxidant effects of *N. jatamansi in vitro* have also shown that the herb extract has significant antioxidant effects (Sharma and Singh, 2012) and can protect C6 glioma cells against hydrogen peroxide-induced oxidative damage (Dhuna et al., 2013).

9.5.2.2 *Ocimum tenuiflorum* in Cerebral Ischemia

The antioxidant effects of *O. tenuiflorum* have made it the subject of cerebral ischemia research. *O. tenuiflorum* has been shown to attenuate the oxidative damage caused by ischemia and reperfusion injury in multiple studies in rats (Jivad and Rabiei, 2015; Mohanty et al., 2006; Yanpallewar et al., 2004) and in mice (Bora et al., 2011; Singh et al., 2017).

O. tenuiflorum can reduce the infarct volume in rats subjected to middle cerebral artery occlusion and reduce oxidative stress by increasing glutathione content of the hippocampus and frontal cortex after middle cerebral artery occlusion (Ahmad et al., 2012).

9.5.2.3 *Nardostachys jatamansi* in Parkinson's Disease

There have been multiple studies regarding *N. jatamansi* and Parkinson's disease models. In a 6-OHDA rat model of Parkinson's disease, *N. jatamansi* (200, 400, and 600 mg/kg body wt.) injected over three weeks dose-dependently improved striatal dopamine content and restored the alterations in locomotor activity and muscle coordination of the 6-OHDA injected rats (Ahmad et al., 2006). This study also showed that a reduction in oxidative stress was significant to *N. jatamansi*'s neuroprotective properties, with dose-dependent increases in superoxide dismutase activity, reduced glutathione (GSH), and catalase activity (Ahmad et al., 2006). Another study utilizing haloperidol-induced catalepsy as a Parkinson's disease model showed a hydroalcoholic extract of *Nardostachys jatamansi* at 30 and 100 mg/kg body wt. could significantly reduce the duration of haloperidol-induced catalepsy. Furthermore, in the same study, the *N. jatamansi* at 30 and 100 mg/kg body wt. reduced tacrin-induced vacuous chewing movements, orofacial bursts, and tongue protrusion. Finally, this study showed *N. jatamansi* could reverse reserpine-induced hypolocomotion (Giri et al., 2020). These effects on haloperidol-induced catalepsy by *N. jatamansi* have been independently confirmed in a separate rat study (Rasheed et al., 2010). Similar research utilizing reserpine-induced orofacial dyskinesia in rats showed that *N. jatamansi* treatment (100 and 300 mg/kg body wt.) could significantly inhibit reserpine-induced catalepsy and induced expression of protective antioxidant enzymes, such as superoxide dismutase and catalase (Patil et al., 2012). In a rotenone-induced Parkinson's model in mice, the extract of *N. jatamansi* root, which contains the active compound nardosinone, effectively reduced Parkinson's disease symptoms in the rotenone-treated mice. Furthermore, this study showed that rotenone reduced dopamine receptor expression, and that *N. jatamansi* root extract could restore the dopamine receptor expression levels (Bian et al., 2021).

9.5.2.4 *Nardostachys jatamansi* in Alzheimer's Disease

Most of the current treatments that have any efficacy for Alzheimer's disease center around increasing acetylcholine in the brain by using acetylcholinesterase inhibitors (McGleenon et al., 1999). Multiple *in vitro* screening studies have identified *N. jatamansi* extracts with acetylcholinesterase inhibitor activity (IC_{50} values 50 – 500 μg/ml) (Mathew and Subramanian, 2014; Mukherjee et al., 2007; Rahman et al., 2011).

The buildup of amyloid-β in the brain and the formation of plaques is a hallmark of Alzheimer's disease (Sadigh-Eteghad et al., 2015); furthermore, amyloid-β has been shown to induce ROS production in neuron cells leading to cell death (Cheignon et al., 2018). In a study using amyloid-β treated SH-SY-5Y neuroblastoma cells, an ethanol extract of *N. jatamansi* root resulted in a reduction of ROS production and increased cell viability (Liu et al., 2018). In the same study, the researchers used Aβ42-expressing drosophila flies as a model and found that the ethanol extract of *N. jatamansi* root treatment resulted in lower neuroinflammation, increased longevity of the flies, and reduced ERK phosphorylation in the brains of the Aβ42-expressing Alzheimer's disease model flies (Liu et al., 2018).

9.5.2.5 *Ocimum tenuiflorum* in Alzheimer's Disease

As with *N. jatamansi* above, multiple studies are investigating the acetylcholinesterase inhibitor activity of *O. tenuiflorum*. In a study using aged mice as an Alzheimer's model, *O. tenuiflorum* at 50 to 100 μg/ml doses significantly decreased acetylcholine esterase activity. Furthermore, in the same study, *O. tenuiflorum* reversed the amnesic effects of scopolamine (Joshi and Parle, 2006a). Further studies have also shown *O. tenuiflorum* to possess anti-acetylcholinesterase activity (Kandhan et al., 2018).

Another study utilizing colchicine or ibotenic acid-induced Alzheimer's showed that *O. tenuiflorum* could improve the cognitive ability of colchicine and ibotenic acid-treated rats measured the Morris' water maze test and learned helplessness test. The study also tested the rats' anxiety using various behavioral tests, including the elevated plus-maze test and Porsolt's swim test, alleviating the depressive-like symptoms in the swim test caused by ibotenic acid and colchicine (Raghavendra et al., 2009). This study showed *O. tenuiflorum* to have little effect on acetylcholinesterase inhibitor activity.

Finally, a study investigating the effect of *O. tenuiflorum* on depressive-like behaviors in rats treated with Aβ1-42 showed that *O. tenuiflorum* reduced immobility time in the forced swim test and increased open arm entries in the elevated maze test, both results suggesting that *O. tenuiflorum* could reduce depressive-like behaviors in this rat model of Alzheimer's disease (Gradinariu et al., 2015).

9.5.3 Nootropic Activity

Nootropic drugs, sometimes colloquially known as smart drugs or cognitive enhancers, are claimed to improve cognitive functions such as memory, creativity, and motivation in healthy individuals (Frati et al., 2015). There is an apparent demand for drugs in the modern, fast-paced society that can improve productivity and ability in the tasks required for learning and creative thinking. Students are under tremendous pressure to succeed with more and more competition in the job market. They feel it is essential to achieve the top grades at all costs. This has created a demand for drugs that can improve cognitive abilities.

One well-known drug that is consumed in some form or other across the world is caffeine. Except for water, tea and coffee are the most consumed drinks globally, both of which contain caffeine. Caffeine is a stimulant that has been used by humankind for thousands of years, with tea being cultivated in China as early as the 10th century BC (Mair et al., 2009), and coffee being consumed since at least the 7th century in parts of North Africa (Weinberg and Bealer, 2004). Caffeine has been shown to have nootropic and psychostimulant properties and improves cognitive performance after exercise (Hogervorst et al., 2008; Hogervorst et al., 1999) as well as improves memory performance in distracted middle-aged (but not younger) subjects (Hogervorst et al., 1998).

Several medicinal herbs have been claimed to possess nootropic activity, such as *Bacopa monnieri* (Brimson et al., 2020; Norman Scholfield et al., 2013), *Panax ginseng* (Zhu and Yueying, 1998), *Ginkgo biloba* (Nathan et al., 2002), and *Centella asiatica* (Kulkarni et al., 2012). However, at least one meta-analysis study for each of *Bacopa monnieri* and *Ginkgo biloba* has found no significant improvements in healthy individuals (Brimson et al., 2021; Canter and Ernst, 2007). There are many claims made of various extracts from medicinal herbs that fail to meet scientific standards of proof. Thus, it is essential to carefully consider the scientific evidence before consuming herbs to improve cognitive performance, especially as there is often a misconception that naturally occurring is equal to healthy and safe.

9.5.3.1 *Nardostachys jatamansi*

N. jatamansi has been shown to improve learning and memory in mice at doses ranging from 50 to 200 mg/kg in both young and old mice. Ethanol extract of *N. jatamansi* root showed memory improvements in the elevated maze test, with a significant reduction in the transfer latency (time taken by the mouse to move into one of the covered maze arms with all four of its legs). Furthermore, 100mg/kg *N. jatamansi* reversed the amnesic effects of diazepam (1mg/kg) and scopolamine (0.4 mg/kg). Moreover, *N. jatamansi* significantly improved the memory and learning of both aged and young mice in the passive avoidance test (where mice remember that stepping down from the platform results in a small electric shock). The time taken for the mice to step down was significantly increased at all doses of *N. jatamansi* in both the young and aged mice, and *N. jatamansi* also reversed the amnesic effect of scopolamine (Joshi and Parle, 2006c). A separate study in rats showed an increase in nootropic activity with *N. jatamansi* extracts and at low doses when combined with lithium carbonate (at higher doses in combination has less effect) (Kasture et al., 2014).

9.5.3.2 *Ocimum tenuiflorum*

With the improvements in spatial learning and memory seen in the Alzheimer's models, researchers have wondered whether improvements can be seen in otherwise healthy animals. A study of the step-down latency in the passive avoidance test showed significant improvements in *O. tenuiflorum*-treated mice (50–200 mg/kg). Furthermore, the amnesia-inducing compound scopolamine shortened the latency (made the mouse memory worse) significantly. At the same time, cotreatment with *O. tenuiflorum* returns the step-down latency to significantly about the saline-treated control. This suggests that *O. tenuiflorum* has a significant benefit to memory and learning. Moreover, the same study investigated aged mice, showing that older mice learned more slowly and had a shorter step-down latency; however, *O. tenuiflorum* significantly improved their spatial learning and memory (Joshi and Parle, 2006b). These improvements are thought to be due to the increase in acetylcholine made available in the brain (Joshi and Parle, 2006a).

9.6 CONCLUSION

Both *N. jatamansi* and *O. tenuiflorum* are medicinal herbs with a long tradition in religious ceremonies and, in the case of *O. tenuiflorum*, in Asian cuisine. Both herbs contain a unique signature of active compounds, and they have a range of potential medicinal uses. A fundamental interest in their pharmacological activities is their cytoprotective activities, in particular their neuroprotective activities. Both herbs are helpful in animal models of Alzheimer's disease, Parkinson's disease, and cerebral ischemia. Furthermore, there has been some interest in these herbs for their nootropic activity. Many of the compounds found in both these herbs have links to antioxidant and anti-inflammatory pathways, which could explain many of the neurological benefits they appear to have. Furthermore, both herbs appear to improve acetylcholine signaling in rodent models' brains of Alzheimer's disease. Both these herbs have the potential to discover new pharmaceuticals aimed at treating neurological disease, either through the herb extract itself or the isolation and modification of compounds found within.

REFERENCES

Ahmad, A., Khan, M.M., Raza, S.S., Javed, H., Ashafaq, M., Islam, F., Safhi, M.M., and Islam, F. (2012). *Ocimum sanctum* attenuates oxidative damage and neurological deficits following focal cerebral ischemia/reperfusion injury in rats. *Neurological Sciences*, 33(6), pp.1239–1247.

Ahmad, M., Yousuf, S., Khan, M.B., Hoda, M.N., Ahmad, A.S., Ansari, M.A., Ishrat, T., Agrawal, A.K., and Islam, F. (2006). Attenuation by *Nardostachys jatamansi* of 6-hydroxydopamine-induced parkinsonism in rats: Behavioral, neurochemical, and immunohistochemical studies. *Pharmacology Biochemistry and Behavior*, 83(1), pp.150–160.

Ali, S., Ansari, K.A., Jafry, M., Kabeer, H., and Diwakar, G. (2000). *Nardostachys jatamansi* protects against liver damage induced by thioacetamide in rats. *Journal of Ethnopharmacology*, 71(3), pp.359–363.

Anandjiwala, S., Kalola, J., and Rajani, M. (2006). Quantification of eugenol, luteolin, ursolic acid, and oleanolic acid in black (Krishna Tulasi) and green (Sri Tulasi) varieties of *Ocimum sanctum* Linn. using high-performance thin-layer chromatography. *Journal of AOAC International*, 89(6), pp.1467–1474.

Arora, R.B., Singh, M., and Kanta, C. (1962). Tranquilising activity of jatamansone: A sesquiterpene from *Nardostachys jatamansi. Life Sciences*, 1(6), pp.225–228.

Atanasov, A.G., Waltenberger, B., Pferschy-Wenzig, E.M., Linder, T., Wawrosch, C., Uhrin, P., Temml, V., Wang, L., Schwaiger, S., Heiss, E.H., and Rollinger, J.M. (2015). Discovery and resupply of pharmacologically active plant-derived natural products: A review. *Biotechnology Advances*, 33(8), pp.1582–1614.

Bagchi, A., Oshima, Y., and Hikino, H. (1990). Spirojatamol, a new skeletal sesquiterpenoid of *Nardostachys jatamansi* roots. *Tetrahedron, 46*(5), pp.1523–1530.

Bagchi, A., Oshima, Y., and Hikino, H. (1991). Neolignans and lignans of *Nardostachys jatamansi* roots. *Planta Medica*, 57(1), pp.96–97.

Bakhru, H. (1992). *Herbs that Heal: Natural Remedies for Good Health.* Orient Paperbacks.

Batra, A., and Sastry, V. (2014). Extraction of ursolic acid from *Ocimum sanctum* and synthesis of its derivatives: Comparative evaluation of anti-oxicant activities. *International Journal of Pharmaceutical Sciences and Research*, 5(10), pp.4486–4492.

Baxendale, I.R., Hayward, J.J., Ley, S.V., and Tranmer, G.K. (2007). Pharmaceutical strategy and innovation: An academics perspective. *ChemMedChem*, 2(6), pp.768–788.

Bian, L.H., Yao, Z.W., Zhao, C.B., Li, Q.Y., Shi, J.L., and Guo, J.Y., (2021). Nardosinone alleviates Parkinson's disease symptoms in mice by regulating dopamine D2 receptor. *Evidence-Based Complementary and Alternative Medicine*, 2021, p.6686965.

Birks, J. (2006). Cholinesterase inhibitors for Alzheimer's disease. *Cochrane Database of Systematic Reviews*, 1, p.Cd005593.

Bora, K.S., Arora, S., and Shri, R. (2011). Role of *Ocimum basilicum* L. in prevention of ischemia and reperfusion-induced cerebral damage, and motor dysfunctions in mice brain. *Journal of Ethnopharmacology*, 137(3), pp.1360–1365.

Bose, B., Tripathy, D., Chatterjee, A., Tandon, P., and Kumaria, S. (2019). Secondary metabolite profiling, cytotoxicity, anti-inflammatory potential and *in vitro* inhibitory activities of *Nardostachys jatamansi* on key enzymes linked to hyperglycemia, hypertension and cognitive disorders. *Phytomedicine*, 55, pp.58–69.

Brimson, J.M., Brimson, S.J., Brimson, C.A., Rakkhitawatthana, V., and Tencomnao, T. (2012). *Rhinacanthus nasutus* extracts prevent glutamate and amyloid-β neurotoxicity in HT-22 mouse hippocampal cells: Possible active compounds include lupeol, stigmasterol and β-sitosterol. *International Journal of Molecular Sciences*, 13(4), pp.5074–5097.

Brimson, J.M., Brimson, S., Prasanth, M.I., Thitilertdecha, P., Malar, D.S., and Tencomnao, T. (2021). The effectiveness of *Bacopa monnieri* (Linn.) Wettst. as a nootropic, neuroprotective, or antidepressant supplement: Analysis of the available clinical data. *Scientific Reports*, 11(1), p. 596

Brimson, J.M., Prasanth, M.I., Plaingam, W., and Tencomnao, T. (2020). *Bacopa monnieri* (L.) wettst. Extract protects against glutamate toxicity and increases the longevity of *Caenorhabditis elegans. Journal of Traditional and Complementary Medicine*, 10(5), pp.460–470

Brownstein, M.J. (1993). A brief history of opiates, opioid peptides, and opioid receptors. *Proceedings of the National Academy of Sciences of the United States of America*, 90(12), pp.5391–5393.

Canter, P.H., and Ernst, E. (2007). *Ginkgo biloba* is not a smart drug: An updated systematic review of randomised clinical trials testing the nootropic effects of *G. biloba* extracts in healthy people. *Human Psychopharmacology: Clinical and Experimental*, 22(5), pp.265–278.

Chadamangsi, K. (2000). *In DoTTaAMaTFaD. Administration (Ed.), Monographs of Selected Thai Materia Medica*, Vol. 1. Bangkok, Thailand: Amarin Printing and Publishing, pp. 79–82.

Chatterjee, A., Basak, B., Saha, M., Dutta, U., Mukhopadhyay, C., Banerji, J., Konda, Y., Harigaya, Y., and Harigaya, Y. (2000). Structure and stereochemistry of nardostachysin, a new terpenoid ester constituent of the rhizomes of *Nardostachys jatamansi. Journal of Natural Products*, 63(11), pp.1531–1533.

Chattopadhyay, R., Sarkar, S., Ganguly, S., Medda, C., and Basu, T. (1992). Hepatoprotective activity of *Ocimum sanctum* leaf extract against paracetamol included hepatic damage in rats. *Indian Journal of Pharmacology*, 24(3), p.163.

Chaudhary, S., Chandrashekar, K.S., Pai, K.S.R., Setty, M.M., Devkar, R.A., Reddy, N.D., and Shoja, M.H. (2015). Evaluation of antioxidant and anticancer activity of extract and fractions of *Nardostachys jatamansi* DC in breast carcinoma. *BMC Complementary and Alternative Medicine*, 15(1), p.50.

Chaudhary, S.K., Mukherjee, P.K., Maity, N., Nema, N.K., Bhadra, S., and Saha, B.P. (2014). *Ocimum sanctum* L. a potential angiotensin converting enzyme (ACE) inhibitor useful in hypertension. *Indian Journal of Natural Products and Resources*, 5(1), pp.83–87.

Cheignon, C., Tomas, M., Bonnefont-Rousselot, D., Faller, P., Hureau, C., and Collin, F. (2018). Oxidative stress and the amyloid beta peptide in Alzheimer's disease. *Redox Biology*, 14, pp.450–464.

Chevalier, A., Marinova, E., and Peña-Chocarro, L. (2014). *Plants and People: Choices and Diversity through Time*. Oxford, UK: Oxbow Books.

Christenhusz, M.J., and Byng, J.W. (2016). The number of known plants species in the world and its annual increase. *Phytotaxa*, 261(3), pp.201–217.

Cuzick, J., Thorat, M.A., Bosetti, C., Brown, P.H., Burn, J., Cook, N.R., Ford, L.G., Jacobs, E.J., Jankowski, J.A., Vecchia, C.L., Law, M., Meyskens, F.L.,Rothwell, P.M., Senn, H.J., and Umar, A. (2015). Estimates of benefits and harms of prophylactic use of Aspirin in the general population. *Annals of Oncology*, 26(1), pp.47–57.

Dhiman, N., and Bhattacharya, A. (2020). *Nardostachys jatamansi* (D.Don) DC.-Challenges and opportunities of harnessing the untapped medicinal plant from the Himalayas. *Journal of Ethnopharmacology*, 246, p.112211.

Dhuna, K., Dhuna, V., Bhatia, G., Singh, J., and Kamboj, S.S. (2013). Cytoprotective effect of methanolic extract of *Nardostachys jatamansi* against hydrogen peroxide induced oxidative damage in C6 glioma cells. *Acta Biochimica Polonica*, 60(1), pp.21–31.

Engels G., and Brinckmann, J. (2013) Holy Basil. In *HerbalGram*, Vol. 98. Austin, TX: American Botanical Council, pp.1–6.

Farnsworth, N.R., Akerele, O., Bingel, A.S., Soejarto, D.D., and Guo, Z. (1985). Medicinal plants in therapy. *Bulletin of the World Health Organization*, 63(6), pp.965–981.

Frati, P., Kyriakou, C., Del Rio, A., Marinelli, E., Vergallo, G.M., Zaami, S., and Busardò, F.P. (2015). Smart drugs and synthetic androgens for cognitive and physical enhancement: Revolving doors of cosmetic neurology. *Current Neuropharmacology*, 13(1), pp.5–11.

Giri, M.A., Bhalke, R.D., Prakash, K.V., and Kasture, S.B. (2020). Comparative evaluation of and as *Nardostachys jatamansi Mucuna pruriens* neuroprotective in Parkinson's disease. *Asian Journal of Pharmacy and Pharmacology*, 6(3), pp.224–230.

Goldberg, D.R. (2009). Aspirin: Turn of the century miracle drug. *Chemical Heritage Magazine*, 27(2), pp.26–30.

Gradinariu, V., Cioanca, O., Hritcu, L., Trifan, A., Gille, E., and Hancianu, M. (2015). Comparative efficacy of *Ocimum sanctum* L. and *Ocimum basilicum* L. essential oils against amyloid beta (1–42)-induced anxiety and depression in laboratory rats. *Phytochemistry Reviews*, 14(4), pp.567–575.

Guerchet, M., Prina, M., and Prince, M. (2013). Policy brief for heads of government: The global impact of dementia 2013–2050. In *Policy Brief for Heads of Government: The Global Impact of Dementia 2013–2050*. London: Published by Alzheimer's Disease International (ADI), pp.1–8.

Hakkim, F.L., Shankar, C.G., and Girija, S. (2007). Chemical composition and antioxidant property of holy basil (*Ocimum sanctum* L.) leaves, stems, and inflorescence and their *in vitro* callus cultures. *Journal of Agricultural and Food Chemistry*, 55(22), pp.9109–9117.

Hayat, S., and Ahmad, A. (2007). *Salicylic Acid: A Plant Hormone*. Berlin Germany: Springer.

Hoerster, H., Ruechker, G., and Joachim T. (1977). Valeranone content in the roots of *Nardostachys jatamansi* and *Valeriana officinalis*. *Phytochemistry*, 16, pp.1070–1071.

Hogervorst, E., Bandelow, S., Schmitt, J., Jentjens, R., Oliveira, M., Allgrove, J., Carter T., Gleeson, M. (2008). Caffeine improves physical and cognitive performance during exhaustive exercise. *Medicine and Science in Sports and Exercise* 40(10):1841–1851.

Hogervorst, E., Riedel, W., Kovacs, E., Brouns, F., and Jolles, J. (1999). Caffeine improves cognitive performance after strenuous physical exercise. *International Journal of Sports Medicine*, 20(06), pp.354–361.

Hogervorst, E., Riedel, W., Schmitt, J., and Jolles, J. (1998). Caffeine improves memory performance during distraction in middle-aged, but not in young or old subjects. *Human Psychopharmacology: Clinical and Experimental*, 13(4), pp.277–284.

Hussain, A.I., Chatha, S.A.S., Kamal, G.M., Ali, M.A., Hanif, M.A., and Lazhari, M.I. (2017). Chemical composition and biological activities of essential oil and extracts from *Ocimum sanctum*. *International Journal of Food Properties*, 20(7), pp.1569–1581.

Ijaz, B., Hanif, M., Mushtaq, Z., Khan, M., Bhatti, I., and Jilani, M. (2017). Isolation of bioactive fractions from *Ocimum sanctum* essential oil. *Oxidation Communications*, 40(1I), pp.158–167.

Irondi, E.A., Agboola, S.O., Oboh, G., and Boligon, A.A. (2016). Inhibitory effect of leaves extracts of *Ocimum basilicum* and *Ocimum gratissimum* on two key enzymes involved in obesity and hypertension in vitro. *Journal of Intercultural Ethnopharmacology*, 5(4), p.396.

Jivad, N., and Rabiei, Z. (2015). Review on herbal medicine on brain ischemia and reperfusion. *Asian Pacific Journal of Tropical Biomedicine*, 5(10), pp.789–795.

Joppa, L.N., Roberts, D.L., Myers, N., and Pimm, S.L. (2011). Biodiversity hotspots house most undiscovered plant species. *Proceedings of the National Academy of Sciences*, 108(32), pp.13171–13176.

Joshi, H., and Parle, M. (2006a). Cholinergic basis of memory improving effect of *Ocimum tenuiflorum* Linn. *Indian Journal of Pharmaceutical Sciences*, 68(3), pp. 364-365.

Joshi, H., and Parle, M. (2006b). Evaluation of nootropic potential of *Ocimum sanctum* Linn. in mice. *Indian Journal of Experimental Biology*, 44(2), pp.133–136.

Joshi, H., and Parle, M. (2006c). *Nardostachys jatamansi* improves learning and memory in mice. *Journal of Medicinal Food*, 9(1), pp.113–118.

Kaewthawee, N., and Brimson, S. (2013). The effects of ursolic acid on cytokine production via the mapk pathways in leukemic T-cells. *Experimental and Clinical Sciences International Online Journal for Advances in Science*, 12, pp.102–114.

Kandhan, T.S., Thangavelu, L., and Roy, A. (2018). Acetylcholinesterase activity of *Ocimum sanctum* leaf extract. *Journal of Advanced Pharmacy Education and Research*, 8(1), 41-44.

Kaprao. (1999). Administration. In TFaD (Ed.), *Herbal Medicines Used in Primary Health Care*. Bangkok Thailand: The Printing Office Agency to Assist Veterans.

Karaöz, E., Çetinalp Demircan, P., Erman, G., Güngörürler, E., and Eker Sarıboyacı, A. (2017). Comparative analyses of immunosuppressive characteristics of bone-marrow, Wharton's Jelly, and adipose tissue-derived human mesenchymal stem cells. [Kemik İliği, Wharton Jölesi ve İnsan Yağ Doku-Kaynaklı Mezenkimal Kök Hücrelerin İmmünsüpresif Özelliklerinin Karşılaştırmalı Olarak İncelenmesi]. *Turkish Journal of Haematology: Official journal of Turkish Society of Haematology*, 34(3), pp.213–225.

Kasture, S.B., Mane-Deshmukh, R.V., and Arote, S.R. (2014). Non-linear dose effect relationship in anxiolytic and nootropic activity of lithium carbonate and *Nardostachys jatamansi* in rats. *Oriental Pharmacy and Experimental Medicine*, 14(4), pp.357–362.

Kelm, M.A., Nair, M.G., Strasburg, G.M., and DeWitt, D.L. (2000). Antioxidant and cyclooxygenase inhibitory phenolic compounds from *Ocimum sanctum* Linn. *Phytomedicine*, 7(1), pp.7–13.

Ko, W., Park, J.S., Kim, K.W., Kim, J., Kim, Y.C., and Oh, H. (2018). Nardosinone-type sesquiterpenes from the hexane fraction of *Nardostachys jatamansi* attenuate NF-κB and MAPK signaling pathways in lipopolysaccharide-stimulated BV2 microglial cells. *Inflammation*, 41(4), pp.1215–1228.

Kot Chadamangsi [Internet]. Department of Pharmaceutical Sciences, Ubon Ratchathani University. [cited 27 Aug 2021]. Available from: http://www.thaicrudedrug.com/main.php?action=viewpage&pid=29#:~:text=%E0%B8%AA%E0%B8%A3%E0%B8%A3%E0%B8%9E%E0%B8%84%E0%B8%B8%E0%B8%93%3A,%E0%B8%9B%E0%B8%B2%E0%B8%81%E0%B9%83%E0%B8%99%E0%B8%84%E0%B8%AD%20%E0%B8%82%E0%B8%B1%E0%B8%9A%E0%B8%A5%E0%B8%A1

Kritikos, P.G., and Papadaki, S. (1967). *The History of the Poppy and of Opium and Their Expansion in Antiquity in the Eastern Mediterranean Area*. New York: UN.

Kulkarni, R., Girish, K., and Kumar, A. (2012). Nootropic herbs (Medhya Rasayana) in Ayurveda: An update. *Pharmacognosy Reviews*, 6(12), pp.147.

Kumar, P., Jyothirmai, N., Hymavathi, P., and Prasad, K. (2013). Screening of ethanolic extract of *Ocimum tenuiflorum* for recovery of atorvastatin induced hepatotoxicity. *International Journal of Pharmacy and Pharmaceutical Sciences*, 5(4), pp.346–349.

Lahon, K., and Das, S. (2011). Hepatoprotective activity of *Ocimum sanctum* alcoholic leaf extract against paracetamol-induced liver damage in Albino rats. *Pharmacognosy Research*, 3(1), p.13.

Lefrak, E.A., Piťha, J., Rosenheim, S., and Gottlieb, J.A. (1973). A clinicopathologic analysis of adriamycin cardiotoxicity. *Cancer*, 32(2), pp.302–314.

Liu, Q.F., Jeon, Y., Sung, Y.-W., Lee, J.H., Jeong, H., Kim, Y.-M., … Cho, K.S. (2018). *Nardostachys jatamansi* ethanol extract ameliorates Aβ42 cytotoxicity. *Biological and Pharmaceutical Bulletin*, 41(4), pp.470–477.

Mair, V.H., Hoh, E., and Hodgson, M. (2009) *The True History of Tea*. London, UK: Thames & Hudson.

Manglik, A., Kruse, A.C., Kobilka, T.S., Thian, F.S., Mathiesen, J.M., Sunahara, R.K., . . . Granier, S. (2012). Crystal structure of the μ-opioid receptor bound to a morphinan antagonist. *Nature*, 485(7398), pp.321–326.

Manov, I., Motanis, H., Frumin, I., and Iancu, T.C. (2006). Hepatotoxicity of anti-inflammatory and analgesic drugs: Ultrastructural aspects. *Acta Pharmacologica Sinica*, 27(3), pp.259–272.

Mathew, M., and Subramanian, S. (2014). *In vitro* screening for anti-cholinesterase and antioxidant activity of methanolic extracts of ayurvedic medicinal plants used for cognitive disorders. *PLoS ONE*, 9(1), p.e86804.

McGleenon, B., Dynan, K., and Passmore, A. (1999). Acetylcholinesterase inhibitors in Alzheimer's disease. *British Journal of Clinical Pharmacology*, 48(4), pp.471.

Mehta, A. (2005). Aspirin. *Chemical & Engineering News*, 83(25), pp.46–47.

Mohanty, I., Arya, D.S., and Gupta, S.K. (2006). *Effect of Curcuma longa and Ocimum sanctum* on myocardial apoptosis in experimentally induced myocardial ischemic-reperfusion injury. *BMC Complementary and Alternative Medicine*, 6(1), pp.1–12.

Mukherjee, P.K., Kumar, V., and Houghton, P.J. (2007). Screening of Indian medicinal plants for acetylcholinesterase inhibitory activity. *Phytotherapy Research: An International Journal Devoted to Pharmacological and Toxicological Evaluation of Natural Product Derivatives*, 21(12), pp.1142–1145.

Nathan, P.J., Ricketts, E., Wesnes, K., Mrazek, L., Greville, W., and Stough, C. (2002). The acute nootropic effects of *Ginkgo biloba* in healthy older human subjects: A preliminary investigation. *Human Psychopharmacology: Clinical and Experimental*, 17(1), pp.45–49.

National Clinical Guideline, C. (2013). *National Institute for Health and Clinical Excellence: Guidance Myocardial Infarction with ST-Segment Elevation: The Acute Management of Myocardial Infarction with ST-Segment Elevation*. London: Royal College of Physicians.

National List of Essential Medicines (NLEM). 2013. *Vol. 2*. Retrieved from http://kpo.moph.go.th/webkpo/tool/Thaimed2555.pdf

Norman Scholfield, C., Dilokthornsakul, P., Limpeanchob, N., Thanarangsarit, P., and Kongkeaw, C. (2013). Meta-analysis of randomised controlled trials on cognitive effects of *Bacopa monnieri* extract. *Journal of Ethnopharmacology*, 151(1), pp.528–535.

Nunn, J.F. (1996). Ancient Egyptian medicine. *Transactions of the Medical Society*, 113, pp.57–68.

Pande, J., Singh, S., Khilnani, G., Khilnani, S., and Tandon, R. (1996). Risk factors for hepatotoxicity from antituberculosis drugs: A case-control study. *Thorax*, 51(2), pp.132–136.

Pandey, M.M., Katara, A., Pandey, G., Rastogi, S., and Rawat, A.K. (2013). An important Indian traditional drug of ayurveda jatamansi and its substitute bhootkeshi: Chemical profiling and antioxidant activity. *Evidence-Based Complementary and Alternative Medicine*, 2013, pp.142517–142517

Patil, R.A., Hiray, Y.A., and Kasture, S.B. (2012). Reversal of reserpine-induced orofacial dyskinesia and catalepsy by *Nardostachys jatamansi*. *Indian Journal of Pharmacology*, 44(3), p. 340.

Pattanayak, P., Behera, P., Das, D., and Panda, S.K. (2010). *Ocimum sanctum* Linn. A reservoir plant for therapeutic applications: An overview. *Pharmacognosy Reviews*, 4(7), pp.95–105.

Raghavendra, M., Maiti, R., Kumar, S., and Acharya, S. (2009). *Role of Ocimum sanctum* in the experimental model of Alzheimer's disease in rats. *International Journal of Green Pharmacy*, 3(1), p.6.

Rahman, H., Muralidharan, P., and Anand, M. (2011). Inhibition of AChE and antioxidant activities are probable mechanism of *Nardostacys jatamansi* DC in sleep deprived Alzheimer's mice model. *International Journal of PharmTech Research*, 3, pp.1807–1816.

Rasheed, A.S., Venkataraman, S., Jayaveera, K.N., Fazil, A.M., Yasodha, K.J., Aleem, M.A., Mohammed, M., Khaja, Z., Ushari,B., Pradeep, H.A., and Ibrahim, M. (2010). Evaluation of toxicological and antioxidant potential of *Nardostachys jatamansi* in reversing haloperidol-induced catalepsy in rats. *International Journal of General Medicine*, 3, pp.127–136.

Rucker G, Paknikar S.K, Mayer R, Breitmaier E, Will G, and Wiehl, L. (1993). Revised structure and stereochemistry of jatamansic acid. *Phytochemistry*, 33, pp.141–143.

Sadigh-Eteghad, S., Sabermarouf, B., Majdi, A., Talebi, M., Farhoudi, M., and Mahmoudi, J. (2015). Amyloid-beta: A crucial factor in Alzheimer's disease. *Medical Principles and Practice*, 24(1), pp.1–10.

Sahu, R., Dhongade, H.J., Pandey, A., Sahu, P., Sahu, V., Patel, D., and Kashyap, P. (2016). Medicinal properties of *Nardostachys jatamansi* (A Review). *Oriental Journal of Chemistry*, 32(2), pp.859–866.

Salim, S., Ahmad, M., Zafar, K.S., Ahmad, A.S., and Islam, F. (2003). Protective effect of *Nardostachys jatamansi* in rat cerebral ischemia. *Pharmacology Biochemistry and Behavior*, 74(2), pp.481–486.

Sastry, S.D., Maheswari, M.L., Chakravarti, K.K., and Bhattacharyya, S.C. (1967). Terpenoids: CVI: The structure of calarenol. *Tetrahedron*, 23(4), pp.1997–2000.

Satyamitra, M., Mantena, S., Nair, C., Chandna, S., Dwarakanath, B., and Uma Devi, P. (2014). The antioxidant flavonoids, orientin and vicenin enhance repair of radiation-induced damage. *SAJ Pharmacy and Pharmacology*, 1(105), pp.1–9.

Schippmann, U., Leaman, D.J., and Cunningham, A. (2002). Impact of cultivation and gathering of medicinal plants on biodiversity: Global trends and issues. In *Biodiversity and the Ecosystem Approach in Agriculture, Forestry and Fisheries*, Chapter 7. Rome, Italy: FAO.

Schläppi, B. (1985). The lack of hepatotoxicity in the rat with the new and reversible MAO-A inhibitor moclobemide in contrast to iproniazid. *Arzneimittelforschung*, 35(5), pp.800–803.

Seethalakshmi, B., Narasappa, A., and Kenchaveerappa, S. (1982). Protective effect of *Ocimum sanctum* in experimental liver injury in albino rats. *Indian Journal of Pharmacology*, 14, p.63.

Shanbhag, S.N., Mesta, C.K., Maheshwari, M.L., Paknikar, S.K., and Bhattacharyya, S.C. (1964). Terpenoids—LII: Jatamansin, a new terpenic coumarin from *Nardostachys jatamansi*. *Tetrahedron*, 20(11), pp.2605–2615.

Sharma, M.K., Kumar, M., and Kumar, A. (2002). *Ocimum sanctum* aqueous leaf extract provides protection against mercury induced toxicity in Swiss albino mice. *Indian Journal of Experimental Biology*, 40(9), pp.1079–82.

Sharma, S.K., and Singh, A.P. (2012). *In vitro* antioxidant and free radical scavenging activity of *Nardostachys jatamansi* DC. *Journal of Acupuncture and Meridian Studies*, 5(3), pp.112–118.

Sharma, V., and Chanda, D. (2018). Ocimum: The holy basil against cardiac anomalies. In *The Ocimum Genome*. New York: Springer, pp. 25–36.

Shyamala, A.C., and Devaki, T. (1996). Studies on peroxidation in rats ingesting copper sulphate and effect of subsequent treatment with *Ocimum sanctum*. *Journal of Clinical Biochemistry and Nutrition*, 20(2), pp.113–119.

Singh, S., Rehan, H., and Majumdar, D. (2001). Effect of *Ocimum sanctum* fixed oil on blood pressure, blood clotting time and pentobarbitone-induced sleeping time. *Journal of Ethnopharmacology*, 78(2–3), pp.139–143.

Singh, V., Krishan, P., Singh, N., Kumar, A., and Shri, R. (2017). Amelioration of ischemia-reperfusion induced functional and biochemical deficit in mice by *Ocimum kilimandscharicum* leaf extract. *Biomedicine and Pharmacotherapy*, 85, pp.556–563.

Suanarunsawat, T., Boonnak, T., Ayutthaya, W.N., and Thirawarapan, S. (2010). Anti-hyperlipidemic and cardioprotective effects of *Ocimum sanctum* L. fixed oil in rats fed a high fat diet. *Journal of Basic and Clinical Physiology and Pharmacology*, 21(4), pp.387–400.

Subashini, R., Yogeeta, S., Gnanapragasam, A., and Devaki, T. (2006). Protective effect of *Nardostachys jatamansi* on oxidative injury and cellular abnormalities during doxorubicin-induced cardiac damage in rats. *Journal of Pharmacy and Pharmacology*, 58(2), pp.257–262.

Suzuki, S., Ohta, S., Takashio, K., Nitanai, H., and Hashimoto, Y. (1990). Augmentation for intratumoral accumulation and anti-tumor activity of liposome-encapsulated Adriamycin by tumor necrosis factor-α in mice. *International Journal of Cancer*, 46(6), pp.1095–1100.

Ubaid, R.S., Anantrao, K.M., Jaju, J., and Mateenuddin, M. (2003). Effect of *Ocimum sanctum* (OS) leaf extract on hepatotoxicity induced by antitubercular drugs in rats. *Indian Journal of Physiology and Pharmacology*, 47(4), pp.465–470.

Uma Devi, P., Ganasoundari, A., Vrinda, B., Srinivasan, K.K., and Unnikrishnan, M.K. (2000). Radiation protection by the ocimum flavonoids orientin and vicenin: Mechanisms of action. *Radiation Research*, 154(4), pp.455–460.

Vetal, M.D., Lade, V.G., and Rathod, V.K. (2012). Extraction of ursolic acid from *Ocimum sanctum* leaves: Kinetics and modeling. *Food and Bioproducts Processing*, 90(4), pp.793–798.

Weinberg, B.A., and Bealer, B.K. (2004). *The World of Caffeine: The Science and Culture of the World's Most Popular Drug*. New York: Routledge.

Westphal, J., Vetter, D., and Brogard, J. (1994). Hepatic side-effects of antibiotics. *Journal of Antimicrobial Chemotherapy*, 33(3), pp.387–401.

World-Health-Organization. (2020). *Global Health Estimates 2020: Deaths by Cause, Age, Sex, by Country and by Region, 2000–2019*. World Health Organization.

Yanpallewar, S., Rai, S., Kumar, M., and Acharya, S. (2004). Evaluation of antioxidant and neuroprotective effect of *Ocimum sanctum* on transient cerebral ischemia and long-term cerebral hypoperfusion. *Pharmacology Biochemistry and Behavior*, 79(1), pp.155–164.

Yoon, C.S., Kim, D.C., Park, J.S., Kim, K.W., Kim, Y.C., and Oh, H. (2018). Isolation of novel sesquiterpeniods and anti-neuroinflammatory metabolites from *Nardostachys jatamansi*. *Molecules*, 23(9), pp. 2367

Zhu, L., and Yueying, G. (1998). Progress in the study of nootropic mechanisms of *Panax ginseng*. Shenyang yao ke da xue xue bao. *Journal of Shenyang Pharmaceutical University*, 15(1), pp.73–76.

10 *Panax quinquefolium* (American Ginseng) and *Physostigma venenosum* (Calabar Bean)

Sushweta Mahalanobish, Noyel Ghosh, and Parames C. Sil

CONTENTS

Abbreviations ... 179
10.1 Introduction ... 180
10.2 Basic Description ... 181
10.2.1 American Ginseng ... 181
10.2.2 Calabar Bean ... 181
10.3 Traditional Knowledge ... 182
10.3.1 American Ginseng ... 182
10.3.2 Calabar Bean ... 182
10.4 Bioactive Compounds ... 182
10.4.1 Bioactive Components of American Ginseng ... 182
10.4.2 Bioactive Components of Calabar Bean ... 183
10.5 Potential Benefits ... 185
10.5.1 Health Benefits of American Ginseng ... 185
10.5.1.1 Neuronal Protection ... 185
10.5.1.2 Cardioprotective Activity ... 186
10.5.1.3 Anticancer Activity ... 186
10.5.1.4 Antidiabetic Activity ... 187
10.5.1.5 Prevention of Obesity ... 188
10.5.1.6 Antiaging Properties ... 189
10.5.1.7 Multiple Sclerosis Prevention ... 190
10.5.1.8 Antimicrobial Activity ... 190
10.5.2 Health Benefits of Calabar Bean ... 191
10.5.2.1 Neuronal Protection ... 191
10.5.2.2 Antidote Agent ... 193
10.5.2.3 Glaucoma Treatment ... 194
10.6 Conclusion ... 194
References ... 195

ABBREVIATIONS

ACC acetyl-CoA carboxylase
AchE acetyl choline esterase
AD Alzheimer's disease
ADAS-Cog Alzheimer's disease Assessment Scale-Cognitive subscale

DOI: 10.1201/9781003205067-10

AGBE	AG berry extract
APP	amyloid precursor protein
BchE	butyryl cholinesterase
Bcl-2	B-cell lymphoma 2
BDNF	brain-derived neurotrophic factor
CGIC	Clinical Global Impression of Change
ChAT	choline acetyltransferase
CNS	central nervous system
ERK	extracellular signal-regulated kinases
ET-1	endothelin-1
FABP4	fatty acid-binding protein 4
FAS	fatty acid synthase
GFAP	glial fibrillary acidic protein
GHB	gamma-hydroxybutyrate
GLUT4	glucose transporter 4
GPx	glutathione peroxidase
iNOS	nitric oxide synthase
IR	ionizing radiation
JNK	c-Jun N-terminal kinase
KGB	konjac-based fiber blend
MAP2	microtubule-associated protein 2
MIC	minimal inhibitory concentration
MMP-9	matrix metallopeptidase-9
MS	multiple sclerosis
NF-kB	nuclear factor kappa-light-chain-enhancer of activated B cells
Nrf2	nuclear factor (erythroid-derived 2)-like 2
OGD	oxygen-glucose deprivation/reperfusion
P. aeruginosa	*Pseudomonas aeruginosa*
PD	Parkinson's disease
PKC	protein kinase C
POF	premature ovarian failure
PPD	protopanaxadiol
PPT	protopanaxatriol
ROS	reactive oxygen species
SAM	senescence-accelerated mice
SERT	serotonin transporter
sICAM-1	soluble intercellular adhesion molecule 1
T2DM	type 2 diabetes mellitus

10.1 INTRODUCTION

To keep up the normal body homeostasis, consumption of a healthy diet rich in natural active ingredients is highly appreciable. Natural compounds are extremely beneficial in protecting our health, and thus, naturally derived compounds have gained crucial attention among the research community. Naturally extracted phytonutrients prevent the onset of several health complications, improve clinical outcomes of a number of diseases, and reduce various side effects associated with chemotherapeutic and radiotherapeutic treatments (Gupta and Prakash, 2014). Phytonutrients, when taken regularly as supplements along with a balanced diet, reduce stress, delay aging, and improve overall health conditions due to their inherent antioxidant, anti-inflammatory, antibacterial, neuroprotective, immunomodulatory activities. Besides, they are chemo-preventive, antiallergic, hypolipidemic, CNS stimulating, antidiabetic, etc., thus, they keep diseases at a bay (Mahalanobish et al.,

2019). Epidemiological studies reveal that people who consume a diet rich in vegetables, fruits, and other plant-derived products have a lower risk of cardiovascular disease, stroke, cancer, etc. (Anand et al. 2015, Aune et al. 2017).

Panax quinquefolium L. (American Ginseng (AG)), is an example of such a nutritive plant that contains bioactive ingredients called "ginsenosides". Ginsenosides exhibit a wide range of health-oriented advantageous activities including anti-inflammatory, hypoglycemic, cardioprotective, anti-cancerous, and antidiabetic effects (He et al., 2015; Li et al., 2010b). It also has shown therapeutic efficacy against chronic obstructive pulmonary disease and diminishes the undesirable effects of menopause (Shergis et al., 2014, Rotem and Kaplan 2007). Another similar bioactive phytochemical, known as physostigmine, has been sourced from *Physostigma venenosum* (calabar bean, (CB)), which is recognized as a powerful, reversible inhibitor of acetylcholine esterase; it is used for treating several central and peripheral nervous system diseases (Houghton et al., 2006). In the following sections, we discuss the two aforesaid plants, their health benefits, phytochemical derivatives, and related activities at the molecular level to fight against several diseases.

10.2 BASIC DESCRIPTION

10.2.1 American Ginseng

The *Panax* genus plays a significant role in human health management. Among 11 species, *Panax quinquefolium* (AG), *Panax ginseng* (Asian ginseng), and *Panax notoginseng* are used widely. All three species are in high demand due to their enormous health-oriented beneficial advantages and used to derive health care products worldwide (Lee et al., 2016; Yang et al., 2016b; Yu et al., 2014). Asian ginseng is abundant in North Asian countries, like Korea, China, and Japan; whereas, notoginseng is mostly found in China (Lee et al., 2017; Pan et al., 2016). AG is solely found in North America, mainly in Oklahoma, Arkansas, and Louisiana in the United States (US) (Cruse-Sanders and Hamrick, 2004). It belongs to the family Araliaceae. These plants are grown under a fully shaded environment, underneath hardwoods. Due to specialized growth requirements and huge demand in the commercial market, AG has been declared an endangered species in some areas, like Maine and Rhode Island in the US. AG is a perennial herb with white flowers in summer (Souther et al., 2012). The plant is generally 15–46 cm in height, with three leaves, each with three to five leaflets. The flower is present on a simple umbel within the major leaf axis. Berry-like red fruit is present that contains up to three seeds. The fleshy white roots of AG are highly branched. With aging, there is a growth of an auxiliary root that acts as a spare in case of damage to the main root. The plant is fertilized by generalist insects like halictid bees. Its reproduction depends solely on seeds and takes place after a pre-reproductive period. The fruits come in July and August and reach maturity and redness in November (Cruse-Sanders and Hamrick, 2004). In cultured AG, root harvesting is possible after 3–4 years, whereas for wildly grown AG, the root is generally harvested after 8 years or later (Lim et al., 2005; Proctor and Shelp, 2014). After harvesting, roots are thermally processed, and based on thermal processing, roots are classified into two varieties: white (dried) and red (fresh). Roots which are kept in the sun for dehydration are known as "white ginseng". Red ginseng is processed at a high temperature that sometimes damages its enzymatic content and cleaves its active ingredients (Kim et al., 2007). Thermal processing of roots increases the shelf life of roots. For medicinal purposes, most of the time, dried roots are preferred over fresh roots.

10.2.2 Calabar Bean

CB, *Physostigma venenosum*, also known as "ordeal bean", is a leguminous, woody climbing plant in nature and is mainly found in the tropical forests of Western Africa. (Proudfoot, 2006). Etymologically, *Physostigma* has been named due to the presence of a beak-like appendage at the end of the stigma within the center of its flower. The plant is an herbaceous, perennial having a woody stem as its base. This is a scarlet runner bean with a height of 50 feet and a diameter of two inches. The flowers are large, heavily veined, pale pink or purplish in color, generally an inch long with axillary peduncles, and grouped in pendulous, fascicled racemes. The seed pods containing two or three seeds are 6–7 inches in length. The beans are similar to horse bean but less flattened and chocolate-brown in appearance (Swain, 1972).

10.3 TRADITIONAL KNOWLEDGE

10.3.1 American Ginseng

Medicinal American Ginseng has been in use in Eastern Europe for over 2000 years. The name of the genus, *Panax*, comes from two Greek words: "*Pan*" meaning "all" and "*anox*" meaning "to treat"; thus, something that treats all diseases. The name was coined by Russian scientist Carl Meyer in the 19th century (Sengupta et al., 2004). Native Americans have also used the roots and leaves of AG for medicinal purposes for centuries. In the 18th century, sang hunters collected the roots of AG and sold them to Chinese traders at a high price. AG can only be grown in temperate regions. Although no evidence has been found regarding the effectiveness of AG against the common cold (Seida et al., 2011), it was reported to reduce sickness. Currently, a wide range of products made from AG is available at the market, in form of pellets, powders, or even teas. Nowadays, AG roots are available either in powder form or in dried shredded slices. Its dried flowers or flakes are also readily available in the market (Jung et al., 2014). In many countries, the inclusion of AG extracts can be noted in beauty products, such as shower gels, lotions, hair conditioners, shampoos, etc. The root and its rhizomes are used as dietary supplements, drugs, and several health products. Today, AG extract-enriched candies and drinks are marketed widely in the US, while in Korea, salads and soups containing AG are very popular. In China, the practice of ginseng extract addition in alcoholic beverages may help to reduce the hangover effect (Ma et al., 2017a; Szczuka et al., 2019).

10.3.2 Calabar Bean

CB or "E-ser-e" was used traditionally by the people of Old Calabar. Efik people, residents of Cross River, and the Ibibio people of Akwa Ibom, used CB as an "ordeal poison". People who were accused of witchcraft or other crimes were punished to intake the white, milky bean. If the accused survived thereafter, they were considered to be honest (Houghton et al., 2006). Although the bean is extremely poisonous, there is no specific taste, smell, or external characteristic feature to distinguish them from their harmless counterparts. Though it is said that the plant was first introduced in Britain in 1840, no exact description of the plant is available before 1861. In 1863, Sir Thomas Richard Fraser investigated its physiological effects and reported the alkaloid nature of this plant. The chief alkaloid of this plant, physostigmine, currently is in use to treat anticholinergic syndromes like dementia, myasthenia gravis, glaucoma, etc.

10.4 BIOACTIVE COMPOUNDS

10.4.1 Bioactive Components of American Ginseng

The bioactive compounds that are present in AG and exert various beneficial effects on human health are known to be ginsenosides or panaxosides. They are basically glycosides in nature consisting of sugar chain along with non-sugar (aglycone) moiety. The chemical structure of ginsenosides contains three types of aglycone – dammarane-type tetracyclic triterpene, pentacyclic oleanolic acid, and tetracyclic ocotillol type. The sugar part of ginsenosides comprises hexoses (glucose, galactose), 6-deoxyhexoses (furanose, rhamnose), pentoses (arabinose, xylose), and uronic acids (glucuronic acid). They are cyclic in nature and connected with aglycone part by hemiacetal bonds

(Kochan et al., 2017; Nag et al., 2012). The nomenclature of ginsenosides is designed as "Rx", where "R" indicates root and "x" indicates the polarity of the molecule in alphabetical order from "a" to "h" index.

Most of the ginsenosides comprise a dammarane skeleton where 17 carbon atoms are arranged in four rings. Based on the number of hydroxyl groups, dammarane-type ginsenosides can be further classified into two subgroups: protopanaxadiol (PPD) and protopanaxatriol (PPT) (Feng et al., 2017). In PPD group, ginsenoside contains sugar moiety at C-3 and/or C-20 position. Here, linear linkages occur between glycosyl chains, and acylation takes place at the 6-OH of the terminal glucose of a three-sugar chain. This subgroup can be exemplified by ginsenosides Rb1, Rb2, Rb3, Rc, Rd, Rg3, and Rh2. Whereas, PPT subgroup covers both ginsenosides such as Re, Rf, Rg1, Rg2, Rh1, F1, F3, and also notoginsenosides like R1, R2 etc. Structurally, compounds of the PPT subgroup contain no more than two glycosyl chains and a linear linkage of saccharide chains (Yang et al., 2014). The presence of hydroxyl (-OH) group at C-6 position in PPTs distinguishes them from PPDs (Wang et al., 2005). Ginsenosides Rb1, Re, Rd, Rg1, and Rb3 are the five major saponins that account for more than 70% of total ginsenosides in AG. The second class includes oleanolic acid as the aglycone moiety, and among its derivatives, ginsenoside Ro is notable (Huang et al., 2018).

The last group is ocotillol-type aglycone, which is signified by a five-membered epoxy ring at C-20 position (Kim, 2012). Pseudoginsenoside F11 (p-F11) is an ocotillol-type panaxoside, found in the roots and leaves of AG (Liu et al., 2017). Discrimination between *P. ginseng* and AG is based upon the presence of two ginsenosides, Rf and p-F11. Ginsenoside Rf is present in Asian ginseng, whereas p-F11 occurs solely in AG (Popovich et al., 2012).

The remainder of the composition includes unsaturated fatty acids, like linolenic acid, which in AG provides protection against various chronic diseases, such as arthritis and arrhythmia (Zhang et al., 2013). Besides that, there are various polysaccharides in ginseng roots that consist of complex chains of monosaccharides containing d-galactose, l-arabinose, l-rhamnose, d-glucuronic acid, d-galacturonic acid, and d-galactosyl residues (Wang et al., 2004). Wang et al. (2015) reported the presence of neutral polysaccharides in AG roots. Analysis of monosaccharide composition showed that it is composed of glucose and galactose in a molar ratio of 1:1:15.

Besides saponins and polysaccharides, terpenes, vitamins, volatile oils, flavonoids, amino acids, phenolic compounds, and minerals are present (Cui et al., 2017; Ludwiczuk et al., 2006). By using trans-Anethole as an elicitor, the extraction amount of triterpene saponins from the hairy root culture of AG can be enhanced (Kochan et al., 2018). Application of trans-anethole triggers the production of 9 distinctive ginsenosides: Rb1, Rb2, Rb3, Rc, Rd, Rg1, Rg2, Re, and Rf among which the production of Re metabolite is highest, i.e., 3.9-fold more compared to the untreated root.

It has been observed that despite the presence of PPDs or PPTs in all groups of ginseng plants, their chemical compositions are different in diverse groups. Asian ginseng has Rb1, Rb2, and Rg1ginsenosides (Wang et al., 2015a); whereas Notoginseng is enriched with ginsenosides Rb1, Rd, Rg1; notoginsenoside has R1; and ginsenosides Rb1, Rd, and Re are present in AG (Liu et al., 2020). Interestingly, it has been noted that based upon geographic location, the chemical composition of active compounds of the same species of AG shows a wide range of variation and also different health benefits (Sun and Chen, 2011; Sun et al., 2012). Along with the advancement of age, the saponin content of AG is also noticed to be increased (Kochan et al., 2004; Xiao et al., 2015). Figure 10.1 depicts the bioactive constituents of AG.

10.4.2 Bioactive Components of Calabar Bean

The active ingredients of CB include physostigmine (eserine), eseridine, eseramine, calabarine, and phytosterin which were first isolated in the 19th century (Swain, 1972). Though a number of other derivatives were identified later, physostigmine is still considered the most important alkaloid of CB (Miller, 1976; Zhao et al., 2004). Physostigmine is highly soluble in chloroform, benzol, alcohol, and carbon disulphide and relatively less soluble in water. Physostigmine is highly unstable, and

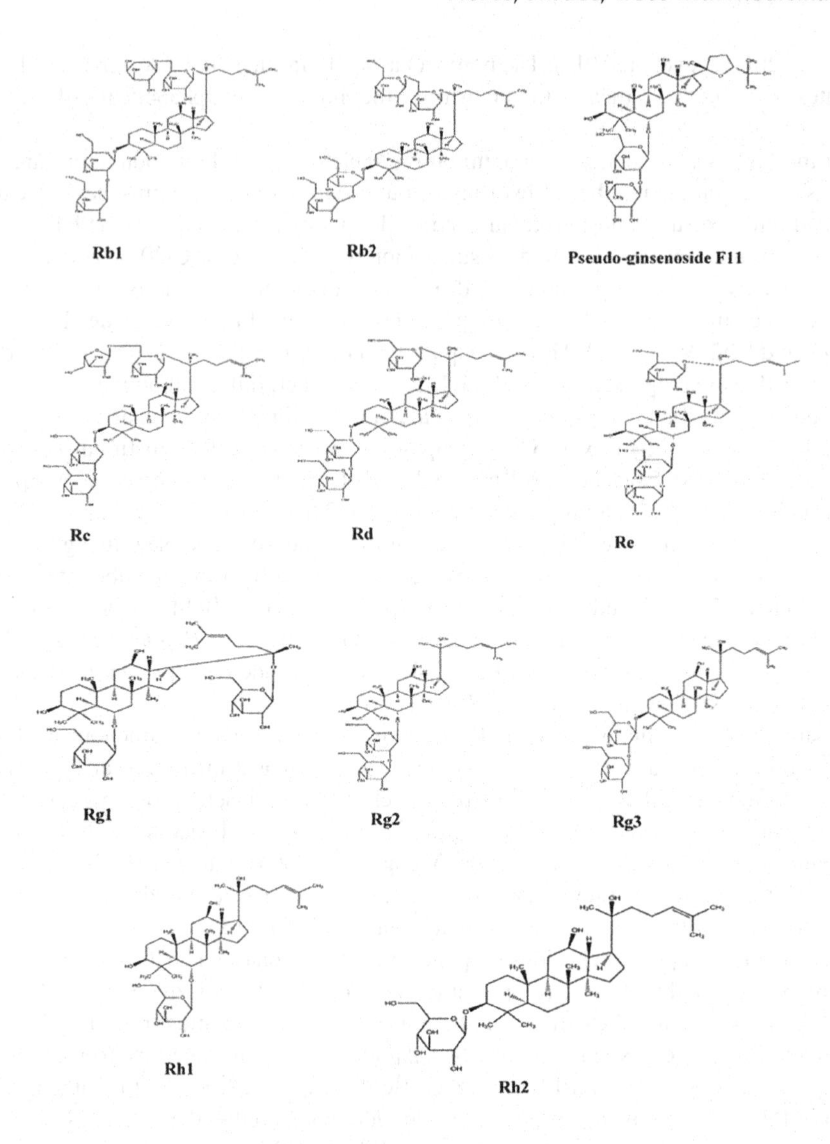

FIGURE 10.1 Bioactive constituents of AG.

light or heat exposure converts it into red color rubreserin (Hemsworth and West, 1970; Rubnov et al., 1999). The interaction between the functional carbamate group of physostigmine at the cholinesterase active site occurs in a way similar to that of the organophosphorus-mediated inhibition of cholinesterase (Coelho and Birks, 2001; Somani and Dube, 1989). Physostigmine contains two stereocenters – the two carbons where the five-membered rings join together. Among 71 derivatives of physostigmine, 33 are racemic mixtures, and 38 are the product of a single enantiomer. In 1935, Julian and Pikl first synthesized total physostigmine. They prepared (*L*)-eseroline by hydrolysis of N-methylcarbamoyl group of physostigmine. (*L*)-eseroline has an effect similar to morphine as it exhibits a high affinity for opioid receptor sites. Calabarine, an ether-insoluble alkaloid is produced as a by-product of physostigmine synthesis and can be converted back to physostigmine by warming with diluted acid (Zhao et al., 2004). Its physiological action has a closer resemblance to strychnine than physostigmine. Other derivatives include epistigmine (heptyl-physostigminetartrate), cymserine, bisnorcymserine (N-demethylated cymserine), and tetrahydrofurobenzofuran cymserine. Phenserine is a phenylcarbamate derivative of physostigmine. Posiphen ((+)-phenserine), is an

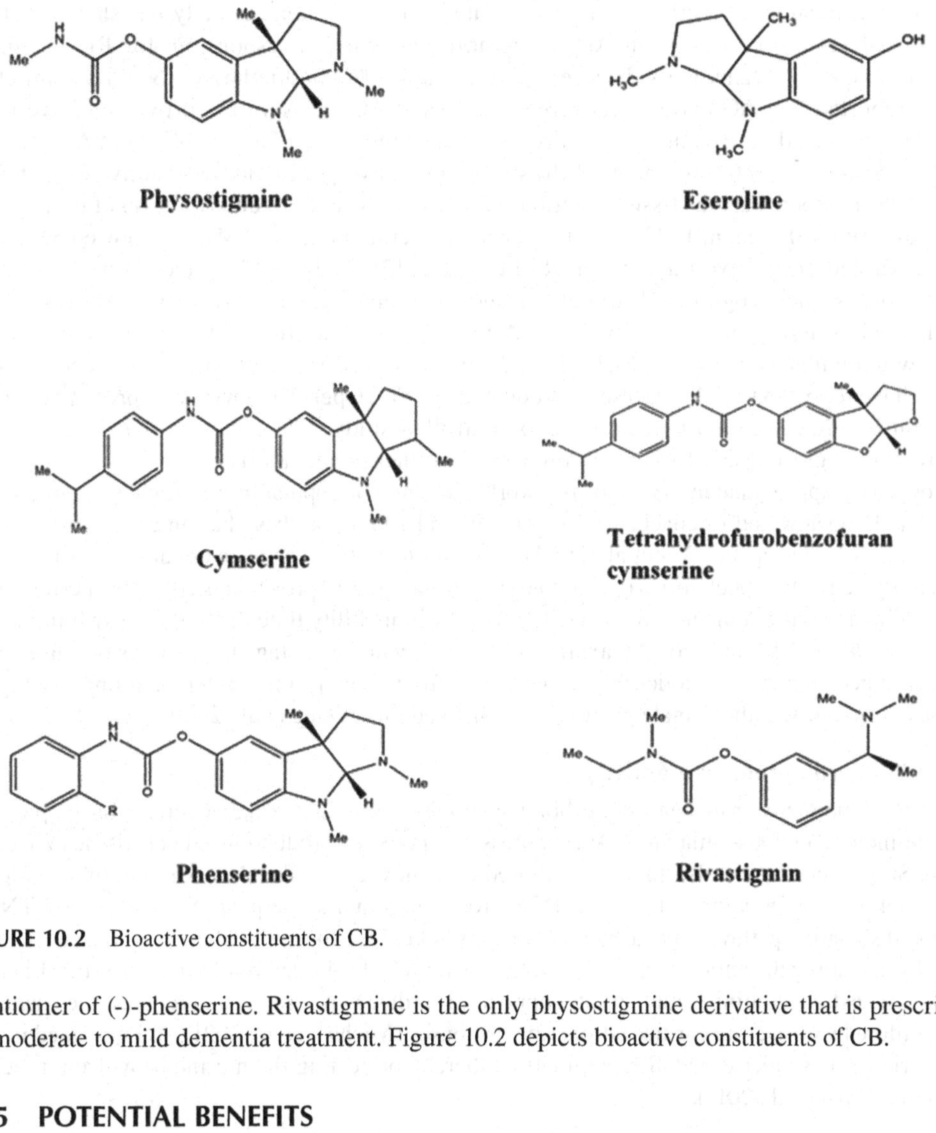

Physostigmine

Eseroline

Cymserine

Tetrahydrofurobenzofuran cymserine

Phenserine

Rivastigmin

FIGURE 10.2 Bioactive constituents of CB.

enantiomer of (-)-phenserine. Rivastigmine is the only physostigmine derivative that is prescribed for moderate to mild dementia treatment. Figure 10.2 depicts bioactive constituents of CB.

10.5 POTENTIAL BENEFITS

10.5.1 Health Benefits of American Ginseng

10.5.1.1 Neuronal Protection

Alzheimer's disease, a degenerative, irreversible neurological disorder, is characterized by the deposition of amyloid-beta (Aβ) peptides in and around brain cells that cause the distortion of the presynaptic cholinergic system and diminishes the choline acetyltransferase (ChAT) activity (Burns and Iliffe, 2009). Administration of AG extract significantly improved cognitive ailment by upregulating ChAT production (Shin et al., 2016). Impaired cognitive function in experimental mice due to intoxication of $A\beta_{1\text{-}42}$ peptide has been shown to improve partially with regular ginseng-extract administration. Rb1 ginsenoside of AG was reported to inhibit synthetic $A\beta_{1\text{-}42}$ peptide-driven cytotoxicity in neuronal stem cells by restoring ChAT gene expression. It is known to restore the normal expression of various biochemical markers like synaptophysin, AChE, microtubule-associated protein 2 (MAP2), whose levels were altered due to $A\beta_{1\text{-}42}$ intoxication. However, the neuroprotective role of AG is not clearly understood; whether AG leads to direct degradation of $A\beta_{1\text{-}42}$ peptides or just ameliorates negative effects of abnormal peptide deposition, is highly debated. In this context, a study has shown that the increased level of neprilysin, an $A\beta_{1\text{-}42}$ degrading enzyme, is responsible for Rg3-mediated protection against Alzheimer's disease (Yang et al., 2009). In another study, Al-Hazmi et al. (2015) reported that AG extract recovered the normal AChE level to improve cognitive function. Therefore, a detailed study is required to resolve the mode of action of AG in Alzheimer's disease. According to the free radical theory of aging, overproduction of reactive oxygen species (ROS) hampers various tissue function that ultimately leads to deterioration of basic physiological activity (Harman, 1992). In senescence-accelerated mice (SAM), AG showed beneficial effects on neurocognitive impairment (Shi et al., 2012). These SAM mice showed advanced aging process where cognitive dysfunctions and the damage of neocortical synapses resulted in intellectual disability in aged individuals. After AG administration, SAM mice were reported to show upregulated insulin and ChAT-mediated improved intellectual ability. Scholey et al. (2010) have reported that AG extract has boosted cognitive performance by improving memory function in an acute, randomized placebo-controlled study. The effect of AG on memory is mainly due to ginsenoside Rb1 which improves cholinergic metabolism.

Nowadays, approximately 33.7% of the worldwide human population is suffering from anxiety disorder (Bandelow and Michaelis, 2015). AG (50 and 100 mg/kg b.w.) has an anxiolytic effect in mice similar to diazepam (Wei et al., 2007). AG administration has been documented to successfully mitigate methamphetamine-induced anxiety in rats and improve their cognitive performance by reducing the error number in the maze test and the immobility time in the forced swimming test (Wu et al., 2003). Rh1 and protopanaxatriol (the primary and end-stage metabolites of ginsenoside Rg1) have been reported to noticeably protect mice from memory impairment and improve hippocampal excitability in the dentate gyrus of anesthetized rats (Wang et al., 2009).

10.5.1.2 Cardioprotective Activity

AG exerts a protective role against cardiomyopathy by inhibiting oxidative stress. In lipopolysaccharide-induced endotoxemia mice, AG suppressed superoxide production in heart tissue (Wu et al., 2016). Similar effect was also found in cultured cardiomyocytes. The AG extract inhibited NOX2 expression (the main culprit of cardiac ROS production) and subsequent NOX2-ERK1/2-TNF-α mediated signaling pathways in cardiomyocytes. AG provided the antioxidant action in cardiomyocytes by activating the antioxidant gene, nuclear factor (erythroid-derived 2)-like 2 (Nrf2) (Li et al., 2010c). AG extract administration in rats suppressed cardiac hypertrophy and reduced the chance of heart failure by mitigating cardiac adrenergic responses (Michael et al., 2021). AG mediated inducible nitric oxide synthase (iNOS) upregulation, thereby protecting the murine heart from reperfusion injury (Wu et al., 2011).

10.5.1.3 Anticancer Activity

Cancer is considered to be one of the deadliest diseases, responsible for worldwide death (Dutta et al., 2019). Implication of phytochemicals in the field of cancer therapy is increasing gradually due to their wide availability and minimal side effects (Dandawate et al., 2016; Zhao et al., 2018). Most of the studies related to the anticancer property of AG have been focused on colorectal cancer. Ginsenoside Rh2, one of the active ingredients of AG, exhibited a potent anticancer effect on the human colorectal cancer cell line HCT116. It has been noted to suppress abnormal HCT116 cell proliferation by inhibiting the increased expression of PDZ-binding kinase/T-LAK cell-originated protein kinase (serine/threonine protein kinase) (Yang et al., 2016a). PPD-mediated cell cycle arrest in G1/S checkpoint in HCT116 cells inhibited cellular proliferation. In a HCT116 xenograft mice model, PPD has been found to suppress nuclear factor kappa-light-chain-enhancer of activated B cells (NF-kB) and c-Jun N-terminal kinase (JNK) signaling pathways to inhibit tumor cell growth (Gao et al., 2013). Additionally, AG-mediated downregulation of pro-inflammatory cytokines has also been shown to provide protection against azoxymethane/dextran sodium sulfate-induced mice colon carcinogenesis (Yu et al., 2015). Application of AG root steamed extract (0.1 mg/mL) has been found to cause mitochondrial damage and exaggerated ROS production to induce apoptotic death in colorectal cancer cells (Li et al., 2010a).

Besides, aqueous extract of AG root (20, 40, 60, and 80 mg/mL) has altered the cellular survival mechanism of SMMC-7721 cells (liver carcinoma) through apoptosis induction (Qu et al., 2018). In MCF-7 cells (breast cancer cells), cellular proliferation was reported to be checked by downregulation of phospho-MEK1/2 and ERK1/2 expression and phospho-Raf-1 upregulation (King and Murphy, 2007). Use of Mitomycin C (MMC) as an anticancer drug results in numerous side effects including DNA alkylation, DNA interstrand crosslinks, chromosomal aberrations, etc. (Blasiak, 2017). Pawar et al. (2007) have shown that the application of AG root extract (50 and 100 mg/kg b.w.) can reduce the number of micro-nucleated polychromatic erythrocyte in MMC pretreated mice. Cis-diamminedichloroplatinum, commonly known as cisplatin, is another widely used anticancer agent which is associated with several side effects including severe nephrotoxicity (Oun et al., 2018). In an experiment, Ma et al. (2017b) found that AG berry extract (AGBE) can reduce cisplatin-triggered histopathological alteration of the kidney as well as effectively lower nitrogen, creatinine, and urea contents in serum. Here, AGBE pretreatment was reported to ameliorate cisplatin-triggered oxidative stress by downregulating pro-apoptotic protein Bax, cytochrome-C, and caspase-3 expression and upregulating antiapoptotic Bcl-2 expression. AG can be used to treat the damage caused by radiotherapy. Use of ionizing radiation (IR) causes direct or indirect effect to induce cellular damage. Direct IR absorption in the target tissue causes cytotoxicity, whereas, indirectly it leads to the radiolysis of water to generate ROS which ultimately causes cellular DNA damage (Desouky et al., 2015). Application of AG extract in human lymphocytes exposed to 1 or 2 Gy γ-irradiation has shown a significant level of protection by reducing oxidative stress, total number of micronuclei, and restoring the total antioxidant capacity (Lee et al., 2010; Lee et al., 2009). Therefore, AG can be used as a post-therapeutic treatment measure to minimalize cytotoxic damage resulting from radiation therapy.

10.5.1.4 Antidiabetic Activity

In type 2 diabetes mellitus (T2DM), impaired glucose metabolism induces the onset of chronic hyperglycemia. In a study, Kan et al. (2017) documented that the application of AG root extract in differentiated adipocytes increases glucose uptake inside the cell and restored insulin sensitivity that cumulatively elicits a hypoglycemic effect reducing the risk associated with T2DM. Combinational application of AG root along with *Morus alba* leaf and *Trigonella foenum graecum* seed extract in 3T3-L1 adipocyte has shown to upregulate glucose transporter 4 (GLUT4) expression to enhance glucose uptake and insulin sensitivity. The abovementioned GLUT4 is a single-pass transmembrane transporter that is responsible for transportation of glucose from blood to tissue; whereas, the expression of GLUT4 is in turn regulated by insulin. In T2DM oxidative stress-mediated excessive ROS production impairs the cellular antioxidant defense mechanism which results in high blood glucose levels and insulin insensitivity. AG consumption for 22 months in a murine model has been studied to decrease ROS production and restore glutathione peroxidase (GPx) activity (Fu and Ji, 2003). Ginseng helps to reduce blood glucose levels (Luo and Luo, 2009). In another study, Yoo et al. (2012) revealed that steamed AG root extract at 150 mg/kg b.w. dose for 30 days can significantly increase hepatic glycogen and plasma insulin levels and decrease blood glucose content in diabetic db/db mice. The presence of cinnamic acid and ferulic acid in AG helped to elevate SOD and GPx production in the liver. AGBE and ginsenoside Re protected hydrogen peroxide-treated MIN-6 pancreatic β cells from chronic oxidative damage (Lin et al., 2008). AG-mediated endothelial nitric oxide (NO) inactivation has been found to impair ROS and NO interaction that otherwise would produce cytotoxic peroxynitrite (Mehendale et al., 2006). It has been found that diabetic glucose fluctuation is strongly correlated with cardiovascular complicacy through endothelial dysfunction and apoptosis induction (Wang et al., 2013a). Treatment for ten weeks with ginseng stem and leaf saponins has been found to lower the level of NO, TNF-α, ET-1 (endothelin-1), sICAM-1 (soluble intercellular adhesion molecule 1) in streptozotocin-induced diabetic mice. Compared to metformin, AG provided more efficient relief from stress in blood vessels and consequent inflammation. In another study, Sen et al. (2011) have shown the beneficiary role of ethanolic extract of AG on both type 1 and type 2 DM. The presence of Rb1 and Re ginsenosides in AG reduced the dysmetabolic condition of diabetic animals and increased heme oxygenase gene expression and various protein expressions related to vessel architecture of retina and heart in diabetic mice. AG root extract supplementation for 3 months (at 3g/day dose) in T2DM patients with hypertension significantly has shown to ameliorate the systolic blood pressure and reduced arterial stiffness (Mucalo et al., 2013). Amin et al. showed that in streptozotocin-induced diabetic mice, AG treatment (300 mg/kg b.w. for 10 days) increased β-cell insulin secretion and C-peptide release from the pancreas (Amin et al., 2011). Their study explored the detailed mechanism of antidiabetic activity of AG with respect to glucose-6-phosphatase (G6Pase) and glycogen phosphorylase inhibition. Vuksan et al. (2000) reported the short-term effect of AG root in controlling postprandial glycemia in nondiabetic and T2DM individuals. It was found that 3 gm of ginseng root preparation administration along with glucose (25 g) had reduced postprandial glycemia by 20% in both normal and diabetic individuals. However, when the preparation was given 40 min before the glucose challenge, it lowered the glucose concentration only in diabetic individuals (Vuksan et al., 2001). But in healthy individuals, 3 gm of alcoholic root extract had no impact on postprandial glycemia and insulin secretion (De Souza et al., 2015). Contrarily, 50% of the ethanolic extract exhibited insulin-sensitizing effects in nondiabetic individuals and boosted metabolism. More importantly, a 12-week treatment of AG root extract (3 gm/day) on T2DM patients did not cause any systemic toxicity (Mucalo et al., 2014). In a clinical trial, root extract application of AG for 4 weeks reduced the systolic blood pressure, fasting blood glucose level, and glycated hemoglobin HbA1c level (Vuksan et al., 2019). Based on the previous reports, an experiment was performed using a combination of konjac-based fiber blend (KGB) and AG (6 gm KGB: 3 gm AG per day) were applied in T2DM patients. This formulation has been able to lower glycated hemoglobin HbA1c levels (0.31%) and serum lipid concentrations (8.3±3.1% of low-density lipoprotein 7.5±2.4% of non-high-density lipoprotein, 5.7±1.9% in total cholesterol) in T2DM patients over a span of 12 weeks (Jenkins et al., 2018).

10.5.1.5 Prevention of Obesity

Lack of physical activity and increased food intake can cause increased fat deposition in the body (mainly in adipose tissue) resulting in weight gain. Obesity is associated with an increased chance of T2DM, cardiovascular problems, and cancers (Zhang et al., 2017). Though the hypoglycemic effect of AG is well investigated, its anti-obesity property has not been explored deeply. Therefore, its effect on lipid metabolism is yet to be understood. During digestion, inhibition of pancreatic lipase secretion is known to cause lower absorption of dietary fat and ultimately leads to weight loss and hyperlipidemia. The presence of saponins in AG stems and leaves acted as a potent lipase inhibitor which was observed to lower the release of oleic acid from triolein (Liu et al., 2008). Daily intake of saponin extract at a dose of 1gm/kg b.w. has been observed to limit the plasma triglyceride buildup and fat deposition in adipose tissue in rats. Liu et al. (2010) showed *in vitro* dose-dependent inhibitory effects of PPD of AG leaves against porcine pancreatic lipase activity. Additionally, a high fat diet intake along with.02 or.05% PPD in mice reduced hepatic triacylglycerol accumulation and total cholesterol level. Adipocyte hypertrophy and preadipocyte hyperplasia are two major events of adipose tissue remodeling. Adipokines released from adipose tissue trigger the preadipocyte differentiation (adipogenesis) that initiates the onset of obesity (Yeo et al., 2011). Application of ethanolic extract of AG containing Rg1, Re, Rb1, Rc, Rb2, and Rd ginsenosides on 3T3L1 mice preadipocytes reduces preadipocyte proliferation and intracellular lipid accumulation in a dose-dependent manner. Adiponectin (a type of adipokines) upregulation is associated with insulin resistance development (Hwang et al., 2009). The presence of Rh2 and Rg3 ginsenosides in AG extract effectively suppressed adipogenesis by activating AMPK (responsible for insulin sensitivity) and inhibiting peroxisome proliferator-activated receptor (decrease the rate of glycolysis). These proteins regulate the process of adipogenesis and lipogenesis by changing the expression of leptin, adiponectin, perilipin, GLUT4, acetyl-CoA carboxylase (ACC), fatty acid synthase (FAS), and fatty acid binding protein 4 (FABP4) (Jenkins et al., 2018). Pseudoginsenoside F11 of AG has been reported to induce adiponectin secretion, fat accumulation, and PPAR activation on in 3T3-L1 cells (Wu et al., 2013). In line, it inhibits obesity related phosphorylation of PPAR to disrupt the interaction between PPAR- and retinoid X receptor RXR and their interaction with DNA. According to Wilson et al. (2013), the biological activity of AG depends upon its extraction process. It was found that aqueous AG extract significantly increased the expression of inflammatory genes (CCL5, TNF-α, MCP1, NOS2, IL6) in adipocytes and other cells cocultured with adipocytes. But the ethanolic extract did not exhibit the same (Garbett et al., 2016). Singh et al. (2017) have shown that ethanolic extract of AG has beneficial property against obesity, insulin resistance, fatty liver, hypertriglyceridemia in Pcyt2 (ethanolamine-phosphate cytidylyl transferase 2) gene-deficient and ETKO/Pcyt2+/-deficient mice. Administration of AG (200 mg/kg b.w.) in obese mice for 24 weeks has been found to improve fatty liver condition and reduce the level of intestinal and hepatic lipoprotein. It has reduced the level of FAS, PPARα, and SREBP1 gene expression in the fatty liver. Such changes seem to be associated solely with AMPK level upregulation without effecting further phosphorylation of ACC participated in malonyl-CoA for fatty acid synthesis. Figure 10.3 shows the mechanism of AG activity against diabetes and obesity.

10.5.1.6 Antiaging Properties

No specific clinical study has been performed to rule out the impact of AG on life expectancy. Most of the studies explored the antioxidant role of AG. AG application (2.25 g/kg) for 14 days gave protection against premature ovarian failure (POF) which resulted due to accelerated aging (Zhu et al., 2015). According to a study by Fernández-Moriano et al. (2017), Rb1, a primary ginsenoside of AG, protects SH-SY5Y cell lines from oxidative stress by scavenging ROS, restoring normal levels of

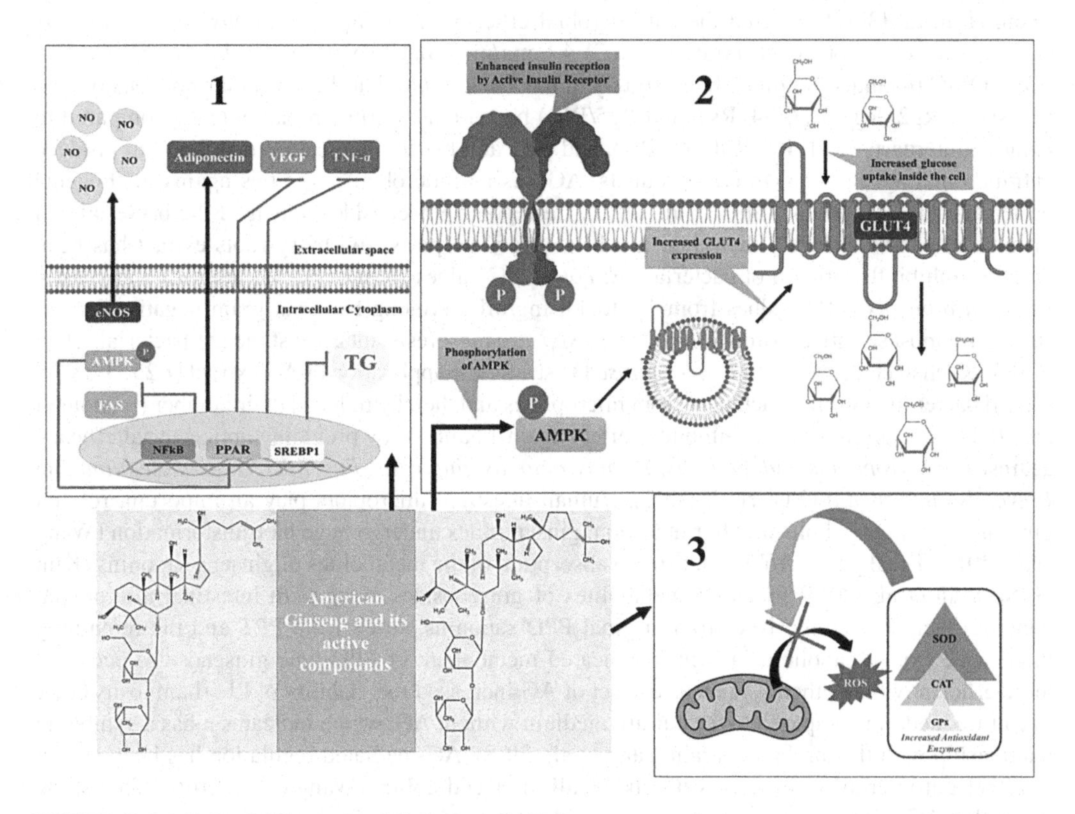

FIGURE 10.3 Role of AG to control diabetes and obesity.

antioxidants glutathione and SOD, and activating the Nrf-2 pathways. Additionally, Rbl-mediated mitochondrial protection against ROS suggests its efficacy against neurodegenerative diseases and mitochondria-dependent aging. Application of AG powdered root (100 mg/kg b.w.) for 14 and 28 days has been noticed to arouse the copulatory behavior of male rats suffering from intromission and ejaculation latencies (Murphy et al., 1998). The chance of ischemic stroke and general stroke is associated with age progression. It is known that Na^+ influx during ischemic stroke can be reversed by using drugs that block voltage-dependent sodium channels. In a study, Liu et al. (2001) showed that application of AG water extract and ginsenoside Rbl blocked the activity of brain (2a) subunits of the voltage-dependent sodium channel. The interaction between Rbl/ ginseng and brain (2a) subunits caused sodium channel inactivation, which had a similar effect to lidocaine, a common Na+ channel blocker.

10.5.1.7 Multiple Sclerosis Prevention

Multiple sclerosis (MS) is a chronic inflammatory neurological disorder where myelinated axons of the central nervous system (CNS) become destroyed due to immune attack. The exact mechanism of MS is not fully understood, but the overproduction of ROS by microglia and other immune cells that attack the CNS, causing demyelination and damage of axons (Ohl et al., 2016). Water extract of ginseng was reported to alter the autoimmune encephalomyelitic symptoms in mice and reduce the demyelination extension (Sloley et al., 1999). By decreasing immune-reactive iNOS and circulating TNF-α level, the anti-oxidative property of AG can suppress such a degenerative neuronal disorder due to its antiaging activity (Yang et al., 2017).

10.5.1.8 Antimicrobial Activity

AG derived ginsenosides have been reported to exhibit anti-staphylococcal properties. The antibiotic resistance property of this bacteria does not obscure the protective effect of AG. In a study, Wang et al. (2013b) discovered the antimicrobial efficacy of AG against *Staphylococcus epidermidis* (mean inhibitory concentration (MIC) 4.1 mg/mL) and two strains of *Propionibacterium acnes* (MIC 64 and 128 g/mL). Interestingly, it has been found that the less polar portions of ginsenosides (Rg2, Rg3, Rg6/F4, Rs3, and Rg5/Rk1) have higher antimicrobial potency compared to polar counterparts (Rgl, Re, Rbl, Rc, Rb2, and Rd), and hence, this feature is utilized to produce antimicrobial agents for skin care products. AG has antimicrobial properties against pathogenic bacterial and yeast strains (Shin et al., 2016). Less polar ginsenosides damaged the bacterial cell membrane more effectively than their much polar counterparts. AG hairy roots extract has been found to inhibit the growth of bacteria – *E. coli* (MIC values from 0.8 to 1.4 mg/mL) and yeast – *Candida albicans* (MIC values from 1.0 to 1.4 mg/mL). Among bacteria, gram negative *E. coli* strains are most sensitive to and *Enterococcus spp.* are most resistant against the antibacterial effect of AG (Kochan et al., 2013). In *P. aeruginosa* O1 strain, the application of AG extract (1.25–5% w/v) caused bacterial biofilm detachment from microplates and thereby reduced their number (Alipour et al., 2011). AG ginsenoside has inhibitory effect against halitosis by providing antibacterial efficacy against *Porphyromonas endodontalis*, *Porphyromonas gingivalis*, *Fusobacterium nucleatum*, and *Prevotella intermedia* (Xue et al., 2017). Human intestinal microbiota play an important role on ginseng saponin metabolism. After ingestion, ginsenosides undergo huge biotransformation (Wang et al., 2012). Fecal gut microbiota uplift the absorption of the metabolites of ginseng saponins (Kim 2018). Wan et al. (2013) found 25 metabolites of ginsenosides along with intestinal microbiota. Among them, 15 were derived from original PPD saponins, seven from PPT and the remaining three were from oleanolic acid, which indicated metabolism of PPD-type ginsenosides occurred more efficiently than others. Aqueous extract of AG increased the viability of Lb. rhamnosus GR-1 in culture medium compared to the culture medium without AG, which indicates it has a symbiotic relationship with this probiotic strain (Jang et al., 2018). AG-mediated regulation has been shown to affect gut microbiota in mice with chemically induced colitis (Wang et al., 2016). This study found that AG inhibited the growth of Bacteroidetes and Verrucomicrobia bacteria and stimulated

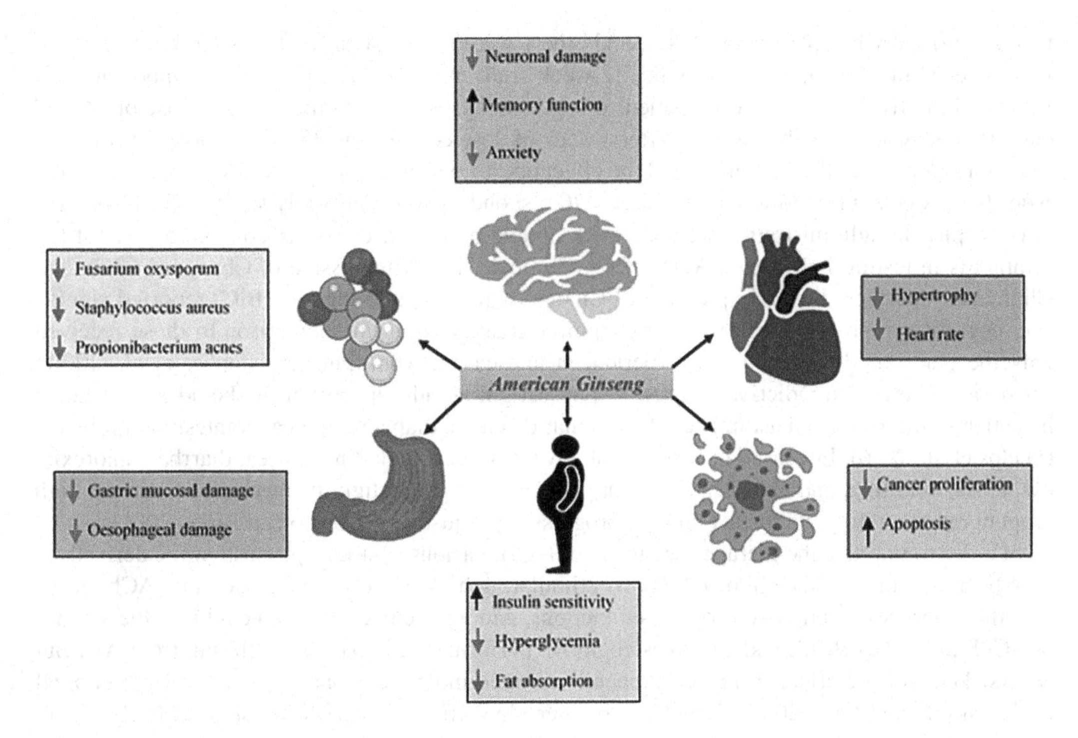

FIGURE 10.4 Health benefits of AG.

the growth of Firmicutes to restore the normal homeostasis of gut microbiota in colitis mice. Figure 10.4 represents the health benefits of AG.

10.5.2 Health Benefits of Calabar Bean

10.5.2.1 Neuronal Protection

Cholinergic dysfunction plays a significant role in the pathogenesis of Alzheimer's disease (AD) and related dementias. Diminished cortical cholinergic neurotransmission causes cognitive and behavioral abnormalities along with altered neurotrophic signaling and aberrant AβP processing, which ultimately leads to AD onset (Ferreira-Vieira et al., 2016; Hampel et al., 2018). The presence of carbamate moiety in CB is responsible for its reversible inhibitory effect against AChE and butyrylcholinesterase (BChE) activity (Singh et al., 2013). It uplifts the amount of active ACh at cholinergic synapses in both central and peripheral nervous systems through AChE and BChE-mediated ACh breakdown. Physostigmine is a tertiary ammonium compound that also can readily cross the blood–brain barrier. This alkaloid has been shown to improve the cognitive ability of rats suffering from scopolamine-induced cognitive impairment (Yoshida and Suzuki 1993). In another study with AD patients, Thal et al. (1999) showed that physostigmine intake at a dose of 30 mg for 24 weeks could improve cognitive behaviors significantly. It impeded both G1 and G4 AChE activity of the mammalian CNS (Grossberg 2003). Administration of physostigmine in mice, has been reported to improve cognitive impairment due to oxygen deficiency. In the same study, it was observed to enhance learning ability in humans (McCaleb, 1990). In a clinical trial, physostigmine was found to significantly enhance the cognitive behaviors in AD patients (Sitaram et al., 1978). Oral physostigmine at a range of doses (0.0, 0.5, 1.0, 1.5, and 2.0 mg) was given every 2 h for 3–5 days in a group of 12 AD patients. Alzheimer's disease Assessment Scale-Cognitive Subscale (ADAS-Cog) measured the symptoms after each dose administration. Later, physostigmine was readministered for every 3–5 days in the placebo and in the "least severe symptoms"-containing group. Out of 10 patients, three patients showed significant recovery with the highest dose of physostigmine in both phases, four were slightly better in both phases, and others were inconsistent to physostigmine-treatment. In another clinical trial, a double-blind, 12-week study was conducted on mild-to-moderate AD patients. Initially, for 1 week, each patient was given physostigmine salicylate at a dose of 24 mg/day, 30 mg/day, and a daily placebo. After a week of dose completion, 35.9% responded, yet 62.4% were unresponsive to the treatment, and the effect upon 1.6 % was unevaluated due to missing data. After the 4-week "placebo-washout phase", 176 responders were randomly selected for physostigmine or placebo administration for a 12-week double-blind phase. Post-treatment analysis of the symptoms in testing individuals ADAS-Cog, Clinical Global Impression of Change (CGIC), and Clinician's Interview-Based Impression of Change with caregiver input (CIBIC) reported significant cognitive improvement in the physostigmine-treated group in comparison to those received only the placebo. However, 47.0% experienced nausea and vomiting (Christopher et al., 2000). There are several contradictions related to physostigmine administration. It should not be taken by patients suffering from asthma, cardiovascular diseases, diabetes, and gastrointestinal problems (Phelps et al., 2016). Increased occurrence of adverse effects including nausea, diarrhea, anorexia, dyspepsia, vomiting, and abdominal pain in patients after physostigmine ingestion, causing a high dropout rate; therefore, its further clinical progress is not justified (Coelho and Birks, 2001).

In order to improve the therapeutic outcome of CB, various synthetic physostigmine derivatives have been developed (Giacobini, 1998a). By stimulating both nicotinic and muscarinic ACh receptors, these derivatives improved cognitive functions. Among them, eptastigmine inhibits the activity of AChE as well as BChE and enhances cognitive performance in patients suffering from AD. But adverse hematologic effects (granulocytopenia) resulted in the postponement of its further clinical trials (Braida and Sala, 2001). Eseroline, another physostigmine derivative, showed fairly potent analgesic effects by interacting through μ-opioid receptors (Agresti et al., 1980). Phenolic ring and pyrrolidine nitrogen of eseroline, which interact with the opioid receptor site, have similar distance like piperidine nitrogen and phenolic ring of morphine molecule (Fürst et al., 1982; Isaksson and Kissinger 1987; Renzi et al., 1980). This eseroline exhibits more toxicity compared to its parent compound physostigmine. Eseroline-mediated ATP depletion has been reported neuronal cell death (Somani et al., 1990). Among other derivatives, bisnorcymserine is a highly potent inhibitor of human serum BChE and also lowered the accumulation of Aβ and amyloid plaque-associated protein that is related to AD (Kamal et al., 2006). Tetrahydrofurobenzofuran also inhibited the BChE activity (Kamal et al., 2008). Phenserine exhibits long-lasting, pseudo-irreversible AChE inhibitor activity. It has been found that phenserine, an amyloid precursor protein (APP) degenerator, exhibited lower toxicity compared to physostigmine and tacrine. It exhibited a neuroprotective effect by minimizing oxidative stress and suppressing glutamate toxicity regulated by protein kinase C (PKC) and extracellular signal-regulated kinase (ERK) pathways (Lilja et al., 2013). In a rat model, focal cerebral ischemia and oxygen–glucose deprivation/reperfusion (OGD/RP) damage in SH-SY5Y neuronal cultures, and (-)-phenserine administration showed neurotrophic and anti-apoptotic activity through ERK-1/2 signaling modulation. (-)-phenserine caused increased brain-derived neurotrophic factor (BDNF) and B-cell lymphoma 2 (Bcl-2) levels and decreased Caspase-3 and Metallopeptidase-9 (MMP-9) expression. It also diminished the expression of APP and glial fibrillary acidic protein (GFAP) (Chang et al., 2017). The noncholinergic (+) chiral enantiomer of phenserine is known as posiphen. Both phenserine enantiomers effectively lowered the α-synuclein expression in SH-SY5Y. They lowered the rate of APP synthesis, both in animals as well as neuronal cell cultures, by controlling the translational regulation APP mRNA 5′UTR (Mikkilineni et al., 2012). It has been found that compared to phenserine, posiphen is a weak AChE inhibitor that can be administered in human and rodent models at a relatively higher amount hence producing a huge amount of its primary metabolites (Klein, 2007; Winblad et al., 2010). Tolserine, a methyl group substitute at C20 position of phenserine, exhibited increased selectivity for AChE compared to BChE (Loizzo et al., 2008). Rivastigmine is the only derivative of physostigmine that has reached the phase IV trial level. It has been prescribed for mild to moderate levels of dementia symptoms and improves cognitive and behavioral abnormalities in AD or Parkinson disease (PD).

Rivastigmine-mediated dual inhibition of AChE and BChE activity prevents advancement of subcortical dementias (PD dementia and subcortical vascular dementia), and it is better in comparison to donepezil and galantamine (it inhibits only acetylcholinesterase activity) (Kandiah et al., 2017; Weinstock, 1999). Pharmaco-interaction between carbamate derivatives and other drugs can be used to explore a new horizon for the development of therapeutic constituents against cognitive impairments and behavioral and psychological symptoms associated with dementia. The combination of physostigmine with propargylamine pharmacophore of selegiline produces dual AChE and monoamine oxidase (MAO) inhibitors. Rivastigmine and fluoxetine combined, and marketed as RS 1259, has been shown to inhibit the activity of AChE and the serotonin transporter (SERT) *in vitro* (Fink et al., 1996). Oral administration of RS-1259 in rodents also inhibited AChE and SERT activity and enhanced cognitive performance in spatial short-term memory tests in aged rats (Toda et al., 2010). Figure 10.5 represents CB action in the nerve terminal.

10.5.2.2 Antidote Agent

Physostigmine is readily absorbed from the gastrointestinal tract, and once it is in the bloodstream, it can easily penetrate the blood–brain barrier. As physosigmine is able to diffuse inside the CNS, it is often used as an antidote against anticholinergic drug overdose. It was first used as an antidote against atropine toxicity (Nickalls and Nickalls, 1988). Currently, physostigmine salicylate is used as an antidote to antihistamines, antipsychotics, benzodiazepines, and tricyclic antidepressants by reversing the action of cholinergic toxicity (Lee et al., 1975; Jastak, 1985; Newton, 1975; Weisdorf et al., 1978). Gamma-hydroxybutyrate (GHB) is a sedative-hypnotic agent used for the treatment of narcolepsy. It is a popular psycho-addictive drug of abuse. Physostigmine acts as an arousal agent to treat coma patients against GHB overdose (Traub et al., 2002). Physostigmine is a powerful ingredient that can stimulate involuntary muscle contraction of the body. Subcutaneous injection of physostigmine can reverse muscle weakness by facilitating acetylcholine-mediated neuromuscular transmission in skeletal muscle: therefore, it is used to treat myasthenia gravis (Nickalls and Nickalls, 1988). The symptoms of myasthenia are similar to those of curare poisoning where

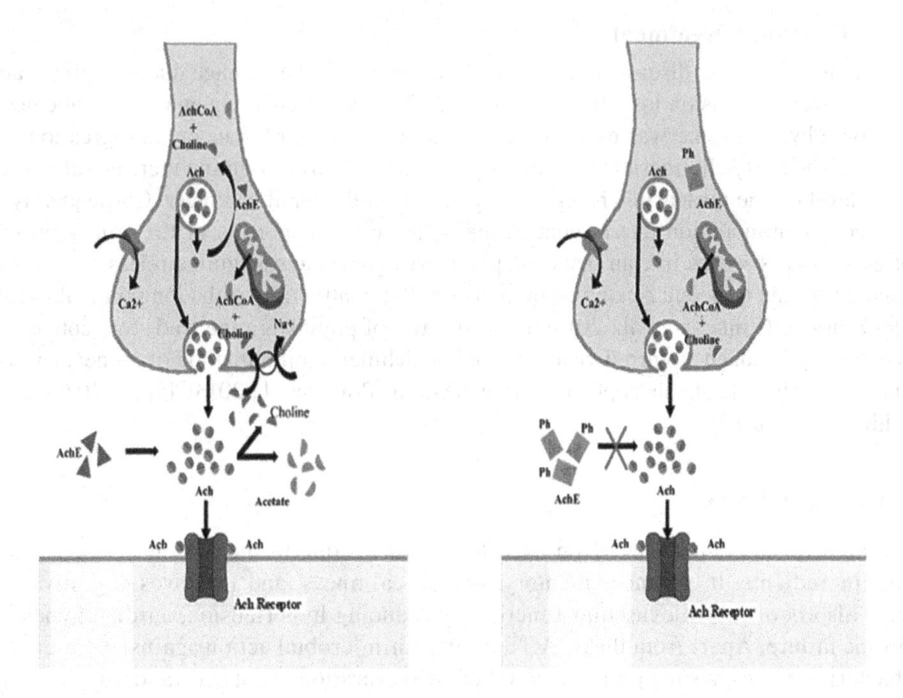

FIGURE 10.5 Mechanism of CB action in the nerve terminal. AcCoA, acetyl-CoA; Ach, acetylcholine; AchE, acetylcholinesterase; Ph, physostigmine; Ca^{+2}, calcium ion.

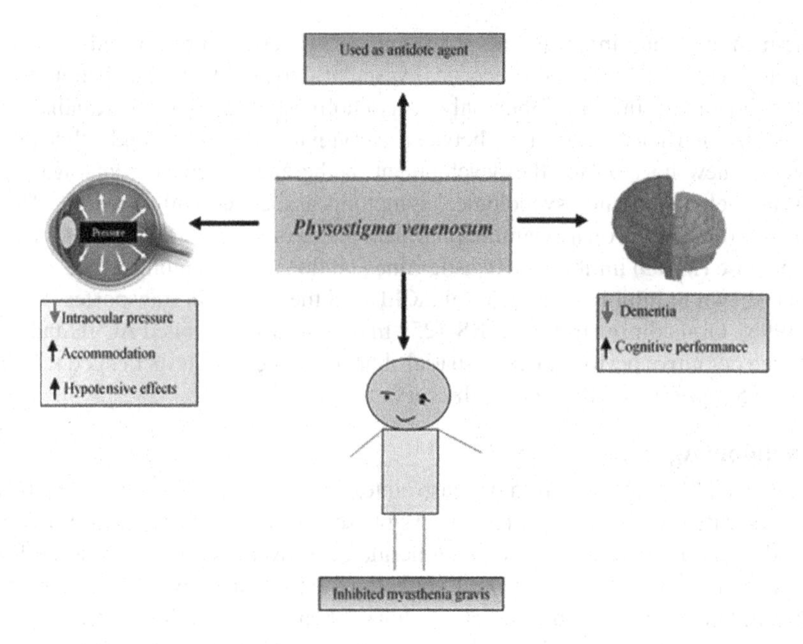

FIGURE 10.6 Beneficial effects of CB.

physostigmine can be used as an effective antidote. It is a reversible neuromuscular blocker which is available in the US market under the trade names Antilirium and Isopto Eserine. Physostigmine also acts as a prophylactic agent against organophosphate intoxication (Miller et al., 1993; Triggle and Filler, 1998). Physostigmine-triggered carbamylation of an enzyme active site of cholinesterase protects the same from organophosphorous attack, which would otherwise bind to the enzyme permanently and cause neuronal damage.

10.5.2.3 Glaucoma Treatment

Glaucoma, an ophthalmic disease, is characterized by optic nerve damage due to high intraocular pressure and resulting vision loss. In 1885, Adolf Weber suggested glaucoma as an optic neuropathy. In 1864, physostigmine was used as the first herbal remedy of glaucoma progression (Orhan et al., 2011). The acetylcholinesterase inhibitory property of physostigmine increases the free acetylcholine level on the muscarinic receptors m_2 and m_3 in the pupil sphincter. Consequently, there is increased accommodation due to contraction of the circular portion of the ciliary muscle. In case of glaucoma, it works like an ocular hypotensive agent. Physostigmine releases excess aqueous humor from the trabecular pathway by inducing the contraction of the longitudinal portion of the ciliary muscle (Pinheiro et al., 2018). However, use of physostigmine leads to a couple of side effects including headache, blurred vision, retinal detachment, inflammation of cornea, iris as well as conjunctiva, which limits its application in glaucoma (Pinho et al., 2013). Figure 10.6 represents the health benefits of CB.

10.6 CONCLUSION

AG as a dietary supplement has been used for centuries due to the presence of ginsenosides as active ingredients. It enhances memory, mental calmness, and improves cognitive performance. It also boosts cardiovascular function by reducing hypertension, cardiac hypertrophy, and cardiac failure. Apart from these, AG exhibits antimicrobial action against different pathogenic bacterial strains. As the proper method of AG separation is not yet standardized, it should be targeted for future research. Though it exhibits a broad range of beneficiary effects in human heath, AG ingestion is not suggested for pregnant and lactating women due to its teratogenic activities in rats and mice embryos (Chan et al., 2003; Liu et al., 2005). AG overdose leads to various side effects like nervousness, depression, insomnia, and gastrointestinal disorders. With respect to its advantageous pro-health activity, probable mechanisms of its mode of action, and potential application in various fields of medicine need further investigation. On the other hand, though the efficacy of CB has been studied widely in the field of neuronal disorders, only rivastigmine has been clinically approved for AD- and PD-associated dementia treatment. Several gastrointestinal issues after ingestion of these phytochemicals have restricted its wide usage in human health. Its application is confined to the field of glaucoma, myasthenia gravis, and anticholinergic drug overdose treatment. The potency of CB still needs to be investigated at both basal and clinical medicinal levels. A safety assessment of this phytochemical should be done very carefully before use.

REFERENCES

Agresti, A., F. Buffoni, J.J. Kaufman, and C. Petrongolo. 1980. Structure-activity relationships of eseroline and morphine: Ab initio quantum-chemical study of the electrostatic potential and of the interaction energy with water. *Molecular Pharmacology*, 18(3), pp.461–467.

Al-Hazmi, M.A., S.M. Rawi, N.M. Arafa, A. Wagas, and A.O. Montasser. 2015. The potent effects of ginseng root extract and memantine on cognitive dysfunction in male albino rats. *Toxicology and Industrial Health*, 31(6), pp.494–509.

Alipour, M., A. Omri, and Z.E. Suntres. 2011. Ginseng aqueous extract attenuates the production of virulence factors, stimulates twitching and adhesion, and eradicates biofilms of *Pseudomonas aeruginosa*. *Canadian Journal of Physiology and Pharmacology*, 89(6), pp.419–427.

Amin, K.A., E.M. Awad, and M.A. Nagy. 2011. Effects of panax quinquefolium on streptozotocin-induced diabetic rats: Role of C-peptide, nitric oxide and oxidative stress. *International Journal of Clinical and Experimental Medicine*, 4(2), pp.136–147.

Anand, S.S., C. Hawkes, R.J. de Souza, A. Mente, M. Dehghan, R. Nugent, M.A. Zulyniak, T. Weis, A.M. Bernstein, R.M. Krauss, D. Kromhout, D.J.A. Jenkins, V. Malik, M.A. Martinez-Gonzalez, D. Mozaffarian, S. Yusuf, W.C. Willett, and B.M. Popkin. 2015. Food consumption and its impact on cardiovascular disease: Importance of solutions focused on the globalized food system: A report from the workshop convened by the world heart federation. *Journal of the American College of Cardiology*, 66(14), pp.1590–1614.

Aune, D., E. Giovannucci, P. Boffetta, L.T. Fadnes, N. Keum, T. Norat, D.C. Greenwood, E. Riboli, L.J. Vatten, and S. Tonstad. 2017. Fruit and vegetable intake and the risk of cardiovascular disease, total cancer and all-cause mortality-a systematic review and dose-response meta-analysis of prospective studies. *International Journal of Epidemiology*, 46(3), pp.1029–1056.

Bandelow, B., and S. Michaelis. 2015. Epidemiology of anxiety disorders in the 21st century. *Dialogues in Clinical Neuroscience* 17(3), p.327.

Blasiak, J. 2017. DNA-Damaging anticancer drugs: A perspective for DNA repair-oriented therapy. *Current Medicinal Chemistry*, 24(15), pp.1488–1503.

Braida, D., and M. Sala. 2001. Eptastigmine: Ten years of pharmacology, toxicology, pharmacokinetic, and clinical studies. *CNS Drug Reviews*, 7(4), pp.369–386.

Burns, D.A., and S. Iliffe. 2009. Enfermedad de Alzheimer. *British Medical Journal* 338, p.b158.

Chan, L., P. Chiu, and T. Lau. 2003. An in-vitro study of ginsenoside Rb1-induced teratogenicity using a whole rat embryo culture model. *Human Reproduction*, 18(10), pp.2166–2168.

Chang, C.F., J.H. Lai, C.C. Wu, N.H. Greig, R.E. Becker, Y. Luo, Y.H. Chen, S.J. Kang, Y.H. Chiang, and K.Y.Chen. 2017. (-)-Phenserine inhibits neuronal apoptosis following ischemia/reperfusion injury. *Brain Research*, 1677, pp.118–128.

Christopher, H.D., N. Paul, E.F. William, and A.M. Jeffrey. 2000. Extended-release physostigmine in Alzheimer disease. A multicenter, double-blind, 12-week study with dose enrichment. *Archives of general psychiatry*, 57(2), pp.157–164.

Coelho, F., and J. Birks. 2001. Physostigmine for Alzheimer's disease. *Cochrane Database System Review*, 2, p.Cd001499.

Cruse-Sanders, J.M., and J.L. Hamrick. 2004. Genetic diversity in harvested and protected populations of wild American ginseng, *Panax quinquefolius* L. (Araliaceae). *American Journal of Botany*, 91(4), pp.540–548.

Cui, S., J. Wu, J. Wang, and X. Wang. 2017. Discrimination of American ginseng and Asian ginseng using electronic nose and gas chromatography–mass spectrometry coupled with chemometrics. *Journal of Ginseng Research*, 41(1), pp.85–95.

Dandawate, P.R., D. Subramaniam, R.A. Jensen, and S. Anant. 2016. Targeting cancer stem cells and signaling pathways by phytochemicals: Novel approach for breast cancer therapy. *Seminars in Cancer Biology*, 40, pp.192–208.

De Souza, L.R., A.L. Jenkins, E. Jovanovski, D. Rahelić, and V. Vuksan. 2015. Ethanol extraction preparation of American ginseng (*Panax quinquefolius* L) and Korean red ginseng (*Panax ginseng* C.A. Meyer): Differential effects on postprandial insulinemia in healthy individuals. *Journal of Ethnopharmacology*, 159, pp.55–61.

Desouky, O., N. Ding, and G. Zhou. 2015. Targeted and non-targeted effects of ionizing radiation. *Journal of Radiation Research and Applied Sciences*, 8(2), pp.247–254.

Dutta, S., S. Mahalanobish, S. Saha, S. Ghosh, and P.C. Sil. 2019. Natural products: An upcoming therapeutic approach to cancer. *Food and Chemical Toxicology*, 128, pp.240–255.

Feng, R., J. Liu, Z. Wang, J. Zhang, C. Cates, T. Rousselle, Q. Meng, and J. Li. 2017. The structure-activity relationship of ginsenosides on hypoxia-reoxygenation induced apoptosis of cardiomyocytes. *Biochemical and Biophysical Research Communications*, 494(3–4), pp.556–568.

Fernández-Moriano, C., E. González-Burgos, I. Iglesias, R. Lozano, and M.P. Gómez-Serranillos. 2017. Evaluation of the adaptogenic potential exerted by ginsenosides Rb1 and Rg1 against oxidative stress-mediated neurotoxicity in an in vitro neuronal model. *PLoS One*, 12(8), p.e0182933.

Ferreira-Vieira, T.H., I.M. Guimaraes, F.R. Silva, and F.M. Ribeiro. 2016. Alzheimer's disease: Targeting the cholinergic system. *Current Neuropharmacology*, 14(1), pp.101–115.

Fink, D.M., M. Palermo, G.M. Bores, F.P. Huger, B.E. Kurys, M.C. Merriman, G.E. Olsen, W. Petko, and G.J. O'Malley. 1996. Imino 1, 2, 3, 4-tetrahydrocyclopent [b] indole carbamates as dual inhibitors of acetylcholinesterase and monoamine oxidase. *Bioorganic & Medicinal Chemistry Letters*, 6(6), pp.625–630.

Fu, Y., and L.L. Ji. 2003. Chronic ginseng consumption attenuates age-associated oxidative stress in rats. *Journal of Nutrition*, 133(11), pp.3603–3609.

Fürst, S., T. Friedmann, A. Bartolini, R. Bartolini, P. Aiello-Malmberg, A. Galli, G.T. Somogyi, and J. Knoll. 1982. Direct evidence that eseroline possesses morphine-like effects. *European Journal of Pharmacology*, 83(3–4), pp.233–241.

Gao, J L., G.Y. Lv, B.C. He, B.Q. Zhang, H. Zhang, N. Wang, C.Z. Wang, W. Du, C.S. Yuan, and T.C. He. 2013. Ginseng saponin metabolite 20 (S)-protopanaxadiol inhibits tumor growth by targeting multiple cancer signaling pathways. *Oncology Reports*, 30(1), pp.292–298.

Garbett, J., S.A. Wilson, J.C. Ralston, A.A. De Boer, E.M. Lui, D.C. Wright, and D.M. Mutch. 2016. North American ginseng influences adipocyte-macrophage crosstalk regulation of inflammatory gene expression. *Journal of Ginseng Research*, 40(2), pp.141–150.

Giacobini, E.1998. Cholinesterase inhibitors for Alzheimer's disease therapy: from tacrine to future applications. *Neurochemistry international*, *32*(5–6), pp.413–419.

Grossberg, G.T. 2003. Cholinesterase inhibitors for the treatment of Alzheimer's disease: Getting on and staying on. *Current Therapeutic Research, Clinical and Experimental*, 64(4), pp.216–235.

Gupta, C., and D. Prakash. 2014. Phytonutrients as therapeutic agents. *Journal of Complementary and Integrative Medicine*, 11(3), pp.151–169.

Hampel, H., M.M. Mesulam, A.C. Cuello, M.R. Farlow, E. Giacobini, G.T. Grossberg, A.S. Khachaturian, A. Vergallo, E. Cavedo, P.J. Snyder, and Z.S. Khachaturian. 2018. The cholinergic system in the pathophysiology and treatment of Alzheimer's disease. *Brain*, 141(7), pp.1917–1933.

Harman, D. 1992. Free radical theory of aging. *Mutation Research*, 275(3–6), pp.257–266.

He, Y.S., W. Sun, C.Z. Wang, L.W. Qi, J. Yang, P. Li, X.D. Wen, and C.S. Yuan. 2015. Effects of American ginseng on pharmacokinetics of 5-fluorouracil in rats. *Biomedical Chromatography*, 29(5), pp.762–767.

Hemsworth, B., and G. West. 1970. Anticholinesterase activity of some degradation products of physostigmine. *Journal of Pharmaceutical Sciences*, 59(1), pp.118–120.

Houghton, P.J., Y. Ren, and M.-J. Howes. 2006. Acetylcholinesterase inhibitors from plants and fungi. *Natural Product Reports*, 23(2), pp.181–199.

Huang, X., Y. Liu, N. Zhang, X. Sun, H. Yue, C. Chen, and S. Liu. 2018. UPLC orbitrap HRMS analysis of *Panax quinquefolium* L. for authentication of panax genus with chemometric methods. *Journal of Chromatographic Science*, 56(1), pp.25–35.

Hwang, J.T., M.S. Lee, H.J. Kim, M.J. Sung, H.Y. Kim, M.S. Kim, and D.Y. Kwon. 2009. Antiobesity effect of ginsenoside Rg3 involves the AMPK and PPAR-gamma signal pathways. *Phytotherapy Research*, 23(2), pp.262–266.

Isaksson, K., and P.T. Kissinger.1987. Metabolism of physostigmine in mouse liver microsomal incubations studied by liquid chromatography with dual-electrode amperometric detection. *Journal of Chromatography*, 419, pp.165–175.

Jang, H.J., J. Jung, H.S. Yu, N.K. Lee, and H.D. Paik. 2018. Evaluation of the quality of yogurt using ginseng extract powder and probiotic lactobacillus plantarum NK181. *Korean Journal for Food Science of Animal Resource*, 38(6), pp.1160–1167.

Jastak, J.T. 1985. Physostigmine: An antidote for excessive CNS depression or paradoxical rage reactions resulting from intravenous diazepam. *Anesthesia Progress*, 32(3), p.87.

Jenkins, A.L., L.M. Morgan, J. Bishop, E. Jovanovski, D.J.A. Jenkins, and V. Vuksan. 2018. Co-administration of a konjac-based fibre blend and American ginseng (*Panax quinquefolius* L.) on glycaemic control and serum lipids in type 2 diabetes: A randomized controlled, cross-over clinical trial. *European Journal of Nutrition*, 57(6), pp.2217–2225.

Jung, J., K.H. Kim, K. Yang, K.H. Bang, and T.J. Yang. 2014. Practical application of DNA markers for high-throughput authentication of *Panax ginseng* and *Panax quinquefolius* from commercial ginseng products. *Journal of Ginseng Research*, 38(2), pp.123–129.

Kamal, M.A., P. Klein, Q.-S. Yu, D. Tweedie, Y. Li, H.W. Holloway, and N.H. Greig. 2006. Kinetics of human serum butyrylcholinesterase and its inhibition by a novel experimental Alzheimer therapeutic, bisnorcymserine. *Journal of Alzheimer's Disease*, 10(1), pp.43–51.

Kamal, M.A., X. Qu, Q.-S. Yu, D. Tweedie, H.W. Holloway, Y. Li, Y. Tan, and N.H. Greig. 2008. Tetrahydrofurobenzofuran cymserine, a potent butyrylcholinesterase inhibitor and experimental Alzheimer drug candidate, enzyme kinetic analysis. *Journal of Neural Transmission*, 115(6), pp.889–898.

Kan, J., R.A. Velliquette, K. Grann, C.R. Burns, J. Scholten, F. Tian, Q. Zhang, and M. Gui. 2017. A novel botanical formula prevents diabetes by improving insulin resistance. *BMC Complementary and Alternative Medicine*, 17(1), p.352.

Kandiah, N., M.-C. Pai, V. Senanarong, I. Looi, E. Ampil, K.W. Park, A.K. Karanam, and S. Christopher. 2017. Rivastigmine: The advantages of dual inhibition of acetylcholinesterase and butyrylcholinesterase and its role in subcortical vascular dementia and Parkinson's disease dementia. *Clinical Interventions in Aging*, 12, p.697.

Kim, D.H. 2012. Chemical diversity of *Panax ginseng, Panax quinquifolium, and Panax notoginseng. Journal of Ginseng Research*, 36(1), pp.1–15.

Kim, D.H. 2018. Gut microbiota-mediated pharmacokinetics of ginseng saponins. *Journal of Ginseng Research*, 42(3), pp.255–263.

Kim, K.T., K.M. Yoo, J.W. Lee, S.H. Eom, I.K. Hwang, and C.Y. Lee. 2007. Protective effect of steamed American ginseng (*Panax quinquefolius* L.) on V79-4 cells induced by oxidative stress. *Journal of Ethnopharmacology*, 111(3), pp.443–450.

King, M.L., and L.L. Murphy. 2007. American ginseng (*Panax quinquefolius* L.) extract alters mitogen-activated protein kinase cell signaling and inhibits proliferation of MCF-7 cells. *Journal of Experimental Therapeutics & Oncology*, 6(2), pp.147–155.

Klein, J. 2007. Phenserine. *Expert Opinion on Investigational Drugs*, 16(7), pp.1087–1097.

Kochan, E., B. Kolodziej, G. Gadomska, and A. Chmiel. 2004. Content of ginsenosides on *Panax quinguefolium* from field cultivation. *Herba Polonica*, 50(1), pp.20–27.

Kochan, E., M. Wasiela, and M. Sienkiewicz. 2013. The production of ginsenosides in hairy root cultures of American Ginseng, *Panax quinquefolium* L. and their antimicrobial activity. *In Vitro Cellular and Developmental Biology. Plant: Journal of the Tissue Culture Association*, 49(1), pp.24–29.

Kochan, E., P. Szymczyk, Ł. Kuźma, A. Lipert, and G. Szymańska. 2017. Yeast extract stimulates ginsenoside production in hairy root cultures of American ginseng cultivated in shake flasks and nutrient sprinkle bioreactors. *Molecules*, 22(6), p.880.

Kochan, E., P. Szymczyk, Ł. Kuźma, G. Szymańska, A. Wajs-Bonikowska, R. Bonikowski, and M. Sienkiewicz. 2018. The increase of triterpene saponin production induced by trans-anethole in hairy root cultures of *Panax quinquefolium. Molecules*, 23(10), p.2674.

Lee, D.G., S.I. Jang, Y.R. Kim, K.E. Yang, S.J. Yoon, Z.W. Lee, H.J. An, I.S. Jang, J.S. Choi, and H.S. Yoo. 2016. Anti-proliferative effects of ginsenosides extracted from mountain ginseng on lung cancer. *Chinese Journal of Integrative Medicine*, 22(5), pp.344–352.

Lee, J.H., H. Turndorf, and P.J. Poppers. 1975. Physostigmine reversal of antihistamine-induced excitement and depression. *Anesthesiology*, 43(6), pp.683–684.

Lee, O.R., N.Q. Nguyen, K.H. Lee, Y.C. Kim, and J. Seo. 2017. *Cytohistological study of the leaf structures of Panax ginseng* Meyer *and Panax quinquefolius* L. *Journal of Ginseng Research*, 41(4), pp.463–468.

Lee, T.K., K.F. O'Brien, W. Wang, C. Sheng, T. Wang, R.M. Johnke, and R.R. Allison. 2009. American ginseng modifies Cs-induced DNA damage and oxidative stress in human lymphocytes. *Open Nuclear Medicine Journal*, 1(1), pp.1–8.

Lee, T.K., K.F. O'Brien, W. Wang, R.M. Johnke, C. Sheng, S.M. Benhabib, T. Wang, and R.R. Allison. 2010a. Radioprotective effect of American ginseng on human lymphocytes at 90 minutes postirradiation: A study of 40 cases. *Journal of Alternative and Complementary Medicine*, 16(5), pp.561–567.

Li, B., C.Z. Wang, T.C. He, C.S. Yuan, and W. Du. 2010b. Antioxidants potentiate American ginseng-induced killing of colorectal cancer cells. *Cancer Letter*, 289(1), pp.62–70.

Li, J., T. Ichikawa, Y. Jin, L.J. Hofseth, P. Nagarkatti, M. Nagarkatti, A. Windust, and T. Cui. 2010c. An essential role of Nrf2 in American ginseng-mediated anti-oxidative actions in cardiomyocytes. *Journal of Ethnopharmacology*, 130(2), pp.222–230.

Lilja, A.M., Y. Luo, Q.S. Yu, J. Röjdner, Y. Li, A.M. Marini, A. Marutle, A. Nordberg, and N.H. Greig. 2013. Neurotrophic and neuroprotective actions of (-)- and (+)-phenserine, candidate drugs for Alzheimer's disease. *PLoS One*, 8(1), p.e54887.

Lim, W., K.W. Mudge, and F. Vermeylen. 2005. Effects of population, age, and cultivation methods on ginsenoside content of wild American ginseng (*Panax quinquefolium*). *Journal of Agricultural and Food Chemistry*, 53(22), pp.8498–8505.

Lin, E., Y. Wang, S. Mehendale, S. Sun, C.Z. Wang, J.T. Xie, H.H. Aung, and C.S. Yuan. 2008. Antioxidant protection by American ginseng in pancreatic beta-cells. *The American Journal of Chinese Medicine*, 36(5), pp.981–988.

Liu, D., B. Li, Y. Liu, A.S. Attele, J.W. Kyle, and C.-S. Yuan. 2001. Voltage-dependent inhibition of brain Na+ channels by American ginseng. *European Journal of Pharmacology*, 413(1), pp.47–54.

Liu, H., X. Lu, Y. Hu, and X. Fan. 2020. Chemical constituents of *Panax ginseng and Panax notoginseng* explain why they differ in therapeutic efficacy. *Pharmacological Research*, 161, p.105263.

Liu, P., Y. Xu, H. Yin, J. Wang, K. Chen, and Y. Li. 2005. Developmental toxicity research of ginsenoside Rb1 using a whole mouse embryo culture model. *Birth Defects Research Part B: Developmental and Reproductive Toxicology*, 74(2), pp.207–209.

Liu, R., J. Zhang, W. Liu, Y. Kimura, and Y. Zheng. 2010. Anti-obesity effects of protopanaxdiol types of Ginsenosides isolated from the leaves of American ginseng (*Panax quinquefolius* L.) in mice fed with a high-fat diet. *Fitoterapia*, 81(8), pp.1079–1087.

Liu, W., Y. Zheng, L. Han, H. Wang, M. Saito, M. Ling, Y. Kimura, and Y. Feng. 2008. Saponins (Ginsenosides) from stems and leaves of *Panax quinquefolium* prevented high-fat diet-induced obesity in mice. *Phytomedicine*, 15(12), pp.1140–1145.

Liu, Y.Y., T.Y. Zhang, X. Xue, D.M. Liu, H.T. Zhang, L.L. Yuan, Y.L. Liu, H.L. Yang, S.B. Sun, C. Zhang, H.S. Xu, C.F. Wu, and J.Y. Yang. 2017. Pseudoginsenoside-F11 attenuates cerebral ischemic injury by alleviating autophagic/lysosomal defects. *CNS Neuroscience and Therapeutics*, 23(7), pp.567–579.

Loizzo, M.R., R. Tundis, F. Menichini, and F. Menichini. 2008. Natural products and their derivatives as cholinesterase inhibitors in the treatment of neurodegenerative disorders: An update. *Current Medicinal Chemistry*, 15(12), pp.1209–1228.

Ludwiczuk, A., T. Wolski, and E. Holderna-Kedzia. 2006. Estimation of the chemical composition and antimicrobial and antioxidant activity of extracts received from leaves and roots of American ginseng [*Panax quinquefolium* L.]. *Herba Polonica*, 52(4), pp.79–90.

Luo, J.Z., and L. Luo. 2009. Ginseng on hyperglycemia: Effects and mechanisms. *Evidence-Based Complementary and Alternative Medicine*, 6(4), pp.423–427.

Ma, Z.N., Y.Z. Li, W. Li, X.T. Yan, G. Yang, J. Zhang, L.C. Zhao, and L.M. Yang. 2017a. Nephroprotective effects of saponins from leaves of *Panax quinquefolius* against cisplatin-induced acute kidney injury. *International Journal of Molecular Sciences*, 18(7), pp.1407–1424.

Ma, Z.N., Z. Liu, Z. Wang, S. Ren, S. Tang, Y.P. Wang, S.Y. Xiao, C. Chen, and W. Li. 2017b. Supplementation of American ginseng berry extract mitigated cisplatin-evoked nephrotoxicity by suppressing ROS-mediated activation of MAPK and NF-κB signaling pathways. *Food and Chemical Toxicology*, 110, pp.62–73.

Mahalanobish, S., S. Saha, S. Dutta, and P.C. Sil. 2019. Mangiferin alleviates arsenic induced oxidative lung injury via upregulation of the Nrf2-HO1 axis. *Food and Chemical Toxicology*, 126, pp.41–55.

McCaleb, R. 1990. Nature's medicine for memory loss. *HerbalGram*, 23, p.15.

Mehendale, S.R., C.Z. Wang, Z.H. Shao, C.Q. Li, J.T. Xie, H.H. Aung, and C.S. Yuan. 2006. Chronic pretreatment with American ginseng berry and its polyphenolic constituents attenuate oxidant stress in cardiomyocytes. *European Journal of Pharmacology*, 553(1–3), pp.209–214.

Michael, S., N.J. Yu, D.E. Sophia, Z. Liu, S. Li, Z. Jing, G. Rui, G. Shan, G. Xiumei, and F. Guanwei. 2021. Ginsenosides for cardiovascular diseases; update on pre-clinical and clinical evidence, pharmacological effects and the mechanisms of action. *Pharmacological Research*, p.105481.

Mikkilineni, S., I. Cantuti-Castelvetri, C.M. Cahill, A. Balliedier, N.H. Greig, and J.T. Rogers. 2012. The anticholinesterase phenserine and its enantiomer posiphen as 5′untranslated-region-directed translation blockers of the Parkinson's alpha synuclein expression. *Parkinsons Disease*, 2012, p.142372.

Miller, R.D. 1976. Antagonism of neuromuscular blockade. *Anesthesiology*, 44(4), pp.318–329.

Miller, S.A., D.W. Blick, S.Z. Kerenyi, and M.R. Murphy. 1993. Efficacy of physostigmine as a pretreatment for organophosphate poisoning. *Pharmacology Biochemistry and Behavior*, 44(2), pp.343–347.

Mucalo, I., E. Jovanovski, D. Rahelić, V. Božikov, Z. Romić, and V. Vuksan. 2013. Effect of American ginseng (*Panax quinquefolius* L.) on arterial stiffness in subjects with type-2 diabetes and concomitant hypertension. *Journal of Ethnopharmacology*, 150(1), pp.148–153.

Mucalo, I., E. Jovanovski, V. Vuksan, V. Božikov, Z. Romić, and D. Rahelić. 2014. American ginseng extract (*Panax quinquefolius* L.) is safe in long-term use in type 2 diabetic patients. *Evidence-Based Complementary and Alternative Medicine*, 2014, p.969168.

Murphy, L.L., R.S. Cadena, D. Chávez, and J.S. Ferraro. 1998. Effect of American ginseng (*Panax quinquefolium*) on male copulatory behavior in the rat. *Physiology and Behavior*, 64(4), pp.445–450.

Nag, S.A., J.J. Qin, W. Wang, M.H. Wang, H. Wang, and R. Zhang. 2012. Ginsenosides as anticancer agents: In vitro and in vivo activities, structure-activity relationships, and molecular mechanisms of action. *Frontiers in Pharmacology*, 3, p.25.

Newton, R.W. 1975. Physostigmine salicylate in the treatment of tricyclic antidepressant overdosage. *Jama*, 231(9), pp.941–943.

Nickalls, R., and E. Nickalls. 1988. The first use of physostigmine in the treatment of atropine poisoning: A translation of Kleinwachter's paper entitled 'observations on the effect of calabar bean extract as an antidote to atropine poisoning'. *Anaesthesia*, 43(9), pp.776–777.

Ohl, K., K. Tenbrock, and M. Kipp. 2016. Oxidative stress in multiple sclerosis: Central and peripheral mode of action. *Experimental Neurology*, 277, pp.58–67.

Orhan, I.E., G. Orhan, and E. Gurkas. 2011. An overview on natural cholinesterase inhibitors-a multi-targeted drug class-and their mass production. *Mini-Reviews in Medicinal Chemistry*, 11(10), pp.836–842.

Oun, R., Y.E. Moussa, and N.J. Wheate. 2018. The side effects of platinum-based chemotherapy drugs: A review for chemists. *Dalton Transactions*, 47(19), pp.6645–6653.

Pan, Y., X. Wang, G. Sun, F. Li, and X. Gong. 2016. Application of RAD sequencing for evaluating the genetic diversity of domesticated *Panax notoginseng* (Araliaceae). *PLoS One*, 11(11), p.e0166419.

Pawar, A.A., D.N. Tripathi, P. Ramarao, and G. Jena. 2007. Protective effects of American ginseng (*Panax quinquefolium*) against mitomycin C induced micronuclei in mice. *Phytotherapy Research*, 21(12), pp.1221–1227.

Phelps, S.J., T.M. Hagemann, K.R. Lee, and A.J. Thompson. 2016. *Pediatric Injectable Drugs: The Teddy Bear Book*. American Society of Health System Pharmacists.

Pinheiro, G.K.L.d.O., I.d. Araújo Filho, I.d. Araújo Neto, A.C.M. Rêgo, E.P.d. Azevedo, F.I. Pinheiro, and A.A.d.S. Lima Filho. 2018. Nature as a source of drugs for ophthalmology. *Arquivos Brasileiros de Oftalmologia*, 81(5), pp.443–454.

Pinho, B.R., F. Ferreres, P. Valentão, and P.B. Andrade. 2013. Nature as a source of metabolites with cholinesterase-inhibitory activity: An approach to Alzheimer's disease treatment. *Journal of Pharmacy and Pharmacology*, 65(12), pp.1681–1700.

Popovich, D.G., C.-R. Yeo, and W. Zhang. 2012. Ginsenosides derived from Asian (*Panax ginseng*), American ginseng (*Panax quinquefolius*) and potential cytoactivity. *International Journal of Biomedical and Pharmaceutical Sciences*, 6(1), pp.56–62.

Proctor, J.T., and B.J. Shelp. 2014. Effect of boron nutrition on American ginseng in field and in nutrient cultures. *Journal of Ginseng Research*, 38(1), pp.73–77.

Proudfoot, A. 2006. The early toxicology of physostigmine: A tale of beans, great men and egos. *Toxicological Reviews*, 25(2), pp.99–138.

Qu, Y., Z. Wang, F. Zhao, J. Liu, W. Zhang, J. Li, Z. Song, and H. Xu. 2018. AFM-detected apoptosis of hepatocellular carcinoma cells induced by American ginseng root water extract. *Micron*, 104, pp.1–7.

Renzi, G., A. Bartolini, A. Galli, R. Bartolini, and P. Malmberg. 1980. The structure and action of eseroline: A new antinociceptive drug. *Inorganica Chimica Acta*, 40, pp.X74–X75.

Rotem, C., and B. Kaplan. 2007. Phyto-female complex for the relief of hot flushes, night sweats and quality of sleep: Randomized, controlled, double-blind pilot study. *Gynecological Endocrinology*, 23(2), pp.117–122.

Rubnov, S., D. Levy, and H. Schneider. 1999. Liquid chromatographic analysis of physostigmine salicylate and its degradation products. *Journal of Pharmaceutical and Biomedical Analysis*, 18(6), pp.939–945.

Scholey, A., A. Ossoukhova, L. Owen, A. Ibarra, A. Pipingas, K. He, M. Roller, and C. Stough. 2010. Effects of American ginseng (*Panax quinquefolius*) on neurocognitive function: An acute, randomised, double-blind, placebo-controlled, crossover study. *Psychopharmacology*, 212(3), pp.345–356.

Seida, J.K., T. Durec, and S. Kuhle. 2011. North American (*Panax quinquefolius*) and Asian Ginseng (*Panax ginseng*) preparations for prevention of the common cold in healthy adults: A systematic review. *Evidence-Based Complementary and Alternative Medicine*, 2011, p.282151.

Sen, S., S. Chen, B. Feng, Y. Wu, E. Lui, and S. Chakrabarti. 2011. American ginseng (*Panax quinquefolius*) prevents glucose-induced oxidative stress and associated endothelial abnormalities. *Phytomedicine*, 18(13), pp.1110–1117.

Sengupta, S., S.A. Toh, L.A. Sellers, J.N. Skepper, P. Koolwijk, H.W. Leung, H.W. Yeung, R.N. Wong, R. Sasisekharan, and T.P. Fan. 2004. Modulating angiogenesis: The yin and the yang in ginseng. *Circulation*, 110(10), pp.1219–1225.

Shergis, J.L., Y.M. Di, A.L. Zhang, R. Vlahos, R. Helliwell, J.M. Ye, and C.C. Xue. 2014. Therapeutic potential of *Panax ginseng* and ginsenosides in the treatment of chronic obstructive pulmonary disease. *Complementary Therapies in Medicine*, 22(5), pp.944–953.

Shi, S., R. Shi, and K. Hashizume. 2012. American ginseng improves neurocognitive function in senescence-accelerated mice: Possible role of the upregulated insulin and choline acetyltransferase gene expression. *Geriatrics and Gerontology International*, 12(1), pp.123–130.

Shin, K., H. Guo, Y. Cha, Y.H. Ban, W. Seo da, Y. Choi, T.S. Kim, S.P. Lee, J.C. Kim, E.K. Choi, J.M. Yon, and Y.B. Kim. 2016. Cereboost™, an American ginseng extract, improves cognitive function via up-regulation of choline acetyltransferase expression and neuroprotection. *Regulatory Toxicology and Pharmacology*, 78, pp.53–58.

Singh, M., M. Kaur, H. Kukreja, R. Chugh, O. Silakari, and D. Singh. 2013. Acetylcholinesterase inhibitors as Alzheimer therapy: From nerve toxins to neuroprotection. *European Journal of Medicinal Chemistry*, 70, pp.165–188.

Singh, R.K., E. Lui, D. Wright, A. Taylor, and M. Bakovic. 2017. Alcohol extract of North American ginseng (*Panax quinquefolius*) reduces fatty liver, dyslipidemia, and other complications of metabolic syndrome in a mouse model. *Canadian Journal of Physiology and Pharmacology*, 95(9): 1046–1057.

Sitaram, N., H. Weingartner, and J.C. Gillin, 1978. Physostigmine: improvement of longterm memory processes in normal humans. *Science*, 201, pp.272–276.

Sloley, B.D., P.K. Pang, B.H. Huang, F. Ba, F.L. Li, C.G. Benishin, A.J. Greenshaw, and J.J. Shan. 1999. American ginseng extract reduces scopolamine-induced amnesia in a spatial learning task. *Journal of Psychiatry & Neuroscience*, 24(5), pp.442–452.

Somani, S., and S. Dube. 1989. Physostigmine: An overview as pretreatment drug for organophosphate intoxication. *International Journal of Clinical Pharmacology, Therapy, and Toxicology*, 27(8), pp.367–387.

Somani, S.M., R.K. Kutty, and G. Krishna. 1990. Eseroline, a metabolite of physostigmine, induces neuronal cell death. *Toxicology and Applied Pharmacology*, 106(1), pp.28–37.

Souther, S., M.J. Lechowicz, and J.B. McGraw. 2012. Experimental test for adaptive differentiation of ginseng populations reveals complex response to temperature. *Annals of Botany*, 110(4), pp.829–837.

Sun, J., and P. Chen. 2011. Differentiation of *Panax quinquefolius* grown in the USA and China using LC/MS-based chromatographic fingerprinting and chemometric approaches. *Analytical and Bioanalytical Chemistry*, 399(5), pp.1877–1889.

Sun, X., P. Chen, S.L. Cook, G.P. Jackson, J.M. Harnly, and P.B. Harrington. 2012. Classification of cultivation locations of *Panax quinquefolius* L. samples using high performance liquid chromatography–electrospray ionization mass spectrometry and chemometric analysis. *Analytical Chemistry*, 84(8), pp.3628–3634.

Swain, T. 1972. *Plants in the Development of Modern Medicine*. Harvard University Press.

Szczuka, D., A. Nowak, M. Zakłos-Szyda, E. Kochan, G. Szymańska, I. Motyl, and J. Blasiak. 2019. American ginseng (*Panax quinquefolium* L.) as a source of bioactive phytochemicals with pro-health properties. *Nutrients*, 11(5), pp.1041–1069.

Thal, L.J., J.M. Ferguson, J. Mintzer, A. Raskin, and S.D. Targum. 1999. A 24-week randomized trial of controlled-release physostigmine in patients with Alzheimer's disease. *Neurology* 52(6): 1146–1152.

Toda, N., T. Kaneko, and H. Kogen. 2010. Development of an efficient therapeutic agent for Alzheimer's disease: Design and synthesis of dual inhibitors of acetylcholinesterase and serotonin transporter. *Chemical and Pharmaceutical Bulletin*, 58(3), pp.273–287.

Traub, S.J., L.S. Nelson, and R.S. Hoffman. 2002. Physostigmine as a treatment for gamma-hydroxybutyrate toxicity: A review. *Journal of Toxicology: Clinical Toxicology*, 40(6), pp.781–787.

Triggle, D.J., and R. Filler. 1998. The pharmacology of physostigmine. *CNS Drug Reviews*, 4, pp.87–136.

Vuksan, V., J.L. Sievenpiper, J. Wong, Z. Xu, U. Beljan-Zdravkovic, J.T. Arnason, V. Assinewe, M.P. Stavro, A.L. Jenkins, L.A. Leiter, and T. Francis. 2001. American ginseng (*Panax quinquefolius* L.) attenuates postprandial glycemia in a time-dependent but not dose-dependent manner in healthy individuals. *American Journal of Clinical Nutrition*, 73(4), pp.753–758.

Vuksan, V., J.L. Sievenpiper, V.Y. Koo, T. Francis, U. Beljan-Zdravkovic, Z. Xu, and E. Vidgen. 2000. American ginseng (*Panax quinquefolius* L.) reduces postprandial glycemia in nondiabetic subjects and subjects with type 2 diabetes mellitus. *Archives of Internal Medicine*, 160(7), pp.1009–1013.

Vuksan, V., Z.Z. Xu, E. Jovanovski, A.L. Jenkins, U. Beljan-Zdravkovic, J.L. Sievenpiper, P. Mark Stavro, A. Zurbau, L. Duvnjak, and M.Z.C. Li. 2019. Efficacy and safety of American ginseng (*Panax quinquefolius* L.) extract on glycemic control and cardiovascular risk factors in individuals with type 2 diabetes: A double-blind, randomized, cross-over clinical trial. *European Journal of Nutrition*, 58(3), pp.1237–1245.

Wan, J.Y., P. Liu, H.Y. Wang, L.W. Qi, C.Z. Wang, P. Li, and C.S. Yuan. 2013. Biotransformation and metabolic profile of American ginseng saponins with human intestinal microflora by liquid chromatography quadrupole time-of-flight mass spectrometry. *Journal of Chromatography A*, 1286, pp.83–92.

Wang, A., C.Z. Wang, J.A. Wu, J. Osinski, and C.S. Yuan. 2005. Determination of major ginsenosides in *Panax quinquefolius* (American ginseng) using high-performance liquid chromatography. *Phytochemical Analysis*, 16(4), pp.272–277.

Wang, C.Z., C. Yu, X.D. Wen, L. Chen, C.F. Zhang, T. Calway, Y. Qiu, Y. Wang, Z. Zhang, S. Anderson, Y. Wang, W. Jia, and C.S. Yuan. 2016. American ginseng attenuates colitis-associated colon carcinogenesis in mice: Impact on gut microbiota and metabolomics. *Cancer Prevention Research*, 9(10), pp.803–811.

Wang, C.Z., G.J. Du, Z. Zhang, X.D. Wen, T. Calway, Z. Zhen, M.W. Musch, M. Bissonnette, E.B. Chang, and C.S. Yuan. 2012. Ginsenoside compound K, not Rb1, possesses potential chemopreventive activities in human colorectal cancer. *International Journal of Oncology*, 40(6), pp.1970–1976.

Wang, C.-Z., Y. Cai, S. Anderson, and C.-S. Yuan. 2015. Ginseng metabolites on cancer chemoprevention: An angiogenesis link. *Diseases*, 3(3), pp.193–204.

Wang, J.S., H.J. Yin, C.Y. Guo, Y. Huang, C.D. Xia, and Q. Liu. 2013a. Influence of high blood glucose fluctuation on endothelial function of type 2 diabetes mellitus rats and effects of *Panax quinquefolius* Saponin of stem and leaf. *Chinese Journal of Integrative Medicine*, 19(3), pp.217–222.

Wang, L., X. Yang, X. Yu, Y. Yao, and G. Ren. 2013b. Evaluation of antibacterial and anti-inflammatory activities of less polar ginsenosides produced from polar ginsenosides by heat-transformation. *Journal of Agricultural and Food Chemistry*, 61(50), pp.12274–12282.

Wang, L., X. Yu, X. Yang, Y. Li, Y. Yao, E.M.K. Lui, and G. Ren. 2015. Structural and anti-inflammatory characterization of a novel neutral polysaccharide from North American ginseng (*Panax quinquefolius*). *International Journal of Biological Macromolecules*, 74, pp.12–17.

Wang, M., L.J. Guilbert, J. Li, Y. Wu, P. Pang, T.K. Basu, and J.J. Shan. 2004. A proprietary extract from North American ginseng (*Panax quinquefolium*) enhances IL-2 and IFN-γ productions in murine spleen cells induced by Con-A. *International Immunopharmacology*, 4(2), pp.311–315.

Wang, Y.-Z., J. Chen, S.-F. Chu, Y.-S. Wang, X.-Y. Wang, N.-H. Chen, and J.-T. Zhang. 2009. Improvement of memory in mice and increase of hippocampal excitability in rats by ginsenoside Rg1's metabolites ginsenoside Rh1 and protopanaxatriol. *Journal of Pharmacological Sciences*, 109(4), pp.504–510.

Wei, X.-Y., J.-Y. Yang, J.-H. Wang, and C.-F. Wu. 2007. Anxiolytic effect of saponins from *Panax quinquefolium* in mice. *Journal of Ethnopharmacology*, 111(3), pp.613–618.

Weinstock, M. 1999. Selectivity of cholinesterase inhibition. *CNS Drugs*, 12(4), pp.307–323.

Weisdorf, D., J. Kramer, A. Goldbarg, and H.L. Klawans. 1978. Physostigmine for cardiac and neurologic manifestations of phenothiazine poisoning. *Clinical Pharmacology & Therapeutics*, 24(6), pp.663–667.

Wilson, S.A., M.H. Wong, C. Stryjecki, A. De Boer, E.M. Lui, and D.M. Mutch. 2013. Unraveling the adipocyte inflammomodulatory pathways activated by North American ginseng. *International Journal of Obesity*, 37(3), pp.350–356.

Winblad, B., E. Giacobini, L. Frölich, L.T. Friedhoff, G. Bruinsma, R.E. Becker, and N.H. Greig. 2010. Phenserine efficacy in Alzheimer's disease. *Journal of Alzheimer's Disease*, 22(4), pp.1201–1208.

Wu, C.F., Y.L. Liu, M. Song, W. Liu, J.H. Wang, X. Li, and J.Y. Yang. 2003. Protective effects of pseudoginsenoside- F11 on methamphetamine-induced neurotoxicity in mice. *Pharmacology Biochemistry and Behavior*, 76(1), pp.103–109.

Wu, G., J. Yi, L. Liu, P. Wang, Z. Zhang, and Z. Li. 2013. Pseudoginsenoside F11, a novel partial PPAR γ agonist, promotes adiponectin oligomerization and secretion in 3T3-L1 adipocytes. *PPAR Research*, 2013, p.701017.

Wu, Y., C. Qin, X. Lu, J. Marchiori, and Q. Feng. 2016. North American ginseng inhibits myocardial NOX2-ERK1/2 signaling and tumor necrosis factor-α expression in endotoxemia. *Pharmacological Research*, 111, pp.217–225.

Wu, Y., X. Lu, F.-L. Xiang, E.M. Lui, and Q. Feng. 2011. North American ginseng protects the heart from ischemia and reperfusion injury via upregulation of endothelial nitric oxide synthase. *Pharmacological Research*, 64(3), pp.195–202.

Xiao, D., H. Yue, Y. Xiu, X. Sun, Y. Wang, and S. Liu. 2015. Accumulation characteristics and correlation analysis of five ginsenosides with different cultivation ages from different regions. *Journal of Ginseng Research*, 39(4), pp.338–344.

Xue, P., Y. Yao, X.S. Yang, J. Feng, and G.X. Ren. 2017. Improved antimicrobial effect of ginseng extract by heat transformation. *Journal of Ginseng Research*, 41(2), pp.180–187.

Yang, J., D. Yuan, T. Xing, H. Su, S. Zhang, J. Wen, Q. Bai, and D. Dang. 2016a. Ginsenoside Rh2 inhibiting HCT116 colon cancer cell proliferation through blocking PDZ-binding kinase/T-LAK cell-originated protein kinase. *Journal of Ginseng Research*, 40(4), pp.400–408.

Yang, L., J. Hao, J. Zhang, W. Xia, X. Dong, X. Hu, F. Kong, and X. Cui. 2009. Ginsenoside Rg3 promotes beta-amyloid peptide degradation by enhancing gene expression of neprilysin. *Journal of Pharmacy and Pharmacology*, 61(3), pp.375–380.

Yang, L., Q.T. Yu, Y.Z. Ge, W.S. Zhang, Y. Fan, C.W. Ma, Q. Liu, and L.W. Qi. 2016b. Distinct urine metabolome after Asian ginseng and American ginseng intervention based on GC-MS metabolomics approach. *Scientific Report*, 6, p.39045.

Yang, Y., C. Ren, Y. Zhang, and X. Wu. 2017. Ginseng: An nonnegligible natural remedy for healthy aging. *Aging and Disease*, 8(6), pp.708–720.

Yang W. Z., Y. Hu, W. Y. Wu, M. Ye, and D. A. Guo. 2014. Saponins in the genus *Panax L.* (Araliaceae): a systematic review of their chemical diversity. *Phytochemistry*, 106, pp.7–24.

Yeo, C.R., S.M. Lee, and D.G. Popovich. 2011. Ginseng (*Panax quinquefolius*) reduces cell growth, lipid acquisition and increases adiponectin expression in 3T3-L1 Cells. *Evidence-Based Complementary and Alternative Medicine*, 2011, p.610625.

Yoo, K.M., C. Lee, Y.M. Lo, and B. Moon. 2012. The hypoglycemic effects of American red ginseng (*Panax quinquefolius* L.) on a diabetic mouse model. *Journal of Food Science*, 77(7), pp.H147–H152.

Yoshida, S., and N. Suzuki. 1993. Antiamnesic and cholinomimetic side-effects of the cholinesterase inhibitors, physostigmine, tacrine and NIK-247 in rats. *European Journal of Pharmacology*, 250(1), pp.117–124.

Yu, C., C.Z. Wang, C.J. Zhou, B. Wang, L. Han, C.F. Zhang, X.H. Wu, and C.S. Yuan. 2014. Adulteration and cultivation region identification of American ginseng using HPLC coupled with multivariate analysis. *Journal of Pharmaceutical and Biomedical Analysis*, 99, pp.8–15.

Yu, C., X.-D. Wen, Z. Zhang, C.-F. Zhang, X.-H. Wu, A. Martin, W. Du, T.-C. He, C.-Z. Wang, and C.-S. Yuan. 2015. American ginseng attenuates azoxymethane/dextran sodium sulfate-induced colon carcinogenesis in mice. *Journal of Ginseng Research*, 39(1), pp.14–21.

Zhang, L., C. Virgous, and H. Si. 2017. Ginseng and obesity: Observations and understanding in cultured cells, animals and humans. *Journal of Nutritional Biochemistry*, 44, pp.1–10.

Zhang, X.-J., L.-L. Huang, X.-J. Cai, P. Li, Y.-T. Wang, and J.-B. Wan. 2013. Fatty acid variability in three medicinal herbs of Panax species. *Chemistry Central Journal*, 7(1), pp.1–8.

Zhao, B., S.M. Moochhala, and S.-y. Tham. 2004. Biologically active components of *Physostigma venenosum. Journal of Chromatography B*, 812(1–2), pp.183–192.

Zhao, Y., X. Hu, X. Zuo, and M. Wang. 2018. Chemopreventive effects of some popular phytochemicals on human colon cancer: A review. *Food and Function*, 9(9), pp.4548–4568.

Zhu, L., J. Li, N. Xing, D. Han, H. Kuang, and P. Ge. 2015. American ginseng regulates gene expression to protect against premature ovarian failure in rats. *BioMed Research International*, 2015, p.767124.

11 *Phytolacca dodecandra* (African Soapberry) and *Picrorhiza kurroa* (Kutki)

K. Meenakshi, Mansi Shah, and Indu Anna George

CONTENTS

11.1 Introduction 203
11.2 *Phytolacca dodecandra* 204
11.2.1 General Description 204
11.2.2 Botanical Aspects and Habitat 204
11.2.3 Traditional Knowledge 205
11.2.4 Bioactive Compounds and Phytochemistry 205
11.2.5 Potential Benefits, Applications, and Uses 206
11.2.5.1 Molluscicidal Property 206
11.2.5.2 Anthelminthic Property 206
11.2.5.3 Antimicrobial Property 206
11.2.5.4 Antimalarial Property 207
11.2.5.5 Antiviral Property 207
11.2.5.6 Hepatoprotective Property 207
11.3 *Picrorhiza kurroa* 207
11.3.1 General Description 207
11.3.2 Botanical Aspects and Habitat 208
11.3.3 Traditional Knowledge 208
11.3.4 Bioactive Compounds and Phytochemistry 209
11.3.5 Potential Benefits, Applications, and Uses 209
11.3.5.1 Nephroprotective Activity 210
11.3.5.2 Antioxidant Activity 210
11.3.5.3 Anti-Inflammatory and Antiallergic Activities 210
11.3.5.4 Immunostimulatory Activity 210
11.3.5.5 Anticarcinogenic and Antineoplastic Activity 210
11.3.5.6 Antidiabetic Activity 210
11.3.5.7 Potential Production of Secondary Metabolites for Medicinal or Commercial Purposes 211
11.4 Conclusion 211
References 211

11.1 INTRODUCTION

Since ancient times, cultural and traditional development was majorly influenced by the local environment and available materials. Local plants were integral components of traditional cultures around the world. The traditional medicinal practices, such as ancient Egyptian medicine, traditional Chinese medicines, traditional Indian medicines (Ayurveda, Unani, Siddha) have depended significantly on plants for the formulation of medicines. Plants have been the driving force for

DOI: 10.1201/9781003205067-11

several industries other than pharmaceuticals, spearheading cultural development from traditional times to the modern era. Essential oils obtained from plants were the sole source of the raw scent for the perfume industry before the dawn of synthetic perfumes. These oils were derived from all parts of the plant. Petals of rose and lavender, leaves of rosemary and lemongrass, seed of nutmeg, peels of orange and lemon, and sandalwood are a few examples (Poucher, 1993). Similarly, the natural dye industries were also predominantly dependent on plants. The traditional natural dyes were derived from indigo and madder wood (Gilbert and Cooke, 2001). The Indian and African continents are abundant in natural plant resources. The culture and daily routine of life have been deeply influenced by the environment and available resources. Neem (*Azadirachta indica*) has been used in India for pest control (Boeke et al., 2004). China berry and *Stellera* have been used as insecticides in China (Yang and Tang, 1988). *Phytolacca dodecandra*, found in Africa and known as "soapberry" for its soap-like properties, and *Picrorhiza kurroa*, found in the alpine belt of the Himalayas and known as "kutki", are examples of two plants that are famous for their medicinal properties in traditional medicines and have influenced the local culture

11.2 *PHYTOLACCA DODECANDRA*

11.2.1 General Description

Phytolacca dodecandra (L'Herit) (synonyms: *P. abyssinica* Hoffm., *Pircunia abyssinica* Moq.) (Figure 11.1), is native to sub-Saharan Africa and Madagascar (Lemma, 1970; Schmelzer and Gurib-Fakim, 2008) and popularly known as "endod" in Ethiopia. In English, it is also referred to as African soapberry, soapberry, or gopo berry (Esser et al., 2003), phytolaque in French, fitolaca in Spanish, and chihakahaka in Tanzania (Legère, 2009). There are two varieties of endod that exist in Ethiopia, namely, arabe (pinkish berries) and ahiyo (greyish berries). Ahiyo means donkey to signify its diminished detergent properties as compared to the arabe variety (Lemma, 1970).

11.2.2 Botanical Aspects and Habitat

P. dodecandra (Family: Phytolaccaceae) is a semi-succulent shrub that is a perennial, dioecious, and woody climber with taproots (Adams et al., 1989; Karunamoorthi et al., 2008). It is a rapid growing plant that attains an average height of 5–8 m (Beressa et al., 2020).

The vegetative parts are usually glabrous or pubescent (less common). The blades of the leaves are 6–15 cm in length, 3–8.5 cm wide, ovate or elliptic in shape, ending in a recurved and mucronate

FIGURE 11.1 *Phytolacca dodecandra.*

tip. They are cuneate, rounded, or cordate and unequal at the base. It is glossy with visible crystals (Plants of the World Online).

The flowers are arranged in bracteolate racemes that are 15–30 cm long. The flowers are scented, dimorphic, and long- and short-staminate (Plants of the World Online). Its fruiting season is generally twice a year (January and July) under favorable climatic conditions in Ethiopia (Lemma, 1970). The berries, 5–8 mm across, are bluntly star-shaped which becomes rounded, juicy, and orange-red on ripening (Plants of the World Online).

P. dodecandra is widely distributed in Africa, Madagascar, Asia, and South America (Karunamoorthi et al., 2008). Its habitats include a range of forests, woodlands, bushlands, thickets, grasslands, old forest lands, plantations, riparian areas, and at elevations of 500–2,400 m (*African Plant Database*).

11.2.3 Traditional Knowledge

P. dodecandra is traditionally used widely for medicinal and nonmedicinal purposes. An aqueous solution of dried powdered berries of *P. dodecandra* is used as a foaming detergent solution for cleaning clothes in Ethiopia, Somalia, and Uganda (Lemma, 1965). Different parts of *P. dodecandra* are widely used as an anthelminthic, laxative, emetic, diuretic, and against diarrhea, in humans; whereas it is used as a purgative for animals in East Africa, Central Africa, and Madagascar (Schmelzer and Gurib-Fakim, 2008). Mashed roots, berries, and leaf sap are conventionally used to treat skin diseases, such as boils, ringworms, leprosy, scabies, and vitiligo (Ogutu et al., 2012). Extracts of berries, roots, and leaves are used to treat rabies, malaria, sore throat, asthma, tuberculosis, and respiratory problems. Soaked root bark or leaves are used for the treatment of epilepsy. Juvenile leaves and shoots are consumed to induce abortion (Nalule et al., 2011). Leaf sap is used to treat ear inflammation (Otitis media) in Rwanda (Schmelzer and Gurib-Fakim, 2008). The leaf or fruit extract is sometimes added to drinks or foods as a stimulant and is used to curdle milk. The plant is used as hedges in Ethiopia. The stems of this climber are used as ties to construct huts and fences. It is generally not used as firewood, as the smoke is believed to reduce male sexual ability.

11.2.4 Bioactive Compounds and Phytochemistry

P. dodecandra is widely used as traditional medicine in Africa despite its toxicity. Various phytochemical compounds of *P. dodecandra* have been identified in different studies.

Polar solvent extract of berries was reported to have saponins, tannins, flavonoids, and genins (Tadeg et al., 2005). Saponins such as 3-*O*-(2′,4′-di-*O*-[β-d-glucopyranosyl]-β-d-glucopyranosyl) 2-β-hydroxyoleanolic acid, 3-*O*-(*O*-alpha-l-rhamnopyranosyl-[1, 2]-*O*-[β-d-galactopyranosyl-[1, 3])-β-d-glucopyranosyl] 2 β-hydroxyoleanolic acid, 3-*O*-(3′-*O*-[β-d-galactopyranosyl]- β-d-glucopyranosyl) 2 β- hydroxyoleanolic acid, and 3-(2,4-di-*O*-[β-d-glucopyranosyl]-β-dglucoprano syl)-olean-12-ene-28-oic acid were identified in the aqueous extract of *P. dodecandra* berries (Birrie et al., 1998; Parkhurst et al., 1974; Thiilborg et al., 1994). Methanol extract of berries contains saponins like *O*-acetyl oleanolate (I) and methyl tri-*O*-acetyl bayogenin (Karunamoorthi et al., 2008), and *n*-butanol extract of berries contains genins, such as hederagenin, 2β-hydroxyoleanolic acid, bayogenin, and oleanolic acid (Stobaeus et al., 1990). Lemmatoxins (triterpenoid saponins) that have molluscicidal activities were named after the Ethiopian scientist, Aklilu Lemma, who isolated it (Domon and Hostettmann, 1984; Ogutu et al., 2012; Parkhurst et al., 1974; Powell and Whalley, 1969; Spengel, 1996; Thiilborg et al., 1993).

The *n*-butanol-extracts of the roots and leaves also contain oleanolic acid, hederagenin, 2β-hydroxyoleanolic acid, bayogenin, phytolaccagenin, and phytolaccagenic acid (Namulindwa et al., 2015). Methanolic extract of the root is composed of phytolaccagenin, phytolaccagenic acid, serjanic acid, dodecandral, and dodecandralol (Lambert et al., 1991). Terpenoids and phenolics are present in the hexane, dichloromethane, ethyl acetate, and methanol solvent extracts of leaves, roots, and stem barks (Lemma, 1965). Glycosides, such as esculentoside and esculentoside L1, were identified in the *n*-butanol extract of callus (Namulindwa et al., 2015).

Matebie et al., (2019) stated that aldehydes and ketones, such as sulcatone, 2-nonanone, benzaldehyde, and phytone, are the major components in the aroma of *P. dodecandra*. Phytone, phytol, and hexadecanoic acid were noticeable components in the essential oils.

11.2.5 Potential Benefits, Applications, and Uses

11.2.5.1 Molluscicidal Property

P. dodecandra is famous in Africa for its molluscicidal activity, which can be used to block the transmission of schistosomiasis (Esser et al., 2003; Lemma, 1970). Schistosoma (blood flukes) are waterborne parasites of humans, and schistosomiasis is an infection that occurs when human skin encounters contaminated freshwater which is home to certain types of snails that carry the blood flukes. Chronic infections may cause an increased risk of liver fibrosis or bladder cancer.

Lemma, the Ethiopian scientist, noticed more dead snails in places along rivers where people washed clothes using the endod powder as the detergent (Goll et al., 1983; Lemma, 1970). His maiden attempt to determine the molluscicidal activity of *P. dodecandra* has been followed by many studies that were conducted to analyze different properties of *P. dodecandra*. The triterpenoid saponins that were responsible for molluscicidal activity isolated by Lemma were named lemma toxins in his honor.

Lemma (1965) stated that the leaves, stem, and bark of the male plant possess higher molluscicidal activity than the female plant, and their potency did not vary with the seasons. The potency of the n-butanol extract was 7–10 times the potency of the aqueous extract. He also observed no change in the molluscicidal property of the arabe and ahiyo varieties of endod berries. The storage stability, checked at regular intervals of six months revealed that the molluscicidal potency of whole and powdered berries remained unchanged even after extended storage up to four years at room temperature (22°C).

Abebe et al. (2005) compared different formulations of *P. dodecandra* for the ease of broadcast and effectiveness in the field and discovered that the spray method is more successful than the drip-feeding method since the spray method displayed 100% mortality of *Biomphalaria pfeifferi* (a freshwater snail).

11.2.5.2 Anthelminthic Property

P. dodecandra is widely used to manage helminthic disease in Uganda (Nalule et al., 2011; Tuwangye and Olila, 2006). Tsehayneh and Melaku, (2019) studied the *in vitro* egg hatchability inhibition effect of *P. dodecandra* against natural infection of ovine GIT nematodes. Nalule et al., (2011), reported the anthelmintic activity of aqueous crude extract of *P. dodecandra* leaves against infected goats (naturally acquired mixed parasites infection) in support their traditional use in veterinary practices.

11.2.5.3 Antimicrobial Property

Extensive studies have been conducted to evaluate the antimicrobial efficacy of *P. dodecandra* (Maatalah et al., 2012; Ogutu et al., 2012; Tadeg et al., 2005; Taye et al., 2011; Tura et al., 2017). The antibacterial activity of *P. dodecandra* leaf extracts was more prominent than its antifungal activity, as opposed to its root extract where the antifungal activity was more pronounced. The methanol extract of its bark was found active against *P. aeruginosa* ATCC 27853 and *Salmonella typhi*. The berry extract (*n*-butanol and aqueous) also showed antifungal activity against *Histoplasma capsulatum* var. *farciminosum* (Mekonnen et al., 2012).

Dichloromethane, ethyl acetate, and aqueous extracts showed activity against *P. aeruginosa* ATCC 27853. The ethyl acetate extract had a mild activity on clinical isolates of *Salmonella typhi*. The aqueous leaf extract of *P. dodecandra* had moderate activity against the clinical isolates of

Microsporum gypseyum KMCC MG 201 and *Trychophyton mentagrophytes* KMCC TM 200 and was mildly active against *Candida albicans* ATCC 90028 (Ogutu et al., 2012). The effect of the aqueous extract of *P. dodecandra* on different strains of dermatophytes was examined by Woldeamanuel et al., (2005) to validate the traditional use of this plant on skin diseases. The minimum inhibitory concentration (MIC) against the dermatophytes studied varied from 0.0195 to 0.156 mg/ml.

11.2.5.4 Antimalarial Property

P. dodecandra possesses antimalarial properties as well and is used for the treatment of malaria in Northwest Ethiopia by tradition (Berhan et al., 2006; Gurmu et al., 2018). Adinew (2014) studied the antimalarial activity of the methanolic extract of *P. dodecandra* leaves against *Plasmodium berghei* in infected Swiss albino mice. The powdered *P. dodecandra* and the aqueous seed extract of *P. dodecandra* at a dose of 50 mg/L in laboratory conditions respectively displayed 100% and 80% mortality of the *Acropora arabensis* larvae (Getachew et al., 2016; Zeleke et al., 2017).

11.2.5.5 Antiviral Property

Koch et al. (1996) attempted to develop protocols to isolate protoplasts from suspension cell cultures of endod and leaves to understand the antiviral mechanism as the cell walls of the *P. dodecandra* leaves sequester ribosome inactivating proteins that display strong antiviral activity.

The antirabies activity was found at higher dose levels in the studies conducted by Admasu et al., (2014), who tested the antirabies activity of hydroethanolic extract of roots and leaves in mice that were infected with rabies virus (CVS-11) and a control group by monitoring their survival rate and period (days).

11.2.5.6 Hepatoprotective Property

Meharie and Tunta, (2021) demonstrated the efficacy of *P. dodecandra* in the management of liver problems by conducting biochemical, histopathological, *in vitro*, and *in vivo* antioxidant studies in mice with CCl4-induced acute liver damage. The root extract of *P. dodecandra* showed significant hepatoprotective and antioxidant activities which are similar to silymarin, an antioxidant compound taken from milk thistle seeds.

11.3 *PICRORHIZA KURROA*

11.3.1 General Description

Picrorhiza kurroa Royle ex Benth is one of the endemic plants found in the alpine region of the Himalayas. It is acknowledged for its medicinal properties in Ayurveda, Siddha, Unani, and folk medicines (Figure 11.2).

P. kurroa is a non-timber forest product (Karki, 2020). It is also used as a substitute for Indian gentian (Kumar and Tewari, 2013). *P. kurroa* is used as raw material in many medicinal preparations. It is traded under the name Katuki or kutaki (Thani, 2021). The annual demand for *P. kurroa* is around 500 tons. Around 90% of this demand is fulfilled through wild plants (Debnath et al., 2020), which leads to overharvesting of the natural habitat. Harvesting of this plant requires long hikes into the forest. Overharvesting the plant has prompted the CITE agreement for its restricted trade across the countries (Mulliken and Crofton, 2008). The updated red list published by the Botanical Survey of Indiahas put *P. kurroa* under the "vulnerable" category (Ayar and Sastry, 1990).

The word "picro" in Greek means bitter and "rhiza" denotes roots as the genus *Picrorhiza* has bitter roots. *P. kurroa* belongs to the plant family Scrophulariaceae (Masood et al., 2015). Various vernacular names for the plants are: Indian gentian, (English), kardi, karoi, karu, karwi, katki, kaur, kuru, kutaki, kutki (Hindi), katukhurohani (Malayalam), katki, kutki, Qusttalakh (Urdu), honlen, hon-len, putsesel (Tibetan), atamvara, katavi, katuka, katukarohini, katuki, katumbhara, katurohani (Sanskrit); Bal kadu, Kali katuki (Marathi); Kadu (Gujarati); Kattki (Bengali); Kaundd (Punjabi);

FIGURE 11.2 *Picrorhiza kurroa.*

Katukarohini (Telgu); Katukarogini (Tamil) (*Indian Medicinal Plants Database*, 2020; Masood et al., 2015; Mulliken and Crofton, 2008; Raina et al., 2021).

11.3.2 Botanical Aspects and Habitat

Picrorhiza kurroa grows naturally at altitudes of 2700–5000 m above sea level in the alpine regions of the Himalaya Mountains. Its cultivation could be achieved at lower altitudes in colder regions. It thrives in the rock crevices of the mountain slopes. It also grows in organic soil (Bhardwaj et al., 2021; Masood et al., 2015; Mulliken and Crofton, 2008; Thani, 2021).

P. kurroa is a dicotyledonous, perennial herb. The plant has a creeping rootstock with only leaves as its aerial part. The leaves of the plant are 5–10 cm long with a coriaceous texture and oval shape. The leaves are basal and alternate. The flowers appear on terminal spikes. The spikes are around 5–10 cm long, sub-calendric, obtuse with brackets that are oblong or lanceolate. The flowers are white or purple, 8 mm long, sessile, zygomorphic, bi-lipped, bisexual, and bloom from June to September. The corolla is actinomorphic, 4–5 mm long, with 4–5 lobes. The stigma is capitated while four slightly didynamous stamens are inserted into the corolla tube. The fruits are 12 mm long, divided into four valves and in the acute capsules tapered at the top. The fruit contains numerous ellipsoid shaped seeds (1mm in size) with a transparent thick seed coat. The rhizome is grayish-brown in color, 25–120 mm long, and 3–10 mm thick. It is straight or slightly curved with longitudinal furrows that look like fish scales. The tip of the rhizome is seen as a bunch of leaves, and it's capable of continued growth. The roots are hard, 0.5–1mm thick, and 50–100 mm long with longitudinal and dotted scars (Bhardwaj et al., 2021; Masood et al., 2015; R. Raina et al., 2010; Sultan et al., 2017; Thani, 2021).

P. kurroa is a self-regenerating plant. Its propagation is through rhizomes or seeds. The seed viability is around 35–60% for six months. The seed prefers a cold climate to germinate (R. Raina et al., 2010). The roots and rhizome of the plants are harvested in September when plants start to wither. The rhizome and rootstock of the plants are bitter in taste and consist of many medicinally important compounds.

11.3.3 Traditional Knowledge

P. kurroa is referred to as kutki in Sanskrit. *Nirukti* means analysis or interpretation of the word in Sanskrit. The *Nirukti* of kutaki is "tikta" and "katuka". Tikta denotes bitterness while katuka refers to increased bile secretion of the human body (D. Raina et al., 2021).

The various formulations of Ayurveda in which kutki is used are Arogyavardhinigutika, Tiktakaghrita, Mahatiktakaghrita, Sarvajvaraharalauha, and Katukadyaghrita. The properties or karma in Ayurveda of Kutaki described in different formulations are Aamhar (carminative), Bhedniya (purgative), Dipana (digestive), Gulmaevum Shoolanashak (pain killer), Hrudya (cardiac tonic), Jwaraghna (antipyretic), Kaas hara (bronchial sedative), Kamlahara (curative of jaundice), Kushtaghna (curative of dermatosis) Lekhaniya (weight reducing), Shvashara (bronchial antspasmodic), and Stanyashodhana (galacto purifying) (D. Raina et al., 2021).

Tibetan folk medicine practitioner, Amchi, uses the rhizome of kutki as a treatment for cough and cold (Masood et al., 2015; Mulliken and Crofton, 2008). Rhizomes and roots of kutki have been used to treat scorpion bites as well as a purgative and laxative agent in Nepal's folk medicine. It is also considered an antiperiodic tonic to stop the reoccurrence of the disease. Kutki is also an ingredient in the medicine for blood, bile, eye and gastric disorders, sore throats, blood disorders, and lung fever in Nepal (Mulliken and Crofton, 2008).

In Pakistan, the herbal essential oil preparations Maajon-e-murravehul-azwah and roghane-qust-talakh, which are used for the treatment of hypothermia, debility, tremors, tetanus, and gout, contain *P. kurroa* as an important ingredient (Masood et al., 2015; Mulliken and Crofton, 2008).

Kutki is used to treat fever and cold in Bhutan as well as in China. The Bhutan National Institute of Medicine uses *P. kurroa* for many medicinal preparations. The Chinese people use kutki for the treatment of malnutrition due to digestive disorders, jaundice, diarrhea, and dysentery (Masood et al., 2015; Mulliken and Crofton, 2008).

11.3.4 Bioactive Compounds and Phytochemistry

The important classes of secondary metabolites reported from *P. kurroa* are iridoids, terpenoids, flavonoids, phenolic acids, and steroids and their glycosides (Bhardwaj et al., 2021). There are more than 50 different compounds that are isolated from various plant parts of *P. kurroa*. The major bioactive compounds reported from *P. kurroa* are iridoids and their glycoside derivatives followed by cucurbitacins and acetophenones (Bhandari et al., 2010; Thani, 2021).

The pharmacologically important iridoids are picrosides or kutikosides. The mixture of picrosides is known as kutkin or picroliv. This comprises of picroside I and II (Upadhyay et al., 2013; P. Verma et al., 2009a). Picrosides III, IV, and V, vernicoside, minecoside, 6-feruloyl catalpol, pikuroside, mussae-nosidic acid, bartsioside, and boschnaloside are other glycoside derivatives of iridoids reported from the plant (Soni and Grover, 2019; Thani, 2021). Biosynthesis of picrosides occurs through a combined biosynthetic route involving non-mevalonate (MEP), mevalonate (MVA), phenylpropanoid, and iridoid pathways (V. Kumar et al., 2017; Shitiz et al., 2015).

Cucurbitacin glycosides B, D, and E are reported from *P. kurroa* (Mallick et al., 2015). The phenolic compounds phenol glycoside picein, apocynin, androsin, and vanillic acid are also reported from the *P. kurroa* (Dorsch et al., 1991; Engels et al., 1992; Krupashree et al., 2014). Carbohydrates such as D mannitol and aromatic compounds cinnamic acid, vanillic acid, and ferulic acid are present in the plant (Masood et al., 2015).

11.3.5 Potential Benefits, Applications, and Uses

The protection reported by Picroliv isolated from *P. kurroa* was similar to that of silymarin in studies using rodent models with hepatic damage induced by galactosamine, paracetamol, thioacetamide, and CCl4 (Verma et al., 2009). The hydroalcoholic extract of the plant inhibited AAPH-induced oxidation of bovine serum albumin and lipid peroxidation of rat hepatic tissues (Krupashree et al., 2014). The ethanolic extract of rhizome and roots inhibited isoniazid and rifampicin induced hepatitis in rats (Jeyakumar et al., 2008). Picroside I and Picroside II isolated from *P. kurroa* showed a significant reduction in the liver lipid content of the galactosamine-induced liver injury in rats (Vaidya et al., 1996).

11.3.5.1 Nephroprotective Activity

Ethanolic extract of the rhizome when an administered at a dose of 600 mg/kg b.w. significantly reduced the elevated serum levels of creatinine and blood urea during Cisplatin (5 mg/kg b.w.i.p.)-induced nephrotoxicity in female Wistar rats (Yamgar et al., 2010). Iridoid glycosides from *P. kurroa* mitigate the cyclophosphamide-induced renal toxicity and peripheral neuropathy in mice via PPAR-γ –mediated pathways (Sharma et al., 2017).

11.3.5.2 Antioxidant Activity

The butanol extracts of *P. kurroa* leaves exhibit antioxidant activity and contain compounds such as luteolin-5-O-glucopyranoside and picein (Kant et al., 2013). The methanolic and aqueous extract of the rhizome showed antioxidant activity (Rajkumar et al., 2011). The levels of antioxidant enzymes, such as catalase, glutathione peroxidase, glutathione-s-transferase, and superoxide dismutase, increased when aqueous extract of the rhizome was administrated to albino mice, treated with 30% carbon tetrachloride for three days, at 250 mg/kg b.w. for ten days (Vinodkumar et al., 2010). Oral administration of ethanolic extract of rhizome and roots prevented adriamycin-induced oxidative stress in rats at 50 mg/kg b.w. for fifteen days (Rajaprabhu et al., 2007). Petroleum ether, dichloromethane, chloroform, ethanol, and aqueous extracts of *P. kurroa* exhibited *in vitro* antioxidant activities (Kalaivani et al., 2010).

11.3.5.3 Anti-Inflammatory and Antiallergic Activities

Extract of *P. kurroa* exhibited anti-inflammatory activity. This is due to the suppression of NF-κB mediated through the suppression of macrophage-derived cytokine signaling (Kumar et al., 2016b). The plant extract reduces edema effectively in albino rats (Kantibiswas et al., 1996). The rhizome extract reduced joint inflammation in adjuvant-induced arthritic rats (Kumar et al., 2016a). Picroliv, derived from *P. kurroa* rhizome, reduced dextran-sulphate-sodium-induced colitis in mice by depressing the expression of IL-1β, TNF-α, and NF-κB p65 (Zhang et al., 2012). Picrosides I, II, and IV, and picrorhizaosides D and E, and minecoside isolated from methanolic extract of rhizome showed hyaluronidase inhibitory activity as good as or better than the widely used antiallergic medicines disodium cromoglycate, ketotifenfumarate, and tranilast (Morikawa et al., 2020).

11.3.5.4 Immunostimulatory Activity

Biopolymeric fractions purified from *P. kurroa* rhizome extract showed a dose-dependent improved immunological response in Balb/c mice (Gupta et al., 2006). Picroliv shielded mice against lipopolysaccharide-induced neuroinflammation by regulating the inflammatory pathway (Li et al., 2018). Alcoholic extract and aqueous extract of the plant rhizome exhibited protective effects against cyclophosphamide-induced immunosuppression in mice (Hussain et al., 2013).

11.3.5.5 Anticarcinogenic and Antineoplastic Activity

The hydro-alcoholic extract of *P. kurroa* inhibited macromolecule damage induced by hydrogen peroxide (Krupashree et al., 2014). The dichloromethane fraction of *P. kurroa* consisting of cucurbitacins B and E, betulinic acid, picrosides 1 and 2, and apocynin displayed potent anticancer activity in Ehrlich ascites carcinoma model in Balb/c mice as well as breast cancer (MCF-7, MDA-MB 231), cervical cancer (HeLa, SiHa) cell lines MDA-MB-435S (human breast carcinoma), Hep3B (human hepatocellular carcinoma), and PC-3 (human prostate cancer) cell lines (Mallick et al., 2015; Rajkumar et al., 2011). Picroside II present in the rhizome of *P. kurroa* inhibited the activity of matrix metalloproteinase 9 (MMP-9) as well as a marker of angiogenesis, the cluster of differentiation 31 (CD31), in MDA-MB-231 cancer cells indicating anticancer and antiangiogenesis activity (Lou et al., 2019)

11.3.5.6 Antidiabetic Activity

Aqueous extract of rhizome of *P. kurroa* increased plasma insulin levels in streptozotocin-induced diabetic rats in 14 days at 100 and 200 mg/kg/day (Husain et al., 2014)

11.3.5.7 Potential Production of Secondary Metabolites for Medicinal or Commercial Purposes

The above biological activities make *P. kurroa* a potential candidate for various drugs. However, its extensive use in traditional medicine has put the natural population at risk. Several strategies have been utilized, both *in vivo* and *in vitro*, to increase its natural population. Micropropagation with the help of different plant regulators has promised an improved plant population (Chandra et al., 2006; Jan et al., 2010). The hairy root cultures have promised high-yield production of picroliv, picrosides, picrolin, and picrotoxins (Mishra et al., 2011; Verma et al., 2015).

11.4 CONCLUSION

Civilizations have depended upon their flora and fauna for managing and curing their ailments and predicaments. This chapter demonstrated the worthiness and the scientific findings of two different plants (*P. dodecandra* and *P. kurroa*) that have been traditionally used as folklore medicine. The traditional knowledge of these plants serves as a fund of resources that can be adapted for medicinal and economic use even in the modern world. Various studies done on these two plants have been discussed briefly in this chapter to throw light on their phytochemical composition and their specific bioactivities. Lemmatoxin and kutkin are the most important phytoconstituents of *P. dodecandra* and *P. kurroa* extracts, respectively. Targeted phytochemicals could be identified, isolated, semi-synthesized, synthesized, or derivatized to form components of formulations from these plants. *P. dodecandra* has proved to be a powerful molluscicide and has also shown several pharmaceutical applications. Validation and clinical trials would be required to commercially realize its potential in medicine. *P. kurroa* is a potential drug (secondary metabolites) source for modern medicine. It is a potential candidate for anticancer drug formulations. Plant biotechnology could be used to conserve its natural population yet produce important metabolites through cell culture. In conclusion, the collective scientific data presented in this chapter on *P. dodecandra* and *P. kurroa* will facilitate the sustainable use of these plants and preservation of traditional knowledge.

REFERENCES

Abebe, F., Erko, B., Gemetchu, T., and Gundersen, S.G. 2005. Control of *Biomphalaria pfeifferi* population and schistosomiasis transmission in Ethiopia using the soap berry endod (*Phytolacca dodecandra*), with special emphasis on application methods. *Transactions of the Royal Society of Tropical Medicine and Hygiene*, 99(10), pp.787–794.

Adams, R.P., Neisess, K.R., Parkhurs, R.M., Makhubu, L.P., and Yohannes, L.W. 1989. *Phytolacca dodecandra* (Phytolaccaceae) in africa: Geographical variation in morphology. *Taxon*, 38(1), pp.17–26.

Adinew, G.M. 2014. Antimalarial activity of methanolic extract of *Phytolacca dodecandra* leaves against *Plasmodium berghei* infected Swiss albino mice. *International Journal of Pharmacology and Clinical Sciences*, 3(3), pp.39–45.

Admasu, P., Deressa, A., Mengistu, A., Gebrewold, G., and Feyera, T. 2014. *In vivo* antirabies activity evaluation of hydroethanolic extract of roots and leaves of *Phytolacca dodecandra*. *Global Veterinaria*, 12(1), pp.12–18.

African Plant Databse. (n.d.). *CJB: African Plant Database.* Retrieved September 5, 2021, from http://www.ville-ge.ch/musinfo/bd/cjb/africa/details.php?langue=an&id=114919

Ayar, M.P., and Sastry, A.R. 1990. *Redlisted Plants.* ENVIS Resource Partner on Biodiversity. http://bsienvis.nic.in/Database/RedlistedPlants_3940.aspx

Beressa, T.B., Ajayi, C.O., Peter, E.L., Okella, H., Ogwang, P.E., Anke, W., and Tolo, C.U. 2020. Pharmacology, phytochemistry, and toxicity profiles of *Phytolacca dodecandra* L'Hér: A scoping review. *Infectious Diseases: Research and Treatment*, 13, pp.1–7.

Berhan, A., Asfaw, Z., and Kelbessa, E. 2006. Ethnobotany of plants used as insecticides, repellents and antimalarial agents in Jabitehnan district, West Gojjam. *SINET: Ethiopian Journal of Science*, 29(1), pp.87–92.

Bhandari, P., Kumar, N., Singh, B., and Ahuja, P.S. 2010. Online HPLC-DPPH method for antioxidant activity of *Picrorhiza kurroa* Royle ex Benth. and characterization of kutkoside by ultra-performance LC-electrospray ionization quadrupole time-of-flight mass spectrometry. *Indian Journal of Experimental Biology*, 48(3), pp.323–328.

Bhardwaj, A., Sharma, A., Cooper, R., Bhardwaj, G., Gaba, J., Mutreja, V., and Chauhan, A. 2021. A comprehensive phytochemical, ethnomedicinal, pharmacological ecology and conservation status of *Picrorhiza kurroa* Royle ex Benth.: An endangered Himalayan medicinal plant. *Process Biochemistry*, 109, pp.72–86.

Birrie, H., Balcha, F., Erko, B., Bezuneh, A., and Gemeda, N. 1998. Investigation into the cercariacidal and miracidiacidal properties of Endod (*Phytolacca dodecandra*) berries (Type 44). *East African Medical Journal*, 75(5), pp.311–314.

Boeke, S.J., Boersma, M.G., Alink, G.M., Van Loon, J.J.A., Van Huis, A., Dicke, M., and Rietjens, I.M.C.M. 2004. Safety evaluation of neem (*Azadirachta indica*) derived pesticides. *Journal of Ethnopharmacology*, 94(1), pp.25–41.

Chandra, B., Palni, L.M.S., and Nandi, S.K. 2006. Propagation and conservation of *Picrorhiza kurrooa* Royle ex Benth.: An endangered himalayan medicinal herb of high commercial value. *Biodiversity & Conservation 15*(7), pp.2325–2338.

Debnath, P., Rathore, S., Walia, S., Kumar, M., Devi, R., and Kumar, R. 2020. *Picrorhiza kurroa*: A promising traditional therapeutic herb from higher altitude of western Himalayas. *Journal of Herbal Medicine* 23, p.100358.

Domon, B., and Hostettmann, K. 1984. New Saponins from *Phytolacca dodecandra* l'HERIT. *Helvetica Chimica Acta*, 67(5), pp.1310–1315.

Dorsch, W., Stuppner, H., Wagner, H., Gropp, M., Demoulin, S., and Ring, J. 1991. Antiasthmatic effects of *Picrorhiza kurroa*: Androsin prevents allergen- and PAF-induced bronchial obstruction in guinea pigs. *International Archives of Allergy and Immunology*, 95(2–3), pp.128–133.

Engels, F., Renirie, B.F., 't Hart, B.A., Labadie, R.P., and Nijkamp, F.P. 1992. Effects of apocynin, a drug isolated from the roots of *Picrorhiza kurroa*, on arachidonic acid metabolism. *FEBS Letters*, 305(3), pp.254–256.

Esser, K.B., Semagn, K., and Wolde-Yohannes, L. 2003. Medicinal use and social status of the soap berry endod (*Phytolacca dodecandra*) in Ethiopia. *Journal of Ethnopharmacology*, 85(2–3), pp.269–277.

Getachew, D., Balkew, M., and Gebre-Michael, T. 2016. Evaluation of endod (*Phytolacca dodecandra*: Phytolaccaceae) as a larvicide against *Anopheles arabiensis*, the principal vector of malaria in Ethiopia. *Journal of American Mosquito Conrol Association*, 32(2), pp.124–129.

Gilbert, K.G., and Cooke, D.T. 2001. Dyes from plants: Past usage, present understanding and potential. *Plant Growth Regulation*, *34*, pp.57–69.

Goll, P.H., Lemma, A., Duncan, J., and Mazengia, B. 1983. Control of schistosomiasis in Adwa, Ethiopia, using the plant molluscicide endod (*Phytolacca dodecandra*). *Tropenmedizin Und Parasitologie*, 34(3), pp.177–183.

Gupta, A., Khajuria, A., Singh, J., Bedi, K.L., Satti, N.K., Dutt, P., Suri, K.A., Suri, O.P., and Qazi, G.N. 2006. Immunomodulatory activity of biopolymeric fraction RLJ-NE-205 from *Picrorhiza kurroa*. *International Immunopharmacology*, 6(10), pp.1543–1549.

Gurmu, A.E., Kisi, T., Shibru, H., Graz, B., and Willcox, M. 2018. Treatments used for malaria in young Ethiopian children: A retrospective study. *Malaria Journal*, 17(1), pp.1–8.

Husain, G.M., Rai, R., Rai, G., Singh, H.B., Thakur, A.K., and Kumar, V. 2014. Potential mechanism of anti-diabetic activity of *Picrorhiza kurroa*. *Tang [Humanitas Medicine]*, 4(4), pp.27.1–27.5.

Hussain, A., Shadma, W., Maksood, A., and Ansari, S.H. 2013. Protective effects of *Picrorhiza kurroa* on cyclophosphamide-induced immunosuppression in mice. *Pharmacognosy Research*, 5(1), p.30.

Indian Medicinal Plants Database. 2020. *National Medicinal Plants Board*. http://www.medicinalplants.in/

Jan, A., Thomas, G., Shawl, A.S., Jabeen, N., and Kozgar, M.I. 2010. Improved micropropagation protocol of an endangered medicinal plant-*Picrorhiza kurroa* Royle ex Benth. promptly through auxin treatments. *Chiang Mai Journal of Science*, 37(2), pp.304–313.

Jeyakumar, R., Rajesh, R., Meena, B., Rajaprabhu, D., Ganesan, B., Buddhan, S., and Anandan, R. 2008. Antihepatotoxic effect of *Picrorhiza kurroa* on mitochondrial defense system in antitubercular drugs (isoniazid and rifampicin)-induced hepatitis in rats. *Journal of Medicinal Plants Research*, 2(1), pp.017–019.

Kalaivani, T., Rajasekaran, C., and Mathew, L. 2010. *In vitro* free radical scavenging potential of *Picrorhiza kurroa*. *Journal of Pharmacy Research*, 3(4), pp.849–854.

Kant, K., Walia, M., Agnihotri, V.K., Pathania, V., and Singh, B. 2013. Evaluation of antioxidant activity of *Picrorhiza kurroa* (leaves) extracts. *Indian Journal of Pharmaceutical Sciences*, 75(3), p.324.

Kantibiswas, T., Marjit, B., and Maity, L.N. 1996. Effect of *Picrorhiza kurroa* Benth. In acute inflammation. *Ancient Science of Life*, 16(1), pp.11–14.

Karki, M.B. 2020. Harnessing the potential of medicinal, aromatic and non-timber forest products for improving the livelihoods of pastoralists and farmers in Himalayan Mountains. In: *Conservation and Utilization of Threatened Medicinal Plants*. Rajasekharan P., Wani S. (eds), Springer, Cham, pp.93–106.

Karunamoorthi, K., Bishaw, D., and Mulat, T. 2008. Laboratory evaluation of Ethiopian local plant *Phytolacca dodecandra* extract for its toxicity effectiveness against aquatic macroinvertebrates. *European Review for Medical and Pharmacological Sciences*, 12(6), pp.381–386.

Koch, P.E., Bonness, M.S., Lu, H., and Mabry, T.J. 1996. Protoplasts from *Phytolacca dodecandra* L'Herit (endod) and *P. americana* L. (pokeweed). *Plant Cell Reports*, 15(11), pp.824–828.

Krupashree, K., Hemanth Kumar, K., Rachitha, P., Jayashree, G.V., and Khanum, F. 2014. Chemical composition, antioxidant and macromolecule damage protective effects of *Picrorhiza kurroa* Royle ex Benth. *South African Journal of Botany*, 94, pp.249–254.

Kumar, R., and Tewari, L. 2013. Studies on natural resources, trade and conservation of Kutki (*Picrorhiza kurroa* Royle ex Benth., Scrophulariaceae) from Kumaun Himalaya. *Scientific Research and Essays*, 8(14), pp.575–580.

Kumar, R., Gupta, Y.K., Singh, S., and Arunraja, S. 2016a. *Picrorhiza kurroa* inhibits experimental arthritis through inhibition of pro-inflammatory cytokines, angiogenesis and MMPs. *Phytotherapy Research*, 30(1), pp.112–119.

Kumar, R., Gupta, Y.K., Singh, S., and Raj, A. 2016b. Anti-inflammatory effect of *Picrorhiza kurroa* in experimental models of inflammation. *Planta Medica*, 82(16), pp.1403–1409.

Kumar, V., Bansal, A., and Chauhan, R.S. 2017. Modular design of Picroside-II biosynthesis deciphered through NGS transcriptomes and metabolic intermediates analysis in naturally variant chemotypes of a medicinal herb, *Picrorhiza kurroa*. *Frontiers in Plant Science*, 8, p.564.

Lambert, J.D.H., Temmink, J.H.M., Marquis, J., Parkhurst, R.M., Lugt, C.B., Lemmich, E., Wolde-Yohannes, L., and de Savigny, D. 1991. Endod: safety evaluation of a plant molluscicide. *Regulatory Toxicology and Pharmacology*, 14(2), pp.189–201.

Legère, K. 2009. Plant Names in the Tanzanian Bantu Language Vidunda: Structure and (Some) Etymology. *Proceedings of the 38th Annual Conference on African Linguistics. Cascadilla Proceedings Project*, pp.217–228.

Lemma, A. 1965. A preliminary report on the molluscicidal property of endod (*Phytolacca dodecandra*). *Ethiopian Medical Journal*, 3(4), pp.187–190.

Lemma, A. 1970. Laboratory and field evaluation of the molluscicidal properties of *Phytolacca dodecandra*. *Bulletin of the World Health Organization*, 42(4), pp.597–612.

Li, L., Jin, X., Zhang, H., and Yin, J. 2018. Protective effect of picroliv against lipopolysaccharide-induced cognitive dysfunction and neuroinflammation by attenuating TLR4/NFκB pathway. *Folia Neuropathologica*, 56(4), pp.337–345.

Lou, C., Zhu, Z., Xu, X., Zhu, R., Sheng, Y., and Zhao, H. 2019. Picroside II, an iridoid glycoside from *Picrorhiza kurroa*, suppresses tumor migration, invasion, and angiogenesis *in vitro* and *in vivo*. *Biomedicine & Pharmacotherapy*, 120, p.109494.

Maatalah, M.B., Bouzidi, N.K., Bellahouel, S., Merah, B., Fortas, Z., Soulimani, R., Saidi, S., and Derdour, A. 2012. Antimicrobial activity of the alkaloids and saponin extracts of *Anabasis articulata*. *E3 Journal of Biotechnology and Pharmaceutical Research*, 3(3), pp.54–57.

Mallick, M.N., Singh, M., Parveen, R., Khan, W., Ahmad, S., ZeeshanNajm, M., and Husain, S.A. 2015. HPTLC analysis of bioactivity guided anticancer enriched fraction of hydroalcoholic extract of *Picrorhiza kurroa*. *BioMed Research International*, 2015, p.513875.

Masood, M., Arshad, M., Qureshi, R., Sabir, S., Amjad, M.S., Qureshi, H., and Tahir, Z. 2015. *Picrorhiza kurroa*: An ethnopharmacologically important plant species of Himalayan region. *Pure and Applied Biology*, 4(3), pp.407–417.

Matebie, W.A., Zhang, W., and Xie, G. 2019. Chemical composition and antimicrobial activity of essential oil from *Phytolacca dodecandra* collected in Ethiopia. *Molecules*, 24(2), p.342.

Meharie, B.G., and Tunta, T.A. 2021. *Phytolacca dodecandra* (Phytolaccaceae) root extract exhibits antioxidant and hepatoprotective activities in mice with CCl4-induced acute liver damage. *Clinical and Experimental Gastroenterology*, 14, pp.59–70.

Mekonnen, N., Makonnen, E., Aklilu, N., and Ameni, G. 2012. Evaluation of berries of *Phytolacca dodecandra* for growth inhibition of *Histoplasma capsulatum* var. *farciminosum* and treatment of cases of epizootic lymphangitis in Ethiopia. *Asian Pacific Journal of Tropical Biomedicine*, 2(7), pp.505–510.

Mishra, J., Bhandari, H., Singh, M., Rawat, S., Agnihotri, R.K., Mishra, S., and Purohit, S. 2011. Hairy root culture of *Picrorhiza kurroa* Royle ex Benth.: A promising approach for the production of picrotin and picrotoxinin. *Acta Physiologiae Plantarum*, 33, pp.1841–1846.

Morikawa, T., Nakanishi, Y., Inoue, N., Manse, Y., Matsuura, H., Hamasaki, S., Yoshikawa, M., Muraoka, O., and Ninomiya, K. 2020. Acylated iridoid glycosides with hyaluronidase inhibitory activity from the rhizomes of *Picrorhiza kurroa* Royle ex Benth. *Phytochemistry*, 169, p.112185.

Mulliken T. & Crofton P. 2008. Review of the status, harvest, trade and management of seven Asian CITES-listed medicinal and aromatic plant species. *Federal Agency for Nature Conservation*, 1(1), pp 61–76.

Nalule, A.S., Karue, C.N., and Katunguka-Rwakishaya, E. 2011. Anthelmintic activity of *Phytolacca dodecandra* and *Vernonia amygdalina* leaf extracts in naturally infected small East African goats. *Livestock Research for Rural Development*, 23(12), p.244.

Nalule, A.S., Mbaria, J.M., Olila, D., and Kimenju, J.W. 2011. Ethnopharmacological practices in management of livestock helminthes by pastoral communities in the drylands of Uganda. *Livestock Research for Rural Development*, 23(2), 1–11.

Namulindwa, A., Nkwangu, D., and Oloro, J. 2015. Determination of the abortifacient activity of the aqueous extract of *Phytolacca dodecandra* (L'Her) leaf in Wistar rats. *African Journal of Pharmacy and Pharmacology*, 9(1), pp.43–47.

Ogutu, A., Lilechi, D., Mutai, C., and Bii, C. 2012. Phytochemical analysis and antimicrobial activity of *Phytolacca dodecandra, Cucumis aculeatus* and *Erythrina excelsa*. *International Journal of Biological and Chemical Sciences*, 6(2), pp.692–704.

Parkhurst, R.M., Thomas, D.W., Skinner, W.A., and Cary, L.W. 1974. Molluscicidal Saponins of *Phytolacca dodecandra*: Lemmatoxin. *Canadian Journal of Chemistry*, 52(5), pp.702–705.

Plants of the World Online. (n.d.). *Phytolacca dodecandra* L'Hér. Kew Science. Retrieved September 5, 2021, from http://www.plantsoftheworldonline.org/taxon/urn:lsid:ipni.org:names:676349-1#source-FTEA

Poucher, W.A. 1993. The production of natural perfumes. In *Perfumes, Cosmetics and Soaps*. Dordrecht: Springer, pp. 16–40.

Powell, J.W., and Whalley, W.B. 1969. Triterpenoid saponins from *Phytolacca dodecandra*. *Phytochemistry*, 8(10), pp.2105–2107.

Raina, D., Raina, S., and Singh, B. 2021. Katuki (*Picrorhiza kurroa*): A promising ayurvedic herb. *Biomedical Journal of Scientific & Technical Research*, 36(1), pp.28238–28242.

Raina, R., Chand, R., and Singh Parmar, Y. 2010. Reproductive biology of *Picrorhiza kurroa*: A critically endangered high value temperate medicinal plant. *Open Access Journal of Medicinal and Aromatic Plants*, 1(2), pp.40–43.

Rajaprabhu, D., Rajesh, R., Jeyakumar, R., Buddhan, S., Ganesan, B., and Anandan, R. 2007. Protective effect of *Picrorhiza kurroa* on antioxidant defense status in adriamycin-induced cardiomyopathy in rats. *Journal of Medicinal Plant Research*, 1(4), pp.80–085.

Rajkumar, V., Guha, G., and Ashok Kumar, R. 2011. Antioxidant and anti-neoplastic activities of *Picrorhiza kurroa* extracts. *Food and Chemical Toxicology*, 49(2), pp.363–369.

Report: *Phytolacca dodecandra* Taxonomic Serial No.: 506583. (n.d.). *Integrated Taxonomic Information System: Report*. Retrieved September 5, 2021, from https://www.itis.gov/servlet/SingleRpt/SingleRpt?search_topic=TSN&search_value=506583#null

Schmelzer, G. & Gurib-Fakim, A. 2008. Plant resources of tropical Africa 11(1). *Medicinal Plants*, 1.

Sharma, S., Sharma, P., Kulurkar, P., Singh, D., Kumar, D., andPatial, V. 2017. Iridoid glycosides fraction from *Picrorhiza kurroa* attenuates cyclophosphamide-induced renal toxicity and peripheral neuropathy via PPAR-γ mediated inhibition of inflammation and apoptosis. *Phytomedicine*, 36, pp.108–117.

Shitiz, K., Sharma, N., Pal, T., Sood, H., and Chauhan, R.S. 2015. NGS transcriptomes and enzyme inhibitors unravel complexity of picrosides biosynthesis in *Picrorhiza kurroa* Royle ex. Benth. *PLOS ONE*, 10(12), p.e0144546.

Soni, D., and Grover, A. 2019. "Picrosides" from *Picrorhiza kurroa* as potential anti-carcinogenic agents. *Biomedicine and Pharmacotherapy*, 109, pp.1680–1687.

Spengel, S.M. 1996. Two pentacyclic triterpenes from *Phytolacca dodecandra* roots. *Phytochemistry*, 43(1), pp.179–182.

Stobaeus, J.K., Heath, G.E., Parkhurst, R.M., Jones, W.O., and Webster, J.E. 1990. A laboratory study of the toxcity of the butanol extract of endod (*Phytolacca dodecandra*) on two species of freshwater fish and two species of aquatic snails. *Veterinary and Human Toxicology*, 32(3), pp.212–216.

Sultan, P., Rasool, S., and Hassan, Q.P. 2017. *Picrorhiza kurroa* Royle ex Benth. A plant of diverse pharmacological potential. *Annals of Phytomedicine: An International Journal*, 6(1), pp.63–67.

Tadeg, H., Mohammed, E., Asres, K., and Gebre-Mariam, T. 2005. Antimicrobial activities of some selected traditional Ethiopian medicinal plants used in the treatment of skin disorders. *Journal of Ethnopharmacology*, 100(1–2), pp.168–175.

Taye, B., Giday, M., Animut, A., and Seid, J. 2011. Antibacterial activities of selected medicinal plants in traditional treatment of human wounds in Ethiopia. *Asian Pacific Journal of Tropical Biomedicine*, 1(5), pp.370–375.

Thani, P.R. 2021. A comprehensive review on *Picrorhiza kurroa* Royle ex Benth. *Journal of Pharmacognosy and Phytochemistry*, 10(3), pp.307–313.

Thiilborg, S.T., Brøgger Christensen, S., Cornett, C., Olsen, C.E., and Lemmich, E. 1994. Molluscicidal saponins from a zimbabwean strain of *Phytolacca dodecandra. Phytochemistry*, 36(3), pp.753–759.

Thiilborg, S.T., Christensen, S.B., Cornett, C., Olsen, C.E., and Lemmich, E. 1993. Molluscicidal saponins from *Phytolacca dodecandra. Phytochemistry*, 32(5), pp.1167–1171.

Tsehayneh, B., and Melaku, A. 2019. In vitro egg hatchability inhibition effect of *Albizia gummifera*, *Phytolacca dodecandra*, and *Vernonia amygdalina* against natural infection of ovine GIT nematodes. *Journal of Medicinal Botany*, 3, pp.4–6.

Tura, G.T., Eshete, W.B., and Tucho, G.T. 2017. Antibacterial efficacy of local plants and their contribution to public health in rural Ethiopia. *Antimicrobial Resistance & Infection Control*, 6(1), pp.1–7.

Tuwangye, I., and Olila, D. 2006. The anthelmintic activity of selected indigenous medicinal plants used by the Banyankole of Western Uganda. In *Journal of Animal and Veterinary Advances*, 5(8), pp.712–717.

Upadhyay, D., Dash, R.P., Anandjiwala, S., and Nivsarkar, M. 2013. Comparative pharmacokinetic profiles of picrosides I and II from kutkin, *Picrorhiza kurroa* extract and its formulation in rats. *Fitoterapia*, 85(1), pp.76–83.

Vaidya, Antarkar, D., Doshi, J., Bhatt, A., Ramesh, V., Vora, P., Perissond, D., and Baxi, A. 1996. *Picrorhiza kurroa* (kutaki) Royle ex. Benth as a hepatoprotective agent: Experimental & clinical studies. *Journal of Postgraduate Medicine*, 42(4), pp.105–108.

Verma, P., Basu, V., Gupta, V., Saxena, G., and Ur Rahman, L. 2009a. Pharmacology and chemistry of a potent hepatoprotective compound picroliv isolated from the roots and rhizomes of *Picrorhiza kurroa* Royle ex Benth. (Kutki). *Current Pharmaceutical Biotechnology*, 10(6), pp.641–649.

Verma, P.C., Singh, H., Negi, A.S., Saxena, G., Rahman, L., and Banerjee, S. 2015. Yield enhancement strategies for the production of picroliv from hairy root culture of *Picrorhiza kurroa* Royle ex Benth. *Plant Signaling & Behavior*, 10(5), pp.1–11.

Vinodkumar, P., Sivraja, A., and Senthilkumar, B. 2010. Hepatoprotective and antioxidant properties of aqueous rhizome extracts of *Picrorhiza kurroa* on CCl 4 induced liver toxicity in albino rats. *Journal Of Pharmacy Research*, 3(6), pp.1280–1282.

Woldeamanuel, Y., Abate, G., and Chryssanthou, E. 2005. *In vitro* activity of *Phytolacca dodecandra* (Endod) against dermatophytes. *Ethiopian Medical Journal*, 43(1), pp.31–34.

Yamgar, S., Sali, L., Salkar, R., Jain, N., and Gadgoli, C.H. 2010. Studies on nephroprotective and nephrocurative activity of ethanolic extract of *Picrorhiza kurroa* royle and arogyawardhinibati in rats. *International Journal Of Pharmacy & Technology*, 2(3), pp.472–489.

Yang, R.Z., and Tang, C.S. 1988. Plants used for pest control in China: A literature review. *Economic Botany*, 42(3), pp.376–406.

Zeleke, A.J., Shimo, B.A., and Gebre, D.Y. 2017. Larvicidal effect of endod (*Phytolacca dodecandra*) seed products against *Anopheles arabiensis* (Diptera: Culicidae) in Ethiopia. *BMC Research Notes*, 10, p.449.

Zhang, D.K., Yu, J.J., Li, Y.M., Wei, L.N., Yu, Y., Feng, Y.H., and Wang, X. 2012. A *Picrorhiza kurroa* derivative, picroliv, attenuates the development of dextran-sulfate-sodium-induced colitis in mice. *Mediators of Inflammation*, 2012, p.751629.

12 *Piper longum* (Long Pepper or Pipli) and *Tinospora cordifolia* (Giloy or Heart-Leaved Moonseed)

Yashashree Pradhan, Hina Alim, Nimisha Patel, Kamal Fatima Zahra, Belkıs Muca Yiğit, Johra Khan, and Ahmad Ali

CONTENTS

12.1 Introduction 218
12.2 Description 218
12.2.1 *Piper longum*, or Long Pepper 219
12.2.2 *Tinospora cordifolia*, Giloy, or Heart-Leaved Moonseed 219
12.3 Traditional Knowledge 220
12.3.1 *Piper longum* 220
12.3.2 *Tinospora cordifolia* 221
12.4 Chemical Derivatives 222
12.4.1 Phytochemicals of *P. longum* 222
12.4.2 Phytochemicals of *T. cordifolia* 223
12.5 Potential Benefits 224
12.5.1 Uses of *P. longum* 224
12.5.1.1 Antiulcer Activity 225
12.5.1.2 Insecticidal Activity 225
12.5.1.3 Antiplatelet Activity 225
12.5.1.4 Anti-Snake Venom Activity 225
12.5.1.5 Coronary Vasodilation and Cardioprotective Activity 225
12.5.1.6 Anti-amebic and Antihelminthic Activity 225
12.5.1.7 Antimicrobial Activity 225
12.5.1.8 Anti-obesity and Hypocholesterolemic Activity 225
12.5.1.9 Antidepressant Activity 226
12.5.1.10 Anticancer and Antitumor Activity 226
12.5.1.11 Radioprotective Activity 226
12.5.1.12 Immunomodulatory and Activity Against COVID-19 226
12.5.1.13 Hepatoprotective Activity 226
12.5.1.14 Anti-inflammatory and Anti-arthritic Activity 226
12.5.1.15 Anti-apoptotic and Antioxidant Activity 226
12.5.1.16 Antiasthmatic and Analgesic Activity 226
12.5.1.17 Antidiabetic activity 227
12.5.1.18 Melanin-Inhibiting and Antifertility Activity 227
12.5.1.19 Antiepileptic and Therapeutic Activity for Alzheimer's Disease 227
12.5.1.20 Antiparkinsonian Activity 227

DOI: 10.1201/9781003205067-12

12.5.2 Uses of *T. cordifolia* 228
12.5.2.1 Anti-ulcer Activity 228
12.5.2.2 Anticancer and Antitumor Activity 228
12.5.2.3 Antimicrobial Activity 228
12.5.2.4 Anti-Inflammatory and Anti-Stress Activity 228
12.5.2.5 Immunomodulatory and Activity against COVID-19 228
12.5.2.6 Hepatoprotective and Anti-Amebic Activity 228
12.5.2.7 Anti-arthritic and Anti-osteoporotic Activity 229
12.5.2.8 Antidiabetic and Hypolipidemic Activity 229
12.5.2.9 Anti-HIV and Wound-Healing Activity 229
12.5.2.10 Antioxidant and Antitoxic Activity 229
12.5.2.11 Antiparkinsonian and Memory-Enhancing Activity 229
12.5.2.12 Against Urinary Calculi and Uremia 229
12.6 Conclusion 229
References 230

12.1 INTRODUCTION

Medicinal plants have been used for the treatment of many diseases from ancient times. These plants are thought to have more value in traditional medicine. After understanding the side effects of synthetic medicines, both common people and researchers again turned their attention toward herbal medicine (Kim, 2005). The studies characterizing nutritional and therapeutic potential of medicinal plants can aid in overcoming the side effects of synthetic drugs as well as immunosuppression. Medicinal plants are rich in phytochemicals, which give a wide range of therapeutic activities. Leaves, bark, fruits, and roots are useful for the treatment of many illnesses (Ali, 2020). Nowadays, modern medicine has commercialized by producing various efficient drugs from medicinal plants. Apart from allopathic practices, herbal product demand has increased to over 5 billion dollars in the US market. Apart from the side effects of long-term use of allopathic medicines, an important reason for the increasing interest in herbal medicines can contribute to factors like cultural acceptability, better compatibility as well as feasibility. The traditional knowledge of many plants is more accessible, reliable, and trusted by common people all over the world (Vidyarthi et al., 2013). Many herbalists formulate herbal medicines by knowing the medicinal properties and effects of plants. The traditional system lacks uniformity, documentation, and maintenance. To evolve traditional medicine as an alternative to modern medicine, a standard scientific protocol is important along with strong clinical validation, analysis, and documentation of the efficacy of herbal products (Kamboj, 2000).

Two of these plants are long pepper or pipli (*Piper longum*) and giloy or heart-leaved moonseed (*Tinospora cordifolia*). These plants have huge importance in Ayurveda due to their properties which give strength and protection to various systems of the body (Gogte, 1982). These two plants contain phytochemicals which act as bioactive compounds. These bioactive compounds show various activities against a number of diseases. Due to increasing use and demand of these plants, sometimes adulteration in these products may also occur. To overcome these malpractices, these plants are cultivated not only in India but also in many countries. Nowadays, modern methods, like micropropagation and plant tissue cultures are used. This chapter discusses traditional knowledge, bioactive compounds, and medicinal properties of *P. longum* and *T. cordifolia* (Chitme et al., 2003). These plants contain various phytochemicals and micronutrients, which aid in fighting against many diseases. As per the report of AYUSH ministry, an immunomodulatory action of these two plants helps in fighting against the current pandemic of novel coronavirus COVID-19 (Ministry of AYUSH, 2020).

12.2 DESCRIPTION

Piper longum and *Tinospora cordifolia* are medicinal herbs that are used to cure many diseases due to their wide spectrum therapeutic activity. These plants are rich in phytochemicals. *P. longum*, or pipli, contains alkaloids, saponins, starch, protein, volatile oils, amygdalin, and many more (Dhanalakshmi et al., 2016); whereas, Giloy, or heart-leaved moonseed, contains steroids, sesquiterpenoids, alkaloids, a mixture of fatty acids, phosphorous, polysaccharides, calcium, etc. (Spandana et al., 2013).

12.2.1 *Piper longum*, or Long Pepper

P. longum (Family: Piperaceae) is an aromatic, perennial, and slender climber. The plant grows into a shrub with large woody roots and numerous creeping and jointed stems which are thickened at the nodes (Choudhary and Singh, 2018). Leaves are wide ovate and cordate. It has cylindrical, pedunculate spike inflorescence. The male flower is larger and slender whereas the female flower is up to 2.5 cm long and 4–5 mm in diameter. The fruits, embedded in fleshy spikes, are small, ovoid berries, and shiny blackish-green in appearance (Zaveri et al., 2010). Fruits are globose with pungent taste and aromatic odor. It is local in the Indo-Malaya region. It is now mainly cultivated in Indonesia, India, Sri Lanka, Nepal, and Malaysia. In India, central Himalayas to Assam, evergreen forests of Western Ghats, Mikir Hills and Khasi, and lower hills of West Bengal (Navneet and Singh, 2018) (Figure 12.1).

12.2.2 *Tinospora cordifolia*, Giloy, or Heart-Leaved Moonseed

T. cordifolia (Family: Menispermaceae) is commonly known as giloy or heart-leaved moonseed. Giloy is a glabrous, succulent, woody, climbing shrub. It is a large deciduous, wide-spreading plant with numerous coiling branches. The stem is fleshy, succulent, and filiform whereas it comprises tetra to penta-arched aerial roots (Raghunathan and Sharma, 1969). It has simple, alternate, long petioled, exstipulate, heart-shaped leaves, and the flowers are greenish-yellow, unisexual where male flowers are clustered, and female flowers exist in solitary inflorescence. Fruits are orange-red in color. Giloy is widely distributed in tropical regions of India, from Kumaon to Assam, Konkan, Kerala, Deccan, Karnataka, Bihar, and West Bengal (Sinha et al., 2004). It is also found in countries like Bangladesh, Pakistan, and Sri Lanka. It is also found in southeast Asian countries like Malaysia, Indonesia, and Tamil Nadu (Spandana et al., 2013) (Figure 12.2).

FIGURE 12.1 *Piper longum.*

FIGURE 12.2 *Tinospora cordifolia.*

12.3 TRADITIONAL KNOWLEDGE

Nowadays, due to increased industry, pollution, deforestation, and changed lifestyles of people, many health-related problems are emerging. The main reasons for these health problems are lack of immunity and generalized debility. For combating such situations, many herbs have proved useful in traditional medicine. The leading two of them in improving immunity and general debility are *P. longum* and *T. cordifolia.*

12.3.1 *PIPER LONGUM*

Piper longum, or long pepper, is mostly used in traditional medicine for treatment of many diseases. It is known as "pipali" in Sanskrit. According to Ayurvedic medicine, it has properties as follows: rasa-katu, i.e., pungent; veerya-Anushnashita, i.e., slightly cold; vipaka-Madhur, i.e., sweet; guna-snigdha, i.e., unctuous; Laghu, i.e., light; tikshna, i.e., sharp; and dosha, i.e., calms *vata* and *kapha* (Zaveri et al., 2010).

Stems and roots of long pepper are used as medicine in Ayurveda and Arabian medicine. A portion of the root is used for parturition, to assist in the expulsion of the placenta. The roots of *P. longum* are used for the management of heart diseases. The fruit is used as an abortifacient, antiasthmatic, stomachic, liver tonic, aphrodisiac, laxative, antidiarrheal, antidysenteric, and anti-bronchitis (Johri and Zutshi, 1992). Fruit also helps in treatment of abdominal complaints, urinary discharges, tumors, diseases of the spleen, pains, inflammation, leprosy, insomnia, jaundice, and hiccups. Roots are also used in the treatment of tuberculosis. The roasted fruits of *P. longum* are beaten up with honey and are used for treatment of rheumatism (Parmar et al., 1997). In the treatment of acute and chronic bronchitis, the decoction of root and dried young fruits are used. The preparation of liniment using long pepper boiled with ginger, mustard oil, buttermilk, and curds is used in cases of paralysis (Navneet and Singh, 2018). The roots and fruits are also used in gout, palsy, and lumbago. In Indian systems of medicine, such as Unani medicine, it is used for the treatment of snakebite, night-blindness, and scorpion sting. Fruits also have stimulant properties, which act as an alternative tonic in paraplegia, enlargement of the spleen, and other abdominal viscera, as well as in chronic cough (Choudhary and Singh, 2018). In Chinese medicine, *P. longum* is used for treatment of stomachache, rhinitis, vomiting, and headache (Gajurel et al., 2021). It is taken as a decoction of three long peppers with honey on the first day, gradually increasing by three every day until the tenth day, at which time 30 long peppers are taken in one dose. Then the number of long peppers gradually decrease by three every day until the last day, when only three long peppers are taken. Then the medicine is stopped (Chauhan et al., 2011). This is also called "vardhaman pippali" (Garde, 1983).

As per the properties described in Ayurveda, long pepper acts on various systems of the body by various mechanisms and actions. Due to its rasa, vipaka, and veerya, it normalizes all three doshas (vata, pitta, kapha); hence, it is called "tridoshaghna". Locally, due to its tikshna guna, it is used as a pain reliever in inflamed conditions. Internally, it acts on various systems. It is also known as rasayana, i.e., nourishing and strengthening all systems (Gogte, 1982).

P. longum acts is nutritious and improves vigor and vitality. Due to its snigdha guna, it acts as a laxative. Katu rasa of long pepper helps to improve appetite, and tikshna guna helps to reduce fullness of the stomach. Being snigdha and ushna, it performs as a peristaltic regulator, pain reliever, and dewormer. All these effects combined helps to improve taste, appetite, indigestion, and reduces piles. It is also used in treating splenomegaly and ascites. *P. longum* is used to treat hyperacidity and gastritis. Pippali, jaggary, honey, and milk are cooked and taken for treatment of gastritis; whereas, honey + pippali – or ghee prepared from pippali decoction taken with honey – helps to treat hyperacidity (Deshpande et al., 2018).

With the help of katu rasa and Madhur vipaka, pippali acts on hematopoiesis, thereby increasing hemoglobin, so it is used in treating anemia and bone marrow suppression. It is also used in many blood-borne diseases. *P. longum* is used in the treatment of pyrexia of various origins. Powdered mixture of long pepper and Indian gooseberry is specifically used in the treatment of anemia. Ghee prepared from pippali reduces loss of taste, headache, cough, breathlessness, and pleural and lung pain. Milk cooked with *P. longum* is useful in chronic fever. Vardhaman pippali is considered the best and permanent treatment for recurrent fever, chronic bronchitis fever, and fever due to tuberculosis (Deshpande et al., 2018).

Due to medhya action, it acts as nervine tonic. It improves the capacity of the brain, thereby increasing grasping and memory. Due to its vataghna action, i.e., pacifying the vata dosha, it acts on joint pain and neurological abnormalities. Due to its pain-relieving action, a decoction of *P. longum* is used for treatment of rheumatoid arthritis. This decoction is also used for treatment of sciatica and hemiplegia (Deshpande et al., 2018). Being nutritious, *P. longum* increases power and the capacity of muscles. In malnourished conditions, ghee cooked with pippali is used (Gogte 1982).

With snigdha, ushna guna, and nutritious properties, *P. longum* improves the capacity of the larynx, thereby helping in treatment of coughs of various etiology. Pippali, black resin, and ghee taken along with honey is used for the treatment of cough due to tuberculosis. Long pepper shows a bactericidal effect against *M. tuberculosis*. One of the known preparations for the treatment of tuberculosis cough is sitopaladi churna. Long pepper is one of the best treatments for bronchitis and bronchial asthma. Powder of pippali, honey, and ghee are used to reduce tachypnea and bronchospasm. It reduces bronchial obstruction by its mucolytic action (Gogte, 1982). *P. longum* acts as an aphrodisiac in males by increasing sperm count and libido. By its action on uterine contractions in females, it is used for the treatment of menstrual disorders and painful deliveries. Powder of the root of long pepper along with honey is used for increasing labor contractions. Due to Madhur vipaka and snigdha guna, it acts as diuretic in urinary diseases (Deshpande et al., 2018).

12.3.2 *Tinospora cordifolia*

T. cordifolia is a broad-spectrum traditional medicine for various diseases. It is known as guduchi or amrita in Sanskrit (Spandana et al., 2013). According to Ayurvedic medicine, it has properties as follows: rasa-tikta, katu, kashay, i.e., bitter, pungent, astringent; veerya-ushna, i.e., worm; vipaka-Madhur, i.e., sweet; guna-snigdha, i.e., unctuous; mrudu, i.e., soft; and dosha-tridoshaghna, i.e., calms vata, kapha, and pitta (Deshpande et al., 2018).

Stem and leaves of giloy are used in Ayurveda for treatment of many diseases. The pills of mixture of stem of the Guduchi (*T. cordifolia*) and the roots of Bhatkatiaya (*Solanum surattense*) are used in the treatment of fever. In North Gujarat, powdered root and bark of giloy with milk is used for treating cancer; whereas, decoction of root is used for the treatment of dysentery and diarrhea. It is also useful in treating asthma, jaundice, gout, and anorexia (Tripathi et al., 2015). According to Ayurvedic literature, *T. cordifolia* also shows properties like bitter tonic, diuretic, astringent, and potential aphrodisiac and curative against skin infections, as well as diabetes. For the treatment of jaundice as well as for tonic action, general debility, and especially in treatment of tuberculosis, guduchi sattva is used (Garde 1983). It is prepared from fresh pieces of giloy, crushed and soaked in water overnight, and then stirred well in the morning, and fibers are removed. Precipitate is removed and repeatedly washed and dried (Joshi and Kaur, 2016). According to Ayurvedic literature, its overall action is appetizer, nutritious, blood purifier, hepatostimulant, and strengthening bones, muscles, and the neurological system. Hence it is called "best rasayana", i.e., nourishing and strengthening all systems (Deshpande et al., 2018).

As per the properties described in Ayurveda, giloy acts on various systems of the body by various mechanisms and actions. Due to its rasa, vipaka, and veerya, it normalizes all three doshas (vata, pitta, kapha); hence, it is called tridoshaghna. Locally, it is used as pain reliever and for dermatological disorders. Internally it acts on various systems (Gogte, 1982).

Due to tikta, katu rasa, and ushna guna it acts as an appetizer, digestive, and dewormer. It reduces the acidic atmosphere in the stomach. Due to ushna guna and Kashaya rasa, it is bulk (feces) forming. It is also useful in thirst, emesis, decreased appetite, pain, liver disorders, jaundice, hyperacidity, dysentery, irritable bowel syndrome (IBS), and worms. In hyperacidity, a decoction of giloy and honey is used. In liver disorders, it reduces liver inflammation and stimulates the liver, thereby reducing hepatitis. For this juice of giloy, the leaves along with rock sugar are advised. In obstructive jaundice, it removes obstructions and improves bile flow. Due to its Kashaya rasa and ushna guna, it absorbs excess liquid in the colon and, hence, it is used for diarrhea and dysentery (Deshpande et al., 2018).

Giloy is mainly used in pyrexia. Due to its tikta rasa and Madhur vipaka, it is used in chronic fever associated with cough, mild pain, and splenomegaly. It is also used in typhoid fever. For this guduchi siddha ghrita, or guduchi sattva, is used which helps to reduce fever along with burning, decreased appetite, and debility. In rheumatoid arthritis, giloy decoction is used along with dry ginger to reduce inflammation, fever, and pain. In gout, ghee or oil cooked in decoction of giloy pulp and milk is used to reduce joint pain, inflammation, and joint deformity (Garde, 1983). In debility, guduchi sattva is used to protect and support the heart, thus acting as hridya. To reduce post-pyrexial fatigue and burning of extremities, guduchi sattva is used along with milk and sugar. Giloy juice is used to reduce burning sensation (Gogte, 1982).

With its snigdha, ushna, and mrudu guna, it reduces debility-induced cough (Deshpande et al., 2018). Juice, decoction, powder, or sattva of *T. cordifolia* along with honey is used to normalize hormonal imbalances, leading to the treatment of diabetes (Garde, 1983). Due to the blood purifying property of giloy, it is used in the treatment of skin diseases and herpes. Decoction of *T. cordifolia, Adhatoda vasica*, and *Azadiracta indica* is used to reduce suppuration and itching in skin disorders. Giloy is also used in the treatment of syphilis (Deshpande et al., 2018). Similar to guduchi sattva used as nutritious (rasayana), another formulation samshamani vati or guduchi ghana vati, i.e., pills of giloy are used as immunomodulators (Garde, 1983).

12.4 CHEMICAL DERIVATIVES

To understand traditional knowledge thoroughly, researchers are finding and extracting phytochemical constituents which show various properties to improve the health of an individual. There are many classes of phytochemicals present in these plants, like alkaloids, amides, terpenoids, volatile oils, lignans, esters, organic acids, steroids, glycosides, and many more.

12.4.1 PHYTOCHEMICALS OF *P. LONGUM*

Root and fruit of *P. longum* are used to treat many diseases because of the various classes of phytochemicals present. Major classes of phytochemicals are alkaloids, amides, volatile oils, lignans, esters, and organic acids (Varughese et al., 2016). Active compounds of *P. longum* are shown in Figure 12.3.

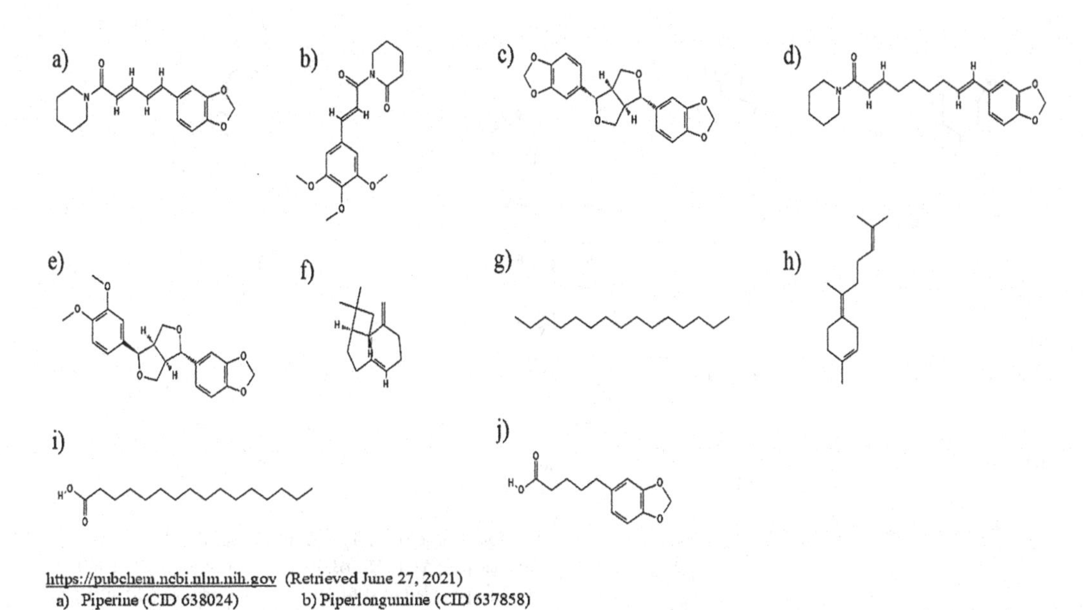

FIGURE 12.3 Major phytochemicals present in *Piper longum*.

The volatile oils present in fruit of *P. longum* are mainly bisabolene (Figure 12.3h), caryophyllene (Figure 12.3f), pentadecane (Figure 12.3g). Along with these oils, it also contains p-methoxy acetophenone, zingiberene, thujine, pcymene, dihydrocarveol, and terpinolene (Zaveri et al., 2010). Sesamin (Figure 12.3c), fargesin (Figure 12.3e), and pulviatilol are lignans present in the fruit of long pepper. The class of alkaloids and amides largely contains piperine (Figure 12.3a), pipernonaline (Figure 12.3d), asarinine, piperundecalidine, N-isobutyl decadienamide, methyl piperine, pellitorine, piperlonguminine, dehydropipernonaline piperidine, piperlongumine (Figure 12.3b), pergumidiene, a dimer of desmethoxypiplartine, brachystine, tetrahydro piperine, piperettine, brachystamide-B, pipercide, retrofractamide A, brachyamide-A, piperderidine, and longamide, which are present in the fruit; whereas; (2E, 4E, 8E) -Nisobutylhenicosa-2,4,8-trienamide, guineesine, 7-epi- eudesm-4(15)-ene-1beta, 6-alpha-diol, biabola-1, 10-diene, 2E,4E-dienamide, 3-(3,4-methylenedioxophenylpropenal, guineesine, piperlongumine (Figure 12.3b), 1-(3,4-methylenedioxyphenyl 1E-tetradecene, piperoic acid, trimethoxy cinnamoyl-piperidine, 3,4-di-hydroxy- biabola-1, piperine (Figure 12.3a), tetrahydropiperlongumine, piperlonguminine are found in the roots of long pepper (Mitra et al., 2007). Tetrahydropiperic acid (Figure 12.3j) and palmitic acid (Figure 12.3i) are the organic acids, whereas Z-12octandecenoic–glycerol-monoester and tridecyl-dihydro-pcoumaarate, eicosanyl-(E)-p-coumarate are esters present in *P. longum* (Dhanalakshmi et al., 2016).

12.4.2 Phytochemicals of *T. cordifolia*

Leaves and stems of *T. cordifolia* are used as therapeutic agents due to the presence of different classes of phytochemicals. These classes include alkaloids, steroids, diterpenoid lactones, glycosides, sesquiterpenoid, aliphatic compounds, essential oils, and lignans (Tripathi et al., 2015). Active compounds of *T. cordifolia* are shown in Figure 12.4.

Steroids present in the shoot of *T. cordifolia* are giloinsterol, makisterone A, 20 δ -hydroxyecdysone, β –sitosterol (Figure 12.4g) and δ-sitosterol. Furanolactone, clerodane derivatives, tinosporon,

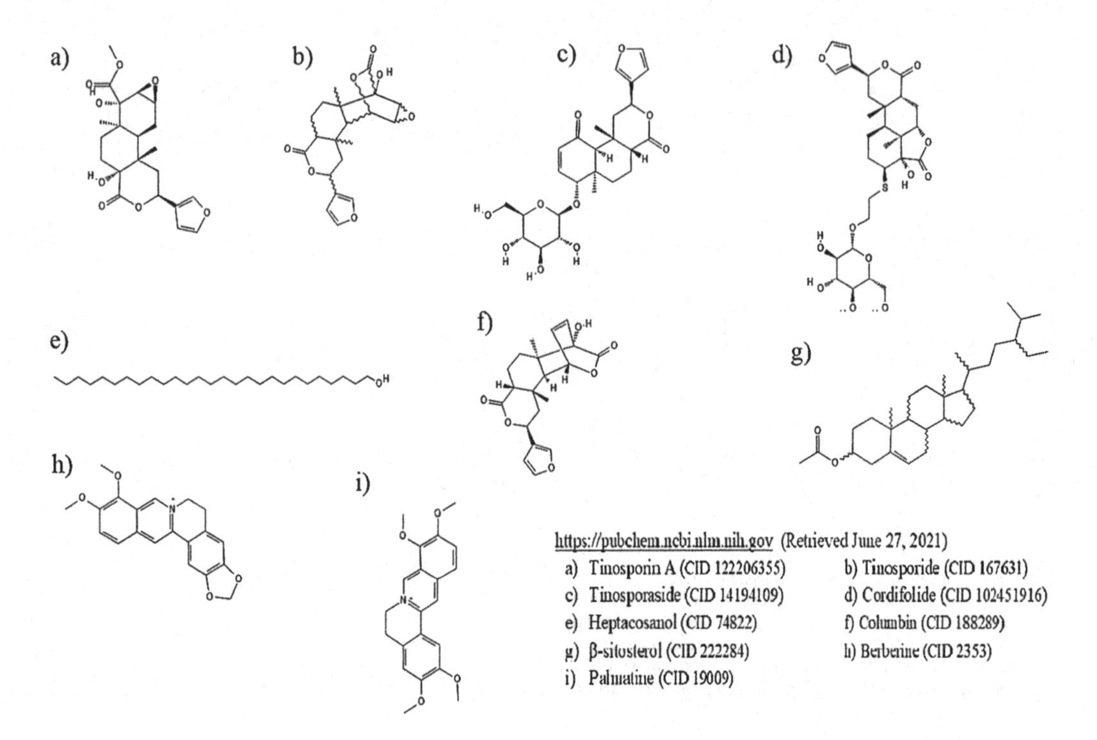

FIGURE 12.4 Major phytochemicals present in *Tinospora cordifolia.*

columbin (Figure 12.4f), jateorine, and tinosporides (Figure 12.4b) are diterpenoid lactones as well as nonacosan-15-one, heptacosanol (Figure 12.4e), octacosanol and dichloromethane are aliphatic compounds present in the whole plant (Spandana et al., 2013). Glycosides present in stem of giloy are tinocordiside, syringinapiosylglycoside, tinocordifolioside, cordifolioside, Syringin, pregnane glycoside, furanoidditerpene glucoside, palmatosides, cordioside, 18-norclerodane glucoside and cordifolide A (Figure 12.4d), B, C, D and E. jatrorrhizine, berberine (Figure 12.4h), magnoflorine, tinosporine (Figure 12.4a), choline, palmatine (Figure 12.4i), tetrahydropalmatine, tembetarine, aporphine alkaloids, and isocolumbin are alkaloids present in the roots and stems, whereas tinocordifolin is a sesquiterpenoid present in the stem only (Bharathi et al., 2018). Other than these chemical derivatives, there are many components like tinosporaside (Figure 12.4c), tetrahydrofuran, N trans-feruloyltyramine, giloinin, cordifol, jatrorrhizine, cordifelone, and tinosporic acid. Giloy also contains potassium, iron, calcium, and chromium (Tripathi et al., 2015).

12.5 POTENTIAL BENEFITS

P. longum and *T. cordifolia* have an abundance of phytochemicals which act as bioactive compounds. These phytochemicals show activities such as antidiabetic, antihyperlipidemic, hepatoprotective, neuroprotective, cardioprotective, antibacterial, aphrodisiac, antifungal, antistress, hypolipidemic, hepatic disorder, anticancer, anti-HIV, antiosteoporotic, antitoxic, wound-healing, anticomplementary, immunomodulating, as well as a cure for many diseases, like gonorrhea, paralysis of the tongue, diarrhea, cholera, chronic bronchitis, asthma, constipation, etc.

12.5.1 Uses of *P. longum*

Phytochemicals of *P. longum* show activities like antiulcer, insecticidal, antiplatelet, anti-snake venom, coronary vasodilation, cardioprotective, anti-amebic, antihelminthic, antimicrobial, anti-obesity, antidepressant, anticancer, antitumor, radioprotective, immunomodulatory, anti-arthritic, antiapoptotic, anti-inflammatory, hepatoprotective, analgesic, antidiabetic, hypocholesterolemic, antiasthmatic, antioxidant, melanin-inhibiting, antifertility, anti-epileptic, antiparkinsonian, and a therapeutic agent for Alzheimer's disease.

12.5.1.1 Antiulcer Activity

P. longum shows activity against ulcer by reducing its severity. It also reduces blood flow in mucosa and cell proliferation, which helps in reducing ulcers (Yadav et al., 2015). Extract from long peppers also reduces the irritative action of ulcers on gastric mucosa (Agrawal et al., 2000).

12.5.1.2 Insecticidal Activity

Pipyahyine, which is an amide extracted from *P. longum,* is used as a larvicidal for mosquitos. Larvae of *C. quinquefasciatus* are also killed by the long pepper extract (Jeon et al., 2019).

12.5.1.3 Antiplatelet Activity

Extracts of *P. longum* show antiplatelet activity by inhibiting the Thromboxane A2, i.e., TXA2 receptor in the hemostatic system (Iwashita et al., 2007). Piperlongumine (Figure 12.3b) is one of the phytochemicals which shows antiplatelet activity. It was tested using collagen, arachidonic acid, and platelet-activating factor, and it was observed that the amides of long pepper showed antithrombogenic effect when checked in rabbits (Jeon et al., 2019).

12.5.1.4 Anti-Snake Venom Activity

Piperine (Figure 12.3a) is present in ethanolic extract of long pepper fruits and shows activity against venom of Russell's viper snake when tested in chicken eggs and mice. Piperine reduces the activity of venom and its lethality by inhibiting necrosis, hemorrhage, and inflammation (Shenoy et al., 2013).

12.5.1.5 Coronary Vasodilation and Cardioprotective Activity

Coronary vasodilation was observed by the usage of *P. longum* extract due to presence of dehydropipernonaline, which is an amide (Shoji et al., 1986). Extracts of fruits of *P. longum* showed efficacy for protection from toxic lesions induced in the cardiac system due to Adriamycin. This was observed in rats given a pretreatment of *P. longum* extract (Wakade et al., 2008). Piperaldehyde also shows protective effect against myocardial ischemia (Dhanalakshmi et al., 2016).

12.5.1.6 Anti-amebic and Antihelminthic Activity

Piperine (Figure 12.3a) present in ethanolic extract of long pepper fruits shows activity against *Entamoeba histolytica* when observed in rats (Sawangjaroen et al., 2004). Glucose uptake in parasite *G. explanatum* was observed whereas the muscular movement of *Fasciola gigantica* was irregular after administration of *P. longum* extract (Singh et al., 2007; Singh et al., 2009). Essential oils from long pepper show paralytic activity in *Ascaris lumbricoides* (Chaudhari et al., 1981).

12.5.1.7 Antimicrobial Activity

P. longum shows activity against microorganisms, such as bacteria and fungi. Piperidine alkaloid, pipernonaline (Figure 12.3d) shows antifungal effect against phytopathogenic fungi viz., *Puccinia recondita*, *Botrytis cinerea*, *Erysiphe graminis*, *Rhizoctonia solani*, *Pyricularia oryzae*, *Phytophthora infestans* (Nigam and Rao, 1976). Antibacterial activity of *P. longum* alcoholic extract was observed against *P. aeruginosa*, *S. albus*, *S. typhi*, *E. coli*, and *B. megaterium* (Khan and Siddiqui, 2007).

12.5.1.8 Anti-obesity and Hypocholesterolemic Activity

By inhibiting the acyl CoA diacylglycerol acyltransferase, long pepper shows activity as an anti-obesity agent (Dhanalakshmi et al., 2016). Hepatic cholesterol is reduced in hypercholesterolemic rats by administration of methyl piperine (Figure 12.3a) extracted from *P. longum* (Wu and Bao, 1992).

12.5.1.9 Antidepressant Activity

Piperine (Figure 12.3a), an alkaloid from long pepper, reversed the corticosterone-induced reduction of brain-derived neuropathic factor as well as it also inhibits monoamine oxidase. These activities suggest the antidepressant activity of *P. longum* (Lee et al., 2008).

12.5.1.10 Anticancer and Antitumor Activity

Piperine (Figure 12.3a) isolated from long pepper inhibits the adhesion of neutrophils to endothelium whereas piperlonguminine (Figure 12.3b) shows inhibitory activity in alpha-MSH-induced tyrosinase synthesis and reduces carcinogenesis in hamsters (Senthil et al., 2007). Piperine (Figure 12.3a) also shows effect as chemopreventive in animals having lung cancer and reduces metastasis (Selvendiran and Sakthisekaran, 2004). Piplartine and piperine (Figure 12.3a) alkaloid amide show cytotoxic activity against tumor and hence act as an antitumor agent (Bezerra et al., 2006).

12.5.1.11 Radioprotective Activity

Hampered glutathione production due to increases in the levels of glutathione pyruvate transaminase, lipid peroxidation, and alkaline phosphatase is normalized by the use of alcoholic extracts of *P. longum* fruits in radiation-treated animals (Kumar et al., 2011).

12.5.1.12 Immunomodulatory and Activity Against COVID-19

Pippali rasayana, which is an Ayurvedic formulation, was tested in mice infected with *Giardia lamblia*. The phagocytic activity and macrophage migration index was seen elevated when tested by hemagglutination titer and phagocytic index. These observations indicate that the long pepper shows immunomodulatory effect (Dhanalakshmi et al., 2016). Piperine (Figure 12.3a) present in *P. longum* also increases the WBC count, α-esterase positive cells, and bone marrow cellularity in mice (Srivastava and Saxena, 2020). Due to these effects of *P. longum*, it is used for boosting immunity. These immune-boosting effects act as preventives against COVID-19 (Srivastava and Saxena, 2020).

12.5.1.13 Hepatoprotective Activity

Lipid peroxidation induced by chemicals, such as carbon tetrachloride and tert-butyl hydroperoxide, and causing hepatotoxicity, is reduced after administration of piperine (Figure 12.3a) extracted from *P. longum*. Liver fibrosis is also inhibited by using *P. longum* extracts (Koul and Kapil, 1993).

12.5.1.14 Anti-inflammatory and Anti-arthritic Activity

Piperine (Figure 12.3a) from long pepper shows anti-inflammatory activity by inhibiting the prostaglandin and leukotrienes COX-1 when tested on carrageenan rat paw edema (Kumar et al., 2005). Freund's Complete Adjuvant-induced arthritis in rats is reduced by the use of aqueous extract of *P. longum* (Dhanalakshmi et al., 2016).

12.5.1.15 Anti-apoptotic and Antioxidant Activity

Gentamicin-induced hair cell loss was reduced by usage of fruit extracts of *P. longum* when checked in neonatal cochlea cultures (Yadav et al., 2014). Petroleum ether extract of *P. longum* fruits shows reduction in lipid peroxide level as well, as it maintains glutathione content due to its antioxidant activity (Navneet and Singh, 2018).

12.5.1.16 Antiasthmatic and Analgesic Activity

Antigen-induced bronchospasm, which can be the cause of asthma was reduced by the usage of extracts of long pepper fruit administered with milk to guinea pigs (Zaveri et al., 2010). Root extracts of long pepper show analgesic activity when tested in mice given chemical and thermal stimulus of pain. Administration of root extract showed delay in the reaction which indicates that it can be used as a pain reliever (Kumar et al., 2011).

12.5.1.17 Antidiabetic activity

Fruits of *P. longum* show anti-hyperglycemic effect which reduces the sugar level when tested in alloxan-induced diabetic rats. Oral administration of ethanolic extract of dried fruits of *P. longum* showed more effect against hyperglycemia (Zaveri et al., 2010).

12.5.1.18 Melanin-Inhibiting and Antifertility Activity

Piperlonguminine (Figure 12.3b) present in *P. longum* exhibits an inhibitory effect on the alpha-melanocyte-stimulating hormone signaling via cAMP responsive element-binding protein and regulates its expression. This interaction leads to inhibition of melanin production (Kumar et al., 2011). Hexane extract of *P. longum* exhibits anti-implantation activity. It also shows contraceptive activity in males and females without interfering with hormones (Srivastava, 2014).

12.5.1.19 Antiepileptic and Therapeutic Activity for Alzheimer's Disease

P. longum extracts show reduction in seizures when administered to rats treated with PTZ for induction of seizures. It also gives protection to 4-aminopyridine and strychnine-induced seizures in mice. Long pepper extract reduces the GABA levels by acting as antiepileptic agent (Yadav et al., 2020). Piperlonguminine and dihydropiperlonguminine regulate the expression of amyloid precursor protein (APP), which further acts as a therapeutic agent for Alzheimer's disease (Srivastava, 2014).

12.5.1.20 Antiparkinsonian Activity

Parkinson's disease is caused due to dopamine accumulation. This accumulation leads to inflammation and damage to neurons. This damage is prevented by *P. longum* extracts and hence shows preventive action against Parkinson's disease (Yadav et al., 2020) (Table 12.1).

TABLE 12.1
Activities of *Piper longum*

Plant Part	Activities	
Root	Analgesic	
Whole Plant	Antimicrobial	
Fruit	Anti-ulcer	Insecticidal
	Antiplatelet	Anti-snake venom
	Coronary vasodilation	Cardioprotective
	Anti-amebic	Antihelminthic
	Anti-obesity	Hypocholesterolemic
	Antidepressant	Radioprotective
	Anticancer	Antitumor
	Immunomodulatory	Activity against COVID-19
	Hepatoprotective	Antidiabetic
	Anti-inflammatory	Antiarthritic
	Antiapoptotic	Antioxidant
	Antiasthmatic	Antiparkinsonian
	Melanin-inhibiting	Antifertility
	Antiepileptic	Therapy for Alzheimer's disease

12.5.2 Uses of *T. cordifolia*

Phytochemicals of *T. cordifolia* show activities like anti-ulcer, antimicrobial, anticancer, antitumor, immunomodulatory, anti-arthritic, anti-inflammatory, anti-stress, hepatoprotective, antidiabetic, hypolipidemic, antioxidant, antiparkinsonian, anti-HIV, antitoxic, anti-osteoporotic, wound-healing, memory-enhancing, anti-amebic, urinary calculi, uremia.

12.5.2.1 Anti-ulcer Activity

Ethanolic extract of root of *T. cordifolia* shows anti-ulcer effect when tested against stress-induced ulceration (Sinha et al., 2004).

12.5.2.2 Anticancer and Antitumor Activity

Methanolic extracts of giloy show increased cell death when tested on HeLa cells. It also reduces the effect of gamma radiation in Swiss albino mice and hence shows anticancer activity. Hexane extract of *T. cordifolia* inhibits the growth of tumor cells, whereas it enhances the expression of the proapoptotic BAX gene (Tiwari et al., 2018). It also reduces weight, occurrence, and carcinogenesis of papillary tumors. Palmatine (Figure 12.4i) and yangambin are the two compounds from giloy that help in the treatment of colon cancer (Bala et al., 2015).

12.5.2.3 Antimicrobial Activity

Proteus vulgaris, *Escherichia coli*, *Pseudomonas aeruginosa*, *Salmonella paratyphi*, *Klebsiella pneumoniae*, *Salmonella typhimurium*, *Enterobacterium aeruginosa*, *Salmonella typhi*, *Shigella flexneri*, *Staphylococcus aureus*, *Enterobacter aerogenes*, and *Serattia marcescens* are gram-positive human pathogens which are inhibited by the antimicrobial action of methanolic extract of *T. cordifolia* (Jabiullah et al., 2018). Ethanolic extract of giloy also shows inhibition of isolates of *Pseudomonas aeruginosa* and *Klebsiella pneumoniae* from the urinary tract (Shanthi and Nelson, 2013). *Aspergillus flavus*, *Aspergillus fumigatus*, and *Aspergillus niger* are the fungi inhibited by *T. cordifolia* extract. *Streptococcus mutans* is one of the dental plaque-forming organisms that is also inhibited by the extract of *T. cordifolia* (Agarwal et al., 2019).

12.5.2.4 Anti-Inflammatory and Anti-Stress Activity

Aqueous extract of *T. cordifolia* exhibits anti-inflammatory activity in acute inflammatory responses in albino rats. Carrageenan-induced hind paw edema in rats was reduced using decoction of *T. cordifolia* (Sinha et al., 2004). Giloy extract also shows a major impact as an anti-stress agent when compared with diazepam (Sarma et al., 1996).

12.5.2.5 Immunomodulatory and Activity against COVID-19

Bioactive compounds of *T. cordifolia* viz., tinocordioside, 11-hydroxy muskatone, cordifolioside A, N-methyl-2-pyrrolidone, N formyl annonain, magnoflorine, and syringin show cytotoxic effects as well as immunomodulatory effects (Mittal et al., 2014). These compounds also increase the phagocytic capacity of macrophage (Tripathi et al., 2015). Extracts of *T. cordifolia* mainly increase the concentration of IgG immunoglobulin which acts as an immunity barrier against infection (Sharma et al., 2019). Water extract of giloy shows production of cytokines and activates the immune effector cells. It also increases the number as well as phagocytic activity of macrophages (Srivastava and Saxena, 2020). These immunomodulatory effects of *T. cordifolia* help to boost the immune system against the COVID-19 pandemic (Srivastava and Saxena, 2020).

12.5.2.6 Hepatoprotective and Anti-Amebic Activity

Increased levels of cholesterol, HDL, gamma-glutamyl transferase, triglyceride, aspartate transaminase, alanine transaminase, and LDL in patients consuming alcohol is significantly decreased after administration of *T. cordifolia* extract, which further acts as a hepatoprotective (Sharma et al., 2019). Ethanolic extract of giloy reduces the SGOT and SGPT levels as well as the weight of the liver which shows its hepatoprotective effect on CCl_4-induced hepatic dysfunction in albino rats (Sinha et al., 2004). Inhibition of enzymes, such as Rnase, Dnase, acid phosphatase, alkaline phosphatase, α-amylase, aldolase, and protease acts as anti-amebic on *Entamoeba histolytica* (Sohni et al., 1995).

12.5.2.7 Anti-arthritic and Anti-osteoporotic Activity

T. cordifolia shows efficiency against rheumatoid arthritis (Chopra et al., 2012). It enhances the differentiation of osteoblast, which further leads to osteoblastic lineage and mineralization of bone. This leads to strengthening of bone. Extracts of giloy plant also increase the thickness of joint cartilage when tested in mice mesenchymal stem cells. Hence, *T. cordifolia* can be used as anti-osteoporotic as well as anti-osteoarthritic agent (Saha and Ghosh, 2012).

12.5.2.8 Antidiabetic and Hypolipidemic Activity

Alkaloids, such as jatrorrhizine, magnoflorine and palmetine (Figure 12.4i) isolated from *T. cordifolia* show effect against gestational diabetes by reduction in birth defects in the fetus. This effect was studied in the streptozotocin-induced diabetic rat. Root extract of giloy reduces urinary glucose levels as well as blood glucose levels. It also reduces creatinine levels (Sharma et al., 2019). Root extract also reduces serum and tissue lipid levels which further helps in the reduction of risk of coronary heart disease (Stanely and Menon, 2003).

12.5.2.9 Anti-HIV and Wound-Healing Activity

Stem extract of *T. cordifolia* shows enhancement of macrophage, lymphocytes, eosinophils, polymorphonuclear leukocytes; it also shows an increase in hemoglobin in HIV-positive patients. It induces immune cells hence acts as anti-HIV agent (Jabiullah et al., 2018). Alcoholic extracts of *T. cordifolia* shows wound-healing activity by increasing collagen synthesis and tensile strength (Shanbhag et al., 2005).

12.5.2.10 Antioxidant and Antitoxic Activity

Antioxidant activity of stem extract of *T. cordifolia* using methanol was seen due to the increase in lipid peroxide and catalase levels. It also increased the activity of superoxide dismutase (SOD), which acts as an antioxidant (Jabiullah et al., 2018). Radical scavenging activity is also present in methanolic extract of giloy, and hence it prevents gamma radiation damage (Tiwari et al., 2018). Alkaloids such as tinosporin (Figure 12.4a), tetrahydropalmatine, palmetine (Figure 12.4i), isocolumbin, choline, and magnoflorine from giloy show antitoxic activity against aflatoxin-induced nephrotoxicity (Tripathi et al., 2015).

12.5.2.11 Antiparkinsonian and Memory-Enhancing Activity

Extracts of the giloy plant suppress dopaminergic inflammation of neurons which further reduces the effect of Parkinson's disease when tested in the MPTP-induced Parkinsonian mouse model (Birla et al., 2019). Due to increases in choline and acetylcholine levels in memory-deficient animals as well as in normal animals, giloy acts as a memory enhancer (Tripathi et al., 2015).

12.5.2.12 Against Urinary Calculi and Uremia

Urinary calculi are dissolved by the usage of aqueous extract of the stem of *T. cordifolia*. Aqueous extracts also reduce blood urea levels when tested in uremic dogs. In a transient fall in blood pressure and in the case of bradycardia, water extract shows an increase in vascular contraction in dogs and diuresis in rats (Singh et al., 1975) (Table 12.2).

12.6 CONCLUSION

P. longum and *T. cordifolia* are two medicinal herbs which are largely used in traditional medicine. These herbs have various properties to cure complaints related to cardiovascular, nervous, digestive,

TABLE 12.2
Activities of *Tinospora cordifolia*

Plant Part	Activities	
Root	Anti-ulcer	Antidiabetic
	Hypolipidemic activity	Anticancer
	Anti-stress	
Stem	Anti-HIV	Wound-healing activity
	Antioxidant	Antitoxic
	Against urinary calculi	Anti-uremic
	Immunomodulatory	Activity against COVID-19
	Hepatoprotective	Antimicrobial
	Antiarthritic	Anti-osteoporotic activity
	Antiparkinsonian	Anti-inflammatory
	Memory-enhancing	Anti-amebic
	Antitumor	

locomotor, reproductive, and many more systems according to traditional medicine. These plants are rich in phytochemicals, like bisabolene, caryophyllene, pentadecane, p-methoxy acetophenone, zingiberene, thujine, pcymene, dihydrocarveol, terpinolene, piperine, methyl piperine piperidine, etc. These phytochemicals act as bioactive compounds that help in the treatment of many diseases. Extracts of *P. longum* and *T. cordifolia* show activities like anti-ulcer, anticancer, antiarthritic, anti-inflammatory, antidiabetic, antioxidant, antimicrobial, anti-amebic, antitumor, hepatoprotective, hypolipidemic, antiparkinsonian, and immunomodulatory. Besides these activities, *P. longum* shows insecticidal, antiplatelet, anti-snake venom, coronary vasodilation, cardioprotective, antihelminthic, anti-obesity, antidepressant, radioprotective, antiapoptotic, analgesic, antiasthmatic, melanin-inhibiting, antifertility, antiepileptic, and as a therapeutic agent for Alzheimer's disease. *T. cordifolia* shows other activities, like anti-stress, against urinary calculi and uremia, anti-HIV, antitoxic, anti-osteoporotic, wound-healing, and memory-enhancing. Due to these activities, the bioactive compounds present in the whole plant as well as in extracts *P. longum* and *T. cordifolia* can be used as good therapeutic agents.

REFERENCES

Agarwal, S., P. Ramamurthy, B. Fernandes, A. Rath, and P. Sidhu. 2019. Assessment of antimicrobial activity of different concentrations of *Tinospora cordifolia* against Streptococcus mutans: An in-vitro study. *Dental Research Journal*, 16(1), pp.24–28.

Agrawal, A., C. Rao, K. Sairam, V. Joshi, and R. Goel. 2000. Effect of *Piper longum* Linn, *Zingiber officianalis* Linn and Ferula species on gastric ulceration and secretion in rats. *Indian Journal of Experimental Biology*, 38, pp.994–998.

Ali, A. 2020. Herbs that heal: The philanthropic behaviour of nature. *Annals of Phytomedicine*, 9, pp.7–17.

Bala, M., K. Pratap, P. Verma, B. Singh, and Y. Padwad. 2015. Validation of ethnomedicinal potential of *Tinospora cordifolia* for anticancer and immunomodulatory activities and quantification of bioactive molecules by HPTLC. *Journal of Ethnopharmacology*, 175, pp.131–137.

Bezerra, D., F. Castro, A. Alves, C. Pessoa, M. Moraes, E. Silveira, M. Lima, F. Elmiro, and L. Costa-Lotufo. 2006. In vivo growth-inhibition of Sarcoma 180 by piplartine and piperine, two alkaloid amides from Piper. *Brazilian Journal of Medical and Biological Research*, 39(6), pp.801–807.

Bharathi, C., A. Reddy, G. Nageswari, B. Sri Lakshmi, M. Soumya, D. Vanisri, and B. Venkatappa. 2018. A Review on medicinal properties of *Tinospora cordifolia*. *International Journal of Scientific Research and Review*, 7(12), pp.585–598.

Birla, H., S. Rai, S. Singh, W. Zahra, A. Rawat, N. Tiwari, R. Singh, A. Pathak, and S. Singh. 2019. *Tinospora cordifolia* Suppresses Neuroinflammation in Parkinsonian Mouse Model. *Neuromolecular Medicine*, 21(1), pp.42–53.

Chaudhari, G., C. Kokate, and A. Nimbkar. 1981. Search for anthelmintics of plant origin: Activities of volatile principles of acorus calamus against ascaris lumbricoides. *Ancient Science of Life*, 1(2), pp.103–105.

Chauhan, K., R. Solanki, A. Patel, C. Macwan, and M. Patel. 2011. Phytochemical and therapeutic potential of *Piper longum* Linn a review. *International Journal of Research in Ayurveda & Pharmacy*, 2(1), pp.157–161.

Chitme, H., R. Chandra, and S. Kaushik. 2003. Studies on antidiarrheal activity of *Calotropis gigantea* R. Br. in experimental animals. *Journal of Pharmacy & Pharmaceutical Sciences*, 7, pp.70–75.

Chopra, A., M. Saluja, G. Tillu, A. Venugopalan, G. Narsimulu, R. Handa, L. Bichile, A. Raut, S. Sarmukaddam, and B. Patwardhan. 2012. Comparable efficacy of standardized Ayurveda formulation and hydroxychloroquine sulfate (HCQS) in the treatment of rheumatoid arthritis (RA): A randomized investigator-blind controlled study. *Clinical Rheumatology*, 31(2), pp.259–269.

Choudhary, N. and V. Singh. 2018. A census of P. *longum*'s phytochemicals and their network pharmacological evaluation for identifying novel drug-like molecules against various diseases, with a special focus on neurological disorders. *PLoS ONE*, 13(1), p.e0191006.

Deshpande, A., R. Javalgekar, and S. Ranade. 2018. *Dravyaguna Vidnyan*. 4th edn. Pune: Proficient Publishing House.

Dhanalakshmi, D., S. Umamaheswari, D. Balaji, R. Santhanalakshmi, and S. Kavimani. 2016. phytochemistry and pharmacology of *piper longum*: A systematic review. *World Journal of Pharmacy and Pharmaceutical Sciences*, 6(1), pp.381–398.

Gajurel, P., S. Kashung, S. Nopi, R. Panmei, and B. Singh. 2021. Can the Ayurvedic pippali plant (*Piper longum* L.) be a good option for livelihood and socio-economic development for Indian farmers. *Current Science*, 120, p.10.

Garde, G. 1983. *Sartha Vagbhata*. 7th edn. Pune: Maharashtra sahakari Mudranalay.

Gogte, V. 1982. *Ayurvedik Materia Medica*. 1st edn. Pune: Continental prakashan.

Iwashita, M., M. Saito, Y. Yamaguchi, R. Takagaki, and N. Nakahata. 2007. Inhibitory effect of ethanol extract of *Piper longum* L. on rabbit platelet aggregation through antagonizing thromboxane A2 receptor. *Biological and Pharmaceutical Bulletin*, 30(7), pp.1221–1225.

Jabiullah, S., J. Battineni, V. Bakshi, and N. Boggula. 2018. *Tinospora cordifolia*: A medicinal plant: A review. *Journal of Medicinal Plants Studies*, 6(6), pp.226–230.

Jeon, H., K. Kim, Y. Kim, and S. Lee. 2019. Naturally occurring Piper plant amides potential in agricultural and pharmaceutical industries: Perspectives of piperine and piperlongumine. *Applied Biological Chemistry*, 62(1), pp.1–7.

Johri, R. and U. Zutshi. 1992. An Ayurvedic formulation "Trikatu" and its constituents. *Journal of Ethnopharmacology*, 37(2), pp.85–91.

Joshi, G. and R. Kaur. 2016. *Tinospora cordifolia*: A phytopharmacological review. *International Journal of Pharmaceutical Sciences and Research*, 7(3), pp.890–897.

Kamboj, V. 2000. Herbal medicine. *Current Science*, 78, pp.35–39.

Khan, M. and M. Siddiqui. 2007. Antimicrobial activity of fruits of *Piper longum*. *Natural product Radiance*, 6, pp.111–113.

Kim, H. 2005. "Do not put too much value on conventional medicines". *Journal of Ethnopharmacology*, 100(1–2), pp.37–39.

Koul, I. and A. Kapil. 1993. Evaluation of the liver protective potential of piperine, an active principle of black and long peppers. *Planta Medica*, 59(5), pp.413–417.

Kumar, S., J. Kamboj, and S. Sharma. 2011. Overview for various aspects of the health benefits of Piper longum linn. fruit. *Journal of Acupuncture Meridian Studies*, 4(2), pp.134–140.

Kumar, S., P. Arya, C. Mukherjee, B. Singh, N. Singh, V. Parmar, A. Prasad, and B. Ghosh. 2005. Novel aromatic ester from *Piper longum* and its analogues inhibit expression of cell adhesion molecules on endothelial cells. *Biochemistry*, 44(48), pp.15944–15952.

Lee, S., S. Hwang, X. Han, C. Lee, M. Lee, S. Choe, S. Hong, D. Lee, M. Lee, and B. Hwang. 2008. Methylpiperate derivatives from *Piper longum* and their inhibition of monoamine oxidase. *Archives of Pharmacal Research*, 31(6), pp.679–683.

Ministry of AYUSH. 2020. *Interdisciplinary Committee for Integration of Ayurveda and Yoga Interventions in the 'National Clinical Management Protocol: COVID-19' First Report and Recommendations.*

Mitra, R., B. Mitchell, S. Agricola, C. Gray, K. Baskaran, and S. Muralitharan. 2007. Medicinal plants of India. *APBN*, 11(11), pp.707–725.

Mittal, J., M. Sharma, and A. Batra. 2014. *Tinospora cordifolia*: A multipurpose medicinal plant- A review. *Journal of Medicinal Plants Studies*, 2, pp.32–47.

Nigam, S. and C. Rao. 1976. Antimicrobial activity of some Indian essential oils. *Indian Drugs*, 14, pp.62–65.

Parmar, V., S. Jain, K. Bisht, R. Jain, P. Taneja, A. Jha, O. Tyagi, A. Prasad, J. Wengel, C. Olsen, and P. Boll. 1997. Phytochemistry of the genus Piper. *Phytochemistry*, 46, pp.597–673.

Raghunathan, K. and P. Sharma. 1969. the aqueous extract of *T. cordifolia* used reduction of blood sugar in alloxan induced hyperglycemic rats and rabbits. *Journal of Research in Indian Medicine*, 3, pp.203–209.

Saha, S. and S. Ghosh. 2012. *Tinospora cordifolia*: One plant, many roles. *Ancient Science of Life*, 31(4), pp.151–159.

Sarma, D., R. Khosa, J. Chansauria, and M. Sahai. 1996. Antistress activity of *Tinospora cordifolia* and *Centella asiatica* extracts. *Phytotherapy Research*, 10(2), pp.181–183.

Sawangjaroen, N., K. Sawangjaroen, and P. Poonpanang. 2004. Antiamoebic effects of *Piper longum* fruit, *Piper sarmentosum* root and *Quercus infectoria* nut gall on caecal amoebiasis in mice. *Journal of Ethnopharmacology*, 91(2–3), pp.357–360.

Selvendiran, K. and D. Sakthisekaran. 2004. Chemopreventive effect of piperine on modulating lipid peroxidation and membrane bound enzymes in benzo(a)pyrene induced lung carcinogenesis. *Biomedicine and Pharmacotherapy*, 58(4), pp.264–267.

Senthil, N., S. Manoharan, S. Balakrishnan, C. Ramachandran, R. Muralinaidu, and K. Rajalingam. 2007. Modifying Effects of *Piper longum* on Cell Surface Abnormalities in 7, 12-dimethylbenz(A)Anthracene Induced Hamster Buccal Pouch Carcinogenesis. *International Journal of Pharmacology, 3*, pp.290–294.

Shanbhag, T., S. Shenoy, and M. Rao. 2005. Wound healing profile of *Tinospora cordifolia. Indian Drugs*, 42, pp.217–221.

Shanthi, V. and R. Nelson. 2013. Antibacterial activity of *Tinospora cordifolia* (Wild) Hook. F. Thoms on urinary tract pathogens. *International Journal of Current Microbiology and Applied Science*, 2, pp.190–194.

Sharma, P., B. Dwivedee, D. Bisht, A. Dash, and D. Kumar. 2019. The chemical constituents and diverse pharmacological importance of *Tinospora cordifolia. Heliyon*, 5(9), p.e02437.

Shenoy, P., S. Nipate, J. Sonpetkar, N. Salvi, A. Waghmare, and P. Chaudhari. 2013. Anti-snake venom activities of ethanolic extract of fruits of *Piper longum* L. (Piperaceae) against Russell's viper venom: Characterization of piperine as active principle. *Journal of Ethnopharmacology*, 147(2), pp.373–382.

Shoji, N., A. Umeyama, N. Saito, T. Takemoto, A. Kajiwara, and Y. Ohizumi. 1986. Dehydropipernonaline, an amide possessing coronary vasodilating activity, isolated from *Piper longum* L. *Journal of Pharmaceutical Science*, 75(12), pp.1188–1189.

Singh, A. and Navneet, N. 2018. Critical review on various ethonomedicinal and pharmacological aspects of piper longum linn.(long pepper or pippali). *International Journal of Innovative Pharmaceutical Science and Research, 6*, pp.48-60.

Singh, K., A. Gupta, V. Pendse, C. Mahatma, D. Bhandari, and M. Mahawar. 1975. Experimental and clinical studies on *Tinospora cordifolia. Journal of Research and education in Indian Medicine*, 10, pp.9–14.

Singh, T., D. Kumar, P. Gupta, and S. Tandan. 2007. Inhibitory effects of alcoholic extracts of *Allium sativum* and *Piper longum* on gross visual motility and glucose uptake of *Fasciola gigantica* and *Gigantocotyle explanatum. Journal of Veterinary Parasitology*, 21(2), pp.121–124.

Singh, T., D. Kumar, S. Tandan, and S. Mishra. 2009. Inhibitory effect of essential oils of *Allium sativum* and *Piper longum* on spontaneous muscular activity of liver fluke, *Fasciola gigantica. Experimental Parasitology*, 123(4), pp.302–308.

Sinha, K., N. Mishra, J. Singh, and S. Khanuja. 2004. *Tinospora cordifolia* (Guduchi), a reservoir plant for therapeutic applications: A Review. *Indian Journal of Traditional Knowledge*, 3(3), pp.257–270.

Sohni, Y., P. Kaimal, and R. Bhatt. 1995. The antiamoebic effect of a crude drug formulation of herbal extracts against Entamoeba histolytica in vitro and in vivo. *Journal of Ethnopharmacology*, 45(1), pp.43–52.

Spandana, U., S. Ali, T. Nirmala, M. Santhi, and S. Sipai. 2013. A Review on *Tinospora cordifolia. International Journal of Current Pharmaceutical Review and Research*, 4(2), pp.61–68.

Srivastava, N., and V. Saxena. 2020. A review on scope of immuno-modulatory drugs in Ayurveda for prevention and treatment of Covid-19. *Plant Science Today*, 7(3), p.417.

Srivastava, P. 2014. Therapeutic potential of *Piper longum* L. for disease management: A review. *International Journal of Pharma Sciences*, 4(4), pp.692–696.

Stanely, P., and V. Menon. 2003. Hypoglycaemic and hypolipidaemic action of alcohol extract of *Tinospora cordifolia* roots in chemical induced diabetes in rats. *Phytotherapy Research*, 17(4), pp.410–413.

Tiwari, P., P. Nayak, S. Prusty, and P. Sahu. 2018. phytochemistry and pharmacology of *Tinospora cordifolia*: A review. *Systematic Reviews in Pharmacy*, 9(1), pp.70–78.

Tripathi, B., D. Singh, S. Chaubey, G. Kour, and R. Arya. 2015. a critical review on *guduchi* (*Tinospora cordifolia* (willd.) miers) and its medicinal properties. *International Journal of Ayurveda and Pharma Research*, 3(5), pp.6–12.

Varughese, T., P. Unnikrishnan, M. Deepak, I. Balachandran, and A. Rema Shree. 2016. Chemical Composition of the Essential Oils from Stem, Root, Fruit and Leaf of *Piper longum Linn. Journal of Essential Oil Bearing Plants*, 19(1), pp.52–58.

Vidyarthi, S., S. Samant, and P. Sharma. 2013. Traditional and indigenous uses of medicinal plants by local residents in Himachal Pradesh, North Western Himalaya, India. *International Journal of Biodiversity Science and Management*, 9, pp.185–200.

Wakade, A., A. Shah, M. Kulkarni, and A. Juvekar. 2008. Protective effect of *Piper longum* L. on oxidative stress induced injury and cellular abnormality in adriamycin induced cardiotoxicity in rats. *Indian Journal of Experimental Biology*, 46, pp.528–533.

Wu, E., and Z. Bao. 1992. Effects of unsaponificable matter of *Piper longum* oil on cholesterol biosynthesis in experimental hypocholestrolaemic mice, *Honggacayano*, 23(4), pp.197–200.

Yadav, M., J. Choi, and J. Song. 2014. Protective effect of hexane and ethanol extract of *Piper longum L.* On gentamicin-induced hair cell loss in neonatal cultures. *Clinical and Experimental Otorhinolaryngology*, 7(1), pp.13–18.

Yadav, V., A. Krishnan, and D. Vohora. 2020. A systematic review on *Piper longum* L. Bridging traditional knowledge and pharmacological evidence for future translational research. *Journal of Ethnopharmacology*, 30(247), p.112255.

Yadav, V., S. Chatterjee, M.Majeed, and V. Kumar. 2015. Long lasting preventive effects of piperlongumine and a Piper longum extract against stress triggered pathologies in mice. *Journal of Intercultural Ethnopharmacology*, 4(4), p.277.

Zaveri, M., A. Khandhar, S. Patel, and A. Patel. 2010. Chemistry and pharmacology of *Piper longum* L. *International Journal of Pharmaceutical Sciences Review and Research*, 5(1), pp.67–76.

13 *Plantago ovata* (Isabgol) and *Rauvolfia serpentina* (Indian Snakeroot)

Ankur Anavkar, Nimisha Patel, Ahmad Ali, and Hina Alim

CONTENTS

13.1 Introduction 236
13.2 Description and Distribution 237
13.2.1 *Plantago ovata* 237
13.2.2 *Rauvolfia serpentina* 239
13.3 Traditional Knowledge 240
13.3.1 *Plantago ovata* 240
13.3.2 *Rauvolfia serpentina* 240
13.4 Chemical Derivatives 241
13.4.1 *Plantago ovata* 241
13.4.2 *Rauvolfia serpentina* 243
13.5 Therapeutic Benefits, Applications, and Side Effects 245
13.5.1 *Plantago ovata* 245
13.5.1.1 Gastrointestinal Functions 245
13.5.1.2 Antibacterial Activity 246
13.5.1.3 Immunomodulatory Actions 246
13.5.1.4 Hypolipidemic Activity 247
13.5.1.5 Anti-obesity Activity 247
13.5.1.6 Anticancer Activity 247
13.5.1.7 Coagulation Activity 247
13.5.1.8 Antileishmanial Effect 247
13.5.1.9 Natural Super Disintegrant 248
13.5.1.10 Anticorrosive Activity 248
13.5.1.11 Used for Making Natural Eye Drops 248
13.5.1.12 Activity as Lead Biosorbent 248
13.5.1.13 Anti-ulcer Activity 248
13.5.1.14 Hepatoprotective Activity 249
13.5.1.15 Antidiabetic Activity 249
13.5.1.16 Anti-inflammatory Activity 249
13.5.1.17 Bioedible Films 249
13.5.1.18 Antioxidant Activity 249
13.5.1.19 Anti-nematode Activity 249
13.5.1.20 Against Parkinson's and Alzheimer's Diseases 250
13.5.1.21 Wound-Healing Activity 250
13.5.1.22 Against Industrial Pollution 250
13.5.1.23 Uses in the Food Industry 250
13.5.1.24 Other Uses 251
13.5.1.25 Side Effects 251

DOI: 10.1201/9781003205067-13

13.5.2 *Rauvolfia serpentina* 251
13.5.2.1 Antidiarrheal Activity 251
13.5.2.2 Anti-Ulcer Activity 251
13.5.2.3 Antidiabetic Activity 252
13.5.2.4 Anti-Alzheimer's Activity 252
13.5.2.5 Antibacterial Activity and Herbal Gels 252
13.5.2.6 Anticancer and Antitumor Activity 253
13.5.2.7 Hypolipidemic Activity 253
13.5.2.8 Anti-SARS Activity 254
13.5.2.9 Anti–Larvicidal Activity 254
13.5.2.10 Anticorrosive Activity 254
13.5.2.11 Anti-oxidative Activity and Anti-Heavy Metal Toxicity 254
13.5.2.12 Nanoparticle Formations 254
13.5.2.13 Associated Microbes 254
13.5.2.14 Side Effects 255
13.6 Conclusion 255
References 256

13.1 INTRODUCTION

"Nature cures, not the physician" are the words by the Father of Medicine, Hippocrates. Since the dawn of the human race, humans have been dependent on nature for basic (food, clothing, shelter) and modern (fertilizers, natural pigments) needs. This reliability on nature over the years has improved traditional knowledge of people toward the plants (Upasani et al., 2017). Over the years the medicinal importance of plants has been realized. Various human civilizations have relied on plant-based medicines for the treatment of diseases. This led to the development of various medicinal systems based on the diversity and availability of plants in their respective geographical regions (Ali, 2020). The documentation of the traditional medicine of Egypt around 2900 BC, known as *Ebers Papyrus*, includes around 700 types of drugs derived from plants. The first millennium recorded the emergence of the two most-known traditional medicinal systems present in the modern age, the Indian Ayurveda and traditional Chinese medicine. Other medicinal systems, like Unani medicine, emerged later. From lichens to higher plants, around 70,000 plant species are proved to have positive therapeutic effect on various diseases. Around 21,000 plants currently have medicinal applications on record, according to the World Health Organization (WHO) (Prasathkumar et al., 2021).

In modern times, there is an increased demand for medicines due to high populations and emergence of various diseases. Synthetic drugs being costly, not completely effective, and often having harmful side effects has forced the world to acknowledge natural or herbal medicines (Malviya and Sason, 2016). Herbal drugs have taken the driver's seat in the modern era. 80% of people are more dependent on herbal drugs, while 25% of prescribed drugs are derived from medicinal plants in developing and developed countries, respectively (Chen et al., 2016). There are a high number of phytonutrients in medicinal plants that have numerous therapeutic effects. The uniqueness of some medicinal plants is such that each part of the plant has a specified active ingredient effective against specific diseases. Some studies have also reported toxic and therapeutic effects shown by the same plant from different parts of the plant. The US alone supports a 5-billion-dollar market for herbal products (Ali, 2020). Chemical substances such alkaloids, flavonoids, and various secondary metabolites all can be used as possible therapeutic drugs. Low cost, fewer side effects, and cultural acceptability are factors which increase the interest of humans toward herbal medicines (Malviya and Sason, 2016).

One-tenth of plant species worldwide are being used in drugs or health products, while the plant distribution is not uniform. India stands second with a rich variety of 7500 species but percentage-wise it stands first with 44% species of medicinal plants (Chen et al., 2016). Indians believe that

Mother Nature itself is the healer. Due to the diversity of cultures in India, each has its own traditional knowledge of medicinal plants for different diseases. It is the belief of practitioners of the traditional Ayurvedic system that every plant has a specific medicinal importance (Upasani et al., 2017). With modern science, the secondary metabolites of these medicinal plants are also being used in other industries, such as food, cosmetics, and fragrances (Malviya and Sason, 2016).

In the current scenario, a proper documentation of the traditional knowledge, bioactive compounds, and secondary metabolites of these plants is needed. The potential benefits and applications of the compounds obtained from these plants need to be investigated. Also, standardizing scientific protocols and validation of products through appropriate clinical trials is equally important (Ali, 2020).

India is the largest producer and exporter of *P. ovata*. thus it has high economic value in India. Developed countries, such as the US import 75% of the total seed husks (Madgulkar et al., 2015). Indian snakeroot has become an endangered species in India due to its over-exploitation for its immense medicinal value (Paturkar and Khobragade, 2016). Thus, this chapter focuses on India's two traditionally important plants *P. ovata* and *R. serpentina*. Commonly, they are known as Isabgol and Indian snakeroot, respectively. Various parts of these plants, such as the roots, leaves, seeds, etc., have various chemical compounds, each of which has been useful in many industries and for many purposes other than medicinal.

13.2 DESCRIPTION AND DISTRIBUTION

13.2.1 *Plantago ovata*

Plantago ovata Forsk. (Family: Plantaginaceae) is one of 300 species belonging to the family Plantaginaceae (Figure 13.1). It is a perennial herb distributed in temperate and tropical regions. *P. ovata* is mainly known for its seed (Tewari et al., 2014). The seed is covered with a husk, a membranous coating on the concave side of the seed. The word *Plantago* is derived from Latin which translates to "sole of the foot", which refers to its leaf, while the word *ovata* refers to the shape of the leaf (Dhar et al., 2005). In India, it is also called Ashwagolam Babka and Spangur, while it is known

FIGURE 13.1 *Plantago ovata* (ICAR-Directorate of Medicinal & Aromatic Plants Research, Anand, Gujarat, n.d.; PubChem. n.d.).

by different names in different regions – Barhanj, Spogel, Ch-chientzu, Obeko, and Buzarqatona in Poland, Iran, China, Japan, and Arabia, respectively (Franco et al., 2020). Blond psyllium is the common English name, but psyllium is more popular; thus, is used for seeds and the complete plant. It is a Greek word meaning "flea" referring to the size, color, and shape of *P. ovata* seeds. Psyllium is the common name for most plant species in the Plantago genus, but in this review, it strictly refers to the *P. ovata* plant. In Sanskrit, *P. ovata* is called "Isabgol"; the root words are "Isab" and "ghol", which translates to "horse" and "flower", referring to the shape of the *P. ovata* seeds (Dhar et al., 2005; Verma and Mogra, 2015).

Psyllium was introduced in India by Muslims in the 16th century. As a medicinal plant, *P. ovata* was used for treatment of chronic dysentery and other gastric problems. The seeds were originally collected from the wilds, but due to high demand and other difficulties, the people began to farm them (Dhar et al., 2005; Verma and Mogra, 2015). Psyllium is native to Mediterranean regions and has high growth rates in agroclimatic regions. Due to low water demand, the plant is limited to arid zones. The cultivation initiated in the districts of Lahore and Multan, and later in India, at Bengal, Mysore, and the Coromandel coast. Psyllium is believed to have originated in Persia (Dhar et al., 2005; Tewari et al., 2014; Verma and Mogra, 2015). Nowadays, it is cultivated in various parts of the world, such as southern Europe, North Africa, West Pakistan, Iran, Arabian countries, and West Asia. India exports 90% gross production of the plant and 93% of the seed husk (Madgulkar et al., 2015; Verma and Mogra, 2015). In Indian states, such as Gujarat, Rajasthan, Maharashtra, Punjab, and Haryana, psyllium is cultivated. Rajasthan and Gujarat are major producers due to their favorable climatic conditions. Gujarat leads with 35% psyllium husk production worldwide (Tewari et al., 2014; Verma and Mogra, 2015).

The crop is cultivated post rainy season, i.e., October to March. *P. ovata* matures in 130 days and positively responds to weather when the temperature is between 15°C and 40°C. The crop develops in a sandy-loam soil even with low fertility. Unusual rains during maturation harms the crop yield (Bemiller et al., 1993; Verma and Mogra, 2015; Franco et al., 2020). *P. ovata* has a short stem, or is stemless, and grows 30–45 cm high. Around 40–86 leaves appear alternately higher on the stem and arranged in rosette fashion close to the ground. Leaves are strap-shaped, linear, filiform having a coat of soft, minute hairs, and are 6–25 cm long and 0.3–1.9 cm wide (Franco et al., 2020). Each spike inflorescence is 0.5–5 cm long and 0.5 cm broad has a cylindrical or ovoidal shape, and can bear 45–70 flowers. Flowers appear roughly after 60 days of plantation, are white in color, 1.25–3.8 cm, and the bracts are 0.4 cm (Dhar et al., 2005; Franco et al., 2020). Flowers favor crossing as they are bisexual, tetramerous, anemophilous, and protogynous. The fruits are 0.8 cm long; an ellipsoid capsule opens up at maturity. They are membranous and obtuse, bearing around 1000 seeds. The seeds are rosy, white, or brown in color. The seeds are ovoid- oblong, boat-shaped, concave-convex, and have a smooth surface (Tewari et al., 2014). A paper covering of the seeds form the seed coat. A secondary deposit in the seed epidermis cells is the source of mucilage.

The husk is separated from seeds through mechanical milling. The husk is odorless, tasteless, white to light pink in color, and has a translucent membranous covering which constitutes 30% of seed weight (Franco et al., 2020). The husk is known for absorbing moisture and forming a mucilaginous substance used in the formulation of various drugs. Both the seeds and husk have medicinal and commercial importance. Psyllium is used in pharmaceuticals, cosmetics, paper- and ink-making, and water-proofing of military explosives (Tewari et al., 2014; Franco et al., 2020). Psyllium has also been incorporated into the food industry, as it was found to be a cheap replacement for agar as a gelling agent in a tissue culture media. It is reported to be a source of fiber, medicinally active, and a completely natural gel-forming polysaccharide. It has high water-holding capacity (up to 80 times its weight) thus having a laxative effect (Dhar et al., 2005; Franco et al., 2020). Psyllium lately has been investigated for various health benefits. Psyllium treats various gastrointestinal problems, obesity, and diabetes. It is used for cancer prevention and also shows prebiotic effects (Franco et al., 2020).

13.2.2 *Rauvolfia serpentina*

Rauvolfia serpentina (L) Bentham ex Kurtz. (Family: Apocynaceae) is one of the 131 species under genus *Rauvolfia* (Figure 13.2) (Malviya and Sason, 2016; Mukherjee et al., 2019). It is a perennial shrub commonly known with different names. Sarpagandha is a common local name for *R. serpentina*, which refers to antidote for snakebites. Due to India's diverse culture, it has various names such as Chandrika in Sanskrit, Chandrabhaga and Chota-chand in Hindi (Keshavan, 2011; Malviya and Sason, 2016). In Sri Lanka, it is known as Acawerya; in Java it is known as Akartikoes or Phoelphandakand; in China it takes the name Lu fu mu (Dey and De, 2010). Other than treating snakebites the herb has also been used to treat mental illness thus known as the "insanity herb" or "PagalkiButi" in Hindi (Paturkar and Khobragade, 2016). Currently, the biggest concern surrounding Sarpagandha is its conservation. Due to various anthropogenic activities leading to erosion of its natural ecosystem and over-exploitation, it is an endangered species. There is a world requirement of 20000 tons/annum of dried Sarpagandha while India itself requires 650 tons for manufacturing processes (Dey and De, 2010; Paturkar and Khobragade, 2016). One report stated that genetic erosion of these species has affected the populations resulting in low alkaloid content. Thus, it has become the only species from Rauvolfia listed in CITES (Convention on International Trade in Endangered Species of wild fauna and flora), Appendix II, and the International Union for Conservation of Nature considers it endangered. It is also endangered in southern Western Ghats, Kanyakumari, and Northeast India. From areas such as south Karnataka and Western Ghats of India, the collection and conservation of the species has been initiated (Dey and De, 2010; Eurlings et al., 2013).

It has been reported that *R. serpentina* originated in Southeast Asia. Now it is distributed worldwide, mainly to tropical regions of Africa, America, Nepal, Sri Lanka, and the Association of Southeast Asian Nations (ASEAN) countries. It is indigenous to India, Bangladesh, and some regions of Asia. In India, it grows at the foothills of the Himalaya range, Gangetic plains, Eastern and Western Ghats, Pune to Cochin and Andaman Island. They can also be seen at sub-Himalayan tracts. The plant thrives best at an altitude from sea level to 1200 m in forests that are moist and deciduous. It also has been associated with bamboo forests in Deccan (Malviya and Sason, 2016; Mukherjee et al., 2019).

R. serpentina mainly flourishes in a tropical climate. The plant prefers the tropical or subtropical regions with temperature around 10–38°C and annual rainfall of 200–250 cm. It also prefers an altitude of 1000–1300 m and soil with high organic matter, rich in nitrogen and humus. *R. serpentina* grows in a variety of soils; when in natural habitat, it prefers clay or clayey loam soil with a pH of 4.6–6.2 (Dey and De, 2010; Paturkar and Khobragade, 2016). It is an evergreen, erect perennial shrub, growing to a height of 60–90 cm. The leaves appear in whorls of 3–5; these are 7–10 cm

FIGURE 13.2 *Rauvolfia serpentina.*

in length and 3.5–5 cm in width. The leaves are lanceolate, brightly colored above, while pale and smooth beneath. The flowers are white with light shades of violet that appear irregularly (Malviya and Sason, 2016; Mukherjee et al., 2019). In Indian conditions, the flowering season is March to May. The inflorescence pedicels and calyx are red, while the corolla is white. The fruit is a drupe, shiny black in color when matured, grows single or in pairs, and 0.5–0.7 in diameter. The root system is tuberous having a soft tap root of 30–0 cm and 1–2.5 cm in length and breadth, respectively for a two-year-old plant (Malviya and Sason, 2016; Paturkar and Khobragade, 2016; Mukherjee et al., 2019). The seeds produced by this plant have been used conventionally for propagation. Due to poor seed viability, germination rate, and loss of habitat the species has declined. Other methods of cultivation, that also help in the conservation of this plant, are root-cutting and *in vitro* culture and micropropagation (Dey and De, 2011; Mukherjee et al., 2019).

The roots and leaves of this plant contain a large number of secondary metabolites. The shrub has been known for its medicinal value since ages ago. This plant has been used to treat high blood pressure and various mental disorders and has sedative properties, which makes it useful for treatment of insomniac patients. In the Ayurvedic medicinal system, this plant has proven to aid various types of illness, and the roots of the plant cure menstrual issues (Malviya and Sason, 2016; Mukherjee et al., 2019). The most important secondary metabolite of this plant is reserpine (Figure 13.4E). Because reserpine is in high demand in the pharmaceutical industry but is cost-prohibitive to synthesize, extraction from roots is preferred (Paturkar and Khobragade, 2016).

13.3 TRADITIONAL KNOWLEDGE

13.3.1 *Plantago ovata*

From ancient times, psyllium has been used throughout the world as a remedy for various diseases. In India, from 1500 BC this plant has been used and mentioned in Ayurvedic literature. In 1927, *Indian Materia Medica,* published by Nadkarni, stated that seeds of psyllium are useful against urinary disorders and dysentery by arresting the pain and flux (Bemiller et al., 1993; Madgulkar et al., 2015). The Chinese also have been using psyllium as traditional medicine from ancient times. *Shennong Bencao* is an ancient Chinese herbal medicine book dated 300 BC–200 AD. The book mentions different parts of the plant having therapeutic effects on coughing, constipation, and hypertension. It also helps in lowering of blood sugar and lipids (Ji et al., 2019). Europeans and Asians have been using psyllium as herbal medicine for treating constipation since the 16th century. At the end of the 20th century, psyllium gained popularity in North America for its healing properties (Madgulkar et al., 2015). In pyretic conditions and urinary tract infections, seeds of this plant are recommended, as they are considered to have cooling and diuretic effects. Seeds are used as a poultice for swelling, boils, ulcers, and also on hairs as a cosmetic. During cold and cough, a tonic prepared from seeds is prescribed. Europeans heat the seeds and use them as a remedy to treat chronic diarrhea. In traditional medicine the seeds are well known to treat dysentery, enteritis, and gastritis. Oral consumption of decoction made from dried seeds is used for treating diarrhea in India. In Thailand and Iran, the water extract from seeds is used for its moisturizing effect on cells. Mucilage obtained from the seeds is applied externally, used as moisturizer for the skin. Seeds are also used for bile-related abnormalities. Gout and rheumatoid arthritis can be treated by application of acetic acid extract obtained from dried seeds. Oral consumption of leaf infusions is advised for treating cold in Spain (Tewari et al., 2014).

13.3.2 *Rauvolfia serpentina*

For Asian countries, this medicinal herb has been of historical significance. Sarpagandha root has been used since pre-Vedic times for treating snakebites and fever. It later gained popularity for its medicinal properties in the Vedic and Ayurvedic periods (Keshavan, 2011). The herb was first mentioned by Sushruta in his book *Sushruta Samhita* in 600 BC. It has also been mentioned in ancient literature by Charaka in 100 BC for its effectiveness against snake and insect bites (Dey and De, 2010; Keshavan, 2011). Ayurveda has been using *R. serpentina* for centuries in the treatment of insanity, insomnia, and snakebites. Sarpagandha ghanavati, Sarpagandha yoga, Sarpagandha churna, and maheshvarivati are some of the preparations made from Sarpagandha in Ayurveda. The roots are valuable and have been used as a remedy for various disorders related to the central nervous system. *R. serpentina* also has great importance in Unani. Pitkriya is an Unani formulation capsule containing Sarpagandha which acts as a sedative, diuretic, and anaesthetic. In Unani medicinal system, *R. serpentina* is used for treatment of urticaria (Dey and De, 2010; Dey and De, 2011). Leaf extract of this plant is used for cataract removal and to prevent eye inflammation (Mukherjee et al., 2019). Crushed roots are consumed for dog bites in some tribal areas. Root extract of the plant has been known to control hypertension. Extracts of leaves, root, and fruit are used by many tribal people for curing gastrointestinal diseases. Stem bark of *R. serpentina* and others trees are given during jaundice. In some districts of Karnataka, the roots of *R. serpentina* have been used for treatment of arthritis (Dey and De, 2011; Bhat et al., 2019). For treatment of respiratory problems root powder of this plant is prescribed, while for treating asthma, leaves and flowers of Sarpagandha are consumed. Roots are also used to treat pain, while root paste is applied on wounds or boils for recovery. Nowadays, it is also incorporated into beverages. For the recovery of livestock from fever and gastrointestinal disorders, roots of Sarpagandha are used. Over the years, tribes have used this herb for treatment of spleen diseases, headache, fever, acquired immuno-deficiency syndrome (AIDS), and menstrual disorders (Dey and De, 2011).

13.4 CHEMICAL DERIVATIVES

13.4.1 *Plantago ovata*

Various investigations have revealed that psyllium produces a decent number of secondary metabolites which include alkaloids, phenols, tannins, saponins, and flavonoids (Table 13.1) (Tlili et al., 2019). Compounds such as phenols, triterpenoids, monoterpenes, and flavonoids play a major role in antioxidant and anticancer activity (Tewari et al., 2014; Franco et al., 2020). A high-performance liquid chromatography (HPLC) analysis of *P. ovata* leaf extract showed the presence of 1027.6 µg/g phenolic, thus leading to high antioxidant activity in the plant responsible for eliminating free radicals (Tlili et al., 2021).

The phytochemicals present in this plant are biologically active which can be used for drug development. Gardoside, 5, 6, 8-Epiloganic acid, and plantamajosides are some of the phytochemicals extracted (Figure 13.3). Plantamajoside is known for its anti-inflammatory activity and antioxidant effect by radical scavenging (Reddy et al., 2018).

TABLE 13.1
Phytochemical Composition of *Plantago ovata*

Phytochemicals		Amount
Alkaloids		0.1±0.01%
Crude saponins		0.8±0.05%
Condensed tannins		5.13±0.11 mg/ml
Total flavonoids	70% methanol	17.63±0.41 mg/g
	70% acetone	16.39±0.26 mg/g
Total polyphenols	70% methanol	23.92±0.021 mg/g
	70% acetone	27.60±0.039 mg/g

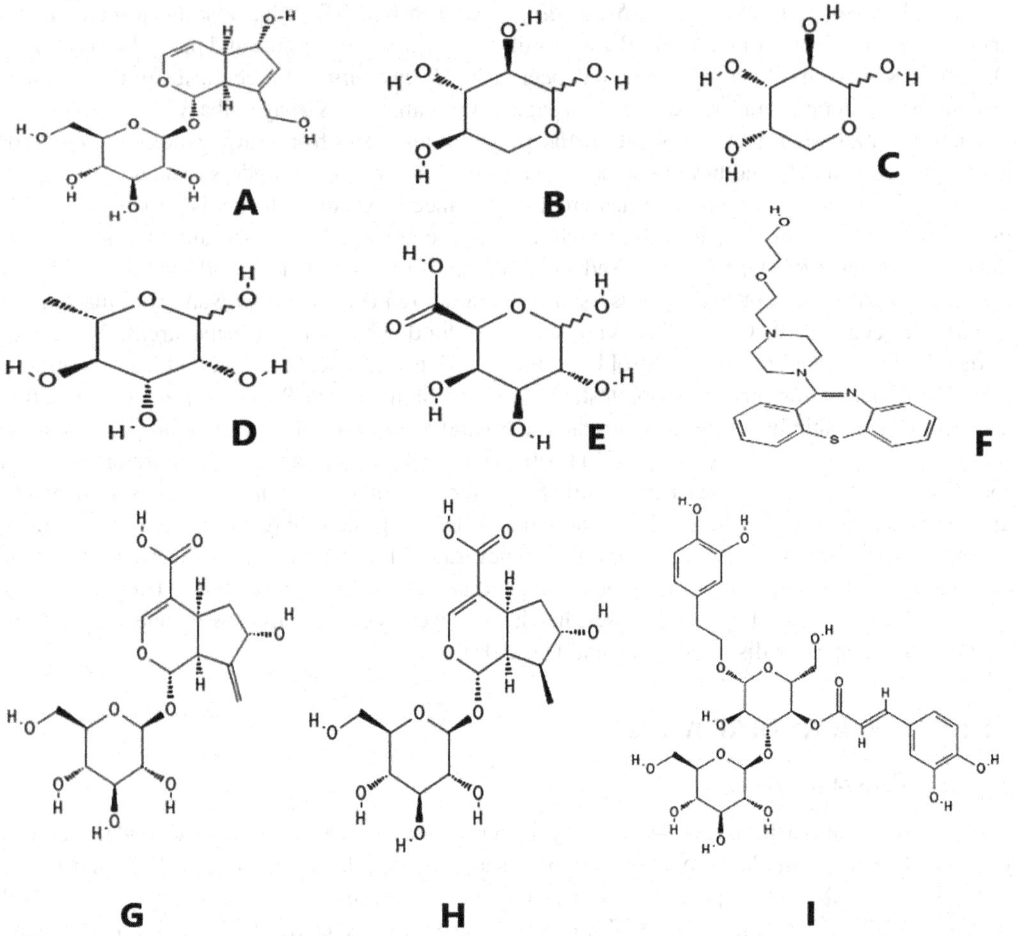

FIGURE 13.3 Phytochemicals and carbohydrates of *Plantago ovata*: A) Aucubin(CID-91458); B) D-xylose (CID-135191); C) L-arabinose (CID-439195); D)Rhamnose (CID 25310); E) D-galacturonic acid (CID 439215); G) Gardoside(CID-46173850); H) 5,6,8-Epiloganic Acid (CID-158144); I) Plantamajoside (CID-5281788) (ICAR-Directorate of Medicinal & Aromatic Plants Research, Anand, Gujarat, n.d.; PubChem. n.d.).

The vitamins and minerals also play an important role in improving immunity and overall health status (Table 13.2) (Sagar et al., 2020). Other than mucilage, 5% of fatty oil is present in *P. ovata* seeds, while acetylcholine-like action is shown by aucubin (Figure 13.3A) and tannins. Aucubin and other chemical constituents show activity against leukemia and lymphoma cell lines. Valine, alanine, glycine, glutamic acid, cystine, lysine, leucine, and tyrosine are amino acids found in the seeds of *P. ovata*. Linoleic acid, oleic acid, and palmitic acid are some of the fatty acids found in psyllium husk. The seeds are rich in starch when the husk is removed. The amount of mucilage is greater in wild species than cultivated ones. The jelly-like mucilage is used for treating patients having constipation. The mucilage is not affected by the gastrointestinal environment. It also absorbs various toxins and prevents them from harming the body (Tewari et al., 2014; Franco et al., 2020).

P. ovata is a good source of natural polysaccharides. These polysaccharides in recent times have been used for drug delivery, removal of flocculants, etc. (Gonçalves and Romano, 2016). The polysaccharide consists of 65% D-xylose, 20% L-arabinose, 6% Rhamnose, and 9% D-galacturonic acid (Figure 13.3). *P. ovata* husk also includes 0.94% of protein, 4.07% of ash, and 6.83% of moisture. Carbohydrates constitute the largest part (84.98%) in the seed husk. Xylose (503.1 μg/g) and arabinose (203.2 μg/g) were the most abundant (Tewari et al., 2014; Franco et al., 2020). The soluble

TABLE 13.2
Mineral and Vitamin Composition of *Plantago ovata*

Vitamin	Amount
Niacin	4.18±0.37 mg/100 g
Pantothenic acid	0.87 mg/100 g
Riboflavin	0.13 mg/100 g
Total Carotenoids	183.86±2.40 μg/100 mg
Total Folates	32.19±0.70 μg/100g
Total Vitamin B6	0.25±0.90 mg/100 g
Minerals	**Amount**
Arsenic	22.53 μg/100 g
Calcium	1500 μg/g
Cobalt	31.20 μg/100 g
Copper	2.39 mg/100 g
Iron	6.75 mg/100 g
Lithium	12.65 μg/100 g
Magnesium	150 μg/g
Manganese	1.06 mg/100 g
Molybdenum	27.14 μg/100 g
Phosphorous	140 μg/g
Potassium	8500 μg/g
Selenium	18.06 μg/100 g
Sodium	640 μg/g
Sulphur	23 μg/g
Zinc	3.15 mg/100 g

(arabinoxylan) and insoluble (cellulose, lignin) polysaccharides are present in the seed husk. The arabinoxylan and water-soluble hemicelluloses constitute around 60% of the husk. Arabinoxylan is a hemicellulose that has xylose and arabinose linked together. Arabinoxylan is known to have antioxidant activity and other health benefits. Both insoluble and soluble fibers help in restoration of gastrointestinal functions and reduction in risk of disorders such as constipation and hemorrhoids (Khan et al., 2021). Due to high fiber content, *P. ovata* degrades slowly causing formation of butyrate and acetate. The butyric acid is known to have antineoplastic activity against colorectal cancer. Thus, the polysaccharides also show anticancer activity (Tewari et al., 2014; Tlili et al., 2019; Franco et al., 2020).

13.4.2 *Rauvolfia serpentina*

R. serpentina has a number of secondary metabolites which include alkaloids, phenols, tannins, and flavonoids (Table 13.3). Alkaloids belong to a large group of organic molecules that contain a nitrogen atom attached to the heterocyclic ring. Alkaloids are secondary metabolites produced for defensive purposes against animals or pathogens (Bunkar, 2017). There are three classes of alkaloids present in *R. serpentina* which include weakly basic indole alkaloids, intermediate basic indole alkaloids, and strong anhydroniumbases. The alkaloids include ajmaline, ajmalimine, ajmalicine, deserpidine, reserpine, reserpiline, rescinnamine, rescinnamidine, sarpagine, serpentine, serpentinine, and yohimbine. (Figure 13.4) (Malviya and Sason, 2016).

Reserpine is the one of the most used alkaloids and has several therapeutic applications. Over the years, it has been useful in the treatment of hypertension, cardiovascular diseases, and neurological diseases. Reserpine acts as a hypotensive agent by depleting the catecholamine. Angiotensin-converting

TABLE 13.3
Phytochemical Composition of *Rauvolfia serpentina*

Phytochemicals	Amount (dry wt.)
Alkaloids	1.48±0.02 mg/100g
Flavonoids	1.72±0.11 mg/100g
Phenols	1.86±0.11 mg/100g
Tannins	0.51±0.20 mg/100g

enzyme (ACE) is inhibited by rescinnamine (Figure 13.4G) to halt the conversion of angiotensin I, resulting in a decrease of plasma angiotensin II, and then lowering the blood pressure. Thus, both reserpine and rescinnamine act as hypotensive agents (Bunkar, 2017; Malviya and Sason, 2016; Shah et al., 2020). Other alkaloids also have a major role to play in various therapeutic activities of *R. serpentina.* Ajmaline (Figure 13.4A) is used as a class 1 anti-arrhythmic and useful in diagnosing a hereditary cardiac disorder known as Bruguda syndrome. Ajmalicine (Figure 13.4C) is useful in treatment of circulatory diseases, while serpentine (Figure 13.4J) has antipsychotic properties. Deserpidine (Figure 13.4D) has both hypotensive and antipsychotic behavior; and Yohimbine (Figure 13.4L), being an alpha-adrenergic antagonist, helps to treat erectile dysfunction (Kumari et al., 2013).

Itoh et al., (2005) reported five new indole alkaloids (*N*b-methylajmaline, *N*b-methylisoajmaline, 3-hydroxysarpagine, yohimbinic acid, isorauhimbinic acid) along with 2 glycosides and 20 compounds in *R. serpentina.* The glycoside 6'-*O*-(3,4,5-trimetoxybenzoyl) glomeratose A is a sucrose derivative, while the 7-epiloganin is an iridoidglucoside. All these compounds are being studied for drug development against various diseases.

In the modern era, phytomedicines have been incorporated into our medical systems. Thus, plant-derived molecules (PDMs) are being constantly used for drug discovering. The complete process of drug discovering can be rationalized by various computational approaches. Approaches such as molecular docking, ligand-based virtual screening, and molecular dynamics are currently used for reducing the cost of drug development (Pathania et al., 2015). Pathania et al., (2015) has structurally compiled 147 PDMs of *R. serpentina* and made a database, Serpentina DB. This database includes plant part source, chemical classification with the International Union of Pure and Applied Chemistry (IUPAC), etc. The database also identifies analogs of natural molecules on the ZINC database. Of the 147 PDMs, 122 are alkaloids, 7 iridoidglucosides, 6 phenols, 4 phytosterols and anhydronium bases each, 3 glycosides, and 1 fatty acid (Pathania et al., 2013).

Flavonoids reduce reactive oxygen species (ROS) and prevent cell damage. Presence of flavonoids also reduces risk of cardiovascular diseases. Flavonoids have antioxidant, anti-inflammatory, and anticancer activities. Thus, *R. serpentina* is currently being studied for drug discovery (Bunkar, 2017; Shah et al., 2020).

Phenols, a secondary metabolite are mostly toxic to pests and pathogens. Thus, it is used as an antimicrobial agent. Antidiabetic, hypolipidemic, and antimicrobial properties of *R. serpentina* extract are due to the phenolic compounds present in high amounts. Saponins have been identified in more than 70 families of plants. These are basically triterpenes or sterols having a glycosidic bond. Saponins have cholesterol-binding ability, help in forming foams, and hemolytic activity. Due to the blood-coagulating properties of saponins, *R. Serpentina* is used to stop bleeding and treat wounds (Bunkar, 2017; Shah et al., 2020).

Most of these compounds are bioactive, thus scientists have been working on their analysis and medicinal importance. The composition of minerals and vitamins is equally important (Table 13.4). The *R. serpentina* has the highest amount of ascorbic acid – a vital vitamin for the body which has diverse roles in physiological functions throughout the body. Other vitamins, such as niacin, riboflavin, and thiamine, are also present (Bunkar, 2017).

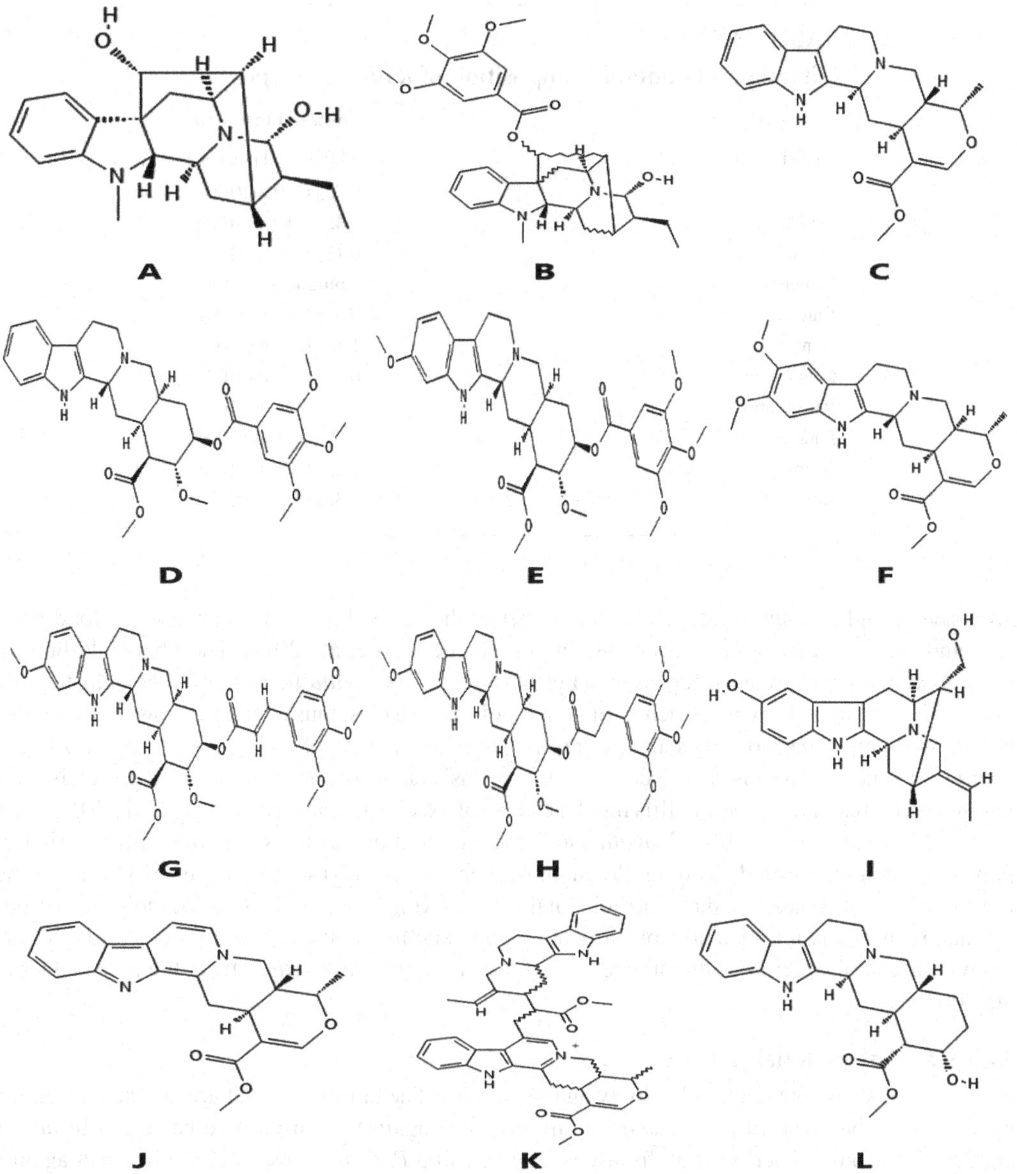

FIGURE 13.4 Chemical structures of alkaloids in *Rauvolfia serpentina*: A) ajmaline (CID-6100671); B) ajmalimine (CID-42621393); C) ajmalicine (CID-441975); D) deserpidine (CID-8550); E) reserpine (CID-5770); F) reserpiline (CID-67228); G) rescinnamine (CID-5280954); H) rescinnamidine (CID-184180); I) sarpagine (CID-12314884); J) serpentine (CID-73391); K) serpentinine (CID-5351576); L) yohimbine (CID-8969) (ICAR-Directorate of Medicinal & Aromatic Plants Research, Anand, Gujarat, n.d.; PubChem. n.d.).

13.5 THERAPEUTIC BENEFITS, APPLICATIONS, AND SIDE EFFECTS

13.5.1 *Plantago ovata*

13.5.1.1 Gastrointestinal Functions

P. ovata has been known for its diverse therapeutic effects. Psyllium is used for treatment of irritable bowel diseases (IBD). One study reported that *P. ovata* seeds cause higher production of fatty acid chains, decrease pro-inflammatory mediators, and improve intestinal cell structure. Chlorigenic

TABLE 13.4
Vitamin and Mineral Composition of *Rauvolfia serpentina*

Vitamins	Amount (dry wt.)
Ascorbic acid	44.03±0.20 mg/100g
Niacin	0.02±0.10 mg/100g
Riboflavin	0.42±0.10 mg/100g
Thiamine	0.18±0.02 mg/100g
Minerals	**Amount** (dry wt.)
Calcium	0.32±0.10 mg/100g
Iron	1.85±0.20 mg/100g
Magnesium	0.10±0.20 mg/100g
Phosphorus	0.18±0.22 mg/100g
Potassium	0.04±0.11 mg/100g
Sodium	0.02±0.10 mg/100g
Zinc	5.38±0.11 mg/100g

properties could be found in the alcoholic extract of the seeds. It causes a decrease in blood pressure and stimulates intestinal movements in animals (Tewari et al., 2014). The fibers add bulk to the stools; thus, they are used for faster relief of dysentery. *P. ovata* helps in both constipated and diarrheal conditions. *P. ovata* is rich in fiber and polysaccharides, thus helping in bowel movements by binding with water molecules and promoting smooth excretion. Another study also reported *P. ovata* supplementation causes 50% decrease for inconsistent stools. *P. ovata* forms a protective lining in gastrointestinal regions, nullifying the effects of excess stomach acids (Soni et al., 2017; Xing et al., 2017; Zhang et al., 2019). *P. ovata* supplementation improves the symptoms of hemorrhoids, preventing progression and bleeding. Pregnant women also are advised to consume psyllium for the prevention of constipation and hemorrhoid anal fissure (Xing et al., 2017). *P. ovata* polysaccharides can act as short chain fatty acids for lubricating and expelling colon contents (Ji et al., 2019). It is known for complete removal of pathogens, such as *Entamoebahistolytica,* from the body (Dhar et al., 2005).

13.5.1.2 Antibacterial Activity

P. ovata also shows antibacterial activity against *E. coli.* The tannins present are less active against entamoeba or bacteria. Ethanolic extracts are effective against gram-positive bacteria (Tewari et al., 2014). A study on periodontal disease was done using *P. ovata* leaves and seed extract against common pathogens *P. gingivalis, F. nucleatum,* and *A. actinomycetemcomitans.* The extract, being effective against periodontal pathogens, has proved to be a good antibacterial agent (Reddy et al., 2018). Another study using *P. ovata* leaf extract on *L. monocytogenes, S. aureus,* and *S. epidermidis* reported antibacterial activity of the extract. The extract also showed antibacterial activity against methicillin-resistant strains, but activity against gram-negative bacteria was not detected (Tlili et al., 2021). A study was conducted using a nanocomposite hydrogel system (rGO-5FU-CMARX) to measure the antibacterial activity of *P. ovata.* The hydrogel systems (1–4) reported antibacterial activity against *S. aureus* and *P. aeruginosa.* The hydrogel system (1) shows the highest antibacterial activity due to less cross-linking in its system (Nazir et al., 2021).

13.5.1.3 Immunomodulatory Actions

The seed extract has been seen to affect the immune system in rabbits by increasing the white blood cells and leucocytes. The extract has also been observed to reduce the anti-HD antibodies (Tewari et al., 2014).

13.5.1.4 Hypolipidemic Activity

Psyllium husk helps in reducing angina attack in patients suffering from cardiovascular diseases. *P. ovata* seed coat has been reported to reduce risk of cardiovascular diseases by increasing the high-density lipid (HDL) cholesterol. A study reported reduction of low-density lipid (LDL) cholesterol and total cholesterol by 8% and 6%, respectively, when seeds were administered to patients suffering from high cholesterol (Tewari et al., 2014; Franco et al., 2020). The seed oil is rich in linoleic acid and acts as a hypocholesterolemic agent, causing reduction of serum cholesterol in animals (Tewari et al., 2014). *P. ovata* husk reduces serum cholesterol by preventing bile acid reabsorption. Various experiments have concluded on *P. ovata* husk to be a hypocholesterolemic agent. For reduction of egg yolk and blood cholesterol levels in layer birds, *P. ovata* is being added to their diet (Xing et al., 2017).

13.5.1.5 Anti-obesity Activity

A study was conducted on the effect of *P. ovata* on appetite. The study reported reduction of hunger and increase in feeling of fullness between meals ((Soni et al., 2017; Sahebkar-khorasani et al., 2019). The seeds of this plant have been proven to reduce insulin, lipids, leptin, oxidative stress, and improve body mass index (BMI) in fat-induced rats. Thus, extracts are used to treat obese patients and many diet supplements. Psyllium has been incorporated in various food products (Tewari et al., 2014; Franco et al., 2020).

13.5.1.6 Anticancer Activity

A study was conducted using a nanocomposite hydrogel system (rGO-5FU-CMARX) for the care and treatment of patients suffering from melanoma skin cancer. The nanocomposite hydrogel systems (1–4) had significantly high anticancer activities against U-87 cell lines. The hydrogel system 1 has high anticancer activity due to low cross-linking of polymers when compared to the other three systems. Thus, the hydrogel systems (1–4) could be useful in drug development for treatment of skin cancer (Nazir et al., 2021). Hot water extract of *P. ovata* seeds and husk, along with alkaline extract of *P. ovata* leaves, has anticancer activities. Polysaccharides extracted from *P. ovata* leaves showed anticancer, antitumor, and antiproliferative activities against hepatocellular carcinoma cell line (Huh-7) and human epithelioid cervix carcinoma cell line (HeLa) (Patel et al., 2018; Patel et al., 2019; Zhang et al., 2019). The ethanolic extract of *P. ovata* also showed anticancer activity against colon carcinoma (CaCo-2) and (myelogenous leukemia (K-52) cell lines (Tlili et al., 2019).

13.5.1.7 Coagulation Activity

A study conducted on water treatment used a bio-coagulant extracted from *P. ovata* seeds using $FeCl_3$. The electron dispersive spectroscopy results concluded that the extract mainly consists of protein and small amounts of phospholipids acids. The FCE ($FeCl_3$ induced crude extract) was successful in water turbidity removal at an optimum dose of 0.25 mg/l and a negligible change in dissolved organic carbon. High bacteriological quality of water was achieved after treatment with FCE (Ramavandi, 2014). Another study used four extracts from *P. ovata* seeds using different solvents, i.e., distilled water, tap water, sodium chloride, and ammonium acetate. All the extracts were less effective in low turbidity. Sodium chloride extract (SCE) showed the highest coagulation activity compared to other extracts. The SCE was also effective in alkaline conditions. The coagulation activity is dependent on temperature, time, pH, etc. (Dhivya et al., 2017). Thus, the extract can be used as a natural coagulant against a range of turbidity in water treatment plants to achieve the standards of drinking water (Ramavandi, 2014).

13.5.1.8 Antileishmanial Effect

A study conducted on *Leishmania major*-infected mice reported that a combination of *P. ovata* powder and white vinegar effectively heals the lesions. The group treated with *P. ovata* powder and water combination healed 3 of the 5 mice treated. Also, the combination of *P. ovata* powder and white vinegar healed six of the ten mice. Though the antileishmanial effect is related to white vinegar, *P. ovata* powder plays a supportive role. Comparatively, treatment with *P. ovata* is less painful and, thus, could play a major role in drug development against leishmaniasis (Moshfe et al., 2011).

13.5.1.9 Natural Super Disintegrant

P. ovata seed husk contains a high proportion of polysaccharides. The extracted polysaccharides have been studied as a super disintegrant for fast-dissolving tablets. Valsartan, used for treating hypertension, was the model drug for the study. Tablets with concentration of 7.5% w/w polysaccharide showed rapid wetting and disintegration time when compared with the same concentration of the control Crospovidone, a synthetic super disintegrant (Pawar and Varkhade, 2014). Another study used *P. ovata* husk powder in the manufacture of orodispersible tablets of the model drug meloxicam. The formulations having lowest wetting and disintegration time with highest water absorption ratio had 16 mg of psyllium husk powder. It outperformed the control and other formulations. Thus, *P. ovata* can be used in orodispersible tablets and its formulations (Draksiene et al., 2019). The *P. ovata* husk polysaccharide is inexpensive and nontoxic and has fewer side effects, good stability, patient compliance; it is also highly bioavailable, and manufacture-friendly. Thus, *P. ovata* could be used as a natural super-disintegrant (Pawar and Varkhade, 2014).

13.5.1.10 Anticorrosive Activity

Arabinosyl (galacturonic acid) rhamnosylxylan (AX), a polysaccharide present in the mucilage of *P. ovata,* is the main constituent responsible for its anticorrosive activity. A study proved AX is an inhibitor of corrosion in carbon steel present in 1M HCL medium. The efficiency of corrosion inhibition increases with high AX concentrations and solution temperatures. The role of AX in carbon steel protection from acid corrosion was confirmed by quality chemical analysis. Various analyses also proved AX to be a green corrosion inhibitor with low environmental risks (Mobin and Rizvi, 2017).

13.5.1.11 Used for Making Natural Eye Drops

A study conducted on *P. ovata* used its mucilage as an ophthalmic drop for the treatment of dry eye disease. The double blind randomized clinical trial reported improvement in symptoms such as light sensitivity, sore eyes, and blurred vision. The hydrating, anti-inflammatory, and antioxidant effects of *P. ovata* play a major role in the treatment. Reports concluded that the mucilage is an inexpensive, natural, safe lubricant having beneficial effects on the ocular symptoms of patients suffering from dry eye disease (Nili et al., 2019).

13.5.1.12 Activity as Lead Biosorbent

Lead is one of the most toxic heavy metals in the environment and poses a threat to humans and animals. It affects the liver, nervous system, and hematopoietic system (mainly in humans). A study investigated *P. ovata* seed mucilage for its lead (II) absorption capability. Administration of 2% psyllium orally caused a decrease in lead absorption and lesions in organs, and an increase in fecal excretion of lead in mice. Thus, it could be an alternative treatment method for lead-exposed or contaminated populations (Basiri et al., 2020).

13.5.1.13 Anti-ulcer Activity

The aqueous extract of *P. ovata* is useful in prevention of lesions in gastrointestinal parts (Tewari et al., 2014). A study was conducted using aqueous extract of *P. ovata* seeds against peptic ulcers in rats induced with indomethacin. A histological and morphological study of the groups reported the reduction of ulcer index in the group administered with extract. This indicates prevention of gastrointestinal lesions by the *P. ovata* seeds. Thus, it could be used for possible drug development against antiulcer activities (Bagheri et al., 2018). Another study was conducted on oral mucositis, a side complication caused during treatment of cancer therapy. The trial included patients suffering from breast cancer who were receiving chemotherapy. They were administered with per dose mouthwash having 500 mg of husk powder and 3 drops of vinegar in 30 ml water. The oral care protocol resulted in reduction in severity of oral mucositis and associated pain. Although it improves quality of life in patients, further studies are needed (Hasheminasab et al., 2020).

13.5.1.14 Hepatoprotective Activity

A study used aqueous extract of *P. ovata* seeds against peptic ulcers in rats induced with indomethacin. Histological analysis of rats administered with aqueous extract showed hepatoprotective effect by maintaining the morphology and functioning integrity (Bagheri et al., 2018). Another study on CCl_4-induced rats reported that administration of *P. ovata* husk mucilage of 100 mg/kg acts as hepatoprotective agent. The mucilage restores the hepatic deterioration in the rat model. The study reported adequate liver protective potential of the *P. ovata* (Wahid et al., 2020).

13.5.1.15 Antidiabetic Activity

A study was conducted using *P. ovata* seeds on streptozotocin-induced diabetic rats having a high fat diet. The study reported that administration of *P. ovata* seeds caused a decrease in blood glucose levels. There was improvement in the lipid profiles, functioning of liver and kidneys, with an increase of antioxidant enzymes in the rats. Thus, it has antidiabetic effects and could be useful in treatment of diabetes and related complications (Abdel-Rahim et al., 2016).

13.5.1.16 Anti-inflammatory Activity

A study was conducted on periodontal disease testing anti-inflammatory activity of *P. ovata* leaf and seed extract against inflammatory mediators. These included matrix metalloproteinase-2 (MMP-2) and matrix metalloproteinase-9 (MMP-9). The extract proved to be a weak anti-inflammatory agent with inhibition of 30% against MMP-2 and 40% against MMP-9 (Reddy et al., 2018). Methanolic extract of *P. ovata* was able to inhibit denaturation of albumin, thus confirming its anti-inflammatory activity (Tlili et al., 2019).

13.5.1.17 Bioedible Films

Non-biodegradable material used in food packaging has been harmful to the environment and humans. A study was conducted that used hydrocolloid of *P. ovata* seeds for formation of biodegradable edible films. Hydrocolloid along with plasticizer (glycerol) was used in formation of edible films. The study reported that *P. ovata* seeds have a good potential for making edible films having various future prospects (Ahmadi et al., 2012). The use of psyllium in biodegradable films reduces its solubility and permeability by water vapor but increases thermostability and tensile strength (Belorio and Gómez, 2022).

13.5.1.18 Antioxidant Activity

Hot water extract of *P. ovata* seed, husk, and leaves and alkaline extract of the plant leaves were reported to have antioxidant activities (Zhang et al., 2019). Antioxidant activity of *P. ovata* leaves, seeds, and husk was confirmed using 2,2'-azino-bis (3-ethylbenzothiazoline-6-sulfonic acid) (ABTS) assay, 2,2-diphenyl-1-picrylhydrazyl (DPPH) assay, hydroxyl radical scavenging, and reducing capacity (Patel et al., 2018; Patel et al., 2019; Wahid et al., 2020). Another study related to genes of phenols and flavonoids in *P. ovata* leaves and ovary tissue reported higher antioxidant activity than seed husk (Kotwal et al., 2019).

13.5.1.19 Anti-nematode Activity

The water seed extract causes death in nematodes, such as *Meloidogyne incognita,* at various concentrations (Franco et al., 2020).

13.5.1.20 Against Parkinson's and Alzheimer's Diseases

The pharmacokinetics of levodopa, a dopamine precursor, is shown to be improved by *P. ovata* husk. This improves the drug response in patients suffering from Parkinson's disease (Goncalves and Romano, 2016). A study reported the inhibiting activity of *P. ovata* against the acetylcholinesterase enzyme responsible for hydrolysis of acetylcholine. Thus, it could be used to overcome acetylcholine deficiency and used in the treatment of Alzheimer's disease (Tlili et al., 2019).

13.5.1.21 Wound-Healing Activity

A study was conducted on albino rats suffering from second-degree burns. The rats when treated with *P. ovata* seeds powder resulted in 60% of wound-healing, fibrous healing, and a significant decrease in wound size. The study concluded with *P. ovata* seeds powder having therapeutic effects for second-degree burn treatment. Despite these positive results, further detailed studies are needed on different components of *P. ovata* having wound-healing activity (Jalilimanesh et al., 2021).

13.5.1.22 Against Industrial Pollution

P. ovata has been useful in the treatment of domestic and industrial wastewater. It helps by removal of solid waste and dye produced from the effluents of the tannery, domestic, and textile industries (Ahmadi et al., 2012).

13.5.1.23 Uses in the Food Industry

Foods with added ingredients that have beneficial effects on human health are on the rise. Thermal analysis conducted on polysaccharide extracted from *P. ovata* leaves confirmed the biopolymer having structural stability and viscosity. This could be used in the food industry as a thickening or gelling agent (Patel et al., 2019). A study conducted on the poultry meat formulations investigated the addition of *P. ovata* husk. The formulation of 40% of water and 3% *P. ovata* husk resulted in high production yields, low hardness, and low lightness – certainly the best results for the producers. Further research for sensory analysis, shelf-life, and stability needs to be conducted (Zajac, 2020). Another study on the effect of *P. ovata* husk on myofibrillar protein gelation was investigated. A 2% addition of *P. ovata* husk gave the best results on water-holding capacity (53.8%), and textural property of the gel. The gel strength and adhesiveness increased by 2 and 6 folds, respectively. Thus, it could potentially be used as an ingredient in meat production (Zhou et al., 2021).

A study used *P. ovata* husk gum as a fat substitute for producing low-fat yogurt gel. At 0.12% of husk gum and 0.63% of fat, the optimum formulation was achieved. Benefits such as optimized pH, total titratable acidity, minimized whey separation, and maximized firmness, viscosity, and stability, was achieved in this formulation of yogurt. These yogurts were more acceptable due to aroma, texture, and all other factors (Ladjevardi et al., 2015). Results from another study showed that addition of 2% *P. ovata* seed mucilage showed positive growth effect and increased survival rate of *L. acidophilus* in probiotic yogurt. Improved water-holding capacity (89%), and low whey separation (6%) were reported. Higher acidity along with low pH was observed when compared to the control. In terms of acceptance of appearance and texture, the probiotic yogurt having 1% mucilage was highly accepted (Mehrinejad Choobari et al., 2021).

A study revealed that *P. ovata* seeds can be used as a substitute for hydrocolloids in the production of gluten-free breads, enhancing the dough yield and textural properties (Ziemichod et al., 2019). Various formulations of *P. ovata* have been used in canola oil, pizza dough, cookies, cakes, and pasta as substitutes for oil or fat. A reduction of glycemic response could be observed in pasta with 15% psyllium. Used as stabilizers and gelling agents in dairy products such yogurt, ice creams, etc. *P. ovata* also improves texture of sugar-free products and reduces syneresis in jam and apple jelly. Additions of *P. ovata* show prebiotic effects in various products (Belorio and Gomez, 2022). *P. ovata* is being used as a substitute for sodium alginate in ice cream manufacturing and in chocolate-making (Dhar et al., 2005)

13.5.1.24 Other Uses

P. ovata has also been used in the cosmetic industry as a base, due to its sizing purpose, and for setting and dressing of hair (Dhar et al., 2005). It is a good substitute for agar-agar and is also used in dyeing. The seed gum is used to produce teeth cleaning powders and germ-killing lubricating gels. It can also be used as cattle and bird feed (Tewari et al., 2014; Franco et al., 2020). *P. ovata* has also been used in treatment of dysphagia by increasing bolus homogeneity and saliva miscibility (Belorio and Gomez, 2022). *P. ovata* husk is used in milk as a replacement supplement to neonatal dairy calves, improving their physiological conditions, performance, and overall health (Xing et al., 2017).

13.5.1.25 Side Effects

13.5.1.25.1 Absorption of Drugs

Psyllium is a common dietary supplement, used as a laxative, and patients do not consider it a normal medication. But these supplements interact with other drugs by slowing down their absorption. Psyllium is known to absorb these drugs on its surface, thereby reducing the drug concentration in the blood. Psyllium, when co-administered with carbamazepine or lithium, has been known to reduce their absorption. Thus, the dose time of psyllium should be separated. Also, the laxative contains hydrocolloid fibers which reduce the carbohydrate absorption rate and gastric emptying (Soni et al., 2017; Mehmood et al., 2019).

13.5.1.25.2 Absorption of Minerals

Trace minerals play an important role in the functioning and good health of the body. A study reported that increases in fecal excretion of trace minerals (Zn, Cu, and Mn) may be due to the consumption of 25g *P. ovata* husk (Soni et al., 2017).

13.5.1.25.3 Bloating

P. ovata is commonly prescribed for people suffering from constipation. Over-consumption of the powder can block gas passage from the gastrointestinal (GI) tract to the rectum. Bloating is caused by this gas retention (Soni et al., 2017).

13.5.1.25.4 Other Effects

A study has reported adverse effects, such as gastrointestinal disturbances, mild headache, along with vomiting and nausea. The fibers present in the husk also cause feelings of fullness, thus delaying the absorption of nutrients (Sahebkar-khorasani et al., 2019). *P. ovata is* responsible for causing allergies in the work environment. A pigment released from powdered seeds can cause injury to the kidney. Consumption of unsoaked or less-soaked seeds can cause gastrointestinal disorders (Dhar et al., 2005).

13.5.2 *Rauvolfia serpentina*

13.5.2.1 Antidiarrheal Activity

Traditionally, *R. serpentina* is used for the treatment of various diseases. One of them is diarrhea. Thus, to verify the antidiarrheal activity a study was conducted on castor-oil induced diarrhea in model organisms. Methanolic extract of *R. serpentina* leaves was used which showed presence of alkaloids, saponins, phenols, etc. All the administered doses of extract showed significant increase in antidiarrheal activity (30–40%), thus supporting its traditional use. In the charcoal meal test, the extract reduced the transit time in comparison to the reference (Ezeigbo et al., 2012).

13.5.2.2 Anti-Ulcer Activity

Peptic ulcer disease has a high mortality rate and is a prominent cause for surgeries worldwide. A major cause is the imbalance of gastrointestinal offensive and defensive factors. Reserpine shows anti-ulcer activity in stress-induced model organisms by increasing mucin secretion (Jain, 2016).

13.5.2.3 Antidiabetic Activity

Today, diabetes affects millions of people worldwide. It is estimated that 435 million individuals will be suffering from diabetes by 2030. Thus, a study was conducted on the effect of methanolic root extract obtained from *R. serpentina* on diabetic mice induced through alloxan (a diabetogenic chemical). The extract was found to increase body weight, regularize the glucose and insulin levels and their ratios, and improve glycosylated and total hemoglobin in test model groups. The test groups also reported decreased levels of total cholesterol, triglycerides, and a very low density of lipoprotein cholesterols. Alanine transferase (ALT) levels were back to normal, and lipolysis with increased amount of glycogenesis was found in the liver tissues. Thus, the methanolic root extract was responsible for improving antiatherogenic, cardioprotective, and glycemic indices in the diabetic mice. The extract thus can be considered an antidiabetic agent (Azmi and Qureshi, 2012). Another study on fructose-induced type 2 diabetic mice concluded that the methanolic root extract plays a vital role in homeostasis of carbohydrates and lipids via inhibiting absorption or decreasing resistance of fructose (Azmi and Qureshi, 2016).

Various studies on the methanolic root extract have proven that it increases glucose tolerance in mice. Methanolic and aqueous methanolic root extract has antioxidant and hematinic properties in type 2 diabetic mice. The aqueous methanolic extract is more effective due to the presence of phenols. Methanolic root extract plays a role in lowering atherogenic dyslipidemia and atherosclerosis in diabetic mice. The *in silico* studies have proved that *R. serpentina* has therapeutic effect by lowering the glycemic index and hypercholesterolemia (Azmi et al., 2021a; Azmi et al., 2021b).

Aldose reductase is a possible target for treatment of diabetes because of being associated with secondary complications in diabetes. Using the SerpentinaDB database 2, plant-derived molecule (PDM) leads, i.e., indobine and indobinine were identified as potential inhibitors of aldose reductase. Further, through structural analogs, 16 leads were determined from the ZINC database. All the leads thus could help in the design of potential drugs that would act as aldose reductase inhibitors with minimal side effects (Pathania et al., 2013).

13.5.2.4 Anti-Alzheimer's Activity

The two important secondary metabolites, reserpine and ajmalicine, were studied as multi-target directed ligands. These indole alkaloids may provide therapeutic relief to patients suffering from Alzheimer's disease. Both the alkaloids showed inhibition of anti-cholinesterase, BACE-1 (β-site amyloid precursor protein cleaving enzyme 1), MAO-B (monoaminoxidase-B enzyme) and were backed by *in silico* analysis. Reserpine and ajmalicine both have anti-aggregational properties and are neuroprotective against Aβ (amyloid beta) toxicity and oxidative stress induced in PC-12 cells. This study proves that both alkaloids can be developed into novel drugs and used for alleviating the symptoms of Alzheimer's disease (Kashyap et al., 2020).

13.5.2.5 Antibacterial Activity and Herbal Gels

A study was conducted to verify the effects of reserpine on *Staphylococcus aureus* biofilm. Reserpine had eradicated up to 72.7% of biofilm with just ½ MIC dosages. It halted the metabolic activity of 50.6% bacterial cells in the biofilm. The alkaloids have shown antibiofilm, antimicrobial, and antiviral effects against the bacteria. Thus, reserpine can help in pharmacological applications individually or in combination for the treatment of *S. aureus* infections (Parai et al., 2020).

The antibacterial activity of *R. serpentina* has been validated by various experiments. Against Acinetobacter bacteria, *R. serpentina* has shown an inhibition of 50–75%. A very high inhibitory zone was produced by ethanolic root extract of the plant against *K. pneumoniae*. The root and leaf extract were tested against *B. subtilis*, *E. coli*, *S. aureus*, and *S. typhi*. The root extract was more effective and the extract showed highest activity against *S. typhi* at 40 mg/ml. The highest activity of methanolic root extract was shown against *S. aureus* with concentration of 6.25 μg/ml when tested against five other types of bacteria (Deshmukh et al., 2012; Anand et al., 2020). Due to the rise of multi-drug-resistant bacteria, phytochemicals are being used as alternatives. A study showed reserpine as an inhibitor of efflux pump in *Bacillus subtilis* and *Streptococcus pneumonia*. This inhibition of efflux pump causes death or reduces drug resistance in the bacteria. Efflux activity of gram-positive bacteria can be evaluated through reserpine assay (Seukep et al., 2020).

Skin is the most exposed part of the body and is exposed to pathogens causing infections or diseases. Acne is the most common in teenagers. Due to side effects of synthetic drugs, herbal alternatives are preferred. *R. serpentina* along with 2 other plants (*Curcuma longa* and *azadirachtaindica)* was used to form a gel system. The gel system had inhibitory activity against *Propionibacterium acnes, Staphylococcus epidermidis*, and *Malassezia furfur*. Thus, the current research may hopefully bring new formulations and lead to advancements in acne treatment (Rasheed et al., 2011).

13.5.2.6 Anticancer and Antitumor Activity

Other than use of reserpine as hypertensive drug it also shows activity against cancer cells both *in vitro* and *in vivo*. Experimental studies have reported an increase in life span of mice suffering from leukemia when administered with reserpine. Many other studies have reported antitumoral activity of reserpine against a variety of mouse sarcomas. A study was conducted using reserpine on drug-resistant tumor cells. Cell analysis reported cell cycle arrest in G1 phase. The reserpine proved effective against tumor cells by increasing intracellular drug accumulation and inhibition of P-glycoprotein. Reserpine improves the activity of anticancer and alkylating drugs. On the other hand, molecular docking showed reserpine binds strongly to receptors when compared to other drugs (Abdelfatah and Efferth, 2015). Reserpine is also an inhibitor of carcinogenesis in the liver caused by FAA (N-2-fluorenylacetamide). Currently, there are very few reports of reserpine being used for liver cancer treatment (Liu et al., 2019). Reserpine also has side effects on the cardiovascular system in patients suffering with just cancer. Thus, reserpine could be beneficial to cancer patients suffering from hypertension but could affect others suffering only from cancer. Thus, further studies are required on antitumor activity of reserpine (Abdelfatah and Efferth, 2015).

A study reported that reserpine downregulates DNA repair, cell proliferation, and invasion in the tumor cells. On the other hand, it also induces apoptosis through TGF-β (Transforming growth factor) signaling (Ramu et al., 2021). The antiproliferative activity of *R. serpentina* against cervical cancer cell lines (HeLa) was reported to be 51.27% and 49.27% by leaf and root extract, respectively. With 196 μg/ml IC_{50} (inhibitory concentration) value, the leaf extract was found to be more effective (Deshmukh et al., 2012).

Few newly isolated alkaloids from *R. serpentina* were tested for inhibitory and cytotoxic activities against topoisomerase I, II, and human promyelocytic leukemia (HL-60) cancer cell lines, respectively. The compounds yohimbinic acid at 30 μM and reserpine at 20μM showed 50% inhibition on human topoisomerase I and II. Also, alkaloids such as etoposide and camptothecin are potent inhibitors of human topoisomerase I and II. The compounds isorauhimbinic acid, reserpine, and yohimbinic acid inhibited the growth of HL-60 cancer cells (Itoh et al., 2005).

13.5.2.7 Hypolipidemic Activity

Hypercholesterolemia has been associated with diabetes and various cardiovascular diseases. The increased amount of cholesterol synthesis is caused by high activity of HMGCR (3-hydroxy-3-methyl-glutaryl CoA reductase) enzyme. Thus, the enzyme is a prominent drug target for inhibition. A study was conducted on the effect of alkaloids from roots of *R. serpentina* toward HMGCR activity using *in silico* investigations. This study concluded that, of the 12 alkaloids tested, ajmalicine, reserpine, indobinine, yohimbine, and indobine are predicted as the potent inhibitors of HMGCR and suppress cholesterol synthesis. Thus the five alkaloids can serve as potential lead compounds for developing new drugs against hyperlipidemia (Azmi et al., 2021). Another study on methanolic extract of *R. serpentina* reported its hypotensive and hypolipidemic effects (Shah et al., 2020).

13.5.2.8 Anti-SARS Activity

In 2019, the emergence of a new strain of Coronavirus caused a pandemic which has affected more than 200 countries. The strain is referred to as SARS-CoV-2 (severe acute respiratory syndrome coronavirus-2) while commonly named COVID-19 (coronavirus disease-2019). Reserpine has EC_{50} (effective concentration) value of 3.4 μM CC_{50} (cytotoxic concentration) value of 25 μM against the virus. Reserpine was tested using ELISA (enzyme-linked immunosorbent assay), flowcytometry, IFA (immunofluorescence assay), and Western blot analysis which confirmed its anti-SARS activity (Wu et al., 2004; Xian et al., 2020).

13.5.2.9 Anti–Larvicidal Activity

Mosquito control has become difficult due to an increase in resistance toward synthetic insecticides. Thus, biological agents, such as phytochemicals, are used against larval mosquitoes. A study was conducted using *Culex quinquefasciatus* mosquito larva to test the larvicidal effect of *R. serpentina* seeds on the larvae. Petroleum ether extract of the seeds with 100 ppm concentration had the highest larval mortality rates on the third instar larval form of the mosquito. This clearly indicates that mature seeds of *R. serpentina* have a larvicidal effect against *Culex quinquefasciatus* and can be used as an eco-friendly and biodegradable plant-based insecticide (Das and Chandra, 2012).

13.5.2.10 Anticorrosive Activity

Aluminum and its alloys have wide applications in various fields. The use of aluminum in Al/air batteries causes corrosion due to its alkaline medium. Anticorrosive chemical compounds are expensive and hazardous to the environment. In a study, plant leaf extract of five different plants were tested for environmentally safe anticorrosive agents. The leaf extract of *R. serpentina* was used as an inhibitor which had the highest efficiency of 97.1% and was achieved at a concentration of 0.2 g L^{-1} (Chaubey et al., 2015).

13.5.2.11 Anti-oxidative Activity and Anti-Heavy Metal Toxicity

Chromium (VI), naturally occurring in drinking water, is harmful when found above permissible limits. It has been known to cause DNA damage, resulting in cancers, reproductive damage, and respiratory and skin ailments. A study reported chromium (VI) toxicity causes oxidative stress in fish (*Channa punctata*). It further reported increased inactivity of liver enzymes, micronuclei formation, chromosomal aberrations, and decreases in the protein number of fish. When administered an ethanolic root extract of *R. serpentina*, improvement could be seen in toxicity symptoms, reduction in DNA damage, and restoration of mitochondrial damage due to its antioxidant properties. Thus, it could be useful for aquatic biodiversity against heavy metal toxicity (Trivedi et al., 2021).

13.5.2.12 Nanoparticle Formations

Plant leaf extract acts as the capping agent of silver nanoparticles (AgNPs). The synthesized nanoparticles have 31.43% of silver weight. The AgNPs have antibacterial activity against both classes of bacteria (gram-positive and gram-negative). The synthesized AgNPs have antifungal activity against pathogenic fungus, e.g., *A. niger* and *C. albicans*. These nanoparticles also have larvicidal activity against *Culex quinquefasciatus*. The AgNPs also showed cytotoxicity toward cell lines of cervical (HeLa) and breast (MCF-7) cancers (Panja et al., 2016). Synthesis of copper oxide nanoparticles (CuO NPs) was done using aqueous leaf extract of the plant. These synthesized CuO NPs showed antibacterial activity. The CuO NPs effectively degraded the carcinogenic dye trypan blue in the presence of UV or sunlight. Thus, the use of plant extracts in synthesizing nanoparticles and further biological applications of the synthesized nanoparticles should be studied (Lingaraju et al., 2015).

13.5.2.13 Associated Microbes

Various bacterial strains are present in the rhizosphere of *R. serpentina*. One such strain isolated was designated "WGR–UOM–BT1". The BT1 strain showed antifungal activities by suppressing foliar or root fungal pathogens and also promotes plant growth in tomato under lab and greenhouse conditions. The strain is positive for rhizosphere colonization in tomato, Indole acetic acid (IAA) production. It can also be used as a biofertilizer due to properties such as phosphate solubilization, 1-aminocyclopropane-1-carboxylate (ACC) deaminase activity, and IAA production which stabilizes various factors of the plants during unstable environmental conditions. The strain produces a completely novel antimicrobial metabolite identified as amino (5-(4-methoxyphenyl)-2-methyl-2-(thiophen-2-yl)-2,3-dihydrofuran-3-yl) methanol (AMTM). It is a heterocyclic compound containing nitrogen (Prasannakumar et al., 2015). Endophyticactino bacteria can be found in medicinal plants such as *R. serpentina*. In the future, these actinobacteria will act as a potent bioresource for the extraction of new bioactive compounds (Gohain et al., 2019).

13.5.2.14 Side Effects

Individuals turn toward herbal supplements known for their natural source and fewer side effects. Although herbal medicines that are used as "self-medications" can cause long-term harm (Mossoba et al., 2015; Rijntjes and Meyer, 2019). Some patients developed symptoms of depression while undergoing reserpine treatment for hypertension. The severity of the symptoms was very high, which ultimately led to the use of antidepressants or hospitalization. Reserpine is responsible for depletion of serotonin and catecholamines in the brain. Serotonin levels in platelets are also affected. Reserpine has also been used for mimicking Parkinson's disease symptoms in model organisms (Fahn, 2015; Minor and Hanff, 2015). Various researchers have reported the side effects caused by reserpine. The norepinephrine in the adrenal medulla also gets depleted. In animals, reserpine causes neuroleptic effects and akinetic effects. A researcher used L-dopa to alleviate the reserpine-induced Parkinson state in animals (Fahn, 2015). A case study reported a person suffering from early parkinsonism and depression for a period of three years. Initially, the hypertension medications were not taken into consideration. But with further investigation, it was confirmed that Rauvolfia tablets had caused such side effects. After three months of without the tablets, the patient recovered fully (Rijntjes and Meyer, 2019).

A study investigated the negative effects of *R. serpentina* on the proximal tubule cells of the human kidney *in vitro*. The study reported highly damaging effects on the cells of the renal system. Thus *R. serpentina* can initiate damage or worsen renal functioning, and prolonged use may cause permanent injury. Further investigations are required on the side effects of the plant on kidney and other organs. Also, dietary supplements need to be investigated (Mossoba et al., 2015). A study conducted by Shah et al., (2020) using methanolic extract of *R. serpentina* on albino rats showed no significant damage to the kidney.

13.6 CONCLUSION

Thus *P. ovata* and *R. serpentina* have been traditionally used to treat various gastrointestinal disorders, bites, wounds, mental illness, respiratory disorders, etc. (Malviya and Sason, 2016; Upasani et al., 2018; Franco et al., 2020; Ji et al., 2019). *P. ovata* and *R. serpentina* have activities such as antibacterial, hypolipidemic, antileishmanial, anticancer, antitumor, anticorrosive, anti-ulcer, antidiabetic, hepatoprotective, anti-inflammatory, antioxidant, anti-nematode, anti-larvicidal, immunomodulatory actions, wound-healing activity, etc. (Xing et al., 2017; Gonçalves and Romano, 2016; Mukherjee et al., 2019; Prasathkumar et al., 2021). *P. ovata* and *R. serpentina* activities are mainly dependent on metabolites, such as polysaccharides, alkaloids, phenols, flavonoids, etc. *P. ovata* showed coagulation activity in turbid water and is used in making tablets for its super-disintegrant activity. But it is also used in making eye drops and bio-edible films. *P. ovata* acts against heavy metals, industrial pollution, and Parkinson's and Alzheimer's diseases. It is used in the food industry as a gelling agent and stabilizer, as well as in the cosmetic industry (Dhar et al., 2005; Ahmadi et al., 2012; Tlili et al., 2019; Franco et al., 2020; Zhou et al., 2021). *R. serpentina* has been confirmed with anti-SARS activity (Xian et al., 2020). *R. serpentina* was also used in nanoparticles having anticancer, antibacterial, antifungal, and larvicidal activity. *R. serpentina* was also used against heavy metal toxicity, Alzheimer's disease, and used in the cosmetic industry (Panja et al., 2016; Trivedi et al., 2021). Although both *P. ovata* and *R. serpentina* have equal amounts of side effects, they are being used in various industries. *P. ovata* affects absorption of drugs and nutrients, causes bloating, gastrointestinal disturbances, headache, vomiting, etc. Meanwhile, side effects of *R. serpentina* include depression, symptoms of Parkinson's disease, and damage to the kidney (Mossoba et al., 2015; Soni et al., 2017; Rijntjes and Meyer, 2019). Thus, more research and future study is required for extraction and utilization of the required chemical constituents while minimizing the side effects from both plants.

REFERENCES

Abdelfatah, S.A., and T. Efferth. 2015. Cytotoxicity of the indole alkaloid reserpine from *Rauwolfia serpentina* against drug-resistant tumor cells. *Phytomedicine: International Journal of Phytotherapy and Phytopharmacology*, 22(2), pp.308–318.

Abdel-Rahim, E.A., M.M. Rashed, Z.M. El-Hawary, M.M. Abdelkader, S.S. Kassem, and R.S. Mohamed. 2016. Anti-diabetic effect of Cichoriumintybus leaves and *Plantago ovata* seeds in high fat diet-streptozotocin induced diabetic rats. *Journal of Food and Nutrition Research*, 4(5), pp.276–281.

Ali, A. 2020. Herbs that heal: The philanthropic behaviour of nature. *Annals of Phytomedicine: An International Journal*, 9(1), pp.7–17.

Ahmadi, R., A. Kalbasi-Ashtari, A. Oromiehie, M.S. Yarmand, and F. Jahandideh. 2012. Development and characterization of a novel biodegradable edible film obtained from psyllium seed (*Plantago ovata* Forsk). *Journal of Food Engineering*, 109(4), pp.745–751.

Anand, U., S. Nandy, A. Mundhra, N. Das, D.K. Pandey, and A. Dey. 2020. A review on antimicrobial botanicals, phytochemicals and natural resistance modifying agents from Apocynaceae family: Possible therapeutic approaches against multidrug resistance in pathogenic microorganisms. *Drug Resistance Updates: Reviews and Commentaries in Antimicrobial and Anticancer Chemotherapy*, 51, p.100695.

Azmi, M.B., and S.A. Qureshi. 2012. Methanolic root extract of *Rauwolfia serpentina* Benth improves the glycemic, antiatherogenic, and cardioprotective indices in alloxan-induced diabetic mice. *Advances in Pharmacological Sciences*, 2012, p.376429.

Azmi, M.B., and S.A. Qureshi. 2016. *Rauwolfia serpentina* improves altered glucose and lipid homeostasis in fructose-induced type 2 diabetic mice. *Pakistan Journal of Pharmaceutical Sciences*, 29(5), pp.1619–1624.

Azmi, M.B., S.A. Qureshi, S. Ahmed, S. Sultana, A.A. Khan, and H.A. Mudassir. 2021a. Antioxidant and haematinic effects of methanolic and aqueous methanolic roots extracts of *Rauwolfia serpentina* Benth in type 2 diabetic mice. *Pakistan Journal of Pharmaceutical Sciences*, 34(2), pp.529–535.

Azmi, M.B., S. Sultana, S. Naeem, and S.A. Qureshi. 2021b. *Insilico* investigation on alkaloids of *Rauwolfia serpentina* as potential inhibitors of 3-hydroxy-3-methyl-glutaryl-CoA reductase. *Saudi Journal of Biological Sciences*, 28(1), pp.731–737.

Bagheri, S.M., F. Zare-Mohazabieh, H. Momeni-Asl,M. Yadegari, A. Mirjalili, and M. Anvari. 2018. Antiulcer and hepatoprotective effects of aqueous extract of *Plantago ovata* seed on indomethacin-ulcerated rats. *Biomedical Journal*, 41(1), pp.41–45.

Basiri, S., S.S. Shekarforoush, S. Mazkour, P. Modabber, and F.Z. Kordshouli. 2020. Evaluating the potential of mucilaginous seed of psyllium (*Plantago ovata*) as a new lead biosorbent. *Bioactive Carbohydrates and Dietary Fibre*, 24, p.100242.

Belorio, M., and M. G6mez. 2022. Psyllium: A useful functional ingredient in food systems. *Critical Reviews in Food Science and Nutrition*, 62(2), pp.527–538.

BeMiller, J.N., R.L. Whistler, D.G., Barkalow, and C.C. Chen. 1993. Aloea, chia, flax seed, okra, psyllium seed, quince seed, and tamarin gums. In R.L.Whistler, and J.N. BeMiller (Eds.), *Industrial Gums*. New York: Academic Press, pp.227–256.

Bhat, S., G.S. Mulgund, and P. Bhat. 2019. Ethnomedicinal practices for the treatment of arthritis in Siddapur region of Uttara Kannada District, Karnataka, India. *Journal of Herbs, Spices & Medicinal Plants*, 25(4), pp.316–329.

Bunkar, A.R. 2017. Therapeutic uses of *Rauwolfia serpentina*. *International Journal of Advanced Science and Research*, 2(2), pp.23–26.

Chaubey, N., D. Yadav, V. Singh, and M. Quraishi. 2015. A comparative study of leaves extracts for corrosion inhibition effect on aluminium alloy in alkaline medium. *Ain Shams Engineering Journal*, 8(4), pp.673–682.

Chen, S.L., H. Yu, H.M. Luo, Q. Wu, C.F. Li, and A. Steinmetz. 2016. Conservation and sustainable use of medicinal plants: Problems, progress, and prospects. *Chinese Medicine*, 11, p.37.

Das, D., and G. Chandra. 2012. Mosquito larvicidal activity of *Rauvolfia serpentina* L. seeds against Culexquinquefasciatus Say. *Asian Pacific Journal of Tropical Medicine*, 5(1), pp.42–45.

Deshmukh S.R., D.S. Ashrit, and B.A. Patil. 2012. Extraction and evaluation of indole alkaloids from *Rauvolfia serpentina* for their antimicrobial and antiproliferative activities. *International Journal of Pharmacy and Pharmaceutical Sciences*, 4(5), pp.329–334

Dey, A., and J.N. De. 2010. Rauvolfia serpentina (L). Benth.Ex Kurz.: A Review. *Asian Journal of Plant Sciences*, 9(6), pp.285–298.

Dey, A., and J.N. De. 2011. Ethnobotanical aspects of *Rauvolfia serpentina* (L). Benth.exKurz. in India, Nepal and Bangladesh. *Journal of Medicinal Plants Research*, 5(2), pp.144–150.

Dhar M.K., S. Kaul, S. Sareen, and A.K. Kou. 2005. *Plantago ovata*: Genetic diversity, cultivation, utilization and chemistry. *Plant Genetic Resources*, 3(2), pp.252–263.

Dhivya, S., S.T. Ramesh, R. Gandhimathi, and P.V. Nidheesh. 2017. Performance of natural coagulant extracted from *Plantago ovata* seed for the treatment of turbid water. *Water, Air, & Soil Pollution*, 228(11), pp.1–11.

Draksiene, G., D.M. Kopustinskiene, R. Lazauskas, and J. Bernatoniene. 2019. Psyllium (*Plantago Ovata* Forsk.) Husk Powder as a natural superdisintegrant for orodispersible formulations: A study on meloxicam tablets. *Molecules*, 24(18), p.3255.

Eurlings, M.C., F. Lens, C. Pakusza, T. Peelen, J.J. Wieringa, and B. Gravendeel. 2013. Forensic identification of Indian snakeroot (*Rauvolfia serpentina* Benth. ex Kurz) using DNA barcoding. *Journal of Forensic Sciences*, 58(3), pp.822–830.

Ezeigbo, I.I., M.I. Ezeja, K.G. Madubuike, D.C. Ifenkwe, I.A. Ukweni, N.E. Udeh, and S.C. Akomas. 2012. Antidiarrhoeal activity of leaf methanolic extract of *Rauwolfia serpentina. Asian Pacific Journal of Tropical Biomedicine*, 2(6), pp.430–432.

Fahn S. 2015. The medical treatment of Parkinson disease from James Parkinson to George Cotzias. *Movement Disorders: Official Journal of the Movement Disorder Society*, 30(1), pp.4–18.

Franco, E.A.N., A. Sanches-Silva, R. Ribeiro-Santos, and N.R. de Melo. 2020. Psyllium (*Plantago ovata* Forsk): From evidence of health benefits to its food application. *Trends in Food Science & Technology*, 96, pp.166–175.

Gohain, A., R.K. Sarma, R. Debnath, J. Saikia, B.P. Singh, R. Sarmah, and R. Saikia. 2019. Phylogenetic affiliation and antimicrobial effects of endophyticactinobacteria associated with medicinal plants: Prevalence of polyketide synthase type II in antimicrobial strains. *Folia microbiologica*, 64(4), pp.481–496.

Gonçalves, S., and A. Romano. 2016. The medicinal potential of plants from the genus *Plantago* (Plantaginaceae). *Industrial Crops and Products*, 83, pp.213–226.

Haji-Ali-Nili, N., F. Khoshzaban, M. Karimi, R. Rahimi, E. Ashrafi, R. Ghaffari, A. Ghobadi, and M.J. Behrouz. 2019. Effect of a natural eye drop, made of *Plantago ovata* mucilage on improvement of dry eye symptoms: A randomized, double-blind clinical trial. *Iranian Journal of Pharmaceutical Research: IJPR*, 18(3), pp.1602–1611.

Hasheminasab, F.S., S.M. Hashemi, A. Dehghan, F. Sharififar, M. Setayesh, P. Sasanpour, M. Tasbandi, and M. Raeiszadeh. 2020. Effects of a Plantago ovata-based herbal compound in prevention and treatment of oral mucositis in patients with breast cancer receiving chemotherapy: A double-blind, randomized, controlled crossover trial. *Journal of Integrative Medicine*, 18(3), pp.214–221.

ICAR-Directorate of Medicinal & Aromatic Plants Research, Anand, Gujarat. n.d. Indian Council of Agricultural Research - Department of Agricultural Research & Education, Government of India. [Retrieved on 25 August 2021] Available from https://dmapr.icar.gov.in//MandCrop/Isabgol.html

Itoh, A., T. Kumashiro, M. Yamaguchi, N. Nagakura, Y. Mizushina, T. Nishi, and T. Tanahashi. 2005. Indole alkaloids and other constituents of *Rauwolfia serpentina. Journal of Natural Products*, 68(6), pp.848–852.

Jain, P. 2016. Secondary metabolites for antiulcer activity. *Natural Product Research*, 30(6), pp.640–656.

Jalilimanesh, M., M. Azhdari, A. Mirjalili, M.A. Mozaffari, and S. Hekmatimoghaddam. 2021. The comparison of clinical and histopathological effects of topical Psyllium (*Plantago ovata*) powder and silver sulfadiazine on second-degree burn wound healing in rats. *World Journal of Plastic Surgery*, 10(1), pp.96–103.

Ji, X., C. Hou, and X. Guo. 2019. Physicochemical properties, structures, bioactivities and future prospective for polysaccharides from *Plantago* L. (Plantaginaceae): A review. *International Journal of Biological Macromolecules*, 135, pp.637–646.

Kashyap, P., V. Kalaiselvan, R. Kumar, and S. Kumar. 2020. Ajmalicine and reserpine: Indole alkaloids as multi-target directed ligands towards factors implicated in Alzheimer's disease. *Molecules*, 25(7), p.1609.

Keshavan, M.S. 2011. The tale of Rauwolfia Serpentina and the contributions of Asian psychiatry. *Asian Journal of Psychiatry*, 4(3), pp.214–215.

Khan, A.W., K. Waseem, S. Safdar, U. Muhammad, A.S. Muhammad, J. Nuzhut, R.P. Jha, M. Baig, S. Shehzadi, Z.K. Muhammad, and K.S. Muhammad. 2021. Nutritional and therapeutic benefits of Psyllium Husk (*Plantago Ovata*). *Acta Scientific Microbiology*, 4(3), pp.43–50.

Kotwal, S., S. Kaul, and M.K. Dhar. 2019. Comparative expression analysis of flavonoid biosynthesis genes in vegetative and reproductive parts of medicinally important plant, *Plantago ovata* Forssk. *Industrial Crops and Products*, 128, pp.248–255.

Kumari, R., B. Rathi, A. Rani, and S. Bhatnagar. 2013. *Rauvolfia serpentina* (L). Benth. Ex Kurz.: phytochemical, pharmacological and therapeutic aspects. *International Journal of Pharmaceutical Sciences Review and Research*, *23*(2), pp.348–355.

Ladjevardi, Z.S., S.M.T. Gharibzahedi, and M. Mousavi. 2015. Development of a stable low-fat yogurt gel using functionality of psyllium (*Plantago ovata* Forsk) husk gum. *Carbohydrate Polymers*, 125, pp.272–280.

Lingaraju, K., H. Raja Naika, K. Manjunath, G. Nagaraju, D. Suresh, and H. Nagabhushana. 2015. *Rauvolfia serpentina*-mediated green synthesis of CuO nanoparticles and its multidisciplinary studies. *ActaMetallurgicaSinica (English Letters)*, 28(9), pp.1134–1140.

Liu, C., S. Yang, K. Wang, X. Bao, Y. Liu, S. Zhou, H. Liu, Y. Qiu, T. Wang, and H. Yu. 2019. Alkaloids from Traditional Chinese Medicine against hepatocellular carcinoma. *Biomedicine & Pharmacotherapy*, 120, p.109543.

Madgulkar, A.R., M.R.P. Rao, and D. Warrier. 2015. Characterization of Psyllium (*Plantago ovata*) Polysaccharide and its uses. In K. Ramawat, and J.M. Mérillon (Eds.), *Polysaccharides*. Switzerland: Springer, pp.871–890.

Malviya, A. and R. Sason. 2016. The phytochemical and pharmacological properties of Sarpagandha: *Rauwolfia Serpentina*. *Ayushdhara*, 3(1), pp.473–478.

Mehmood, Z., M.S. Khan, F.A. Qais, and I. Ahmad. 2019. Herb and modern drug interactions: Efficacy, quality, and safety aspects. In *New Look to Phytomedicine*. London: Elsevier, pp.503–520.

MehrinejadChoobari, S.Z., A.A.Sari, and A. DaraeiGarmakhany. 2021. Effect of *Plantago ovata* Forsk seed mucilage on survivability of *Lactobacillus acidophilus*, physicochemical and sensory attributes of produced low-fat set yoghurt. *Food Science & Nutrition*, 9(2), pp.1040–1049.

Minor, T.R., and T.C. Hanff. 2015. Adenosine signaling in reserpine-induced depression in rats. *Behavioural Brain Research*, 286, pp.184–191.

Mobin, M., and M. Rizvi. 2017. Polysaccharide from Plantago as a green corrosion inhibitor for carbon steel in 1 M HCl solution. *Carbohydrate Polymers*, 160, pp.172–183.

Moshfe, A., M. Bahmani, M. Naghmachi, S. Askarian, A. Rezaei, and R. Zare. 2011. PP-180 Anti leishmanial effect of *Plantago psyllium* (ovata) and white vinegar on *Leishmania* major lesions in BALB/c mice: 1.3-024. *International Journal of Infectious Disease*, 15, p. S95.

Mossoba, M.E., T.J. Flynn, S. Vohra, P.L. Wiesenfeld, and R.L. Sprando. 2015. Human kidney proximal tubule cells are vulnerable to the effects of *Rauwolfia serpentina*. *Cell Biology and Toxicology*, 31(6), pp.285–293.

Mukherjee, E., S. Gantait, S. Kundu, S. Sarkar, and S. Bhattacharyya. 2019. Biotechnological interventions on the genus *Rauvolfia*: Recent trends and imminent prospects. *Applied Microbiology and Biotechnology*, 103(18), pp.7325–7354.

Nazir, S., M.U.A. Khan, W.S. Al-Arjan, S.I. AbdRazak, A. Javed, and M.R.A. Kadir. 2021. Nanocomposite hydrogels for melanoma skin cancer care and treatment: In-vitro drug delivery, drug release kinetics and anti-cancer activities. *Arabian Journal of Chemistry*, 14(5), p.103120.

Panja, S., I. Chaudhuri, K. Khanra, and N. Bhattacharyya. 2016. Biological application of green silver nanoparticle synthesized from leaf extract of *Rauvolfia serpentina* Benth. *Asian Pacific Journal of Tropical Disease*, *6*(7), pp.549–556.

Parai, D., M. Banerjee, P. Dey, and S.K. Mukherjee. 2020. Reserpine attenuates biofilm formation and virulence of *Staphylococcus aureus*. *Microbial Pathogenesis*, 138, p.103790.

Patel, M.K., B. Tanna, A. Mishra, and B. Jha. 2018. Physicochemical characterization, antioxidant and anti-proliferative activities of a polysaccharide extracted from psyllium (P. ovata) leaves. *International Journal of Biological Macromolecules*, 118, pp.976–987.

Patel, M.K., B. Tanna, H. Gupta, A. Mishra, and B. Jha. 2019. Physicochemical, scavenging and anti-proliferative analyses of polysaccharides extracted from psyllium (*Plantago ovata* Forssk) husk and seeds. *International Journal of Biological Macromolecules*, 133, pp.190–201.

Pathania, S., S.M. Ramakrishnan, V. Randhawa, and G. Bagler. 2015. SerpentinaDB: A database of plant-derived molecules of *Rauvolfia serpentina. BMC Complementary and Alternative Medicine*, 15, p.262.

Pathania, S., V. Randhawa, and G. Bagler. 2013. Prospecting for novel plant-derived molecules of *Rauvolfia serpentina* as inhibitors of Aldose Reductase, a potent drug target for diabetes and its complications. *PloS One*, 8(4), p.e61327.

PaturkarM., and P. Khobragade. 2016. Mass cultivation of Sarpagandha (*Rauwolfia Serpentina* Benth. Ex Kurz) in consideration with environmental factors and cultivation techniques. *International Journal of Ayurveda and Pharma Research*, 4(10), pp.58–62.

Pawar, H., and C. Varkhade. 2014. Isolation, characterization and investigation of *Plantago ovata* husk polysaccharide as superdisintegrant. *International Journal of Biological Macromolecules*, 69, pp.52–58.

Prasannakumar, S.P., H.G. Gowtham, P. Hariprasad, K. Shivaprasad, and S.R. Niranjana. 2015. Delftiatsuruhatensis WGR-UOM-BT1, a novel rhizobacterium with PGPR properties from *Rauwolfia serpentina* (L.) Benth.exKurz also suppresses fungal phytopathogens by producing a new antibiotic-AMTM. *Letters in Applied Microbiology*, 61(5), pp.460–468.

Prasathkumar, M., S. Anisha, C. Dhrisya, R. Becky, and S. Sadhasivam. 2021. Therapeutic and pharmacological efficacy of selective Indian medicinal plants: A review. *Phytomedicine Plus*, 1(2), p.100029.

PubChem. n.d. National Centre for Biotechnology, National Library of Medicine, National Institute of Health. [Retrieved on 26 June 2021].Available from https://pubchem.ncbi.nlm.nih.gov.

Ramavandi, B. 2014. Treatment of water turbidity and bacteria by using a coagulant extracted from *Plantago ovata. Water Resources and Industry*, 6, pp.36–50.

Ramu, A.K., D. Ali, S. Alarifi, M.H. Syed Abuthakir, and B.A. Ahmed Abdul. 2021. Reserpine inhibits DNA repair, cell proliferation, invasion and induces apoptosis in oral carcinogenesis via modulation of TGF-β signaling. *Life Sciences*, 264, p.118730.

Rasheed, A., G. Avinash Kumar Reddy, S. Mohanalakshmi, and C.K. Ashok Kumar. 2011. Formulation and comparative evaluation of poly herbal anti-acne face wash gels. *Pharmaceutical Biology*, 49(8), pp.771–774.

Reddy, P.R.T., K.V. Vandana, and S. Prakash. 2018. Antibacterial and anti-inflammatory properties of *Plantago ovata* Forssk. leaves and seeds against periodontal pathogens: An in vitro study. *Ayu*, 39(4), p.226.

Rijntjes, M., and P.T. Meyer 2019. No free lunch with herbal preparations: Lessons from a case of Parkinsonism and depression due to herbal medicine containing reserpine. *Frontiers in Neurology*, 10, p.634.

Sagar, S., G. Goudar, M. Sreedhar, A. Panghal, and P. Sharma. 2020. Characterization of nutritional content and in vitro-antioxidant properties of Plantago ovata seeds. *International Journal of Food and Nutritional Sciences*, 9(2), pp.27–31.

Sahebkar-Khorasani, M., L. Jarahi, H. Cramer, M. Safarian, H. Naghedi-Baghdar, R. Salari, P. Behravanrad, and H. Azizi. 2019. Herbal medicines for suppressing appetite: A systematic review of randomized clinical trials. *Complementary Therapies in Medicine*, 44, pp.242–252.

Seukep, A.J., V. Kuete, L. Nahar, S.D. Sarker, and M. Guo. 2020. Plant-derived secondary metabolites as the main source of efflux pump inhibitors and methods for identification. *Journal of Pharmaceutical Analysis*, 10(4), pp.277–290.

Shah, S., S. Naqvi, N. Munir, S. Zafar, M. Akram, and J. Nisar. 2020. Antihypertensive and antihyperlipidemic activity of aqueous methanolic extract of *Rauwolfia Serpentina* in albino rats. *Dose-response: A Publication of International Hormesis Society*, 18(3), p.1559325820942077.

Soni A., H. Thakur, S. Goyal, and S. Shivali. 2017. Isolation and characterization of mucilage from Plantago ovata. *World Journal of Pharmaceutical and Life Sciences*, 3(7), pp.285–291.

Tewari, D., N. Anjum, and Y.C. Tripathi. 2014. Phytochemistry and pharmacology of Plantago ovate: A natural source of laxative medicine. *World Journal of Pharmaceutical Research*, 3(9), pp.361–372.

Tlili, H., A. Marino, G. Ginestra, F. Cacciola, L. Mondello, N. Miceli, and A. Nostro. 2021. Polyphenolic profile, antibacterial activity and brine shrimp toxicity of leaf extracts from six Tunisian spontaneous species. *Natural Product Research*, 35(6), pp.1057–1063.

Tlili, H., N. Hanen, A. Ben Arfa, M. Neffati, A. Boubakri, D. Buonocore, M. Dossena, M. Verri, and E. Doria. 2019. Biochemical profile and in vitro biological activities of extracts from seven folk medicinal plants growing wild in southern Tunisia. *Plos One* 14(9), p.e0213049.

Trivedi, S.P., V. Kumar, S. Singh, and M. Kumar. 2021. Efficacy evaluation of Rauwolfia serpentina against Chromium (VI) toxicity in fish, Channapunctatus. *Journal of Environmental Biology*, 42(3), pp.659–667.

Upasani, S.V., V.G. Beldar, A.U. Tatiya, M.S. Upasani, S.J. Surana, and D.S. Patil. 2017. Ethnomedicinal plants used for snakebite in India: A brief overview. *Integrative Medicine Research*, 6(2), pp.114–130.

Upasani, M.S., S.V. Upasani,V.G. Beldar, C.G. Beldar, and P.P. Gujarathi. 2018. Infrequent use of medicinal plants from India in snakebite treatment. *Integrative Medicine Research*, 7(1), pp.9–26.

Verma, A., and R. Mogra. 2015. Psyllium (Plantago ovata) husk: A wonder food for good health. *International Journal of Science and Research*, 4(9), pp.1581–1585.

Wahid, A., S.M.N. Mahmoud, E.Z. Attia, A.S. Yousef, A.M.M. Okasha, and H.A. Soliman. 2020. Dietary fiber of psyllium husk (*Plantago ovata*) as a potential antioxidant and hepatoprotective agent against CCl4-induced hepatic damage in rats. *South African Journal of Botany*, 130, pp.208–214.

Wu, C.Y., J.T. Jan, S.H. Ma, C.J. Kuo, H.F. Juan, Y.S. Cheng, H.H. Hsu, H.C. Huang, D. Wu, A. Brik, F.S. Liang, R.S. Liu, J.M. Fang, S.T. Chen, P.H. Liang, and C.H. Wong. 2004. Small molecules targeting severe acute respiratory syndrome human coronavirus. *Proceedings of the National Academy of Sciences of the United States of America*, 101(27), pp.10012–10017.

Xian, Y., J. Zhang, Z. Bian, H. Zhou, Z. Zhang, Z. Lin, and H. Xu. 2020. Bioactive natural compounds against human coronaviruses: A review and perspective. *Acta Pharmaceutica Sinica B*, 10(7), pp.1163–1174.

Xing, L.C., D. Santhi, A.G. Shar, and M. Saeed. 2017. Psyllium husk (Plantago ovata) as a potent hypocholesterolemic agent in animal, human and poultry. *International Journal of Pharmacology*, 13(7), pp.690–697.

Zając, M.H. 2020. The properties of poultry batters depending on the amount of water and Plantago ovata husk. *Acta Scientiarum Polonorum Technologia Alimentaria*, 19(4), pp.475–482.

Zhang, J., C. Wen, H. Zhang, and Y. Duan. 2019. Review of isolation, structural properties, chain conformation, and bioactivities of psyllium polysaccharides. *International Journal of Biological Macromolecules*, 139, pp.409–420.

Zhou, Y., H. Dai, L. Ma, Y. Yu, H. Zhu, H. Wang, and Y. Zhang. 2021. Effect and mechanism of psyllium husk (*Plantago ovata*) on myofibrillar protein gelation. *LWT*, 138, p.110651.

Ziemichód, A., M. Wójcik, and R. Różyło. 2019. Seeds of Plantago psyllium and Plantago ovata: Mineral composition, grinding, and use for gluten-free bread as substitutes for hydrocolloids. *Journal of Food Process Engineering*, 42(1), p.e12931.

14 *Saussurea costus* (Kust) and *Senna alexandrina* (Senna)

Amita Dubey, Soni Gupta, Mushfa Khatoon, and Anil Kumar Gupta

CONTENTS

14.1 Introduction 262
14.2 ***Saussurea costus*** (Kust) 262
14.2.1 Description 264
14.2.1.1 Taxonomic Hierarchy 264
14.2.1.2 Botanical Description 264
14.2.1.3 Traditional Knowledge 265
14.2.2 Phytochemistry 265
14.2.2.1 Bioactive Phytoconstituents of Costus Roots 265
14.2.2.2 Costus Oil 266
14.2.3 Potential Benefits, Applications, and Uses 266
14.2.3.1 Anti-Inflammatory 266
14.2.3.2 Anticancer and Antitumor 267
14.2.3.3 Hepatoprotective 268
14.2.3.4 Anti-Ulcerogenic and Cholagogic Activity 268
14.2.3.5 Immunomodulatory Activity 268
14.2.3.6 Respiratory Diseases and Asthma 269
14.2.3.7 Miscellaneous Activities 269
14.2.3.8 Other Uses 270
14.3 *Senna alexandrina* Mill 271
14.3.1 Description 276
14.3.1.1 Botanical Description 276
14.3.1.2 Traditional Knowledge 276
14.3.2 Chemical Derivatives (Bioactive Compounds – Phytochemistry) 278
14.3.3 Potential Benefits, Applications, and Uses 278
14.3.3.1 Senna as a Laxative Drug 279
14.3.3.2 Antibacterial Activity 280
14.3.3.3 Antifungal Activity 280
14.3.3.4 Antioxidant Activity 281
14.3.3.5 Antihelminthic Activity 281
14.3.3.6 Anticancer 281
14.3.3.7 Anti-Obesity and Antidiabetic Activities 281
14.4 Conclusion 281
Acknowledgments 282
References 282

DOI: 10.1201/9781003205067-14

14.1 INTRODUCTION

Since ancient times, the cure and treatment of diseases is one of the priorities for the well-being of mankind. In the traditional medicine system of many countries, including India, plant-based products are the main component of drug formulations (Arif et al., 2009). The use of medicinal plants as drugs has gained more importance these days due to fewer side effects in comparison to synthetic medicine (Kala et al., 2006). Out of the total population of vascular plants worldwide, approx. 10% have medicinal significance (Salmerón-Manzano et al., 2020). In India, there are about 17,000 species of higher plants, out of which 7500 are used for medicinal purposes (Kala et al., 2006). According to an estimate reported by the World Health Organization (WHO), about 25% of the total medicines used these days are derived from plant sources (Singh et. al., 2019). Due to the presence of numerous types of bioactive compounds, medicinal plants are highly explored for their therapeutic potential (Raina et al., 2014). Medicinal plants are used for treatment of various ailments including hypertension, liver disorders, respiratory problems, immunodeficiency, cancer, kidney-related problems, and bacterial and fungal infections, etc.

From this vast array of medicinal plants, some have gained immense popularity in traditional pharmacopeias. Two such plants inhabiting different ecological zones are described here. These have not only been extensively used traditionally, but they are also traded in the international market earning foreign exchange for India. The first plant described here is *Saussurea costus* (Kust, Family: Asteraceae). It is the most important species of genus *Saussurea* found in the lap of the Himalayas, and its medicinal significance has been mentioned in the traditional systems of medicine of various countries. The second plant is *Senna alexandrina* Mill. (Senna, Family: Fabaceae), which is now cultivated in about 25,000 hectares in arid and semiarid regions of the country. It is one of the most traded medicinal plants of India, mainly due to its laxative potential. The medicinal significance, botanical description, phytochemistry, and potential uses of both *S. costus* and *S. alexandrina* are discussed in the following sections of this chapter.

14.2 *SAUSSUREA COSTUS* (KUST)

The Asteraceae family is one of the largest families of plants, which include about 1000 genera and 30,000 species. In India approx. 177 genera and 1052 species are reported that belong to this family (Rao et al., 1988.)

The *Saussurea* genus of this family includes roughly 400 species of medicinal plants in the high-altitude Himalayan region, of which approx. 27 species have medicinal significance (Butola and Samant, 2010; Lipschitz, 1979). The genus is named after Horace Benedict de Sassure, a Swiss philosopher (Dhar et al., 1984; Nadda et al., 2020). Plants of this genus are distributed throughout the Holarctic regions which cover the area from the Arctic to Southeast Asia but are highly populated in Asian regions (Kita et al., 2004). Different species of this genus are found in lowlands and in the zones between 3500 m and 5000 m of the Himalayas (Haffner, 2000; Kita et al., 2004). Different species of genus *Saussurea* are used for various purposes, like medicine, food, fuel, fodder, ornamental, and religious causes (Butola and Samant, 2010). Some of the important species of genus *Saussurea* and their medicinal uses are listed in Table 14.1.

Among numerous species of genus *Saussurea*, *S. costus* is the most important species both commercially and medicinally. The medicinal significance of *S. costus* has been mentioned in Ayurvedic, Tibetan, and Chinese systems of medicine as the major ingredient in about 175 herbal formulations (Butola and Samant, 2010). *Saussurea costus* Lipschitz (Falc.), synonymous with *Saussurea lappa* C.B. Clarke and *Aucklandia costus* Falc. is one of the most important species of genus *Saussurea*. *S. costus* also known as costus in English. There are other names in India *viz.* Kur (in Bengal), Kut (in Gujrat), Postkhai (in Kashmir), Kot (in Punjab), Kushta (in Sanskrit), Kostum (in Tamil), Kushta (in Maharashtra), and Kuth (in Hindi) (Kirtikar and Basu, 2001). *S. costus* is extensively used in indigenous systems of medicine to cure a variety of diseases and has emerged as an important medicinal

TABLE 14.1
List of Important Species of *Saussurea* with Their Distribution and Uses

Species Name	Common Name	Uses	Distribution	Reference
S. costus (Syn. *lappa*)	Kust	Treatment of cholera, Cough, skin diseases, rheumatism, Toothache, vermifuge for intestinal worm, tuberculosis, and epilepsy.	3200 to 5000 m (in the subalpine zones); Jammu and Kashmir, Uttaranchal, and Himachal Pradesh.	Pandey *et al.*, 2007; Ajaib *et al.*, 2021
S. involucrate	Snow lotus	Accelerate blood circulation, rheumatoid arthritis, anticancerous, gynecological problems.	High latitude region of the western Tianshan Mountains.	Gong *et al.*, 2020
S. ceratocarpa	Pashka	Treatment of headache, lumbar pain, renal pain, and menorrhea.	Western Himalayas (3000 to 5000 m); Himachal Pradesh and Jammu and Kashmir.	Butola and Samant, 2010
S. laniceps	Cotton-headed snow lotus	Anti-nociceptive efficacy, analgesic, antioxidant, anti-microbial.	3500 to 5700 m (in open places); Uttaranchal, Himachal Pradesh Jammu and Kashmir, and Sikkim.	Chen *et al.*, 2016
S. affinis	Ganga Mula	Leaves and young shoots are edible, gynecological problem.	East Himalaya (Assam), East Asia, China, Japan.	Butola and Samant, 2010
S. auriculata	PachakKut	Antisyphilitic, purgative.	2000 to 4300 m; Jammu and Kashmir, Himachal Pradesh.	Singh and Hajra, 1997
S. bracteata	PrerakMul	Boils, cough, headache, cold, lung infection, fever, and also as a good soil binder.	3500 to 5600 m Himachal Pradesh, Jammu and Kashmir.	Hajra *et. al.*, 1995
S. gossypiphora	Kasturi Kamal	As an essential oil in perfumery, gynecological disorders, hysteria, and menstrual disorders.	Himalayas (3500 to 5700 m); Himachal Pradesh, Jammu and Kashmir, Sikkim and Uttaranchal.	Dhar and Kachroo, 39 1983
S. obvallata	Brahm Kamal	Boils, cuts, and wounds, applied to bruises, and also used as antiseptic and nerve tonic.	3800 to 4600 m (on rocky slopes); Uttaranchal, Jammu and Kashmir, Himachal Pradesh, Sikkim.	Samant *et. al.*, 1998
S. medusa	Saw-wort	high blood pressure, headache, menstrual problems, regulate menstrual cycles. Also used as remedy for arthritis and tonic for weakness.	China, India (Kashmir), Nepal, East Tibet, North Pakistan.	Dhar and Kachroo, 1983

plant in the international drug market (Fan et al., 2014). The plant is mainly the inhabitant of the subalpine zones of Jammu and Kashmir, Uttaranchal, and Himachal Pradesh of India, usually at an altitude of 3200–3800 m. Due to overexploitation of the plant for various medicinal and commercial purposes, it has been listed as a critically endangered species of medicinal plants in Appendix I of CITES (Convention on International Trade in Endangered Species) of Wild Fauna and Flora (Pandey et al., 2007). The plant has also been listed among the Himalayan medicinal plants that are selected for *ex situ* and *in situ* conservation (Pandey et al., 2007; Kuniyal et al., 2005). *S. costus* has been listed among the 29 medicinal and aromatic plants whose export has been regulated by the Ministry of Commerce Government of India (Pandey et al., 2007). Dried roots of costus and its oil are the main products of costus which are traded in the herbal drug market (Zahara et al., 2014). India was the second largest exporter of costus after China from 1983 to 2009 with 266 tons of export. However, the largest importer of the plant was France. Drugs formulated using costus are often reported to be adulterated with *Inula racemosa*, a plant in the northwestern Himalayan region (Sastry et al., 2013; Sharma et al., 2017).

14.2.1 Description

14.2.1.1 Taxonomic Hierarchy

The taxonomic account of the family Asteraceae is notably given by C.B. Clarke and Sir J.D. Hooker. The taxonomic statuses of about 608 Asteraceae plant species are dealt with in *Flora of British India* by Hooker (1881). Many other workers have also discussed morphological features of different Asteraceae species including *S. costus*. In *S. costus*, parameters like plant size, leaves, habit, capitula, etc. have been described in detail (Pandey et al., 2007; Chaudhary and Rao, 2000). Further details of taxonomic hierarchy of the plant are described by Zahara et al. (2014)

14.2.1.2 Botanical Description

S. costus is a perennial plant which grows to 1–2 meters in height (Figure 14.1). The plant is upright, strong, with hairs on its surface. Roots are 40–60 cm long, dark brown, and possess specific odor that is described by some as like "wet dog" (Hajra et al., 1995). Leaves of the plant are membranous, glabrate, auricled, and irregularly toothed. Upper leaves are small and often sub-sessile, but the basal leaves are very big up to 0.50–1.25 m long and have a winged petiole. Upper layer of the basal leaves is rough, but the lower layer is smooth with auriculate base.

Flowers of the plant are hard, bluish purple in color, spherical, and grouped in 2–5 at the axils or terminals of leaves. Many hairless, purple-colored bracts are present which appear in whorls (involucral) and are long and pointed (Pandey et al., 2007). A tubular corolla, approx. 2 cm long,

FIGURE 14.1 *Saussurea costus* (A) in field and (B) floral part.

blue-purple, with 1.7 cm long double feathery pappus is present on the fruiting flower head. Fruits are cupped, hairy, curved achenes which are compressed and about 8 mm long (Pandey et al., 2007).

Upadhyay et al. (1993) and Bruchhausen et al. (1994) have depicted both micro and macroscopic characteristics of *S. costus* roots. Fruit morphology is discussed by Saklani et al. (2000).

14.2.1.3 Traditional Knowledge

S. costus has been described as a drug in various indigenous systems of medicine worldwide. In India, the use of plant alone or in combination with other drugs is mentioned in the Ayurveda, Siddha, and Unani systems of medicine (Madhuri et al., 2012). In Ayurveda, Kust roots are prescribed for curing various ailments, like itching, vomiting, headache, scabies, epilepsy, and leukoderma, and enhancing skin complexion. Its use is also reported in treating tridoshas, that is, vata, kapha, and pitta (Madhuri et al., 2012). In Unani medicine system, plant formulations are used as an aphrodisiac, tonic, carminative, and a cure for various liver, blood, and kidney-related disorders. The medicinal potential of this plant is also documented in traditional medicine systems of China and Tibet. Out of 171 formulations described in the *Handbook of Traditional Tibetan Drugs*, the principal constituent of 71 formulations is *S. costus* (Tsarong, 1986) which are mainly used for curing diseases like chest congestion, lung disorders, cough, etc.

Roots of the Kust, with its unique odor and bitter taste, are the most important medicinal part of the plant which is used alone or in combination for the treatment of bronchial asthma, paralysis, fever, cholera, deafness, and various skin-related disorders (Kirtikar and Basu, 1975; Dhar et al., 1984; Chopra et al., 1956). The roots contain alkaloids, resinoids, essential oil, and minor quantities of tannins, sugars, etc. (Ahmed et al., 2016). Butola and Samant (2010) described that the seeds of many plants belonging to this genus are used as animal fodder and their stems as fuelwood. Milled roots of costus along with *Foeniculum vulgare* (saunf), *Piper oblongum* (kalimirch), *Bunium persicum* (kala jeera), *Cuminum cyminum* (jeera), *Angelica glauca* (chaura), milk, and water are boiled and sieved to prepare an extract. This extract is used to treat various disorders like cold, stomachache, cough, and fever (Rawat et al., 2004). Costus is often used as an antiseptic – a medicine to treat helminthic infection – and a treatment for convulsions, tuberculosis, leprosy, chest pain, toothache, and gout. It is also used to promote spermatogenesis (Shah, 1982; Malik et al., 2011).

14.2.2 Phytochemistry

Roots of the costus are mainly responsible for attributing medicinal properties to the plant. Extracts of roots in different solvents and oil extracted from its roots possess various biologically active phytochemicals. First report on phytochemicals present in basic fraction of Kust was given by Salooja et al. (1950). Afterward other works reported a vast array of bioactive constituents in Kust. A detailed description of various bioactive phytoconstituents is discussed in the following sections.

14.2.2.1 Bioactive Phytoconstituents of Costus Roots

From the extracts of fresh roots of costus phytochemicals like monoterpenes, sesquiterpenes and their derivatives, triterpenes, lignans, glycosides, flavonoids, steroids, etc. have been reported.

Costus roots are especially rich in sesquiterpenes and their derivatives. Sesquiterpene lactones which include costunolides, sulfocostunolide A and B (Wang et al., 2010), 13-sulfo-dihydro reynosin and 13-sulfodihydrosantamarine (Yin et al., 2005), etc. are the main bioactive components of the costus. Saussurea lactone (a crystalline lactone) and costunolide were first isolated from the oil extracted from costus roots by Rao and his group (Rao and Verma, 1951, Rao et al., 1960).

Govindan and Bhattacharaya (1977) reported dehydrocostuslactone, palmitic acid, costic acid, linoleic acid, β-sitosterol, isoalantolactone, and other constunolide derivatives in costus. α-amorphenic acid and 12-methoxy dihydrodehydrocostuslactone were also isolated from Kust oil (Rijke et al., 1978, Dhillon et al., 1987).

Other chemical constituents present in roots include germacrenes *viz.* (+)-germacrene A, lappadilactone, cynaropicrin, guainolides viz. isozaluzanin C and isodehydrocostus lactone (Kraker et al., 2001; Taniguchi et al., 1995; Kalsi et al. 1984).

Saussureal and 4α, 4β-methoxydehydrocostus lactone were isolated from the petroleum ether extract of roots (Talwar et al., 1991, Singh et al., 1992).

β-cyclocostunolide and dihydro costunolide were reported from hexane extract, five new saussuramines A–E (amino acid-sesquiterpene adducts) and (–)-massoniresinol 4′′-O-β-D-glucopyranoside (a lignin glycoside) were isolated from methanolic extract, and later these components were synthesized in the laboratory (Yoshikawa et al., 1993; Matsuda et al., 2000). A baccharane type triterpene, α-amyrin, 3-β-acetoxy-9(11)-baccharene, a new sesquiterpene aldehyde, chlorogenic acid, and saussurine were also separated from costus (Yang et al., 1997; Lalla et al., 2002; Pandey et al., 2004). Moko lactone, betulinic acid methyl ester, and betulinic acid were characterized by activity dependent fractionation (Choi et al., 2009). Costus has been reported as a new source of inulin by Viswanathan and Kulkarni (1995).

The percentage of different constituents in costus roots is 5% resinoids (both solid and liquid), 1.5% essential oil, and 0.05% alkaloid (Ansari, 2019).

14.2.2.2 Costus Oil

The essential oil isolated from the plant roots is famous as costus oil. It is light yellow to brown in color and reported to be used for different purposes. It is often used as a component in high-quality perfumes and in the production of hair oils. Different methods of oil extraction from costus roots have been reported by Singh et al. (1957).

The quality of the oil and the quantity of the different constituents in the oil may vary, which depends upon the factors related to chemotype, genetic background, ecotype, and variations in conditions of environment *viz.,* temperature, relative humidity, irradiance, and photoperiod. Chief components of the oil are reported to be dehydrocostus lactone, β-costol, δ-elemene, costunolide, 8-cedren-13-ol, α-curcumene, α-selinene, β-selinene, α-costol, 4-terpinol, and 22-dihydrostigmasterol (Dhillon et al., 1987; Maurer and Grieder, 1997; Jain and Banks, 1968).

14.2.3 Potential Benefits, Applications, and Uses

Kust is an important medicinal plant and an array of compounds have been isolated from the plant which are responsible for its medicinal attributes. Several biological activities are reported to be present in costus extracts as well as in the costus oil. Besides its medicinal significance, the plant also possesses many other potential benefits, as it is utilized in the production of hair oils, perfumes, etc.

Biological activities *viz.*, anti-inflammatory, anti-ulcerogenic, antitumor, anticancer, hepatoprotective, immunomodulatory, antiviral, antiarthritic, etc. of different costus extracts and oil are reported by many researchers. A detailed description of pharmacological activities has been given in the following section (Figure 14.2).

14.2.3.1 Anti-Inflammatory

S. costus is often used as a medicine for inflammatory disease in the traditional Korean medicine system. The methanolic extracts of *S. costus* at a concentration of 0.1mg/ml exhibited about 50% reduction in cytokine induced neutrophil chemotactic factor (CINC) induction (Lee et al., 1995), and the sesquiterpene lactones present in methanolic extracts inhibited the induction of CINC-1 in lipopolysaccharide-stimulated NRK-52-E rat kidney cells (Ha et al., 1997). The purification of methanolic extract yielded 3 sesquiterpene lactones, i.e., reynosin, cynaropicrin, and santamarine which played a significant role in the inhibition of TNF-alpha production. The IC_{50} values of reynosin, cynaropicrin, and santamarine were found to be 8.24 μM, 87.4 μM, and 105 μM, respectively. The inhibitory effect of cynaropicrin is reported to be abolished on TNF-alpha

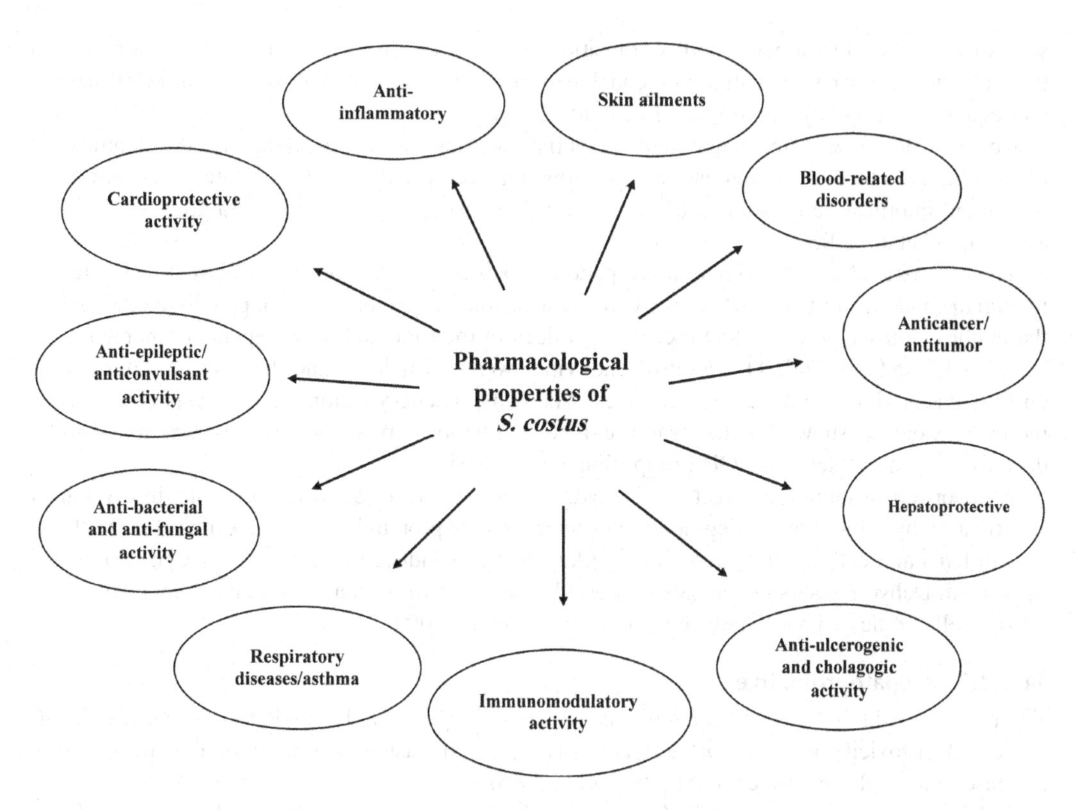

FIGURE 14.2 Pharmacological properties of *Saussurea costus.*

generation when it was treated with sulfhydryl (-SH) compounds like dithiothreitol, L-cysteine, and 2-mercaptoethanol. These results indicated that the major inhibitory constituent of the *S. costus* is cynaropicrin and its suppressive activity is facilitated by combining with the SH-group of desired protein (Cho et al., 1998).

The anti-inflammatory activity of cynaropicrin was further inspected *in vitro* on TNF-α, release of nitric oxide and lymphocyte production, in which it mainly suppressed the TNF- α release from lipopolysaccharide-stimulated RAW 264.7 cells (murine macrophage), and U937 cells (differentiated human macrophage) which were reported to produce significant amounts of TNF-alpha. It also strongly decreased the amount of Nitric oxide (NO) produced by lipopolysaccharide and IFN-ϒ-activated RAW246.7 cells in a dose-dependent mode. Cynaropicrin also inhibited the proliferation of B and T-lymphocytes from splenocyte by MTT assay in the presence of mitogens. These findings suggested the role of cynaropicrin in the inflammatory response by suppressing the proliferation of inflammatory cells and lymphocyte production (Cho et al., 2000).

The suppression of the NO production in lipopolysaccharide stimulated RAW 264.7 cells caused by *S. costus*-derived dehydrocostus lactone, was due to the decrease in expression of inducible nitric oxide synthase (iNOS) (Lee et al., 1999). Later, it was reported by Jin et al. (2000) that the decrease in iNOS gene expression was due to the inactivation of the NF-Kappa B.

14.2.3.2 Anticancer and Antitumor

The roots of *S. costus* contain costunolide, which was analyzed for its action on the initiation of apoptosis in human leukemia cells (HL-60) and its possible mode of action. It was found that costunolide was a potential initiator of apoptosis which acted *via* generation of reactive oxygen species (ROS) and thus inducing cytochrome C release in the cytosol as well as the mitochondrial permeability transition (MPT). However, successive apoptotic cell death, ROS production, and MPT was blocked by an antioxidant N-acetylcysteine (NAC) in costunolide-treated cells. Apoptosis and MPT were also repressed by a permeability transition inhibitor, cyclosporin A. Thus, the first report on the anticancer property of costunolide concluded that it induced the ROS-dependent MPT and the consequent release of cytochrome C (Lee et al., 2001).

Costunolide showed anti-angiogenic potential by substantially delaying the development of endothelial cells by the vascular endothelial growth factor (VEGF). It also inhibited the chemotaxis of human umbilical vein endothelial cells (HUVECs) caused by VEGF (Thara and Zuhra, 2012; Mohammad et al., 2013).

The cytotoxic effect of seven sesquiterpene lactones and lappadilactone of *S. costus* was tested in human cancer cell lines in which dehydrocostuslactone, costunolide, and lappa dilactone showed the major cytotoxic potential and their CD_{50} values in the time and dose-dependent manner were between 1.6 and 3.5 g/ml. The cytotoxic activities were nonspecific and displayed similar results on OVCAR-3, HeLa, and HepG2 cell lines. The α-methylene-γ-lactone moiety was essential for the cytotoxicity as shown by the structure–activity relationship, and its activity was found to be decreased by the presence of -OH group (Sun et al., 2003).

A major root sesquiterpene lactone, Dehydrocostus lactone, of *S. costus* prevents the NF-kappa B activation by inhibiting the degradation of its inhibitory protein known as I-kappa B α in HL-60 human leukemia cells. The degradation of I-kappa B α is induced by phosphorylation carried out by TNF-α. Dehydrocostus lactone also reported to increase the caspase-3 and caspase-8 activity in HL-60 cells prone to TNF-α-induced apoptosis (Oh et al., 2004).

14.2.3.3 Hepatoprotective

The role of costus in hepatoprotection was ascertained by the study carried out in deltamethrin-induced liver toxicity. It was reported that costus increased plasma liver biomarkers with very little increase in total plasma proteins (Alnahdi et al., 2016).

A decrease in expression of HBsAg (hepatitis B surface antigen) in human hepatoma cell line Hep 3B was shown by the treatment of two major constituents of *S. costus,* i.e., dehydrocostus lactone and costunolide in a dose-dependent manner with very little impact on cell viability. The decrease in HBsAg gene expression by dehydrocostus lactone and costunolide at the mRNA level was confirmed by northern blotting. Besides this, the human hepatoma cells, HepA2, which were derived from HepG2 cell line by transfection of a tandemly repeated hepatitis B virus DNA, were also suppressed by the costunolide. Similarly, these two components also suppressed the mRNA level of HBsAg in HepA2 cells. These research findings established that costus is a valuable source for the development of anti-HBV drugs in the future (Chen et al., 1995).

14.2.3.4 Anti-Ulcerogenic and Cholagogic Activity

Both cholagogic and inhibitory effects of costunolide and acetone extract of *S. costus* were observed on the gastric ulcers formed in water restraint mice (Yamahara et al., 1985). Chen et al. (1994) reported that the decoction made from costus could enhance the endogenous motilin release and gastric emptying. The main component in the herbal formulation known as UL409 having an anti-ulcer activity is *S. costus* (Venkataranganna et al., 1998), which causes the improvement in gastric cytoprotection by the modification of defensive factors (Mitra et al., 1996). The dried roots of *S. costus* have an anti-ulcerogenic effect on ethanol/hydrochloric acid-initiated abrasion in rats owing to the presence of saussureamines A, B, and C, amino acid–sesquiterpene adducts. Yoshikawa et al. (1993) reported that the formation of a stress-induced ulcer is decreased by Saussureamine A. The gastroprotective activity on mucosal abrasion induced by acidified ethanol in rats was reported in dehydrocostus lactone and costunolide along with saussureamines (Matsuda et al., 2000).

14.2.3.5 Immunomodulatory Activity

The *S. costus* extract containing dehydrocostus lactone and costunolide inhibited the killing potential of cytotoxic T lymphocytes (9 CTLs). Costunolides prevented the rise in tyrosine phosphorylation due to the cross-linking of T-cell receptors in $CD8^+$ CTL clone OE4 (Taniguchi et al., 1995).

Cynaropicrin isolated from Kust was investigated for immunomodulatory activities, immunosuppression effects, nitric oxide production, and the release of cytokine (Martel et al., 2017; Ghasham et al., 2017; Alnahdi et al., 2017; Zahara et al., 2014). KM1608, an herbal formulation containing costus as one of the principal constituents, was found to stimulate the expression of immune cytokines in macrophages (Trinh et al., 2020).

The structure–activity relationship of dehydrocostus lactone and mokko lactone from costus and other guainolides were tested for the inhibition activity of the killing potential of CTL and stimulation of intercellular adhesion molecule-1 (ICAM-1). It was reported that the guaianolides, having α-methylene-γ-lactone moiety, showed strong inhibitory action on the killing potential of CTL and the ICAM-1 stimulation (Yuuya et al., 1999).

14.2.3.6 Respiratory Diseases and Asthma

Roots of costus are used for treating cold, cough, and throat infections traditionally. Studies also showed that constituents present in costus are important against inflammation caused during asthma (Nadda et al., 2020).

Several tests were performed to study the activity of different *S. costus* extracts on asthma and chronic bronchitis. The alkaloid extract of costus was found to be nontoxic and has had minimal impact on breathing and blood pressure of the rabbit and cat (Pandey et al. 2007). Spasmolytic effect of costus was observed on the tracheal and smooth intestinal muscles of the histamine-stimulated guinea pig (Dutta et al., 1960).

Several experiments were performed on Tincture Saussurea obtained from extraction of the defatted costus roots, petroleum ether extract, and in other root extracts. The tincture, obtained from defatted roots and other extracts, did not initiate the broncho-constriction in the guinea pig. However, Tincture Saussurea and petroleum ether extract showed the broncho-constriction effect, thus establishing that Tincture Saussurea lacking petroleum ether soluble part can be used in the treatment of asthma and chronic bronchitis (Sastry and Dutta, 1961). These results justified the traditional use of costus as a medicine for asthma or other respiratory diseases.

14.2.3.7 Miscellaneous Activities

14.2.3.7.1 Potential for Treating Skin Ailments

Traditionally, root powder of *costus* is applied on the skin for diseases like scalp scabies, rashes, skin eruption, and skin allergy. Cytotoxicity assay was carried out on human cultured keratinocyte (HaCaT) cells showed that no cytotoxicity was seen after the 24 h treatment with methanol extract of *S. costus* (synm. *S. lappa*) at a concentration range of 7.8–31.3 μg/mL. However, significant inhibition of IFN-γ and TNF-α-induced MDC, TARC, and RANTES mRNA expression levels, which are atopic dermatitis markers in the HaCaT cells, was observed at the 5, 10 and 30 μg/mL extract concentrations. The *S. lappa* extract inhibited the production of histamine in MC/9 cells stimulated by phorbol myristate acetate (PMA) and A24187 (ionophore). However, an increase in the liver weight was observed when the mice were administered 500 mg/kg of *S. lappa* extract, which indicated liver toxicity (Lim et al., 2015). Boistir-AD (an extract of house dust mites)-induced skin lesions and dermatitis were found to be decreased in an *in vivo* experiment carried out on *S. lappa* treated 10-week-old Nc/Nga mice. Treatment of *S. lappa* caused a decrease in symptoms, such as hemorrhage, scarring, edema, atopic dermatitis, erosion, and erythema. *S. lappa* was also reported to reduce the histopathological indications of atopic dermatitis in mast cells (Lim et al., 2014).

14.2.3.7.2 Potential for Treating Blood-Related Disorders

Blood samples from 20 healthy volunteers were examined for *in vitro* thrombolytic activity using *S. lappa* bark extract, and a 14.85% clot lysis was detected by determining the weight of clot before and after treatment which is comparable with the streptokinase-created effect (Chaudhary et al., 2015). Some of the inherent properties of bark were lost due to its cleaning and drying in the hot air oven for the removal of moisture.

14.2.3.7.3 Antibacterial and Antifungal Activity

The effectiveness of several costus extracts against *Pseudomonas aeruginosa*, *Bacillus subtilis*, *Klebsiella pneumoniae*, *Staphylococcus aureus*, and *Escherichia coli* was tested by the cup plate diffusion technique, and the chloroform extract was found to be most active showing the highest activity (Alaagib and Ayoub, 2015). The ethanol extract of *S. costus* root showed remarkable inhibitory effect on several microorganisms such as *E. coli*, *Acinetobacter baumannii*, *K. pneumoniae*, *P. aeruginosa*, and *S. aureus*, having the inhibitory concentration ranging from 2.0 μg/μL to 12.0 μg/μL (Hasson et al., 2013). Due to the presence of various bioactive components, the ethanol extract of costus roots showed the highest antibacterial activity against the gram-positive bacteria, *Staphylococcus aureus*, but the extract did not show any effect against the gram-negative bacteria, *Salmonella* species (Omer et al., 2019). The antibacterial activity was detected in root extracts of costus prepared in cold water, hot water, and methanol with a zone of inhibition between 9 and 14 mm against *B. subtilis*, *Enterococcus faecalis*, *P. aeruginosa*, *S. aureus*, and *S. typhi* (Khalid et al., 2011).

Flavonoids present in costus oil (KSR1–KSR4) were investigated for antifungal activity against different species of *Aspergillus*, *Penicillium*, *Trichoderma viride*, *Alternaria alternata*, and *Cladosporium cladosporioides* and KSR4 was found to be more effective than the other three flavonoids (Rao et al., 2007).

14.2.3.7.4 Antiepileptic or Anticonvulsant Activity

To analyze the anticonvulsant activity of alcoholic, petroleum ether, and aqueous extracts of costus, convulsion induced by pentylenetetrazol (PTZ) and picrotoxin and maximal electroshock (MES) were assessed. It was found that maximum anticonvulsant activity was shown by petroleum ether extract (Ambavade et al., 2009).

Antiepileptic activity was also shown by the alcoholic extract of costus roots when administered for MES- and PTZ-induced seizures (Harish et al., 2010; Gupta et al., 2009).

The distress caused by the toxicity of thorium in male albino rats was decreased due to the protective effect of costus extract which inhibited the deposition of thorium in different parts of the brain, provided antioxidant activity, and modulated thyroid gland. The anticonvulsant activity and decrease in convulsion was attributed to the monoterpenes, essential oils, alkaloids and 4-hydroxybenzaldehyde present in costus extract (Rahman et al., 2020).

14.2.3.7.5 Cardioprotective activity

The cardioprotective activity of costus aqueous extract was investigated in rats administered with 85mg/kg isoproterenol, which was responsible for causing myocardial disease and scarring. A considerable decrease in oxidative stress was observed when 200 mg/kg of costus extract was given. This decrease was found to be due to the modulation of myocardial activity *via* lowering the concentration of thiobarbituric acid reactive substances (TBARs) (Saleem et al., 2013).

Costus serves as a cardiac tonic because the cardiac glycosides and flavonoids present in costus roots act as a cardiac stimulant which reduce the risk of cardiovascular diseases (Abdallah et al., 2017; Akbar, 2020; Ahmed et al., 2016; Ansari, 2019). Presence of calcium channel blockers, cholinergic components, sesquiterpene lactones, and flavonoids are also found to be responsible for cardiotonic activity of costus extract (Akhtar et al., 2013).

14.2.3.8 Other Uses

14.2.3.8.1 As Insect and Pest Repellent

Due to deteriorating effect of synthetic pesticides on our environment, the world is facing ecological and economical challenges in the fight against insects and pests. Therefore, plant-based products are frequently used for the management of insects and pests (Hossain et al., 2017). Additionally, herbal products can be used as ovicides, egg-laying deterrents, and antifeedants (Niroumand et al., 2016;

Hikal et al., 2017). The essential oil, costunolide, and dehydrocostus lactone obtained from costus showed significant larvicidal activity, which can be used for the preparation of larvicides in future (Liu et al., 2012).

Alkaloids and essential oil obtained from costus roots can reliably be used as a component of insect repellent. In countries like India and China, the costus roots are burnt to produce smoke which repel different kinds of insects. Costus roots are also kept in woollen fabrics to protect them from insects (Nadda et al., 2020).

14.2.3.8.2 Perfumery

The oil obtained from the roots, called "Costus oil", is used in the production of hair oil and top-grade perfumes. Sesquiterpene lactones isolated from the costus roots are highly valuable in the perfumery business (Gwari et al., 2013). Due to its strong and pleasant fragrance, it is widely used in the preparation of perfumes and mixed with other perfumes (Bhattacharya, 2001; Butola and Samant, 2010). But, the sesquiterpenoid-methylene lactones present in the oil causes allergies, which could restrict its use in the perfume industry. So, a synthetic reconstitution devoid of these lactones would be ideal for it to be widely used in the perfume industry (Mitchell, 1974).

14.3 *SENNA ALEXANDRINA* MILL.

Fabaceae is the third largest angiosperm family accounting for almost 811 genera and 24,840 species (GBIF: The Global Biodiversity Information Facility (2021)). Senna (Fabaceae, subfamily Caesalpinioideae, tribe Cassieae) is a large genus of **the** legume family and contains 250–300 accepted species mostly in tropical and temperate regions (Oladeji et al., 2021). They are mostly found in Asia, Africa, Europe, and Latin America. Several members of this genus are medicinally useful and find application in traditional systems of medicine worldwide. They have been popularly used in Africa, Brazil, China, India, Malaysia, Mexico, and Thailand. These species have been used in folklore for the treatment of ailments, such as abdominal pain, inflammation, typhoid, gonorrhea, diabetes, malaria, measles, and sexually transmitted diseases (STDs) (Silva et al., 2008). About 147 genera, 805 species, 33 subspecies, 155 varieties, and 14 forma of legumes are there in India (Gore and Gaikwad, 2015). Some of the important species of the genus Senna found in India are listed in Table 14.2.

A huge range of structurally diverse bioactive molecules exist in these species. The species have been reported as a good source of anthraquinones, flavonoids, mucilage, and polysaccharides (Sanghi et al., 2006). Recently, the genus has been explored for these molecules and subsequently followed by testing for biological and pharmacological activities of various plant parts. More than 120 structurally distinct phytochemicals, such as alkaloids, anthraquinones, anthrones, flavonoids, glycosides, piperidine, polyphenol saponins, steroids, terpenoids, and tannins have been reported (Hennebelle et al., 2009; Oladeji et al., 2021). The genus Senna has several pharmacological activities, such as analgesic, anti-inflammatory, antidiabetic, antimalarial, antimicrobial, antioxidant, antitumor, antinociceptive, and anticancer as detected in their crude extracts, fractions, or individually isolated metabolites of the plant/plant parts (Ibrahim and Islam, 2014).

One of the medicinally popular species of this genus, *Senna alexandrina* Mill., is used as a stimulant laxative (Figure 14.2). The main constituents in the leaves and fruit pods responsible for purgative activity are anthraquinones, which are found to be quite safe and effective in adults. Known as Alexandrian Senna (sourced from African countries) and Tinnevelly Senna (sourced from India), it is extensively used in habitual constipation and as a bowel preparation before medical procedures, such as colonoscopy (Radaelli and Minoli, 2002) and as relief from drug-induced constipation (Ulbricht et. al., 2011). It is a part of traditional systems of medicine as well as several modern pharmacopeias worldwide. Senna is approved by Food and Drug Administration (FDA) as a non-prescription drug too. The anthraquinones are structurally similar to anthracene and have basic structure of 9,10-dioxoanthracene. The glycosidic derivatives of these anthraquinones are

TABLE 14.2
List of Important Senna Species Reported from India

Name of Species	Common Name	Distribution (Global/ Indian)	Medicinally Important Plant Part	Medicinal or Other Uses	References
Senna siamea (Lam.) H.S. Irwin & Barneby	Siamese Cassia,	South East Asia; India (Assam, Maharastra, Gujarat, Rajasthan)	Leaf, stem, root, heartwood	Antioxidant, stomach complaints, mild purgative, anti-tumor, chemopreventive	Kaur et al., 2006
Senna auriculata (L.) Roxb.	Tanner's Cassia	Asia; India (Karnataka, Kerela, Tamil Nadu)	Flower, leaves	Antidiabetic, leprosy, asthma, gout, rheumatism, antipyretic, antiulcer, skin infections	Pari and Latha, 2002; Ayyanar and Ignacimuthu, 2008
Senna alata (L.) Roxb.	Candle bush	Asia, Africa, America, the Caribbean, and Oceania; India (Assam, Kerala, Uttar Pradesh)	Flower, root, leaves, seed, and bark	Used to treat ringworm, skin ailments. stomach problems, fever, asthma, and snakebite	
Senna septemtrionalis (Viv.) H.S.Irwin&Barneby	Smooth Senna	Java, Ethiopia, Madagascar Guatemala, India	Seeds, leaves	Unripe seeds, young shoots, and the leaves are cooked and eaten	Sosef and Maesen, 1997; Du Puy et al., 2002; Hanelt et al., 2001
Senna polyphylla (Jacq.) H.S.Irwin&Barneby	Desert Cassia	Caribbean – Puerto Rico, Virgin Islands, India (Andhra Pradesh)	None known	Used as fuel wood	Tropical Plants Database
Senna tora (L.) Roxb.	Sickle Senna	Asia; India (Kerala, Assam)	Seeds, leaves	Bronchitis, colic, cough, cardiac disorders, dyspepsia, leprosy, ringworm, flatulence constipation	Deore et al., 2009
Senna uniflora (Mill.) H.S. Irwin & Barneby	Oneleaf Senna	Brazil, Venezuela, through Central America to Mexico and the Caribbean; South India	Leaves, roots	Poultices for wounds; combating dropsy	Tropical Plants Database
Senna occidentalis (L.) Link	Coffee Senna,	Tropical regions; India (Andaman and Nicobar Island, Assam, Bihar, Kerala, Maharastra, Rajasthan, tamil Nadu)	Seeds, leaves	Hepatitis, inflammation, constipation, liver disorders, fungal infections, ulcers, respiratory infections, snakebite	Nuhu and Liyu, 2008; Aragao et al., 2009

(*Continued*)

TABLE 14.2 (CONTINUED)
List of Important Senna Species Reported from India

Name of Species	Common Name	Distribution (Global/ Indian)	Medicinally Important Plant Part	Medicinal or Other Uses	References
Senna sophera (L.) Roxb.	Pepper-Leaved Senna	Australia, Tropical America, S.E. Asia, Malaysia, China; India (Assam, Meghalaya, Kerela)	Root, seed, and leaves	Reduce fevers, anthelmintic, expectorant and febrifuge; relieve painful menstruation, treat epilepsy, diabetes	Tropical Plants Database
Senna spectabilis (DC) H.S. Irwin & Barneby	Spectacular cassia, golden wonder tree	Brazil, Ecuador; India (Western Ghats)	Leaves and flower-buds	Seeds are source of commercial gum, antifungal activity	Dave and Ledwani, 2012; Tropical Plants Database
Senna hirsuta (L.) H.S. Irwin & Barneby	Hairy Senna, Stinking Cassia	Asia, America, West Indies India (hilly tracks of South India, Andaman Island, Assam, Bihar, Kerela, Tamil Nadu)	Leaves	Liver ailments, stomach troubles, dysentery, abscesses, rheumatism, hematuria, fever	Joshua and Nwodo, 2010
Senna surattensis (Burm.f.) H.S. Irwin & Barneby	Glaucous Cassia	South East Asia, Australia	Roots, leaves, flowers	Gonorroea, dysentery, purgative	Tropical Plants Database
Senna pallida (Vahl) H.S. Irwin & Barneby	Twin-flowered cassia	South America, Central America; Caribbean	None known	branches used for making rough brooms or brushes	Tropical Plants Database
Senna obtusifolia (L.) H.S. Irwin & Barneby	Sicklepod	Asia	Seeds	Laxative, removes visual acuity and heat from liver, antiseptic, diuretic, diarrheal, antioxidant, antimutagenic	Dave and Ledwani, 2012
Senna pendula (Willd.) H.S. Irwin & Barneby	Easter cassia, climbing cassia, winter senna and valamuerto	South America, Australia	Leaves	Antioxidant	Monteiro et al., 2018
Senna alexandrina Mill.	Alexandrian Senna, Tinnevelly Senna	Africa, Middle East, India, Pakistan	Seeds, pods and leaves	Laxative, hepatoprotective	Silva et al., 2008

(Continued)

TABLE 14.2 (CONTINUED)
List of Important Senna Species Reported from India

Name of Species	Common Name	Distribution (Global/ Indian)	Medicinally Important Plant Part	Medicinal or Other Uses	References
Senna didymobotrya (Fresen.) H.S. Irwin & Barneby	Candelabra Tree	Tropical Africa	Leaves, stems and roots	Laxative, antifungal, antibacterial infections, anti-hypertension, used for the treatment of hemorrhoids, sickle cell anemia, gynecological diseases	Tropical Plant database
Senna corymbosa (Lam.) H.S. Irwin & Barneby	Argentine senna, buttercup bush	South America	Leaves and pods	Cathartic and laxative properties	Bianco and Kraus, 1997
Senna racemosa (Mill.) H.S. Irwin & Barneby		Mexico, Venezuela, Cuba	Leaves, roots, stem bark	Diarrhea, eye infections	Moo-Pucc et al., 2007
Senna atomaria (L.) H.S. Irwin & Barneby		South and Central America, Caribbean	Leaves	Purgative, treatment of skin itch, insect bites	Tropical Plants Database
Senna timoriensis (DC) H.S.Irwin&Barneby		South east Asia, Australia	Seedpods, bark	Anthelmintic properties, treatment against scabies	Tropical Plants Database
Senna floribunda (Cav.) H.S. Irwin & Barneby		Eastern Australia	Seeds	Malnutrition, purgative	Vadivel and Janardhanan, 2001
Senna bicapsularis (L.) Roxb.	Rambling senna, Winter cassia, Christmas bush, Money bush, and Yellow candlewood	northern South America	Seedpod	Cathartic, heals sores, rashes, eczema, cathartic	Tropical Plants Database
Senna multiglandulosa (Jacq.) H.S. Irwin & Barneby	-	Africa, Asia, Australia, Central America; Western and Eastern Ghats of India	-	-	Dave and Ledwani, 2012, https://indiabiodiversity.org/

(*Continued*)

TABLE 14.2 (CONTINUED)
List of Important Senna Species Reported from India

Name of Species	Common Name	Distribution (Global/ Indian)	Medicinally Important Plant Part	Medicinal or Other Uses	References
Senna italica Mill.	Italian Senna	Africa, Arabia, Iraq, Iran, Indian subcontinent	Root, stem, leaves, pods	Antimicrobial, antitumor, for curing stomach complaints, fever, jaundice, venereal diseases, and biliousness	Tropical Plants Database
Senna fruticosa (Mill.) H.S. Irwin & Barneby	English Christmas bush	South America, West Africa	Flowers	Treatment of pulmonary troubles	Wong, 1976
Senna reticulata (Willd.) H.S. Irwin & Barneby	Maria mole (Portuguese)	South America	Leaves, flowers	Antimicrobial	Parolin, 2001
Senna multijuga (Rich.) H.S. Irwin & Barneby	November shower	Mexico to southern Brazil and Bolivia	Leaves	Sedative: seeds are used as a source of industrial gum	Singh, 1981

used as laxatives. They are also known for the treatment of fungal skin diseases and slimming agents. Although valued for their cathartic and purifying activities, they may cause abdominal cramps, nausea, and diarrhea upon prolonged use or overdose.

Senna alexandrina Mill. is native to Sudan and is also under cultivation in the upper Nile region. It is also found in Southern Arabia and the Indian subcontinent. In India, it is found and cultivated in the districts of Madurai, Tinnevelly, and Tiruchirapally of Tamil Nadu; Mysore, Rajasthan, and Gujarat. Cultivation expenses of Senna are quite low as it does not require special requirements in terms of irrigation, manure, pesticides, etc. It can be cultivated in arid and semiarid regions which receive bright sunlight and have well-drained soil.

Earlier literature suggested that the Alexandrian Senna (*Cassia acutifolia*) is different from its Indian counterpart, Tinnevelly Senna (*Cassia angustifolia* Vahl). These were believed to be different species, as they showed some distinct morphological features and different naphthalene glycosides (hydroxymusicin in Alexandrian Senna; and Tinnevellin glucoside in Tinnevelly Senna) (Franz, 1993). However, the chromosome number is the same, i.e., 28 in both Alexandrain and Tinnevelly Senna. Recently, *Cassia acutifolia* and *Cassia angustifolia* Vahl. are considered synonyms, and the accepted scientific name is *Senna alexandrina* Mill. according to World Flora Online (http://www.worldfloraonline.org/taxon/wfo-0000164723. Accessed September 9, 2021).

The leaves and fruit pods are economically useful parts of this plant. The pod shells accumulate a higher amount of anthraquinone glycosides (sennoside A and sennoside B) than the leaves. The leaves and pods are collected from wild or cultivated plants, dried in the sun, and graded according to quality. Unbroken leaves and pods are exported to the international market, whereas damaged or broken leaves are considered inferior and are used by the extraction/herbal industry in the domestic market. India exports 75% of the Senna produced worldwide while the rest is sourced from African countries. The three Indian states of Tamil Nadu, Rajasthan, and Gujarat are primarily involved in the production and global trade of Senna and its products. About two-thirds of India's export volume of senna is imported by the following 10 countries: Egypt, China, Germany, Japan, Mexico, Poland, Spain, Thailand, the United States, and Vietnam (Brinckmann and Smith, 2018).

14.3.1 Description

14.3.1.1 Botanical Description

The plants of *Senna alexandrina* Mill. are woody perennial herbs or shrubs of approximately 0.5–2 m height with low branches. Leaves are alternate, pinnate, 5–9 pairs, leaflets are narrow acute lanceolate, glabrous and pale green in color. Flowers are ebracteate, axillary, and terminal racemes 10–22 cm in length with 10 mm long pedicels. Sepals are 7–9 mm long, obtuse, and glabrous. Petals are bright yellow, 1–1.5 cm long. Total number of stamens is 10, out of which the upper 3 are reduced to staminodes. The numbers of fertile stamens are 7 out of which 2 anthers are ~ 1 cm long and falcate, one is ~4.5 mm long and straight; and remaining four are 3–4 mm long and straight. The ovary is strigose and stigma is punctiform. The fruits are dehiscent pods of about 4–5.2 × 2 cm oblong, straight or upwardly slightly falcate, flat, shallowly elevated over the seeds and transversely septate. The seeds are 5–8 in number per pod, each 5–6 × 3 mm, obovate-oblong, and narrowed at hilum end (Figure 14.3).

14.3.1.2 Traditional Knowledge

Traditionally, infusions of Senna taken in the form of tea or swallowed in powdered form have been used as a laxative for hundreds of years in both the East and the West. The first medicinal records of Senna dates back to the 9th century by Arab physicians (Abulafatih, 1987). In the 9th century, the famous Christian Arab physician, Mesue, the Elder, brought Senna leaves native to Africa to treat the constipation of the Caliph Harun al-Rashid (Khan, 2020). The name of

FIGURE 14.3 *Senna alexandrina* plant in flowering and fruiting stages.

the plant can be traced to two words: "sena" (Arabic) and "cassia" (Hebrew), which refers to its peelable bark.

Known as Swarnapatri in Sanskrit language (Ramchander and Middha, 2017), *Cassia angustifolia* is documented as "pitta shodhaka", i.e., removes the pitta dosha from body and "vataanulomaka", i.e., removes vata through the anal route (*The Ayurvedic Pharmacopoeia of India*, Part I, Volume I, 2001).

It is a popular drug in various traditional systems of medicine including Unani, Ayurvedic, homeopathic, Siddha, Chinese, etc. Some of the well-known compounds made by using Senna are PanchSakaara Churna, Shtshakaar Churna, and Yashtyaadi Churna (Ayurvedic); Safoof-e-Mulaiyia, Majoon-e Senaai, Itrifal Ustukhudus (Unani); and Nilaavarrai Choornam (Siddha) (Ramchander and Middha, 2017). Popular Ayurvedic formulations using Senna are: Sarivadyasava (for the treatment of diabetes, skin diseases, and gout); Ayulax (for the treatment of distention of abdomen and constipation); Kultab (for the treatment of piles and hemorrhoids); Pylend (for the treatment of constipation and piles); and Raktansoo (blood purifier). The leaves of the plant are used in several ailments such as constipation, indigestion, loss of appetite, anemia, malaria, jaundice, spleenomegaly, and hepatomegaly. Purgation therapy termed as "Virechna" has been advised in Ayurveda in cases of jaundice, spleenomegaly, and hepatomegaly so as to remove excess pitt from the body (Ramchander and Anil, 2017). Similar therapeutic effects have also been seen in the pods of the plant, although it shows a gentler effect (Tripathi, 1999). In the Unani system of medicine, Senna leaf (sanamakki) is used for treating gout, acne, cardiac asthma, colic, scabies, sciatica, pimples, backache, joint pain, and hip pain (*The Unani Pharmacopoeia of India*, Part I, Volume I, 2007). The decoctions and infusions are used in folk remedies for the treatment of bacterial infections and several skin ailments like acne, psoriasis, and wounds. It is a popular remedy for anemia, bronchitis, jaundice, and skin problems in the Indian Ayurvedic system of medicine. A paste prepared in vinegar from powdered leaves has been documented to treat skin problems such as pimples (Tripathi, 1999). An ointment made by mixing powdered seeds of *C. fistula* and senna with curd is useful for curing ringworm (Tripathi, 1999).

The Chinese traditional medicine also documents Senna for the treatment of atherosclerosis; to remove "heat" arising from constipation and abdominal pain; and edema (Brinckmann and Smith, 2018). Senna was known as "Purging Cassia" in the medieval age due to its use as a purgative in an Italian medical school.

14.3.2 Chemical Derivatives (Bioactive Compounds – Phytochemistry)

The pioneering work in characterizing chemical constituents of Senna was done by Stoll in 1941 (Franz, 1993). The two crystalline glycosides isolated by him were named sennoside A and sennoside B. Later work by different investigators helped in the identification of other constituents, sennoside C and sennoside D, along with other active molecules (Franz, 1993). Hydroxyanthracene glycosides (HAGs) are basically the bioactive constituents in Senna. The HAGs are present in the form of dianthrone glycosides (sennosides A, A1, B, C, and D) and anthraquinone glycosides (rhein-8-O-glucoside and rhein-8-sophoroside) (Meier et. al., 2017; Wichtl, 2004). The anthranoids identified in the leaf are dianthrones (75–80%) and anthrones (20–25%) which are the active constituents in the plant. The leaves and pods of *S. alexandrina* contain ~2.5% of anthraquinone glycosides (sennosides A and B), that are basically derived from rhein and aloe-emodin.

The leaves of *Senna alexandrina* Mill. contain aloe-emodin, its 8-glucoside, aloe-emodin dianthraone, chrysophanol, emodin 8-O-sophoroside, rhein, rheum-emodin glycoside, aloe-emodin dianthraone-diglucoside, sennoside A, sennoside B, sennoside C and sennoside D, sennoside G, III, A_1, anthranoids of the emodin and aloe-emodin (The Wealth of India, 1992; Rastogi and Mehrotra, 1990; 1998).

The pods are reported to contain aloe-emodin, chrysophanol, rhein and their glucosides, emodin, anthranoids of the emodin and aloe emodin, sennoside A, B and sennoside $A_{1.}$

Chrysophanol, physcion, rheum emodin, aloe-emodin, and rhein had been identified in callus culture from cotyledons (Friedrich and Baier, 1973). The various constituents identified from seedlings and roots are: several mono- and di-glucosides of anthrones, chrysophanol, physcion, emodin, aloe emodin, rhein, chrysophanein, physcionin, gluco-aloe-emodin, emodin-8-O-β-glucoside, gluco-rhein, and sennosides A, B, C (Kinjo et al., 1994; Ilavarasan et al., 2001; Werner and Merz, 2007; Dave and Ledwani, 2012). The formation of hydroxyanthraquinone has been reported in cell cultures of *C. angustifolia* Vahl and *C. senna* Linn (Dave and Ledwani, 2012).

The leaves of *Senna alexandrina* Mill. contain seven pharmacologically active compounds: aloe emodin, emodin-8-O-beta-D-glucopyranoside, apigenin-6,8-di-C-glycoside, kaempferol, isorhamnetin-3-O-beta-gentiobioside, D-3-O-methylinositol, and tinnevellin glycoside (Wu et al. 2007). Ahmed et al. 2016reported quercimeritrin, scutellarein, and rutin as the main phenolic constituents of the plant.

Coumarins, flavonoids, and phenolic acids were the three groups of secondary metabolites detected in pod extract of *Senna alexandrina* by Elansary et al. (2018). Relatively high content (369.15 mg/ 100 g DW) of a precursor of the phenolic acids, benzoic acid, was detected along with seven phenolic acids: caffeic (13.34 mg/ 100 g DW); gallic (15.94 mg/ 100 g DW); gentisic (363.21 mg/ 100 g DW); neochlorogenic (65.15 mg/ 100 g DW); protocatechuic (31.52 mg/ 100 g DW); syringic (5.88 mg/ 100 g DW); and vanillic (5.06 mg/ 100 g DW). They also identified seven flavonoids: cynaroside (73.28 mg/ 100 g DW); isoquercetin (195.15 mg/ 100 g DW); isorhamnetin (51.31 mg/ 100 g DW); kaempferol (137.74 mg/ 100 g DW); luteolin (22.81 mg/ 100 g DW); quercetin (80.06 mg/ 100 g DW); and rhamnetin (291.30 mg/ 100 g DW). The two coumarins identified in their study were 6-hydroxy-4-methylcoumarin (62.72 mg/ 100 g DW) and psoralene (20.93 mg/ 100 g DW).

Two new prenyloxyanthraquinones (Madagascin (3-isopentenyloxyemodin) and 3- geranyloxyemodine) were isolated and reported from the fruit and leaves of *Senna alexandrina* Mill. by Epifano et al., 2015.

14.3.3 Potential Benefits, Applications, and Uses

Senna is approved by the FDA as an over-the-counter (OTC) laxative and its purchase does not require a prescription. Senna is also used for anal or rectal surgery, irritable bowel syndrome (IBS), hemorrhoids, tears in the lining of the anus (anal fissures), and weight loss.

In addition to being well known as a natural laxative (Rama Reddy et al., 2015) the senna leaf has been found to exhibit antioxidant, antibacterial, antifungal, antitumor, and antidiabetic activities

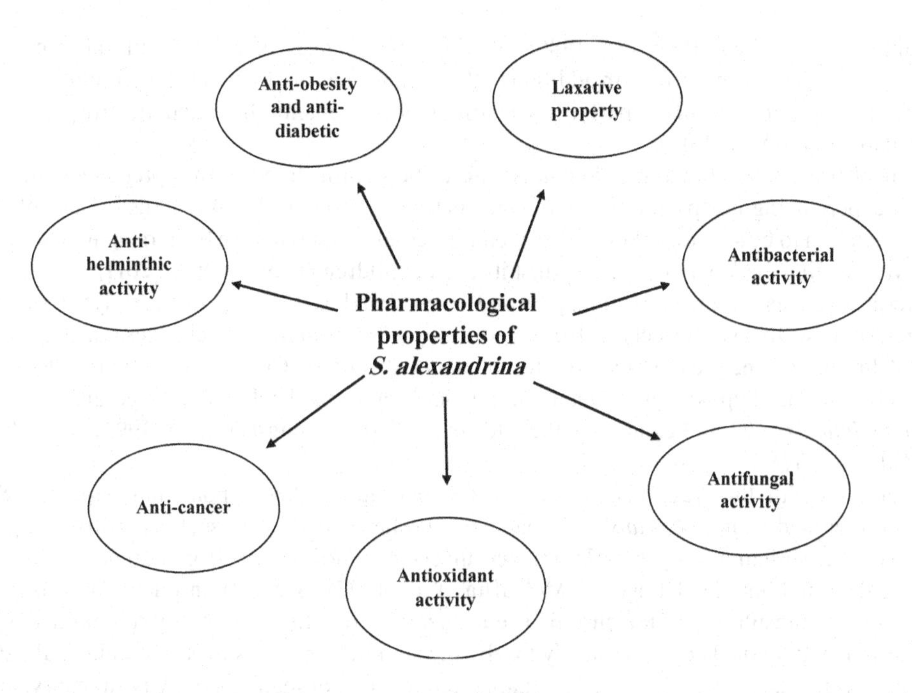

FIGURE 14.4 Pharmacological properties of *Senna alexandrina*.

(Elansary et al., 2018). Some of its applications and potential benefits are described in the following sections (Figure 14.4).

14.3.3.1 Senna as a Laxative Drug

Senna is considered one of the most widely used anthraquinone laxatives, as it is relatively cheaper and more effective (Ulbricht et al., 2011). The phytoconstituents of *S. alexandrina*, aloe emodin, sennoside, and dianthrone glycoside (anthraquinone derivatives and glycosides), are the reason for its laxative effect (Monkheang et al., 2011).

Scientifically, the use of Senna for the treatment of chronic constipation or constipation induced by childbirth or pharmaceutical drugs is well-documented (Ulbricht et al., 2011). It is also used in the majority of cases of terminally ill cancer patients who are on opioids for pain relief. It was seen that Senna is equally effective as lactulose in these patients. Patients on drugs (tricyclic antidepressants, opioids, and phenothiazines) who suffer from constipation as a side effect have an option to use Senna.

The mechanism of action of Senna has been reviewed by several researchers (Leng-Peschlow, 1992a, 1992b; Lemli, 1986). After the oral intake of the drug, sennosides neither get chemically modified in the upper part of the gastrointestinal tract nor get absorbed by the epithelial cells of the gut due to their hydrophilic nature (Lemli, 1988a; Kobashi et al., 1980). They move to the large intestine unabsorbed where the active form of aglycon gets released by bacterial hydrolysis of the sugar (de Witte and Lemli, 1990). The metabolization of sennosides occurs by the beta-glucosidases of the microflora (*Bifidobacterium adolescentis, Clostridium perfringens, Clostridium sphenoides, Eubacterium spp., Lactobacillus brevis, and Peptostreptococcus intermedius*) to form their corresponding glucuronide and sulfate derivatives) (de Witte, 1993; Lemli, 1988; Ulbricht et al., 2011). After the splitting of sennosides into sennidins and anthrones in the intestine by intestinal bacteria, these are oxidized to form anthraquinones outside the large intestine (Krumbiegel and Schulz, 1993). Downstream metabolism of the resulting rheinanthrones does not occur by bacteria but these are absorbed by intestinal epithelial cells. The anthrone is structurally similar to danthron, a known hepatoxic laxative. The colon transit is accelerated by sennosides directly effecting colon (large intestine) motility (Leng-Peschlow, 1986a, 1986b). The changes in colonic fluid absorption and motility cause the laxative effect. In addition to this, sennosides are known to be secretory in action. The laxative action of Senna is partially via stimulation of colonic fluid and electrolyte secretion (Ramchander and Anil, 2017).

Clinical trials have shown that Senna is safe to be administered during pregnancy and those women experiencing postpartum-induced constipation. Although, WHO regards the use of Senna during lactation to be safe for infants, it still remains controversial for use by nursing mothers. Some negative scientific evidence presents against its use in children (Ulbricht et al., 2011).

Overdose causes griping abdominal pain and severe diarrhea, resulting in electrolyte loss. *In vivo* research suggests that side effects such as griping may arise from motility changes (Leng-Peschlow, 1980). Chronic use may lead to electrolyte imbalance disorders. Chronic use may result in hematuria, albuminuria, hypokalemia, metabolic alkalosis, and renal tubular damage and may cause *Pseudomelanosis coli* (pigmentation of the intestinal mucosa) (Leng-Peschlow, 1992c; Ramchander and Middha, 2017)

When used with diuretics and other laxatives Senna causes additive potassium depletion. Senna has been regarded as *possibly unsafe* in patients who have conditions, such as obstructed gastrointestinal tract, inflammatory bowel, stomach ulcers, hemorrhoids, and gall stones or are using anticoagulant and antiplatelet agents (McGuffin et al., 1997). A condition known as "lazy-bowel syndrome" may result due to the chronic use of laxatives in which the ability to contract without being stimulated by the laxative is slowly lost by the stomach and intestines (Ulbricht et al., 2011).

In vivo trials lacked any evidence of lethality, teratogenic, or fetotoxic activity in embryo due to sennosides (Mengs, 1986). Also, the corelation between colon carcinogenesis and long-term use of anthracene drugs is still controversial. The scientific evidences support Senna as a laxative drug and its widespread use remains undisputed.

14.3.3.2 Antibacterial Activity

Several reports show antibacterial activity of leaves and pods of *S. alexandrina* (Selim et al., 2013; Al Saiym et al., 2015). The extracts of pods of *S. alexandrina* showed high antibacterial activity against gram-positive and gram-negative bacteria, including *Listeria monocytogenes* (clinical isolate), *Staphylococcus aureus* (ATCC 6538), *Bacillus cereus* (ATCC 14579), *Micrococcus flavus* (ATCC 10240), *Pseudomonas aeruginosa* (ATCC 27853), and *Escherichia coli* (ATCC 35210). The minimum inhibitory (MIC) and bactericidal concentration (MBC) ranged from 0.02 to 0.12 mg/mL and 0.04 to 0.29 mg/mL, respectively (Elansary et al., 2018). Ahmed et al. (2016) tested the antibacterial activity of Senna aqueous and organic extracts against *Acinetobacter junii* IARS2, *Enterobacter cloacae* IARS7, *Pseudomonas aeroginosa* IARS8, *Salmonella typhi* ATCC 14079, and *Serratia mercescens* IARS6 and found the highest bactericidal activity against *S. mercescens* in the ethyl acetate extract. Methanol extract showed inhibitory effects against all bacterial strains tested whereas aqueous extract was ineffective against *A. junni*, *E. cloacae*, and *P. aeroginosa*. Significant antibacterial activity against some human pathogenic bacteria, *E. coli*, *Klebsiella pneumoniae, and Shigella shinga* has been reported by Bameri et al., 2013. Antibacterial potential of n-butanol extract of Senna against *S. aureus* and *typhi* was also reported by Gnanavel et al. (2012).

14.3.3.3 Antifungal Activity

Significant antifungal activity of *S. alexandrina* leaf extracts against some plant pathogens has been reported by Rizwana et al. (2021), but the Senna extracts were ineffective against human pathogenic fungus *Candida*. However, antifungal activity of leaf extracts against some species of *Candida* and *Aspergillus* have also been reported (Hossain et al., 2012; Ahmed et al., 2016; Al Marzoqi et al., 2016). Fungicidal properties of *S. alexandrina* pods against pathogenic fungi are mainly attributed due to the activity of phenols and flavonoids as reported by earlier research (Ahmed et al., 2016; Vijaysekhar et al., 2016). The methanolic extracts of pods of *S. alexandrina* showed high antifungal activities against several fungi: *Aspergillus flavus* (ATCC 9643); *A. ochraceus* (ATCC 12066); *A.*

niger (ATCC 6275); *Penicillium ochrochloron* (ATCC 48663); *P. funiculosum* (ATCC 56755); and *Candida albicans* (ATCC 12066) (Elansary et al., 2018).

14.3.3.4 Antioxidant Activity

The antioxidant activity of the methanolic extracts of Senna pods was evaluated by 2,2′-diphenylpicrylhydrazyl (DPPH) and β-carotene-linoleic acid assays. Relatively high antioxidant activities were detected when expressed as IC_{50}, *μ*g/mL for *S. alexandrina* pods (DPPH = 2.6; *β*-carotene = 2.4) (Elansary et al., 2018). The presence of benzoic acid in high amount in the extract was one of the probable reasons for high antioxidant activity. Also, flavonoids, such as kaempferol, also are strong antioxidants that limit free radical accumulation and scavenge them. Ahmed et al. (2016) also evaluated the Senna aqueous and organic (methanol, ethanol, acetone, ethyl acetate) extracts for antioxidant potential by the DPPH radical scavenging assay. Maximum DPPH free radical scavenging activity (93%) was seen in the methanol, whereas poor DPPH scavenging activity (68%) was seen in the aqueous extract. The other organic extracts of *Senna* exhibited moderate DPPH scavenging activities.

14.3.3.5 Antihelminthic Activity

Crude ethanolic extract of Senna has reported to possess anti-helminthic activity on *Hymenolepis diminuta* (Kundu et al., 2017). Senna leaf extracts (*Senna alata*, *S. alexandrina*, and *S. occidentalis*) individually and in combination showed potential anti-trematocidal activity against rumen fluke, *Paramphistomum gracile*, by damaging body tegument and neural propagation and proved to be a potential trematocidal drug candidate in livestock (Roy and Lyndem, 2019).

14.3.3.6 Anticancer

MTT colorimetric assay was employed by Ahmed et al. (2016) to evaluate the cytotoxicity potential of aqueous and organic Senna extracts and explore anticancer potential against Hep2, HeLa, MCF-7, and normal HCEC cell lines. Anticancerous activity against Hep2 cell line (28% cell death) was exhibited by methanol extract; HeLa cells (33 and 23 % cell death) by methanol and ethanol extracts, respectively. Also, 43 and 23% cell death was observed in MCF-7 cell lines by methanol and ethanol extracts, respectively. Further, normal HCEC cells were 100% viable against different concentrations of methanol and ethanol extracts. It was concluded that the antioxidant, antimicrobial, and anticancer activities of Senna extracts are probably due to the isolated flavonoids to a certain extent.

14.3.3.7 Anti-Obesity and Antidiabetic Activities

Yuniarto et al., 2018 tested the effect of Senna and pomegranate (extracts and fractions) through *in vitro* studies such as alpha-glucosidase inhibitory activity, alpha-amylase inhibitory activity, pancreatic lipase inhibitory activity, and antioxidant activity by DPPH method and concluded that these can be useful in strategies for treatment of obesity and type-2 diabetes mellitus. Folium Sennae Ethanolic (FSE) extract mediates uptake of glucose in L6 myoblasts by increasing Ca^{2+} concentration followed by the fusion of glucose transporter 4 (GLUT4) vesicles with plasma membrane, allowing glucose uptake. Therefore, FSE may be a potential drug in the area of natural therapies for improving Type 2 diabetes mellitus (Zhao et al., 2018).

14.4 CONCLUSION

The two important medicinal plants, *S. costus* and *S. alexandrina*, are highly recommended in various traditional medicine systems and have been summarized. *S. costus* has been recommended for *ex-situ* and *in-situ* conservation whereas *S. alexandrina* M. is under wide cultivation covering about 25, 000 hectares in India. *S. costus* has been widely used as a cure for numerous ailments due to the presence of a variety of bioactive compounds. The use of plant as drug is mentioned in indigenous medicine systems of Tibet, India, Korea, and China. Costus is used traditionally for the treatment of asthma, lung disorders, inflammation, ulcer, cold, rheumatism, cough, etc. in many parts of the world. Costunolides, a sesquiterpene lactone is one of the main phytoconstituent of the plant. Dehydrocostus lactone, cynaropicrin, reynosin, and saussuramines are some other important bioactive components reported from the plant. Due to its overexploitation for commercial and medicinal uses, the plant has been listed under the endangered category in India. The essential oil extracted from costus has an intense fragrance and is used in the manufacture of oil and good quality perfumes. Anthraquinones (both natural and synthetic) have acquired a large number of applications in health care as well as industry, impacting human populations indirectly and directly. Anthraquinones in the plant extracts from genus Senna are finding newer applications in the cosmetics, dye, food, and pharmaceutical sectors. The demand of *Senna alexandrina* Mill. globally is foreseen to increase as Senna enjoys the status of an effective and safe drug to relieve constipation. It has several therapeutic potentials, such as purgative, antibacterial, antifungal, antioxidant, anticancerous, antidiabetic, and anti-obesity, which make it useful in different pharmacopoeias worldwide. The three largest suppliers of Senna raw material and value-added products to the international market are India followed by Sudan and Egypt. Focus on research toward development of varieties with higher active constituents will help in maintaining leadership position in the supply chain. Many of the secondary metabolic pathway steps still remain to be elucidated which requires biotechnological interventions to understand the genes involved for anthraquinone biosynthesis. Also, adulteration of the drug with similar species in the trade is yet another undesirable factor which needs to be strictly controlled by the concerned authorities.

ACKNOWLEDGMENTS

Financial assistance under the Women Scientist Scheme (WOS A), Department of Science and Technology, Government of India to Dr. Soni Gupta to carry out the research as Project Investigator is duly acknowledged. Authors also acknowledge the support and resources provided by Integral University, Lucknow in writing this manuscript.

REFERENCES

Abdallah, E.M., K.A. Qureshi, A.M.H. Ali and G.O. Elhassan. 2017. Evaluation of some biological properties of *Saussureacostus* crude root extract. *Bioscience Biotechnology Research Communications*, 10(4), pp.601–611.

Abulafatih, H.A. 1987. Medicinal plants of Southwestern Saudi Arabia. *Economic Botany*, 41(3), pp.354–360.

Ahmed, A., S. Ahmad, K. Soni, et al. 2016. Suitable solvent and drying condition to enhance phenolics and extractive value of *Saussurea costus*. *Journal of Ayurvedic and Herbal Medicine*, 2(5), pp.165–170.

Ahmed, S.I., M.Q. Hayat, Tahir, M. et al. 2016. Pharmacologically active flavonoids from the anticancer, antioxidant and antimicrobial extracts of *Cassia angustifolia* Vahl. *BMC Complementary and Alternative Medicine*, 16, p.460.

Ajaib, M., M. Ishtiaq, K.H. Bhatti, et al. 2021. Inventorization of traditional ethnobotanical uses of wild plants of Dawarian and RattiGali areas of District Neelum, Azad Jammu and Kashmir Pakistan. *PLoS ONE*, 16(7), p.e0255010.

Akbar, S. 2020. *Saussurea lappa* (Falc.) Lipsch. (Asteraceae/Compositae). In *Handbook of 200 Medicinal Plants*. Cham: Springer, pp 1609–1617.

Akhtar, M.S., S. Bashir, M.H. Malik and R. Manzoor.2013. Cardiotonic activity of methanolic extract of *Saussurea lappa* Linn roots. *Pakistan Journal of Pharmaceutical Sciences* 26(6), pp.1197–1201.

Alaagib, R.M.O. and S.M.H. Ayoub. 2015. On the chemical composition and antibacterial activity of Saussurea lappa (Asteraceae). *Pharma Innovation Journal* 4(2), pp.73–76.

Al-Marzoqi, A.H., M.Y. Hadi and I.H. Hameed. 2016. Determination of metabolites products by Cassia angustifolia and evaluate antimicrobial activity. *Journal of Pharmacognosy Phytotherapy* 8, pp.25–48.

Alnahdi, H.S., E.N Danial, M.E.A.E Elhalwagy and N.O. Ayaz. 2017. Phytochemical studies, antioxidant properties and antimicrobial activities of herbal medicinal plants costus and cidir used in Saudi Arabia. *International Journal of Pharmacology* 13(5), pp.481–487.

Alnahdi, H.S., N.O. Ayaz and M.E. Elhalwagy. 2016. Prophylactic effect of cousts Saussurea lappa against liver injury induced by deltamethrin intoxication. *International Journal of Clinical and Experimental Pathology* 9(1), pp.387–394.

Al-Saiym, R.A., H.H. Al-Kamali and A.Z. Al-Magboul. 2015. Synergistic antibacterial interaction between *Trachyspermum ammi*, *Senna alexandrina* Mill and *Vachellia nilotica* spp. Nilotica extract and antibiotics. *Pakistan Journal of Biological Sciences* 18, pp.115–121.

Ambavade, S.D., N.A. Mhetre, A.P. Muthal and S.L. Bodhankar. 2009. Pharmacological evaluation of anticonvulsant activity of root extract of *Saussurea lappa* in mice. *European Journal of Integrative Medicine* 1(3), pp.131–137.

Ansari, S. 2019. Ethnobotany and pharmacognosy of Qust/Kut (*Saussurea lappa*, C. B. Clarke) with special reference to Unani medicine. *Pharmacognosy Reviews* 13(26), pp.71–76.

Aragao, T.P., M.M. Lyra and M.G. Silva. 2009. Toxicological reproductive study of *Cassia occidentalis* L. in female Wistar rats. *Journal of Ethnopharmacology* 123(1), pp.163–166.

Arif, T., J.D. Bhosale, N. Kumar, T.K. Mandal, R.S Bendre, G.S. Lavekar and R. Dabur. 2009. Natural products: Antifungal agents derived from plants. *Journal of Asian Natural Products Research* 11(7), pp.621–638.

Ayurvedic Pharmacopoeia Committee. 2001. *The Ayurvedic Pharmacopoeia of India, Part I*, Vol. I, 1st edn. New Delhi, India: The Controller of Publications.

Ayyanar, M. and S. Ignacimuthu. 2008. Pharmacological action of *Cassia auriculata* L. and *Cissus quadrangularis* Wall.: A short review. *Journal of Pharmacology and Toxicology* 3(3), pp.213–221.

Bameri, Z., A.B. Negar, S. Saeide and B. Saphora. 2013. Antibacterial activity of C. angustifolia extract against some human pathogenic bacteria. *Journal of Novel Applied Science* 2, pp.584–586.

Bhattacharya, S.K. 2001. *Handbook of Medicinal Plants*. Jaipur, India: Aavishkar Publishers, p.504.

Bianco, C.A. and T.A. Kraus. 1997. Observation on the *Senna* (Leguminodae-Caesalpinioideae) species in the south of the province of Córdoba. *Multequina* 6, pp.33–47.

Brinckmann, J. and T.Smith. 2018. *Cassia angustifolia* and *Cassia senna* (syn. *Cassia acutifolia*, *Senna alexandrina*) Family: Fabaceae (Leguminosae). *Herbal Gram* 120, pp.6–13.

Bruchhausen, F.Y., G. Dannhardt, S. Ebel, et al. 1994. HagersHandbuch der Pharmazeutischen Praxis. Band 9, Stoffe P-Z Bandwerk Hager: Hdbpharmaz.Praxis (5.Aufl.) 5., vollst¨andigneubearb. Aufl., XXX, 1255S., Geb.

Butola JS and S.S. Samant. 2010. *Saussurea* species in Indian Himalayan Region: diversity, distribution and indigenous uses. *International Journal Plant Biology* 1, pp.43–51.

Chaudhary, H.J. and R.R. Rao. 2000. Trans-Himalaya: A vast genetic resource centre of less known economic plants. *Indian Journal of Forestry* 23, pp.446–456.

Chaudhary, S., P.K. Godatwar and R. Sharma. 2015. In vitro thrombolytic activity of dhamasa (*Fagonia arabica* inn.), kushta (*Saussurea lappa* Decne.), and guduchi (*Tinospora cordifolia*thunb.). *AYU* 36, pp.421–424.

Chen, H.C., C.K. Chou, S.D. Lee, J.C. Wang and S.F. Yeh. 1995. Active compounds from *Saussurea lappa* Clarke that suppress hepatitis B virus surface antigen gene expression in human hepatoma cells. *Antiviral Research* 27(1–2), 99–109.

Chen, Q.L., X.Y. Chen, L. Zhu, et al. 2016. Review on *Saussurea laniceps*, a potent medicinal plant known as "snow lotus": Botany, phytochemistry and bioactivities. Springer, *Nature* 15, pp. 537–565.

Chen, S.F., Y.Q. Li and F.Y. He. 1994. Effect of *Saussurea lappa* on gastric functions. *Chinese Journal of Integrated Traditional and Western Medicine* 14, pp.406–408.

Cho, J.Y., J. Park, E.S. Yoo, et al. 1998. Inhibitory effect of sesquiterpene lactones from *Saussurealappa* on tumor necrosis factor-alpha production in murine macrophage-like cells. *Planta Medica* 64, pp.594–597.

Cho, J.Y., K.U. Baik, J.H. Jung and M.H. Park. 2000. *In vitro* anti-inflammatory effect of cyanaropicrin, a sesquiterpene lactone, from *Saussurea lappa*. *European Journal of Pharmacology* 398, pp.399–407.

Choi, E.M., G.H. Kim and Y.S. Lee. 2009. Protective effects of dehydrocostus lactone against hydrogen peroxide-induced dysfunction and oxidative stress in osteoblastic MC3T3-E1 cells. *Toxicology In Vitro* 23, pp.862–867.

Chopra, R.N., S.L. Nayar and I.C. Chopra. 1956. *Glossary of Indian Medicinal Plants*. New Delhi: Publication and Information Directorate.

Dave, H. and L. Ledwani. 2012. A review on anthaquinones isolated from *Cassia* species and their applications. *Indian Journal of Natural Products and Resources* 3, pp.291–319.

de Witte, P. and L. Lemli. 1990. The metabolism of anthranoid laxatives. *Hepatogastroenterology* 37, pp.601–605.

de Witte, P. 1993. Metabolism and pharmacokinetics of anthranoids. *Pharmacology* 47(1) Supplement 1, pp.86–97.

Deore, S.L., S.S. Khadabadi, K.S. Kamdi, et al. 2009. In vitro anthelmintic activity of *Cassia tora*. *International Journal of ChemTech Research* 1(2), pp.177–179.

Dhar, G.H., J. Virjee, P. Kachroo and G.M. Buth. 1984. Ethnobotany of Kashmir: I. Sind Valley. *Journal of Economic Taxonomic Botany* 5, pp.668–675.

Dhar, U. and P. Kachroo. 1983. *Alpine Flora of Kashmir, Himalaya*. Jodhpur, India: Scientific Publishers.

Dhillon, R.S., P.S. Kalsi, W.P. Singh, V.K. Gautam and B.R. Chhabra.1987. Guaianolide from *Saussurea lappa*. *Phytochemistry* 26, pp.1209–1210.

Du Puy, D.J., N.-N. Labat, R. Rabevohitra, J.-F. Villiers, J. Bosser andJ. Moat. 2002. *The Leguminosae of Madagascar.* Kew: Royal Botanic Gardens, pp.1–737.

Dutta, N.K., M.S. Sastry and R.G. Tamhane. 1960. Pharmacological actions of an alkaloidal fraction isolated from *Saussurea lappa* (Clarke). *Indian Journal of Pharmacy* 22, pp.6–7.

Elansary, H.O., A. Szopa, P. Kubica, et al. 2018. Bioactivities of traditional medicinal plants in Alexandria. *Evidence-Based Complementary and Alternative Medicine*, Jan 31, 2018, p.e1463579. doi: 10.1155/2018/1463579

Epifano, F., S. Fiorito, M. Locatelli, V.A. Taddeo and S. Genovese. 2015. Screening for novel plant sources of prenyloxyanthraquinones: *Senna alexandrina* Mill. and *Aloe vera* (L.) Burm. F. *Natural Product Research* 29(2), pp.180–184.

Fan, H., F. Liu, S.W. Bligh, S. Shi and S. Wang. 2014. Structure of a homofructosan from *Saussurea costus* and anti-complementary activity of its sulfated derivatives. *Carbohydrate Polymers* 105, pp.152–60.

Franz, G. 1993. The senna drug and its chemistry. Pharmacology, Suppl 1 Suppl 1, pp.2–6.

Friedrich, H. and S. Baier. 1973. Anthracen-derivate in kalluskulturen aus *Cassia angustifolia*. *Phytochemistry* 12(6), pp.1459–1462.

Ghasham, A.A., M.A. Muzaini, K.A. Qureshi, et al. 2017. Phytochemical screening, antioxidant and antimicrobial activities of methanolic extract of Ziziphus mauritiana Lam. leaves collected from Unaizah, Saudi Arabia. *International Journal of Pharmaceutical Research and Allied Sciences* 6(3), pp.33–46.

Gnanavel, S., R. Harathidasan, R. Mahalingam, P. Madhanraj and A. Panneerselvam. 2012. Antimicrobial Activity of *Strychnos nux-vomica* Linn and C. *angustifolia* Linn. *Asian Journal of Pharmacy and Technology* 2, pp.8–11.

Gong, G., J. Huang, Y. Yang, et al. 2020. *Saussureae involucratae* Herba (Snow Lotus): Review of chemical compositions and pharmacological properties. *Journal Frontiers in Pharmacology* 10, p. 1549

Gore, R. and S. Gaikwad. 2015. Checklist of fabaceaelindley in balaghat ranges of maharashtra, India. *Biodiversity Data Journal* 3, p.e4541.

Govindan, S.V. and S.C. Bhattacharaya. 1977. Alantolides and cyclocostunolides from *Saussurea lappa*. *Indian Journal of Chemistry* 15, p.956.

Gupta, P.S., S.S. Jadhav, M.M. Ghaisas and A.D. Deshpande. 2009. Anticonvulsant activity of *Saussurea lappa*. *Pharmacology Online* 3, pp.809–814.

Gwari, G., U. Bhandari, H.C. Andola, H. Lohani and N. Chauhan. 2013. Volatile constituents of Saussurea costus roots cultivated in Uttarakhand Himalayas, India. *Pharmacognosy Research* 5(3), pp.179–182.

Ha, J.Y., B. Min, J.H. Jung, et al. 1997. Sesquiterpene lactone from *Saussurea lappa* with inhibitory effect on IL-8/CINC-1 induction of LPS-stimulated NRK-52E cells. *Phytomedicine* 3, p.178.

Haffner, E. 2000. On the phylogeny of the subtribe Carduineae (tribe Cardueae, Compositae). *Englera* 21, pp.1–209.

Hajra, P.K., R.R. Rao, D.K. Singh and B.P. Uniyal. 1995. *Flora of India. Vols. 12 & 13*: *Asteraceae*. Calcutta, India: Botanical Survey of India.

Hanelt, P., R. Buttner and R. Mansfeld. 2001. *Mansfeld's Encyclopedia of Agricultural and Horticultural Crops (Except Ornamentals).* Berlin, Germany: Springer.

Harish, B.B., L.S. Mohana and K.A. Saravana. 2010. A review on the traditional system of medicine for treating epilepsy. *International Journal of Biological & Pharmaceutical Research* 1, pp.1–6.

Hasson, S.S.A., M.S. Al-Balushi, J. Al-Busaidi et al. 2013. Evaluation of anti–resistant activity of Aucklandia (Saussurea lappa) root against some human pathogens. *Asian Pacific Journal of Tropical Biomedicine* 3(7), pp.557–562.

Hennebelle, T., B. Weniger, H. Joseph, S. Sahpaz and F. Bailleul. 2009. Senna alata. *Fitoterapia* 80(7), pp.385–393.

Hikal, W.M., R.S. Baeshen and H.A.H. Said-Al Ahl. 2017. Botanical insecticide as simple extractives for pest control. *Cogent Biology* 3(1), pp.1–16.

Hooker, J.D., 1881. *Flora of British India*, vol. 3. London: L. Reeve and Co.

Hossain, K., M. Hassan, N. Parvin, M. Hasan, S. Islam and A. Haque. 2012. Antimicrobial, cytotoxic and thrombolytic activity of *Cassia senna* leaves (family: Fabaceae). *Journal of Applied Pharmaceutical Science* 2, pp.186–190.

Hossain, L., R. Rahman and M.S. Khan. 2017. In Khan, M.S., Rahman, M.S., (Eds.), Pesticide Residue in Foods. Alternatives of Pesticides, 1st ed. New York: Springer, pp.147–165.

Ibrahim, M.A. and Islam, M.S., 2014. Anti-diabetic effects of the acetone fraction of *Senna singueanastem* bark in a type 2 diabetes rat model. *Journal of Ethnopharmacology* 153, pp.392–399.

Ilavarasan, R., S. Mohideen, L.M. Vijay and G. Manonmani. 2001. Hepatoprotective effect of *Cassia angustifolia* Vahl. *Indian Journal of Pharmaceutical Sciences* 63(6), pp.504–507.

Jain, T.C. and C.M. Banks. 1968. 22-Dihydrostigmasterol from *Saussurea lappa. Canadian Journal of Chemistry* 46, pp.2325–2327.

Jin, M., H.J. Lee, J.H. Ryu and K.S. Chung. 2000. Inhibition of LPS-induced NO Production and NF-kappa B activation by a sesquiterpene from *Saussurea* lappa. *Archives of Pharmacological Research* 23, pp.54–58.

Joshua, P.E. and O.F.C. Nwodo. 2010. Hepatoprotective effect of ethanolic leaf extract of *Senna hirsuta* (*Cassia hirsuta*) against carbon tetrachloride (CCl4) intoxication in rats. *Journal of Pharmacy Research* 3(2), pp.310–316.

Kala, C.P., P.P. Dhyani, and B.S. Sajwan. 2006. Developing the medicinal plants sector in northern India: Challenges and opportunities. *Journal of Ethnobiology and Ethnomedicine* 2, p.32.

Kaur, G., M.S. Alam, Z. Jabbar, K. Javed, M. Athar. 2006. Evaluation of antioxidant activity of Cassia siamea flowers. *Journal of Ethnopharmacology*, 108 (3), pp.340–348.

Kalsi, P.S., G. Kaur, S. Sharma and K.K. Talwar. 1984. Dehydrocostuslactone and plant growth activity of derived guaianolides. *Phytochemistry* 23, pp.2855–2862.

Khalid, A., U. Rehman, A. Sethi et al. 2011. Antimicrobial activity analysis of extracts of Acacia modesta, Artemisia absinthium, Nigella sativa and Saussurea lappa against Gram positive and Gram-negative microorganisms. *African Journal of Biotechnology* 10(22), pp.4574–4580.

Khan, M.S.A. 2020. A review on Senna: An excellent prophetic herbal medicine. *World Journal of Pharmaceutical and Medical Research* 6 (7), pp.113–118.

Kinjo, J., T. Ikeda, K. Watanabe and T. Nohara. 1994. An anthraquinone glycoside from *Cassia angustifolia* leaves. *Phytochemistry* 37(6), pp.1685–1687.

Kirtikar, K.R. and B.D Basu. 1975. *Indian Medicinal Plants*, 2nd ed. Oriental Enterprises, pp.1961–1965.

Kirtikar, K.R. and B.D. Basu. 2001. *Indian Medicinal Plants*. 2nd Edition. Uttaranchal: Oriental Enterprises, Volume 8, p.2604.

Kita, Y., K.Fujikawa,M. Ito, H.Ohba and M. Kato. 2004. Molecular phylogenetic analysis and systematics of the genus *Saussurea* and related genera (Asteraceae, Cardueae). *Taxon* 53, pp.679–690.

Kobashi, K., Nishimura, T., Kusaka, M., Hattori, M. and Namba, T. 1980. Metabolism of sennosides by human intestinal bacteria. *Planta Medica* 40, pp.225–236.

Kraker, D.J.W., M.C. Franssen, A. De Groot, T. Shibata and H.J. Bouwmeester. 2001. Germacrene from fresh costus roots. *Phytochemistry* 58(3), pp.481–487.

Krumbiegel, G. and H.U Schulz. 1993. Rhein and aloe-emodin kinetics from senna laxatives in man. *Pharmacology* 47(1) Supplement 1, pp.120–124.

Kundu, S., S. Roy, S. Nandi, B. Ukil and L.M. Lyndem. 2017. *Senna alexandrina* Mill. induced ultrastructural changes on Hymenolepisdiminuta. *Journal of Parasitic Diseases: Official Organ of the Indian Society for Parasitology* 41(1), pp.147–154.

Kuniyal, C.P., Y.S. Rawat, S.S. Oinam, J.C. Kuniyal and S.C.R. Vishvakarma. 2005. Kuth (*Saussurea lappa*) cultivation in the cold desert environment of the Lahaul valley, north-western Himalaya, India: Rising threats and need to revive socio-economic value. *Biodiversity and Conservation* 14, pp.1035–1045.

Lalla, J.K., P.D. Hamrapurkar, S.A. Mukherjee and U.R. Thorat. 2002. Sensitivity of HPTLC v/s HPLC for the analysis of alkaloid-Saussurine. In 54th Indian Pharmaceutical Congress, Pune, India, p.293.

Lee, G.I., J.Y. Ha, K.R. Min, et al. 1995. Inhibitory effects of oriental herbal medicines on IL-8 induced in lipopolysaccharide activated rat macrophages. *Planta Medica* 61, pp.26–30.

Lee, H.J., N.Y. Kim, M.K. Jang, et al. 1999. A sesquiterpene, dehydrocostus lactone, inhibits the expression of inducible nitric oxide synthase and TNF-alpha in LPS-activated macrophages. *Planta Medica* 65, pp.104–108.

Lee, M.G., K.T. Lee, S.G. Chi and J.H. Park. 2001. Costunolide induces apoptosis by ROS-mediated mitochondrial permeability and cytochrome C release. *Biological and Pharmaceutical Bulletin* 24, pp.303–306.

Lemli J. 1988a. Metabolism of sennosides: An overview. *Pharmacology* 36(1) Supplement 1, pp.126–128.

Lemli J. 1986. Senna: Chemistry and pharmacology. *Verh K Acad Geneeskd Belg* 48, pp.51–62.

Leng-Peschlow, E. 1980. Inhibition of intestinal water and electrolyte absorption by senna derivatives in rats. *Journal of Pharmacy and Pharmacology* 32, pp.330–335.

Leng-Peschlow, E. 1986a. Acceleration of large intestine transit time in rats by sennosides and related compounds. *Journal of Pharmacy and Pharmacology* 38, pp.369–373.

Leng-Peschlow, E. 1986b. Dual effect of orally administered sennosides on large intestine transit and fluid absorption in the rat. *Journal of Pharmacy and Pharmacology* 38, pp.606–610.

Leng-Peschlow, E. 1992a. Modes of action of senna. *Pharmacology* 44(1), pp.16–19.

Leng-Peschlow, E. 1992b. Senna and habituation. *Pharmacology* 44(1), pp.30–32.

Leng-Peschlow, E. 1992c. Site of senna action. *Pharmacology* 44(1), pp.10–15.

Lim, H.S., H. Ha, H. Shin and S Jeong. 2015. The genome-wide expression profile of *Saussurea lappa* extract on house dust mite-induced atopic dermatitis in Nc/Nga mice. *Molecular Cell* 38 (9), pp.765–772.

Lim, H.S., H. Ha, M.Y. Lee, S.E. Jin, S.J. Jeong, W.Y. Jeon, N.R. Shin, D.E. Sok and H.K. Shin. 2014. *Saussurea lappa* alleviates inflammatory chemokine production in HaCaT cells and house dust mite-induced atopic-like dermatitis in Nc/Nga mice. *Food and Chemical Toxicology* 63, pp.212–220.

Lipschitz, S. 1979. *The genus Saussurea DC. (Asteraceae).* Leningrad: Nauka Publishing, pp.1–281. [In Russian].

Liu, Z.L., Q. He, S.S. Chu, et al. 2012. Essential oil composition and larvicidal activity of Saussurea lappa roots against the mosquito Aedes albopictus (Diptera: Culicidae). *Parasitology Research* 110(6), pp.2125–2130.

Madhuri, K., K. Elango and S. Ponnusankar. 2012. *Saussurea lappa* (Kuth root): A review of its traditional uses, phytochemistry, and pharmacology. *Oriental Pharmacy and Experimental Medicine* 12, pp.1–9.

Malik, A.H., A.A. Khuroo, G.H. Dar and Z.S. Khan. 2011. Ethnomedicinal uses of some plants in the Kashmir Himalaya. *Indian Journal of Traditional Knowledge* 10(2), pp.362–366.

Martel, J., Y.F. Ko, D.M. Ojcius, et al. 2017. Immunomodulatory properties of plants and mushrooms. *Trends in Pharmacological Sciences* 38(11), pp.967–981.

Matsuda H, T. Kageura, Y. Inoue, T. Morikawa and M. Yoshikawa.2000. Absolute stereo structures and syntheses of saussureamines A, B, C, D and E, amino acid-sesquiterpene conjugates with gastroprotective effect from the roots of *Saussurea lappa*. *Tetrahedron* 56, pp.7763–7777.

Maurer, B. and A. Grieder. 1997. Sesquiterpenoids from Costus root oil (*Saussurealappa* Clarke). *Helvetica Chimica Acta* 60(7), pp.2177–2190.

McGuffin M, Hobbs C, Upton R and Goldberg A. 1997. *American Herbal Products Association's Botanical Safetly Handbook.* Boca Raton, FL: CRC Press.

Meier, N., B. Meier, S. Peter and E. Wolfram. 2017. High-performance thin-layer chromatographic fingerprint method for the detection of sennosides in Cassia senna L. and Cassia angustifolia Vahl. *Journal of Planar Chromatography: Modern TLC* 30(4), pp.238–244.

Mengs, U. 1986. Reproductive toxicological investigations with sennosides. *Arzneimittel forschung* 36, pp.1355–1358.

Mitchell, J.C. 1974. In Asakawa, Y., Ourisson, G., Aratani, T. (Eds.), Archives of Dermatology. *Tetrahedron Letters* 109, p.572.

Mitra, S.K., S. Gopumadhavan, T.S. Hemavathi, T.S. Muralidhar and M.V. Venkatarangan. 1996. Protective effect of UL-409, a herbal formulation against physical and chemical factors induced gastric and duodenal ulcers in experimental animals. *Journal of Ethnopharmacology* 52(3), pp.165–169.

Mohammad, A.K., A. Anzar, H. Sadique, et al. 2013. Qust (Saussurea lappa Clarke.): A potent herb of unani medicine: A review. *International Journal of Current Pharmaceutical Research* 5(4), pp.213–219.

Monkheang P, Sudmoon R, Tanee T, Noikotr K, Bletter N and Chaveerach A. 2011. Species diversity, usages, molecular markers and barcode of medicinal Senna species (Fabaceae, Caesalpinioideae) in Thailand, *Journal of Medicinal Plants Research* 5(26), pp.6173–6181.

Monteiro, J.A., J.M. Ferreira, I.R. Oliveira, et al. 2018. Bioactivity and toxicity of *Senna cana* and *Senna pendula* extracts. *Biochemistry Research International* 2

Moo-Pucc, R.E., G.J. Mena-Rejona, L. Quijanob and R. Cedillo Rivera. 2007. Antiprotozoal activity of *Senna racemose. Ethnopharmacology* 112(2), pp.415–416.

Nadda, R.K., A. Ali, R.C. Goyal, P.K. Khosla and R. Goyal. 2020. *Aucklandia costus* (syn. *Saussurea costus*): Ethnopharmacology of an endangered medicinal plant of the Himalayan region. *Journal of Ethnopharmacology* 263, p.113199.

Niroumand, M.C., M.H. Farzaei, E.K. Razkenari et al. 2016. An evidence-based review on medicinal plants used as insecticide and insect repellent in traditional Iranian Medicine. *Iranian Red Crescent Medical Journal* 18(2), pp.1–8.

Nuhu, A.A. and R. Iiyu. 2008. Effects of *Cassia occidentalis* aqueous leaf extract on biochemical markers of tissue damage in rats. *Tropical Journal of Pharmaceutical Research* 7(4), pp.1137–1142.

Oh, G.S., H.O. Pae, H.T. Chung, et al. 2004. Dehydrocostus lactone enhances tumor necrosis factor-alpha-induced apoptosis of human leukemia HL-60 cells. *Immunopharmacology and Immunotoxicology* 26, pp.163–175.

Oladeji, O.S., F.E. Adelowo and A.P. Oluyori. 2021.The genus Senna (Fabaceae): A review on its traditional uses, botany, phytochemistry, pharmacology and toxicology. *South African Journal of Botany* 138, pp. 1–32.

Omer, R.E.E., F.H.M. Koua, I.M. Abdelhag and A.M. Ismail. 2019. Gas chromatography/mass spectrometry profiling of the costus plant *Saussurea lappa* (Decne.) C.B. Clarke root extracts and their antibacterial activity. *Journal of Applied Pharmaceutical Science* 9(5), pp.73–81.

Pandey, M.M., R. Govindarajan, A.K.S. Rawat, Y.P.S. Pangtey and S. Mehrotra. 2004. High performance liquid chromatographic method for quantitative estimation of an antioxidant principle chlorogenic acid in *Saussurea costus* and *Arctium lappa*. *Natural Product Science* 10, pp.40–42.

Pandey, M.M., R. Subha and A.K. Singh Rawat. 2007. *Saussurea costus*: Botanical, chemical and pharmacological review of an ayurvedic medicinal plant. *Journal of Ethnopharmacology* 110, pp.379–390.

Pari, L. and M. Latha. 2002. Effect of *Cassia auriculata* flowers on blood sugar levels, serum and tissue lipids in streptozotocin diabetic rats. *Singapore Medical Journal* 43(12), pp.617–621.

Parolin, P. 2001. *Senna reticulata*, a pioneer tree from Amazonian Vårzea floodplains. *Botanical Review* 67 (2) pp.239–254.

Radaelli, F. and G. Minoli. 2002. Colonoscopy preparation: Is there still room for senna? *Gastrointestinal Endoscopy* 56, p.463.

Rahman, A.M., M.M. Rezk, O.A. Ahmed-Farid, S. Essam and A.E.A. Moneim. 2020. *Saussurea lappa* root extract ameliorates the hazards effect of thorium induced oxidative stress and neuroendocrine alterations in adult male rats. *Environmental Science and Pollution Research* 27, pp.13237–13246.

Rama Reddy, N.R., R.H. Mehta, P.H. Soni, J. Makasana, N.A. Gajbhiye, M. Ponnuchamy and J. Kumar. 2015. Next generation sequencing and transcriptome analysis predicts biosynthetic pathway of sennosides from Senna (*Cassia angustifolia* Vahl.), a non-model plant with potent laxative properties. *PLoS ONE* 10(6), p.e0129422.

Ramchander, P.J. and A. Middha. 2017. Recent advances on senna as a laxative: A comprehensive review. *Journal of Pharmacognosy and Phytochemistry* 6(2), pp.349–353.

Ramchander, P.J. and M. Anil. 2017. Recent advances on senna as a laxative: A comprehensive review. *International Journal of Pharmacognosy and Phytochemical Research* 6, pp.349–353.

Raina, H., G. Soni, N. Jauhari, N. Sharma and N. Bharadvaja. 2014. Phytochemical importance of medicinal plants as potential sources of anticancer agents. *Turkish Journal of Botany* 38, pp.1027–1035.

Rao, A.S., G.R. Kelkar and S.C. Bhattacharyya. 1960. Terpenoids—XXI, The structure of costunolide, a new sesquiterpene lactone from costus root oil. *Tetrahedron* 9, pp.275–283.

Rao, K.S., G.V. Babu and Y.V. Ramnareddy. 2007. Acylated flavone glycosides from the roots of *Saussurea lappa* and their antifungal activity. *Molecules* 12(3), pp.328–344.

Rao, P.S. and B.S. Verma. 1951. Isolation of a new lactone from the resinoid of costus roots. *Journal of Scientific and Industrial Research* 10, p.166.

Rao, R.R., H.J. Chowdhery, P.K. Hajra, et al. 1988. *FloraiIndicae Enumeratio-Asteraceae*. Calcutta: BSI.

Rastogi, R.P. and B.N. Mehrotra. 1990. *Compendium of Indian Medicinal Plants*, Vol. I. New Delhi, India: Publication and Information Directorate, CSIR, pp.81–83.

Rastogi, R.P. and B.N. Mehrotra. 1998. *Compendium of Indian Medicinal Plants*, Vol. V. New Delhi, India: Publication and Information Directorate, CSIR, pp.173–180.

Rawat, Y.S., S.S. Oinam, S.C.R. Vishvakarma and J.C. Kuniyal. 2004. *Saussurea costus* (falc.) Lipsch.: A promising medicinal Crop under cold desert agroecosystem in north western Himalaya. *Indian Journal of Forestry* 27(3), pp.297–303.

Rijke, D.D., P.C. Traas, R.T. Heide, H. Boelen and H.J. Takken. 1978. Acidic components in essential oils of costus root, patchouli and olibanum. *Phytochemistry* 17, pp.1664–1666.

Rizwana H, Fatimah A, Alharbi RA, Albasher G, Moubayed NMS and Alqusumi R. 2021. Morphology and ultrasructure of some pathogenic fungi altered by leaf extracts of *Senna alexandrina* Mill. *Pakistan Journal of Agricultural Sciences* 58(1), pp.389–408.

Roy, S. and L.M. Lyndem. 2019. An in vitro confirmation of the ethono-pharmacological use of Senna plants as anthelmintic against rumen fluke Paramphistomum gracile. *BMC Veterinary Research* 15, p.360.

Saklani, A., R.R. Rao and L.B. Chaudhary. 2000. SEM characterizations of achene morphology towards the taxonomy of Indian sp. of *Saussurea* D.C. *Rheedea* 10, pp.1–18.

Saleem, T.M., N. Lokanath, A. Prasanthi, et al. 2013. Aqueous extract of *Saussurealappa* root ameliorate oxidative myocardial injury induced by isoproterenol in rats. *Journal of Advanced Pharmaceutical Technology and & Research* 4(2), pp.94–100.

Salmerón-Manzano, E., Garrido-Cardenas, J.A., Manzano-Agugliaro, F. 2020. Worldwide research trends on medicinal plants. *International Journal of Environmental Research and Public Health* 17, 3376. https://doi.org/10.3390/ijerph17103376

Salooja, K.C., V.N. Sharma and S. Siddiqi. 1950. Chemical examination of the roots of S. lappa Part I, on the reported isolation of the alkaloid "Saussurine". *Journal of Scientific and Industrial Research* 9, p.1.

Samant, S.S., U. Dhar and L.M.S. Palni.1998. *Medicinal Plants of Indian Himalaya: Diversity, Distribution Potential Values.* Gyan. Prakash., Nainital: HIMAVIKAS Publ. No.13, 163.

Sanghi R, B Bhattacharya and V Singh. 2006. Use of *Cassia javanika* seed gum and gum-g-polyacrylamide as coagulant aid for the decolorization of textile dye solutions, *Bioresource Technology* 97(10), pp.1259–1264.

Sastry, J.L.N., T.M. Nesari and H. Rajendra. 2013. Comparative study on the substitutes used in Ayurveda with special reference to substitutes for Pushkaramula (*Inula racemose* Hook. f.) viz., Kushta (*Saussurea lappa* Decne.) and Erandamula (*Ricinus communis* L.): A clinical study. *International Research Journal of Pharmacy* 4(5), pp.92–100.

Sastry, M.S. and N.K. Dutta. 1961. A method for preparing tincture *Saussurea. Indian Journal of Pharmacy* 23, pp.247–249.

Selim, S.A.H and S.M. El Alfy 1. 2013. M.H. AbdelAziz, M. Mashait and M.F. Warrad. Antibacterial activity of selected Egyptian ethnomedicinal plants. *Malaysian Journal of Microbiology* 9, pp.111–115.

Senna alexandrina Mill. in GBIF Secretariat. 2021. *GBIF Backbone Taxonomy.* Checklist dataset https://doi.org/10.15468/39omei accessed via GBIF.org on 2021-09-11.

Shah, N.C. 1982. Herbal folk medicines in northern India. *Ethnopharmacology* 6, pp. 293–301.

Sharma, P., A. Malav and P. Dubey. 2017. A review on the micro propogation of an important medicinal plant Inula racemosa Hook. F. *International Journal of Development Research* 7 (10), pp.16369–16372.

Silva, C.R., M.R. Monteiro, H.M. Rocha, et al. 2008. Assessment of antimutagenic and genotoxic potential of Senna (*Cassia angustifolia* Vahl.) aqueous extract using *in vitro* assays. *Toxicology in vitro* 22(1), 212–218.

Singh, A., R. Hart, S. Chandra, M.C. Nautiyal, and A.K. Sayok. 2019. Traditional herbal knowledge among the inhabitants: A case study in urgam valley of Chamoli Garhwal, Uttarakhand, India. *Evidence-Based Complementary and Alternative Medicine.*

Singh, D.K. and P.K. Hajra.1997. Floristic diversity. In Gujral (Ed.), *Biodiversity Status in the Himalaya.* New Delhi, India: British Council, pp.23–38.

Singh, H., T. Singh and K.L. Handa. 1957. A note on costus oil from Kashmir costus roots. *Indian Forester* 83, p.606.

Singh, I.P., K.K. Talwar, J.K. Arora, B.R. Chhabra and P.S. Kalsi. 1992. A biologically active guanolide from *Saussurea lappa. Phytochemistry* 31, pp.2529–2531.

Singh, J. 1981. Phytochemical investigation of *Cassia multijuga* seeds. *Planta Medica* 41(4), pp.397–399.

Sosef, M.S.M., L.J.G. Maesen van der. 1997. Senna septemtrionalis (Viv.) Irwin &Barneby. In Faridah Hanum, I., Maesen, L.J.G. van der (Eds.), *Plant Resources of South-East Asia (PROSEA) No. 11: Auxiliary Plants.* Leiden, Netherlands: Backhuys Publisher.

Sun, C.M., W.J. Syu, M.J. Don, J.J. Lu and G.H. Lee. 2003. Cytotoxic sesquiterpene lactones from the root of *Saussurea lappa. Journal of Natural Product* 66, pp.1175–1180.

Talwar, K.K., I.P. Singh and P.S. Kalsi. 1991. A sesquiterpenoid with plant growth regulatory activity from *Saussurea lappa. Phytochemistry* 31, pp.1336–1338.

Taniguchi, M., T. Kataoka, H. Suzuki et al. 1995. Costunolide and dehydrocostus lactone as inhibitors of killing function of cytotoxic T lymphocytes. *Bioscience Biotechnology and Biochemistry* 59, pp.2064– 2067.

Thara, K.M. and K.F. Zuhra. 2012. Comprehensive in-vitro pharmacological activities of different extracts of *Saussurea lappa. European Journal of Experimental Biology* 2(2), pp.417–420.

The Wealth of India. 1992. *A Dictionary of India Row materials and Industrial Products Raw Materials, Revised Ser,* Vol. 3 (Ca-Ci). New Delhi, India: Publications and Information Directorate, CSIR, pp.327–331.

Trinh, T.A., J. Park, J.H. Oh, et al. 2020. Effect of herbal formulation on immune response enhancement in RAW 264.7 macrophages. *Biomolecules* 10(424), pp.1–16.

Tripathi, Y.C. 1999. *Cassia angustifolia*, a versatile medicinal crop. *International Tree Crops Journal* 10, pp.2, 121–129, DOI: 10.1080/01435698.1999.9752999

Tropical Plants Database, Ken Fern. tropical.theferns.info. 2021-09-07.

Tsarong, TJ. 1986. *Handbook of Traditional Tibetan Drugs, Their Nomenclature, Composition, Use and Dosage*. India: Tibetan Medical Publications 1st edition, p.101.
Ulbricht, C., J. Conquer, D. Costa, et al. 2011. An Evidence-Based Systematic Review of Senna (*Cassia senna*) by the Natural Standard Research Collaboration. *Journal of Dietary Supplements* 8(2), pp.189–238.
Unani Pharmacopoeia Committee. 2007. *The Unani Pharmacopoeia of India, Part I*, Vol I. New Delhi, India: Department of Ayurveda, Yoga & Naturopathy, Unani, Siddha and Homoeopathy (AYUSH), Ministry of Health & Family Welfare, Government of India.
Upadhyay, O.P., J.K. Ojha and S.K. Datta.1993. Pharmacognostic study of the root of Saussurea lappa C.B. Clarke. *Sachitra Ayurveda* 8, pp.608–612.
Vadivel, V. and K. Janardhanan. 2001. Nutritional and antinutritional attributes of the under-utilized legume, *Cassia floribunda* Cav. *Food Chemistry* 73(2), pp.209–215.
Venkataranganna, M.V., S. Gopumadhavan, R. Sundaram and S.K. Mitra. 1998. Evaluation of possible mechanisms of antiulcerogenic activity of UL-409 a herbal preparation. *Journal of Ethnopharmacology* 63, pp.187–192.
Vijayasekhar, V.E.V., M.S. Prasad, D.S.D.Suman Joshi, K. Narendra, A.K. Satya and K.R.S. Sambasiva Rao. 2016. Assessment of phytochemical evaluation and in-vitro antimicrobial activity of *Cassia angustifolia. International Journal of Pharmacognosy and Phytochemical Research* 8, pp.305–312.
Viswanathan, P. and P.R. Kulkarni. 1995. *Saussurea lappa* as a new source of inulin for fermentative production of inulinase in a laboratory stirred fermenter. *Bioresource Technology* 52, pp.181–184.
W.F.O. 2021. *Senna alexandrina* Mill. Published on the Internet. http://www.worldfloraonline.org/taxon/wfo-0000164723.
Wang, Y.F., Z.Y. Ni and M. Dong. 2010. Secondary metabolites of plants from the Genus *Saussurea*: Chemistry and biological activity. *Chemistry and Biodiversity* 7(11), pp.2633–2807.
Werner, C. and B. Merz. 2007. *Assessment Report on Cassia senna L. and Cassia angustifolia Vahl, folium.* London: European Medicines Agency, pp.1–32.
Wichtl, M. 2004. *Herbal Drugs and Phytopharmaceuticals: A Handbook for Practice on a Scientific Basis.* 3rd ed. Stuttgart: Medpharm Scientific Publishers.
Wong, W. 1976. Some folk medicinal plants from trinidad. *Journal of Economic Botany* 30, pp.103–142
Wu, Q.P., Z.J. Wang, M.H. Fu, L.Y. Tang, Y. He, J. Fang and Q.F. Gong. 2007. Chemical constituents from the leaves of *Cassia angustifolia. Zhong Yao Cai* 30(10), pp.1250–2. Chinese.
Yamahara, J., M. Kobayashi, K. Miki, et al. 1985. Cholagogic and antiulcer effect of *Saussurea radix* and its components. *Chemical and Pharmaceutical Bulletin* 33, pp.1285–1288.
Yang, H., J. Xie and H. Sun. 1997. Study on chemical constituents of *Saussurea lappa* I. *Acta Botanica Yunnanica* 19, pp.85–91.
Yin, H.Q., H.W Fu and H.M. Hua. 2005. Two new sesquiterpene lactones with the sulfonic acid group from *Saussurea lappa. Chemical and Pharmaceutical Bulletin* 53(7), pp.841–842.
Yoshikawa, M., S. Hatakeyama, Y. Inoue and J. Yamahara. 1993. Saussureamines A, B, C, D and E, new antiulcer principles from Chinese Saussurea radix. *Chemical and Pharmaceutical Bulletin* 41.
Yuniarto, A., E.Y. Sukandar, I. Fidrianny, F. Setiawn and I.K. Adnyana. 2018. Antiobesity, antidiabetic and antioxidant activities of senna (*Senna alexandrina* Mill.) and pomegranate (*Punica granatum* L.) leaves extracts and its fractions. *International Journal of Pharmaceutical and Phytopharmacological Research* 8(3), pp.18–24.
Yuuya, S., H. Hagiwara, T. Suzuki, et al. 1999. Guaianolides as immunomodulators, synthesis and biological activities of dehydrocostus lactone, mokko lactone, eremanthin, and their derivatives. *Journal of Natural Product* 62, pp.22–30.
Zahara, K., Tabassum, S., Sabir, S., Arshad, M., Qureshi, R., Amjad, M.S. and Chaudhari, S.K., 2014. A review of therapeutic potential of *Saussurea lappa*: An endangered plant from Himalaya. *Asian Pacific Journal of Tropical Medicine* 7(1), pp.S60–S69.
Zhao, P., Ming, Q., Qiu, J., Tian, D., Liu, J., Shen, J., Liu, Q.-H. and Yang, X. 2018. Ethanolic extract of folium sennae mediates the glucose uptake of L6 cells by GLUT4 and Ca2+. *Molecules* 23, p.2934

15 *Swertia chirata* (Chirata) and *Withania somnifera* (Ashwagandha)

Suchita V. Jadhav, Pankaj S. Mundada, Mahendra L. Ahire, Devashree N. Patil, and Swati T. Gurme

CONTENTS

15.1 Introduction 291
15.2 *Swertia chirata* 292
15.2.1 Botanical Features and Habitat 292
15.2.2 Importance and Uses 292
15.2.3 Functional in Mild to Moderate Cases of COVID-19 293
15.2.4 Traditional Knowledge of *S. chirata* 293
15.2.5 Chemical Constituents 293
15.2.6 Pharmacological Activity of *S. chirata* 294
15.3 *Withania somnifera* 295
15.3.1 Botanical Description 295
15.3.2 Habitat and Cultivation 296
15.3.3 Traditional Knowledge 296
15.3.4 Chemical Constituents Present in *W. somnifera* 296
15.3.5 Potential Benefits, Applications,and Uses of *W. somnifera* 297
15.4 Conclusion 297
References 298

15.1 INTRODUCTION

Plants have been used from ancient times as the dominant source of traditional medicines and also for the discovery of new medicinal drugs. The majority of the world's population still depends on medicinal plants for the treatment of many diseases (Muthuet al.,2006). The main advantage of using medicinal plants for therapeutic purposes is their safety. It has fewer side effects compared to chemically derived drugs. Besides this, they are economical, effective, and easily available (Atal et al., 1989; Siddiqui, 1993). In many countries, especially Asia,different plants are used traditionally for the treatment of many diseases and disorders, like kidney failure, inflammation, fever, hypertension, immune scarcity, and cancer. The treatment and results rely on the constituents of photochemical and bioactive compounds that are present in those medicinal plants. The extracts of these bioactive compounds are also used in formulations of new drugs (Cragg and Newman, 2013). Numbers of plants from different families are used as medicinal plants. Among them, *Swertia chirata* (chirata) and *Withania somnifera* (Ashwagandha) is the most popular herb. In the current chapter, we focus on traditional knowledge, botanical description in addition to the natural bioactive elements present in them, and their health support and disease prevention.

DOI: 10.1201/9781003205067-15

15.2 *SWERTIA CHIRATA*

15.2.1 Botanical Features and Habitat

Swertia chirata (Family: Gentianaceae), an ancient herb,was first introduced to Europe in 1839(Aleem and Kabir, 2018). The genus *Swertia* is morphologically dissimilar while taxonomically distinct. Currently this taxon contains 150 species, and it is an annual, biennial, or perennial herb of height ranging from 2–4 cm to 1.5 m. Out of these, 40 species of *Swertia* are found in India. The habitat of the genus is alpine or temperate of Asia, Africa, and North America. Most of the plants from this genus are medicinally useful. *Swertia chirata* is one of them, which is found in temperate Himalayas, between 1200 m and 3000 m altitudes spread out from Kashmir to Bhutan (Duke, 2002). *S. chirata* is commonly grown in friable, well-drained soil, in areas with mild rainfall in the rainy season and cold winters. Morphologically, *S. chirata* is an erect, annual, branched herb. The brownish-purple stem is strong and cylindrical. Leaves are generally 10 cm long and sharp at the tip with five nerved, lanceolate, contradictory, and sessile in nature. Flowers of *S. chirata* are abundant and purple in color; it contains four sepals and petals (Figure 15.1). The seeds are small, smooth, and many-angled. *S. chirata* can be intercropped with potato but prefers to be a pure crop for cultivation. The plant can be propagated generatively (through seeds); mature seeds can be collected in the autumn (Kumar and Staden,2016).

15.2.2 Importance and Uses

Swertia chirata is a medically important herb well-known for its antipyretic, antihelmintic activity. It has usability in cathartic, antiperiodic situations. In Ayurveda, it is useful for the treatment of asthma, leukorrhea, stomachic, analeptic, inflammation, uterine discomfort in pregnant women, and fever (Kirtikar and Basu, 1984). *S. chirata* can be used as a purifier of breast milk, as a laxative, and acarminative(Tabassum et al., 2012). *Chirata* is a good medication for gastrointestinal problems, ulcers, liver, kidney diseases, cough, and fever. (Garg and Parasar,1965; Sharma and Karanikar,1985). Oral intake of *chirata* has proven to be a good source for the treatment of constipation, loss of appetite, upset stomach, intestinal worms, and stomach and liver inflammation. Furthermore, *chirata*is

FIGURE 15.1 Morphology of *Swertia chirata* (chirata): A) habit; B) flowering branch; and C) branch with buds.

used in the treatment of high blood pressure, hiccups, and scorpion bites in combination with other medicines (Kumar and Staden, 2016). In India, *S. chirata* has been used as malaria treatment. Also, *chirata* can be used in the manufacturing of alcoholic and nonalcoholic liquors. Owing to its strong medicinal value, *chirata* is utilized in Ayurvedic industries for the formation of skin tonic "Safi", which is mainly used by females for blood purification during the menstrual cycle, specifically in India. Over and above it also used in skin products,like soaps and cosmetics (Phoboo and Jha, 2010). Besides all these human health applications, *chirata* has recently been used as an effective insecticidal to kill the larvae of *Aedys aegypti* mosquito (Mallikarjun et al., 2010).

15.2.3 Functional in Mild to Moderate Cases of COVID-19

The polyherbal drug AYUSH-64 originally developed by India in 1980 for the treatment of malaria, has now been repurposed for the treatment of Covid-19. It is formulated from different herbs and medicinal plants. *Swertia chirata* is one of the main constituents of the formulation used in the treatment of Covid-19 (Rastogi et al., 2020). The Central Council for Research in Ayurvedic Science (CCRAS), a research institute under the Ministry of Ayush and devoted to Ayurveda,conducted clinical trials of AYUSH-64 in collaboration with the Council of Scientific and Industrial Research (CSIR) and many other research organizations. The trials resulted in notable antiviral, immunomodulatory, and antipyretic activities of AYUSH-64. This is also found to be useful in the treatment and clinical recovery of mild and moderate Covid-19 infected patients. According to Ayush, the tablets can be taken at any stage of Covid-19 by those who have symptoms like fever, malaise, cough, headache, body pain, and nasal congestion and discharge. Ingestion of AYUSH-64 significantly improves the patient's anxiety, fatigue, stress, appetite, and sleep too (https://www.ayush.gov.in/docs/Ayush123.pdf).

15.2.4 Traditional Knowledge of *S. chirata*

Swertia chirata, also known as chirayita, is one of the important medicinal plants from the Asian region. About 45% of the total volume is exported by Nepal alone (Barakoti, 2004). The entire *chirata* plant is commonly used for various treatments. The plant is also used in combination with different medicinal plants for the treatment of various disorders and diseases in Ayurveda, Siddha, Unani, Tibetan, and Chinese traditional medicine. In India,*chirata* is commonly known as "bitter tonic" due to its bitter taste. This is caused by the presence of the amarogentin compound,which mainly has hypoglycemic, anti-inflammatory, anti-helmintic, and hepatoprotective activities (Phoboo and Jha, 2010). In Ayurveda,the *chirata* plant is described as bitter in taste, having a cooling property, and is easily digestible (Joshi and Dhawan, 2005). Traditionally, the whole plant is used in the treatment of fever, malaria, hepatitis, skin diseases, worms, blood purification, constipation, anemia, asthma, and diabetes (Phoboo and Jha, 2010; Banerjee et al., 2000). It is taken as a concentrated infusion, tincture, or extract, and in powder form.The decoction of *chirata* species is traditionally used for cardiostimulation, antifatigue, antiaging, antidiarrheal, antibacterial, antifungal, and blood pressure and blood sugar control (Schimmer and Mauthner, 1996; Seher et al., 2020). Widespread use of *S.chirata* in traditional medication in many countries resulted in over-exploitation of the plant from its natural habitat, constricted ecological availability, unsolved inherited seed germination, and viability issues ultimately took the plant to the brink of extinction. More research related to conservation and alternate propagation methods development is urgently needed to avoid the probable disappearance of this medicinally valuable species.

15.2.5 Chemical Constituents

Chemical constituents of *S. chirata* vary, from one cultivar to another, in fresh or dry plants and in different plant parts. More than twenty polyhydroxylatedxanthones and some other phytochemicals, like swertinin, swerchirin, mangeiferin, and many more have been estimated (Figure 15.2) (Bhattacharya

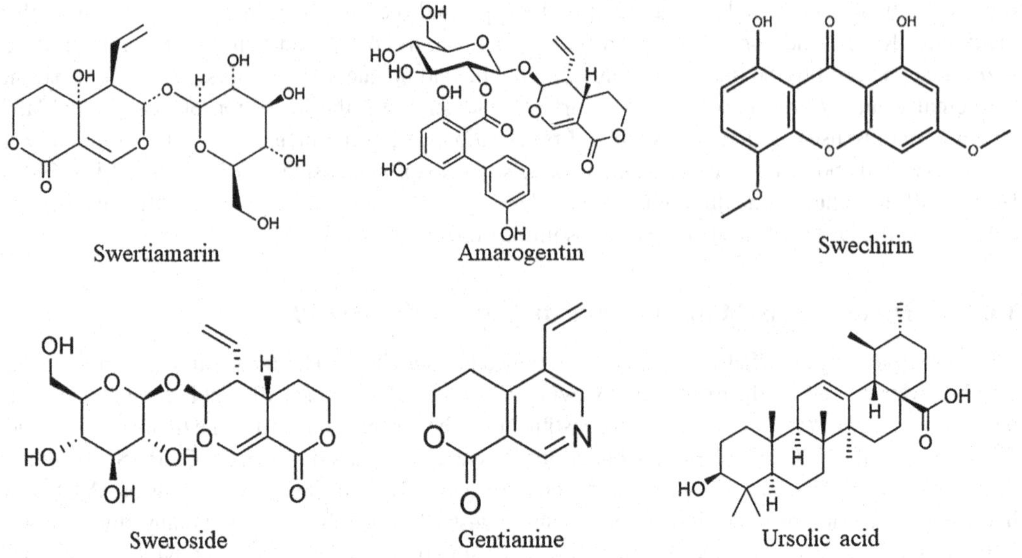

FIGURE 15.2 Major phytoconstituents in *Swertia chirata*.

et al., 1976). Mangiferin is one of the important constituents of chirata, commonly used for its antiviral, anti-inflammatory, antioxidant, and immunomodulatory activities. It shows downregulation of TNF-alpha, IL-lbeta, IL-6, and IFN-gamma and upregulated the IL-10 in joint homogenated of mice (Kumar et al., 2003). Chirata is also reported as an anti-HIV, chemopreventive, antiparkinsonian (Yoshimi et al., 2001). Swertiamarin, a secoiridoid glycoside, is a photochemical specifically found in *S. Chirata* and mostly useful for its cardio-protective, antidiabetic, antiarthritic, anticholinergic, and CNS depressive activities. Swerchirin is a xanthone found in many plants of Gentianaceae family, including *S. chirata.* It has hepatoprotective, antimicrobial, hypoglycemic, and chemo preventive activities and is also used as a blood glucose-lowering agent (Hirakawa, 1987). Swerosode is another bioactive compound found only in *Swertia,* which could be used for hyperpigmentation, hepatoprotective activity, osteoporosis, and bacterial issues (Kumar and Staden,2016).

Amarogentin, also called chirantin, is a glycoside and bitter in taste. It has antileishmanial, anticancer, anti-diabetic, antihelmintic, gastroprotective, and topoisomerase inhibition activities (Saha and Das,2005). Amaroswerin, found in *S. chirata,* proved to have gastroprotective activity (Niiho et al., 2006). Antimalarial, antiamebic, antibacterial, diuretic, lenitive, antipsychotic activities are possessed by Gentianinea monoterpene, an alkaloid obtained from several species of Gentianaceae including *S. chirata.* It also possesses anti-inflammatory, anesthetic, anticonvulsant properties. *S. chirata* consists of many terpenoids (namely,swertanone, swetenol, episwertinol, gammacer-16-en-3β-ol, 21-a-H-hop-22(29)- en-3ß-ol, taraxerol, oleanolic acid, ursolic acid, swerta-7, 9(11)-dien-3ß-ol, and pichierenol) and pentacyclictriterpenoids, like ß-amyrin, friedlin, chiratenol, kairatenol,oleanolic acid, and ursolic acid. Among these terpenoids and pentylcyclictriterpenoids, ursolic acid has anti-inflammatory, antimicrobial, and chemoprotective activity, while kairatenol is well-known for its hypoglycemic activity (Chatterjee and Pakrashi, 1995; Rastogi and Mehrotra, 1998).

15.2.6 Pharmacological Activity of *S. chirata*

Biological activity of *S. chirata* from different plant parts was evaluated using different solvent extracts in *in-vivo* and *in-vitro* conditions. Ethanolic whole-plant extract shows *in-vitro* antibacterial activity against *E. coli, Klebsiella pneumonia, Pseudomonas aeruginosa,* and *Proteus vulgaris* (Rehman et al., 2011). Antibacterial activity of methanol extract of the whole plant has been studied by Laxmi et al. (2011) against *Bacillus subtilis, E. coli, Satphylococcus aureus, Vibria cholera*, *Salmonella typhi*, *S. pyogenes*, *Pseudomonas aeruginosa*, and others. *In vitro* antibacterial activity has been proved from the stem (methanol, ethanol extract) against *Bacillus*, *Pseudomonas*, *Salmonella*,and *Staphylococcus* species (Sultana et al., 2007; Khalid et al., 2011). Similarly, antifungal and antiviral activities were also depicted by whole-plant methanol extract and leave or stem water extract, respectively (Laxmi et al., 2011; Verma et al., 2008). The *chirata* plant has *in vitro* antileishmanial, antihelmintic, and antimalarial activities, which were proven by methanol, ethanol, and water extracts (Kumar and Staden,2016). The anti-hepatitis B activity of whole plant extract in 50% ethanol has been demonstrated by Zhou et al. (2015) by using HepG 2.2.15 cell lines. Along with these biological activities *S. chirata* has been shown to have anti-inflammatory, hypoglycemic, antidiabetic, and analgesic activities in a variety of solvent extracts (Kumar and Staden,2016).

15.3 *WITHANIA SOMNIFERA*

15.3.1 Botanical Description

Withania (Family: Solanaceae) is a genus that contains nearly 26 species native to North Africa, southern Europe, and the Asian continent – mostly flowering plants. Out of these, *W. somnifera* and *W. coagulans* are found in India and are economically and medicinally important (Chadha, 1976). *Withania somnifera*,an evergreen annual shrub growing in Indian, Middle Eastern, and African habitat is alternatively known as Ashwagandha, Indian ginseng, winter cherry, or poison gooseberry. Ashwagandha is one of the most important medicinal plants in India, and it ranked in the top 32 prime concerned plants according to the National Medicinal Plant Board of India (http://www.nmpb.nic.in). Not only in India, but worldwide, it has recognition as an important medicinal plant, and it is also included in the list of the World Health Organization (WHO) (Namdeo and Ingawale, 2021). In Sanskrit, Ashwagandha means "horse smell" and the root of the plant probably smells like a sweaty horse, and it is used for medication.

Ashwagandha plant is an erect, branched plant; the branches grow radially in a star pattern (stellate) and are fully covered with woolly hairs. It grows to a height of 30–150 cm with long tuberous brown-colored roots. The leaves of somnifera are 4–10 cm long and 2–7 cm wide with a simple structure and covered with a continual grayish tomentum on the side. The orientation of leaves on the vegetative shoot is alternate (Figure 15.3). The pale green flowers of somnifera are bisexual and

FIGURE 15.3 Morphology of *Withania somnifera* (Ashwagandha): A) branch with fruit; B) habitat; and C) branch with buds.

bloom from March to July (Mir et al., 2012). The ripened fruits are orange-red and have a milk-coagulating property (Mirjalili et al., 2009).

15.3.2 Habitat and Cultivation

Some species of *Withania* that have medicinal properties and are grown in different parts of the world are *W. coagulens*, *W. simonii*, *W. adunensis*, and *W. riebeckii*. In India, *W. somnifera* is commercially cultivated on a large scale, specifically in dry region. But in some mountainous regions of Himachal Pradesh and Jammu & Kashmir, Ashwagandha have been nurtured (Kandalkar et al., 1993). Besides India, *W. somnifera* have been reported from different countries like Pakistan, Afghanistan, Palestine, Morocco, Jordan, Egypt, Canary Islands, eastern Africa, Madagascar, and South Africa, etc. (Dymock et al., 1981). It is commonly cultivated in red, black, sandy soil with a pH range 6.5–8.0 at a temperature range of 20–35°C. Propagation is normally completed by seedin the early spring or using vegetative propagation (greenwood cutting/ asexual) in the later spring.

15.3.3 Traditional Knowledge

Ashwagandha has a long history of use in traditional medicine in Ayurveda and native medicine for more than 3000 years. Some herbalists call it Indian ginseng (Bharti et al., 2016). In Ayurveda, it is considered a Rasayana, meaning it helps to maintain the mental and physical health of a person. Traditionally, different parts (leaves, stem, flowers, roots, seeds, bark, and whole plant) are used for different purposes (Verma and Kumar,2011). Specifically, Ashwagandha is used for children who are extremely thin or under nourished in combination with milk as a tonic. It is also applied as treatment for hindrance from old age, leucoderma, constipation, insomnia, rheumatism, goiter, etc. The somnifera species is specifically known for its sleep-inducing property (Verma and Kumar, 2011). The root paste mixed with water is known to treat joint inflammation (Bhandari, 1970). It is also applied to reduce ulcers, painful swelling, worms, piles, cough, respiratory problems,and flatulent colic (Mishra,2004). In Ayurveda, roots are used as a sedative and hypnotic in anxiety neurosis, to treat tumors, scrofula, and rheumatism. Traditionally, leaves of Ashwagandha are advised to reduce fever, painful swellings, and antibacterial issues. Root bark infusion has been used for asthma,while the fruits and seeds are diuretic. Ashwagandha is also reported for stimulating and increasing sperm count (Singh et al., 2011).

15.3.4 Chemical Constituents Present in *W. somnifera*

Phytochemicals of *Withania* have been studied extensively by different research groups. Ashwagandha contains various alkaloids, saponins, and steroidal lactones and has been used for different purposes. It contains steroidal lactones called withanolides (Elsakka et al., 1990). The roots of *Withania* contain numerous alkaloids, including withanine, withananine, withananinine, pseudo-withanine, somnine, somniferinine; additionaly, roots have about 18 fatty acids, β-sitesterol, polyphenols, and phytosterols (Figure 15.4). Amino acids, like aspartic acid, glycine, tyrosine, alanine, proline, tryptophan, glutamic acid, and cystine are specifically found in roots of Ashwagandha (Krutika et al., 2016).

The leaves of *W. somnifera* contain secondary metabolites, like Withanolide A and withaferin A. Besides these, it has12-deoxywithastramonolide, 12-deoxy withanstramonolide, withanoside-IV, withanoside-Vashwagandhanolide, withanone, withanolide-B, withanolide-D, withanolide-E, pseudo-withanine27-hydroxywithanone, somnine, 20-deoxywithanolide A, somniferinine, withastramonolide, somniferine, and tropine (Sivanandhan et al., 2011; Supe et al., 2006).The healing action of Withania is mainly due to the presence of withaferin A in the roots and leaves. It possesses antibacterial, antitumor, and antiarthritic properties.A bitter alkaloid, somniferine,is present in the roots and has hypnotic activity. Other alkaloids, such as isopelletierine, anaferine, cuseohygrine, and anahygrine are also reported from *W. somnifera* (Supe et al., 2006).

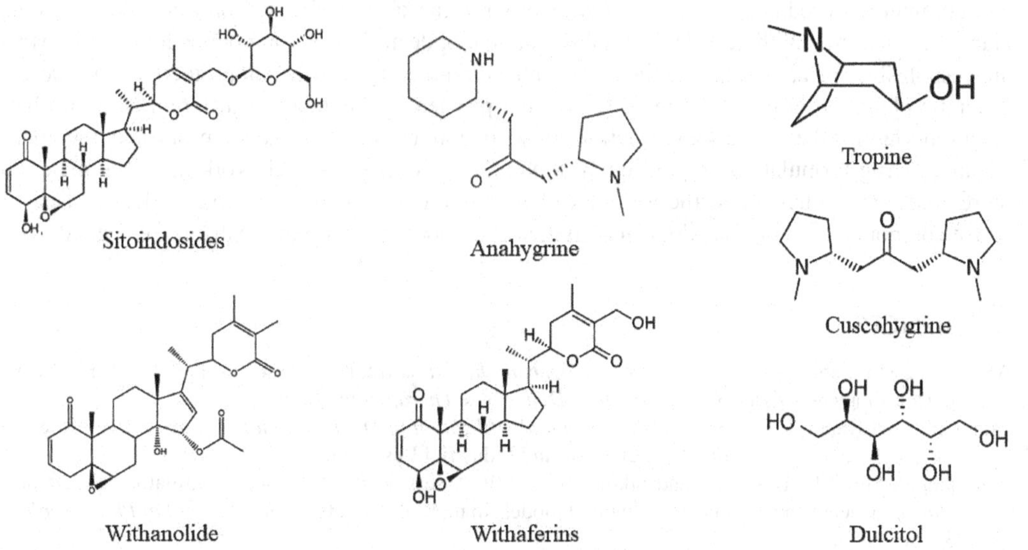

FIGURE 15.4 Major phytoconstituents in *Withania somnifera*.

15.3.5 Potential Benefits, Applications,and Uses of *W. somnifera*

The plant extracts of *W. somnifera* have many bioactive compounds that exhibit antioxidant, anti-inflammatory, and immunomodulatory activities. The drugs produced from leaves of Ashwagandha are useful in the treatment of fever and to kill intestinal worms. Also, it is used in the treatment of bowel movement, swollen hands, feet, and sore eyes. Treatment of bedsores and wounds is also possible using the ointment formulated by boiling the leaves in fat (Chaurasia et al., 2013). The root of Ashwagandha is the most regularly used and more beneficial than the other plant parts. It contributes to boosting immunity, increasing white blood cell count, and specifically to promoting sound sleep. The alkaloids and steroidal lactones present in the roots are effective in herbal supplements to treat weight loss and cancer (Jayaprakasam et al., 2003). In India, Ashwagandha is popular as a male sex tonic that helps to treat erectile dysfunction and improve male sperm count (Chaurasia et al., 2013). The aqueous extract of *W. somnifera* is a good antioxidant, which also helps in prevention of radiation-induced hepatotoxicity in rats (Mansour and Hafez, 2012) andiron-induced hepatotoxicity in chickens. Some studies reported the applicability of Ashwagandha in the treatment of neurodegenerative diseases, like Parkinson's and Alzheimer's. It is also useful for reducing stress, epilepsy, and promotes the general physical and mental health of humans. Additionally, it inhibits aging and catalyzes the anabolic reaction of the body (Jesbergerand Richardson, 1991; Sehgal et al., 2012). *W. somnifera* has antimicrobial activity proven by many researchers against many bacterial and fungal strains too. The first antibacterial activity was reported by Kurup (1956) against *Salmonella aurens*.

15.4 CONCLUSION

The present chapter highlights the potential benefits and traditional knowledge of two medicinally important plants *Swertia chirata* (chirata) and *Withania somnifera* (Ashwagandha), respectively. The potential herbal therapy of *S. chirata,* with no serious side effects,has generated increased demand for the plant at both the national and international levels. The promising results against various illness like malaria, asthma, kidney, liver, neurological disorder, skin diseases, cough, cancer, and Covid-19 treatment prove its wide usability. Despite these good aspects, over-exploitation and habitat destruction of *S. chirata* have turned the focus of researchers toward studying methods of conservation and micropropagation and utilizing more biotechnological interventions and commercial production of *S. chirata*. Similarly, the chapter focused on a second medicinal plant, *W.somnifera*. Withaferin A and Ashwagandhanolide are major compounds found in it, which are responsible for antioxidant, anticancer, anti-inflammatory, and hepatoprotective activities. In India,the plant is useful to treat male sex-related problems. A number of different activities have been reported for the extracts and fractions of withanolides, which make it a prominent memberin advanced drug formulation. Regarding *W. somnifera*,extensive research work has been done, but more study with reference to the isolation of bioactive compounds on a large scale, synthesis of active compounds,its pharmacological activities with mode of action of the drug, are needed.

REFERENCES

Aleem, A., and Kabir, H. (2018). Review on *Swertia chirata* as traditional uses to its pyhtochemistry and phrmacological activity. *Journal of Drug Delivery & Therapeutics*, 8, pp.73–78.

Atal, C.K., and Kapur, B.M. (1982). *Cultivation and Utilization of Medicinal plants*. Jammu-Tawi: Regional Research Laboratory Council of Scientific and Industrial Research, p. 28.

Banerjee, S., Sur, T.P., Das, P.C., and Sikdar, S. (2000). Assessment of the anti-inflammatory effects of *S. chirata* in acute and chronic experimental models in male albino rats. *Indian Journal of Pharmacology*, 32, pp.21–24.

Barakoti, T.P. (2004). *Attempts Made for Domestication, Conservation and Sustainable Development of Chiretta (Swertia Chirayita)*. Dhankuta, Nepal: Nepal Agriculture Research Centre (NARC), 2 pp.

Bhandari, C. (1970). *Ashwagandha (Withaniasomnifera) "VanaushadhiChandroday" (An Encyclopedia of Indian Herbs)*, vol 1. Varanasi: CS Series of Varanasi Vidyavilas Press, pp. 96–97.

Bharti, V.K., Malik, J.K., and Gupta, R.C. (2016). Ashwagandha: Multiple health benefits. In: Gupta RC. (ed.), *Nutraceuticals: Efficacy, Safety and Toxicity*. London: Academic Press, pp. 717–733.

Bhattacharya, S.K., Reddy, P.K.S.P., Ghosal, S., Singh, A.K., and Sharma, P.V. (1976). Chemical constituents of gentianaceae XIX: CNS-depressant effects of swertiamarin. *Journal of Pharmaceutical Sciences*, 65, pp.1547–1549.

Chadha, Y.R. (1976). *The Wealth of India*, Vol. 10. New Delhi, India: Council of Scientific and Industrial Research, p.164.

Chatterjee, A., and Pakrashi, S.C. (1995). *The Treatise on Indian Medicinal Plants used in Ayurveda*, Vol. 4. New Delhi, India: Publication and Information Directorate, p.92.

Chaurasia, P., Bora, M., and Parihar, A. (2013). Therapeutic properties and significance of different parts of Ashwagandha: A medicinal plant. *International Journal of Pure & Applied Bioscience*, 1, pp.94–101.

Cragg, G.M., and Newman, D.J. (2013). Natural products: A continuing source of novel drug leads. *BiochimBiophys Acta*, 1830, pp.3670–3695.

Duke, J.A. (2002). *Handbook of Medicinal Herbs*. Washington, DC: CRC Press, p.190.

Dymock, W., Warden, C.J.H., and Hopper, D. (1981). *Pharmacographia Indica: A History of the Principal Drugs of Vegetable Origin*. London: Kegan Paul, Trench, Trubner and Co., Ltd., pp. 120–121.

Elsakka, M., Grigoreseu, E., Stanescu, U., and Dorneanu, V. (1990). New data referring to chemistry of *Withania somnifera* species. *Revista medico-chirurgicala a Societatii de Medici si Naturalisti din Iasi*, 94, pp.385–387.

Garg, L.C., and Parasar, G.C. (1965). Effect of Withania *somnifera* on reproduction in mice. *Planta Medica*, 13, pp.46–47.

Hirakawa, K. (1987). Chemo preventive action of xanthone derivatives on photosensitized DNA damage. *Photochemistry and Photobiology*, 81, pp.314–319.

Jayaprakasam, B., Zhang, Y., Seeram, N., and Nair, M. (2003). Growth inhibition of tumor cell lines by withanolides from *Withaniasomnifera* leaves. *Life Sciences*, 74, pp.125–132.

Jesberger, J.A., and Richardson, J.S. (1991). Oxygen free radicals and brain dysfunction. *International Journal of Neuroscience*, 57, pp.1–17.

Joshi, P., and Dhawan, V. (2005). *Swertia chirayita*: An overview. *Current Science*, 89, pp.635–640.

Kandalkar, V.S., Patidar, H., and Nigam, K.B. (1993). Genotypic association and path coefficent analysis in Ashwagandha. *Indian Journal of Genetics and Plant Breeding*, 53, pp.257–260.

Khalid, A., Waseem, A., Saadullah, M., Rehman, U., Khiljee, S., Sethi, A., et al. (2011). Antibacterial activity analysis of extracts of various plants against gram -positive and -negative bacteria. *African Journal of Pharmacy and Pharmacology*, 5, pp.887–893.

Kirtikar, K.R., and Basu, B.D. (1984). *Indian Medicinal Plants*, Vol. III. Allahabad: Lalit Mohan Basu, pp. 1664–1666.

Krutika, J., Swagata, T., Kalpesh, P., Praveen, K.A., and Nishteswar, K. (2016). Studies of Ashwagandha (*Withaniasomnifera* Dunal). *International Journal of Pharmaceutical & Biological Archives*, 7, pp.1–11.

Kumar, I.V., Paul, B.N., Asthana, R., Saxena, A., Mehrotra, S., and Rajan, G. (2003). *Swertia chirayita* mediated modulation of interleukin-1beta, interleukin-6, interleukin-10, interferon-gamma, and tumor necrosis factor-alpha in arthritic mice. *Immunopharmacology and Immunotoxicology*, `25, pp.573–583.

Kumar, V., and Van Staden, J. (2016). A review of *Swertia chirayita* (Gentianaceae) as a traditional medicinal plant. *Frontiers in Pharmacology*, 6, p.308.

Kurup, P.A. (1956). Antibiotic principals of the leaves of *Withania somnifera*. *Current Science*, 25, pp.57–60.

Laxmi, A., Siddhartha, S., and Archana, M. (2011). Antimicrobial screening of methanol and aqueous extracts of *Swertia chirata*. *International Journal of Pharmacy and Pharmaceutical Sciences*, 3, pp.142–146.

Mallikarjun, N., Mesta, S.C., PrashithKekuda, T.R., Sudharshan, S.J., and Vinayak, K.S. (2010). Mosquito (Insecticidal) activity of extracts of *Hemidesmus indicus* and *Sweriachirata* against *Aedys aegypti* mosquito larvae: A comparative study. *Drug Invention Today*, 2, pp.106–108.

Mansour, H.H., and Hafez, F. (2012). Protective effect of *Withaniasomnifera* against radiation-induced hepatotoxicity in rats. *Ecotoxicology and Environmental Safety*, 80, pp.14–19.

Mir, B.A., Khazir, J., Mir, N.A., and Koul, T.H.S. (2012). Botanical, chemical and pharmacological review of *Withania somnifera* (Indian ginseng): an ayurvedic medicinal plant. *Indian Journal of Drugs and Diseases*, 1, pp.147–160.

Mirjalili, M.H., Moyano, E., Bonfill, M., Cusido, R.M., and Palazón, J. (2009). Steroidal Lactones from *Withania somnifera*, an Ancient Plant for Novel Medicine. *Molecules*, 14, pp.2373–2393.

Mishra, B. (2004). *Ashwagandha: BhavprakashNigantu (Indian Materia Medica)*. Varanasi: Chaukhambha Bharti Academy, pp.393–394.

Muthu, C., Ayyanar, M., Raja, N., and Ignacimuthu, S. (2006). Medicinal plants used by traditional healers in Kancheepuram district of Tamil Nadu, India. *Journal of Ethnobiology and Ethnomedicine*, 2, p.43.

Namdeo, A.G., and Ingawale, D.K. (2021). Ashwagandha: Advances in Plant Biotechnological approaches for propagation and production of bioactive compounds. *Journal of Ethnopharmacology*, 271, p.113709.

Niiho, Y., Yamazaki, T., Nakajima, Y., Yamamoto, T., Ando, H., Hirai, Y., Toriizuka, K., and Ida, Y. (2006). Gastroprotective effects of bitter principles isolated from Gentian root and Swertia herb on experimentally-induced gastric lesions in rats. *Journal of Natural Medicines*, 60, pp.82–88.

Phoboo, S., and Jha, P.K. (2010). Trade and sustainable conservation of *Swertia chirayita* (Roxb. ex Fleming) h. Karst in Nepal. *Nepal Journal of Science and Technology*, 11, pp.125–132.

Rastogi, R.P., and Mehrotra, B.N.. (1993). *Compendium of Indian Medicinal Plants*, Vol. 3 New Delhi, India: CDRI, Lukhnow and National Institute of Science Communication.

Rastogi, S., Pandey, D.N., and Singh, R.H.2020. COVID-19 pandemic: A pragmatic plan for ayurveda intervention. *Journal of Ayurveda and Integrative Medicine*, 13, pp. 100312.

Rehman, S., Latif, A., Ahmad, S., and Khan, A.U. (2011). *In vitro* antibacterial screening of *Swertia chirayita* Linn. against some gram negative pathogenic strains. *International Journal of Pharmaceutical Research and Development*, 4, pp.188–194.

Saha, P., and Das, S. (2005). Bitter fraction of Swertia chirata prevent carcinogenic risk due to DMBA Exposure. *Indian Journal of Medical Research*, pp.22–27, Poster Presentation.

Schimmer, O., and Mauthner, H. (1996). Polymetoxylated xanthones from the herb of Centaurium erythraea with strong antimutagenic properties in *Salmonella typhimurium*. *Planta Medica*, 62, pp.561–564.

Seher, A., Hanif, M.A., Hanif, M., and Hanif, A., 2020. Chirayita. *Medicinal Plants of South Asia*, Elsevier, 10, pp.25–134.

Sehgal, N., Gupta, A., Khader, R., Shanker, V., Joshi, D., Mills, J.T., Hamel, E., Khanna, P., Jain, S.C., Thakur, S.S., and Ravindranath, V. (2012). *Withania somnifera* reverses Alzheimer's disease pathology by enhancing low-density lipoprotein receptor-related protein in liver. *Proceedings of the National Academy of Sciences*, 109, pp.3510–3515.

Sharma, S.D., and Karanikar, S.M. (1985). Effects of long-term administration of the roots of Ashwagandha and shatavari in rats. *Indian Drugs*, 29, pp.133–139.

Siddiqui, H.H. (1993). Safety of herbal drugs-an overview. *Drugs News Views*, 1, pp.7–10.

Singh, N., Bhalla, M., de Jager, P., and Gilca, M. (2011). An overview on ashwagandha: a Rasayana (rejuvenator) of Ayurveda. *African Journal of Traditional, Complementary and Alternative Medicines*, 8(Supplement), pp.208–213.

Sivanandhan,G.,Arun, M., Mayavan, S., Rajesh, M., Mariashibu, T.S., Manickavasagam, M., Selvaraj, N., and Ganapathi, A. (2011). Chitosan enhances withanolides production in adventitious root cultures of *Withania somnifera* (L.) Dunal. *Industrial Crops and Products*, 37, pp.124–129.

Sultana, M.J., Molla, M.T.H., Alam, M.T., and Ahmed, F.R.S. (2007). Investigation of antimicrobial activities of the plant *Swertia chirayita* ham. *Journal of Life and Earth Science*, 2, pp.31–34.

Supe, U., Dhote, F., and Roymon, M.G. (2006). In vitro Plant Regeneration of *Withania somnifera. Plant Tissue Culture & Biotechnology*, 16, pp.111–115.

Tabassum, S., Mahmood, S., Hanif, J., Hina, M., and Uzair, B., 2012. An overview of medicinal importance of Swertia chirayita. *International Journal of Applied Science and Technology*, 2, pp.298–304.

Verma, H., Patil, P.R., Kolhapure, R.M., and Gopalkrishna, V. (2008). Antiviral activity of the Indian medicinal plant extract, *Swertia chirata* against herpes simplex viruses: a study by *in-vitro* and molecular approach. *Indian Journal of Medical Microbiology*, 26, pp.322–326.

Verma, S.K., and Kumar, A. (2011). Therapeutic uses of Withania somnifera (Ashwagandha) with a note on withanolides and its pharmacological actions. *Asian Journal of Pharmaceutical and Clinical Research*, 4, pp.1–4.

Yoshimi, N., Matsunaga, K., Katayama, M., Yamada, Y., Kuno, T., Qiao, Z., Hara, A., Yamahara, J., and Mori, H. (2001). The inhibitory effects of mangiferin, a naturally occurring glucosylxanthone, in bowel carcinogenesis of male F344 rats. *Cancer Letter*, 163, pp.163–170.

Zhou, N.J., Geng, C.A., Huang, X.Y., Ma, Y.B., Zhang, X.M., and Wang, J.L., et al. (2015). Anti-hepatitis B virus active constituents from *Swertia chirayita. Fitoterapia*100, pp.27–34.

16 *Vinca rosea* (Madagascar Periwinkle) and *Adhatoda vesica* (Malabar Nut)

Rajib Hossain, Md Shahazul Islam, Dipta Dey, and Muhammad Torequl Islam

CONTENTS

16.1 Introduction 302
16.1.1 *Vinca rosea* 302
16.1.2 *Adhatoda vasica* 303
16.2 Description 303
16.2.1 *Vinca Rosea* 303
16.2.2 *Adhatoda vasica* 305
16.3 Traditional Knowledge 305
16.3.1 *Vinca rosea* 305
16.3.2 *Adhatoda vasica* 306
16.4 Chemical Derivatives 310
16.4.1 *Vinca rosea* 310
16.4.2 *Adhatoda vasica* 310
16.5 Potential Benefits, Application, and Uses 316
16.5.1 *Vinca rosea* 316
16.5.1.1 Antioxidant Effect 316
16.5.1.2 Anticancer Effect 317
16.5.1.3 Cytotoxic Effect 321
16.5.1.4 Antidiabetic Effect 321
16.5.1.5 Antidiarrheal Effect 322
16.5.1.6 Antihelminthic Effect 322
16.5.1.7 Hypolipidemic Effect 322
16.5.1.8 Anti-HIV Effect 323
16.5.1.9 Antihypertensive Effect 323
16.5.1.10 Antimicrobial Effect 323
16.5.1.11 Antimycobacterium Tuberculosis Effect 324
16.5.1.12 Neuroprotective Effect 324
16.5.1.13 Antiplatelet Aggregation Effect 324
16.5.1.14 Anti-Ulcer Effect 324
16.5.1.15 Wound-Healing Effect 324
16.5.1.16 Other Effects 324
16.5.1.17 Safety Aspects 325
16.5.2 *Adhatoda vasica* 325
16.5.2.1 Antioxidant Effect 325
16.5.2.2 Antibacterial Effect 325
16.5.2.3 Antimicrobial Effect 325
16.5.2.4 Anti-Inflammatory Effect 326

DOI: 10.1201/9781003205067-16

16.5.2.5 Antitussive Effect ... 326
16.5.2.6 Hepatoprotective Effect ... 326
16.5.2.7 Antiviral Effect ... 326
16.5.2.8 Thrombolytic Effect ... 327
16.5.2.9 Uterine Effect ... 327
16.5.2.10 Antifungal Effect ... 327
16.5.2.11 Anthelmintic Effect ... 327
16.5.2.12 Antidiabetic Effect ... 327
16.5.2.13 Antituberculosis Effect ... 328
16.5.2.14 Anticestodal Effect ... 328
16.5.2.15 Hepato-Protective Effect ... 328
16.5.2.16 Radio-Modulatory Effect ... 328
16.5.2.17 Immunomodulatory Effect ... 329
16.5.2.18 Effect on Reproductive Organs ... 329
16.6 Conclusion ... 329
Acknowledgment ... 329
References ... 329

16.1 INTRODUCTION

16.1.1 *VINCA ROSEA*

Numerous naturally produced herbs that can be utilized for therapeutic purposes can be found all around us. *Vinca rosea* is one of them, and it's a plant that's located all over the world in tropical places. V. rosea (also known as *Catharanthus roseus*) is a perennial plant that may be found in tropical areas. This genus includes seven species unique to Madagascar and one member indigenous to South Asia (Johnson et al., 1963; Joshi et al., 1992; Sharma, 1998). *Catharanthus roseus,* Madagascar periwinkle, brilliant eyes, Cape periwinkle, cemetery plant, old maid, pink periwinkle, and rose periwinkle myrtle are some of the popular names for *V. rosea.* (Atal, 1980; Moudi et al., 2013) (Figure 16.1).

FIGURE 16.1 (A) leaves, (B) flower, and (C) root of *Vinca rosea.*

Furthermore, it is also named Madagascar periwinkle. In Malaysia, "Kemunting Cina" is local name of *V. rosea*. The periwinkle logo is used by Malaysia's National Cancer Council (Majlis Kanser Nasional, MAKNA) as a sign of hope for cancer sufferers. This plant breeds lovely blooms in a range of colors, including purple, pink, and white, and is frequently used as an ornamental plant. Madagascar periwinkle has been utilized for a variety of ailments in the past, including diabetes, hypertension, inflammation, and infection. The milky fluid secreted by the shoot of *V. rosea* contains around 70 distinct indole alkaloids (Avula et al., 2008). Vincristine and vinblastine (Vinca alkaloid) are the two most pronounced anticancer agents which are biosynthesized from this plant. The vinca alkaloids are the earliest category of botanical alkaloids that have already been prescribed in cancer medication (Moudi et al., 2013). Vincristine is being utilized in the treatment of Hodgkin's lymphoma, whereas vinblastine is employed in the treatment of pediatric leukemia. These vinca alkaloids attach to tubulin dimers, blocking cellular microtubule formations and therefore preventing mitosis from reaching metaphase in the cell cycle (Chattopadhyay and Das, 1990; Chattopadhyay, 1999; Loh, 2008). Peripheral neuropathy, hair loss, hyponatremia, and constipation are the most common adverse effects of these medications (Johnson et al., 1963). They are mostly used to treat hypertension, diabetes, blood cancer, malaria, non-small-cell lung cancer, Hodgkin's lymphoma, and cognitive impairment. It also has antibacterial, antioxidant, antidiarrheal, antifeedant, hypolipidemic, and wound-healing properties (Cordall et al., 1974; Dymock et al., 1980).

16.1.2 *Adhatoda vasica*

Vasaka is the popular name for *Adhatoda vasica* (Malabar nut) Nees, a well-known medicinal herb. It is frequently utilized in the Unani and Ayurvedic medical systems. Since ancient times, all parts of the plant have been utilized for medicinal purposes, notably in upper respiratory tract illnesses including bronchitis and asthma. It has been utilized in the Indian subcontinent's traditional medical system for over 2000 years (Claseon et al., 2000; Iyengar et al., 1994; Adnan et al., 2010).

Most prominent Unani scholars have described it as an effective drug for a variety of respiratory ailments, including bronchitis, bronchial asthma, and fever. It is described as Munaffis-e-balgham, an expectorant in Unani medicine, and this property makes it an elixir for respiratory ailments like bronchitis. The entire plant can help with colds, coughs, asthma, and chronic bronchitis. It provides reliable alleviation in the acute phases of bronchitis, especially if the sputum is thick and sticky; it liquefies sputum to make it easier to expel. The leaves are primarily employed in the treatment of chest illnesses, notably as an expectorant and bronchial antiseptic. They are also thought to be beneficial in the treatment of TB. Asthmatics smoke dried leaves that are made into cigarettes. It generates ammoniacal vapor (El–Merzabani et al., 1979; El-Sawi et al., 1979), which helps asthma sufferers breathe easier. Experimental and clinical investigations have demonstrated its hypotensive, bronchodilator, expectorant, hypoglycemic, antibacterial, antitubercular, and uterine properties (Figure 16.2) (Jaleel, 2009; Hassan et al., 2011).

Plants are directly responsible for human and animal survival. Plant diversity is made up of about 5,00,000 botanical species. Plants are an important part of biodiversity since they help to preserve the earth's environmental balance and ecological stability. Herbal medicine is well-known for its therapeutic properties. It dates back to 1600 BC1 and comes from ancient Greek. Herbal plant renaissances are happening all over the world, and therapeutic herbs are making a huge comeback. According to ethnobotanical data from the Indian subcontinent, more than 6000 higher plant species, or around 40% of the higher plant variety, are utilized in codified and traditional medicine (Ved and Goraya, 2007; Jha et al., 2012).

16.2 DESCRIPTION

16.2.1 *Vinca Rosea*

Catharanthus (from the Greek katharos (pure) and anthos (fruit) was named after Linneaus (flower). The scientific name of vinca was established to be *Cartharantus roseus* by Scottish botanist George

FIGURE 16.2 (A) leaves, (B) flower, and (C) root of *Adhatoda vesica*.

Don, which has been the subject of several disputes and contentious arguments concerning its classification. Carl von Linneanus, a Swedish scientist, was named after the first of his species, *V. rosea*, in 1759. Heinrich Gottlieb Ludwig Reichenbach, a German botanist, suggested the name Lochnera in 1828. In 1838, an Austrian botanist called Stephan Ladislaus Endlicher christened the plant *Lochnera rosea* (Kirtikar and Basu, 1918; Karthikeyan et al., 2009). The scientific name *Catharanthus roseus* was verified by William Stearn as the proper name for the Madagascan periwinkle. Furthermore, in his chapter "Synopsis of the Genus Catharanthus," Stearn pointed out that even the genus name Lochnera is invalid since it is too similar to the naming of some other genus, Lochneria, established in 1777 by scientist Giovanni Antonio Scopoli. Further details of taxonomical classification are presented by Paarakh et al. (2019). In India, periwinkle is a popular blooming plant. Its scientific name is *V. rosea*, and it belongs to the Apocynaceae family. It is a year-round blooming shrub with smooth, glossy, dark green leaves that grows to a height of 1–3 feet. Blue, purple, violet, pink, and white periwinkle flowers are among the many colors available. North America, Bangladesh, Europe, India, and China are all habitats to these plants. Medicinal qualities may be found in almost every section of the Madagascar periwinkle (Mondal and Mandal, 2009; Nisar et al., 2016). Periwinkle is an herbal medicinal plant with a wide range of uses. Although all plant parts are therapeutic, alkaloids are predominant in the roots and bark. Ajmalicine, reserpine, and serpentine are three of the most significant Rauvolfia alkaloids. Vindoline, vincristine, and vinblastine are some of the other significant alkaloids discovered. *Vinca rosea* is utilized in both Ayurvedic and Western medicine. Synonyms of *V. rosea* are *Ammocallis rosea* (L.) Small, *Lochnera rosea* (L.) Rchb, *Lochnera rosea* var. flava Tsiang, *Pervinca rosea* (L.) Gaterau, *Vinca gulielmi*-waldemarii Klotzsch, *Vinca rosea* L., *Vinca rosea* var. albiflora Bertol., *Vinca speciosa* Salisb.

It is an evergreen sub-shrub or herbaceous plant that grows to a height of 32 cm to 80 cm. It flowers all summer long and has a glistening dark green hue. The blossoms are pale pink with a purple "eye" in the middle in their natural state. Suffrutex, up to 1 m tall, upright or accumbent, with white latex. Purple or red tinges on green stems. Decussate, petiolate oval leaves (1–2in long); lamina varied, elliptic, obovate, or narrowly obovate, apex mucronate. Flowers are average 4–5 cm, classy, white or pink, with a purple, red, pale yellow, or white center. Follicle 1.2–3.8 × 0.2–0.3 cm, susceptible on the axial side. Seeds 1–2 mm, are numerous and grooved on one side. Climate, soil,

TABLE 16.1
Vernecular Names of *Vinca rosea*

Country	Name	Reference
Malaysia	*Kemuning cina*	Siddiqui et al., 2010; Paarakh et al., 2019
Bangladesh	Nayantara	
India	Sadabahar,Sadaphuli,Sadasuhagi, Sadaphuli, Billaganneru, Nithyakalyaani,	
Srilanka	Shayam Naari	
Others	Savanari, Periwinkle, Nithyakalyani, Cape Periwinkle.	
English	Madagascar Periwinkle	

and propagation. Throughout the year in equatorial conditions, and from spring to late autumn, in warm temperate climates is the flowering periods (Sharma, 1998) (Table 16.1).

16.2.2 *Adhatoda vasica*

A. vasica Nees is a member of the Acanthaceae family of medicinal plants. It reaches a height of 1.5–2.0 meters, with leaves that are 10–15 centimeters long and 5.0 centimeters broad, white or purple blooms, and four-seeded fruits. The leaves are dark green on top and light yellow on the bottom. The flowers are white and placed in a pedunculated spike. Adhatoda leaves have been utilized for over 2000 years in Ayurvedic medicine, especially to treat respiratory problems. Tender stem cuttings are used to propagate it. Plant stem cuttings with 3–4 nodes and a length of 15–20 cm are excellent. Adhatoda is gathered from open fields or purchased from commercial sources. Seeds are the most common method of propagation; however, cuttings can also be used in the early spring (Bjaj and Williams, 1995). Medicinal trees are important to the world's health care systems (Pandey and Mishra, 2010). Further details of taxonomical classification are presented by U.S. Department of Agriculture (Nosálová et al., 1993; Paikara et al., 2017). The *A. vasica* (Malabar nut) tree is a widespread tiny evergreen sub-herbaceous bush found across India, particularly in the lower Himalayas (up to 1300 meters above sea level), India, Sri Lanka, Burma, and Malaysia. It is recognized as "vasaca" (Prajapati et al., 2003) in Ayurveda, the traditional system of Indian medicine.

16.3 TRADITIONAL KNOWLEDGE

16.3.1 *Vinca rosea*

Periwinkle has a variety of traditional and folklore applications that have been time-tested and verified by people's beliefs. The paste made from the leaves is a great wound healer that also helps to ease the discomfort of wasp stings. It can halt bleeding, allowing the mending process to proceed more quickly. Periwinkle is also said to help with depression, headaches, and exhaustion, according to some sources. Traditionally, *V. rosea* has been used in several diseases and disorders, such as bee sting, wasp sting, diabetes, stomach cramps, intestinal parasitism, menorrhagia, cancer, dysentery, vomitive, purgative, vermifuge, depurative, hemostatic and toothache, indigestion and dyspepsia, eye wash in infants, asthma, tuberculosis and flatulence, hypertension, insomnia, cancer, bleeding, sore throats, chest ailments, laryngitis, and rheumatism in India, Philippines, Madagascar, Mauritius, West Indies, Nigeria, Cuba Jamaica, Bahamas, Malaysia, Hawaii, America, and Africa (Patel et al., 2011; Patil et al., 2014).

Chinese alternative medicine employs leaves, flowers, and roots. Diabetes, malaria, leukemia, and Hodgkin's disease are all treated using the plant's extraction in Western medicine. Wasp stings are treated with fresh leaf juices, sore mouths are treated with a mouthwash, and babies' eyes are washed with flower extracts. Diabetes and cough are treated with periwinkle tea. Alkaloidal metabolites with antitumor and anticancer effects are found in the leaves, branches, and shoots. People use leaves to control diabetes and hypertension. Sedative and tranquilizing effects are also provided by the alkaloids found in *V. rosea* (Ramya, 2008; Rajput et al., 2011). It also reduces muscular pain and sadness which is used to treat wasp stings due to its detoxifying and poison-counteracting properties. Nasal bleeds, gum disease, oral ulcers, and sore throats are all treated with this herb. When used internally, it can help with gastritis, cystitis, enteritis, diarrhea, diabetes, and other conditions.

The *V. rosea* plant is good for the brain. It contains substances that enhance blood flow to the brain and boost the amount of oxygen available to the brain. It also inhibits irregular blood-clotting and increases levels of serotonin. Vincamine is an alkaloid that helps to maintain blood thinness and improves cognitive function. As a result, it is beneficial in the prevention of dementia, particularly vascular dementia. If taken orally, periwinkle can be harmful. Pregnant women should stay away from the plant (Singh et al., 1991, 2017). In folk and traditional medicine, the alkaloids from *V. rosea* are used to treat a range of non-malignant illnesses; while in Africa, herbal medicine is used to treat menorrhagia and rheumatism (Waltz, 2004). *V. rosea* has been employed to treat diabetes for many decades, and its hypoglycemic effect has been demonstrated via inducing diabetes in mice (Cuellar and Lorincz, 1975). Because of the rise in hyperglycemia as well as the stimulation of insulin synthesis, it has hypoglycemic properties. (Singh et al., 2001; Kumari and Gupta, 2013).

Wasp bites are healed using the leaves and stem in India and used to cure diabetes, internal bleeding, and wound repair in Brazil. In Hawaii, a boiling plant extract has been recommended to control hemorrhage. It is used to treat throat discomfort and laryngitis in Central and South America. The crude white flower extract is frequently used as an eye rinse for newborns and as compresses for relieving ophthalmic problems in Cuba, Puerto Rico, Jamaica, and other islands (Waltz, 2004; Amirjani, 2013).

Many nations have utilized this herb as a folk and traditional remedy (Table 16.2). In Northeast India, the Cook Islands, Australia, England, Thailand, Natal, Mozambique, Philippines, Dominican Republic, Jamaica, Northern Europe, and Vietnam, the dried leaves or complete shrub is cooked with water and afterward, the extract is given orally to control hyperglycemia (Swanston-Flatt, et al., 1989; Holdsworth, 1990; Marles and Farnsworth, 1995; Khan, 2010; Vo, 2012). Peoples of Cook Island and Vietnam utilize the aqueous leaf extract or the entire plant (Holdsworth, 1990; Vo, 2012), while Kenyans are using adjunctive medicines to treat malignancies of the mouth, abdomen, and esophagus (Ochwang'i, et al., 2014). People in the Kancheepuram District of Tamil Nadu, India, combine the powder of the entire *V. rosea* plant with cow's milk to cure diabetes (Muthu et al., 2006). In South Africa, *V. rosea* root provides for the prevention of urogenital infections, gonorrhea (Semenya and Potgieter, 2013; Fernandes et al., 2008), and abdominal pain in Zimbabwe (Chigora et al., 2007).

16.3.2 *Adhatoda vasica*

The leaves, flower, root, and other parts of the *A. vasica* plant are widely used as an expectorant in traditional Indian sub-continental medicine in various forms such as decoction, infusion, powder, and fresh juice to treat colds, flu, and chronic respiratory diseases, such as whooping cough, asthma, and chronic bronchitis (Ahmad et al., 2009). It helps by reducing mucus production and cleaning the airways. The herb's powder is cooked with sesame oil to treat ear infections and stop bleeding. Boiled leaves are used to alleviate the discomfort of rheumatoid arthritis and urinary tract infections. (*Medicinal Plants in Orrisa*) It's also thought to have anti-abortive effects. It's used in certain areas of India to speed up delivery by stimulating uterine contractions. (Gupta et al., 1978). The following are some of the most important Unani and Ayurvedic formulations that include Syrup Basakarista, Basadik Wath, Basaboleho, Sarbat Ejaz, Sarbat Tulsi, Sarbat Sadar, and Sarbat Vasac.

Vasica is a common medication in the Unani and Ayurvedic medical systems. It is used to treat respiratory illnesses as well as some stomach and intestinal problems. The drug's leaves and blossoms are used as medicine in the form of powder or decoction; the entire plant is also used as medication on occasion. The following is a broad description of the drug (Yadav and Tangpu, 2008; Wang et al., 2012c) (Table 16.3).

TABLE 16.2
Traditional Uses of *Vinca rosea*

Part Used	Disease	Preparation	Mode of Administration	Country	References
The whole plant, leaf	Diabetes, Bee sting/ wasp sting	The whole plant is powdered and mixed with cow's milk.	Oral intake	India	Muthu et al., 2006
	Diabetes	The leaf is boiled with water.			Khan and Yadava, 2010
Leaf	Diabetes mellitus	The dried leaf is decocted.	Oral intake	Northern Europe	Swanston-Flatt et al., 1989
Leaf of purple or white-flowered varieties	Diabetes, hypertension, and cancer	Eighteen leaves are boiled in a kettle of water. The cool solution is drunk daily		Cook Island	Holdsworth, 1990
Whole plant	Throat, stomach, oesophageal cancer	The whole plant is boiled with water. Pound	Oral intake. Usually taken together with Sesbania sesban whole plant. Applied topically	Kenya	Ochwang'I et al., 2014
Root	Urogenital infections	The root is air-dried, ground, and decocted	Oral intake	Venda region, South Africa	Fernandes et al., 2008
Root	Gonorrhea, menorrhagia, and rheumatism	The root is boiled for 20 min.		Limpopo Province, South Africa	Semenya and Potgieter, 2013
Root	Stomach	Crushed roots are mixed with a cup of water		Mutirikwi area of Zimbabwe	Chigora et al., 2007
Whole plant	Diabetes, hypertension, dysentery, cancer	The whole plant is boiled with water		Vietnam	Vo, 2012

The medication is categorized as a mucolytic and expectorant drug by *Charaka Samhita*. The plant's active components, roots, leaves, and flowers, have a variety of pharmacological characteristics and are used to treat cough, chronic bronchitis, rheumatism, asthma, and bronchial asthma. The majority of medicinal plant species are high in biomolecules that can combat health risks, and many species of plants have lately been shown to have antibacterial activity. The leaves and roots contain several alkaloids (the most important of which is quinazoline alkaloid but also vasicine, vasicinone, vasicinolone, and vasicol), which may have bronchodilator properties. These alkaloids are believed to occur in conjunction with adhatoda acid. It has sedative, expectorant, antispasmodic, antihelmintic, bronchial antiseptic, and bronchodilator properties. For many years, leaf extract has been used to treat bronchitis and asthma. It helps with coughing and shortness of breath. It's also used to treat idiopathic thrombocytopenic purpura, local bleeding caused by peptic ulcer, piles, and menorrhagia. Its local use provides relief from pyorrhea and bleeding gums. Because the alkaloid content of the plant varies by genotype, vegetative propagation of the *A. vasica* plant is advised.

TABLE 16.3
Ethnopharmacological Uses of *A. vasica* Reported in Field Studies

Country	Part Used	Indication or Use	Preparation	References
India, Myanmar, Nepal, Pakistan, Sri Lanka, Nepal, Thailand	Leaves	Respiratory diseases, cough, colds, asthma, expectorant, liquefy sputum, antispasmodic, bronchodilator, bronchitis, bronchial catarrh, tuberculosis, phthisis, consumption	Juice, fluid extract, decoction, infusion, electuary, powder, cigarettes	Agarwal, 1986; Atta et al., 1986; Kapoor, 1990; Salalamp et al., 1992, 1996; Pushpangadan et al., 1995
India, Myanmar, Nepal, Pakistan		Rheumatism, rheumatic and painful inflammatory swellings, neuralgias	Decoction, infusion, poultice	Watt, 1972; Kirtikar et al., 1975; Nadkarni, 1976; Ikram and Hussain, 1978; Jayaweera, 1981; Agarwal, 1986; Atta et al., 1986.
India, Myanmar, Nepal, Pakistan, Sri Lanka		Fever, malarial fever	Juice, decoction, infusion, electuary, powder	Watt, 1972; Kirtikar et al., 1975; Nadkarni, 1976; Jayaweera, 1981; Atta et al., 1986.
India, Myanmar, Pakistan, Sri Lanka:		Gastrointestinal disorders: diarrhea, dysentery, colic	Juice	Watt, 1972; Jayaweera, 1981; Atta et al., 1986.
India, Nepal, Pakistan, Sri Lanka		Insecticide; toxic to the lower life, flies, flees	Infusion, alcoholic extract	Nadkarni, 1976; Ikram and Hussain, 1978; Agarwal, 1986; Atta et al., 1986.
India, Myanmar, Nepal, Pakistan, Thailand		Bleeding, hemorrhage, hemoptysis	Juice, decoction, infusion, crushed leaves	Jayaweera, 1981; Atta et al., 1986; Salalamp et al., 1992, 1996.
India, Pakistan, Sri Lanka		Antiseptic	Leave extract, Juice, decoction, poultice	Perry and Metzger, 1980; Agarwal, 1986; Atta et al., 1986.
India, Myanmar, Pakistan, Sri Lanka		Skin diseases, wounds		Jayaweera, 1981Atta et al., 1986.
Myanmar, Pakistan		Headache	Infusion	Jayaweera, 1981; Atta et al., 1986.
India, Sri Lanka		Pain	Decoction, fomentation	Roberts, 1931; Rafael and Barreto, 1967
India		Emmenagogue	NS	Kirtikar et al., 1975.
Pakistan		Gonorrhoea	Infusion	Kirtikar et al., 1975; Atta et al., 1986.
India, Sri Lanka		Snake-bites	Bruised fresh leaves	Roberts, 1931
India, Sri Lanka		Liver congestion, jaundice, biliousness	Extracts, syrups	
Pakistan		Vomiting, Leprosy	Infusion	Atta et al., 1986

(*Continued*)

TABLE 16.3 (CONTINUED)
Ethnopharmacological Uses of *A. vasica* Reported in Field Studies

Country	Part Used	Indication or Use	Preparation	References
Myanmar		Wealthy persons suffering from certain humors	Spirit	Kirtikar et al., 1975
India		Diuretic	Juice	Watt, 1972
Nepal		Vegetable	NS	Malla et al., 1982
India, Nepal, Pakistan, Sri Lanka, Thailand	Roots	Respiratory diseases, cough, colds, asthma, expectorant, liquefy sputum, antispasmodic, bronchitis, tuberculosis, consumption, phthisis, lung tonic, diphtheria	Powder, decoction, infusion, electuary, powder, paste	Nadkarni, 1976; Ikram and Hussain, 1978; Khory and Katrak, 1981; Agarwal, 1986; Atta et al., 1986; Kapoor, 1990; Salalamp et al., 1992, 1996.
India, Pakistan, Sri Lanka		Fever, malarial fever	Decoction, infusion, electuary, powder,	Nadkarni, 1976; Jayaweera, 1981; Atta et al., 1986; Kapoor, 1990.
India, Pakistan		Gonorrhea, Antiseptic, Rheumatism, Antiperiodic, Anthelmintic, Facilitates the expulsion of	Powder	Atta et al., 1986; Kirtikar et al., 1975
India, Sri Lanka		Eye diseases	NS	Kirtikar et al., 1975;
India, Nepal		Strangury, Leucorrhoea, Bilious vomiting, Diuretic, Abnormal labor,	NS	Pathak, 1970; Kirtikar et al., 1975
Sri Lanka	Roots, Bark	Wounds, Haemoptysis, Heart diseases, Catarrh	Decoction, bruised	Kirtikar et al., 1975; Jayaweera, 1981
India, Sri Lanka	Bark	Chest diseases, asthma, expectorant, antispasmodic, phthisis	Decoction, bruised,	Kirtikar et al., 1975; Jayaweera, 1981
India, Nepal, Pakistan, Sri Lanka	Flowers	Colds, consumption, phthisis, asthma, bronchitis, cough, antispasmodic	Infusion, electuary	Watt, 1972; Kirtikar et al., 1975; Ikram and Hussain, 1978; Khory and Katrak, 1981; Atta et al., 1986; Kapoor, 1990.
India, Pakistan, Sri Lanka		Ophthalmia, Fever, Gonorrhoea, Antiseptic, Hectic heat of blood, improve blood circulation	Fresh flowers, Infusion, electuary	Kirtikar et al., 1975; Atta et al., 1986;
India, Pakistan, Sri Lanka	Fruit	Colds, antispasmodic, bronchitis	NS, hung round the children's necks	Kirtikar et al., 1975; Watt, 1972; Atta et al., 1986
India, Pakistan Sri Lanka		Diarrhea, dysentery, Fever, Excessive phlegm, Menorrhagia, Laxative	Unripened fruit boiled with milk	Kirtikar et al., 1975; Atta et al., 1986;

NS, not stated.

16.4 CHEMICAL DERIVATIVES

16.4.1 *Vinca rosea*

V. rosea has a wide variety of alkaloids (nitrogen-containing chemical substances apart from amino acids, peptides, purines and derivatives, amino sugars, and antimicrobials) (Wansi et al., 2013) (Table 16.4). Based on their chemical constitution, alkaloids are categorized as heterocyclic or non-heterocyclic. Heterocyclic alkaloids have a nitrogen atom in the ring system and are classified as pyrrole, pyrrolizidine, pyridine, piperidine, quinoline, isoquinoline, norlupinane, or indole alkaloids, depending on their size. Non-heterocyclic alkaloids, also known as proto-alkaloids or biological amines, are much less abundant in nature than heterocyclic alkaloids. These compounds, like ephedrine, cathinone, and colchicine, have a nitrogen atom that is not a component of any ring structure. (Wansi et al., 2013).

Alkaloids, tannins, flavonoids, saponins, phenolic compounds, triterpenoids, amino acids, protein, carbohydrate, reducing sugars, and phlorotannins were all found in *V. rosea* shoot extracts, whereas aromatic acids, and xanthoprotein were not. The shoot and flower and root extract of *V. rosea* showed negative results for terpenoids and xanthoprotein (Prakash and Jain, 2011). The shoot extract showed positive results for terpenoids, alkaloids, saponins, and sterols, phenols as mentioned in *C. roseus* (Paarakh et al., 2019) (Table 16.5 and Figure 16.3).

16.4.2 *Adhatoda vasica*

The presence of alkaloids, phytosterols, polyphenolics, and glycosides as a significant class of chemicals was shown by a phytochemical study of different sections of Adhatodavasic. Its main components are quinazoline alkaloids, the most prominent of which being vasicine. Vitamin C and carotene are abundant in the leaves, which also produce essential oil. Essential oils, lipids, resins, sugar, gum, amino acids, proteins, and vitamin C are among the chemical compounds found in the leaves and roots of this plant (Dymock, 1972). A tiny quantity of crystalline acid is also present in the leaves. The seeds contain 25.8% of a deep yellow oil comprised of glycerides of arachidic 3.1%, behenic 11.2%, lignoceric 10.7%, cerotic 5%, oleic 49.9%, and linoleic acids 12.3%, according to a study published in India in 1956 (Paarakh et al., 2019). *A. vasica* includes mainly K, Na, Ca, and Mg, with traces of Zn, Cu, Cr, Ni, Co, Cd, Pb, Mn, and Fe, according to atomic absorption

TABLE 16.4
Qualitative Phytocompound Analysis Data for the Shoot, Flower, and Root Extracts of *V. rosea*

Phytochemical class	Result
Alkaloids	+
Triterpenoids	+
Tannins	+
Flavonoids	+
Amino acid	+
Saponin	+
Philobatinins	+
Phenolic compounds	+
Aromatic acids	-
Xantho proteins	-
Proteins	+
Reducing sugar	+
Carbohydrate	+

TABLE 16.5
Phytochemical Analysis of *Vinca rosea*

Plant Parts	Class	Compounds	References
plant extract and leaf	Alkaloid	β-carboline, apparicine, ammocalline, anthirine, akuammicine, sitsirikine, lochrovicine, pericyclivine, cavincine, rosicine, catharanthine, tabersonine, perividine, perivine, lochneridine, N-oxide fluorocarpamine, serpentine, cathenamine, ajmalicine, 19-epi,3-iso ajmalicine, 3-epi ajmalicine, O-deacetyl akuammiline, lochnericine, minovincine, rosamine, tetrahydroalstonine, Nb-oxide vindolinine, 19,20-cis16(R)-isositsirikine, 19,20-trans16(R)-isositsirikine, 19,20-trans 16(S)-isositsirikine, yohimbine, dihydro-sitsirikine, perimivine, 11-methoxy tabersonine, vincoline, vindolinine, 19-epi vindolinine, vincolidine, akuammine, lochnerinine, lochrovidine, 19-hydroxy-11-methoxy tabersonine, lochrovine, akuammiline, vincarodine, deacetoxy-vindoline, 19-acetoxy-11-hydroxy tabersonine, deacetyl-vindoline, lochnerinine, cathovaline, vindolidine, vindoline, strictosidine, bannucine, leurosivine, 17-deacetoxy- leurosine, 4-deacetoxy-vinblastine, deacetyl vinblastine, leurosinine, 4'-deoxy-vinblastine, vinosidine, N-demethyl-vinblastine, catharanthamine, leurosine, roseadine, vincathicine, roseamine, 5'-oxo leurosine, carosine, Nb'-oxide leurosine, vinamidine, vincristine, Nb-oxide leurosidine, 14'-hydroxy-vinblastine, 15'-hydroxy-vinblastine, neoleurocristine, vindolidine, leurosinone, neoleurosidine, Nb-oxide neoleurosidine, vindolicine, Ajmalicine, Anhydrovinblastine, Vindoline, Catharanthine, Vincamine, Serpentine, Vindolicine, Catharanthine, Vindolidine, Vindolicine, Vindolinine, Catharanthamine, 14',15'-didehydrocyclovinblastine, 17-deacetoxycyclovinblastine, 17-deacetoxyvinamidine	El-Sayed and Cordell, 1981; Tikhomiroff et al., 2002; Sottomayor et al., 2005; Hedhili et al., 2007; Taha et al., 2009; Siddiqui et al., 2010; Mu et al., 2012; Tiong et al., 2013; Jacobs et al., 2004
Leaf	Triterpenoids	Loganic acid, Ursolic acid, Oleanolic acid, Perivine, Vindolinine, Vinblastine, Vincristine, Reserpine, Serpentine, Horhammericine, Alstonine, Tabersonine, Vindogentianine, Vinpocetine	Huang etal., 2012; Zahari et al., 2018; Ragasa et al., 1998; Tiong et al., 2013; Sottomayor et al., 2005; Retna and Ethalsa, 2013; Almagro et al., 2015; Mishra et al., 2001; Zenk et al., 1977; Elisabetsky and Costa-Campos, 2006; Zhang et al., 2018; Tiong et al., 2015; Kiss and Karpati, 1996.

(*Continued*)

TABLE 16.5 (CONTINUED)
Phytochemical Analysis of *Vinca rosea*

Plant Parts	Class	Compounds	References
	Phenolic Flavonoid	Kamauritianin, quercetin 3-O-α-L-rhamnopyranosyl-(1→2)-α-L-rhamnopyranosyl-(1→6)-β-D-galactopyranoside and 2,3-dihydroxybenzoic acid, vincoside and chlorogenic acid, 3-O-caffeoylquinic acid, Kaempferol-3-O-(2,6-di-O-rhamnosyl-galactoside)-7-O-hexoside, 4-O-caffeoylquinic acid, 5-O-caffeoylquinic acid, Quercetin-3-O-(2,6-di-O-rhanmosyl-galactoside), kaempferol-3-O-(2,6-di-O-rhanmosyl-galactoside), vinpocetine	Nishibe et al., 1996; Daneshtalab, 2008; Ferreres et al., 2008; Pereira et al., 2009;Yogesh, 2011
	Alkaloid	Catharanthine, Vindolidine, Vindolicine, Vindolinine, Catharanthamine,	Mu et al., 2012; Tikhomiroff and Jolicoeur, 2002; Tiong et al., 2013; El-Sayed and Cordell, 1981
	Miscellaneous	Vincamicine, vinaspine, vinaphamine, rovindine, perosine, maandrosine, cavincidine, cathindine, 20'-epi-vinblastine,	Jacobs et al., 2004
Root and hairy root	Alkaloid	Ammocalline, akuammicine, cavincine, catharanthine, perivine, venalstonine, 21-hydroxy cyclolochnerine, lochneridine, alstonine, serpentine, ajmalicine, tetrahydroalstonine, yohimbine, dihydro-sitsirikine, perimivine, 19(S)-epimisiline, vinosidine, lochnerinine, strictosidine lactam, strictosidine, Vincristine, Vinblastine, Vinpocetine, Reserpine, Ajmalicine, Ajmaline, Yohimbine, Vindesine, Serpentine, Vindoline,	Kumar et al., 2018; Tiong et al., 2013; Tikhomiroff and Jolicoeur, 2002; Jacobs et al., 2004
	Miscellaneous	ammorosine, cathalanceine Cathindine,	Jacobs et al., 2004
Flower	Alkaloid	Apparicine, catharanthine, perivine, coronaridine, ajmalicine, tetrahydroalstonine, 11-methoxy tabersonine, mitraphyllline, vindoline, vinblastine, catharicine, carosine,	
	Phenolic	4-O-caffeoylquinic acid, Quercetin-3-O-(2,6-di-O-rhamnosyl-galactoside); Kaempferol-3-O-(2,6-di-O-rhamnosyl-galactoside), Kaempferol-3-O-(2,6-di-O-rhamnosyl-glucoside), Kaempferol-3-O-(6-O-rhamnosyl-galactoside), Kaempferol-3-O-(6-O-rhamnosyl-glucoside), Isorhamnetin-3-O-(6-O-rhamnosyl-galactoside), Isorhamnetin-3-O-(6-O-rhamnosyl-glucoside)	Ferreres et al., 2008; Pereira et al., 2009
Seedling And Seed	Alkaloid	Catharanthine, tabersonine, serpentine, ajmalicine, preakuammicine, tetrahydroalstonine, yohimbine, deacetoxy-vindoline, vindoline, strictosidine, vinsedine, vinsedicine, vingramine, methyl-vingramine, vinblastine,	Jacobs et al., 2004

(Continued)

TABLE 16.5 (CONTINUED)
Phytochemical Analysis of *Vinca rosea*

Plant Parts	Class	Compounds	References
	Phenolic	Kaempferol-3-O-(2,6-di-O-rhamnosyl-galactosid e)-7-O-hexoside, Quercetin-3-O-(2,6-di-O-r hanmosyl-galactoside), kaempferol-3-O-(2,6-di-O-rhanmosyl-galactoside); kaempferol -3-O-(2,6-di-O-rhanmosyl-glucoside); Isorhamnet in-3-O-(2,6-di-O-rhamnosyl-glucoside); Kaemp ferol-3-O-(6-O-rhamnosyl-galactoside); Isorh amnetin-3-O-(6-O-rhamnosyl-glucoside)	Ferreres et al., 2008; Pereira et al., 2009
Stem	Alkaloid	Akuammicine, sitsirikine, catharanthine, 21-hydroxy cyclolochnerine, serpentine, ajmalicine, tetrahydroalstonine, hörhammericine, vindolinine, 11- methoxy hörhammericine, strictosidine lactam, vindoline, 3',4'-anhydro-vinblastine, leurosine, Catharine,	Jacobs et al., 2004
	Phenolic	3-O-caffeoylquinic acid, 4-O-caffeoylquinic acid, 5-O-caffeoylquinic acid, Quercetin-3-O-(2,6-d i-O-rhanmosyl-galactoside), kaempferol -3-O-(2,6-di-O-rhanmosyl-galactoside), isorh amnetin-3-O-(2,6-di-O-rhanmosyl-galactoside),	Ferreres et al., 2008; Pereira et al., 2009
Whole plant	Alkaloid	Catharoseumine, 14',15' -Didehydr ocyclovinblastine, 17-deacetoxycy Cycloleurosine, Cycloleurosine, 17-deacetoxyvinamidine, Vinamidine, Leurosine, Leurosidine, Catharine, cathachunine, Vinblastine, vinorelbine, vincristine and vindesine, Vindoline, Ajmalicine, Tetrahydroalstonine, Serpentine, Lochnerine, Reserpine, Akuammine, Catharanthine, Lochnericine, Leurosine Vincaleu- koblastine Perivine, Virosine, leurosine and vincaleukoblastine	Svoboda et al., 1959; Wang et al., 2012a,b; Moudi et al., 2013.; Jacobs et al., 2004; Wang et al., 2016
	Miscellaneous	Vincadifformine, vallesiachotamine, isovallesiachotamine, akuammigine, 19-hydroxy-tabersonine, 19-epi-N-oxide vindolinine, minovincinine, 7-hydroxyindolenine ajmalicine, pseudo-indoxyl ajmalicine, 10-hydroxydeacetyl akuammiline, O-deacetyl-vindolidine, 19-acetoxy-11- methoxy tabersonine, xylosyloxy-akuammicine,	Jacobs et al., 2004

spectrophotometry (Jabeen et al., 2010). Gulfraz et al. conducted a chemical study of different bioactive components extracted from *A. vasica* leaves and roots (Gulfraz et al., 2004). They found protein (8.5%), vasicine (7.5%), vitamin C (5.2%), and lipids (2.5%) in *A. vasica* root samples, according to the findings. Except for sugar (16.4%), fiber (5.2%), vasicinone (3.5%), Zn (0.6%), S (1.3%), and Fe (1.3%), the number of such compounds was low in leaves (1.2%) (Tables 16.6, 16.7 and Figure 16.4).

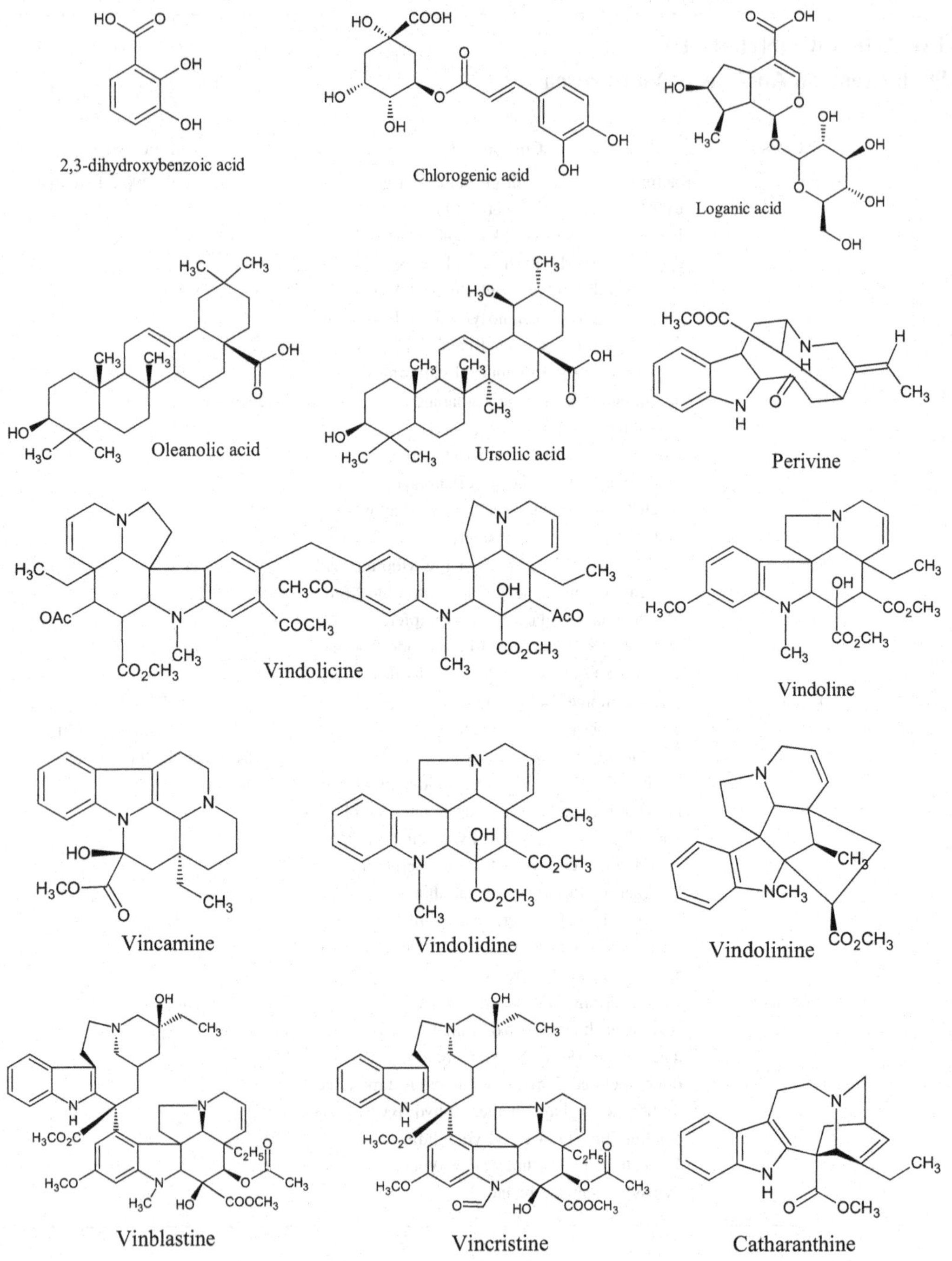

2,3-dihydroxybenzoic acid

Chlorogenic acid

Loganic acid

Oleanolic acid

Ursolic acid

Perivine

Vindolicine

Vindoline

Vincamine

Vindolidine

Vindolinine

Vinblastine

Vincristine

Catharanthine

FIGURE 16.3 Bioactive compounds present in *Vinca rosea*.

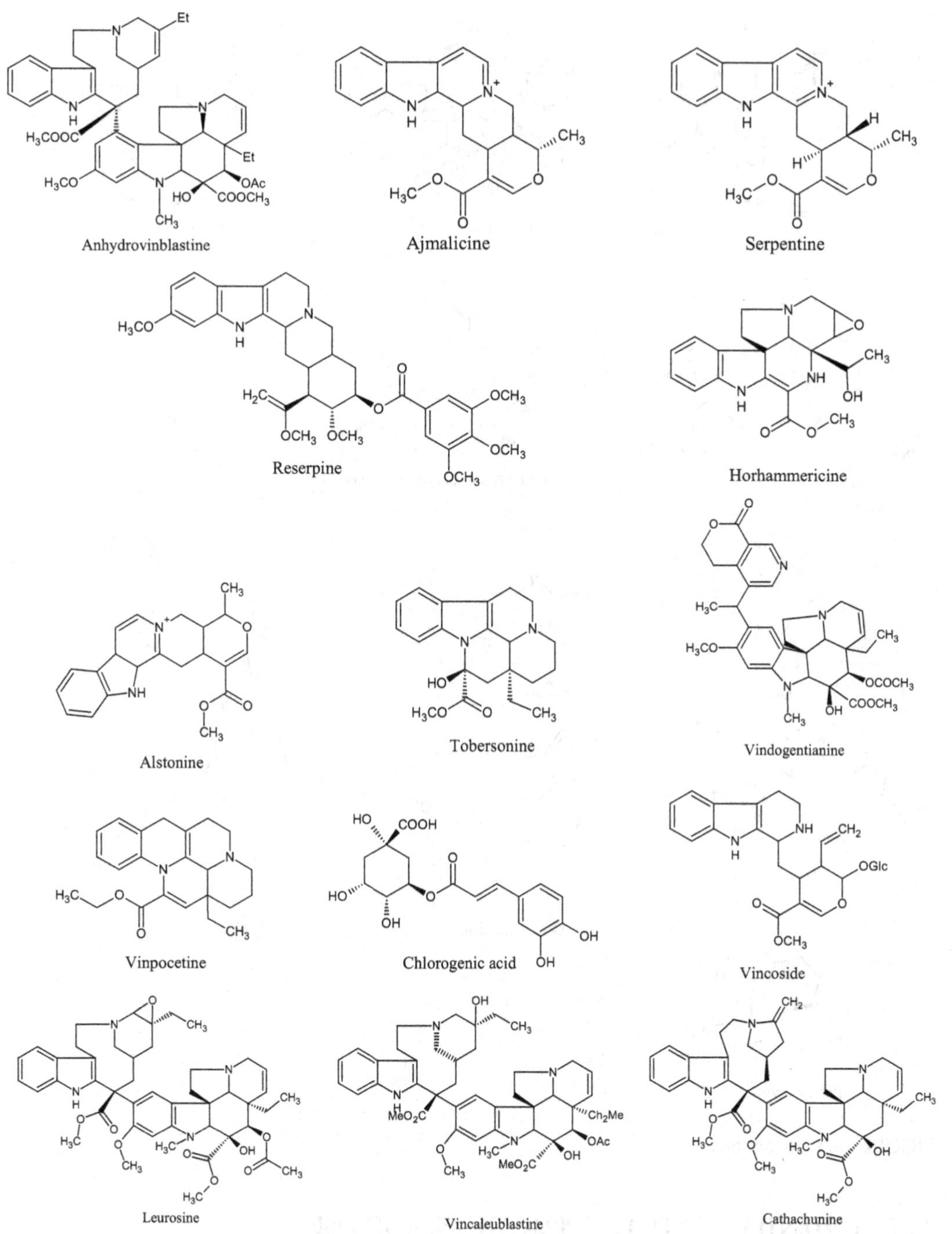

Anhydrovinblastine

Ajmalicine

Serpentine

Reserpine

Horhammericine

Alstonine

Tobersonine

Vindogentianine

Vinpocetine

Chlorogenic acid

Vincoside

Leurosine

Vincaleublastine

Cathachunine

FIGURE 16.3 (Continued)

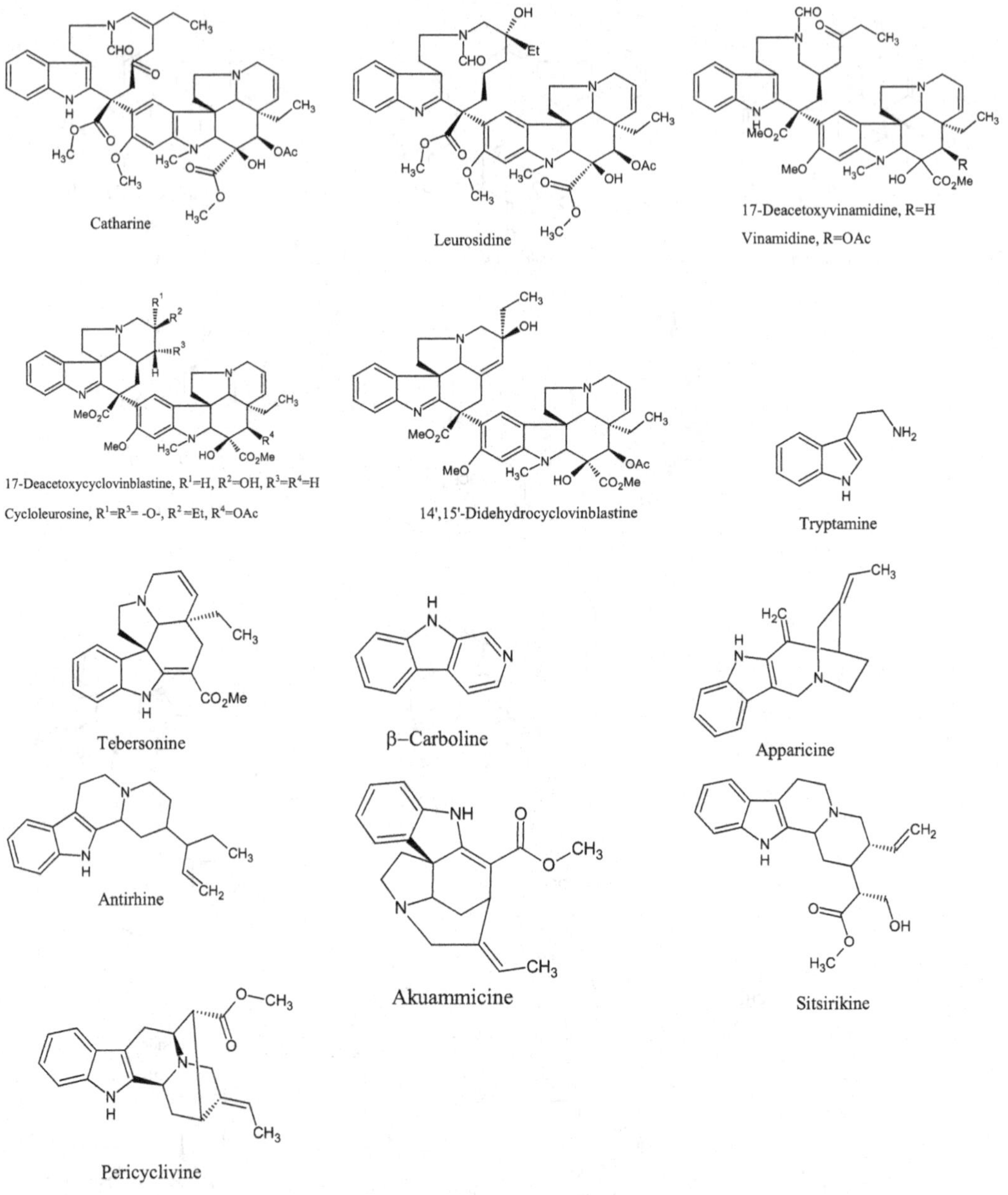

Catharine

Leurosidine

17-Deacetoxyvinamidine, R=H
Vinamidine, R=OAc

17-Deacetoxycyclovinblastine, R^1=H, R^2=OH, R^3=R^4=H
Cycloleurosine, R^1=R^3= -O-, R^2=Et, R^4=OAc

14',15'-Didehydrocyclovinblastine

Tryptamine

Tebersonine

β–Carboline

Apparicine

Antirhine

Akuammicine

Sitsirikine

Pericyclivine

FIGURE 16.3 (Continued)

16.5 POTENTIAL BENEFITS, APPLICATION, AND USES

16.5.1 *Vinca rosea*

16.5.1.1 Antioxidant Effect

The antioxidant potential is mostly associated with the roots of white and pink blooms that contain extracts, as determined by several assays, such as hydroxyl radical, peroxide radical, DPPH radical, and nitric oxide radical suppression (Alba and Bhise, 2011). 2,3-dihydroxybenzoic acid demonstrated significant radical-scavenging action, which is linked to reduced cancer risk (Nishibe, 1997). The antioxidant properties of enzymes such as superoxide dismutase (SOD), peroxidase (POX), and

TABLE 16.6
Phytochemical Analysis of Methanol Extract of *A. vasica* (Prathiba & Giri, 2018)

Phytochemicals	Result
Carbohydrate	+
Terrapins	+
Triterpenoids	+
Flavonoids	+
Saponins	-
Tannins	-
Amino acids	+
Glycosides	+
Alkaloids	+
Steroids	+

Key: + Present, - Absent

catalase (CAT) were studied in particular. According to the results of the aforementioned investigation, *V. rosea* is a suitable plant for salt-affected locations, and we may acquire plants with increased antioxidant and therapeutic properties (Jayakumar et al., 2010). Kumar et al. (2012) investigated the antioxidant properties of *V. rosea*. They determined that the habitat's heat would have a direct influence on antioxidant capacity, allowing them to see that the superoxide dismutase and polyphenol oxidase enzymes, in contrast to catalase, had more antioxidant properties whenever the temperature rose. (Kumar et al., 2012). The investigation was guided by a methanolic extract of the leaves. Rasool et al. conducted numerous *in vitro* antioxidant experiments in 2011 to investigate the influence of the solution on the extract and overall anti-radical capability of various *V. rosea* extracts. The extraction and fractionation were shown to be an excellent source of antioxidants in the studies. The shoots of *V. rosea* demonstrated excellent antioxidant potential in a 100% methanolic extract and a portion of 100% ethyl acetate (Rasool et al., 2011).

16.5.1.2 Anticancer Effect

Carcinoma, or cancer, is a hereditary disease marked by uncontrolled cell growth of a specific cell type in the body. When the regulation of cell proliferation fails, the cells begin to proliferate and divide needlessly; when the new cells inherited the tendency to reproduce without management, the outcome is a duplication that grows forever, eventually becoming cancer. These malignancies can be benign or malignant, depending on their ability to infiltrate and spread to other parts, a process known as metastasis (Barrales et al., 2012).

Alkaloids with antitumor and anticancer effects are derived from the leaves and branches of *V. rosea*. Cancer is inhibited by alkaloids. Vinblastin is being used to treat chorio carcinoma, a kind of Hodgkin's disease tumor. Vincristine is a drug that is used to treat childhood leukemia. Oncovin is a medication that contains vinblastin, which is marketed as Velban or Vincristine. Moderate vinca alkaloids, like vinorelbine and vinflunine, have been created to enhance the treatment efficacy. Vinorelbine and vinflunine interact with microtubules and hence have cytotoxic effects. The alkaloids, commonly known as mitotic spindle poisons, prevent microtubules from entering the cell cycle, preventing mitosis cell division in the cell cycle. As a result, vinca alkaloids aid in the prevention of cancer (Moudi et al., 2013). Various fractions of raw methanolic extract of vinca have exhibited anticancer efficacy against such a range of cell types (Moreno-Valenzuela et al., 1998) and exert a pronounced effect on multi-drug-resistant malignancies (Wang et al., 2004). Anhydrovinblastine in the lung and cervicouterine cancer cell lines exhibited significant effects (Tiong et al., 2013). In a variety of cancer cells,

TABLE 16.7
Important Phytoconstituents of *Adhatoda vasica*

Plant Part	Class	Compounds	References
Leaves	Alkaloid	Vasnetine, vasicinol, vasicinolone, adhatonine, adhavasinone, anisotine, deoxyvasicine, 3-hydroxyanisotine, l-vasicinone, l-vasicol, maiontone, 5-methoxyvasicinone, 7-methoxyvasicinone hydrate, 7-methoxy-3*R*-hydroxy-1,2,3,9-tet rahydropyrrolo-[2,1-b]-quinazolin-9-one(7-methoxy-vasicinone); 1,2,3,9-tetrahydro-5 -methoxy-pyrrolo[2,1-b]quinazoline-3-Ol, vasicine,	Johne et al., 1971; Bhartiya and Gupta, 1982; Jain and Sharma, 1982; Chowdhury, and Bhattacharyya, 1987; Ram et al., 1991; Iyenger et al., 1997; Shinwari and Khan, 1998;
	Flavonoid	Vitexin, isovitexin, violanthin, 2”-*O*-xylosylvitexin, rhamnosylvitexin	Jain and Sharma, 1982; Mueller et al., 1993
	Steroid	epitaraxerol	Mueller et al., 1993; Atta et al., 1997
	Pyrroloquinazolines	desmethoxyaniflorine	
Flower	Flavonoid	Vitexin, isovitexin, violanthin, astragalin, kaempferol, 2”-*O*-xylosylvitexin, quercetin, rhamnosylvitexin,	Jain and Sharma, 1982; Mueller et al., 1993
	Triterpenoid	α-amyrin	Huq et al., 1967
	Alkaloid	Betaine	
	Chalcone	2’,4-dihydroxychalcone-4-O-β-D-glucopyra noside	Bhartiya and Gupta, 1982
	Steroid	Epitaraxerol	Mueller et al., 1993; Atta et al., 1997
	Alkyl Ketone	4-heptanone	Ahmed et al., 1999
	Alkanone	3-methylheptanone	
Pollen	Amino acid	amino-*n*-butyric acid, glycine, proline, serine, valine,	Amal et al., 2009
Inflorence	Alkaloid	Vasicinolone, adhatodine, vasicoline, vasicolinone, vasicol	Johne et al., 1971; Singh, 1997
Aerial Parts	Alkaloid	Vasicinolone, N-oxides and glycosides of vasicine, N-oxides and glycosides of vasicinone, peganidine, vasicinone, vasicinolone	Jain and Sharma, 1982, Atta et al., 1997; Ahmed et al., 1999
	Flavonoid	apigenin	Mueller et al., 1993
	Triterpenoid	3α-hydroxy-D-friedoolean-5-ene, 3α-hydroxy-oleanane-5-ene	Singh, 1997
	Steroid	Epitaraxerol, sitosterol	Mueller et al., 1993; Atta et al., 1997
	Alkanol	29-methyltriacontan-1-ol	Sultana et al., 2001
	Alkyl Hydroxyketones	37-hydroxy-hexatetracont-1-en-15-one, 37-hydroxy-hentetracontan-19-one	Ram et al., 1991
Part Not Specified	vitamin	vitamin C	Suthar et al., 2009; Srivastava et al., 2001
	Alkaloid	Vasicine, vasicinone, vasinol	
	Organic acid	adhatodic acid	Singh, 1997
	Terpenoid	β-carotene	Wealth of India, 1989
	Glucoside	β-sitosterol-D-glucoside	Ahmed et al., 1999
	Chalcone	hydroxyl oxychalcone	
	Fatty acid	Vasakin	Wealth of India, 1989

(Continued)

TABLE 16.7 (CONTINUED)
Important Phytoconstituents of *Adhatoda vasica*

Plant Part	Class	Compounds	References
Root	Alkaloid	Vasicinol, vasicinolone, 9-acetamido-3,4-dihydropyrido-(3,4-b)-indole, adhatodine, anisotine, deoxyvasicine, deoxyvasicinone, l-vasicinone, maiontone, vasicine, vasicinone, vasicol, vasicinolone, vasicinol	Johne et al., 1971; Jain et al., 1980; Singh, 1997; Jain and Sharma, 1982, Iyenger et al., 1997, Bhartiya and Gupta, 1982, Shinwari and Khan, 1998,
	Glucoside	β-glucoside-galactose, sitosterol-β-D-glucoside	Jain et al., 1980; Iyenger et al., 1997
	Galactoside	O-ethyl-α-D-galactoside	Jain et al., 1980;
	Sterol	Daucosterol, sitosterol	Jain and Sharma, 1982; Iyenger et al., 1997
	Glucose	D-galactose, D-glucoside	Jain et al., 1980; Jain and Sharma, 1982
	Chalcone	2'-glucosyl-4-hydroxyl - oxychalcone	Jain and Sharma, 1982
	Hydrocarbon	Tritriacontane	Johne et al., 1971
Young Plant	Alkaloid	Adhatodine, anisotine, vasicoline, vasicolinone,	Huq et al., 1967; Johne et al., 1971; Jain and Sharma, 1982

ursolic acid has been demonstrated to exhibit antiproliferative properties (Neto, 2007). Ursolic acid upregulates apoptosis cell death which was initiated through using Fas activation and caspase-3/8, and PARP breakage, which was accompanied by increasing Bax and decreasing Bcl-2, as well as the production of cytochrome c from mitochondria with the reduction in MMP (Kim et al., 2011).

Vinblastine, vincristine, vindoline, vindolidine, vindolicine, vindolinine, and vindogentianine are a class of alkaloids in *V. rosea* that have already been discovered to have chemotherapeutic effects (Cragg and Newman, 2003; Tiong et al., 2013; 2015). They exhibited antiproliferation characteristics via altering microtubular dynamics and enhancing apoptosis (Almagro et al., 2015). The first plant-derived chemotherapeutic agent to be used in clinical trials were vinblastine and vincristine (Cragg and Newman, 2003). Vinblastine sulfate is used to treat acute and chronic leukemia, Hodgkin's disease, lymphosarcoma, choriocarcinoma, neuroblastoma, and carcinomas. Vincristine sulfate is an oxidized version of vinblastine (Almagro et al., 2015; Aslam et al., 2010).

Other *V. rosea*-derived indole alkaloids drastically demonstrated cytotoxic properties against a wide range of cancer. At IC_{50} of 6.28 μM concentration, catharoseumine, a novel monoterpenoid indole alkaloid isolated from *V. rosea*, exerted a suppressive effect against leukemia (HL-60) cell line (Wang et al., 2012a). Wang et al. (2012b) discovered that three novel indole alkaloids, 14',15'-didehydrocyclovinblastine, 17-deacetoxycyclovinblastine, and 17-deacetoxyvinamidine – when combined with active compounds like vinamidine, leurosine, Catharine, cycloleurosine, and leurosidine – had an anticancer effect on breast cancer at IC50 concentrations of 0.73–10 μM (Wang et al., 2012b).

Interestingly, cathachunine (bisindole alkaloid) attenuates tumor growth in leukemia cells while causing considerably less cytotoxicity in normal endothelium cells, showing that such action preferentially suppressed targeting leukemia and other cancer cell line (Pham et al., 2018; 2019). Furthermore, Fernández-Pérez et al. (2013) revealed that *V. rosea* has the potential antitumor agents which are indole alkaloid (Fernández-Pérez et al., 2013). These findings indicate a synergistic response and constructive contact between the active compounds identified in *V. rosea* and cancerous cells, that have previously been discovered in other organic materials (Barth et al., 2007; Bhuyan

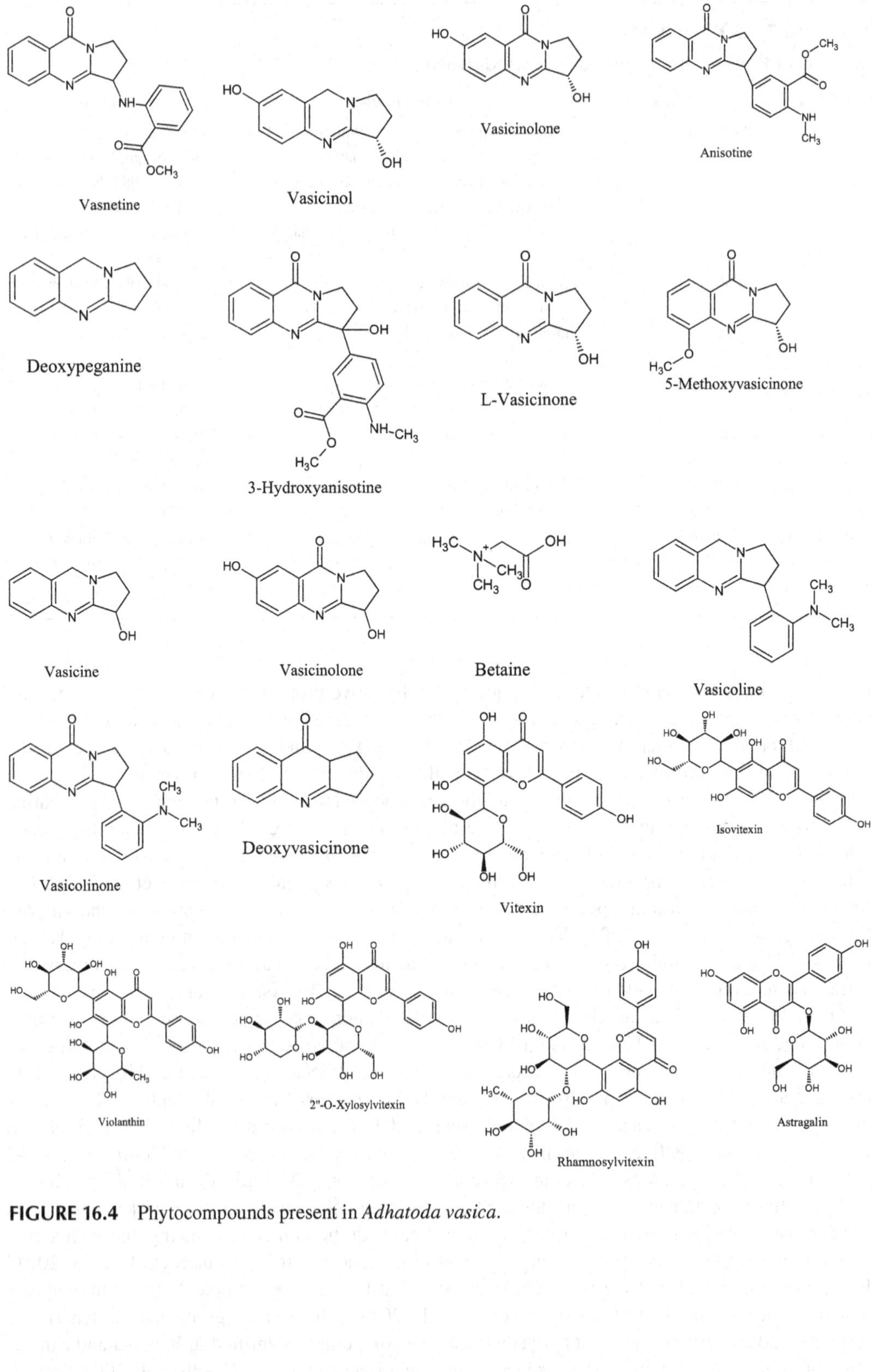

Vasnetine

Vasicinol

Vasicinolone

Anisotine

Deoxypeganine

3-Hydroxyanisotine

L-Vasicinone

5-Methoxyvasicinone

Vasicine

Vasicinolone

Betaine

Vasicoline

Vasicolinone

Deoxyvasicinone

Vitexin

Isovitexin

Violanthin

2"-O-Xylosylvitexin

Rhamnosylvitexin

Astragalin

FIGURE 16.4 Phytocompounds present in *Adhatoda vasica*.

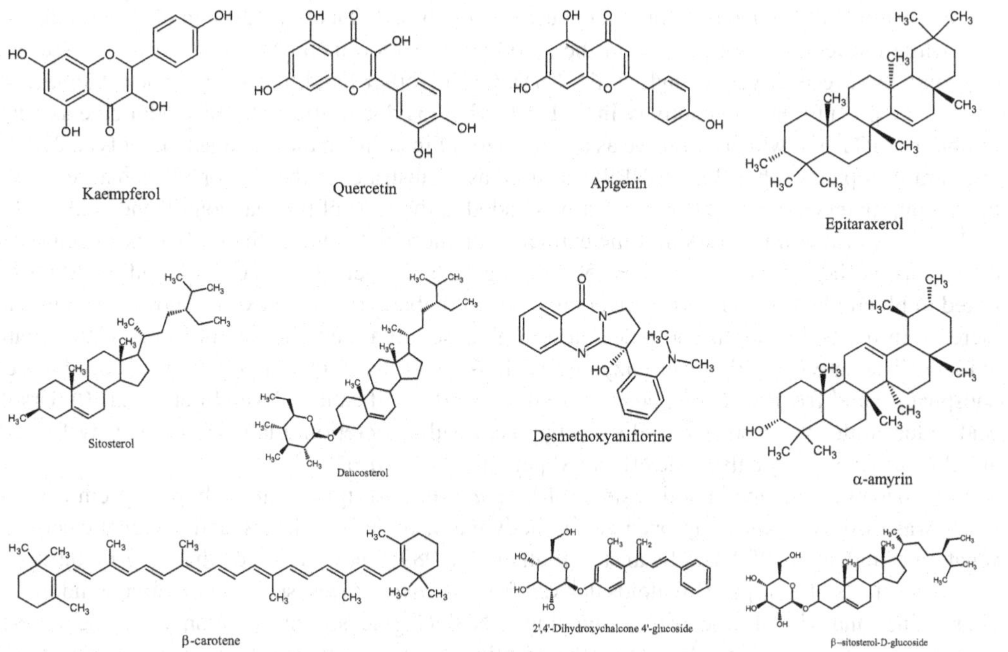

Kaempferol

Quercetin

Apigenin

Epitaraxerol

Sitosterol

Daucosterol

Desmethoxyaniflorine

α-amyrin

β-carotene

2',4'-Dihydroxychalcone 4'-glucoside

β-sitosterol-D-glucoside

FIGURE 16.4 (Continued)

et al., 2018) as well as being studied as a cancer therapy method (Jiang et al., 2010). This occurrence is conceivable owing to the aid of substances with very little activity to the primary chemical components by enhancing absorption, reducing metabolism and excretion, modes of action, reversing resistance, and regulating adverse effects of the bioactive constituents (Rasoanaivo et al., 2011).

16.5.1.3 Cytotoxic Effect

In cell culture, the alkaloid component of dried leaves was potent in CA-9KB, with the median effective dosage (ED50) of 0.0435 g/mL (El-Sayed and Cordell, 1981). At the dose of 50 mg, plant extract exerts a cytotoxic effect. The leaf extract of *V. rosea* has been shown to inhibit human cancer cell proliferation (Wong et al., 2011). Catharanthine produces cytotoxic action on the HCT-116 colorectal carcinoma cell line at 200 μg/mL (Siddiqui et al., 2010).

16.5.1.4 Antidiabetic Effect

Flower and leaf extract of *V. rosea* exerts an indistinguishable hypoglycemic effect similar to a standard drug, glibenclamide. Because of the increased glucose consumption inside the liver, hypoglycemic action has developed (Chattopadhyay et al., 1991; Ghosh and Suryawanshi, 2001; Singh et al., 2001). Methanolic extract of leaves and twigs of vinca along with dichloromethane at a 1:1 ratio exhibited hypoglycemic effect in a streptozotocin-induced diabetic rat model at 500 mg/kg dose orally. On the other hand, a 57.6% hypoglycemic effect has been observed. *V. rosea* extract enhances liver enzymes, such as glycogen synthase, glucose 6-phosphate dehydrogenase, succinate dehydrogenase, and malate dehydrogenase. These enzymes are essential for glucose storage in the liver. Another study suggested that the methanolic extract of *V. rosea* drastically exerted an antihyperglycemic effect in a dose-dependent manner (300 and 500 mg/kg), improving body weight and lipid profile. Furthermore, it can regenerate β-cells of the pancreas in diabetic rats (Ahmed et al., 2010). Several studies in animals have shown that the ethanol extracts of leaves and flowers of *C. roseus* decrease the levels of glucose in the blood (Ghosh RK and Gupta, 1980; Chattopadhyay et al., 1991).

Furthermore, the aqueous extracts can decrease the blood glucose by 20% in trials with diabetic rats, while the reduction of glucose in the blood with dichloromethane and methanol extracts are 49% and 58%, respectively (Singh et al., 2011). On the other hand, the vindolicine alkaloid demonstrated a potent inhibition activity in PTP-1B, which is due to the effect of vindolicine as a new inhibitor of PTP-1B, which can serve as a "sensitizer of insulin" in the management of type 2 diabetes. Para-nitrophenyl phosphate (pNPP) was used as a substrate for the trial on phosphatase activity (the recombinant enzyme PTP-1B) and it was added at the start of the reaction (Tiong et al., 2013).

The daily oral administration of the extract of dichloromethane: methanol (1:1) has been evaluated by using the leaves of *C. roseus* (500 mg/kg of body weight) for 20 days, and its effect was tested in blood glucose and normal liver enzymes in diabetic rats. The extract showed a significant increase in the body weight and a decrease in glucose, urea, and cholesterol levels of the treated animals. The activity of the liver enzymes, such as hexokinase, increased while that of glucose-6 phosphatase and fructose 1,6-biphosphatase decreased significantly (Jayanthi et al., 2010). Ursolic acid reduces glucose absorption, glucose uptake, insulin secretion, diabetic vascular dysfunction and this provides a hypoglycemic effect (Alqahtani et al., 2013).

In diabetic rats, the methanolic extract of *V. rosea* showed significant antihyperglycemic action, which was associated with improvements in body mass, cholesterol levels, and pancreatic cell rejuvenation (Singh et al., 2001). A Swanston-Flatt et al. (1989) study suggested that *in vitro* antihyperglycemic effects of four pure alkaloids derived from *V. rosea* leaves, such as vindoline, vindolidine, vindolicine, and vindolinine leaf, employing 2-NBDG glucose consumption tests and PTP-1B enzyme suppression, adversely affects the insulin signaling pathway. Improved glucose absorption in pancreatic or muscle cells might help type 2 diabetes patients with hyperglycemia. The four alkaloids were discovered to enhance glucose absorption in mice -TC6 pancreas and mice myoblast C2C12 cells while inhibiting PTP-1B. Vindolicine was the most active of the bunch. The findings backed up the traditional use of *C. roseus* for diabetes therapy, indicating that it is a promising candidate for additional research into antidiabetic medicines (Swanston-Flatt et al., 1989).

16.5.1.5 Antidiarrheal Effect

In addition to the pretreatment of the vinca plant extract, the *in vivo* antidiarrheal efficacy of the ethanol extract of leaves was evaluated in Male Wistar rats utilizing castor oil as the study diarrhea-producing drug. The usual medications were loperamide and atropine sulfate. The ethanolic leaf extracts of *V. rosea* exert an antidiarrheal effect in a dose-dependent inhibition of castor oil-induced diarrhea. At dosages of 200 and 500 mg/kg, gastrointestinal propulsion of charcoal meal and diarrheal symptoms are both inhibited in a dose-dependent manner. This evidence backs up the use of vinca in the prevention and treatment of diarrhea (Hassan et al., 2011).

16.5.1.6 Antihelminthic Effect

Infections caused by helminths are long-term illnesses that afflict humans. In medicine, *V. rosea* is utilized as an antihelminthic agent. Humans and livestock are much more susceptible to helminthic infectious diseases, which are chronic. Vinca contains antihelminthic properties, which were tested using a *Pherithema postuma* experiment model and piperazine citrate as a reference standard. The ethanolic extract of *V. rosea* at 250 mg/ml concentration provides antihelminthic effects (Swati et al., 2011). Furthermore, Agarwal et al. used the same methods for testing the antihelminthic property of the *C. roseus* experimental model. At 250 mg/ml *V. rosea* ethanolic extract significantly attenuated helminthic activity at 46.3 min. This ethnomedical study looks at *V. rosea* as a powerful antihelminthic agent (Agarwal et al., 2011).

16.5.1.7 Hypolipidemic Effect

The leaf juice of vinca reduces blood serum levels, such as total cholesterol, triglycerides, LDL-c, and VLDLc, indicating anti-atherosclerotic action. As a consequence, the flavonoid, a vinpocetine-like molecule found in vinca leaf extract has antioxidant properties (Patel et al., 2011).

16.5.1.8 Anti-HIV Effect

Ursolic acid is isolated from *V. rosea*, *Prosopis glandulosa* (Leguminosae), *Phoradendron juniperinum* (Loranthaceae), *Syzygium claviflorum* (Myrtaceae), and *Hyptis capitata* (Lamiaceae). The anti-HIV effect of ursolic acid was determined at an EC50 value of 4.4 μM and was found to exhibit a level of activity similar to that shown by oleanolic acid (Babalola et al., 2013). Thus, some ursolic acid derivatives (3-O-acyl ursolic acid derivatives) exert anti-HIV activity. (Kashiwada et al., 2000).

16.5.1.9 Antihypertensive Effect

The leaf extract, which contains 150 valuable alkaloids as well as other biologically active molecules, has hypotensive properties. The ethanolic extract has been shown to exhibit hypoglycemic and hypotensive effects in experimental animals (Pillay et al., 1959). An indole alkaloidal compound isolated from *V. rosea* plant extract, serpentine, has hypotension potential properties (Hedhili et al., 2007). In adrenaline-induced hypertension rats, the leaf extract of *V. rosea* was tested for hypotensive and hypolipidemic effects. Animals given *C. roseus* leaf extract showed hypotensive responses at 30 mg/kg b.w. dose (Ara et al., 2009). Several bioactive compounds are isolated from *V. rosea*. Among them, alstonine, horhammericine, serpentine, reserpine, and ajmalicine possess antihypertensive properties in *in vivo* and *in vitro* test systems by reducing blood volume, cardiac output, and vasodilation (Mishra et al., 2001; Elisabetsky and Costa-Campos, 2006). After the study, the *V. rosea* leaves extract caused substantial alterations in each cardiovascular parameter. The bioactive constituents of *V. rosea* are responsible for hypotensive activity in animal models.

16.5.1.10 Antimicrobial Effect

Vinca has a wide range of medicinal properties and particularly aids in the development of innovative medicines because most bacterial infections improve resistance to numerous antimicrobial treatments. Organic therapeutic compounds are also found in plants, which imply a broad range of activity with a focus on prevention (Patil and Ghosh, 2010). Muhammad et al. discovered antibacterial properties in the plant extract of several sections of *V. rosea* (leaves, stalks, roots, and flowers) against therapeutically relevant strains of bacteria (Muhammad et al., 2009). The antimicrobial effect of *V. rosea* leaf extract was evaluated toward microbes including *Pseudomonas aeruginosa*, *Salmonella typhimurium*, and *Staphylococcus aureus*, and it has been discovered that this extract could be used as a prophylactic agent in the treatment of a variety of infections (Patil and Ghosh, 2010).

Kumari and Gupta investigated the potentiality of *V. rosea* plant extract in a variety of hazardous bacteria finding that, at a concentration of 50 mg/mL, *V. rosea* was effective against *Bacillus fusiformis*, while at a concentration of 20 mg/mL, it was selective against *Aspergillus fumigatus, Candida albicans, Escherichia coli,* and *Bacillus fusiformis* (Kumari and Gupta, 2013).

In an experimental rat model, ethanol *V. rosea* flower extract was found to have therapeutic effects (Nayak and Pereira, 2006). The extract improves wound-healing, raises hydroxyproline levels, and has antimicrobial activity against *P. aeruginosa* and *S. aureus*. Furthermore, *V. rosea* extracts from the leaves, stems, flowers, and roots exhibit antibacterial action against *E. coli, Streptococcus pyogenes*, *Streptococcus agalactiae*, *Salmonella typhi*, and *Aeromonas hydrophila* (Muhammad et al., 2009). 2,3-dihydroxybenzoic acid, vincoside, and chlorogenic acid possess defensive mechanisms against fungal pathogens by blocking 1,3-β-glucan synthase enzyme (Moreno et al., 1994; Daneshtalab, 2008). Moreover, loganic acid also has antifungal properties (Huang et al., 2012).

Oleanolic and ursolic acids showed antimicrobial activity against vancomycin-resistant enterococci at minimum inhibitory concentrations (MICs) 8 and 4 μg/ml, respectively (Huang et al., 2012). Furthermore, both of them possess bactericidal effects against *Streptococcus pneumoniae* and methicillin-resistant *Staphylococcus aureus* (Horiuchi et al., 2007; Kim et al., 2012; Kurek et al., 2012). Ursolic acid effectively controls the growth of *Rigidoporus microporous*, *Ganoderma philippii*, and *Phellinus noxious* fungi (Zahari et al., 2018).

16.5.1.11 Antimycobacterium Tuberculosis Effect

Several evaluations of antimycobacterial alternative remedies have mentioned the efficacy of plant-derived terpenoids toward mycobacterium tuberculosis (Newton et al., 2000; Cantrell et al., 2001; Kanokmedhakul et al., 2005). Ursolic acid has been exhibited to suppress tuberculosis H37Rv growth at MIC of 41.9gl/ml. Antimycobacterial action has also been documented in a wide range of triterpenes. Moreover, ursolic acid derived from various sources was shown to have a MIC of 12.5gml-1 and 32m against mycobacterium tuberculosis. (Newton et al., 2000; Cantrell et al., 2001).

16.5.1.12 Neuroprotective Effect

Ajmalicine and indole terpene alkaloids isolated from *V. rosea* have a potential effect on depression and stress (Taha et al., 2008). Vas and Gulyas (2005) demonstrated that Vincamine protects the brain from cerebral disorders and insufficiencies (Vas and Gulyas, 2005). Another isolated compound, serpentine, attenuated anxiety disorder (Hedhili et al., 2007). Vinoceptine is an alkaloid that can increase cognitive abilities and memories, which is advantageous to Alzheimer's disease. In clinical studies of dementia and stroke, vinpocetine at a well-tolerated dosage of up to 60 mg/d showed no notable side effects (Sekar, 1996).

Alzheimer's disease (AD) is a neurological disorder of the brain and nervous system that causes 50–60% of dementia in people. Similarly, serpentine, an alkaloid found in the leaves, stems, and roots of *V. rosea* showed significant anti-AchE action with a low IC50 value (0.775 M). These results indicate that this plant might be a source of useful chemicals for the treatment of neurological diseases, such as Alzheimer's disease.

16.5.1.13 Antiplatelet Aggregation Effect

The effects of pentacyclic triterpenes ursolic, oleanolic acids on platelet aggregation have been studied *in vivo*.

According to a Babalola et al. (2013) study, oleanolic acid and ursolic acid exhibit antiplatelet aggregation characteristics against thrombin, ADP, and epinephrine (Babalola et al., 2013).

16.5.1.14 Anti-Ulcer Effect

Vincamine and Vindoline are two *V. rosea* alkaloids that have anti-ulcer properties. The plant leaves contain vincamine, which possesses cerebrovasodilatory and neuroprotective properties; however, they also cause stomach injury in animals (Babulova et al., 2003). Furthermore, Vindoline has anti-ulceration properties in several experimental models (Taha et al., 2008).

16.5.1.15 Wound-Healing Effect

The healing process is the procedure of repairing the epidermis as well as other body tissues after an injury, infection, trauma, or lesion. An inflammatory reaction happens after an injury, and cells underneath the basement (profound skin layer) start to produce more collagen (connective tissue). The epidermis (outer skin) regenerates subsequently later (Stadelmann et al., 1998; Iba et al., 2004). Wound-healing is divided into three phases: inflammation, proliferation, and remodeling (Nayak et al., 2007). The EtOH extract of *V. rosea* leaf drastically improves the wound in the animal experimental model compared with the standard drug. At a concentration of 100 mg/kg/day, *V. rosea* leaf extract decreases the wound surface area and time of the epithelization process in a rat model, which was manifested by a boost in dry weight and hydroxyproline levels. Together with hydroxyproline, there seems to be an improvement in tensile strength that aids in tissue repair maintenance (Nayak et al., 2007).

16.5.1.16 Other Effects

Ajmalicine exerted an antispasmodic effect (Taha et al., 2008). Furthermore, serpentine can reduce spasmodic properties (Hedhili et al., 2007). Oleanolic acid is a pentacyclic triterpenoid that can reduce inflammation in carrageenan and dextran-induced edema in rats model (Huang et al., 2012). By inhibiting COX-2 catalyzed prostaglandin biosynthesis process, ursolic acid provides an anti-inflammatory effect at 50 μg/ml in human mammary epithelial cells (Subbaramaiah et al., 2000; Huang et al., 2009). Ursolic acid further blocks the NF-kB pathway via suppressing several proinflammatory cytokines and chemokines (Shanmugam et al., 2013). According to Fischer et al. (1990), DHBA has a significant antifeedant effect against the Mexican bean beetle (*Epilachna varivestis*). The biological properties of *V. rosea* against larvae of the gram pod borer, *Helicoverpa armigera*, was investigated (Lepidoptera: Noctuidae). Ethyl acetate leaf extract of *V. rosea* has been discovered to be an effective biopesticide. Deshmukh et al. also discovered that *V. rosea* had insecticidal capabilities (2010).

16.5.1.17 Safety Aspects

It's possible that a medication, when used in recommended doses, is safe. Vinblastine has an LD50 of 17 mg/kg iv and vincristine has an LD50 of 5.2 mg/kg i.p. in rodents (Paarakh, et al., 2019). In adult rats, continuous oral treatment of 0.1 g/kg of methanol leaves extracts of *V. rosea* was shown to be nontoxic, with no substantial hepatic or kidney failure (Kevin et al., 2012).

16.5.2 *Adhatoda vasica*

16.5.2.1 Antioxidant Effect

Two significant medicinal plants endemic to India are *A. vasica* Nees and *Sesbania Grandiflora* (L.) Pers. The DPPH radical-scavenging activity, hydroxyl radical-scavenging activity in Fe^{3+}/ascorbate/EDTA/H_2O_2 system, prevention of lipid peroxidation caused by $FeSO_4$ in egg yolk, and metal-chelating activity of the aqueous leaf extracts of these two plants have all been investigated *in vitro*. Standard antioxidants, such as butylated hydroxytoluene (BHT), ascorbic acid, and EDTA were used to compare free radical-scavenging capabilities. The reduction of Mo (VI) to Mo (V) by the extraction and subsequent production of green phosphate/Mo (V) complex at acid pH, as well as reducing power via Fe^{3+}–Fe^{2+} transition in the presence of extracts, were used to determine total antioxidant activity. Total phenolics (measured in milligrams of gallic acid equivalents per gram) and total flavonoids (measured in milligrams of quercetin equivalents per gram), as well as antioxidant enzymes, were measured. *In vitro*, the antioxidant activity of *A. vasica* and *S. Grandiflora* was shown to be considerable. The antioxidant and radical-scavenging properties of *A. vasica* were found to be more significant than those of *S. Grandiflora*. (Padmaja et al., 2011).

16.5.2.2 Antibacterial Effect

The crude extracts produced from the leaf of *A. vasica* were subjected to preliminary phytochemical and antibacterial studies utilizing polarity-varying solvents. The tests revealed the presence of phenols, tannins, alkaloids, anthraquinones, saponins, flavonoids, amino acids, and reducing sugars. On *Staphylococcus aureus*, *Staphylococcus epidermidis*, *Bacillus subtilis*, *Enterococcus faecalis*, *Escherichia coli*, *Pseudomonas aeruginosa*, *Proteus vulgaris*, *Klebsiella pneumoniae*, and *Candida albicans*, the effects of ethanol, petroleum ether, and water extracts were investigated. For different species, the lowest inhibitory concentration of crude extracts was found (Jayakumar et al., 2010).

16.5.2.3 Antimicrobial Effect

The phytochemical content and antioxidant activity of *A. vasica* leaves, total antioxidant activity, 2, 2 diphenyl-1-picrylhydrazyl radical-scavenging activity, reducing power potential, and iron-chelating activity were used to determine the antioxidant activity of *A. vasica* methanol extract. The agar well diffusion technique was used to test antimicrobial activity. Total phenolic content was determined using the Folin-Ciocalteu reagent technique, and total flavonoid content was determined using the aluminum chloride method. Saponins, oils and lipids, phytosterol, phenolic compounds, tannins, glucose, alkaloids, flavonoids, and proteins were discovered in the leaves of *A. vasica*. Various antioxidant tests revealed that the extract has a strong antioxidant activity. The presence of significant quantities of polyphenolic substances (phenolic compounds and flavonoids) in the extract of *A. vasica* may be the cause of the plant's antioxidant action. Furthermore, the extract exhibited moderate antibacterial and cytotoxic action (lethality of brine shrimp) (Kotakadi et al., 2013).

16.5.2.4 Anti-Inflammatory Effect

A. vasica (L.) Nees is a well-known Ayurvedic and Unani medicinal plant. It's been used to treat a variety of ailments, most notably inflammatory and cardiovascular disorders. The scientific reason and methods by which it works in certain illnesses, however, are unknown. The goal of this research was to investigate the inhibitory effect of *A. vasica* aqueous and butanolic fractions on arachidonic acid metabolism in the *Journal of Medicinal Plants Studies.* Aqueous and butanolic fractions of *A. vasica* were tested for activity against arachidonic acid (AA) metabolites, and their efficacy was further assessed by investigating platelet aggregation caused by AA, adenosine diphosphate (ADP), platelet-activating factor (PAF), and collagen. A thin layer chromatography system was used to study AA metabolism, while a dual channel Lumiaggrego meter was used to detect platelet aggregation. Through the COX (TXB2) and LOX pathways, the aqueous fraction of *A. vasica* suppressed the AA metabolites, but not the butanolic fraction (LP1 and 12-HETE). However, platelet aggregation tests revealed that butanolic extract of *A. vasica* inhibited AA, PAF, and collagen-induced aggregation but not ADP (Ahmed et al., 2013).

16.5.2.5 Antitussive Effect

In contrast to the control group, the impact of ethanol extracts *A. vasica* on SO_2 gas-produced cough in experimental animals has highly significant effects at the level of P0.01 in suppressing the cough reflex at doses of 800 mg/kg and 200 mg/kg b.w. p.o. Within 60 minutes of the trial, mice treated with *Glycyrrhiza glabra* exhibited a 35.62% reduction in cough and 43.02% reduction in cough when treated with Adhatodaasica. The extract's antitussive efficacy was similar to that of codeine sulfate (10, 15, 20 mg/kg b.w.), a common antitussive. At doses of 10 mg/kg, 15 mg/kg, and 20 mg/kg, codeine sulfate caused 24.80%, 32.98%, and 45.73% inhibition in cough, respectively, while codeine sulfate (20 mg/kg) generated maximal 45.73% (P0.001) inhibition at 60 minutes of the trial. (Jahan and Siddiqui, 2012).

16.5.2.6 Hepatoprotective Effect

In Swiss albino rats, the hepatoprotective effect of ethyl acetate extract of *A. vasica* was studied against CCl4-induced liver injury. CCl4 caused liver injury in rats when given at a dosage of 1ml/kg, as shown by statistically significant increases in blood alanine aminotransferase (ALT), aspartate aminotransferase (AST), alkaline phosphatase (ALP), and bilirubin. The three blood level enzymes, as well as bilirubin, were statistically reduced when rats were given the ethyl acetate extract of *A. vasica* (100mg/kg and 200mg/kg) prior to the CCl4 dosage of 1ml/kg. Histopathological findings corroborated the aforementioned findings, although the 200mg/kg dosage was shown to be more active. According to current findings, *A. vasica* ethyl acetate extract exhibits a strong hepatoprotective effect against CCl4-induced liver injury. (Ahmad et al., 2013).

16.5.2.7 Antiviral Effect

Influenza viruses are important etiologic agents of human respiratory diseases, and they cause significant health and economic harm. The current research investigated the antiviral activity of *A. vasica* crude extracts against influenza virus *in vitro* by reducing hemagglutination (HA) in two distinct layouts of simultaneous and post-treatment assays. Antiviral activity in the noncytotoxic range was tested using aqueous and methanolic extracts. At a dosage of 10 mg/ml, the methanolic extract reduced HA by 100% in both simultaneous and posttreatment tests. In a simultaneous test, aqueous extracts at doses of 10 mg/ml and 5 mg/ml decreased HA to 33% and 16.67%, respectively. These findings indicate that extracts have potent anti-influenza virus action, inhibiting viral attachment and/or replication, and therefore may be utilized as a viral prophylactic. (Chavan and Chowdhary, 2014).

16.5.2.8 Thrombolytic Effect

The extractives of *A. vasica* were tested for thrombolytic action as part of the discovery of cardio-protective medicines from natural sources, and the findings are given in Table 16.1. The addition of 100l SK, a positive control (30,000 I.U.) to the clots, and subsequent incubation at 37°C for 90 minutes resulted in 80.65% clot lysis. At the same time, distilled water was used as a negative control, resulting in little clot lysis (4.08%). The methanolic fraction (MF) had the greatest thrombolytic activity in this research (53.23%) (Shahriar, 2013).

16.5.2.9 Uterine Effect

The uterotonic action of vasicine, a quinazoline alkaloid found in *A. vasica*, was investigated in-depth both *in vitro* and *in vivo*, using uteri from various species of animals with diverse hormonal effects. Similar to oxytocin and methyl ergometrine, uterotonic action was observed. Under the priming impact of estrogens, the abortifacient action of vasicine, as well as its uterotonic effect, was more pronounced. Vasicine-induced abortion in rats, guinea pigs, hamsters, and rabbits was investigated. According to the findings, vasicine works by releasing PGs. *In vitro* experiments showed that synthesized vasicine and vasicinone derivatives have oxytocic action at doses greater than 1 mg/ml (Rao et al., 1983).

16.5.2.10 Antifungal Effect

Plants have been shown to be potential sources of novel and physiologically active natural compounds with great therapeutic efficacy. Natural goods and active plant extracts have been more popular in recent years, and new medicines are being found thanks to technical advances. The current study investigates the phytochemical components of *A. vasica* and their human pathogenic fungal effective agent. The phytochemical extract's lowest inhibitory activity was determined. The extract will be further investigated for partial characterization by thin liquid chromatography (TLC) as well as antifungal activity as assessed by agar disc diffusion and germ tube formation inhibition activities. The goal of this research on the impact of *A. vasica* on pathogenic fungi *A. ruber* and *T. rubrum* was to determine the antifungal activity of *A. vasica* (Ramachandran and Sankaranarayanan, 2013).

16.5.2.11 Anthelmintic Effect

The primary goal of this research was to see whether *A. vasica* (Acanthaceae) has any anthelmintic action *in vitro* against sheep gastrointestinal nematodes. Egg-hatching and larval development tests were used to assess the aqueous and ethanolic extracts of *A. vasica* aerial parts. The aqueous and ethanolic extracts had ovicidal and larvicidal (P0.05) action against gastrointestinal nematodes at doses of 25–50 mg/ml. The plant extracts inhibited the cells in a dose-dependent manner (P0.05). Inhibiting egg-hatching and larval development of gastrointestinal nematodes were more successful with the ethanolic extract at a dosage of 50.0 mg/ml. The effective dosage (ED50) of aqueous and ethanolic extracts was visually calculated using a linear regression equation with a probit scale, y = 5. *A. vasica* extracts may be helpful in the management of gastrointestinal nematodes in sheep, according to the findings of this research (Al-Shaibani et al., 2008).

16.5.2.12 Antidiabetic Effect

Diabetes-related cognitive decline (DACD), which includes oxidative nitrosative stress, inflammation, and cholinergic dysfunction, is referred to as "diabetic encephalopathy". The goal of this research was to see how *A. vasica*, a well-known anti-inflammatory, antioxidant, anticholinesterase, and antihyperglycemic herb, affected diabetic encephalopathy. *A. vasica* leaves ethanolic extract (AVEE) was given to diabetic Wistar rats caused by streptozotocin (STZ) for six weeks at doses of 100, 200, and 400 mg/kg/day. Learning and memory were examined in a single Y-maze and passive avoidance test during the fifth week of therapy. Biochemical markers, such as acetylcholinesterase

(AchE) activity, nitrite levels, tumor necrosis factor-alpha (TNF-α), and oxidative stress were evaluated in the cerebral cortex and hippocampal areas of the brain after the research. In the cerebral cortex of diabetic rats, AchE activity was found to be 70% higher. The levels of lipid peroxidation (LPO) in the cerebral cortex and hippocampus of diabetic rats were raised by 100% and 94%, respectively. Nonprotein thiol levels, as well as superoxide dismutase and catalase enzymatic activity, were shown to be lower in the cerebral cortex and hippocampus regions of diabetic rats. Nitrite levels rose by 170% and 137%, respectively, in both diabetic brain areas. In diabetic rats, TNF-α, a pro-inflammatory cytokine, was shown to be substantially higher. Animals given AVEE, on the other hand, showed a substantial reduction in these behavioral and biochemical abnormalities. The findings point to *A. vasica* Nees having a preventive function against diabetic encephalopathy, which may be due to its antioxidant, anticholinesterase, anti-inflammatory, and glucose-lowering properties (Patil et al., 2014).

16.5.2.13 Antituberculosis Effect

Phytochemical analysis was used to validate the extraction and detection of alkaloids. In the leaf of *A. vasica*, six distinct quinazoline alkaloids (vasicoline, vasicolinone, vasicinone, vasicine, triterpenes, and anisotine) were discovered (*A. vasica*). The existence of the HPLC peaks revealed the leaf's varied type of alkaloid. The enzyme -ketoacyl-acyl carrier protein synthase III, which catalyzes the first stage of fatty acid biosynthesis (FabH) through a type II fatty acid synthase, has a distinctive structural profile and is found in all Mycobacterium TB strains (M. tuberculosis). As a result, it was chosen as a target for the development of antituberculosis drugs. Docking simulations were performed on the aforementioned alkaloids generated from *A. vasica*. The combination of docking and scoring revealed new information on the binding and activity of many inhibitors. These findings will aid in the development of *M. tuberculosis* inhibitors as well as provide a solid foundation for natural plant-based medicinal chemistry (Kumar et al., 2012).

16.5.2.14 Anticestodal Effect

Studies confirm that the EPG count was decreased by 79.57%, and the percentage of worm recovery rate was reduced by 16.60%, indicating that an 800 mg/kg double dosage of *A. vasica* extract had significant effectiveness against adult worms. These results were superior to the usual therapy of 5 mg/kg praziquantel in a single dosage. The extract exhibited a substantial decrease in worm recovery rate (from 100% in control to 20.00% at 800 mg/kg dosage of extract) when used against immature worms. (Yadav and Tangpu, 2008).

16.5.2.15 Hepato-Protective Effect

In an evaluated study on traditional medicine, the Indian Council of Medical Research, New Delhi, identified liver disease to be one of the frequent thrust areas. *A. vasica* has been shown to protect rats from CCl4-induced liver impairment to different degrees. The goal of this study was to see whether *A. vasica* whole plant powder might help prevent Wister rats from liver damage caused by CCl4. The hepatoprotective effect was assessed using blood and tissue biochemical indicators of the liver: GOT, GPT, alkaline phosphate, glucose, bilirubin, triglycerides, GT, cholesterol, DNA, RNA, total protein. *A. vasica* whole plant powder was compared to Silymarin in a conventional procedure and shown to have a better hepatoprotective effect, suggesting that *A. vasica* may have a promising impact against liver diseases. As a result, it may be used as a liver tonic in humans (Shirish and Pingale, 2009).

16.5.2.16 Radio-Modulatory Effect

At different post-irradiation intervals between 6 h and 30 days, the effects of *A. vasica* Nees leaf extract on radiation-induced hematological changes in Swiss albino mice peripheral blood were investigated. Animals pretreated with *A. vasica* leaf extract, on the other hand, survived for 81.25% of the time until 30 days following exposure, and hematological parameters gradually improved. Even after 30 days, however, these hematological levels remained considerably below normal. In control mice, there was a substantial reduction in blood glutathione (GSH) concentration and an increase in lipid peroxidation (LPO) (Radiation alone). Pretreated irradiation mice with *A. vasica* leaf extract had significantly higher GSH levels and lower LPO levels. During the whole research period, *A. vasica* leaf extract pretreatment irradiated animals had a substantial increase in blood alkaline phosphatase activity and a significant reduction in acid phosphatase activity. (Wasserman and Kuo, 1991).

16.5.2.17 Immunomodulatory Effect

Leaves of the Indian medicinal plant *A. vasica* Linn were extracted using methanol, chloroform, and diethyl ether. In experimental animals, they were pharmacologically verified for their immunomodulatory effects. The percentage of neutrophil adherence to nylon fibers increased substantially after oral administration of extracts at a dosage of 400 mg/kg to mature male Wister rats (P 0.001). In comparison to the control groups, the observed outcomes at various dosages were significant. These results indicated that *A. vasica* Linn extracts favorably regulate the host's immunity (Vinothapooshan and Sundar, 2011).

16.5.2.18 Effect on Reproductive Organs

An extract of *A. vasica* leaf spissum was tested in rats for its potential abortive impact. Vasicine (0.85 0.03%) was the main alkaloid found in the extract. Between the first and ninth days of pregnancy, five pregnant females were given the extract (325 mg/kg/day) via a stomach cannula. Between the first and ninth days of pregnancy, 9 pregnant women were given 0.25 and 2.5% *A. vasica* in their water. The administration of *A. vasica* did not result in abortion in any of the treatment groups, according to the findings (Burgos et al., 1997).

16.6 CONCLUSION

V. rosea is high in phytochemicals such as alkaloids and phenolics, which provide a wide variety of biological activities, notably as anticancer, antidiabetic, antioxidant, antibacterial, and antihypertensive abilities. Several alkaloids and phenolics have also been found in this substance, but many others still have to be discovered. As a result, work on identifying and isolating novel phytoconstituents isolated from *V. rosea* should proceed. Furthermore, prospective possibilities of phytochemical compounds generated from this product in the cosmeceutical, nutraceutical, and pharmaceutical sectors must be explored thoroughly. The phytochemical and pharmacological properties of *A. vasica* have been extensively studied in the literature. Currently, *A. vasica* belongs to a class of herbal medicines having a solid conceptual or traditional foundation. It's a good source of vasicine, vasicinone, and vasicolone, as well as a few other alkaloids. Antibacterial, antifungal, hepatoprotective, antitussive, radio-modulation, anti-inflammatory and anti-ulcer, abortifacient, antiviral, thrombolytic, antimutagenic, cardiovascular protection, hypoglycemic, antitubercular, antioxidant characteristics are all attributed to these components.

ACKNOWLEDGMENT

We would like to express our sincere gratitude to the Department of Pharmacy, Life Science Faculty, Bangabandhu Sheikh Mujibur Rahman Science and Technology University, Gopalganj (Dhaka)-8100, and International Centre for Empirical Research and Development, Bangladesh.

REFERENCES

Adnan, M., Hussain, J., Tahir Shah, M., Shinwari, Z.K., Ullah, F., Bahader, A., Khan, N., Latif Khan, A. and Watanabe, T., 2010. Proximate and nutrient composition of medicinal plants of humid and sub-humid regions in North-west Pakistan. *Journal of Medicinal Plants Research*, *4*(4), pp.339–345.

Agarwal, V.S., 1986. *Economic Plants of India.* Kailash Prakashan, Calcutta, p. 8.

Agarwal, S., Jacob, S., Chettri, N., Bisoyi, S., Tazeen, A., Vedamurthy, A.B., Krishna, V. and Hoskeri, H.J., 2011. Evaluation of in-vitro anthelminthic activity of Catharanthus roseus extract. *International Journal of Pharmaceutical Sciences and Drug Research, 3*(3), pp.211–3.

Ahmad S, Madhukar G, Maksood A, Mhaveer S, Tanwir AM, Husain AS. 2009, A Phytochemical Overview On AadhatodaZyelanicaMedica A Vasica(Linn.) Nees. *Natural Product Radiance.* 8(5), pp.549–554.

Ahmad, R., Raja, V. and Sharma, M., 2013. Hepatoprotective activity of ethyl acetate extract of Adhatoda vasicain Swiss albino rats. *International Journal of Current Research and Review, 5*(6), p.16.

Ahmed, M.F., Kazim, S.M., Ghori, S.S., Mehjabeen, S.S., Ahmed, S.R., Ali, S.M. and Ibrahim, M., 2010. Antidiabetic activity of Vinca rosea extracts in alloxan-induced diabetic rats. *International Journal of Endocrinology, 2010*, p. 841090.

Ahmed, S., Gul, S., Gul, H. and Bangash, M.H., 2013. Dual inhibitory activities of Adhatoda vasica against cyclooxygenase and lipoxygenase. *International Journal of Endorsing Health Science, 1*, pp.14–17.

Alba Bhutkar, M.A. and Bhise, S.B., 2011. Comparative Studies on Antioxidant Properties of Catharanthus Rosea and Catharanthus. *International Journal of Pharmaceutical Techniques, 3*(3), pp.1551–1556.

Almagro, L., Fernández-Pérez, F. and Pedreño, M.A., 2015. Indole alkaloids from Catharanthus roseus: bio-production and their effect on human health. *Molecules, 20*(2), pp.2973–3000.

Alqahtani, A., Hamid, K., Kam, A., Wong, K.H., Abdelhak, Z., Razmovski-Naumovski, V., Chan, K., Li, K.M., Groundwater, P.W. and Li, G.Q., 2013. The pentacyclic triterpenoids in herbal medicines and their pharmacological activities in diabetes and diabetic complications. *Current Medicinal Chemistry, 20*(7), pp.908–931.

Al-Shaibani, I.R.M., Phulan, M.S., Arijo, A. and Qureshi, T.A., 2008. Ovicidal and larvicidal properties of Adhatoda vasica (L.) extracts against gastrointestinal nematodes of sheep in vitro. *Pakistan Veterinary Journal, 28*(2), pp.79–83.

Amirjani, M.R., 2013. Effects of drought stress on the alkaloid contents and growth parameters of Catharanthus roseus. *Journal of Agricultural and Biological Science, 8*(11), pp.745–750.

Ara, N., Rashid, M. and Amran, S., 2009. Comparison of hypotensive and hypolipidemic effects of Catharanthus roseus leaves extract with atenolol on adrenaline induced hypertensive rats. *Pakistan Journal of Pharmaceutical Sciences, 22*(3).

Aslam, J., Khan, S.H., Siddiqui, Z.H., Fatima, Z., Maqsood, M., Bhat, M.A., Nasim, S.A., Ilah, A., Ahmad, I.Z., Khan, S.A. and Mujib, A., 2010. Catharanthus roseus (L.) G. Don. An important drug: it's applications and production. *Pharmacie Globale (IJCP), 4*(12), pp.1–16.

Atal, C.K., 1980. *Chemistry and Pharmacology of Vasicine: A New Oxytocic and Abortifacient.* Regional Research Laboratory.

Atta-Ur-Rahman, Said, H.M., Ahmad, V.U., 1986. *Pakistan Encyclopaedia Planta Medica*, vol. 1. Hamdard Foundation Press, Karachi, pp.181–187.

Atta-Ur-Rahman, Sultana, N., Akhter, F., Nighat, F. and Choudhary, M.I., 1997. Phytochemical studies on Adhatoda vasica Nees. *Natural Product Letters, 10*(4), pp.249–256.

Avula, B., Begum, S., Ahmed, S., Choudhary, M.I. and Khan, I.A., 2008. Quantitative determination of vasicine and vasicinone in Adhatoda vasica by high performance capillary electrophoresis. *Die Pharmazie-An International Journal of Pharmaceutical Sciences, 63*(1), pp.20–22.

Babalola, I.T., Shode, F.O., Adelakun, E.A., ROpoku, A. and Mosa, R.A., 2013. Platelet-aggregation inhibitory activity of oleanolic acid, ursolic acid, betulinic acid, and maslinic acid. *Journal of Pharmacognosy and Phytochemistry, 1*(6).

Babulova A, Machova J, and Nosalova V, 2003. Protective action of vinpocetine against experimentally induced gastric damage in rats. *Arzneimittel forschung*, 43, pp.981–985.

Bajaj, M. and Williams, J.T., 1995. *Healing forests, healing people: report of a Workshop on Medicinal Plants held on 6–8 Feb., 1995, Calicut, India.* IDRC, New Delhi, India.

Barrales, C.H.J., De la Rosa, M.C.R. and Villegas, O.S., 2012. Hacia una genética celular del cáncer. *Revista la Ciencia y el Hombre, 25*(2), pp.1–6.

Barth, S.W., Faehndrich, C., Bub, A., Watzl, B., Will, F., Dietrich, H., Rechkemmer, G. and Briviba, K., 2007. Cloudy apple juice is more effective than apple polyphenols and an apple juice derived cloud fraction in a rat model of colon carcinogenesis. *Journal of Agricultural and Food Chemistry, 55*(4), pp.1181–1187.

Bhartiya, H.P. and Gupta, P.C., 1982. A chalcone glycoside from the flowers of Adhatoda vasica. *Phytochemistry, 21*(1), p.247.

Bhuyan, D.J., Vuong, Q.V., Bond, D.R., Chalmers, A.C., Bowyer, M.C. and Scarlett, C.J., 2018. Eucalyptus microcorys leaf extract derived HPLC-fraction reduces the viability of MIA PaCa-2 cells by inducing apoptosis and arresting cell cycle. *Biomedicine & Pharmacotherapy, 105*, pp.449–460.

Burgos, R., Forcelledo, M., Wagner, H., Müller, A., Hancke, J., Wikman, G. and Croxatto, H., 1997. Non-abortive effect of Adhatoda vasica spissum leaf extract by oral administration in rats. *Phytomedicine*, *4*(2), pp.145–149.

Cantrell, C.L., Franzblau, S.G. and Fischer, N.H., 2001. Antimycobacterial plant terpenoids. *Planta medica*, *67*(08), pp.685–694.

Chattopadhyay, S.P. and Das, P.K., 1990. Evaluation of Vinca rosea for the treatment of warts. *Indian Journal of Dermatology Venereology and Leprology*, *56*(2), pp.107–108.

Chattopadhyay, R.R., Sarkar, S.K., Ganguly, S., Banerjee, R.N. and Basu, T.K., 1991. Hypoglycemic and anti-hyperglycemic effect of leaves of Vinca rosea linn. *Indian Journal of Physiology and Pharmacology*, *35*(3), pp.145–151.

Chattopadhyay, R.R., 1999. A comparative evaluation of some blood sugar lowering agents of plant origin. *Journal of Ethnopharmacology*, *67*(3), pp.367–372.

Chavan, R. and Chowdhary, A., 2014. In vitro inhibitory activity of Justicia adhatoda extracts against influenza virus infection and hemagglutination. *International Journal of Pharmaceutical Sciences Review and Research*, *25*(02), pp.231–236.

Chigora, P., Masocha, R. and Mutenheri, F., 2007. The role of indigenous medicinal knowledge (IMK) in the treatment of ailments in rural Zimbabwe: the case of Mutirikwi communal lands. *Journal of Sustainable Development in Africa*, *9*(2), pp.26–43.

Chowdhury, B.K. and Bhattacharyya, P., 1987. Vasicine and related-compounds. 4. Adhavasinone: A New Quinazolone Alkaloid from Adhatoda-Vasica Nees. *Chemistry & Industry*, 1, pp.35–36.

Claeson, U.P., Malmfors, T., Wikman, G. and Bruhn, J.G., 2000. Adhatoda vasica: a critical review of ethnopharmacological and toxicological data. *Journal of Ethnopharmacology*, *72*(1–2), pp.1–20.

Cordall, G.A., Weiss, S.G. and Farnsworth, N.R., 1974. Structure elucidation and chemistry of Catharanthus alkaloids. XXX. Isolation and structure elucidation of vincarodine. *The Journal of Organic Chemistry*, *39*(4), pp.431–434.

Cragg, G.M. and Newman, D.J., 2003. Plants as a source of anti-cancer and anti-HIV agents. *Annals of Applied Biology*, *143*(2), pp.127–133.

Cuellar, A. and Lorincz, C., 1975. Catharanthus roseus (L.) G. Don que crece en Cuba. Aislamiento y caracterizaci6n de vinblastina y leurosina, dos alcaloides con propiedades citostáticas. *Rev Cubana Farm*, *9*(3), pp.183–99.

Daneshtalab, M., 2008. Discovery of chlorogenic acid-based peptidomimetics as a novel class of antifungals. A success story in rational drug design. *Journal of Pharmacy & Pharmaceutical Sciences*, *11*(2), pp.44s–55s.

Deshmukhe, P.V., Hooli, A.A. and Holihosur, S.N., 2010. Bioinsecticidal potential of Vinca rosea against the tobacco caterpillar, Spodoptera litura Fabricius (Lepidoptera: Noctuidae). *Recent Research in Science and Technology*, *2*(2), pp.1–5.

Dymock, W., 1972. India Pharmacographia of Plants. Hamdard National Foundation Pak, *3*, pp.343–344.

Dymock, W., Warden, C.J.H. and Hooper, D., 1890. *Pharmacographia Indica: A History of the Principal Drugs of Vegetable Origin*. Trubner & Company, ld, K. Paul, Trench.

Elisabetsky, E. and Costa-Campos, L., 2006. The alkaloid alstonine: a review of its pharmacological properties. *Evidence-based Complementary and Alternative Medicine*, *3*(1), pp.39–48.

El–Merzabani, M.M., El–Aaser, A.A., Attia, M.A., El–Duweini, A.K. and Ghazal, A.M., 1979. Screening system for egyptian plants with potential anti–tumour activity. *Planta Medica*, *36*(06), pp.150–155.

El-Sayed, A. and Cordell, G.A., 1981. Catharanthus alkaloids. XXXIV. Catharanthamine, a new antitumor bisindole alkaloid from Catharanthus roseus. *Journal of Natural Products*, *44*(3), pp.289–293.

El-Sawi, A.S., El-Megeed Hashem, A.F. and Ali, A.M., 1999. Flavonoids and antimicrobial volatiles from Adhatoda vasica Nees. *Pharmaceutical and Pharmacological Letters*, *9*(2), pp.52–56.

Fernandes, L., Van Rensburg, C.E.J., Hoosen, A.A. and Steenkamp, V., 2008. In vitro activity of medicinal plants of the Venda region, South Africa, against Trichomonas vaginalis. *Southern African Journal of Epidemiology and Infection*, *23*(2), pp.26–28.

Fernández-Pérez, F., Almagro, L., Pedreño, M.A. and Gomez Ros, L.V., 2013. Synergistic and cytotoxic action of indole alkaloids produced from elicited cell cultures of Catharanthus roseus. *Pharmaceutical Biology*, *51*(3), pp.304–310.

Ferreres, F., Pereira, D.M., Valentão, P., Andrade, P.B., Seabra, R.M. and Sottomayor, M., 2008. New phenolic compounds and antioxidant potential of Catharanthus roseus. *Journal of Agricultural and Food Chemistry*, *56*(21), pp.9967–9974.

Fischer, D.C., Kogan, M. and Paxton, J., 1990. Deterrency of Mexican bean beetle (Coleoptera: Coccinellidae) feeding by free phenolic acids. *Journal of Entomological Science*, *25*(2), pp.230–238.

Ghosh, R.K. and Gupta, I., 1980. Effect of Vinca rosea and Ficus racemosus on hyperglycaemia in rats. *Indian Journal of Animal Health*, 19(2), pp.145–149.

Ghosh, S. and Suryawanshi, S.A., 2001. Effect of Vinca rosea extracts in treatment of alloxan diabetes in male albino rats. *Indian Journal of Experimental Biology*, 39(8), pp.748–759.

Gulfraz, M., Arshad, M., Nayyer, N., Kanwal, N. and Nisar, U., 2004. Investigation for Bioactive Compounds of Berberis Lyceum Royle and Justicia Adhatoda L. *Ethnobotanical Leaflets, 2004*(1), p.5.

Gupta, O.P., Anand, K.K., Ghatak, B.J. and Atal, C.K., 1978. Vasicine, alkaloid of Adhatoda vasica, a promising uterotonic abortifacient. *Indian Journal of Experimental Biology, 16*(10), pp.1075–1077.

Hassan, K.A., Brenda, A.T., Patrick, V. and Patrick, O.E., 2011. In vivo antidiarrheal activity of the ethanolic leaf extract of Catharanthus roseus Linn.(Apocyanaceae) in Wistar rats. *African Journal of Pharmacy and Pharmacology, 5*(15), pp.1797–1800.

Hedhili, S., Courdavault, V., Giglioli-Guivarc'h, N. and Gantet, P., 2007. Regulation of the terpene moiety biosynthesis of Catharanthus roseus terpene indole alkaloids. *Phytochemistry Reviews, 6*(2–3), pp.341–351.

Holdsworth, D.K., 1990. Traditional medicinal plants of Rarotonga, Cook Islands part I. *International Journal of Crude Drug Research, 28*(3), pp.209–218.

Horiuchi, K., Shiota, S., Hatano, T., Yoshida, T., Kuroda, T. and Tsuchiya, T., 2007. Antimicrobial activity of oleanolic acid from Salvia officinalis and related compounds on vancomycin-resistant enterococci (VRE). *Biological and Pharmaceutical Bulletin, 30*(6), pp.1147–1149.

Huang, L., Li, J., Ye, H., Li, C., Wang, H., Liu, B. and Zhang, Y., 2012. Molecular characterization of the pentacyclic triterpenoid biosynthetic pathway in Catharanthus roseus. *Planta, 236*(5), pp.1571–1581.

Huang, Y., Nikolic, D., Pendland, S., Doyle, B.J., Locklear, T.D. and Mahady, G.B., 2009. Effects of cranberry extracts and ursolic acid derivatives on P-fimbriated Escherichia coli, COX-2 activity, pro-inflammatory cytokine release and the NF-κβ transcriptional response in vitro. *Pharmaceutical Biology, 47*(1), pp.18–25.

Huq, M.E., Ikram, M. and Warsi, S.A., 1967. Chemical composition of Adhatoda vasica Linn. II. *Pakistan Journal of Scientific and Industrial Research, 10*, pp.224–225.

Iba, Y., Shibata, A., Kato, M. and Masukawa, T., 2004. Possible involvement of mast cells in collagen remodeling in the late phase of cutaneous wound healing in mice. *International Immunopharmacology, 4*(14), pp.1873–1880.

Ikram, M., Fazal Hussain, S., 1978. *Compendium of Medicinal Plants.* Pakistan Council of Scientific and Industrial Research, Peshawar, p. 1.

Iyengar, M.A., Jambaiah, K.M., Kamath, M.S. and Rao, M.N.A., 1994. Studies on an antiastham kada: A proprietary herbal combination. Part I: Clinical study. *Indian Drugs, 31*(5), pp.183–186.

Jabeen, S., Shah, M.T., Khan, S. and Hayat, M.Q., 2010. Determination of major and trace elements in ten important folk therapeutic plants of Haripur basin, Pakistan. *Journal of Medicinal Plants Research, 4*(7), pp.559–566.

Jacobs, D.I., Snoeijer, W., Hallard, D. and Verpoorte, R., 2004. The Catharanthus alkaloids: pharmacognosy and biotechnology. *Current Medicinal Chemistry, 11*(5), pp.607–628.

Jahan, Y. and Siddiqui, H.H., 2012. Study of antitussive potential of Glycyrrhiza glabra and Adhatoda vasica using a cough model induced by sulphur dioxide gas in mice. *International Journal of Pharmaceutical Sciences and Research, 3*(6), p.1668.

Jain, M.P. and Sharma, V.K., 1982. Phytochemical investigation of roots of Adhatoda vasica. *Planta Medica, 46*(12), pp.250–250.

Jain, M.P., Koul, S.K., Dhar, K.L. and Atal, C.K., 1980. Novel nor-harmal alkaloid from Adhatoda vasica. *Phytochemistry, 19*(8), pp.1880–1882.

Jaleel, C.A., 2009. Changes in non enzymatic antioxidants and ajmalicine production in Catharanthus roseus Twith different soil salinity regimes. *Molecules*, 2, p.2.

Jayakumar, D., Mary, S.J. and Santhi, R.J., 2010. Evaluation of antioxidant potential and antibacterial activity of Calotropis gigantea and Vinca rosea using in vitro model. *Indian journal of Science and Technology, 3*(7), pp.720–723.

Jayanthi, M., Sowbala, N., Rajalakshmi, G., Kanagavalli, U. and Sivakumar, V., 2010. Study of anti hyperglycemic effect of Catharanthus roseus in alloxan induced diabetic rats. *International Journal of Pharmacy and Pharmaceutical Sciences*, 2(4), pp.114–116.

Jayaweera, M.A., 1981. *Medicinal Plants (Indigenous and Exotic) Used in Ceylon, Part I.* National Science Council, Colombo, pp.4–5.

Jha, D.K., Panda, L., Lavanya, P., Ramaiah, S. and Anbarasu, A., 2012. Detection and confirmation of alkaloids in leaves of Justicia adhatoda and bioinformatics approach to elicit its anti-tuberculosis activity. *Applied Biochemistry and Biotechnology, 168*(5), pp.980–990.

Jiang, H., Shang, X., Wu, H., Huang, G., Wang, Y., Al-Holou, S., Gautam, S.C. and Chopp, M., 2010. Combination treatment with resveratrol and sulforaphane induces apoptosis in human U251 glioma cells. *Neurochemical Research*, *35*(1), pp.152–161.

Johne, S., Gröger, D. and Hesse, M., 1971. Neue alkaloide aus Adhatoda vasica Nees. *Helvetica Chimica Acta*, *54*(3), pp.826–834.

Johnson, I.S., Armstrong, J.G., Gorman, M. and Burnett, J.P., 1963. The vinca alkaloids: a new class of oncolytic agents. *Cancer Research*, *23*(8 Part 1), pp.1390–1427.

Kanokmedhakul, K., Kanokmedhakul, S. and Phatchana, R., 2005. Biological activity of Anthraquinones and Triterpenoids from Prismatomeris fragrans. *Journal of Ethnopharmacology*, *100*(3), pp.284–288.

Kapoor, L.D., 1990. *Handbook of Ayurvedic Medicinal Plants.* CRC Press, Boca Raton, FL, p. 216.

Karthikeyan, A., Shanthi, V. and Nagasathaya, A., 2009. Preliminary phytochemical and antibacterial screening of crude extract of the leaf of Adhatoda vasica. L. *International Journal of Green Pharmacy (IJGP)*, *3*(1).

Kashiwada, Y., Nagao, T., Hashimoto, A., Ikeshiro, Y., Okabe, H., Cosentino, L.M. and Lee, K.H., 2000. Anti-AIDS agents 38. Anti-HIV activity of 3-O-acyl ursolic acid derivatives. *Journal of Natural Products*, *63*(12), pp.1619–1622.

Kevin, L.Y.W., Hussin, A.H., Zhari, I. and Chin, J.H., 2012. Sub–acute oral toxicity study of methanol leaves extract of Catharanthus roseus in rats. *Journal of Acute Disease*, *1*(1), pp.38–41.

Khan, M.H. and Yadava, P.S., 2010. *Antidiabetic Plants used in Thoubal district of Manipur, Northeast India.*

Khory, N.R., Katrak, N.N., 1981. *Materia Medica of India and Their Therapeutics.* Neeraj Publishing House, Delhi, p. 464.

Kim, K.H., Seo, H.S., Choi, H.S., Choi, I., Shin, Y.C. and Ko, S.G., 2011. Induction of apoptotic cell death by ursolic acid through mitochondrial death pathway and extrinsic death receptor pathway in MDA-MB-231 cells. *Archives of Pharmacal Research*, *34*(8), pp.1363–1372.

Kim, S.G., Kim, M.J., Jin, D.C., Park, S.N., Cho, E.G., Freire, M.O., Jang, S.J., Park, Y.J. and Kook, J.K., 2012. Antimicrobial effect of ursolic acid and oleanolic acid against methicillin-resistant Staphylococcus aureus. *Korean Journal of Microbiology*, *48*(3), pp.212–215.

Kirtikar, K.R. and Basu, B.D., 1918. Indian medicinal plants. *Indian Medicinal Plants.*

Kirtikar, K.R., Basu, B.D., An, L.C.S., 1975. *Indian Medicinal Plants*, vol. 3, second ed. Bishen Singh Mahendra Pal Singh, Delhi, pp.1899–1902.

Kiss, B. and Karpati, E., 1996. Mechanism of action of vinpocetine. *Acta Pharmaceutica Hungarica*, *66*(5), pp.213–224.

Kotakadi, V.S., Rao, Y.S., Gaddam, S.A., Prasad, T.N.V.K.V., Reddy, A.V. and Gopal, D.S., 2013. Simple and rapid biosynthesis of stable silver nanoparticles using dried leaves of Catharanthus roseus. Linn. G. Donn and its anti microbial activity. *Colloids and Surfaces B: Biointerfaces*, *105*, pp.194–198.

Kumar, A., Singhal, K.C., Sharma, R.A., Vyas, G.K. and Kumar, V., 2012. Analysis of antioxidant activity of Catharanthus roseus (L.) and its association with habitat temperature. *Asian Journal of Experimental Biological Sciences*, *3*(4), pp.706–713.

Kumar, S., Singh, A., Kumar, B., Singh, B., Bahadur, L. and Lal, M., 2018. Simultaneous quantitative determination of bioactive terpene indole alkaloids in ethanolic extracts of Catharanthus roseus (L.) G. Don by ultra high performance liquid chromatography–tandem mass spectrometry. *Journal of Pharmaceutical and Biomedical Analysis*, *151*, pp.32–41.

Kumari, K. and Gupta, S., 2013. Phytopotential of Catharanthus roseus L.(G.) Don. var."Rosea" and "Alba" against various pathogenic microbes in vitro. *International Journal of Research in Pure and Applied Microbiology*, *3*(3), pp.77–82.

Kurek, A., Nadkowska, P., Pliszka, S. and Wolska, K.I., 2012. Modulation of antibiotic resistance in bacterial pathogens by oleanolic acid and ursolic acid. *Phytomedicine*, *19*(6), pp.515–519.

Loh, K.Y., 2008. Know the medicinal herb: Catharanthus roseus (Vinca rosea). *Malaysian family physician: the official journal of the Academy of Family Physicians of Malaysia*, *3*(2), p.123.

Malla B.S., Rajbhandari B.S., Shrestha B.T., Adhikari M.P., Adhikari R.S., 1982. Wild edible plants of Nepal. Department of Medicinal Plants, Nepal, Bulletin No. 9. Kathmandu, pp.3–4.

Marles, R.J. and Farnsworth, N.R., 1995. Antidiabetic plants and their active constituents. *Phytomedicine*, *2*(2), pp.137–189.

Mishra, P., Uniyal, G.C., Sharma, S. and Kumar, S., 2001. Pattern of diversity for morphological and alkaloid yield related traits among the periwinkle Catharanthus roseus accessions collected from in and around Indian subcontinent. *Genetic Resources and Crop Evolution*, *48*(3), pp.273–286.

Mondal, A.K. and Mandal, S., 2009. The free amino acids of pollen of some angiospermic taxa as taxonomic markers for phylogenetic interrelationships. *Current Science*, pp.1071–1081.

Moreno, P.R., van der Heijden, R. and Verpoorte, R., 1994. Elicitor-mediated induction of isochorismate synthase and accumulation of 2, 3-dihydroxy benzoic acid in Catharanthus roseus cell suspension and shoot cultures. *Plant Cell Reports*, *14*(2), pp.188–191.

Moreno-Valenzuela, O.A., Galaz-Avalos, R.M., Minero-García, Y. and Loyola-Vargas, V.M., 1998. Effect of differentiation on the regulation of indole alkaloid production in Catharanthus roseus hairy roots. *Plant Cell Reports*, *18*(1), pp.99–104.

Moudi, M., Go, R., Yien, C.Y.S. and Nazre, M., 2013. Vinca alkaloids. *International Journal of Preventive Medicine*, *4*(11), p.1231.

Mu, F., Yang, L., Wang, W., Luo, M., Fu, Y., Guo, X. and Zu, Y., 2012. Negative-pressure cavitation extraction of four main vinca alkaloids from Catharanthus roseus leaves. *Molecules*, *17*(8), pp.8742–8752.

Mueller, A., Antus, S., Bittinger, M., Dorsch, W., Kaas, A., Kreher, B., Neszmelyi, A., Stuppner, H. and Wagner, H., 1993. Chemistry and Pharmacology of the Antiasthmatic Plants Galphimia glauca, Adhatoda vasica, and Picrorhiza kurrooa. *Planta Medica*, *59*(S 1), pp.A586–A587.

Muhammad, L.R., Muhammad, N., Tanveer, A. and Baqir, S.N., 2009. Antimicrobial activity of different extracts of Catharanthas roseus. *Clinical and Experimental Medical Journal*, *3*, pp.81–85.

Muthu, C., Ayyanar, M., Raja, N. and Ignacimuthu, S., 2006. Medicinal plants used by traditional healers in Kancheepuram District of Tamil Nadu, India. *Journal of Ethnobiology and Ethnomedicine*, *2*(1), pp.1–10.

Nadkarni, K.M., 1976. *Indian Materia Medica, With Ayurvedic, Unani-Tibbi, Siddha, Allopathic, Homeopathic, Naturopathic & Home Remedies, Appendices & Indexes*. Popular Prakashan, Bombay, pp.40–43.

Nayak, B.S. and Pereira, L.M.P., 2006. Catharanthus roseus flower extract has wound-healing activity in Sprague Dawley rats. *BMC Complementary and Alternative Medicine*, *6*(1), p.41.

Nayak, B.S., Anderson, M. and Pereira, L.P., 2007. Evaluation of wound-healing potential of Catharanthus roseus leaf extract in rats. *Fitoterapia*, *78*(7–8), pp.540–544.

Neto, C.C., 2007. Cranberry and its phytochemicals: a review of in vitro anticancer studies. *The Journal of Nutrition*, *137*(1) Supplement, pp.186S–193S.

Newton, S.M., Lau, C. and Wright, C.W., 2000. A review of antimycobacterial natural products. *Phytotherapy Research*, *14*(5), pp.303–322.

Nisar, A., Mamat, A.S., Hatim, M.I., Aslam, M.S. and Syarhabil, M., 2016. An updated review on Catharanthus roseus: phytochemical and pharmacological analysis. *Indian Research Journal of Pharmacy and Science*, *3*(2), pp.631–653.

Nishibe, S., Takenaka, T., Fujikawa, T., Yasukawa, K., Takido, M., Morimitsu, Y., Hirota, A., Kawamura, T. and Noro, Y., 1996. Bioactive phenolic compounds from Catharanthus roseus and Vinca minor. *Natural Medicines 生r學雜誌*, *50*(6), pp.378–383.

Nishibe, S., 1997. Bioactive phenolic compounds for cancer prevention from herbal medicines. In *Food Factors for Cancer Prevention*. Springer, Tokyo, pp.276–279.

Nosálová, V., Machova, J. and Babulová, A., 1993. Protective action of vinpocetine against experimentally induced gastric damage in rats. *Arzneimittel-forschung*, *43*(9), pp.981–985.

Ochwang'i, D.O., Kimwele, C.N., Oduma, J.A., Gathumbi, P.K., Mbaria, J.M. and Kiama, S.G., 2014. Medicinal plants used in treatment and management of cancer in Kakamega County, Kenya. *Journal of Ethnopharmacology*, *151*(3), pp.1040–1055.

Oshi, S.K., Sharma, B.D., Bhatia, C.R., Singh, R.V. and Thakur, R.S., 1992. *The Wealth of India Raw Materials*. Council of Scientific and Industrial Research Publication, New Delhi, *3*, pp.270–271.

Paarakh, M.P., Swathi, S., Taj, T., Tejashwini, V. and Tejashwini, B., 2019. Catharanthus roseus Linn-Review. *Acta Scientific Pharmaceutical Sciences*, *3*(10), pp.19–24.

Padmaja, M., Sravanthi, M. and Hemalatha, K.P.J., 2011. Evaluation of antioxidant activity of two Indian medicinal plants. *Journal of Phytology*, *3*(3), pp.86–91.

Paikara, D., Pandey, B. and Singh, S., 2017. Phytochemical analysis and antimicrobial activity of Catharanthus roseus. *Indian Journal of Scientific Research*, *12*(2), pp.124–127.

Pandey, R. and Mishra, A., 2010. Antibacterial activities of crude extract of Aloe barbadensis to clinically isolated bacterial pathogens. *Applied Biochemistry and Biotechnology*, *160*(5), pp.1356–1361.

Patel, Y., Vadgama, V., Baxi, S. and Tripathi, C.B., 2011. Evaluation of hypolipidemic activity of leaf juice of Catharanthus roseus (Linn.) G. Donn. in guinea pigs. *Acta Poloniae Pharmaceutica*, *68*(6), pp.927–935.

Pathak, R.R.P., 1970. *Therapeutic Guide to Ayurvedic Medicine (A Handbook on Ayurvedic Medicine)*. Shri Ramdayal Joshl Memorial Ayurvedic Research Institute. Publ. Series No. 1, pp.121, 208–209.

Patil, P.J. and Ghosh, J.S., 2010. Antimicrobial activity of Catharanthus roseus–a detailed study. *British Journal of Pharmacology and Toxicology*, *1*(1), pp.40–44.

Patil, M.Y., Vadivelan, R., Dhanabal, S.P., Satishkumar, M.N., Elango, K. and Antony, S., 2014. Anti-oxidant, anti-inflammatory and anti-cholinergic action of Adhatoda vasica Nees contributes to amelioration of diabetic encephalopathy in rats: Behavioral and biochemical evidences. *International Journal of Diabetes in Developing Countries*, *34*(1), pp.24–31.

Pereira, D.M., Ferreres, F., Oliveira, J., Valentão, P., Andrade, P.B. and Sottomayor, M., 2009. Targeted metabolite analysis of Catharanthus roseus and its biological potential. *Food and Chemical Toxicology*, *47*(6), pp.1349–1354.

Perry, L.M., Metzger, J., 1980. *Medicinal Plants of East and Southeast Asia*. The MIT Press, Cambridge, p. 1.

Pham, H.N.T., Sakoff, J.A., Van Vuong, Q., Bowyer, M.C. and Scarlett, C.J., 2018. Screening phytochemical content, antioxidant, antimicrobial and cytotoxic activities of Catharanthus roseus (L.) G. Don stem extract and its fractions. *Biocatalysis and Agricultural Biotechnology*, *16*, pp.405–411.

Pham, H.N.T., Sakoff, J.A., Van Vuong, Q., Bowyer, M.C. and Scarlett, C.J., 2019. Phytochemical, antioxidant, anti-proliferative and antimicrobial properties of Catharanthus roseus root extract, saponin-enriched and aqueous fractions. *Molecular Biology Reports*, *46*(3), pp.3265–3273.

Pillay, P.P., Nair, C.P.M. and Santi Kumari, T.N., 1959. Lochnera rosea as a potential source of hypotensive and other remedies. *Bulletin of Research Institute of the University of Kerala*, *1*, pp.51–54.

Pingale, S.S., 2009. Hepatosuppression by Adhatoda vasica against CCl4 Induced Liver Toxicity in Rat. *Pharmacologyonline*, *3*, pp.633–639.

Prajapati, N.D., Purohit, S.S., Sharma, D.D. and Tarun, K., 2003. *A Handbook of Medicinal Pants*. 1st edn, agrobiaos, Jodhpur, India, pp.13–14.

Prakash, S. and Jain, A.K., 2011. Antifungal activity and preliminary phytochemical studies of leaf extract of Solanum nigrum Linn. *International Journal of Pharmaceutical Sciences*, *3*, pp.352–355.

Prathiba, M. and Giri, R., 2018. Pharmacognostic Study on *Adhatoda vasica*. *En. Asian Journal of Innovative Research*, *3*(1), pp.48–54.

Pushpangadan, P., Nyman, U., George, V. (Eds.), 1995. *Glimpses of Indian Ethnopharmacology*. Tropical Botanic Garden and Research Institute, Kerala, p. 309, 383.

Rafael, J., Barreto R., 1967. Plantas Medicinais de Goa. *Tipografia Rangel, Bastorá*, pp 8–9.

Rajput, M.S., Nair, V., Chauhan, A., Jawanjal, H. and Dange, V., 2011. Evaluation of antidiarrheal activity of aerial parts of Vinca major in experimental animals. *Middle-East Journal of Scientific Research*, *7*(5), pp.784–788.

Ramachandran, J. and Sankaranarayanan, S., 2013. Antifungal Activity and the mode of Action of Alkaloid Extract from the Leaves of Adhatoda vasica. *International Journal of Ethnomedicine and Pharmacological Research*, *1*(1), pp.80–87.

Ramya, S., 2008. In Vitro Evaluation of Antibacterial Activity Using Crude Extracts of Catharanthus roseus L.(G.) Don. *Ethnobotanical Leaflets*, *2008*(1), p.140.

Rao, M.N.A., Krishnan, S., Jain, M.P. and Anand, K.K., 1983. Synthesis of vasicine and vasicinone derivatives for oxytocic and bronchodilatory activity. *Chemischer Informationsdienst*, *14*(39).

Rasoanaivo, P., Wright, C.W., Willcox, M.L. and Gilbert, B., 2011. Whole plant extracts versus single compounds for the treatment of malaria: synergy and positive interactions. *Malaria Journal*, *10*(1) Supplement 1, pp.1–12.

Rasool, N., Rizwan, K., Zubair, M., Naveed, K.U.R., Imran, I. and Ahmed, V.U., 2011. Antioxidant potential of different extracts and fractions of Catharanthus roseus shoots. *International Journal of Phytomedicine*, *3*(1), p.108.

Retna, A.M. and Ethalsa, P., 2013. A review of the taxonomy, ethnobotany, chemistry and pharmacology of Catharanthus roseus (Apocyanaceae). *International Journal of Engineering Research & Technology*, *2*(10), pp.3899–3912.

Roberts E, 1931.*Vegetable materiamedica of India and Ceylon*. Plate Limited, Colombo, pp.16–17.

Salalamp, P., Temsiririrkkul, R., Chuakul, W., Riewpaiboon, A., Prathanturarug, S., Charuchinda, C., Pongcharoensuk, P., 1992. *Medicinal Plants in Siri Ruckhachati Garden*. Amarin Printing Group Co., Bangkok, p. 37.

Salalamp, P., Chuakul, W., Temsiririrkkul, R., Clayton, T., 1996. *Medicinal Plants in Thailand*, vol. 1. Amarin Printing and Publishing Public Co, Bangkok, p. 21.

Sekar, P., 1996. Vedic clues to memory enhancer. *The Hindu*, March, 21.

Semenya, S.S. and Potgieter, M.J., 2013. Catharanthus roseus (L.) G. Don.: Extraordinary Bapedi medicinal herb for gonorrhoea. *Journal of Medicinal Plants Research*, *7*(20), pp.1434–1438.

Shahriar, M., 2013. Phytochemical screenings and thrombolytic activity of the leaf extracts of Adhatoda vasica. *International Journal of Sciences and Technology. The Experiment*, *7*(4), pp.438–441.

Shanmugam, M.K., Dai, X., Kumar, A.P., Tan, B.K., Sethi, G. and Bishayee, A., 2013. Ursolic acid in cancer prevention and treatment: molecular targets, pharmacokinetics and clinical studies. *Biochemical Pharmacology*, *85*(11), pp.1579–1587.

Sharma, S.K., 1998. *Medicinal Plants Used in Ayurveda National Academy of Ayurveda*. Ministry of Health and Family Welfare, Govt. of India, New Delhi, India.

Shinwari, M.I. and Khan, M.A., 1998. Indigenous use of medicinal trees and shrubs of Margalla Hills National Park, Islamabad. *Pakistan Journal of Forestry*, *48*(1/4), pp.63–90.

Siddiqui, M.J., Ismail, Z., Aisha, A.F.A. and Abdul Majid, A.M.S., 2010. Cytotoxic activity of Catharanthus roseus (Apocynaceae) crude extracts and pure compounds against human colorectal carcinoma cell line. *IJP-International Journal of Pharmacology*, *6*(1), pp.43–47.

Singh, R.S., Misra, T.N., Pandey, H.S. and Singh, B.P., 1991. Aliphatic hydroxyketones fromAdhatoda vasica. *Phytochemistry*, *30*(11), pp.3799–3801.

Singh, A., 1997. *Therapeutic monograph-Adhatoda vasica*. Ind-Swift Ltd, Mohali, Chandigarh, pp.25–45.

Singh, S.N., Vats, P., Suri, S., Shyam, R., Kumria, M.M.L., Ranganathan, S. and Sridharan, K., 2001. Effect of an antidiabetic extract of Catharanthus roseus on enzymic activities in streptozotocin induced diabetic rats. *Journal of Ethnopharmacology*, *76*(3), pp.269–277.

Singh, R., Kharb, P. and Rani, K., 2011. Rapid micropropagation and callus induction of Catharanthus roseus in vitro using different explants. *World Journal of Agricultural Sciences*, *7*(6), pp.699–704.

Singh, S.K., Patel, J.R., Dangi, A., Bachle, D. and Kataria, R.K., 2017. A complete over review on Adhatoda vasica a traditional medicinal plants. *Journal of Medicinal Plants Studies*, *5*(1), pp.175–180.

Srivastava, S., Verma, R.K., Gupta, M.M., Singh, S.C. and Kumar, S., 2001. HPLC determination of vasicine and vasicinone in Adhatoda vasica with photo diode array detection. *Journal of Liquid Chromatography & Related Technologies*, *24*(2), pp.153–159.

Stadelmann, W.K., Digenis, A.G. and Tobin, G.R., 1998. Physiology and healing dynamics of chronic cutaneous wounds. *The American Journal of Surgery*, *176*(2), pp.26S–38S.

Subbaramaiah, K., Michaluart, P., Sporn, M.B. and Dannenberg, A.J., 2000. Ursolic acid inhibits cyclooxygenase-2 transcription in human mammary epithelial cells. *Cancer Research*, *60*(9), pp.2399–2404.

Sultana, N., Afza, N., Ali, Y. and Anwar, M.A., 2005. Phytochemical studies on Adhatoda vasica. *Biological Sciences-PJSIR*, *48*(3), pp.180–183.

Suthar, A.C., Katkar, K.V., Patil, P.S., Hamarapurkar, P.D., Mridula, G., Naik, V.R., Mundada, G.R. and Chauhan, V.S., 2009. Quantitative estimation of vasicine and vasicinone in Adhatoda vasica by HPTLC. *Journal of Pharmacy Research*, *2*(12), pp.1893–1899.

Svoboda, G.H., Neuss, N. and Gorman, M., 1959. Alkaloids of Vinca rosea Linn.(Catharanthus roseus G. Don.) V. Preparation and characterization of alkaloids. *Journal of the American Pharmaceutical Association*, *48*(11), pp.659–666.

Swanston-Flatt, S.K., Day, C., Flatt, P.R., Gould, B.J. and Bailey, C.J., 1989. Glycaemic effects of traditional European plant treatments for diabetes. Studies in normal and streptozotocin diabetic mice. *Diabetes Research*, *10*(2), pp.69–73.

Swanston-Flatt, S.K., Day, C., Bailey, C.J. and Flatt, P.R., 1989. Evaluation of traditional plant treatments for diabetes: studies in streptozotocin diabetic mice. *Acta diabetologia latina*, *26*(1), pp.51–55.

Swanston-Flatt, S.K., Day, C., Flatt, P.R., Gould, B.J. and Bailey, C.J., 1989. Glycaemic effects of traditional European plant treatments for diabetes. Studies in normal and streptozotocin diabetic mice. *Diabetes Research*, *10*(2), pp.69–73.

Swathi, K., Ks, P.D. And Sangeetha, A.,2011. Herbal Antioxidants-A Review. International Journal Of Engineering Sciences & Research Technology, 2(6), pp.115–123

Taha, H.S., El-Bahr, M.K. and Seif-El-Nasr, M.M., 2008. In vitro studies on egyptian Catharanthus roseus (L.) G. Don.: 1-calli Production, direct shootlets Regeneration and alkaloids determination. *Journal of Applied Science and Research*, *4*(8), pp.1017–22.

Tikhomiroff, C. and Jolicoeur, M., 2002. Screening of Catharanthus roseus secondary metabolites by high-performance liquid chromatography. *Journal of Chromatography A*, *955*(1), pp.87–93.

Tiong, S.H., Looi, C.Y., Hazni, H., Arya, A., Paydar, M., Wong, W.F., Cheah, S.C., Mustafa, M.R. and Awang, K., 2013. Antidiabetic and antioxidant properties of alkaloids from Catharanthus roseus (L.) G. Don. *Molecules*, *18*(8), pp.9770–9784.

Tiong, S.H., Looi, C.Y., Arya, A., Wong, W.F., Hazni, H., Mustafa, M.R. and Awang, K., 2015. Vindogentianine, a hypoglycemic alkaloid from Catharanthus roseus (L.) G. Don (Apocynaceae). *Fitoterapia*, *102*, pp.182–188.

Vas, A., and Gulyas, B., 2005, Eburnamine derivatives and the brain. *Medicinal Research Reviews*, 25(6): pp.737–757.

Ved, D.K. and Goraya, G.S., 2007. *Demand and Supply of Medicinal Plants in India*. NMPB, New Delhi & FRLHT, Bangalore, India, 18.

Vinothapooshan, G. and Sundar, K., 2011. Anti-ulcer activity of Adhatoda vasica leaves against gastric ulcer in rats. *Journal of Global Pharma Technology*, *3*, pp.7–13.

Vo, V.C., 2012 Dictionary of Vietnamese medicinal plants, Medical Publishing House, Ha Noi. *American Journal of Plant Sciences*, 4, pp.210–215.

Waltz, L.R., 2004. *The Herbal Encyclopedia: A practical Guide to the Many Uses of Herbs*. iUniverse.

Wang, S., Zheng, Z., Weng, Y., Yu, Y., Zhang, D., Fan, W., Dai, R. and Hu, Z., 2004. Angiogenesis and anti-angiogenesis activity of Chinese medicinal herbal extracts. *Life Sciences*, *74*(20), pp.2467–2478.

Wang, L., He, H.P., Di, Y.T., Zhang, Y. and Hao, X.J., 2012a. Catharoseumine, a new monoterpenoid indole alkaloid possessing a peroxy bridge from Catharanthus roseus. *Tetrahedron Letters*, *53*(13), pp.1576–1578.

Wang, C.H., Wang, G.C., Wang, Y., Zhang, X.Q., Huang, X.J., Zhang, D.M., Chen, M.F. and Ye, W.C., 2012b. Cytotoxic dimeric indole alkaloids from Catharanthus roseus. *Fitoterapia*, *83*(4), pp.765–769.

Wang, L., He, H.P., Di, Y.T., Zhang, Y. and Hao, X.J., 2012c. Catharoseumine, a new monoterpenoid indole alkaloid possessing a peroxy bridge from Catharanthus roseus. *Tetrahedron Letters*, *53*(13), pp.1576–1578.

Wang, X.D., Li, C.Y., Jiang, M.M., Li, D., Wen, P., Song, X., Chen, J.D., Guo, L.X., Hu, X.P., Li, G.Q. and Zhang, J., 2016. Induction of apoptosis in human leukemia cells through an intrinsic pathway by cathachunine, a unique alkaloid isolated from Catharanthus roseus. *Phytomedicine*, *23*(6), pp.641–653.

Wansi, J.D., Devkota, K.P., Tshikalange, E. and Kuete, V., 2013. Alkaloids from the medicinal plants of Africa. In *Medicinal Plant Research in Africa*. Elsevier, pp.557–605.

Wasserman, H.H. and Kuo, G.H., 1991. The chemistry of vicinal tricarbonyls. An efficient synthesis of (±)-vasicine. *Tetrahedron Letters*, *32*(48), pp.7131–7132.

Watt G., 1972. *The Economic Products of India*, Periodical Experts, Delhi, pp.108–110.

Wong, S.K., Lim, Y.Y., Abdullah, N.R. and Nordin, F.J., 2011. Antiproliferative and phytochemical analyses of leaf extracts of ten Apocynaceae species. *Pharmacognosy Research*, *3*(2), p.100.

Yadav, A.K. and Tangpu, V., 2008. Anticestodal activity of Adhatoda vasica extract against Hymenolepis diminuta infections in rats. *Journal of Ethnopharmacology*, *119*(2), pp.322–324.

Zahari, R., Halimoon, N., Ahmad, M.F. and Ling, S.K., 2018. Antifungal compound isolated from Catharanthus roseus L.(Pink) for biological control of root rot rubber diseases. *International Journal of Analytical Chemistry*, *2018*, p. 8150610.

Zenk, M.H., El-Shagi, H., Arens, H., Stöckigt, J., Weiler, E.W. and Deus, B., 1977. Formation of the indole alkaloids serpentine and ajmalicine in cell suspension cultures of Catharanthus roseus. In *Plant Tissue Culture and Its Bio-technological Application*. Springer, Berlin, Heidelberg, pp. 27–43.

Zhang, D., Li, X., Hu, Y., Jiang, H., Wu, Y., Ding, Y., Yu, K., He, H., Xu, J., Sun, L. and Qian, F., 2018. Tabersonine attenuates lipopolysaccharide-induced acute lung injury via suppressing TRAF6 ubiquitination. *Biochemical Pharmacology*, *154*, pp.183–192.

17 *Aegle marmelos* (Bael) and *Annona squamosa* (Sugar Apple)

Abhidha Kohli, Taufeeq Ahmad, and Sachidanand Singh

CONTENTS

17.1 Introduction 340
17.2 Traditional Knowledge 342
17.2.1 Nutritional 342
17.2.2 Medicinal 343
17.2.3 Commercial 344
17.3 Chemical Derivatives of *Aegle Marmelos* and *Annona Squamosa* 344
17.3.1 Phytochemicals Associated with *Aegle marmelos* 344
17.3.2 Phytochemicals Associated with *Annona squamosa* 345
17.4 Potential Benefits of *Aegle marmelos* (Bael) and *Annona squamosa* (Custard apple or Sugar apple) 346
17.4.1 Antioxidant Activity 346
17.4.1.1 *Aegle marmelos* 346
17.4.1.2 *Annona squamosa* 349
17.4.2 Antimicrobial Activity 350
17.4.2.1 *Aegle marmelos* 350
17.4.2.2 *Annona squamosa* 350
17.4.3 Anticancer Agents 351
17.4.3.1 *Aegle marmelos* 351
17.4.3.2 *Annona squamosa* 351
17.4.4 Antimalarial, Antidiabetic Activities 352
17.4.4.1 *Aegle marmelos* 352
17.4.4.2 *Annona squamosa* 352
17.4.5 Hepatoprotective and Cardioprotective Activities 353
17.4.5.1 *Aegle marmelos* 353
17.4.5.2 *Annona squamosa* 353
17.4.6 Antipyretic, Anti-Inflammatory and Analgesic Activities 354
17.4.6.1 *Aegle marmelos* 354
17.4.6.2 *Annona squamosa* 354
17.5 COVID-19 Perspective of *Aegle marmelos* and *Annona squamosa* 354
17.6 Conclusion 355
Acknowledgments 355
References 355

DOI: 10.1201/9781003205067-17

17.1 INTRODUCTION

Nature provides bountiful resources that become numerous remedial measures to cure ailments in mankind. According to the World Health Organization (WHO), traditional medicine which includes peculiar practices based on theories and beliefs of indigenous cultures, is widely used for prevention, diagnosis, and treatment of physical or mental illness as well as maintaining health. Traditional medicine meets primary health care needs for approximately 80% of the worldwide population: 80% in India, 85% in Burma, and 90% in Bangladesh (Gangadhar et al., 2012). Developing countries have immense popularity and higher consumption of herbal medicines (WHO, 2015). Ayurveda, herbal, Siddha or Unani medicines act as major remedial approaches in traditional medical systems. They have been in effect for thousands of years and have made excellent contributions to human health (Manoharachary and Nagaraju, 2016). Owing to a rich heritage and knowledge of plants and plant-based drugs, there lies a great potential to explore and address potentially beneficial plant species. Medicinal plants are generally rich sources of secondary metabolites (potential sources of drugs) and essential oils, which are all of great therapeutic significance and are safe, effective, easily available, and economical (Ezzat et al., 2019). Two of the commercially competent and potent therapeutic plants are *Aegle marmelos* and *Annona squamosa* as shown in Figure 17.1 and Figure 17.2 and possessing herbal, phytochemical, and edible properties. This chapter discusses the traditional knowledge, chemical derivatives, and potential benefits of the fruit-bearing plants *Aegle marmelos* (bael) and *Annona squamosa* (sugar apple).

A. marmelos (Family: Rutaceae) is also known as bael, Bengal quince, golden apple or stone apple (Bhardwaj and Nandal, 2015). The tree is worshiped in India and has mythological importance. It is one of the most important medicinal plants in Ayurveda and all its parts, except the poisonous leaves, have medicinal attributes (Kintzios, 2006). However, some studies did show that leaves extract of *A. marmelos* have potential benefits owing to presence of bioactive compounds such as coumarin (Sahare et al., 2008). Fruits, bark, leaves, seeds, and roots are used in folk and Ayurvedic medicine systems to treat various ailments (Kintzios, 2006; Baliga et al., 2011). *A. marmelos* fruits and leaves are used to treat dysentery, rheumatism, malabsorption, dyspepsia, edema, vomiting, and neurological diseases (Chanda et al., 2008). A glance at the mythological importance and nutritional, phytochemical, and medicinal value of *A. marmelos* is depicted in Figure 17.3.

A. squamosa (Family: Annoaceae) is a plant species which bears a fruit commonly known as "custard apple" or "sugar apple". It is a small tropical tree which is native to South America and distributed throughout India and other tropical Asian countries. The ripe fruit pulp contains around

FIGURE 17.1 *Aegle marmelos* plant featuring its fruit.

FIGURE 17.2 *Annona squamosa* plant featuring its fruit.

88.9–95.7 g calories, and the sugar content is 14.58%, amino acid lysine (54–69 mg), carotene (5–7 IU), and ascorbic acid (34.7–42.2 mg) (Morton, 1987). Various chemical constituents isolated from leaves, stems, and roots of *A. squamosa* include anonaine, aporphine, coryeline, isocorydine, norcorydine, and glaucine (Pandey and Barve, 2011b) to which their great medicinal benefits may be attributed. Folkloric reports suggest its use as an insecticidal, antitumor (Vikas et al., 2017), antidiabetic (Shirwaikaret al., 2004a, c), antioxidant, antilipidemic (Gupta et al., 2008), and anti-inflammatory agent (Yang et al., 2008) which may be attributed to the presence of the cyclic peptides (Gajalakshmi et al., 2011; Ramalingum et al., 2014). A glance at the edible, phytochemical, and pharmacological profile of *A. squamosa* is depicted in Figure 17.4.

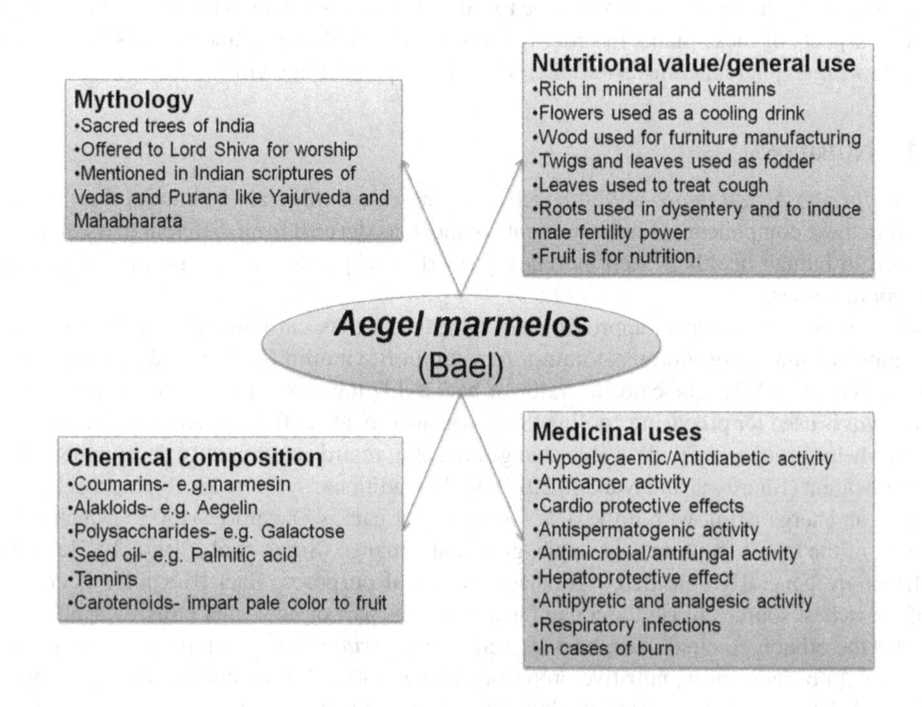

FIGURE 17.3 An overview of *Aegle marmelos* mythological significance, chemical composition, and nutritional and medicinal value.

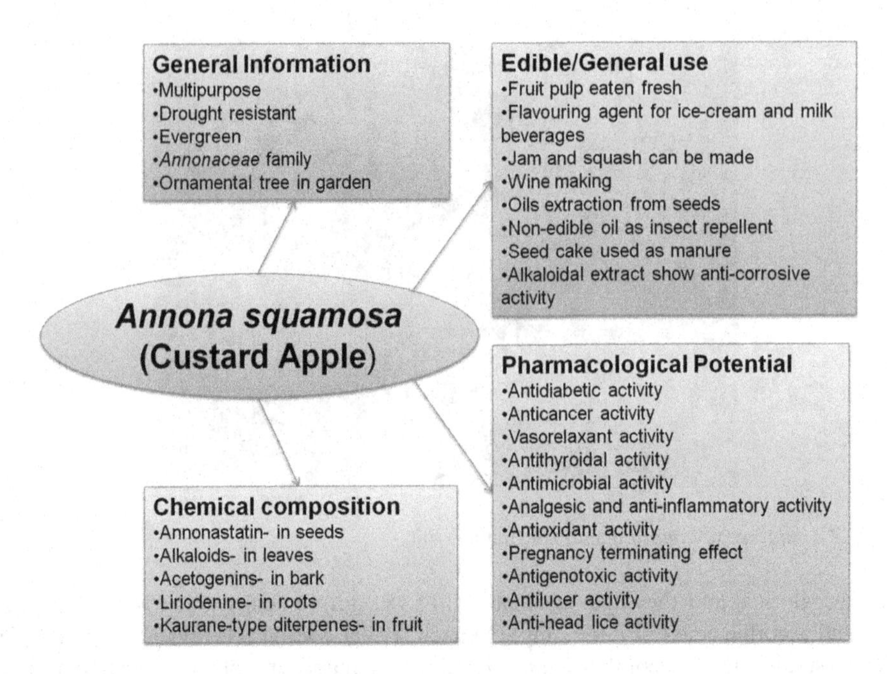

FIGURE 17.4 An overview of *Annona squamosa* chemical composition, edible uses, and pharmacological potential.

17.2 TRADITIONAL KNOWLEDGE

The conventional knowledge about the nutritional, medicinal, and commercial aspects of *A. marmelos* and *A. squamosa* plants get developed, sustained and passed on to subsequent generations in the Indian community. It is of great significance to study and have a holistic view of existing potentially beneficial aspects of these plants in order to conserve the biodiversity and harness the therapeutic potential owing to phytochemicals and their derivatives present in them.

17.2.1 Nutritional

A. marmelos and *A. squamosa* are both rich in vitamins and minerals and attain their nutritional value from these components. All the nutrient components derived from different parts of the plants are useful in human health as well as retain properties of protection against various diseases in ethnic communities.

Bael fruit is rich in water (approximately 64.2%) and has carbohydrates, proteins, minerals (potassium, calcium, phosphorous), vitamin A, vitamin B, vitamin C, fiber, and a small amount of fat (Sharma et al., 2007). The calorific value of bael is 137 (Gurjar et al., 2015). Flower infusion of *A. marmelos* is used for preparing cooling drinks (Sharma et al., 2007). Presence of vitamins in bael fruit juice helps protect any undesirable changes in color, retard any rancidity development, and act as an antioxidant (Bhardwaj and Nandal, 2015). In the traditional system of medicine, bael fruit pulp is used as an energy drink with milk. The food value of bael is 88 calories $100g^{-1}$, which is higher than most of the major fruits such as apple, guava, and mango (Singh et al., 2014a, b). Half-ripened bael fruits are generally considered good for medicinal purposes. Bael fruit juice is considered one of the richest sources of riboflavin and is a nutritious part of one's diet (Bhardwaj and Nandal, 2015). Of the ethnomedicinal uses, bael is used to treat *Tridosha.e.vata* (air), *pitta* (phlegm), and *kapha* (cough) because of its nutritive properties (Kala, 2006). Ethnic communities get enormous nutritional benefits of *A. marmelos* attributed to components derived from its leaves, fruit, seeds, and other plant parts (Dutta et al., 2014).

Phytochemical analysis of leaf extracts of *A. squamosa* reveal the presence of proteins, carbohydrates, vitamins, alkaloids, flavonoids, and minerals, such as magnesium, calcium, iron, phosphorous, sodium, potassium, copper, zinc, and selenium. Vitamins such as A, B1 (thiamine), B2 (riboflavin), B3 (niacin) and B9 (folic acid), C (Ascorbic acid), and E, are present in appreciable concentrations in *A. squamosa* leaves (Shukry et al., 2019; Kumar et al., 2021). These nutritional components help maintain healthy teeth, bones, muscles, prevent blood-clotting, regulate blood pressure, maintain nerve function, immune system health, and energy metabolism (Shukry et al., 2019). *A. squamosa* fruit pulp showed good nutritional value in normal and induced diabetic rabbits which conferred controlled body weight; an improved protein efficiency ratio, digestibility coefficient, and biological value and net protein utilization in normal, healthy and diabetic subjects (Gupta et al., 2005b).

17.2.2 Medicinal

All traditional systems of medicine consider plants to be among the rich sources of chemicals that prove to be of great medicinal and health advantage for humans. According to Ayurvedic and herbal systems of medicine, plants constitute phytochemical compounds, giving them tremendous abilities to be a source of prevention as well as cure for many diseases, such as respiratory, digestive, heart, and metabolic ailments. Among various important medicinal plant species *A. marmelos* and *A. squamosa* draw specific attention because of their extensive medicinal coverage of treatment of ailments from diabetes, microbial infections, oxidative stress related ailments to fungal infections, liver ailments, and cancer.

Bael fruit has been in use since time immemorial as a traditional medicine to treat constipation, diarrhea, peptic ulcer, dysentery, and respiratory infections. Owing to extensive research various medicinal properties of *A. marmelos* have been established, such as antidiabetic, antimicrobial, anti-inflammatory, analgesic, antipyretic, cardioprotective, anticancer, antispermatogenic, radioprotective (Sharma et al., 2007), antifungal, hepatoprotective, and healing (Bhardwaj and Nandal, 2015). Almost all parts of the bael plant are used for the preparation of herbal medicines. Fruits are used for treatment of maximum number of ailments. Bael is a highly regarded plant in Ayurvedic medicine by ethnic communities, as it is used in treating digestive disorders, diabetes, heart diseases, and cholera due to its digestive and carminative properties. Fruit, leaf, bark, root, and flowers of bael in combination are assumed to be a cure for mental disorders (Kala, 2006). The root decoction from *A. marmelos* is reported to treat melancholy, intermittent fevers, and palpitation (Dutta et al., 2014). As mentioned by Gurjar et al. (2015), *A. marmelos* has tremendous medicinal advantage whose properties are mentioned in *Charaka Samhita*, an ancient medical treatise in Sanskrit and listed in *Ayurvedic*, *Unani*, and *Siddha* medicine systems.

Leaves, shoots, bark, roots, and fruit of *A. squamosa* contain most of its medicinal properties. The unripe and dried sugar apple is used as an antidysenteric agent (Kumar et al., 2021) and, along with bark can be used as an astringent. The leaves have anti-fertility activity and are used as vermicide for treating cancerous tumors and are also useful in skin ailments. Powdered seeds need to be taken with care so that they do not come in contact with eyes and are useful in killing head lice and fleas. *A. squamosa* components show various advantages against commonly occurring and rather serious diseases by activities such as antidiabetic, antitumor (Vyas et al., 2012), anti-HIV, antioxidant, anti-ulcer, hepatoprotective, anthelmintic, anti-arthritic, anti-inflammatory, analgesic, cytotoxic, and wound-healing (Saha, 2011). The leaves, bark, fruit, and seed of *A. muricata* have also been the focus of countless medicinal uses (Badrie and Schauss, 2010). The most extensively used preparation in traditional medicine is the decoction of bark, root, seed, leaf, and its applications are varied. According to TRAMIL, a program of applied research about traditional medicinal plant resources in the Caribbean islands, leaves are used for treatment of skin ailments (Boulogne et al., 2011). While in Mauritius (Sreekeesoon and Mahomoodally, 2014), New Guinea (WHO, 2009), and Ecuador (Tene et al., 2007), the application of leaves is used locally on the pain site. The fruit is not only treasured as food, but its juice is also used as galactogogue and to treat diarrhea, heart and liver diseases (Badrie and Schauss, 2010; Hajdu and Hohmann, 2012). Bioactive compounds are the source of many traditional medicines from all around the world. Moreover, these medicinal properties have played a major role in providing the remedies against all sorts of inflammatory conditions. Annonaceae family is very famous in tropical regions due to its extensive use in traditional medicine. *A. muricata* is used in Brazil for weight loss treatment, hypertension, snakebite, inflammation, and dizziness (Cercato et al., 2015). Moreover, in various parts of Indonesia the dried leaves are orally ingested for its strong analgesic effect (Badrie and Schauss, 2010; Bele et al., 2011).

17.2.3 Commercial

The economical and commercial benefit is an additional aspect that puts plants like *A. marmelos* and *A. squamosa* on a pedestal. Besides all the medicinal utilities these wild plant species have, they also have been found to be beneficial for the environment and industry concerns which includes crop improvement, dealing with infertile land spaces, and preparation of food products like jam, syrup, etc.

One of the most important commercial advantages of *A. marmelos* is that it can be cultivated on waste and unproductive land and thus can be beneficial for the upliftment of farmers (Sharma et al., 2007). It is used as a rescue for infertile marginal lands and is considered good for reforestation purposes (Pathirana et al., 2020). Bael fruit pulp is extensively used for the preparation of products like jam, syrup, and pudding (Baliga et al., 2011). Its seeds produce unique fatty acid (12-hydroxyoctadec-cis-9-enoic acid or ricinoleic acid) which can be converted to produce biodiesel fuel (Katagi et al., 2011). It is among the richest plant sources for the preparation of herbal medicines. All these properties and uses makes *A. marmelos* an incredible economical and industrially vital indigenous plant species (Pathirana et al., 2020).

The fruit pulp of *A. squamosa* is used as a flavoring agent in ice creams (Kumar et al., 2021) and juice preparations. There can be much health-related and medicinal products derived from sugar apple, owing to the presence of bioactive compounds such as alkaloids, carbohydrates, fixed oils, tannins, and phenolics. It is also grown for its fruits and ornamental value (Pandey and Barve, 2011b).

17.3 CHEMICAL DERIVATIVES OF *AEGLE MARMELOS* AND *ANNONA SQUAMOSA*

It is important to understand the reason behind the valuable medicinal properties of *A. marmelos* and *A. squamosa* plants so as to reach out to the underlying molecular players. Thus, it becomes obligatory to review the phytochemicals or chemical derivatives from these plants. The most common chemical derivatives belong to groups of alkaloids, flavonoids, steroids, among others, and their potential functions are reviewed in this chapter.

17.3.1 Phytochemicals Associated with *Aegle marmelos*

A. marmelos is constituted of a variety of phytochemicals which form the basis of its medicinal properties. These chemical constituents, or bioactive compounds, mainly belong to the coumarin, alkaloid, polysaccharide, and tannin groups (Dutta et al., 2014). They are isolated from different parts of the plant such as leaves, roots, bark, flowers, and fruits. The phytoconstituents mainly include marmenol, marmin, marmelosin, marmelide, psoralen, alloimperatorin, rutaretin, scopoletin, α-phellandrene, betulinic acid, aegelin, marmelin, fagarine, anhydromarmelin, limonene, marmesin, imperatorin, marmelosin, luvangentin, auroptene (Bansal and Bansal, 2011), tannins, such as skimmianine, riboflavin (Yadav et al., 2011), and other organic acids, including oxalic, tartaric, malic and ascorbic acids, chlorogenic acid, ellagic acid, ferulic acid, gallic acid, protocatechuic acid, and quercetin (Rahman and Parvin, 2014). The pale color of the fruit is due to the presence of carotenoids (Dutta et al., 2014). It has been reported that methanolic extract of *A. marmelos* leaves showed anti-microfilarial activity against *Brugiamalayi microfilariae* which potentially might be attributed to the co-presence of the coumarin bioactive component (Sahare et al., 2008). oxazoline derivative aeglemarmelosine isolate whose structure was characterized as mentioned in a short report by Laphookhieo et al., (2011) were observed in the roots and twigs of *A. marmelos*. Another derivative, called "carboxymethylated fruit gum", has enhanced mucoadhesive potential (Srivastava et al., 2015). Essential oil constituted of β- terpinyl acetate, 2,3-pinanediol, and 5- isopropenyl-2- methyl-7-oxabicyclo (4.1.0) hepten-2-ol derived from the leaves of the bael plant showed significant insecticidal properties against *Aedes aegypti* and *Culex quinquefasciatus* (Sarma et al., 2017). Antihistaminic constituents present in the alcoholic extract of *A. marmelos* have been reported to cause relaxation in the guinea pig isolated ileum and tracheal chain, thus supporting its potential to treat asthmatic ailments (Arul et al., 2004). Phenylethyl cinnamides derived from *A. marmelos* leaves act as α-glucosidase inhibitors and can help treat diabetes mellitus (Phuwapraisirisan et al., 2008). Various studies have revealed through the phytochemical analysis of extracts from *A. marmelos* plant parts, its utility for therapeutic purposes which have been tabulated in Table 17.1.

17.3.2 Phytochemicals Associated with *Annona squamosa*

More than 200 bioactive components have been derived from different genuses of the Annonaceae family. *A. squamosa* constitutes phytochemicals isolated from all plant parts, such as leaves, stems,

TABLE 17.1
Chemical Derivatives, Their Source, and Function from *Aegle marmelos*

		Aegle marmelos		
S. No	Phytochemical	Source	Function	References
1	Coumarin	Leaves extract of *Aegle marmelos*	Antifilarial effect by loss of motility of microfilariae	Sahare et al., 2008
2	Aegelin	Leaves extract of *Aegle marmelos*	Cardioactive, antihyperglycemic activity	Maity et al., 2009
3	Marmelosin	Fruit of *Aegle marmelos*	Antibacterial, antihelminthic	Maity et al., 2009
4	Tannin	Unripened fruit	Anti-diarrhea, astringent	Maity et al., 2009
5	Oxazoline derivative	Roots and twigs	Lack antimalarial activity	Laphookhieo et al., 2011
6	Carboxymethylated derivative of bael fruit gum	Fruit of *Aegle marmelos*	Improved mucoadhesive potential	Srivastava et al., 2015
7	Essential oil (contituting p- terpinyl acetate, 5- isopropenyl-2- methyl-7-oxabicyclo (4.1.0) hepten-2- ol and 2,3-pinanediol)	Leaf extract of *Aegle marmelos*	Might play role in insecticidal activities	Sarma et al., 2017
8	Phenylethyl cinnamides	Leaf extract of *Aegle marmelos*	α-glucosidase inhibitors	Phuwapraisirisan et al., 2008

roots, and fruits. The chemical derivatives belong to groups of glycosides, alkaloids, saponins, flavonoids, tannins, carbohydrates, proteins, phenolic compounds, phytosterols, and amino acids. They include anonaine, aporphine, coryeline, isocorydine, norcorydine, glaucine, 4-(2-nitro-ethyl 1)-1-6-((6-o-β-Dxylopyranosyl-β-D-glucopyranosyl)-oxy) benzene, benzyl tetrahydroisoquinoline, camphene, borneol, camphor, car-3-ene, β caryphyllene, carvone, eugenol, geraniol, farnesol, 16-hetriacontanone, higemamine, hexacontanol, isocorydine, linalool acetate, limonine, menthone, rutin, methyl anthranilate, methylsalicylate, methylheptenone, p-(hydroxybenzyl)-6,7-(2-hydroxy ,4-hydro), α-pinene, n-octacosanol, b-pinene, stigmasterol, β-sitosterol, thymol, and n-triacontanol; while leaf extracts lack alkaloids, proteins, and amino acids (Pandey and Barve, 2011b; Fofana et al., 2012). Acetogenins are the most common chemical compounds derived from plants of the Annoaceae family besides alkaloids and phenols (Coria-Téllezet al., 2018; Attiq et al., 2017; Alali et al., 1999). Bullatacin, bullatacinone, and squamone are the types of acetogenins derived from bark of *A. squamosa* and are found to be cytotoxic against breast carcinoma cell line MCF-7 (Li et al., 1990). Leaf extract derivative rutin (quercetin-3-rhamnosyl glucoside) is a flavonoid that has been found to be antioxidant, thus rendering it antiviral, anti-inflammatory, anticancer properties (Soni et al., 2018). Various studies through the phytochemical analysis of extracts from *A. squamosa* plant parts revealed its utility for therapeutic purposes and have been tabulated in Table 17.2.

17.4 POTENTIAL BENEFITS OF *AEGLE MARMELOS* (BAEL) AND *ANNONA SQUAMOSA* (CUSTARD APPLE OR SUGAR APPLE)

Bael and custard apple both have promising benefits regarding nutritional, environmental, industrial, and, most importantly, medicinal aspects. Medicinal benefits are varied and cover a plethora of diseases and symptoms from diarrhea, dysentery, inflammation, microbial, fungal, viral infection, heart disease, respiratory disorders, to severe illness, like cancer. The applications and potential uses of both these plants are attributed to various properties such as antioxidant, antimicrobial, anticancer, antidiabetic, antipyretic, anti-inflammatory, hepatoprotective, cardioprotective, etc. (Rahman and Parvin, 2014; Kumar et al., 2021). These potential benefits have been discussed and reported by many scientists and open an arena of deeper understanding of the molecular underpinnings of the phytochemicals rendering such medicinal benefits. Out of the various potential benefits, some of the most common ones have been discussed in this chapter and compiled in Table 17.3 for *A. marmelos* and Table 17.4 for *A. squamosa.*

17.4.1 Antioxidant Activity

The antioxidant activity of bael and sugar apple plants has been extensively studied and renders potential benefit for oxidative stress involving diseases and symptoms such as diabetes and cancer (Sekar et al., 2011; Kaleem et al., 2006). Both *A. marmelos* and *A. squamosa* have reported antioxidant activities owing to the phytochemical constituents present in them.

17.4.1.1 *Aegle marmelos*

It has been reported that *A. marmelos* leaves (AML) extract when administered on alloxan-induced diabetic rats proved to be beneficial in managing diabetes mellitus and showed decreased blood glucose levels (Upadhya et al., 2004). Alloxan-induced male rats have shown improvement in lipid profile, glucose, insulin, malondialdehyde (MDA) and glutathione (GSH) levels and superoxide dismutase (SOD) activity (Abdallah et al., 2017). *A. marmelos* leaf extract and water extract from fruits were examined for treatment of diabetic rats and showed a dose-related increase in antioxidative activity parameters like reduced glutathione, glutathione reductase, glutathione peroxidase, catalase, and SOD and a decrease in lipid peroxidation, conjugated diene, and hydroperoxide levels (Sabu and Kuttan, 2004; Kamalakkannan and Stanely, 2003). Natural antioxidants from fruits have gained interest due to their protective effect against oxidative stress-derived free radicals involved

TABLE 17.2
Chemical Derivatives, Their Source and Function from *Annona squamosa*

Annona squamosa				
S. No	**Phytochemical**	**Source**	**Function**	**References**
1	Bullatacin, Bullatacinone and Squamone	Bark of *Annona squamosa*	Increased cytotoxicities to MCF-7 human breast carcinoma cell lines	Li et al., 1990
2	Rutin (quereetin-3-rhamnosyl glucoside)	Leaves of *Annona squamosa*	Antioxidant activity with beneficial effects such as anti-inflammatory, antiallergic, antiviral, anticancer activity. Treatment of diabetes, hepatotoxicity	Soni et al., 2018
3	Saturated and unsaturated fatty acids or their esters	Ethanolic seed extract of *Annona squamosa L.*	May act as natural insecticide against subterranean termites	Acda, 2014
4	ent-kaurane diterpenes	Pericarp of *Annona squamosa L.*	Some diterpenes showed cytotoxic effect on SMMC-7721 and HepG2 hepatocarcinoma cell lines	Chen et al., 2020
5	Kaurane derivative- 16β,17-dihydroxy-ent-kauran-19-oic acid	Fruits of *Annona squamosa*	Activity against HIV replication in H9 lymphocyte cells	Wu et al., 1996
6	Water soluble polysaccharide- ASPW80-1 and its sulphated derivative ASPW80-M1	Fruit pulp of *Annona squamosa*	Inhibition of DPPH and hydroxyl radicals, promote proliferation of mouse splenocytes	Tu et al., 2016
7	Flavonoid compound- AS-1	Ethanolic leaf extract of *Annona squamosa*	Free radical scavenging activity	Chandrashekhar and Kulkarni, 2011
8	Cyclopeptide- Squamtin A	Seeds of *Annona squamosa*	Functional studies were not carried out owing to limited natural availability	Jiang et al., 2003
9	Squamocin- trihydroxy-bis-tetra hydro furan fatty acid y lactone	Petroleum ether seed extract of *Annona squamosa*	cytotoxic activity	Fujimoto et al., 1988

in the development of diseases such as cancer, cardiovascular affections, arthritis, degenerative illnesses such as Parkinson's and Alzheimer's disease (Almeida et al., 2011). Bael fruit pulp has been found to be beneficial against gastrointestinal infections in humans, probably attributed to the antioxidative potential of pytochemicals present therein, such as alkaloids, flavonoids, steroids, terpenoids, tannins, lignins, etc. in aqueous and alcoholic extracts (Rajan et al., 2011; Raja and Khan, 2017). *A. marmelos* keeps a high potential of being an antioxidant because its extracts show selective pH and temperature stability at the same time increased efficiency of scavenging free radicals (Reddy and Urooj, 2013). The antioxidative potential of *A. marmelos* has been reported to be profound in dealing with Dalton's lymphoma in transplanted mice by revealing a decrease in hepatic lipid peroxidation and an increase in antioxidant hepatic enzymes, such as SOD and catalase (Chockalingam et al., 2012). According to (Siddique et al., 2011), the greater radical-scavenging effect of *A. marmelos* as an antioxidant lies in greater phenolic compound constituents enriched in its methanolic leaf extracts.

TABLE 17.3
***Aegle marmelos* Potential Benefits and Their Attributions**

		Aegle marmelos		
S. No	**Plant Part/Species**	**Potential Benefits**	**Benefits Attributed To**	**References**
1	Bael (*Aegle marmelos L. Correa*)	Economically valuable tree, fruit pulp for preparing jam, syrup, pudding, medicinal values as antidiabetic, anticancerous, antifertility, antimicrobial, immunogenic, insecticidal, useful for reforestation of infertile land	Bioactive compounds render medicinal properties	Pathirana et al., 2020
2	*Aegle marmelos L.*	Health benefits such as radio-protective effects, peroxidation, antibacterial, lipid inhibition, antidiarrheal, gastroprotective, antiviral, anti- ulcerative colitis, cardioprotective, free-radical scavenging (antioxidant), and hepatoprotective effects	Fibers, carotenoids, phenolics, terpenoids, coumarins, flavonoids, and alkaloids	Venthodika et al., 2021
3	*Aegle marmelos* fruit	Higher economic potential; Engage people in small scale cottage industries	Jam, Jelly, Juice like edible products from wild fruit	Maikhuri et al., 2004
4	Bhopal variety of *Aegle marmelos L. Correa*	Good antioxidant, cyto/DNA protective properties, anti-ulcer, and anticancer activities	by inhibiting H+, K+ - ATPase (anti-ulcer), and tyrosinase (anticancer)	Hasitha et al., 2018
5	Ethanolic extract of *Aegle marmelos* leaves	Therapeutic benefit against induced chronic fatigue syndrome in animal model	Antioxidant nature of bael extracts helps combat oxidative stress in CFS	Lalremruta and Prasanna, 2012
6	Silver nanoparticles of methanolic extract of fruit of *Aegle marmelos*	Antimicrobial potential	Secondary metabolites such as tannins, terpenes etc. help plants protect against microbial invasions	Devi et al., 2020
7	Bael (*Aegle marmelos L. Correa*) fruit juice	Nutritional and therapeutic importance	Vitamins (Riboflavins), minerals, trace elements (Nutritional); Bioactive compounds like tannins, steroids, coumarins (Therapeutic)	Bhardwaj and Nandal, 2015
8	Bael (*Aegle marmelos L. Correa*)	Radioprotective effects	Attributed to the property of scavenging free radicals and antioxidant activity	Baliga et al., 2010
9	Leaf extract of *Aegle marmelos*	Chloroform extract shows antibacterial and antifungal activity against *E. coli* and *Fusarium oxysporum*	Traditional medicinal value and potential to be used as herbal medicine	Abirami et al., 2014
10	Leaves of *Aegle marmelos*	Antioxidant and antidiabetic agent	Bioactive compounds and traditional medicinal advantage	Nigam and Nambiar, 2015

TABLE 17.4
***Annona squamosa* Potential Benefits and Their Attributions**

		Annona squamosa		
S. No	**Plant Part/ Species**	**Potential Benefits**	**Benefits Attributed To**	**References**
1	Custard apple (*Annona squamosa Linn.*)	Nutraceutical and medicinal benefit	Higher vitamin C, potassium, dietary fiber (nutraceutical), and chemical constituents like alkaloids, acetogenin, squamone (medicinal)	Singh et al., 2019
2	Fruit extract of *Annona squamosa*	Antioxidant activity	Free-radical scavenging action	Boakye et al., 2015
3	Leaves of *Annona squamosa L.*	Human health benefits and biological activities such as antidiabetic, anticancer, antioxidant, antiobesity, antimicrobial, lipid-lowering, and hepatoprotective functions	Various phytochemicals such as phenol-based compounds e.g.- proanthocyanidins	Kumar et al., 2021
4	Extracts of Thailand sugar apple (*Annona squamosa*) cultivar peels	Antioxidant activities	Phenolic compounds play a major role in antioxidant activity	Manochai et al., 2018
5	*Annona squamosa* plant	Treatment of epilepsy, dysentery, cardiac problem, warm infection, constipation, hemorrhage, antibacterial infection, dysuria, fever, and ulcers	Phytochemicals such as acetogenins, cyclopeptides, and alkaloids	Srivastava et al., 2011
6	Seed, peel, and pulp of *Annona squamosa* (*Annona b, Annona h*)	Antioxidant, antimicrobial, and in vitro anticancer activity	Anticancer activity might be attributed to downregulation in mRNA expression of Bcl-2 and upregulation in p53 in tested colon (Caco-2), prostate (PC3), liver (HepG-2), and breast (MCF-7) cell lines	Shehata et al., 2021
7	Fruit pulp of *Annona squamosa*	Nutraceutical benefit and functional food source	Phenolic compounds, copper, manganese (nutraceutical); 2-octanol, D-mannitol, Heptanoic acid (functional foods)	Bhardwaj et al., 2014

17.4.1.2 *Annona squamosa*

Oral administration of *A. squamosa* leaf extracts on streptozotocin-induced diabetic rats has shown hypoglycemic effect and significant reduction in lipid peroxidation while observing increases in activities of plasma insulin and antioxidant enzymes, such as SOD and catalase (Kaleem et al., 2006). Methanol, chloroform, and aqueous extracts of *A. squamosa* have shown concentration-dependent antioxidant activity as depicted by free radical-scavenging activity probably owing to the presence of phytoconstituents, such as glycosides, saponins, tannins, flavonoids, and phenols

(Kalidindi et al., 2015). Ethanolic extracts of *A. squamosa Linn* bark have shown significantly high antioxidant activity as result interpreted by DPPH, hydroxyl, and superoxide radical-scavenging activity. This can be used as a natural antioxidant, which is easily accessible and can be used as a food supplement (Pandey and Barve, 2011a). Water extract of *A. squamosa* seeds has shown higher radical-scavenging activity indicating the possibility of nonedible parts of fruits having promising therapeutic implications (Kothari and Seshadri, 2010; Verma et al., 2021). According to a preliminary study, a wide variety of phytochemicals were determined in *A. squamosa* ethanolic leaf extract, such as alkaloids, coumarins, tannins, phenols, flavonoids, etc. out of which phenol and flavonoid contents were correlated with the antioxidant activities (Nguyen et al., 2020). Phenolic content of the *A. squamosa* leaf extract has shown about 89% contribution to antioxidant activity and about 43–54% of hydrogen peroxide-scavenging capability (El-Chaghaby et al., 2014). *In vitro* studies on rat brain homogenates for antioxidant potential of *A. squamosa* leaves have justified its therapeutic potential and value in the traditional medicine system (Shirwaikar et al., 2004b). Methanolic extracts of *A. squamosa* fruit pulp have demonstrated higher antioxidative potential as compared to aqueous extract in *in vitro* models owing to higher phenols and flavonoid constituents (Nandhakumar and Indumathi, 2013). Previous studies have also suggested that phenolic compounds extracted and derived from roots and bark of *A. squamosa* can be utilized as natural antioxidants as displayed by their higher antioxidant activities (Mariod et al., 2012). Published reports on the potential of antioxidant activity have opened novel potential on the uses of phytoconstituents derived from medicinal plants as drug formulations for the treatment of diseases centering on oxidative stress.

17.4.2 Antimicrobial Activity

Antimicrobial activity is one of the potential benefits of medicinal plants in order to treat infectious diseases and symptoms. Both *A. marmelos* and *A. squamosa* have been reported in various original studies to be effective against standard pathogenic microorganisms as well as certain fungal species. Since infectious diseases are an important concern to be addressed, it is imperative to understand the antimicrobial activities of bael and sugar apple plants.

17.4.2.1 *Aegle marmelos*

Previous reports have shown that methanol, chloroform, and aqueous extracts from the leaves, bark, and fruits of *A. marmelos* have antibacterial activity against *Bacillus subtilis*, *Staphylococcus aureus*, *Klebsiella pneumoniae*, *Proteus mirabilis*, *Escherichia coli*, *Salmonella paratyphi A*, and *Salmonella paratyphi B* (Poonkothai and Saravanan, 2008). *In vitro* antibacterial and antifungal activity tested against various bacterial and fungal species have shown promising antimicrobial activities with potential for sourcing bioactive agents for treating microbial infections (Kothari et al., 2011; Sivaraj et al., 2011). Methanolic extracts from different plant parts of *A. marmelos* have shown inhibitory effect on both gram-positive and gram-negative clinical pathogens and have the potential to deal with infectious diseases caused by resistant microorganisms (Suresh et al., 2009). On comparing the antimicrobial effect of the leaf extract of *A. marmelos* with conventional antibiotic penicillin, it was revealed that *A. marmelos* has a relatively higher zone of inhibition and can be explored for phytomedicine potential (Venkatesan et al., 2009b). Phytochemical constituents that were responsible for antimicrobial activity against *Bacillus subtilis*, *Staphylococcus aureus*, *Escherichia coli*, and *Pseudomonas aeruginosa* were identified as alkaloids, cardiac glycosides, terpenoids, saponins, tannins, flavonoids, and steroids (Venkatesan et al., 2009a; Mujeeb et al., 2014).

17.4.2.2 *Annona squamosa*

Microbial contamination and pesticidal residue are serious issues with post-harvest pulses, which can potentially be mitigated by flavonoids present in the aqueous extract of *A. squamosa* as reported in Kotkar et al. (2002). *A. squamosa* seed extract has shown inhibitory effect on both gram-negative and gram-positive standard pathogenic bacteria, for example, *Staphylococcus aureus*, *Klebsiella pneumoniase*, *Escherichia coli*, *Salmonella typhi*, etc. and found to be synergistically antimicrobial when combined with the seed extract of dates (*Phoenix dactylifera*) (Aamir et al., 2013). Different solvent extracts of *A. squamosa* have shown similar antibacterial effects on sensitive pathogenic bacterial strains *in vitro*, whereas the highest inhibition was observed with mathanolic extract (Aher et al., 2012). Antibacterial potential of *A. squamosa* from Similipal Biosphere Reserve, Orissa, India has also shown higher inhibition of pathogenic bacteria species (Padhi et al., 2011). Isomeric hydroxy ketones from leaf cuticular waxes have shown comparable antifungal activity as compared to palmitone in *A. squamosa*. The essential oil derived from *A. squamosa* bark is reported to contain certain volatile constituents that can be useful against the microbes as depicted by the antimicrobial activity of the oil against *Bacillus subtilis* and *Staphylococcus aureus* (Chavan et al., 2006). According to various antibacterial evaluation studies of *A. squamosa*, growth inhibition has been shown against various selective gram-negative and gram-positive pathogenic microorganisms (Saha, 2011). The higher antibacterial activity of *A. squamosa* leaf extracts is attributed to the presence of various primary and secondary metabolites, and phytochemical analysis revealed the presence of such constituents, like glycosides, alkaloids, oils, saponins, and flavonoids (Neethu Simon et al., 2016).

17.4.3 Anticancer Agents

According to WHO, cancer is the second leading cause of death worldwide responsible for approximately 9.6 million deaths annually, reported as of 2018 (WHO). Some tumor types are considered epidemic, for example lung cancer, owing to the rise in smoking (Addario, 2015). Since the cost of allopathic medication is becoming alarmingly high, scientists have started to look for more economic alternatives in herbal medicines, and thus, medicinal plants are being investigated for potential anticancer properties. According to several published reports, *A. marmelos* and *A. squamosa* have shown anticancer properties, thus proving beneficial for interventions in cancer therapy advancements.

17.4.3.1 *Aegle marmelos*

It is reported that hydroalcoholic extract of *A. marmelos* leaves shows an anticancer effect on a mice model of transplanted Ehrlich ascites carcinoma (Jagetia et al., 2005). Ethanolic extract of *A. marmelos* leaves has shown strong antitumor effect against Dalton's lymphoma ascites-bearing mice (Chockalingam et al., 2012). Chemical derivatives from *A. marmelos* butyl p-tolyl sulfide, 6-methyl-4-chromanone, and butylated hydroxyanisole as identified by gas-chromatography mass-spectrometry (GC/MS) analysis have shown inhibitory effect on the growth of several cancer cells, such as human leukemic K562, melanoma Colo38, B-lymphoid Raji, T-lymphoid Jurkat, erythroleukemic HEL, and breast cancer MCF7 and MDAMB-231 cell lines (Lampronti et al., 2003). Skimmianine, an alkaloid derived from the *A. marmelos* leaf and immature bark of the tree has also shown antitumor activity against A2780 human ovarian cancer cell line (Maity et al., 2009). Toxicity assays like brine shrimp lethality assay, hemolysis assay, sea urchin-egg assay, and MTT assay were conducted on tumor cell lines and *A. marmelos* extract along with other medicinal plants used in Bangladeshi folk medicine showed toxicity effect (Costa-Lotufo et al., 2005; Yadav and Chanotia, 2009). An *in vitro* investigation of cytotoxicity of Bangladeshi medicinal plants on breast tumor cell lines reported the cytotoxic effect of *A. marmelos* extract on MCF-7 and MDA-MB-231 cells (Lambertini et al., 2004; Manandhar et al., 2018). *A. marmelos* extract in combination with decoy, a transcription factor that regulates the human ER-α gene, showed modulatory effect on cell proliferation in MDA-MB-231 ER-α negative breast cancer cell lines (Lambertini et al., 2005; Manandhar et al., 2018).

17.4.3.2 *Annona squamosa*

Approximately 150 chemical derivatives of *A. squamosa* derived from all plant parts such as fruit, bark, stem, and seeds have been collectively reviewed in (Ma et al., 2017) and show anticancer activity against several cancer cell lines, such as A2780 ovarian cancer, MDR MCF-7 A breast cancer, A549 lung cancer, HepG2 liver cancer, L1210 leukemia cancer, and PC-3 prostate cancer cells. Epidermoid carcinoma cell lines KB-3-1 and colon cancer cell lines HCT-116 have shown sensitivity against crude extract and ethyl acetate extract of *A. squamosa* leaves (Wang et al., 2014). Assessment of antitumor activity of *A. squamosa* seed oil showed promising potential for the treatment of cancer (Chen et al., 2016). Acetogenin compound, a constituent of *A. squamosa* seed extract, has shown antitumor potential in the *in vitro* and *in vivo* H_{22} hepatoma cell-transplanted mice model (Chen et al., 2012). Mechanistically, the antitumor activity of *A. squamosa* seed extract acts through promotion of apoptosis with enhanced caspase-3 enzyme activity, downregulation of Bcl-2 and Bcl_{XL} anti-apoptotic genes, and increase in intracellular ROS in the rat histocytic tumor cell line, AK-5 (Pardhasaradhi et al., 2004). Annonareticin, a lactone compound from the seeds of *A. squamosa*, is used in the treatment of lung and breast cancer and has shown higher antitumor activity as compared to the conventional 5-fluorouracil drug (Chen et al., 2011). *A. squamosa* leaf extract-derived silver nanoparticles showed cytotoxicity on mammalian (HEK-293) and cancerous (HeLa) cells, revealing the potential to be used as nanomedicine (Ruddaraju et al., 2019).

17.4.4 Antimalarial, Antidiabetic Activities

Malaria and diabetes are important diseases of concern as the former has identified resistance to antimalarial drugs (Dhankar et al., 2011; Dutta et al., 2014) and the latter is a slow killing disease with limitation of blood sugar regulation. Some reports have also uncovered potential crosstalk between the two diseases and thus make both of them appealing to scientific fraternity. *A. marmelos* and *A. squamosa* have shown antimalarial and antidiabetic properties according to various published reports and opens potential for identification of novel therapeutic molecules.

17.4.4.1 *Aegle marmelos*

Bael seed alcoholic extract was found to kill the mature malaria parasite when examined in *in vitro* and *in vivo* systems against Nk-65 strain of *Plasmodium berghei* (Misra et al., 1991; Dhankar et al., 2011; Rahman and Parvin, 2014). *A. marmelos* root extract showed 50% decrease in parasite growth when tested against multidrug variant K1, *Plasmodium falciparum* (Bhar et al., 2019). The leaf methanol extract of *A. marmelos* sourced from Dharmapuri region, Tamil Nadu, India showed antiplasmodial activity against *Plasmodium falciparum*, thus validating traditional usage of *A. marmelos* against malaria (Kamaraj et al., 2012).

Among Ayurvedic medicines, *A. marmelos* leaf extract is considered one of the best for diabetes mellitus. According to a study on alloxan induced diabetic rats, ethanolic extracts of *A. marmelos* were found to reduce blood sugar levels and histopathologically regenerate the β-cells thus promising antidiabetic activity (Bhavani, 2014). A report on streptozotocin diabetic rabbits when administered with methanol extract from leaf and callus of *A. marmelos* showed significant decrease in blood sugar levels and a potential in diabetes management (Arumugam et al., 2008). The aqueous seed extract of *A. marmelos* has shown hypolipidemic and antidiabetic effects in diabetic rats (Kesari et al., 2006). A mechanistic study on streptozotocin-induced diabetic rat model also showed antihyperglycemic activity when administered with Aege-line 2, an alkaloidal-amide from the leaves of *A. marmelos* (Narender et al., 2007). Even clinical examination of bael leaves has observed decreases in blood sugar levels in patients in combination with oral hypoglycemic agents, suggesting a potential benefit for patients with uncontrolled diabetes or those with adverse effects on dose increments (Ismail, 2009a). A combination of 20 g of fenugreek powder and decoction of 5 g of bael leaf powder showed marked antidiabetic activity on clinical evaluation in diabetic patients (Ismail, 2009b).

17.4.4.2 *Annona squamosa*

A. squamosa along with other African medicinal plants were examined for their antimalarial activity and were found promising using their crude extracts (Kaou et al., 2008). Multidrug resistant strain of *Plasmodicum falciparum* has been found to be sensitive to leaf extract of *A. squamosa* (Oo and Khine, 2017; Bagavan et al., 2011), and N-Nitrosoxylopine, roemerolidine, and duguevalline alkaloids are observed responsible for antimalarial properties (Johns et al., 2011). According to a previous report, essential oil extracts of *A. squamosa Linn.* leaves have also been identified having antimalarial property, thus causing parasitic growth inhibition (Meira et al., 2015). *A. squamosa* alkaloids were identified as larvicidal and chemosterilant against malaria parasite *Anopheles stephensi* (Saxena et al., 1993).

Evaluation of antidiabetic activity of *A. squamosa* is found to be beneficial for reduction of blood sugar levels in streptozotocin diabetic rats as compared to untreated groups (Tomar and Sisodia, 2012). The ethanol and aqueous extract of *A. squamosa* leaves were found to be inhibiting high blood sugar levels in streptozotocin-induced diabetic rats and alloxan-induced diabetic rabbits (Gupta et al., 2005a; Gupta et al., 2005c; Mohd et al., 2009; Sangala et al., 2011). Aqueous extracts of *A. squamosa* leaves were also found antidiabetic in streptozotocin-nicotinamide type 2 diabetic rats (Shirwaikaret al., 2004c). Oral administration of leaf alcohol extract of *A. squamosa* revealed antidiabetic activity with reduction in plasma glucose levels, and significant difference in serum lipid profiles was observed in NIDDM (non-insulin dependent diabetes mellitus induced Wistar rats (Shirwaikaret al., 2004a).

17.4.5 Hepatoprotective and Cardioprotective Activities

Any injury, disease, toxicity, or malfunctioning in the liver or heart remains a grave concern, especially as there are so many alarming hepato and cardiac associated disorders. Therefore, it is inevitable for scientists to come up with understanding of hepatoprotective and cardioprotective properties of medicinal plants. Both *A. marmelos* and *A. squamosa* have been reported in the literature to contribute to hepatoprotective and cardioprotective activities.

17.4.5.1 *Aegle marmelos*

Ethanolic and aqueous extracts of *A. marmelos* are have shown to be potential hepatoprotective agents against carbon tetrachloride-induced liver damage in mice subjects as evident by moderate activity of enzymes serum glutamate oxaloacetate transaminase (SGOT), serum glutamate pyruvate transaminase (SGPT), and alkaline phosphatase (ALP) as compared to silymarin as a control (Kalaivani et al., 2009). Fruit pulp/seed extracts of *A. marmelos* have also shown significant reduction in CCl_4-induced elevation in plasma enzymes and bilirubin concentration in liver-damaged rats (Singh and Rao, 2008). According to studies published by Arun and Balasubramanian (2011)and Singanan et al., (2007), *A. marmelos* leaves have enormous hepatoprotective value as observed by significant variations in biochemical parameters such as increased thiobarbituric acid reactive substances (TBARS), decreased glutathione reductase (GSH), superoxide dismutase (SOD) and Catalase (CAT), increased glutathione peroxidase (GPx), decreased vitamin E and C, and increased serum iron and copper levels in ethanol-induced albino rats as compared to control animals. Methanolic extracts of *A. marmelos* have also shown hepatoprotective activity in CCl_4-intoxicated rats (Siddique et al., 2011).

The aqueous extracts of *A. marmelos* leaves have shown significant cardioprotective effect as experimentally evident in isoproterenol-induced myocardial injury in rats (Ramachandra et al., 2012). Doxorubicin-induced cardiotoxic mice were protected by the hydroalcoholic extract of bael (Jagetia and Venkatesh, 2015). Aurapten, a phytoconstituent of the bael fruit, imparts cardioprotective ability to *A. marmelos* (Jhajhria and Kumar, 2016). Periplogenin, a cardenolide phytochemical constituent of *A. marmelos*, is effective against doxorubicin-induced cardiac disorder and hepatotoxicity in rats (Panda and Kar, 2009).

17.4.5.2 *Annona squamosa*

A. squamosa leaf extract has shown experimental evidence of hepatoprotective efficiency as observed by the increase in enzymes and bilirubin, triglycerides, and cholesterol and the decrease in protein content in paracetamol-induced liver toxicity in rats (Rajeshkumar et al., 2015). Hydroalcoholic seeds, leaves, and bark extract of *A. squamosa* have shown hepatoprotective effects against CCl_4-induced liver toxicity in rats (Mehta and Paliwal, 2017; Sonkar et al., 2016). Liver injury induced by alcohol in sprague dawley rats and diethylnitrosamine (DEN) in Swiss albino mice showed hepatoprotective activity when administered with seed and alcoholic leaf extract of *A. squamosa*, respectively (Zahid et al., 2020; Raj et al., 2009).

A. squamosa leaf extract was examined for its phytochemical constitution, and it was found that Kaempferol one of the flavonoid components showed cardioprotective applications along with other pharmacological activities (Varadharajan et al., 2012). Since *A. squamosa* plants have hypocholesterolemic and hypolipidemic effect, it becomes a potential cardioprotective agent (Sharma et al., 2019).

17.4.6 Antipyretic, Anti-Inflammatory and Analgesic Activities

Pain and inflammation occur as a first line of defense mechanism against any injuries. Medicinal plants in general have shown promising results in dealing with pain, inflammation, and fever-like symptoms. *A. marmelos* and *A. squamosa* keep high potential of delivering therapy to deal with such symptoms.

17.4.6.1 *Aegle marmelos*

Unripe fruit ethanolic extract of *A. marmelos* shows anti-inflammatory, analgesic, and anti-ulcerogenic activities (Rao et al., 2003), while water extract of dried flower of *A. marmelos* possesses anti-inflammatory property as identified in Wistar rats (Kumari et al., 2014). In experimental acute and chronic inflammatory animal models, aqueous root bark extract of *A. marmelos* has also shown anti-inflammatory effects (Benni et al., 2011; Arulmozhi et al., 2018). The results of these various studies reveal that leaf extract of *A. marmelos* is effective as antipyretic, analgesic, and anti-inflammatory agents in rats (Arul et al., 2005).

17.4.6.2 *Annona squamosa*

A chemical derivative 18-acetoxy-ent-kaur-16-ene from *A. squamosa* exhibits significant analgesic and anti-inflammatory activity (Chavan et al., 2011). Caryophyllene oxide in the methanolic extract of the bark of *A. squamosa* exhibited anti-inflammatory and analgesic activity at comparable levels with the standard anti-inflammatory drug Pentazocine and analgesic drug aspirin (Singh et al., 2014c; Chavan et al., 2011). Phytochemical analysis of alcoholic extracts of *A. squamosa* revealed they have anti-inflammatory components such as sterols, triterpenoids, flavonoids, and tannins, thus can be useful for therapy against inflammation (Hemalatha and Satyanarayana, 2009). When Wistar mice and rats were subjected to ethanolic extract of *A. squamosa* leaves, the results showed significant analgesic and anti-inflammatory activity comparable to Pentozocine and Aceclofenac (Singh et al., 2012).

17.5 COVID-19 PERSPECTIVE OF *AEGLE MARMELOS* AND *ANNONA SQUAMOSA*

There is a need to address the impending applications of the medicinal plants *A. marmelos* and *A. squamosa* as a prophylactic measure against COVID-19 disease caused by SARS-CoV-2 virus. According to previous reports, it is factual that consumption of bael fruits is effective against respiratory diseases and in enhancing immunity, thus it can be beneficial for minimizing COVID-19 (Yadav et al., 2021). Ayurvedic medicine system is largely based on medicinal plants that provide active compounds, such as scopoletin, gallic acid, curcumine, etc. with potential antiviral properties in Ayurvedic medicines such as *Indukantham Kwatham* (IK), *MukkamukkatuvadiGulika* (MMG), and *Vilvadi Gulika* (VG) (Sulaiman et al., 2021). In order to have a preliminary idea about bioactive compounds from medicinal plants for their effectiveness as drugs and SARS-Cov2 components as target site, *in silico* molecular docking analysis was conducted. Phytocompounds EBDGp ((2S)-Eriodictyol 7-O-(6''-O-galloyl)-beta-D-glucopyranoside) from *Phyllanthus emblica* has shown high stability with SARS-Cov2 RdRp (RNA-dependent RNA-polymerase) (Pandey et al., 2021), and Seselin from *A. marmelos* has shown inhibitory potential against receptor SARS-CoV-2S spike protein, free enzyme of the SARS-CoV-2 (2019-nCoV) main protease and COVID-19 main protease (Nivetha et al., 2020). Both studies need experimental validation before considering them as potentially effective drug molecules. *In silico* analysis of *A. squamosa* also revealed the importance of quercitin-3glucoside, kaempferol-3-rutinoside, and rutin as bioactive phytocomponents which may bind to receptor binding domain of spike protein of SARS-CoV-2 and, thus, can have anti-COVID-19 activity (Muna et al., 2021).

17.6 CONCLUSION

It can be concluded that *Aegle marmelos* and *Annona squamosa* hold immense potential to be used for deriving chemical constituents which can be used as drug formulations for treating various diseases and syndromes. They can be used to treat ailments as mild as fever, pain, and diarrhea to as severe as diabetes, hepatic disorders, and cancer. According to several published reports, it can be concluded that fruits, leaves, flowers, roots, seeds, and bark of *Aegle marmelos* and *Annona squamosa* plants are used as phytochemical resources and thus display various medicinal properties like antioxidant, antimicrobial, anticancer, antidiabetic, anti-inflammatory, analgesic, hepatoprotective, and cardioprotective. Collective knowledge about *Aegle marmelos* and *Annona squamosa* nutritional, medicinal, and commercial value indicates that they are rich sources of phytochemicals which make them potent for herbal and traditional forms of medicine. More than 200 phytochemicals which constitute *Aegle marmelos* and *Annona squamosa* put them at a pedestal for exploring their potential to produce pharmaceuticals which eventually can be used for treatment of diseases of concern in *in vivo* and in clinical settings. Seselin from bael and quercitin and rutin derivatives from sugar apple have also shown preliminary importance as protective agents against the ongoing pandemic of COVID-19 disease. Conclusively, their medicinal importance can help to bring up prophylactic as well as curable therapeutic advancements.

ACKNOWLEDGMENTS

We are grateful to the Department of Biotechnology, IMS Engineering College, Ghaziabad-201015, Uttar Pradesh and Vignan's Foundation for Science, Technology, and Research (deemed a university), and Vadlamudi, Guntur, Andhra Pradesh, for their continuous encouragement during the subject writing.

REFERENCES

Aamir, J., Kumari, A., Khan, M.N. and Medam, S.K. 2013. Evaluation of the combinational antimicrobial effect of *Annona squamosa* and *Phoenix dactylifera* seeds methanolic extract on standard microbial strains. *International Research Journal of Biological Sciences*, 2(5):68–73.

Abdallah, I.Z., Salem, I., El-Salam, A. and Nayrouz, A.S. 2017. Evaluation of antidiabetic and antioxidant activity of *Aegle marmelos L. Correa* fruit extract in diabetic rats. *Egyptian Journal of Hospital Medicine*, 67(2):731–741.

Abirami, S.G., Vivekanandhan, K., Hemanthkumar, R., Prasanth, S. and Kumar, J.R. 2014. Study of antimicrobial potential of *Aegle marmelos*. *Journal of Medicinal Plants*, 2(2): 113–116.

Acda, M.N. 2014. Chemical composition of ethanolic seed extract of *Annona squamosa L.* and *A. muricata L.* (Annonaceae) using GC-MS analysis. *Philippine Agricultural Scientist*, 97(4):422–426.

Addario, B.J. 2015. Lung cancer is a global epidemic and requires a global effort. *Annals of Translational Medicine*, 3(2):26.

Aher, P.S., Shinde, Y.S. and Chavan, P.P. 2012. In vitro evaluation of antibacterial potential of *Annona squamosa L.* against pathogenic bacteria. *International Journal of Pharmaceutical Sciences and Research*, 3(5):1457.

Alali, F.Q., Liu, X.X. and McLaughlin, J.L. 1999. Annonaceous acetogenins: Recent progress. *Journal of Natural Products*, 62(3):504–540.

Almeida, M.M.B., de Sousa, P.H.M., Arriaga, Â.M.C., do Prado, G.M., de Carvalho Magalhães, C.E., Maia, G.A. and de Lemos, T.L.G. 2011. Bioactive compounds and antioxidant activity of fresh exotic fruits from northeastern Brazil. *Food Research International*, 44(7):2155–2159.

Arul, V., Miyazaki, S. and Dhananjayan, R. 2004. Mechanisms of the contractile effect of the alcoholic extract of *Aegle marmelos Corr.* on isolated guinea pig ileum and tracheal chain. *Phytomedicine*, 11(7–8):679–683.

Arul, V., Miyazaki, S. and Dhananjayan, R. 2005. Studies on the anti-inflammatory, antipyretic and analgesic properties of the leaves of *Aegle marmelos Corr. Journal of Ethnopharmacology*, 96(1–2):159–163.

Arulmozhi, S., Steffi, M.M. and Singh, S. 2018. Unraveling the inhibitory activities of Gentiana derivatives against inflammatory response of rheumatoid arthritis using molecular modeling approach. *International Journal of Pharmaceutical and Clinical Research*, 10(1):1–15.

Arumugam S, Kavimani S, Kadalmani B, Ahmed ABA, Akbarsha MA, Rao MV. 2008. Antidiabetic activity of leaf and callus extracts of *Aegle marmelos* in rabbit. *Science Asia*, 34:317–321.

Arun, K. and Balasubramanian, U. 2011. Comparative study on Hepatoprotective activity of *Aegle marmelos* and *Eclipta alba* against alcohol induced in albino rats. *International Journal of Environmental Sciences*, 2(2):389–402.

Attiq, A., Jalil, J. and Husain, K. 2017. Annonaceae: Breaking the wall of inflammation. *Frontiers in Pharmacology*, 8:752.

Badrie, N. and Schauss, A.G. 2010. Soursop (*Annona muricata L.*): Composition, nutritional value, medicinal uses, and toxicology. In *Bioactive Foods in Promoting Health*. New York: Academic Press, 621–643.

Bagavan, A., Rahuman, A.A., Kaushik, N.K. and Sahal, D. 2011. In vitro antimalarial activity of medicinal plant extracts against Plasmodium falciparum. *Parasitology Research*, 108(1):15–22.

Baliga, M.S., Bhat, H.P., Joseph, N. and Fazal, F. 2011. Phytochemistry and medicinal uses of the Bael fruit (*Aegle marmelos Correa*): A concise review. *Food Research International*, 44(7):1768–1775.

Baliga, M.S., Bhat, H.P., Pereira, M.M., Mathias, N. and Venkatesh, P. 2010. Radioprotective effects of *Aegle marmelos (L.) Correa* (Bael): A concise review. *Journal of Alternative and Complementary Medicine*, 16(10):1109–1116.

Bansal, Y. and Bansal, G. 2011. Analytical methods for standardization of *Aegle marmelos*: A review. *Journal of Pharmaceutical Education and Research*, 2(2):37–44.

Bele, M.Y., Focho, D.A., Egbe, E.A. and Chuyong, B.G. 2011. Ethnobotanical survey of the uses Annonaceae around mount Cameroon. *African Journal of Plant Science*, 5(4):237–247.

Benni, J.M., Jayanthi, M.K. and Suresha, R.N. 2011. Evaluation of the anti-inflammatory activity of *Aegle marmelos* (Bilwa) root. *Indian Journal of Pharmacology*, 43(4):393–397.

Bhar, K., Mondal, S. and Suresh, P. 2019. An eye-catching review of *Aegle marmelos L.*(golden apple). *Pharmacognosy Journal*, 11(2):207–224.

Bhardwaj, A., Satpathy, G. and Gupta, R.K. 2014. Preliminary screening of nutraceutical potential of *Annona squamosa*, an underutilized exotic fruit of India and its use as a valuable source in functional foods. *Journal of Pharmacognosy and Phytochemistry*, 3(2):172–180.

Bhardwaj, R.L. and Nandal, U. 2015. Nutritional and therapeutic potential of Bael (*Aegle marmelos Corr.*) fruit juice: A review. *Nutrition & Food Science*, 45:895–919.

Bhavani, R. 2014. Antidiabetic activity medicinal plant *Aegle marmelos* (*linn.*) on alloxan induced diabetic rats. *International Research Journal of Pharmaceutical and Biosciences (IRJPBS)*, 1(1):36–44.

Boakye, A.A., Wireko-Manu, F.D., Agbenorhevi, J.K. and Oduro, I. 2015. Antioxidant activity, total phenols and phytochemical constituents of four underutilised tropical fruits. *International Food Research Journal*, 22(1):262–268.

Boulogne, I., Germosén-Robineau, L., Ozier-Lafontaine, H., Fleury, M. and Loranger-Merciris, G. 2011. TRAMIL ethnopharmalogical survey in Les Saintes (Guadeloupe, French West Indies): A comparative study. *Journal of Ethnopharmacology*, 133(3):1039–1050.

Cercato, L.M., White, P.A., Nampo, F.K., Santos, M.R. and Camargo, E.A. 2015. A systematic review of medicinal plants used for weight loss in Brazil: Is there potential for obesity treatment?.*Journal of Ethnopharmacology*, 176:286–296.

Chanda, R., Ghosh, A., Mitra, T., Mohanty, J.P., Bhuyan, N. and Pawankar, G. 2008. Phytochemical and pharmacological activity of *Aegle marmelos* as a potential medicinal plant: An overview. *Internet Journal of Pharmacology*, 6(1):3.

Chandrashekar, C. and Kulkarni, V.R. 2011. Isolation, Characterizations and Free radical scavenging activity of *Annona squamosa* leaf. *Journal of Pharmacy Research*, 4(3):610–611.

Chavan, M.J., Shinde, D.B. and Nirmal, S.A. 2006. Major volatile constituents of *Annona squamosa L.* bark. *Natural Product Research*, 20(8):754–757.

Chavan, M.J., Wakte, P.S. and Shinde, D.B. 2011. Analgesic and anti-inflammatory activities of 18-acetoxy-ent-kaur-16-ene from *Annona squamosa L.* bark. *Inflammopharmacology*, 19(2):111–115.

Chen, J., Chen, Y. and Li, X. 2011. Beneficial aspects of custard apple (*Annona squamosa L.*) seeds. In *Nuts and Seeds in Health and Disease Prevention*. Cambridge, MA: Academic Press, 439–445.

Chen, Y., Chen, Y., Shi, Y., Ma, C., Wang, X., Li, Y., Miao, Y., Chen, J. and Li, X. 2016. Antitumor activity of *Annona squamosa* seed oil. *Journal of Ethnopharmacology*, 193:362–367.

Chen, Y., Xu, S.S., Chen, J.W., Wang, Y., Xu, H.Q., Fan, N.B. and Li, X. 2012. Anti-tumor activity of *Annona squamosa* seeds extract containing annonaceousacetogenin compounds. *Journal of Ethnopharmacology*, 142(2):462–466.

Chen, Y.Y., Ma, C.Y., Wang, M.L., Lu, J.H., Hu, P., Chen, J.W., Li, X. and Chen, Y. 2020. Five new ent-kaurane diterpenes from *Annona squamosa L.* pericarps. *Natural Product Research*, 34(15):2243–2247.

Chockalingam, V., Kadali, S.S. and Gnanasambantham, P. 2012. Antiproliferative and antioxidant activity of *Aegle marmelos* (*Linn.*) leaves in Dalton's Lymphoma Ascites transplanted mice. *Indian Journal of Pharmacology*, 44(2):225.

Coria-Téllez, A.V., Montalvo-Gónzalez, E., Yahia, E.M. and Obledo-Vázquez, E.N. 2018. *Annona muricata*: A comprehensive review on its traditional medicinal uses, phytochemicals, pharmacological activities, mechanisms of action and toxicity. *Arabian Journal of Chemistry*, 11(5):662–691.

Costa-Lotufo, L.V., Khan, M.T.H., Ather, A., Wilke, D.V., Jimenez, P.C., Pessoa, C., de Moraes, M.E.A. and de Moraes, M.O. 2005. Studies of the anticancer potential of plants used in Bangladeshi folk medicine. *Journal of Ethnopharmacology*, 99(1):21–30.

Devi, M., Devi, S., Sharma, V., Rana, N., Bhatia, R.K. and Bhatt, A.K. 2020. Green synthesis of silver nanoparticles using methanolic fruit extract of *Aegle marmelos* and their antimicrobial potential against human bacterial pathogens. *Journal of Traditional and Complementary Medicine*, 10(2):158–165.

Dhankar, S., Ruhil, S., Balhara, M., Dhankhar, S. and Chhillar, A.K. 2011. *Aegle marmelos* (*Linn.*) *Correa*: A potential source of Phytomedicine. *Journal of Medicinal Plants Research*, 5(9):1497–1507.

Dutta, A., Lal, N., Naaz, M., Ghosh, A. and Verma, R. 2014. Ethnological and Ethno-medicinal importance of *Aegle marmelos* (*L.*) *Corr* (Bael) among indigenous people of India. *American Journal of Ethnomedicine*, 1(5):290–312.

El-Chaghaby, G.A., Ahmad, A.F. and Ramis, E.S. 2014. Evaluation of the antioxidant and antibacterial properties of various solvents extracts of *Annona squamosa L.* leaves. *Arabian Journal of Chemistry*, 7(2):227–233.

Ezzat, S.M., Jeevanandam, J., Egbuna, C., Kumar, S. and Ifemeje, J.C. 2019. Phytochemicals as sources of drugs. In *Phytochemistry: An in-silico and in-vitro Update*. Singapore: Springer, 3–22.

Fofana, S., Keita, A., Balde, S., Ziyaev, R. and Aripova, S.F. 2012. Alkaloids from leaves of *Annona muricata*. *Chemistry of Natural Compounds*, 48(4):714–714.

Fujimoto, Y., Eguchi, T., Kakinuma, K., Ikekawa, N., Sahai, M. and Gupta, Y.K. 1988. Squamocin, a new cytotoxic bis-tetrahydrofuran containing acetogenin from *Annona squamosa. Chemical and Pharmaceutical Bulletin*, 36(12):4802–4806.

Gajalakshmi, S., Divya, R., Divya, V., Mythili, S. and Sathiavelu, A. 2011. Pharmacological activities of *Annona squamosa*: A review. *International Journal of Pharmaceutical Sciences Review and Research*, 10(2):24–29.

Gangadhar, M., Shraddha, K. and Ganesh, M. 2012. Antimicrobial screening of garlic (*Allium sativum*) extracts and their effect on glucoamylase activity in-vitro. *Journal of Applied Pharmaceutical Science*, 2(1):16.

Gupta, R.K., Kesari, A.N., Diwakar, S., Tyagi, A., Tandon, V., Chandra, R. and Watal, G. 2008. In vivo evaluation of anti-oxidant and anti-lipidimic potential of *Annona squamosa* aqueous extract in Type 2 diabetic models. *Journal of Ethnopharmacology*, 118(1):21–25.

Gupta, R.K., Kesari, A.N., Murthy, P.S., Chandra, R., Tandon, V. and Watal, G. 2005a. Hypoglycemic and antidiabetic effect of ethanolic extract of leaves of *Annona squamosa L.* in experimental animals. *Journal of Ethnopharmacology*, 99(1):75–81.

Gupta, R.K., Kesari, A.N., Watal, G., Murthy, P.S., Chandra, R. and Tandon, V. 2005b. Nutritional and hypoglycemic effect of fruit pulp of *Annona squamosa* in normal healthy and alloxan-induced diabetic rabbits. *Annals of Nutrition and Metabolism*, 49(6):407–413.

Gupta, R.K., Kesari, A.N., Watal, G., Murthy, P.S., Chandra, R., Maithal, K. and Tandon, V. 2005c. Hypoglycaemic and antidiabetic effect of aqueous extract of leaves of Annona squamosa (L.) in experimental animal. *Current Science*, 88(8):1244–1254.

Gurjar, P.S., Lal, N., Gupta, A.K. and Marboh, E.S. 2015. A review on medicinal values and commercial utility of Bael Pawan Singh Gurjar, Narayan Lall, Alok Kumar Guptal, evening stone Marbohl. *International Journal of Life Sciences Scientific Research*, 1(1):5–7.

Hajdu, Z. and Hohmann, J. 2012. An ethnopharmacological survey of the traditional medicine utilized in the community of Porvenir, Bajo Paraguá Indian Reservation, Bolivia. *Journal of Ethnopharmacology*, 139(3):838–857.

Hasitha, P., Ravi, S.K., Shivashankarappa, A. and Dharmesh, S.M. 2018. Comparative morphological, biochemical and bioactive potentials of different varieties of bael [*Aegle marmelos* (*L.*) *Corrêa*] of India. *Indian Journal of Traditional Knowledge*, 17(3):584–591.

Hemlatha, K. and Satyanarayana, D. 2009. Anti-inflammatory activity of *Annona squamosa Linn. Biomedical and Pharmacology Journal*, 2(1):17–20.

Ismail, M.Y.M. 2009a. Clinical evaluation of antidiabetic activity of Bael leaves. *World Applied Sciences Journal*, 6:1518–1520.

Ismail, M.Y.M. 2009b. Clinical evaluation of antidiabetic activity of Trigonella seeds and *Aegle marmelos* leaves. *World Applied Sciences Journal*, 7(10):1231–1234.

Jagetia, G.C. and Venkatesh, P. 2015. An indigenous plant Bael (*Aegle marmelos* (*L.*) *Correa*) extract protects against the doxorubicin-induced cardiotoxicity in mice. *Biochemphysiol*, 4(163):2.

Jagetia, G.C., Venkatesh, P. and Baliga, M.S. 2005. *Aegle marmelos* (*L.*) *CORREA* Inhibits the Proliferation of Transplanted Ehrlich Ascites Carcinoma in Mice. *Biological and Pharmaceutical Bulletin*, 28(1):58–64.

Jhajhria, A. and Kumar, K. 2016. Tremendous pharmacological values of *Aegle marmelos. International Journal of Pharmaceutical Sciences Review and Research*, 36(2):121–127.

Jiang, R.W., Lu, Y., Min, Z.D. and Zheng, Q.T. 2003. Molecular structure and pseudopolymorphism of squamtin A from *Annona squamosa. Journal of Molecular Structure*, 655(1):157–162.

Johns, T., Windust, A., Jurgens, T. and Mansor, S.M. 2011. Antimalarial alkaloids isolated from *Annona squamosa. Phytopharmacology*, 1(3):49–53.

Kala, C.P. 2006. Ethnobotany and ethnoconservation of *Aegle marmelos (L.) Correa. Indian Journal of Traditional Knowledge*, 5(4):537–540.

Kalaivani, T., Premkumar, N., Ramya, S., Siva, R., Vijayakumar, V., Meignanam, E., Rajasekaran, C. and Jayakumararaj, R. 2009. Investigations on hepatoprotective activity of leaf extracts of *Aegle marmelos* (*L.*) *Corr.* (Rutaceae). *Ethnobotanical Leaflets*, 2009(1):4.

Kaleem, M., Asif, M., Ahmed, Q.U. and Bano, B. 2006. Antidiabetic and antioxidant activity of *Annona squamosa* extract in streptozotocin-induced diabetic rats. *Singapore Medical Journal*, 47(8):670.

Kalidindi, N., Thimmaiah, N.V., Jagadeesh, N.V., Nandeep, R., Swetha, S. and Kalidindi, B. 2015. Antifungal and antioxidant activities pf organic and aqueous extracts of *Annona squamosa Linn.* leaves. *Journal of Food and Drug Analysis*, 23(4):795–802.

Kamalakkannan, N. and Stanely, M.P.P. 2003. Effect of *Aegle marmelos Correa.* (Bael) fruit extract on tissue antioxidants in streptozotocin diabetic rats. *Indian Journal of Experimental Biology*, 41:1285–1288.

Kamaraj, C., Kaushik, N.K., Rahuman, A.A., Mohanakrishnan, D., Bagavan, A., Elango, G., Zahir, A.A., Santhoshkumar, T., Marimuthu, S., Jayaseelan, C. and Kirthi, A.V. 2012. Antimalarial activities of medicinal plants traditionally used in the villages of Dharmapuri regions of South India. *Journal of Ethnopharmacology*, 141(3):796–802.

Kaou, A.M., Mahiou-Leddet, V., Hutter, S., Aïnouddine, S., Hassani, S., Yahaya, I., Azas, N. and Ollivier, E. 2008. Antimalarial activity of crude extracts from nine African medicinal plants. *Journal of Ethnopharmacology*, 116(1):74–83.

Katagi, K.S., Munnolli, R.S. and Hosamani, K.M. 2011. Unique occurrence of unusual fatty acid in the seed oil of *Aegle marmelos Corre*: Screening the rich source of seed oil for bio-energy production. *Applied Energy*, 88(5):1797–1802.

Kesari, A.N., Gupta, R.K., Singh, S.K., Diwakar, S. and Watal, G. 2006. Hypoglycemic and antihyperglycemic activity of *Aegle marmelos* seed extract in normal and diabetic rats. *Journal of Ethnopharmacology*, 107(3):374–379.

Kintzios, S.E. 2006. Terrestrial plant-derived anticancer agents and plant species used in anticancer research. *Critical Reviews in Plant Sciences*, 25(2):79–113.

Kothari, S., Mishra, V., Bharat, S. and Tonpay, S.D. 2011. Antimicrobial activity and phytochemical screening of serial extracts from leaves of *Aegle marmelos* (*Linn.*). *Acta Poloniae Pharmaceutica*, 68(5):687–692.

Kothari, V. and Seshadri, S. 2010. Antioxidant activity of seed extracts of *Annona squamosa* and Carica papaya. *Nutrition & Food Science*: 403–408.

Kotkar, H.M., Mendki, P.S., Sadan, S.V.G.S., Jha, S.R., Upasani, S.M. and Maheshwari, V.L. 2002. Antimicrobial and pesticidal activity of partially purified flavonoids of *Annona squamosa*. *Pest Management Science: Formerly Pesticide Science*, 58(1):33–37.

Kumar, M., Changan, S., Tomar, M., Prajapati, U., Saurabh, V., Hasan, M., Sasi, M., Maheshwari, C., Singh, S., Dhumal, S. and Thakur, M. 2021. Custard apple (*Annona squamosa L.*) Leaves: Nutritional Composition, Phytochemical Profile, and Health-Promoting Biological Activities. *Biomolecules*, 11(5):614.

Kumari, K.D.K.P., Weerakoon, T.C.S., Handunnetti, S.M., Samarasinghe, K. and Suresh, T.S. 2014. Anti-inflammatory activity of dried flower extracts of *Aegle marmelos* in Wistar rats. *Journal of Ethnopharmacology*, 151(3):1202–1208.

Lalremruta, V. and Prasanna, G.S. 2012. Evaluation of protective effect of *Aegle marmelos Corr.* in an animal model of chronic fatigue syndrome. *Indian Journal of Pharmacology*, 44(3):351.

Lambertini, E., Lampronti, I., Penolazzi, L., Khan, M.T.H., Ather, A., Giorgi, G., Gambari, R. and Piva, R. 2005. Expression of estrogen receptor α gene in breast cancer cells treated with transcription factor decoy is modulated by Bangladeshi natural plant extracts. *Oncology Research Featuring Preclinical and Clinical Cancer Therapeutics*, 15(2):69–79.

Lambertini, E., Piva, R., Khan, M.T.H., Lampronti, I., Bianchi, N., Borgatti, M. and Gambari, R. 2004. Effects of extracts from Bangladeshi medicinal plants on in vitro proliferation of human breast cancer cell lines and expression of estrogen receptor α gene. *International Journal of Oncology*, 24(2):419–423.

Lampronti, I., Martello, D., Bianchi, N., Borgatti, M., Lambertini, E., Piva, R., Jabbar, S., Choudhuri, M.S.K., Khan, M.T.H. and Gambari, R. 2003. In vitro antiproliferative effects on human tumor cell lines of extracts from the Bangladeshi medicinal plant *Aegle marmelos Correa*. *Phytomedicine*, 10(4):300–308.

Laphookhieo, S., Phungpanya, C., Tantapakul, C., Techa, S., Tha-in, S. and Narmdorkmai, W. 2011. Chemical constituents from *Aegle marmelos*. *Journal of the Brazilian Chemical Society*, 22:176–178.

Li, X.H., Hui, Y.H., Rupprecht, J.K., Liu, Y.M., Wood, K.V., Smith, D.L., Chang, C.J. and McLaughlin, J.L. 1990. Bullatacin, bullatacinone, and squamone, a new bioactive acetogenin, from the bark of *Annona squamosa*. *Journal of natural products*, 53(1):81–86.

Ma, C., Chen, Y., Chen, J., Li, X. and Chen, Y. 2017. A review on *Annona squamosa L.*: Phytochemicals and biological activities. *American Journal of Chinese medicine*, 45(05):933–964.

Maikhuri, R.K., Rao, K.S. and Saxena, K.G. 2004. Bioprospecting of wild edibles for rural development in the central Himalayan mountains of India. *Mountain Research and Development*, 24(2):110–113.

Maity, P., Hansda, D., Bandyopadhyay, U. and Mishra, D.K. 2009. Biological activities of crude extracts and chemical constituents of Bael, *Aegle marmelos* (*L.*) Corr. *Indian Journal of Experimental Biology*, 47:849–861.

Manandhar, B., Paudel, K.R., Sharma, B. and Karki, R. 2018. Phytochemical profile and pharmacological activity of *Aegle marmelos Linn*. *Journal of Integrative Medicine*, 16(3):153–163.

Manochai, B., Ingkasupart, P., Lee, S.H. and Hong, J.H. 2018. Evaluation of antioxidant activities, total phenolic content (TPC), and total catechin content (TCC) of 10 sugar apple (*Annona squamosa L.*) cultivar peels grown in Thailand. *Food Science and Technology*, 38:294–300.

Manoharachary, C. and Nagaraju, D. 2016. Medicinal plants for human health and welfare. *Annals of Phytomedicine*, 5(1):24–34.

Mariod, A.A., Abdelwahab, S.I., Elkheir, S., Ahmed, Y.M., Fauzi, P.N.M. and Chuen, C.S. 2012. Antioxidant activity of different parts from *Annona squamosa*, and *Catunaregam nilotica* methanolic extract. *Acta Scientiarum Polonorum Technologia Alimentaria*, 11(3):249–258.

Mehta, S.D. and Paliwal, S., 2017. Hepatoprotective activity of hydroalcohilic extract of *Annona squamosa* seeds. *International Journal of Pharmaceutical and Phytopharmacological Research*, 9:997–1000.

Meira, C.S., Guimarães, E.T., Macedo, T.S., da Silva, T.B., Menezes, L.R., Costa, E.V. and Soares, M.B. 2015. Chemical composition of essential oils from Annona vepretorum Mart. and *Annona squamosa L.* (Annonaceae) leaves and their antimalarial and trypanocidal activities. *Journal of Essential Oil Research*, 27(2):160–168.

Misra, P., Pal, N.L., Guru, P.Y., Katiyar, J.C. and Tandon, J.S. 1991. Antimalarial activity of traditional plants against erythrocytic stages of Plasmodium berghei. *International Journal of Pharmacognosy*, 29(1):19–23.

Mohd, M., Alam, K.S., Mohd, A., Abhishek, M. and Aftab, A. 2009. Antidiabetic activity of the aqueous extract of *Annona squamosa* in streptozotocin induced hyperglycemic rats. *Pharmaceutical Research*, 2:59.

Morton, J. 1987. *Sugar Apple in Fruits of Warm Climates*. Miami, FL: Julia F. Morton, 69–72.

Mujeeb, F., Bajpai, P. and Pathak, N. 2014. Phytochemical evaluation, antimicrobial activity, and determination of bioactive components from leaves of *Aegle marmelos*. *BioMed Research International*, 2014: 497606.

Muna, I.N., Suharti, S., Muntholib, M. and Subandi, S. 2021, May. Powder preparation of sugar apple (*Annona squamosa L.*) and analyzing its potencies as anti-gout and anti-COVID-19. In *AIP Conference Proceedings*, 2353(1):030096. AIP Publishing LLC.

Nandhakumar, E. and Indumathi, P. 2013. In vitro antioxidant activities of methanol and aqueous extract of *Annona squamosa* (*L.*) fruit pulp. *Journal of Acupuncture and Meridian Studies*, 6(3):142–148.

Narender, T., Shweta, S., Tiwari, P., Reddy, K.P., Khaliq, T., Prathipati, P., Puri, A., Srivastava, A.K., Chander, R., Agarwal, S.C. and Raj, K. 2007. Antihyperglycemic and antidyslipidemic agent from *Aegle marmelos*. *Bioorganic & Medicinal Chemistry Letters*, 17(6):1808–1811.

Neethu Simon, K., Santhoshkumar, R. and Neethu, S.K. 2016. Phytochemical analysis and antimicrobial activities of *Annona squamosa* (*L*) leaf extracts. *Journal of Pharmacognosy and Phytochemistry*, 5(4):128–131.

Nguyen, M.T., Nguyen, V.T., Le, V.M., Trieu, L.H., Lam, T.D., Bui, L.M., Nhan, L.T.H. and Danh, V.T. 2020. Assessment of preliminary phytochemical screening, polyphenol content, flavonoid content, and antioxidant activity of custard apple leaves (*Annona squamosa Linn.*). In *IOP Conference Series: Materials Science and Engineering*, 736(6):062012. IOP Publishing.

Nigam, V. and Nambiar, V.S. 2015. Therapeutic potential of *Aegle marmelos* (*L.*) *Correa* leaves as an antioxidant and anti-diabetic agent: A review. *International Journal of Pharma Sciences and Research*, 6(3):611–621.

Nivetha, R., Bhuvaragavan, S. and Janarthanan, S., 2020. Inhibition of multiple SARS-CoV-2 proteins by an antiviral biomolecule, seselin from *Aegle marmelos* deciphered using molecular docking analysis. Preprint:https://doi.org/10.21203/rs.3.rs-31134/v1

Oo, W.M. and Khine, M.M. 2017. Pharmacological Activities of *Annona squamosa*: Updated. *Chemistry*, 3(6):86–93.

Padhi, L.P., Panda, S.K., Satapathy, S.N. and Dutta, S.K. 2011. In vitro evaluation of antibacterial potential of *Annona squamosa L.* and *Annona reticulata L.* from Similipal Biosphere Reserve, Orissa, India. *Journal of Agricultural Technology*, 7(1):133–142.

Panda, S. and Kar, A. 2009. Periplogenin-3-O--D-Glucopyranosyl-(1→ 6)--D-Glucopyaranosyl--(1→ 4)-D-Cymaropyranoside, Isolated from *Aegle marmelos* Protects Doxorubicin Induced Cardiovascular Problems and Hepatotoxicity in Rats. *Cardiovascular Therapeutics*, 27(2):108–116.

Pandey, K., Lokhande, K.B., venkateswara Swamy, K., Nagar, S. and Dake, M. 2021. In silico exploration of phytoconstituents from Phyllanthus emblica and *Aegle marmelos* as potential therapeutics against SARS-CoV-2 RdRp. Preprint: https://doi.org/10.21203/rs.3.rs-225174/v1

Pandey, N. and Brave, D. 2011a. Antioxidant activity of ethanolic extract of *Annona squamosa Linn* Bark. *International Journal of Biomedical and Pharmaceutical Sciences*, 2:1692–1697.

Pandey, N. and Barve, D. 2011b. Phytochemical and pharmacological review on *Annona squamosa Linn*. *International Journal of Research in Pharmaceutical and Biomedical Sciences*, 2(4):1404–1412.

Pardhasaradhi, B.V.V., Reddy, M., Ali, A.M., Kumari, A.L. and Khar, A. 2004. Antitumour activity of *Annona squamosa* seed extracts is through the generation of free radicals and induction of apoptosis. *Indian Journal of Biochemistry and Biophysics*, 41:167–172.

Pathirana, C.K., Madhujith, T. and Eeswara, J. 2020. Bael (*Aegle marmelos L. Corrêa*): A medicinal tree with immense economic potentials. *Advances in Agriculture*, 2020: 8814018.

Phuwapraisirisan, P., Puksasook, T., Jong-Aramruang, J. and Kokpol, U. 2008. Phenylethyl cinnamides: A new series of α-glucosidase inhibitors from the leaves of *Aegle marmelos*. *Bioorganic & Medicinal Chemistry Letters*, 18(18):4956–4958.

Poonkothai, M. and Saravanan, M. 2008. Antibacterial activity of *Aegle marmelos* against leaf, bark and fruit extracts. *Ancient Science of Life*, 27(3):15.

Rahman, S. and Parvin, R. 2014. Therapeutic potential of *Aegle marmelos* (L.)-An overview. *Asian Pacific Journal of Tropical Disease*, 4(1):71–77.

Raj, D.S., Vennila, J.J., Aiyavu, C. and Panneerselvam, K. 2009. The hepatoprotective effect of alcoholic extract of *Annona squamosa* leaves on experimentally induced liver injury in Swiss albino mice. *International Journal of Integrative Biology*, 5(3):182–186.

Raja, W.W. and Khan, S.H., 2017. Estimation of some phytoconstituents and evaluation of antioxidant activity in *Aegle marmelos* leaves extract. *Journal of Pharmacognosy and Phytochemistry*, 6(1):37–40.

Rajan, S., Gokila, M., Jency, P., Brindha, P. and Sujatha, R.K. 2011. Antioxidant and phytochemical properties of *Aegle marmelos* fruit pulp. *International Journal of Current Pharmaceutical Research*, 3(2):65–70.

Rajeshkumar, S., Tamilarasan, B. and Sivakumar, V. 2015. Phytochemical screening and hepatoprotective efficacy of leaves extracts of *Annona squamosa* against paracetamol induced liver Toxicity in rats. *International Journal of Pharmacognosy*, 22:178–185.

Ramachandra, Y., Ashajyothi, C. and Padmalatha, R. 2012. Cardio protective effect of *aegle marmelos* on isoproterenol induced myocardial infarction in rats. *International Journal of Biology, Pharmnacy and Applied Sciences*, 1:718–729.

Ramalingum, N. and Mahomoodally, M.F. 2014. The therapeutic potential of medicinal foods. *Advances in Pharmacological and Pharmaceutical Sciences Advances*, 2014:354264.

Rao, C.V., Ojha, S.K., Amresh, G., Mehrotra, S. and Pushpangadan, P. 2003. Analgesic, antiinflammatory and antiulcerogenic activity of unripe fruits of *Aegle marmelos. Acta Pharmaceutica Turcica*, 45:85–91.

Reddy, V.P. and Urooj, A. 2013. Antioxidant properties and stability of *Aegle marmelos* leaves extracts. *Journal of Food Science and Technology*, 50(1):135–140.

Ruddaraju, L.K., Pallela, P.N.V.K., Pammi, S.V.N., Padavala, V.S. and Kolapalli, V.R.M. 2019. Synergetic antibacterial and anticarcinogenic effects of *Annona squamosa* leaf extract mediated silver nano particles. *Materials Science in Semiconductor Processing*, 100:301–309.

Sabu, M.C. and Kuttan, R. 2004. Antidiabetic activity of *Aegle marmelos* and its relationship with its antioxidant properties. *Indian Journal of Physiology and Pharmacology*, 48(1):81–88.

Saha, R. 2011. Pharmacognosy and pharmacology of *Annona squamosa. International Journal of Pharmacy and Life Sciences*, 2:1183–1189.

Sahare, K.N., Anandhraman, V., Meshram, V.G., Meshram, S.U., Reddy, M.V.R., Tumane, P.M. and Goswami, K. 2008. Anti-microfilarial activity of methanolic extract of *Vitex negundo* and *Aegle marmelos* and their phytochemical analysis. *Indian Journal of Experimental Biology*, 46:128–131.

Sangala, R., Kodati, D., Burra, S., Gopu, J. and Dubasi, A. 2011. Evaluation of antidiabetic activity of *Annona squamosa Linn* Seed in alloxan–induced diabetic rats. *Diabetes*, 2(1):100–106.

Sarma, R., Mahanta, S. and Khanikor, B. 2017. Insecticidal activities of the essential oil of *Aegle marmelos* (Linnaeus, 1800) against *Aedes aegypti* (Linnaeus, 1762) and *Culex quinquefasciatus* (*Say*, 1823). *Indian Journal of Agricultural Research*, 5(5):304–311.

Saxena, R.C., Harshan, V., Saxena, A., Sukumaran, P., Sharma, M.C. and Kumar, M.L. 1993. Larvicidal and chemosterilant activity of *Annona squamosa* alkaloids against Anopheles stephensi. *Journal-American Mosquito Control Association*, 9:84–84.

Sekar, D.K., Kumar, G., Karthik, L. and Rao, K.B. 2011. A review on pharmacological and phytochemical properties of *Aegle marmelos* (*L.*) *Corr. Serr.*(Rutaceae). *Asian Journal of Plant Science and Research*, 1(2):8–17.

Sharma, P.C., Bhatia, V., Bansal, N. and Sharma, A. 2007. A review on Bael tree. *Indian Journal of Natural Products and Resources*, 6:171–178.

Sharma, S.K., Gupta, M.L. and Kumar, B. 2019. Hypocholesterolemic Efficacy of *Annona squamosa* (*L.*) Extract In Mice Diabetic Models. *Journal of Biochemistry and Biotechnology*, 5:41–45.

Shehata, M.G., Abu-Serie, M.M., Abd El-Aziz, N.M. and El-Sohaimy, S.A. 2021. Nutritional, phytochemical, and in vitro anticancer potential of sugar apple (*Annona squamosa*) fruits. *Scientific Reports*, 11(1):1–13.

Shirwaikar, A., Rajendran, K. and Kumar, C. 2004a. Oral antidiabetic activity of *Annona squamosa* leaf alcohol extract in NIDDM rats. *Pharmaceutical Biology*, 42(1):30–35.

Shirwaikar, A., Rajendran, K. and Kumar, C.D. 2004b. In vitro antioxidant studies of *Annona squamosa Linn.* leaves. *Indian Journal of Experimental Biology*, 42(8):803–807.

Shirwaikar, A., Rajendran, K., Kumar, C.D. and Bodla, R. 2004c. Antidiabetic activity of aqueous leaf extract of *Annona squamosa* in streptozotocin–nicotinamide type 2 diabetic rats. *Journal of Ethnopharmacology*, 91(1):171–175.

Shukry, W.M., Galilah, D.A., Elrazek, A.A. and Shapana, H.A. 2019. Mineral Composition, Nutritional Properties, Vitamins, and Bioactive Compounds in *Annona squamosa L.* grown at different sites of Egypt. Ser. Bot. *Environmental Science*, 1:7–22.

Siddique, N.A., Mujeeb, M., Najmi, A.K., Aftab, A. and Aslam, J. 2011. Free radical scavenging and hepatoprotective activity of *Aegle marmelos* (*Linn.*) *corr* leaves against carbon tetrachloride. *International Journal of Comprehensive Pharmacy*, 2(08):1–6.

Singanan, V., Singanan, M. and Begum, H. 2007. The hepatoprotective effect of Bael leaves (*Aegle marmelos*) in alcohol induced liver injury in albino rats. *International Journal of Science & Technology*, 2(2):83–92.

Singh, A., Sharma, H.K., Kaushal, P. and Upadhyay, A. 2014a. Bael (*Aegle marmelos Correa*) products processing: A review. *African Journal of Food Science*, 8(5):204–215.

Singh, A.K., Chakraborty, I. and Chaurasiya, A.K. 2014b. Bael preserve-syrup as booster of human health as a health drink. *Bioscan*, 9(2):565–569.

Singh, D.P., Mishra, B. and Mishra, R. 2012. Anti-nociceptive and anti-inflammatory activity of *Annona squamosa L.* leaf extract in mice and rats. *Research Journal of Pharmacognosy and Phytochemistry*, 4(3):182–185.

Singh, R. and Rao, H.S. 2008. Hepatoprotective effect of the pulp/seed of *Aegle marmelos correa* ex Roxb against carbon tetrachloride induced liver damage in rats. *International Journal of Green Pharmacy (IJGP)*, 2(4): 232.

Singh, T.P., Singh, R.K. and Malik, P. 2014c. Analgesic and anti-inflammatory activities of *Annona squamosa Linn* bark. *Journal of Scientific and Innovative Research*, 3:60–64.

Singh, Y., Bhatnagar, P. and Thakur, N. 2019. A review on insight of immense nutraceutical and medicinal potential of custard apple (*Annona squamosa Linn.*). *International Journal of Chemical Studies*, 7(2):1237–1245.

Sivaraj, R., Balakrishnan, A., Thenmozhi, M. and Venckatesh, R. 2011. Antimicrobial activity of *Aegle marmelos, Ruta graveolens, Opuntia dellini, Euphorbia royleena* and *Euphorbia antiquorum. Journal of Pharmacy Research*, 4(5):1507.

Soni, H., Malik, J., Yadav, A.P. and Yadav, B. 2018. Characterization of rutin isolated by leaves *Annona squamosa* by modern analytical techniques. *European Journal of Biomedical and Pharmaceutical Sciences*, 5(6):484–489.

Sonkar, N., Yadav, A.K., Mishra, P.K., Jain, P.K. and Rao, C.V. 2016. Evaluation of hepatoprotective activity of *Annona squamosa* leaves and bark extract against carbon tetrachloride liver damage in wistar rats. *World Journal Pharmacy and Pharmaceutical Science*, 5:1353–1360.

Sreekeesoon, D.P. and Mahomoodally, M.F. 2014. Ethnopharmacological analysis of medicinal plants and animals used in the treatment and management of pain in Mauritius. *Journal of Ethnopharmacology*, 157:181–200.

Srivastava, A., Gowda, D.V., Hani, U., Shinde, C.G. and Osmani, R.A.M. 2015. Fabrication and characterization of carboxymethylated Bael fruit gum with potential mucoadhesive applications. *RSC Advances*, 5(55):44652–44659.

Srivastava, S., Lal, V.K. and Pant, K.K. 2011. Medicinal potential of *Annona squamosa*: At a glance. *Journal of Pharmaceutical Research*, 4:4596–4598.

Sulaiman, C.T., Deepak, M., Ramesh, P.R., Mahesh, K., Anandan, E.M. and Balachandran, I. 2021. Chemical profiling of selected Ayurveda formulations recommended for COVID-19. *Beni-Suef University Journal of Basic and Applied Sciences*, 10(1):1–5.

Suresh, K., Senthilkumar, P.K. and Karthikeyan, B. 2009. Antimicrobial activity of *Aegle marmelos* against clinical pathogens. *Journal of Phytology*, 1(5):323–327.

Tene, V., Malagon, O., Finzi, P.V., Vidari, G., Armijos, C. and Zaragoza, T. 2007. An ethnobotanical survey of medicinal plants used in Loja and Zamora-Chinchipe, Ecuador. *Journal of Ethnopharmacology*, 111(1):63–81.

Tomar, R.S. and Sisodia, S.S. 2012. Antidiabetic activity of *Annona squamosa L.* in experimental induced diabetic rats. *International Journal of Pharmaceutical and Biological Archive*, 3:1492–1495.

Tu, W., Zhu, J., Bi, S., Chen, D., Song, L., Wang, L., Zi, J. and Yu, R. 2016. Isolation, characterization and bioactivities of a new polysaccharide from *Annona squamosa* and its sulfated derivative. *Carbohydrate Polymers*, 152:287–296.

Upadhya, S., Shanbhag, K.K., Suneetha, G., Balachandra Naidu, M. and Upadhya, S. 2004. A study of hypoglycemic and antioxidant activity of *Aegle marmelos* in alloxan induced diabetic rats. *Indian Journal of Physiology and Pharmacology*, 48(4):476–480.

Varadharajan, V., Janarthanan, U.K. and Krishnamurthy, V. 2012. Physicochemical, phytochemical screening and profiling of secondary metabolites of *Annona squamosa* leaf extract. *World Journal of Pharmaceutical Research*, 1(4):1143–1164.

Venkatesan, D., Karrunakarn, C.M., Kumar, S.S. and Swamy, P. 2009a. Identification of phytochemical constituents of *Aegle marmelos* responsible for antimicrobial activity against selected pathogenic organisms. *Ethnobotanical Leaflets*, 2009(11):4.

Venkatesan, D., Karunakaran, M., Kumar, S.S., Palaniswamy, P.T. and Ramesh, G. 2009b. Antimicrobial activity of *Aegle marmelos* against pathogenic organism compared with control drug. *Ethnobotanical Leaflets*, 2009(8):1.

Venthodika, A., Chhikara, N., Mann, S., Garg, M.K., Sofi, S.A. and Panghal, A. 2021. Bioactive compounds of *Aegle marmelos L.*, medicinal values and its food applications: A critical review. *Phytotherapy Research*, 35(4):1887–1907.

Verma, N., Sao, P., Srivastava, A. and Singh, S., 2021. Physiological and Molecular Responses to Drought, Submergence and Excessive Watering in Plants. In *Harsh Environment and Plant Resilience: Molecular and Functional Aspects*. Cham: Springer, 305–321.

Vikas, B., Akhil, B.S., Remani, P. and Sujathan, K. 2017. Free radical scavenging properties of *Annona squamosa*. *Asian Pacific Journal of Cancer Prevention: APJCP*, 18(10):2725–2731.

Vyas, K., Manda, H., Sharma, R.K. and Singhal, G. 2012. An update review on *Annona squamosa*. *IJPT*, 3(2):107–118.

Wang, D.S., Rizwani, G.H., Guo, H., Ahmed, M., Ahmed, M., Hassan, S.Z., Hassan, A., Chen, Z.S. and Xu, R.H. 2014. *Annona squamosa Linn*: Cytotoxic activity found in leaf extract against human tumor cell lines. *Pakistan Journal of Pharmaceutical Sciences*, 27(5) Spec no:1559–1563.

World Health Organization 2009. *Medicinal Plants in Papua New Guinea: Information on 126 Commonly Used Medicinal Plants in Papua New Guinea*. Manila: WHO Regional Office for the Western Pacific, 26–27.

World Health Organization 2015. *General Guidelines for Methodologies on Research and Evaluation of Traditional Medicine (WHO/EDM/TRM/2000.1)*. http://apps.who.int/iris/bitstream/handle/10665/66783/WHO_EDM_TRM_2000.1.pdf;jsessionid=FC6651BE55523C7818CAC02248D53ED5?sequence=1

Wu, Y.C., Hung, Y.C., Chang, F.R., Cosentino, M., Wang, H.K. and Lee, K.H. 1996. Identification of ent-16β, 17-dihydroxykauran-19-oic acid as an anti-HIV principle and isolation of the new diterpenoids annosquamosins A and B from *Annona squamosa*. *Journal of Natural Products*, 59(6):635–637.

Yadav, N., Tyagi, G., Jangir, D.K. and Mehrotra, R. 2011. Rapid determination of polyphenol, vitamins, organic acids and sugars in *Aegle marmelos* using reverse phase-high performance liquid chromatography. *Journal of Pharmacy Research*, 4(3):717–719.

Yadav, N.P. and Chanotia, C.S. 2009. Phytochemical and pharmacological profile of leaves of *Aegle marmelos Linn*. *Pharmaceutical Reviews*, 11:144–150.

Yadav, V.K., Singh, G. and Jha, R.K., Kaushik, P. 2021. Visiting Bael (*Aegle marmelos*) as a protective agent against COVID-19: A review. *Indian Journal of Traditional Knowledge (IJTK)*, 19:S-153–S-157.

Yang, Y.L., Hua, K.F., Chuang, P.H., Wu, S.H., Wu, K.Y., Chang, F.R. and Wu, Y.C. 2008. New cyclic peptides from the seeds of *Annona squamosa* L. and their anti-inflammatory activities. *Journal of Agricultural and Food Chemistry*, 56(2):386–392.

Zahid, M., Arif, M., Rahman, M.A. and Mujahid, M. 2020. Hepatoprotective and antioxidant activities of *Annona squamosa* seed extract against alcohol-induced liver injury in Sprague Dawley rats. *Drug and Chemical Toxicology*, 43(6):588–594.

18 *Azadirachta indica* (Neem) and *Berberis aristata* (Indian Barberry)

Swati T. Gurme, Devashree N. Patil, Suchita V. Jadhav, Mahendra L. Ahire, and Pankaj S. Mundada

CONTENTS

18.1 Introduction 365
18.2 Description 366
18.3 Traditional Knowledge 368
18.4 Chemical Derivatives (Bioactive Compounds – Phytochemistry) 369
18.5 Potential Benefits, Applications and Uses 370
18.5.1 Neem 370
18.5.1.1 Medical Applications 370
18.5.1.2 Agronomic Applications 371
18.5.1.3 Food 372
18.5.2 *Berberis aristata* 372
18.6 Conclusion 373
Acknowledgments 373
References 373

18.1 INTRODUCTION

Medicinal plants are known to be the primary source of drugs from ancient times. The widespread use of herbal remedies and health care preparations from traditional herbs and medicinal plants has been traced for their medicinal properties (Sharma et al., 2020). Though humans developed modern sophisticated pharmaceutical chemicals to treat illness, medicinal plants remain as important tools for treating illness. Human beings have been utilizing medicinal plants for basic preventive and curative health care properties. *A. indica* (Family: Meliaceae) has a long history of use as a traditional treatment against various illnesses due to its prominent medicinal properties (NRC, 1992; Mafara and Bakura, 2021). The abundant biological, chemical, and pharmacological properties have been reported by many scientific groups over the time. The remarkable therapeutic nature of neem is its antiviral, antibacterial, antifungal, anti-inflammatory, antihelminthic, antitumor, antigastric ulcer, antipyretic, and antihyperglycemic properties (Arun and Sivaramkrishnan, 1990). The medicinal properties are seen from most parts of the neem plant, for example, the hypoglycemic action can be seen by extracts of leaves, stem, bark, and seeds (Biswas et al., 2002; Ebong et al., 2008). Neem is reported to have a tremendous ability to regulate numerous biological processes *in vitro* and *in vivo*, and the wide range of medicinal properties which are attributed to bioactive compounds such as nimbolide, azadirachtin, and gedunin (Sarkat et al., 2021). There are a total of eight genes expressed in neem, which correspond to the synthesis of sesquiterpenes and triterpenes leading to azadirachtin-A (Krishnan et al., 2012). Normally, the neem plant is known for its drought resistance. It can grow in several diverse soil types but grows best in sandy soil. The neem tree can

DOI: 10.1201/9781003205067-18

grow at temperature range of 21–32°C. It can tolerate high temperature but cannot grow at low temperature (below 5°C). It is also considered a weed in several regions.

Berberis aristata, or Indian barberry, is one of the plants used in different medicinal systems like Ayurveda, homeopathy, Unani, Chinese, and allopathy for treatment of several ailments. It is a native plant of Himalaya and widely distributed from the Himalayan region to Sri Lanka, Bhutan, and hilly regions of Nepal. It is commonly called "Daru haldhi and Chitra" or tree turmeric (Mazumder et al., 2011). Nearly each part of this medicinal plant has medicinal importance. The *Berberis* has many useful applications in Indian and European systems for treatment of fever, eye diseases, jaundice, rheumatism, and kidney and gall bladder stones. Methanolic root extract shows the anti-inflammatory effect too. It is also used in snake and scorpion bite (Srivastava et al., 2015). Different alkaloids, flavonoids, phenols, lignin, steroids, vitamins, proteins, lipids, and carotenoids have been isolated from the *Berberis* plant species. The major bioactive compounds present in *Berberis aristata* are berberine, berbamine, palmatine, magnoflorine, jatrorrhizine, caffeic acid, quercetin, chlorogenic acid, meratin, and rutin (Chander et al., 2017; Basera et al., 2021).

Despite the many medicinal properties and the broad range of bioactive compounds, the masses, especially in the urban areas, are not taking advantage of these plants or they are not being utilized substantially. Sin these plants are so useful in the fields of medicine, agriculture, industry, and many others as confirmed by numerous researchers, this chapter aims to compile information on their chemical derivatives and therapeutic values along with other benefits.

18.2 DESCRIPTION

Azadirachta indica (neem, Family: Meliaceae) is a perennial tree that is most commonly found on the Indian subcontinent and Southeast Asia. It has different common names in different countries: English – Bastard tree, bead tree, Indian cedar, neem, paradise tree, Persian lilac; Arabic – azad-daraknun-hind; Brazil – neem, nim; Ethiopia – azaddarakhtihindi, nib; French – azadirac de l'Inde, margosier, margousier; Indonesia – imba, membha, mimba; India – limbo, nim, nimbi, medusa, vempu, nimani, olle, nindbetain, nimgachh; Iran – azad-darakhat-hindi; Malaysia – baypay, mambu, sadu, veppam; Myanmar – bowtamaka, tambabin, tamaka; Pakistan – nim; Sri Lanka – Kohomba; Spanish – margosa, mim.

This evergreen plant has broad leaves, grows to a height of 35 m with extended branches scattering 12 m across (Yamazaki et al., 2008). The flower and fruits are found on the axillary clusters; during the maturity of fruit, the color of drupes becomes greenish-yellow and consists of a single seed encircled by sweet pulp (Figure 18.1). The single seed contains a shell and 1–3 kernels; these are specifically containing the active compound azadirachtin and its analogs (Rupani and Chavez, 2018). The approximate amount of azadirachtin is 4–6 gm/kg seeds which can differ with the genotype and environmental interactions (Allan et al., 1994). The presence of azadirachtin and other biologically active molecules is also recorded in the bark and leaves of neem in small amounts. The neem tree is distributed widely in the subtropical regions of the world and used mostly for providing shade, alley cropping, and other agroforestry activities, including in many cases to produce pesticides, antifeedant, and insect repellant. The azadirachtin extracted from seeds mainly shows the antifeedant effect, contributing to the toxic effects on the insects (Allan et al., 1999).

Neem has two closely related species that are present everywhere. The *A. indica* A. Juss and *Melia azedarach*. The former is popularly known as Indian neem (margosa tree) or Indian lilac; the latter is called the Persian lilac (Benelli et al., 2017). Because of its traditional medicinal property, it is called the "village pharmacy" or "village dispensary" in Asia. Generally, all the parts of the tree like leaves, bark, twigs, flower, seeds, roots, sap, and gum are prescribed as medication by Indian Ayurvedic, Siddha, and herbal practitioners to treat various human diseases.

The neem plant parts contain microelements of chlorophyll, calcium, phosphorus, iron, thiamine, riboflavin, niacin, vitamin C, carotene, oxalic acid, sodium potassium, and salts in varied ratios (Graz et al., 2018). The neem tree contains other chemicals which can contribute to therapeutics, such as limonoids, terpenoids, steroids, tetranortriterpenoids, and fatty acid derivatives,

FIGURE 18.1 Morphology of *Azadirachta indica.* A) habit; B) flowering branch; C) branch with fruits.

like margosinone and margosinolone; coumarins like scopoletin, dihydrosococumarins; hydrocarbons like docosane, pentacosane, hetacosane, octacosane, etc.; sulphur compounds, phenolics, flavonoglycosides, tannins (Tembe-Fokunang et al., 2019). Approximately 135 varied structural compounds are identified and reported from various parts of the neem tree (Ahmad et al., 2019). These compounds are categorized into two types: (a) Isoprenoids containing limonoids, azadirone, vilasinin, nimbin, azadirechtin, etc. (b) Nonisoprenoids containing amino acids, proteins, sulfur-containing compounds, polyphenols, polysaccharides, glycosides, tannins, coumarin, flavonoids, aliphatic compounds, etc. (Ahmad et al., 2019). The neem plant has about 20,000 genes with an average transcript length of 1.69 kbp (Neeraja et al., 2012). According to Neeraja et al. (2012) the abundance of particular genes in specific organs correlate their functions very well; for example, highly expressed genes involved in the function of ion homeostasis, photosynthesis, ion transport, and ATPase activity are observed in the root, leaf, stem, and flower of *A. indica*, respectively.

Berberis aristata (Family: Berberidaceae) is one of the most primitive family of Angiosperm. The English name of *B. aristata* is Indian barberry or tree turmeric. In India it is popular with different names in different states, like darhaldi (Bengal); rasontkashmal (Himachal Pradesh); chitra (Nepal); daruhalad (Maharashtra); suvarnavarna (Sanskrit); and radaut, kashmal (Hindi) (Mazumder et al., 2011). Berberis spiny shrub grows up to height of 1.5–2.0 m and consists of a thick woody root covered by thin brittle yellow bark. The leaves having cylindrical shape with smooth spines are about 1.5 cm long. The flowers of Berberis are numerous – 11–16 per cluster and yellow in color (Figure 18.2); these plants start blooming in March and remain in progress to the end of April. The fruits are small berries and smooth, they ripen from mid-May to the end of June (Mazumder et al., 2011). The *Berberis aristata* and other *Berberis* varieties contain yellow berberine in their roots and woods, which are a bitter molecule commonly dissolved in acids and form alkaloid salts (Mazumder et al., 2011).

FIGURE 18.2 Morphology of *Berberis aristata.* A) habit; B) flowering branch; C) branch with fruits.

18.3 TRADITIONAL KNOWLEDGE

Many plant remedies are known in traditional medicine and used for treatment and management of many diseases or illnesses (Aktar and Ali, 1984), and some of them have been authorized by scientific studies to apply for biological action against diseases or their complications. Traditionally, neem is used as a household remedy from ancient times. The concoction is normally consumed in the early morning on an empty stomach for a minimum of 15 days, and it will probably protect the human body from most infections throughout the year. Hair loss can be controlled by washing hair regularly with neem decoction. It also promotes hair growth. The application of neem oil can stop the growth of hair bugs (Tembe-Fokunang et al., 2019). The leaves of *A. indica* are traditionally and commonly used as an antidiabetic. High amount of protein and fiber and comparatively low-fat content has positively been concerned in the management of diabetes and post-prandial hyperglycemia. Traditional blood purification mixture also contains mainly neem in it and is used for detoxification purposes as well (Aktar and Ali, 1984). Oil extract from neem seed is used for the preparation of soaps and cosmetics, while the twigs of the plant are used for toothbrushing (Brahmachari, 2004), and flowers are used for tonic and stomachic purposes.

Beside these traditional uses, neem or neem extracts are used to treat fungal infections, painful joints, and muscle treatments. It is well known for the treatment of diseases like malaria, arthritis, jaundice, intestinal worms, tuberculosis, and many skin diseases. Neem oil is used regularly to reduce acne and pimples (Ahmad et al., 2019; Kausik et al., 2002). The neem leaves are a famous remedy for skin disease treatment, and they are also safe to take internally on a regular basis. Fresh leaf juice mixed with salt is best to treat intestinal worms, and mixing this neem leaf juice with honey is useful for jaundice and skin diseases. It is also used as an antiseptic; while a hot blend of leaves can be used for swollen glands, bruises, and sprains. Neem leaves are also used against viral infections including cold, herpes, influenza, and chickenpox. By eating neem leaves, fever related to viral infection can be stopped. The fruits of neem (dry and fresh) are useful to control intestinal worms, piles, and urinary tract infections. The seed oil is also important as an ingredient in medical product formation, as it has strong antiseptic properties. As this oil saponifies easily, it is mostly used for medical soap formulation (Ahmad et al., 2019). Some studies also reported that neem extract is beneficial to control the kissing bugs that spread the Chagas disease. During treatment, the neem extract does not actually kill the insect but imbalances the immune system of the parasite living inside the bug, thereby curing the symptoms of Chagas disease (Martinez and van Emden, 1999).

B. aristata has a long history of healing many illnesses. It has different uses in different countries. For instance, in India it is more commonly used to treat dysentery, aphthous sore abrasions, and ulcerations of the skin. In Egypt, it was used for plague prevention. In European traditional medicine, it is more commonly used for treatment of liver and gallbladder ailment (Chakarvarti et al., 1950). Russian herbalists use neem against inflammation, high blood pressure, and unusual uterine bleeding. Complexions prepared from *Berberis* are used as a bitter tonic, stomachic, cholagogue, antiperiodic.; It is extremely helpful in intermittent fevers and periodic neuralgia and menorrhagia. The plant is useful for the treatment of jaundice, enlargement of the spleen, constipation, leprosy, and as an antiseptic (Ray and Roy, 1941).

18.4 CHEMICAL DERIVATIVES (BIOACTIVE COMPOUNDS – PHYTOCHEMISTRY)

The qualitative proximate composition analysis of *A. indica* leaves performed by Atangwho et al. (2009) shows that neem contains approximately 13.42 ± 0.12% crude protein, 20.11± 0.45% crude fibers, 5.17 ± 0.09% fat, and 11.93 ± 0.09% ash. The vitamin composition shows the presence of vitamins A, E, C, riboflavin, thiamin, and niacin in different concentrations. Leaves of *A. indica* contain minerals like Mn, Se, Zn, Fe, Cu, Mg, Cr, etc. in varied concentrations. According to Atangwho et al. (2009) the presence of these metals, vitamins, and phytochemicals, like flavonoids, polyphenols, and alkaloids, in the neem leaves may also support higher effectiveness in antidiabetic action.

The bioactive compounds in the neem plant that have multiple uses have been identified through different studies as: nimbin – anti-inflammatory; nimbidin – antibacterial, anti-ulcer; nimbidol – antitubercular, anti-protozoan; gedunin – antimalaria, antifungal; sodium nimbinate – diuretic, anti-arthritic; and salannim – repellent (Figure 18.3) (Subapriya and Nagini, 2005). The volatile organic compounds like di-n-propyl-disulfide and as many as 25 volatile compounds have been reported from fresh seeds of neem (Ahmad et al., 2019). The neem

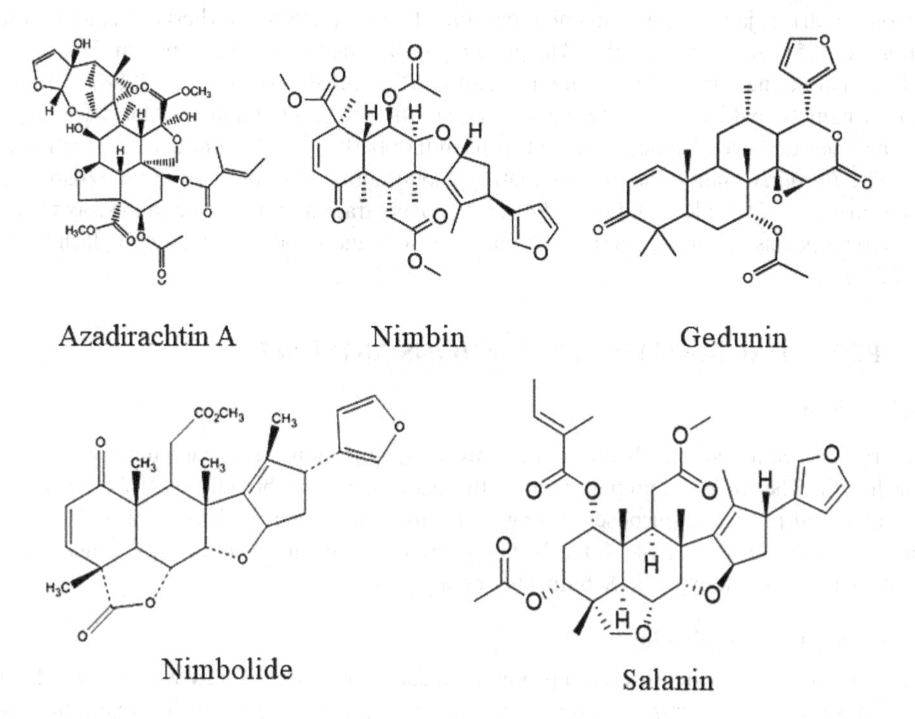

FIGURE 18.3 Major phytoconstituents in *Azadirachta indica.*

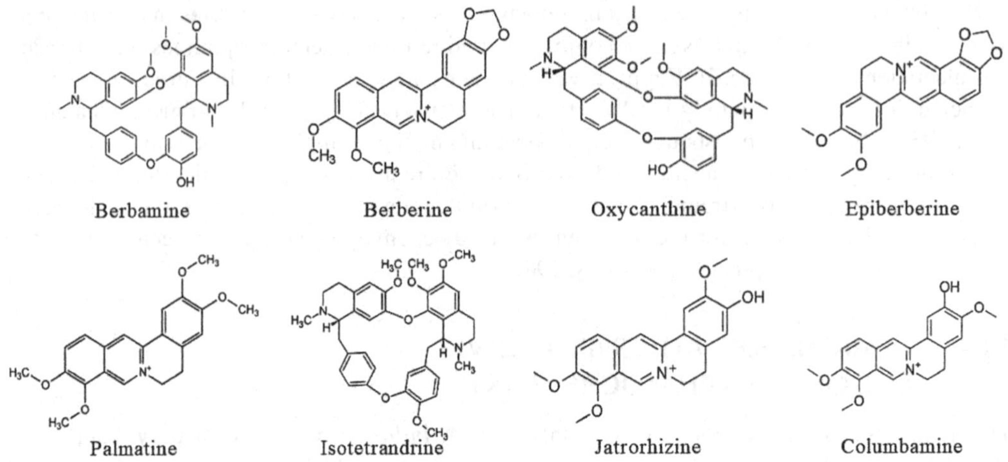

Berbamine Berberine Oxycanthine Epiberberine

Palmatine Isotetrandrine Jatrorhizine Columbamine

FIGURE 18.4 Major phytoconstituents in *Berberis aristata*.

extracts exhibit a high amount of limonoids, especially azadirachtin A and its analogues, which are most commonly considered biopesticides. Azadirachtin concentration varies from one part of the neem plant to another and depends upon the concentration of the active principle compound, so the formulation of pesticide and its effects may vary. Song et al. (2018) demonstrated the presence of five different azadirachtins (A, B, D, H, and I) in seed and leaf samples of the neem plant.

The *Berberis* have a main constituent of alkaloid, i.e., berberine, in its roots and wood. This plant contains more alkaloids in the roots, namely, karachine and taxalamine. Along with these two alkaloids, roots contain berbamine, berberine, oxycanthine, epiberberine, palmatine, dapehydrocaroline, isotetrandrine, jatrorhizine and columbamine, dihyrokarachine, oxyberberine, and aromoline (Chatterjiee, 1951; Saied et al., 2007). The plant also contains oxyberberine, berbamine, aromoline, karachine, palmatine, oxyacanthine, and taxilamine (Figure 18.4) (Ambastha, 1988). Another four alkaloids, namely, pakistanine, 1-O methyl pakistanine, pseudopalmatine chloride, and pseudoberberine chloride have also been isolated from Indian barberry (Bhakuni et al., 1968; Lect et al., 1983). The *Berberis* plant contains palamatine, isoquinoline alkaloid (secobisbenzlisoquinoline) (Chakarvarti et al., 1950; Ray and Roy, 1941). The concentration of berberine in the roots and stems is affected by potassium and the moisture content of soil, and varies depending on altitude (Andola et al., 2010).

18.5 POTENTIAL BENEFITS, APPLICATIONS AND USES

18.5.1 Neem

The evergreen neem plant has been used for diverse applications, as it has numerous benefits for human health. The different preparations of the neem plant are broadly used for agrochemical, insecticidal, and pesticidal purposes. Along with this, it has been used for thousands of years as alternative and modern medicine. It has been reported to treat various infections, cancerous conditions, etc. with different extracts (Schumacher et al., 2011).

18.5.1.1 Medical Applications

Application of neem or neem-derived products in cancer research is now well known. Different studies showed success against inhibition of various cancer cell growth; for example, prostate cancer, human choriocarcinoma cells, murine Ehrlich's carcinoma, melanoma cells, leukemia cells, head-and-neck squamous cells, etc. Similarly, the work done by Sharma et al. (2014) also shows antiproliferative activity of neem ethanolic extract alone or in combination with cisplatin against human breast (MCF-7) and cervical (HeLa) cells. As cancerous is related to the imbalance in antiapoptotic and proapoptotic mechanism of rapid and uncontrolled proliferation of cancer cells, inducing cell death has become a major goal in cancer treatment. These types of results are seen during the experimentation on cancerous cells. This proves the usability of neem extract for the stimulation of cancer cell death following the apoptotic pathway (Sharma et al., 2014). This induction is mainly due to the presence of neem and nimbolide which upregulate Box gene expression in human prostrate and colon cancer (Kumar et al., 2009; Gunadharini et al., 2011); while downregulation of the cyclin D1 gene is also responsible for the antiproliferative action (Kumar et al., 2010; Priyadarsini et al., 2010). The benefit of neem extract in preventing cyclophosphamide, cisplatin, and 5-fluorouracil-induced hematological complications was also reported (Ghosh et al., 2006; Ezz-Din et al., 2011). Fauziah et al. (2012) have reported the proliferation suppression of cancerous cells, particularly the c-Myc mRNA expression after treatment of mice with neem leaf extract, and it ultimately induces the apoptosis against 4T1 cancer cells. Likewise, increases in antioxidant levels in gastric cancer were reported after the pretreatment of neem leaf extract (ethanol) (Subapriya et al., 2003). This increased antioxidant enzyme ultimately resulted in inhibition of chemically-induced hepatocarcinogenesis in rats after consumption of 5% neem leaf aqueous extract (Manal et al., 2007). The antioxidant activity of neem leaf and bark extract was reported by Ghimeray et al. (2009), and the significant central nervous system depressant activity of leaf extract (acetone) was also proved on rat model by Paul et al. (2011).

Various skin diseases such as ringworm, eczema, and scabies can be treated with neem extracts. Neem leaf-derived lotion is also beneficial for dermatological treatment: it can cure the acute stage or chronic situation within 3–4 days of application (Allan et al., 1994). Neem and turmeric paste is another effective skin treatment (Tuchinda et al., 2008) and is used to treat malaria (specifically in India) (Gupta et al., 2017).

Neem leaf oil and neem extracts have antimicrobial activity against a wide spectrum of gram-positive and gram-negative bacteria. The more prominent bacterial strains are *S. aureus*, *S. pyogenes*, *Cornebacterium*, *E. coli*, *S. typhimurium*, *M. tuberculosis*, and streptomycin-resistant strains. Growth inhibition of *Vibrio cholera*, *Klebsiella pneumonia*, *M. tuberculosis*, and *M. pyogenes* was also reported. Neem extract shows antifungal activity against some human fungal genera like *Trichophyton*, *Epidermophyton*, *Microsporum*, and *Geotrichum* (Tembe-Fokunang et al., 2019; Reed and Majumdar, 1998). Besides these uses, different neem extracts are used to treat eczema, acne, dry skin, itchy scalp, skin ulcer, and leprosy.

18.5.1.2 Agronomic Applications

Neem has agricultural applications, such as a pesticide and treatment of livestock illnesses, making neem oil and its limonoids, like azadirachtin, nimbin, and salannin economically valuable (Brahmachari 2004). The effectiveness of azadirachtin was proven against many insect species and shows lower toxicity to mammals (Ley, 1994). Other benefits of neem-derived azadirachtin and limonoids are as anti-proliferative, cytotoxic, larvicidal, and anti-inflammatory agents (Schumacher et al., 2011; Harish et al., 2009; Akudugu et al., 2001; Okumu et al., 2007; Denardi et al., 2011; Thoh et al., 2010). The applicability of botanical products as pest control is a growing trend nowadays, as people are more aware of organic and sustainable farming. The azadirachtin of neem gives antifeedant and toxic effects on the larvae of *Spodoptera littoralis* (Boisd), *S. gregaria*, and *Oncopeltus fasciatus* Dallas (milkweed bug) (Tembe-Fokunang et al., 2019). However, antifeedant activity is varied between species, insect order, and within orders. *A. indica* provides a sustainable and eco-friendly pesticide option for farmers by utilizing neem seed-derived azadirachtin. Another major economic benefit of neem is essential oils, synthesized by terpene synthase (Neeraja et al., 2012). Another potential benefit of neem and neem extracts is their herbicidal or allelopathic activity.

TABLE 18.1
Neem: A Medicinal Plant and Its Potential Benefits

Plant Parts	Uses	References
Leaf	Antioxidant; central nervous system depressant activity; Anticancerous; antiproliferative; antimicrobial; skin treatment; antidiabetic	Reed and Majumdar (1998); Tuchinda et al. (2008); Paul et al. (2011)
Bark	Antioxidant	Ghimeray et al. (2009)
Seeds	Antifeedant; pesticide; herbicidal; soaps and cosmetics	Sindhu et al. (2005)
Tender shoot and flower	Vegetable	Chowdhury and Mukherjee (2012)

Neem is considered a promising source for synthetic herbicidal production because of its low toxicity to mammals (Sindhu et al., 2005).

18.5.1.3 Food

Various part of the neem plant is applicable as food for humans and animals too. Plant parts, like the tender shoots and flowers are used as vegetables or to make soup. Also, neem leaves are used to make some dishes with eggplant (brinjal) and rice, which act as an appetizer in the Indian kitchen. Leaves of neem can be occasionally used as feed for ruminants and rabbits (Table 18.1).

18.5.2 *Berberis aristata*

Berberis aristata is beneficial medication against sun-blindness, ophthalmia, and other eye diseases. The root extract of *Berberis* is used as an effective antibacterial agent against bacteria like *Bacillus cereus*, *Escherichia coli*, and *Staphylococcus aureus*, while the stem extract acts against *Bacillus cereus* and *Streptococcus pneumoniae*. It has prominent activity against fungal strains, like *Aspergillus terreus* and *Aspergillus flavus*, other viruses, protozoans, helminthes, and chlamydia. The *B. aristata* antimicrobial activity may be due to the presence of its major alkaloid, berberine (Mazumder et al., 2011). Another potential use of *B. aristata* is its antidepressant activity. It is well known that berberine inhibits monoamine oxidase-A (MOA), an enzyme that catalyzes oxidative deamination of amines, like dopamine, norepinephrine, and serotonin. This proves the activity of berberine in the central nervous system. The antidepressant effect of berberine was proved on mice using the forced swim test and tail-suspension test. Acute administration of berberine resulted in a 31% increase in level of dopamine and norepinephrine, while serotonin increased up to 47% in the whole brain. Similarly, 15 days of chronic administration showed an increase of up to 52% in level of dopamine. Locomotor activity is not affected at lower concentrations of berberine, whereas it confirms hypothermic actions in rats and demonstrates analgesic effect in mice. The antidepressant effect of berberine involved the nitric oxide pathway and sigma receptors (Sabnis, 2006).

Some studies also prove antidiabetic activity with type 2 diabetes mellitus. Oral administration for 1–3 months along with proper diet resulted in the disappearance of major diabetic symptoms. Animal testing of berberine in a diabetic model suggested this may be possible due to the functional recovery and promoting regeneration of β-cells (Mazumder et al., 2011). Leaf and fruit extracts of *Berberis* display hepatoprotective activity by preventing induction of alkaline phosphatase, amino-transaminase, and ALT levels. Alkaloid from root and bark shows enhanced proliferative response of spleen cells toward the mitogens, like concanavalin A and phytohemagglutinin. In Chinese folk medicine, this plant is commonly used to enhance leukogenesis and act as anti-arrhythmics, and antihypertensives (Mazumder et al., 2011) (Table 18.2).

TABLE 18.2
Tree Turmeric: A Medicinal Plant and Its Potential Benefits

Plant Part	Uses	References
Root	Antimicrobial; enhance proliferation of spleen cells; antidepressant	Mazumder et al. (2011); Sabnis (2006)
Leaves	hepatoprotective	Mazumder et al. (2011)
Fruits	hepatoprotective	Mazumder et al. (2011)

18.6 CONCLUSION

Neem, *Azadirachta*, a traditional medicinal perennial tree has been widely used for agricultural and therapeutic purposes, the extracts and biological active molecules of neem showing numerous applications for different human diseases. The nontoxic extracts and presence of multiple bioactive compounds make neem an eco-friendly, cost-effective, and sustainable source for different human and agricultural treatments. *Berberis aristata,* or Indian barberry, exposed its usability as an antimicrobial, hepatoprotective, antidepressant, and immunomodulatory agent. The traditional knowledge, wide availability, and great demand for herbal medicine in the developing world make the neem and tree turmeric plants two of the easiest sources for the treatment of many diseases and infections. These characteristics also make them excellent alternatives to modern drugs, as they are reasonably priced, have wide cultural acceptance, and better parallelism with the human body – with fewer side effects.

ACKNOWLEDGMENTS

Authors are grateful to Yashavantrao Chavan Institute of Science, Satara for financial assistance and providing necessary laboratory facilities. Financial assistance to the faculty under self-funded project is gratefully acknowledged. The DST-INSPIRE meritorious fellowship to Ms. Patil D.N. from DST, New Delhi, India, is gratefully acknowledged.

REFERENCES

Ahmad, S., Maqbool, A., Srivastava, A., Gogoi, S. 2019. Biological detail and therapeutic effect of *Azadirachta indica* (neem tree) products: A review. *Journal of Evidence Based Medicine and Healthcare* 6:1607–1612.

Akhtar, F. M., Ali, M. R. 1984. Study of antidiabetic effect of a compound medicinal plant prescription in normal and diabetic rabbits. *Journal of the Pakistan Medical Association* 34(8):239–244.

Akudugu, J., Gade, G., Bohm, L. 2001. Cytotoxicity of azadirachtin A in human glioblastoma cell lines. *Life Science* 68:1153–1160.

Allan, E.J., Eeswara, P., Johnson, S., Mordue, A.J., Luntz, E.D., Morgan, Stuchbury, T. 1994. The production of azadirachtin *in-vitro* tissue culture of neem, *Azadirachta indica. Pesticide Science* 42:147–152.

Allan, E.J., Stuckbury, T., Mordue, A.J. (Luntz). 1999. *Azadirachta indica* A. Juss. (Neem tree): *In vitro* culture, micropropagation, and the production of azadirachtin and other secondary metabolites. *Medicinal and Aromatic Plants XI*, Springer, Berlin, Heidelberg43:11–41.

Ambastha, S.P. 1988. *The Wealth of India, vol. 2B. Publication and Information Directorate*. New Delhi: CSIR, p.118.

Andola, H.C., Gaira, K.S., Singh, R.R., Rawat, M.S.M., Bhatt, I.D. 2010. Habitat-dependent variations in berberine content of *Berberis asiatica Roxb.* ex. in Kumaon, Western Himalaya. *Chemistry and Biodiversity* 7:415–420.

Arun, K.S., Sivaramkrishnan, V.M. 1990. Plant products as protective agents against cancer. *Indian Journal of Experimental Biology* 28:1008–1011.

Atangwho, I.J., Ebong, P.E., Eyong, E.U., Williams, I.O., Eteng, M.U., Egbung, G. E. 2009. Comparative chemical composition of leaves of some antidiabetic medicinal plants: *Azadirachta indica, Vernonia amygdalina and Gongronema latifolium. African Journal of Biotechnology* 8:4685–4689.

Basera, I.A., Girme, A., Bhatt, V.P., Shah, M.B. 2021. A validated high-performance thin-layer chromatography method for the simultaneous estimation of berberine, berbamine, palmatine, magnoflorine and jatrorrhizine from *Berberis aristata*. *Journal of Planar Chromatography–Modern TLC* 23:1–9.

Benelli, G., Buttazzoni, L., Canale, A., D'Andrea, A., Del Serrone, P., Delrio, G., Foxi, C., Mariani, S., Savini, G., Vadivalagan, C., Murugan, K., Toniolo, C., Nicoletti, M., Serafini, M. 2017. Bluetongue utbreaks: Looking for effective control strategies against *Culicoides* vectors. *Research in Veterinary Science* 115:263–270.

Bhakuni, D.S., Shoheb, A., Popali, S.P. 1968. Medicinal plants: Chemical constituent of *Berberis aristata*. *Indian Journal of Chemistry* 6:123.

Biswas, K., Chattepadhya, I., Banergee, R.K., Bandyopadhyayi, U. 2002. Biological activities and medicinal properties of neem (*Azadirachta indica*). *Current Science* 82:1336–1346.

Brahmachari, G. 2004. Neem–an omnipotent plant: A retrospection. *Chembiochem: A European Journal of Chemical Biology* 5:408–421.

Chakarvarti, K.K., Dhar, D.C., Siddhiqui, S. 1950. Alkaloidal constituent of the bark of *Berberis aristata*. *Journal of Scientific and Industrial Research* 9b:161–164.

Chander, V., Aswal, J.S., Dobhal, R., Uniyal, D.P. 2017. A review on Pharmacological potential of Berberine; an active component of Himalayan *Berberis aristata*. *JPHYTO* 6:53–58.

Chatterjiee, R.P. 1951. Isolation of new phytoconstituents from the plants of *Berberidaceae* family. *Journal of the Indian Chemical Society* 28:225.

Chowdhury, M. and Mukherjee, R. 2012. Wild edible plants consumed by local communities of Maldah district of West Bengal, India. *Indian Journal of Scientific Research* 3(2):163–170.

Denardi, S.E., Bechara, G.H., de Oliveira, P.R., Camargo, M.M.I. 2011. Inhibitory action of neem aqueous extract (*Azadirachta indica* A. Juss) on the vitellogenesis of Rhipicephalus sanguineus (Latreille, 1806) (Acari: Ixodidae) ticks. *Microscopy Research and Technique* 74:889–899.

Ebong, P.E., Atangwho, I.J., Eyong, E.U., Egbung, G.E. 2008. The antidiabetic efficacy of combined extracts from two continental plants: *Azadirachta indica* (A. Juss) (Neem) and *Vernonia amygdalina* (Del.) (African bitter leaf). *American Journal of Biochemistry and Biotechnology* 4:239–244.

Ezz-Din, D., Gabry, M.S., Farrag, A.R.H., Abdel Moneim, A.E. 2011. Physiological and histological impact of *Azadirachta indica* (neem) leaves extract in a rat model of cisplatin-induced hepato and nephrotoxicity. *Journal of Medicinal Plant Research* 5:5499–5506.

Fauziah, O., Gholamreza, M., Sally, L.T.P., Asmah, R., Rusliza, B., Chong, P.P. 2012. Effect of neem extract (*Azadirecta indica*) on c-Myc oncogene expression in 4T1 breast cancer cells of BALB/c mice. *Cell Journal* 14:53–60.

Ghimeray, A.K., Jin, C., Ghimire, B.K., et al. 2009. Antioxidant activity and quantitative estimation of azadirachtin and nimbin in Azadirachta Indica A. Juss grown in foothills of Nepal. *African Journal of Biotechnology* 8:3084–3091.

Ghosh, D., Bose, A., Haque, E., Baral, R. 2006. Pretreatment with neem (*Azadirachta indica*) leaf preparation in Swiss mice diminishes leukopenia and enhances the antitumor activity of cyclophosphamide. *Phytotherapy Research* 20:814–818.

Graz, H., D'Souza, V.K., Alderson, D.E.C., Graz, M. 2018. Diabetes-related amputations create considerable public health burden in the UK. *Diabetes Research and Clinical Practice* 135:158–165.

Gunadharini, D.N., Elumalai, P., Arunkumar, R., Senthilkumar, K., Arunakaran, J. 2011. Induction of apoptosis and inhibition of PI3K/Akt pathway in PC-3 and LNCaP prostate cancer cells by ethanolic neem leaf extract. *Journal of Ethnopharmacology* 134:644–650.

Gupta, S.C., Prasad, S., Tyagi, A.K., Kunnumakkara, A.B., Aggarwal, B.B. 2017. Neem (*Azadirachta indica*): An Indian traditional panacea with modern molecular basis. *Phytomedicine* 34:14–20.

Harish, K.G., Chandra, M.K.V., Jagannadha, R.A., Nagini, S. 2009. Nimbolide a limonoid from Azadirachta indica inhibits proliferation and induces apoptosis of human choriocarcinoma (BeWo) cells. *Investigational New Drugs* 27:246–252.

Kausik, B., Chattopadhyay, I., Banerjee, R.K., et al. 2002. Biological activities and medicinal properties of neem (*Azadirachta indica*). *Current Science* 82:1336–1345.

Krishnan, N. M., Pattnaik, S., Jain, P., Gaur, P., Choudhary, R., Vaidyanathan, S., ... Panda, B. 2012. A draft of the genome and four transcriptomes of a medicinal and pesticidal angiosperm Azadirachta indica. *BMC Genomics* 13(1): 1–13.

Kumar, G.H., Chandra Mohan, K.V.P., Jagannadha R.A., Nagini, S. 2009. Nimbolide a limonoid from *Azadirachta indica* inhibits proliferation and induces apoptosis of human choriocarcinoma (BeWo) cells. *Investigational New Drugs* 27(3) pp. 246–252.

Kumar, G.H., Vidya, P.R., Vinothini, G., Vidjaya, L.P., Nagini, S. 2010. The neem limonoids azadirachtin and nimbolide inhibit cell proliferation and induce apoptosis in an animal model of oral oncogenesis. *Investigational New Drugs* 28:392–401.

Lect, E.J., Elango, V., Hussain, F.S., Sharma, M. 1983. Secobisbenzlisoquinoline or simple isoquinoline dimmer. *Heterocycle* 20:425–429.

Ley, S.V. 1994. Synthesis and chemistry of the insect antifeedant azadirachtin. *Pure and Applied Chemistry* 66:2099–2102.

Mafara, S. M., Bakura, T. L. (2021). Multiple potentials of neem tree (*Azadirachta indica*); A review. *Bakolori Journal of General Studies* 11(1), 1–2.

Manal, M.E.T., Hanachi, P., Patimah, I., Siddig, I.A., Fauziah, O. 2007. The effect of neem (*Azadirecta indica*) leaves extract on Alpha-fetoprotein, Serum concentration, glutathione S-transferase and glutathione activity in hepatocarcinogenesis induced rats. *International Journal of Cancer Research* 3:111–118.

Martinez, S.S., van Emden, H.F. 1999. Sublethal concentrations of azadirachtin affect food intake, conversion efficiency and feeding behaviour of *Spodoptera littoralis* (Lepidoptera: Noctuidae). *Bulletin of Entomological Research* 89:65–71.

Mazumder, P.M., Das, S., Das, S., Das, M.K. 2011. Phyto pharmacology of *Berberis aristata* DC: A review. *Journal of Drug Delivery & Therapeutics* 1:46–50.

National Research Council (NRC). (1992). *Neem: A Tree for Solving Global Problems.* Washington, DC: National Academy Press, pp. 23–39.

Neeraja, M.K., Swetansu, P., Prachi, J., Prakhar, G., Rakshit, C., Srividya, V., Sa Deepak, Arun, K.H., Bharath, K.P.G., Jayalakshmi, N., Linu, V., Naveen, K.V., Kunal, D., Krishna, R., Binay, P. 2012. A draft of the genome and four transcriptomes of a medicinal and pesticidal angiosperm *Azadirachta indica. BMC Genomics* 13:464–477.

Okumu, F.O., Knols, B.G., Fillinger, U. 2007. Larvicidal effects of a neem (*Azadirachta indica*) oil formulation on the malaria vector Anopheles gambiae. *Malaria Journal* 6:63.

Paul, R., Prasad, M., Sah, N.K. 2011. Anticancer biology of *Azadirachta indica* L (neem): A mini review. *Cancer Biology & Therapy* 12:467–476.

Priyadarsini, R.V., Murugan, R.S., Sripriya, P., Karunagaran, D., Nagini, S. 2010. The neem limonoids azadirachtin and nimbolide induce cell cycle arrest and mitochondria-mediated apoptosis in human cervical cancer (HeLa) cells. *Free Radical Research* 44:624–634.

Ray and Roy. 1941. Folkloric uses of *Berberis aristata. Science and culture* 6:b13.

Reed, E., Majumdar, S.K. 1998. Differential cytotoxic effects of azadirachtin on *Spodoptera frugiperda* and mouse cultured cells. *Entomologia Experimentalis et Applicata* 89:215221.

Rupani, R., Chavez, A. 2018. Medicinal plants with traditional use: Ethnobotany in the Indian subcontinent. *Clinics in Dermatology* 36:306–309.

Sabnis, M. 2006. *Chemistry and Pharmacology of Ayurvedic Medicinal Plants.* Varanasi: Chaukhambha Surabharati Prakashana.

Saied, S., Batool, S., Naz, S. 2007. Phytochemical studies of *Berberis aristata. Journal of Basic and Applied Sciences* 3:1–4.

Sarkar, S., Singh, R. P., Bhattacharya, G. 2021. Exploring the role of *Azadirachta indica* (neem) and its active compounds in the regulation of biological pathways: An update on molecular approach. *3 Biotech* 11:4, 1–12.

Schumacher, M., Cerella, C., Reuter, S., Dicato, M., Diederich, M. 2011. Antiinflammatory, pro-apoptotic, and anti-proliferative effects of a methanolic neem (*Azadirachta indica*) leaf extract are mediated via modulation of the nuclear factor-kappa B pathway. *Genes and Nutrition* 6:149–160.

Sharma, C., Vas, A.J., Goala, P., Gheewala, T.M., Rizvi, T.A., Hussain, A. 2014. Ethanolic neem (*Azadirachta indica*) leaf extract prevents growth of MCF-7 and HeLa cells and potentiates the therapeutic index of cisplatin. *Journal of Oncology* http://dx.doi.org/10.1155/2014/321754

Sharma, M., Thakur, R., Sharma, M., Sharma, A.K. 2020. Changing scenario of medicinal plants diversity in relation to climate change: A review. *Plant Archives* 20:4389–4400.

Sindhu, A., Kumar, S., Sindhu, G., Ali, H., Abdulla, M.K. 2005. Effect of neem (*Azadirachta indica* A. Juss) leachates on germination and seedling growth of weeds. *Allelopathy Journal* 16:329–334.

Song, L., Wang, J., Gao, Q., Ma, X., Wang, Y., Zhang, Y., Xun, H., Yao X., Tang F. 2018. Simultaneous determination of five azadirachtins in the seed and leaf extracts of *Azadirachta indica* by automated online solidphase extraction coupled with LC–QTOF–MS. *Chemistry Central Journal* 12:85 https://doi.org/10.1186/s13065-018-0453-y

Srivastava, S., Srivastava, M., Misra, A., Pandey, G., Rawat, A.K.S. 2015. A review on biological and chemical diversity in *Berberis* (Berberidaceae). *EXCLI Journal* 14:247–267.

Subapriya, R., Nagini, S. 2005. Medicinal properties of neem leaves: A review. *Current Medicinal Chemistry: Anti-Cancer Agents* 5:149–156.

Subapriya, R., Kumaraguruparan, R., Chandramohan, K.V., Nagini, S. 2003. Chemoprotective effects of ethanolic extract of neem leaf against MNNG-induced oxidative stress. *Pharmazie* 58:512–517.

Tembe-Fokunang, E.A., Fokunang, C., Nubia, K., Gatsing, D., Agbor, M., Ngadjui, B. 2019. The Potential Pharmacological and Medicinal Properties of Neem (*Azadirachta indica* A. Juss) in the Drug Development of Phytomedicine. *Journal of Complementary and Alternative Medical Research.* 7:1–18.

Thoh, M., Kumar, P., Nagarajaram, H.A., Manna, S.K. 2010. Azadirachtin interacts with the tumor necrosis factor (TNF) binding domain of its receptors and inhibits TNF-induced biological responses. *Journal of Biological Chemistry* 285:5888–5895.

Tuchinda, P., Pohmakou, M., Korsalkulkam. 2008. The dicpetalin triterpenoid and lignans from aerial parts of *Phyllanthus acutissima. Journal of Natural Products* 71:655–663.

Yamazaki, M., Nakamura, A., Hanai, R., Hirota, H. 2008. Chemical constituents and the diversity of *Ligularialan kongensis*in Yunnan Province of China. *Journal of Natural Products* 71:520–524.

19 *Cinchona officinalis* (Cinchona Tree) and *Corylus avellana* (Common Hazel)

Sawsan A. Oran, Arwa Rasem Althaher, and Mohammad S. Mubarak

CONTENTS

19.1 Introduction 378
19.2 *Corylus avellana* 378
19.2.1 Botanical Aspects 378
19.2.2 Taxonomy 379
19.2.3 Habitat and Ecology 380
19.2.4 Importance and Usage 380
19.2.5 Uses of Hazelnuts in Traditional Medicine 380
19.2.6 Chemical Composition of *C. avellana* 381
19.2.6.1 Phenolics 381
19.2.6.2 Flavonoids 381
19.2.6.3 Tannins and Proanthocyanidins 382
19.2.6.4 Diarylheptanoids 382
19.2.6.5 Lignans 382
19.2.6.6 Taxanes 383
19.2.6.7 Volatile Compounds 383
19.2.7 Biological Activities 383
19.3 The Genus *Cinchona* 384
19.3.1 Botanical Description 384
19.3.2 Etymology and Common Names 384
19.3.3 History 384
19.3.4 Taxonomy 384
19.3.4.1 *Cinchona calisaya* Wedd 385
19.3.4.2 *Cinchona ledgeriana* Moons 386
19.3.4.3 *Cinchona officinalis* Linn. 386
19.3.4.4 *Cinchona pubescens* Vahl. (syn. *Cinchona succirubra* Pav. ex klotzsch) 386
19.3.5 Cultivation 386
19.3.6 Medicinal Uses 386
19.3.7 Chemical Constituents of *Cinchona* 387
19.3.8 Toxicology of *Cinchona* 388
19.3.9 COVID-19 Treatment with Chloroquine and Hydroxychloroquine 388
19.4 Conclusions and Future Perspectives 389
References 389

DOI: 10.1201/9781003205067-19

19.1 INTRODUCTION

Humans have been using medicinal plants and phytochemicals derived from these plants since ancient times. According to the World Health Organization (WHO), more than 80% of the world population still relies on traditional plant-based remedies for primary health care services (Wangchuk et al., 2011). These plants represent important sources of modern drugs. In this respect, approximately, 50,000 plant species are used in the traditional medicine system worldwide, although the majority of them are part of the Asian medicines (Wangchuk, 2004; Gewali and Awale, 2008). These medicinal plants contain chemicals that could serve as leads for new drug discovery. Furthermore, phytochemicals extracted from medicinal plants have been widely used in various countries to treat disorders including inflammation, hypertension, kidney problems, immune deficiency, and cancer (Cragg and Newman, 2013). Accordingly, the consumption of fruits and vegetables is necessary to maintain good health. Similarly, with increased awareness regarding health, phytochemicals and bioactive molecules are also becoming popular as promising options against various diseases. A number of plant species belonging to diverse plant families have been validated to have some sort of biological activity. Among these medicinal plants are *Corylus avellana* (hazelnut), which is one of the most popular tree nuts on a worldwide basis, and the *Cinchona* (genus *Cinchona*) *officinalis*. In this chapter, we will discuss the natural compounds present in these plants as well as their botanical descriptions and traditional uses for health promotion and disease prevention.

19.2 *CORYLUS AVELLANA*

19.2.1 Botanical Aspects

Corylus avellana, common hazel, is one of many hazel species in the Betulaceae family. Although *C. avellana* is native to Europe and Asia, ranging from the British Isles south to Iberia, Italy, Greece, Turkey, and Cyprus, north to central Scandinavia, and east to the central Ural Mountains, the Caucasus, and northwestern Iran, its natural distribution until recently has been limited to the northern hemisphere. Turkey and Italy produce most of the world's hazelnuts (80%), but hazelnut farming has moved to new growing locations in recent years, including the southern hemisphere (Torello Marinoni et al., 2018). Common hazel is a shrub that usually grows to 3–8 m tall but can grow to be 15 m tall (Figure 19.1A). The deciduous leaves are rounded, 6–12 cm long, slightly hairy on both surfaces, and have a double-serrated border (Masullo et al., 2015) (Figure 19.1B). The nut of commerce, the kernel (Figures 19.1C and D), is encased in a dark brown perisperm and shielded by a woody shell (Fanali et al., 2018). Hazelnuts grow in clusters of one to 12, each encased in a cup of green leafy cover that shades about three-quarters of the nut (Cerulli et al., 2017). Moreover, the flowers (Figures 19.1E and F) appear before the leaves in early spring and are monoecious with single-

sex catkins. Male catkins are pale yellow and 5–12 cm long, while female catkins are small and largely hidden in the buds, with only the bright-red, 1–3 mm long styles visible (Masullo et al., 2016; Bottone et al., 2019).

C. avellana has an average life span of 80–90 years (Enescu et al., 2016) and can be propagated both vegetatively and generatively (via seeds). Shoot and root suckers and cuttings are the most common methods of vegetative propagation (Contessa et al., 2011; Enescu et al., 2016). *C. avellana* can sprout and spread quickly – after fires, in particular (Tinner et al., 1999). The hazelnut has been extensively used in breeding programs due to its larger nuts and thinner shells than other hazelnut species (Erdogan and Mehlenbacher, 2000). There are over 400 described cultivars of *C. avellana* (Enescu et al., 2016). Unfortunately, the hazelnut has a well-known disadvantage: its pollen and nuts cause significant allergic reactions in sensitive individuals (Nikolaieva et al., 2014).

FIGURE 19.1 (A and B) Hazelnut tree (David Fenwick, 2006) © Hazelnut leaves (Robert Read / WTML) (D) Hazelnut young fruit (kernel) (Ben Lee / WTML) (E) Hazelnut, mature fruit (Ben Lee/ WTML) (F) *Corylus avellana* female flower (David Fenwick, 2006) (G) male flowers (catkin).

19.2.2 Taxonomy

The hazelnut tree gets its name from the Italian town of Avella, where Carl Linnaeus (1707–78) described it as "Avellana nux sylvestris", which means "Avellana's wild nut" (Celenk et al., 2020). The genus contains approximately 15–20 species disjunctly distributed throughout the Northern Hemisphere, with high species diversity in Eastern Asia (mainly in China) including the tree hazels: *C. chinensis*, *C. fargesii* Schneid., and *C. ferox*; and the shrub hazels: *C. heterophylla* and *C. sieboldiana*. Around ten species are found in China, one in Korea and Japan, three in North America (*C. americana* and *C. cornuta* in the east and *C. californica* in the west), one in the Himalayas (*C. jacquemontii*), and two in Europe and the Mediterranean (*C. avellana* and *C. colurna*) (Whitcher and Wen, 2001). Although the distribution of *Corylus* species shows a distinct disjunct pattern, the biogeographical study by Whitcher and Wen (2001) is the only one that examined this pattern by calculating the substitution rate of the internal transcribed spacer (ITS) region. Furthermore, despite the abundance of fossil records, no age estimation of *Corylus* has been conducted (Crane, 1989; Chen et al., 1999). As a result, the origin of the genus *Corylus* and its biogeographic patterns have not been addressed.

According to various authors, the number of *Corylus* species has varied. While several classification treatments have been limited to taxa on a regional scale (Yang et al., 2018), even among classifications treating the same species, taxa inclusion within each section and subgenus differs significantly. In *Corylus*, species identification is also contentious. In this respect, two species complexes have been subjected to differing taxonomic interpretations: the *Corylus heterophylla* Fisch. and the *C. cornuta* Marsh. complexes (Li and Cheng, 1979). The *C. heterophylla* complex includes three leafy-husked shrubs (*C. heterophylla* Fisch, *C. kweichowensis* Hu, and *C. yunnanensis* Fisch) that are widely distributed in Eastern Asia. *C. cornuta* Marshall, *C. californica* Marshall, *C. sieboldiana* Blume, and *C. mandshurica* Maxim are members of the *C. cornuta* complex, which includes four bristle-husked shrubs that thrive in East Asia and North America. The species within each complex have been grouped and separated in various ways. In this regard, several molecular studies have provided important insights into the phylogeny and taxonomy of *Corylus* in recent decades (Bassil et al., 2013; Yang et al., 2018). However, due to incomplete taxa sampling and low resolution in species delimitation, the above studies did not achieve meaningful results.

19.2.3 Habitat and Ecology

The hazelnut tree often grows as an understory species in mixed deciduous forests (Kull and Niinemets, 1993). It can grow both in the sun and shade due to some adaptations at the leaf level (Catoni et al., 2015). This tree thrives best on fertile and nutrient-rich soil that is only slightly acidic, or neutral, and can additionally thrive on dry calcareous soils (Pfeifer, 2019). Furthermore, hazelnut develops in locations with sufficiently high temperatures throughout the growing season and can withstand cold temperatures or even frost (Pfeifer, 2019). In Turkey, the climate is ideal for hazelnut cultivation, where the average temperature is 13–16 °C with rainfall of more than 700 mm (Ustaoglu and Karaca, 2014).

19.2.4 Importance and Usage

The wood of the hazelnut was traditionally used for fencing, barrel hoops, and wattle plasterwork, while the leaves were utilized for cattle fodder (Pfeifer, 2019). On the other hand, the fruit is the most valuable part of this species. Hazelnuts have been used for foods since ancient times (Kubiak-Martens, 1999). Therefore, people grow the tree for its nuts, and it is one of the most economically significant tree nut crops on the planet (Köksal et al., 2006). Additionally, nuts are high in protein and have high vitamin E, thiamine, and magnesium levels. In 2012, Turkey, Italy, the United States, Azerbaijan, and Georgia were the top five hazelnut producers and exporters. Also, in 2012 Turkish hazelnut production reached 660,000 tons, accounting for more than 75% of global production (Nakai, 2018).

During the cold season, the nuts are an important food source for several deer species, the edible dormouse, squirrels, and birds (Vander Wall, 2001). In addition, several animals, including invertebrates like Lepidoptera spp., feed on the leaves during the growing season. The findings showed that hazel could be a viable alternative for producing *taxol*, one of the most expensive anti-cancer drugs on the market (Wang et al., 2007). Furthermore, hazelnut is valued as an ornamental shrub, particularly the form with an unusual leaf morphology namely cutleaf hazelnut (*C. avellana* L. f. *heterophylla* (Loud.) Rehder) (Mehlenbacher and Smith, 2006).

19.2.5 Uses of Hazelnuts in Traditional Medicine

Corylus avellana (hazelnut) is one of the most popular tree nuts globally due to its nutritional value as food and the health-promoting benefits of its contents. Therefore, there is a growing interest in hazelnut by-products, including hazelnut skin, hard hazelnut shell, and hazelnut green leafy cover, as well as the hazelnut tree leaf. The phytochemicals derived from the roasting, cracking, shelling/hulling, and harvesting processes have exhibited enjoyable biological activities. This medicinal tree has been used in traditional medicine in different cultures. Along this line, Persian traditional medicine used hazelnut in treating cognitive illnesses, such as amnesia and dementia. Hazelnut and almond are mentioned in several Persian medicinal works of literature to protect brain tissue, prevent brain atrophy, and boost memory (Gorji et al., 2018). In addition, the leaves of *C. avellana*, known as fandogh, are widely consumed in the form of infusion as an effective liver tonic in Iranian traditional medicine due to their astringency and vasoprotective and anti-edema properties (Akbarzadeh et al., 2015). Furthermore, they are used in folk medicine to treat hemorrhoids, varicose veins, phlebitis, and lower limb edema. Hazelnut leaf galenic preparations have been used to treat ulcers and oropharyngeal infections. There have also been reports of antidysenteric, antifungal, and cicatrizant activities (Amaral et al., 2010). The leaves and bark of the hazelnut are used to relieve pain in Swedish traditional medicine (Tunon et al., 1995).

19.2.6 Chemical Composition of *C. avellana*

The chemical composition of *C. avellana* varies between cultivars. As a result, the bioactive metabolites found in the hazelnut kernels and other biomasses of *C. avellana* are categorized based on chemical characteristics. The phenolic compounds are the most prevalent in kernels and other processing by-products, with various derivatives and quantities of these substances recorded (Bottone et al., 2019).

19.2.6.1 Phenolics

The following compounds in the hazelnut kernel are considered phenolic acids belonging to the hydroxybenzoic acid class: *p*-Hydroxybenzoic acid, salycilic acid, 4-hydroxysalicylic acid, gallic acid, vanillic acid, and syringic acid (Prosperini et al., 2009; Pelvan et al., 2018). Gallic acid, which is found in hazelnut kernels, green leafy cover (Alasalvar et al., 2006), skin, tree leaves (Shahidi et al., 2007), and shells (Yuan et al., 2018), is the primary constituent of *C. avellana.* On the other hand, protocatechuic acid has been isolated from kernels (both fresh and roasted) and hazelnut skin (Pelvan et al., 2018), and has been recognized as the predominant ingredient in the brown skin and shells of hazelnuts produced in the United States (Alasalvar et al., 2006; Yuan et al., 2018). Furthermore, vanillic acid, methyl gallate, and veratric acid have all been found in hazelnut shells (Esposito et al., 2017; Yuan et al., 2018).

p-coumaric acid, caffeic acid, ferulic acid, and sinapic acid are hydroxycinnamic derivatives found in hazelnut kernels, green leafy cover (Alasalvar et al., 2006), skin, and tree leaves (Shahidi et al., 2007). Only the hazelnut kernel contained *m*-hydroxycinnamic acid, *o*-coumaric acid, and isoferulic acid (Prosperini et al., 2009; Pelvan et al., 2018). 3-, 4-, and 5-Caffeoylquinic acids, *p*-coumaroyltartaric acid, caffeoyltartaric acid (Oliveira et al., 2007), rosmarinic acid, and a caffeoyl-hexoside derivative (Riethmüller et al., 2013) were all found in the hazelnut leaves; whereas 3- and 5-caffeoylquinic acids were found in the leaf cover (Masullo et al., 2016). Additionally, published research indicates that hazelnut shells contain galloylquinic acid, coumaroylquinic acid, feruloylquinic acid, a pentose ester of coumaric acid, and a hexose ester of syringic acid (Yuan et al., 2018). In contrast, *p*-coumaric acid, caffeic acid, ferulic acid, and sinapic acid are hydroxycinnamic derivatives found in hazelnut kernels, green leafy cover (Alasalvar et al., 2006), skin, and leaves (Shahidi et al., 2007). The following compounds, however, are only found in the hazelnut kernel: *m*-Hydroxycinnamic acid, *o*-coumaric acid, and isoferulic acid (Prosperini et al., 2009; Pelvan et al., 2018); whereas 3-, 4-, and 5-caffeoylquinic acids, *p*-coumaroyltartaric acid, caffeoyltartaric acid (Oliveira et al., 2007), rosmarinic acid, and a caffeoyl-hexoside derivative (Riethmüller et al., 2013) were all found in the hazelnut leaves and 3- and 5-caffeoylquinic acid in the leaf cover (Masullo et al., 2016). Additionally, hazelnut shells included galloylquinic acid, coumaroylquinic acid, feruloylquinic acid, a pentose ester of coumaric acid, and a hexose ester of syringic acid (Yuan et al., 2018).

19.2.6.2 Flavonoids

Flavonoids are a class of polyphenolic secondary metabolites present in plants and often used in the human diet. These plant-derived compounds possess a wide range of health benefits. Several flavonoids have been identified in *C. avellana,* including aglycone quercetin, which was identified in the hazelnut kernels and shells (Pelvan et al., 2018; Yuan et al., 2018) and myricetin in the hazelnut kernels, shells, and skins (Del Rio et al., 2011; Yuan et al., 2018). Also, quercetin 3-rhamnoside and myricetin 3-rhamnoside were detected in kernels, shells, leaves, and bark (Oliveira et al., 2007; Jakopic et al., 2011; Riethmüller et al., 2013), and quercetin 3-rhamnoside in the skins (Del Rio et al., 2011). Similarly, kaempferol 3-rhamnoside was found in the leaves and bark (Oliveira et al., 2007; Riethmüller et al., 2013), quercetin-3-glucoside in kernels and leaves (Amaral et al., 2005; Pelvan et al., 2018), and kaempferol 3-glucoside in kernels (Cerulli et al., 2018).

A published study revealed hyperoside (Peev et al., 2007) in hazelnut leaves and rutin (Yuan et al., 2018; Solar et al., 2009) in kernels and shells. In addition, the hazelnut kernel has been recognized as containing quercetin glucuronide and isorhamnetin-3-O-rutinoside (Pelvan et al., 2018). Besides, eriodictyol was tentatively detected in the hazelnut kernel (Prosperini et al., 2009), taxifolin in the shells (Yuan et al., 2018), and naringin in the shells and skins of hazelnuts (Mattonai et al., 2017), while phloretin 2′-O-glucoside, a dihydrochalcone glucoside, has been found in hazelnut kernels and shells (Jakopic et al., 2011; Yuan et al., 2018). Catechin, epicatechin, and epigallocatechin have been found in hazelnut kernels (Fanali et al., 2018), shells (Yuan et al., 2018), and skins (Monagas et al., 2009), gallocatechin in the skins (Del Rio et al., 2011).

19.2.6.3 Tannins and Proanthocyanidins

Tannins (also known as tannic acid) are water-soluble polyphenols found in many plant diets, whereas proanthocyanidins (also known as condensed tannins) are plant-derived flavonoid polymers with a variety of health benefits (Xie et al., 2006). Hazelnut contains five hydrolyzable tannins and related compounds: Ellagic acid hexoside isomer, ellagic acid pentoside isomer, flavogallonic acid dilactone isomer, *bis*(hexahydroxydiphenoyl)-glucose (HHDP-glucose) isomer, and valoneic acid dilactone/sanguisorbic acid dilactone (Prosperini et al., 2009). The condensed tannins oproanthocyanidins are oligomers or polymers found in hazelnut and its by-products that are classified as procyanidins, propelargonidins, or prodelphinidins based on the flavan-3-ol unit (epi)catechin, (epi)afzelechin, or (epi)gallocatechin, respectively (Jakopic et al., 2011). Procyanidins A2, B1, and B2 (ZEPPA and GERBI, 2010; Fanali et al., 2018) and other procyanidin dimers and trimers, were found in hazelnut kernels (Jakopic et al., 2011). Furthermore, four isomers of B-type procyanidin have been identified in hazelnut shells (Yuan et al., 2018). A-type PAs were found as minor compounds, while (epi)-gallocatechin and gallate derivatives were identified as monomer units (Piccinelli et al., 2016).

19.2.6.4 Diarylheptanoids

Diarylheptanoids are a type of plant phenolic that comes in a variety of structural forms. As the wide variety of pharmacological actions of diarylhepatanoids and nutraceutical formulations, with relevance to human health, has just been revealed, this class has been employed in traditional medicines and homemade cures to treat many disorders (Ganapathy et al. 2019). There are three types of diarylheptanoids: linear, cyclic, and cyclic diaryletherheptanoids. The following linear diarylheptanoids were identified primarily in *C. avellana* leaves (1,7-Bis(3,4-dihydroxyphenyl)-4-hepten-3-one (hirsutenone), 1,7-bis(4-hydroxyphenyl)-4,6-heptadien-3-one, 5-hydroxy-1-(4-hydroxyphenyl)-7-(3,4-dihydroxyphenyl)-heptan-3-one, 5-hydroxy-1-(3,4-dihydroxyphenyl)-7-(4-hydroxyphenyl)-heptan-3-one, 7-bis(3,4-dihydroxyphenyl)-4-hepten-3-one-hexoside,1,7-bis(4-hydroxyphenyl)-4,6-heptadien-3-one-hexoside, and 1,7-bis-(4-hydroxyphenyl)-4-hepten-3-one-hexoside) (Riethmüller et al., 2013). Giffonins I, L– P, T, and U, as well as carpinontriol B, were isolated from hazelnut leaves, giffonins I, T, U, and carpinontriol B from hazelnut leaf covers, and giffonin P, carpinontriol B, and giffonin V were found in hazelnut shells (Masullo et al., 2015; Cerulli et al., 2017; Masullo et al., 2017). Furthermore, giffonin I and alnusone were identified in hazelnut flowers (Masullo et al., 2016).

19.2.6.5 Lignans

Lignans are secondary metabolites comprised of two phenylpropanoid units joined by a C-C bond between the C8 and C8′ carbon atoms of the side chain, whereas neolignans are made up of two phenilpropanoid units linked by a connection other than a C8-C8 bond. Erythro-(7S,8R)-guaiacylglycerol—O-4′-dihydroconiferyl alcohol, erythro-(7S,8R)-guaiacylglycerol—coniferyl aldehyde ether, erythro-(7S,8R)-guaiacylglycerol (7R,8S)-guaiacylglycerol—O-4′-dihydroconiferyl alcohol, threo-1,2-bis (4-hydroxy-3-methoxyphenyl) -1,3-propandiol, ceplignan, ficusal, ent-cedrusin, dihydrodehydrodiconiferyl alcohol, and balanophonin, were described in the hazelnut shells (Masullo et al., 2017).

19.2.6.6 Taxanes

Hazelnut leaves and shells contain various phytochemicals, including paclitaxel, 10-deacetylpaclitaxel, 10-deacetylbaccatin III, baccatin III, 7-xylosylpaclitaxel, 10-deacetyl-7-xylosylpaclitaxel, 7-epipaclitaxel, 10-deacetyl-7-epipaclitaxel, 10-deacetyl-7-xylosylcephalomannine, cephalomannine, paclitaxel C, 10-deacetyl-7-xylosylpaclitaxel C, and taxinine M. Furthermore, paclitaxel, 10-deacetyl baccatin III, baccatin III, and cephalomannine were also isolated in small amounts from Tombul hazelnut hard shells, green leafy covers, and leaves (Hoffman and Shahidi, 2009).

19.2.6.7 Volatile Compounds

Hazelnuts are well-known for their distinct flavor and aroma. Fresh and roasted hazelnuts comprise ketones, aldehydes, pyrazines, alcohols, aromatic hydrocarbons, furans, pyrroles, terpenes, and acids (Alasalvar et al., 2003; Marzocchi et al., 2017; Cordero et al., 2019). Furthermore, among the various volatile compounds, 5-methyl-(E)-2-hepten-4-one (filbertone) has been identified as the primary odorant (nutty-roasty and hazelnut-like) of roasted hazelnuts.

19.2.7 Biological Activities

There is a growing interest in hazelnut by-products such as hazelnut skin, hard shell, hazelnut green leafy cover, and hazelnut tree leaf. These roasting, cracking, shelling/hulling, and harvesting by-products have been identified as a source of important phytochemicals that possess interesting and versatile biological activities. The total antioxidant activities of extracts obtained from raw and roasted hazelnut kernels, with or without skins, from different geographical origins, varieties, and storage conditions, were identified (Bottone et al., 2019). The polar extracts of shells, leaves, and green leafy covers of the Italian cultivars such as *C. avellana* and Tonda di Giffoni showed a significant reduction of the number of cancer cell lines, including the human lung adenocarcinoma (A549), human epithelioid cervix carcinoma (Hela), human skin fibroblasts (HaCat), human B lymphoma (DeFew), human osteosarcomas (U2OS and Saos-2) (Cerulli et al., 2017; Masullo et al., 2015; Masullo et al., 2017). Moreover, the methanol extract, neolignans balanophonin and ent-cedrusin, and the phenol derivative gallic acid obtained from hazelnut shells of two other Italian varieties (Mortarella and Lunga San Giovanni) showed a low inhibitory effect on the growth of human cancer cell lines of primary and metastatic melanoma (A375 and SK-Mel-28, respectively) and on human epithelioid cervix carcinoma cell line (HeLa) (Esposito et al., 2017). On the other hand, the leaf and stem extracts of three different Spanish hazel trees significantly reduced the viability of three human-derived cancer cell lines; HeLa, liver hepatocellular cells (HepG2), and human breast adenocarcinoma cell (MCF-7) (Gallego et al., 2017).

The antimicrobial capacity of hazelnut kernels and by-products has been investigated. Cerulli and coworkers (2017) reported that the methanolic extract of *C. avellana* leaves produced two new cyclic diarylheptanoids, giffonins T and U, in addition to two known cyclic diarylheptanoids, quinic acid, flavonoid, and citric acid derivatives. Moreover, the activity of the methanol extract and isolated compounds against the gram-positive strains *Bacillus cereus* and *Staphylococcus aureus* and the gram-negative strains *Escherichia coli* and *Pseudomonas aeruginosa* have been studied. The results showed that carpinontriol B and giffonin U are the most effective against tested strains (Cerulli et al., 2017). While the aqueous extracts of three different *C. avellana* cultivars grown in Portugal including cv. Daviana, Fertille de Coutard, and M. Bollwiller demonstrated antimicrobial activity against the gram-positive bacterial strains *B. cereus*, *B. subtilis*, *S. aureus*, and the gram-negative bacteria *P. aeruginosa*, *E. coli*, *Klebsiella pneumoniae*, as well as the fungi *Candida albicans*, *Cryptococcus neoformans*. The results also revealed that hazelnut extracts have strong antibacterial action against gram-positive bacteria, with a MIC of 0.1 mg/mL. (Oliveira et al., 2008).

The *ex vivo* studies demonstrated that Turkish and Italian hazelnut varieties could mitigate neurochemical alterations in Alzheimer's neurodegeneration in rodents by consistently increasing norepinephrine levels and decreasing dopamine levels (Bahaeddin et al., 2017; Mollica et al., 2018). Furthermore, the aqueous extract of hazelnut leaves collected in Ukraine exhibited intense anticonvulsant activity, resulting in a significant increase in the latency period of the first seizure, as well as a reduction in lethality and duration of the convulsive period in mice (Blyznyuk et al., 2016).

19.3 THE GENUS *CINCHONA*

19.3.1 Botanical Description

Cinchona (genus *Cinchona*) is a genus of flowering plants, belonging to the family Rubiaceae (coffee family). It includes about 23 tropical evergreen trees, shrubs, or small tree species, reaching up to 20 m high with a diameter between 15–20 cm (Júnior et al., 2012). Furthermore, it is native to the Andes of western South America (Andersson, 1997). The Cinchona grows best in temperatures ranging from 13.5 to 21°C, with a minimum daily relative humidity of 68–97%. In addition, the soil should be fertile, loose, not rocky, rich in organic matter, and have a pH of 4.6–6.5, with an optimal pH of 5.8, and have 2000–3000 mm of annual rainfall.

Cinchonopsis, Jossia, Ladenbergia, Remijia, Stilpnophyllum, and Ciliosemina are all members of the Cinchoneae tribe (Andersson and Antonelli, 2005). *Cinchona* plant has opposite, lanceolate leaves that are elliptical to ovate and have the entire edge (Figure 19.2A). The flowers are small pink, creamy, or brown with a small lobed calyx united at the base. The tubular corolla comprises five spreading lobes (Figure 19.2B). The fruit is an oblong capsule with 40–50 winged seeds inside it (Figure 19.2C). The bark (Figure 19.2D and E) includes several alkaloids, the most well-known being quinine, an antipyretic compound (fever antidote) used to treat malaria. Moreover, only *C. officinalis* and *C. pubescens* (syn. *C. succirubra*) manufacture quinine.

19.3.2 Etymology and Common Names

The Swedish botanist, LINNAEUS Sin 1742 named the genus *Cinchona* after the Countess, wife of the Viceroy of Peru. According to well-cited legend, the countess recovered after all other treatments failed in 1638 when she was given cinchona bark. It was then widely used for malaria treatment, with the Jesuits propagating it all across the world. *Cinchona* was then designated as Ecuador's national tree and was depicted on Peru's coat of arms (Jäger, 2004).

19.3.3 History

For more than four centuries, the bark of the *Cinchona* tree was the primary treatment of malaria. According to studies, a few *Cinchona* species including, *C. calisaya*, *C. ledgeriana*, *C. officinalis*, and *C. pubescens* (*C. succirubra*), have a high quinine level and cinchonine alkaloids (Kaufman and Ruveda, 2005). Jesuit monks discovered the medical potential of *Cinchona* bark at Loxa (now Loja, Ecuador) in the 17th century, and shipments of different types of *Cinchona pubescens* Vahl (red bark) from South America to Europe quickly reached half a million kg bark each year (Roersch van der Hoogte and Pieters, 2015). Because imports were insufficient to fulfill demand, the British, Dutch, and French empires embarked on a search for the most productive source of *Cinchona* trees in order to create plantations. *Cinchona calisaya* Wedd. (Yellow bark) from Bolivia is proven to be the most productive species to date (Greenwood, 1992; Nair, 2010).

19.3.4 Taxonomy

In 1738, Charles Marie de La Condamine was the first to define *Cinchona* L. as a genus (Lee, 2002). Because this genus is medicinally significant, taxonomists have historically paid close attention to

FIGURE 19.2 (A) *Cinchona pubescens* leaves (Jäger, 2004) (B) *Cinchona pubescens* flower (Jäger, 2004) (C) *Cinchona pubescens* fruits (Jäger, 2004) (D) *Cinchona pubescens* tree trunk showing the bark (Jäger, 2004). (E) Quinine bark.

it. Many alternative names were in the 19th century (Popenoe, 1949). Andersson (1997) evaluated over 330 names (many at the variety level) and designated 23 species in his review of the genus *Cinchona*. *C. pubescens* is particularly difficult to classify because it frequently hybridizes with other *Cinchona* species in the wild (Camp, 1949; Andersson, 1997). In 1808, Von Humboldt noted this ambiguity, stating that if one did not have the opportunity to examine the branch in the wild, one would consider *Cinchona* leaves from the same branch to be specimens of different species. *C. pubescens* was discovered to hybridize with four species in 1946 (Humboldt, 1808), and seven species in 1998 (Jäger, 2004) (*C. barbacoensis*, *C. calisaya*, *C. lancifolia*, *C. lucumifolia*, *C. macrocalyx*, *C. micrantha*, *C. officinalis*). *C. pubescens*, and *C. calisaya* hybrids are the most abundant in nature, and they appear to have been cultivated as well (Andersson, 1997).

19.3.4.1 *Cinchona calisaya* Wedd

It is a large bushy tree with a straight trunk found in the lower portions of Bolivia and southeastern Peru and prefers lower elevations (400–1000 m above mean sea level). The leaves are thick, oblong to lanceolate with smooth surfaces. A huge panicle of pale pink flowers forms the inflorescence. Capsules are oblong and 8–17 mm in length. The bark is thick (2–5 mm) with a grayish outer layer, a broad longitudinal fissure, and a few transverse fissures that peel off in places. The bark contains 3.89–7.24 % of the total alkaloids, with quinine concentration ranging from 0.78 to 5.57% (Nair, 2021). *C. calisaya* is one of 23 tree species in the genus *Cinchona* that has varying amounts of alkaloids. Although the four primary Cinchona alkaloids (quinine, quinidine, cinchonine, and cinchonidine) exhibit antimalarial action, their pharmacological profiles differ (Maldonado et al., 2017).

19.3.4.2 *Cinchona ledgeriana* Moons

It is a fast-growing, densely branched tree that reaches a height of 6–16 m at maturity. In the Darjeeling hills, the species thrives between 1000 and 1900 m above sea level. The leaves are light green, oval, acuminate, and have a slight axil curve. The flowers are tiny and pale yellow, the flowering period from May through October. The capsule is oval-lanceolate in shape and is 15–19 mm in length. Moreover, the bark is similar to *C. calisaya* in thickness (2–5 mm), but there are more cracks and fewer in deep sections. The root and trunk (10–12 years old) had an average total alkaloid content of 7.21 and 6.01 %, respectively, with quinine at 5.4 and 1.98% (Kurian and Sankar, 2007).

19.3.4.3 *Cinchona officinalis* Linn.

It is a slow-growing tree, slender, 6–10 m tall, with dark green leaves. It prefers a cooler temperature at 1200–2000 m elevation above sea level in the Nilgiri hills. The leaves are small, smooth, ovate-lanceolate with reddish petioles, measuring 4–10 cm in length. Flowers are 1.4–1.6 cm long, deep pink to rosy in hue, and borne in tiny terminal panicles. The capsule is 1.5–2.0 cm in length and ovoid-oblong in shape. The bark is rough and brown on the outside and yellow on the inside, with a thickness of 1.5 mm and numerous transverse fissures with recurved edges. The trunk bark yields 4–6% total alkaloids on average, with quinine accounting for up to half (Chatterjee et al. 1988).

19.3.4.4 *Cinchona pubescens* Vahl. (syn. *Cinchona succirubra* Pav. ex klotzsch)

It is evergreen, massive, sturdy, actively growing tree with a straight trunk. It is native to Peru and Ecuador, grows to around 10–25 m high, and thrives at elevations between 1200 and 2000 m above mean sea level in Tamil Nadu's milder temperature. The species is noteworthy for its capacity to resist both high humidity and drought. The leaves are huge (40–50 cm × 30–40 cm), thin, light green in color, and elliptical. Flowers are rosy, pink in color, bloom all year, 1–2 cm long, and the upper surface of the corolla is white with pink stripes. The capsule is oblong and is 2–3 cm in length. The bark has a dark brown color with longitudinal wrinkles and a few transverse fissures, 2–6 mm thick. It contains much cinchonine but not a lot of quinine (Chatterjee, 1993; Jäger, 2004). The total alkaloids in the bark of the root, trunk, and twigs are 7.21, 6.09, and 4.0%, respectively, with quinine content of 0.76–1.42, 1.1–1.74, and 0.8–1.16%, respectively.

19.3.5 Cultivation

Cinchona has been cultivated for its quinine-rich bark and roots in numerous tropical places across the world. Quinine was used to cure malaria and had an enormous economic impact from the seventeenth to the twentieth centuries. Cinchona's importance waned when quinine was synthesized in 1944, although natural quinine is still used in places where the synthetic is unavailable and for other therapeutic purposes (Jäger, 2004).

19.3.6 Medicinal Uses

Cinchona is well-known for its medicinal properties in traditional and modern medicine. It is a traditional medicinal plant whose extracts are used in Indian Ayurvedic formulations to treat various fevers, including malaria (Bohórquez et al., 2012). Apart from that, in many parts of the world people have used *Cinchona* barks as medicinal agents for various purposes. In Brazil, people used the plant as a digestive stimulant and fever reducer whereas people in South America and Europe employed it in the treatment of cancer (Li and Tian, 2016). In addition, people in the United States used the *Cinchona* extracts as anti-protozoal and anti-spasmodic, digestive tonic, and treatment cardiac diseases. *Cinchona* contains organic compounds that have been isolated from the bark of roughly 40 different types of *Cinchona* trees all over the world. Quinine, quinidine, cinchonine, and cinchonidine are the main chemical elements of these alkaloids. Quinine is the most important alkaloid found in *Cinchona* bark, and it is used to prevent and treat malaria. Quinine hydrochloride is used to treat varicose veins and internal hemorrhoids as a sclerosing agent, while quinine sulfate has long been used in European medicines to relieve night cramps, but it has largely been phased out due to thrombocytopenia problems (El-Tawil et al., 2015). Quinine water (100 ppm) is also used to treat sore throats and unpleasant odors as a gargle. The majority of *Cinchona* extracts have been used as bitter additives in the food and beverage industry. Several *Cinchona* species are currently widely cultivated in various regions due to their rising commercial importance (Li and Tian, 2016).

19.3.7 Chemical Constituents of *Cinchona*

Cinchona bark is high in quinoline alkaloids, accounting for 6–10% of total quinoline alkaloids. Quinine, quinidine, cinchonine, cinchonidine, and 30 other minor bases related to quinine, are the significant elements of alkaloids (Dayrit et al., 1994). Quinoline and quinuclidine rings contain a vinyl group in these alkaloids. The bark also contains coloring matter (>10%), flavonoids, an essential oil, and polyphenols in addition to alkaloids. The alkaloids are mainly presented as quinine and cinchonic acid salts, and their relative amounts differ between species. Because these alkaloids are formed during the sap's descent, their percentage concentration is lowest in the twigs, and highest in the trunk (bark) and the root (bark). The collar part (30–45 cm in length, at the base) has the highest concentration of quinine. Depending on the species, the alkaloid content of trees increases with age (1–12 years). Quinine is separated as quinine sulfate from the total alkaloids. It is a white, crystalline, odor-free substance with a strong bitter flavor, and highly soluble in organic solvents. In most *Cinchona* barks, quinidine is found in minor amounts (0.2 %), although it is found in more significant concentrations in *C. calisaya* and at its highest concentration in *C. tayansis*. The chemical conversion of quinine through oxidation is used to make it commercially (Dayrit et al., 1994). Up to 1% of total alkaloids are found in leaves, with younger ones holding more (Keene et al., 1983). Moreover, *C. ledgeriana* leaves contain five monomeric indole, quinomine, aricine, and 3-epi quinine, quasidimeric indole alkaloids (Dayrit et al., 1994; Nair, 2021).

Quinine ($C_2OH_{24}N_2O_2$) (Figure 19.3): *Cinchona*'s most important and distinctive alkaloids constitute 16% quinine in the bark (Kacprzak, 2013), while the quantity fluctuates (6–10%) according to the species variety. It is commonly used as an anti-malarial medication, as well as a flavoring component in carbonated beverages (Gurung and De, 2017), a skeletal muscle relaxant, a treatment for hemorrhoids, and used as oxytoxic agent (Kacprzak, 2013). Quinine's antimalarial (Achan et al., 2011) characteristics are derived from its interference with DNA synthesis in the merozoite phase of *Plasmodium* protozoa (Gurung and De, 2017). Quinine, sometimes described as "a broad protoplasmic toxin" (Goldenberg and Wexler, 1988) has a wide range of biological effects. The foundation for its therapeutic usage in man for muscular cramps and malaria is its curare-like action on skeletal muscle and toxic effects on bacteria and unicellular organisms such as *Plasmodium* (Goldenberg and Wexler, 1988).

Cinchonine ($C_{19}H_{22}N_2O$) (Figure 19.3) is the second most major alkaloid found in *Cinchona* after quinine, and it is also employed as an anti-malarial drug (Tracy, 1996). Cinchonine has lower toxicity than quinine and higher activity than other quinine-related compounds (Genne et al., 1994). However, the exact percentage of cinchonine in *Cinchona* is controversial. Moreover, it is mainly used as an antimicrobial agent, and broadly used for schizonticide, amoebiasis, flu, dysentery, and fever. It also serves as a moderate stimulant to the mucosa of the stomach (Kacprzak, 2013).

Another major alkaloid is quinidine (Figure 19.3), found in *Cinchona* bark in concentrations ranging from 0.25 to 3.0% (Gurung and De, 2017). It is a dextrorotatory quinine stereoisomer. Its primary role is as an anti-malarial drug, although it is also effective as an anti-arrhythmic agent (Noujaim et al., 2011). Cinchonidine (C_{19} H_{22} N_2O) (Figure 19.3) is found in most *Cinchona* bark species, particularly in the bark of *C. pusescensval* and *C. pitayensis* and is primarily employed as

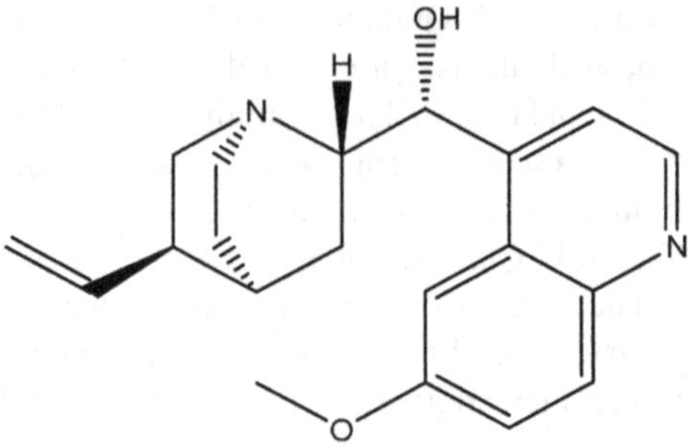

Quinine

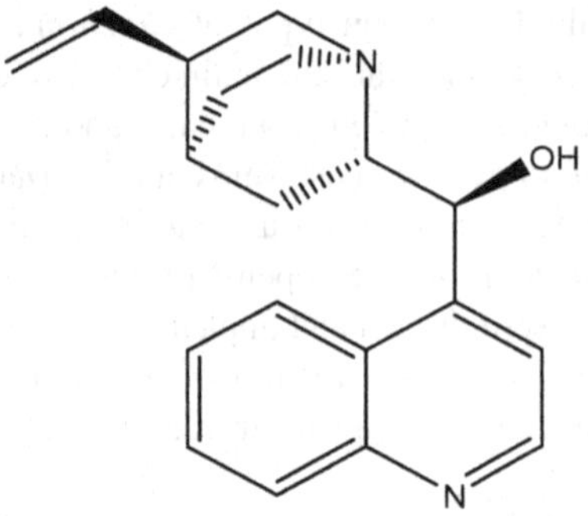

Cinchonine

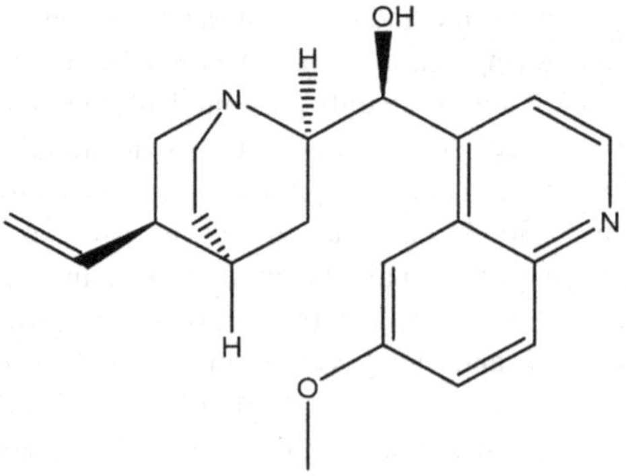

Quinidine

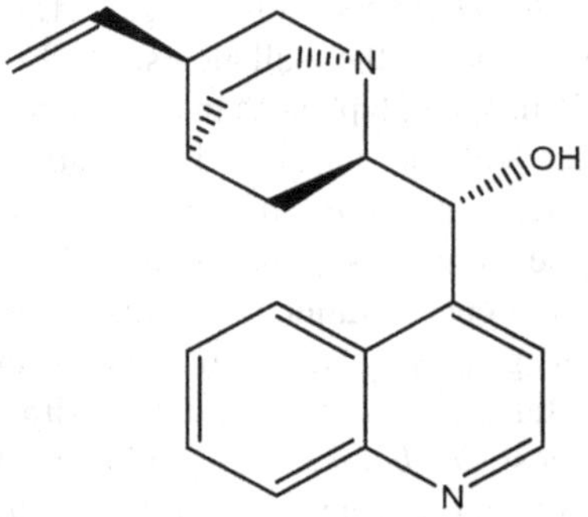

Cinchonidine

FIGURE 19.3 Chemical structures of the major alkaloids found in *Cinchona*.

an anti-malarial drug. It is a stereoisomer and pseudo-enantiomer of cinchonine, and it is mainly employed as a quinine substitute. Furthermore, epicinchonidine is used chiefly as an antimalarial (Gurung and De, 2017).

19.3.8 Toxicology of *Cinchona*

Both ground cinchona and quinine can provoke urticaria, contact dermatitis, and other hypersensitive reactions in humans. When these alkaloids are consumed, they can cause a chemical illness known as "cinchonism", which is characterized by severe headache, abdominal discomfort, convulsions, visual problems, blindness, and auditory disturbances such as ringing in the ears, paralysis, and even collapse (Leung, 1980). Quinidine and similar alkaloids are absorbed through the gastrointestinal tract, and a single oral dose of 2–8 g of quinidine can be deadly to an adult (Nair, 2021). Because of its fetal and abortifacient properties, quinine is not recommended for usage during pregnancy.

19.3.9 COVID-19 Treatment with Chloroquine and Hydroxychloroquine

Quinine is a compound found in the Peruvian bark of *Cinchona* trees and was previously used to treat malaria (Permin et al., 2016). Chloroquine (CQ) is an amino acid tropic form of quinine, developed by Bayer in Germany in 1934 (Parhizgar and Tahghighi, 2017). Chloroquine (CQ) and hydroxychloroquine (HCQ) have been proposed as promising agents against the novel coronavirus

SARS-CoV-2, which causes COVID-19, and as a potential therapy for reducing the duration of the viral disease (Colson et al., 2020). In addition to its antiviral effects, early reports on CQ/HCQ have demonstrated its potential to lower immunological reactions to infection, suppress pneumonia exacerbation, reduce fever duration, and enhance lung health (Devaux et al., 2020; Wang et al., 2020). Furthermore, an *in vitro* investigation revealed that CQ and HCQ inhibit the virus's passage from early endosomes to endolysosomes, preventing the virus's DNA from being released and disrupting its reproductive cycle (Liu et al., 2020).

CQ demonstrated several activities, one of which is to raise the pH of intracellular vesicles, interfering with pH-dependent processes of viral replication such as endosome and lysosome maturation and fusion, as well as virus uncoating (Wang et al. 2020). CQ also caused changes in angiotensin-converting enzyme 2 (ACE2), which hinders S-protein binding, phagocytosis, and subsequent release into the cytoplasm, where viral replication occurs (Yang et al. 2004; Vincent et al. 2005). CQ reduces cytokine synthesis, aiding the inflammatory consequences of viral infections, in addition to its direct antiviral impact (Karres et al. 1998; Savarino et al. 2003). For enveloped viruses, post-translational modification of membrane glycoproteins occurs in the vesicles of the endoplasmic and trans-Golgi apparatus.

19.4 CONCLUSIONS AND FUTURE PERSPECTIVES

In summary, this chapter has summarized and highlighted the recent published research work dealing with two important medicinal plants that have been used by different traditional medicines. Research conducted on *C. avellana*, the common hazel, native to Europe and -Asia, Greece, Turkey, and Cyprus indicated that the tree contains significant amounts of thiamine and vitamin B 6, as well as smaller amounts of other B vitamins, and is an excellent source of vitamins E and B6. Extracts obtained from different parts of *C. avellana* exhibited several biological activities including antioxidant, anti-proliferative, antimicrobial along with neuroprotective effects. Thus, this chapter highlights *C. avellana* as a source of bioactive chemicals that could be responsible for the biological activity of extracts from this plant. However, more work is still needed to address the health benefits of *C. avellana* extracts and their possible use as food supplements. Similarly, this chapter has summarized the recent work related to *Cinchona* (genus *Cinchona*), a genus of flowering plants belonging to the family Rubiaceae (coffee family). Plants of this genus have been used in the traditional to treat several illnesses including fever, cancer, digestive stimulant, as anti-protozoal and anti-spasmodic, digestive tonic, in addition to cardiac diseases. This bioactivity could be attributed to the presence of bioactive compounds such as quinine, quinidine, cinchonine, and cinchonidine. We hope this chapter will benefit people who work in the field.

REFERENCES

Achan, J., Talisuna, A.O., Erhart, A., Yeka, A., Tibenderana, J.K., Baliraine, F.N., Rosenthal, P., & D'Alessandro, U. (2011). Quinine, an old anti-malarial drug in a modern world: role in the treatment of malaria. *Malaria Journal*, 10(1), 1–12.

Akbarzadeh, T., Sabourian, R., Saeedi, M., Rezaeizadeh, H., Khanavi, M., & Ardekani, M.R.S. (2015). Liver tonics: review of plants used in Iranian traditional medicine. *Asian Pacific Journal of Tropical Biomedicine*, 5(3), 170–181.

Alasalvar, C., Karamać, M., Amarowicz, R., & Shahidi, F. (2006). Antioxidant and antiradical activities in extracts of hazelnut kernel (*Corylus avellana* L.) and hazelnut green leafy cover. *Journal of Agricultural and Food Chemistry*, 54(13), 4826–4832.

Alasalvar, C., Shahidi, F., & Cadwallader, K.R. (2003). Comparison of natural and roasted Turkish tombul hazelnut (*Corylus avellana* L.) volatiles and flavor by DHA/GC/MS and descriptive sensory analysis. *Journal of Agricultural and Food Chemistry*, 51(17), 5067–5072.

Amaral, J.S., Ferreres, F., Andrade, P.B., Valentão, P., Pinheiro, C., Santos, A., & Seabra, R. (2005). Phenolic profile of hazelnut (*Corylus avellana* L.) leaves cultivars grown in Portugal. *Natural Product Research*, 19(2), 157–163.

Amaral, J.S., Valentão, P., Andrade, P.B., Martins, R.C., & Seabra, R.M. (2010). Phenolic composition of hazelnut leaves: influence of cultivar, geographical origin and ripening stage. *Scientia Horticulturae*, 126(2), 306–313.

Andersson, L. (1997). A new revision of Joosia (Rubiaceae-Cinchoneae). *Brittonia*, 49(1), 24–44.

Andersson, L., & Antonelli, A. (2005). Phylogeny of the tribe Cinchoneae (Rubiaceae), its position in Cinchonoideae, and description of a new genus, Ciliosemina. *Taxon*, 54(1), 17–28.

Bahaeddin, Z., Yans, A., Khodagholi, F., Hajimehdipoor, H., & Sahranavard, S. (2017). Hazelnut and neuroprotection: improved memory and hindered anxiety in response to intra-hippocampal Aβ injection. *Nutritional Neuroscience*, 20(6), 317–326.

Bassil, N., Boccacci, P., Botta, R., Postman, J., & Mehlenbacher, S. (2013). Nuclear and chloroplast microsatellite markers to assess genetic diversity and evolution in hazelnut species, hybrids and cultivars. *Genetic Resources and Crop Evolution*, 60(2), 543–568.

Blyznyuk, N.A., Prokopenko, Y.S., Georgiyants, V.A., & Tsyvunin, V.V. (2016). A comparative phytochemical and pharmacological analysis of the extracts from leaves of the Ukrainian flora shrubs. *News of Pharmacy*, 1 (85), 29–32.

Bohórquez, E.B., Chua, M., & Meshnick, S.R. (2012). Quinine localizes to a non-acidic compartment within the food vacuole of the malaria parasite *Plasmodium falciparum*. *Malaria Journal*, 11(1), 1–6.

Bottone, A., Cerulli, A., D'Urso, G., Masullo, M., Montoro, P., Napolitano, A., & Piacente, S. (2019). Plant specialized metabolites in hazelnut (Corylus avellana) kernel and byproducts: an update on chemistry, biological activity, and analytical aspects. *Planta Medica*, 85(11/12), 840–855.

Camp, W.H. (1949). *Cinchona* at high altitudes in Ecuador. *Brittonia*, 6(4), 394–430.

Catoni, R., Granata, M.U., Sartori, F., Varone, L., & Gratani, L. (2015). *Corylus avellana* responsiveness to light variations: morphological, anatomical, and physiological leaf trait plasticity. *Photosynthetica*, 53(1), 35–46.

Celenk, V.U., Argon, Z.U., & Gumus, Z.P. (2020). Cold pressed hazelnut (*Corylus avellana*) oil. In *Cold Pressed Oils*. Cambridge, MA and London, UK, Academic Press, pp. 241–254.

Cerulli, A., Lauro, G., Masullo, M., Cantone, V., Olas, B., Kontek, B., Nazzaro, F., Bifulco, G., & Piacente, S. (2017). Cyclic diarylheptanoids from *Corylus avellana* green leafy covers: determination of their absolute configurations and evaluation of their antioxidant and antimicrobial activities. *Journal of Natural Products*, 80(6), 1703–1713.

Cerulli, A., Napolitano, A., Masullo, M., Pizza, C., & Piacente, S. (2018). LC-ESI/LTQOrbitrap/MS/MSn analysis reveals diarylheptanoids and flavonol O-glycosides in fresh and roasted hazelnut (*Corylus avellana* cultivar "Tonda di Giffoni"). *Natural Product Communications*, 13(9), 1934578X1801300906.

Chatterjee, S.K. (1993). Domestication studies of some medicinally important exotic plants growing in India. *Acta Horticulturae*, 331, 151–160.

Chatterjee, S.K., Bharati, P., Chhetri, K.B., & Ramsong, A.F. (1988). Agrotechnology of cultivating cinchona and ipecac in Darjeeling hills. In Proceedings of the 5th ISHA Symposium of Medicinal, Aromatic and Spice Plants, 188-A:Special Issue, pp.25–27.

Chen, Z.D., Manchester, S.R., & Sun, H.Y. (1999). Phylogeny and evolution of the Betulaceae as inferred from DNA sequences, morphology, and paleobotany. *American Journal of Botany*, 86(8), 1168–1181.

Colson, P., Rolain, J.M., Lagier, J.C., Brouqui, P., & Raoult, D. (2020). Chloroquine and hydroxychloroquine as available weapons to fight COVID-19. *International Journal of Antimicrobial Agents*, 55(4), 105932.

Contessa, C., Valentini, N., Caviglione, M., & Botta, R. (2011). Propagation of *Corylus avellana* L. by means of semi-hardwood cutting: rooting and bud retention in four Italian cultivars. *European Journal of Horticultural Science*, 76(5), 170.

Cordero, C., Guglielmetti, A., Bicchi, C., Liberto, E., Baroux, L., Merle, P., Tao, Q., & Reichenbach, S.E. (2019). Comprehensive two-dimensional gas chromatography coupled with time of flight mass spectrometry featuring tandem ionization: challenges and opportunities for accurate fingerprinting studies. *Journal of Chromatography A*, 1597, 132–141.

Cragg, G.M., & Newman, D.J. (2013). Natural products: a continuing source of novel drug leads. *Biochimica et Biophysica Acta*, 1830(6), 3670–3695.

Crane, P.R. (1989). Early fossil history and evolution of the Betulaceae. *Evolution, Systematics, and Fossil History of the Hamamelidae. Volume 2." Higher Hamamelidae." Systematic Association*, 40, 87–116.

Dayrit, F.M., Guldotea, A., Generalao, M.L., & Serna, C. (1994). Determination of the quinine content in the bark of the *Cinchona* tree grown in Mt. *Kitangland, Bukindon. Philippine Journal of Science*, 123(3), 52–57.

Del Rio, D., Calani, L., Dall'Asta, M., & Brighenti, F. (2011). Polyphenolic composition of hazelnut skin. *Journal of Agricultural and Food Chemistry*, 59(18), 9935–9941.

Devaux, C.A., Rolain, J.M., Colson, P., & Raoult, D. (2020). New insights on the antiviral effects of chloroquine against coronavirus: what to expect for COVID-19?. *International Journal of Antimicrobial Agents*, 55(5), 105938.

El-Tawil, S., Al Musa, T., Valli, H., Lunn, M.P., Brassington, R., El-Tawil, T., & Weber, M. (2015). Quinine for muscle cramps. *Cochrane Database of Systematic Reviews*, 5(4): CD005044.

Enescu, C.M., Durrant, T.H., de Rigo, D., & Caudullo, G. (2016). *Corylus avellana* in Europe: distribution, habitat, usage and threats. *European Atlas of Forest Tree Species*. Luxembourg: EU Publication Office, 86–87.

Erdogan, V., & Mehlenbacher, S.A. (2000). Interspecific hybridization in hazelnut (*Corylus*). *Journal of the American Society for Horticultural Science*, 125(4), 489–497.

Esposito, T., Sansone, F., Franceschelli, S., Del Gaudio, P., Picerno, P., Aquino, R.P., & Mencherini, T. (2017). Hazelnut (*Corylus avellana* L.) shells extract: phenolic composition, antioxidant effect and cytotoxic activity on human cancer cell lines. *International Journal of Molecular Sciences*, 18(2), 392.

Fanali, C., Tripodo, G., Russo, M., Della Posta, S., Pasqualetti, V., & De Gara, L. (2018). Effect of solvent on the extraction of phenolic compounds and antioxidant capacity of hazelnut kernel. *Electrophoresis*, 39(13), 1683–1691.

Gallego, A., Metón, I., Baanante, I.V., Ouazzani, J., Adelin, E., Palazon, J., Bonfill, M., & Moyano, E. (2017). Viability-reducing activity of *Coryllus avellana* L. extracts against human cancer cell lines. *Biomedicine & Pharmacotherapy*, 89, 565–572.

Ganapathy, G., Preethi, R., Moses, J.A., & Anandharamakrishnan, C. (2019). Diarylheptanoids as nutraceutical: a review. *Biocatalysis and Agricultural Biotechnology*. 19(5), 101109.

Genne, P., Duchamp, O., Solary, E., Pinard, D., Belon, J.P., Dimanche-Boitrel, M.T., & Chauffert, B. (1994). Comparative effects of quinine and cinchonine in reversing multidrug resistance on human leukemic cell line K562/ADM. *Leukemia*, 8(1), 160–164.

Gewali, M. B., & Awale, S. (2008). Aspects of traditional medicine in Nepal. *Japan: Institute of Natural Medicine University of Toyama*, 140–142.

Goldenberg, A.M., & Wexler, L.F. (1988). Quinine overdose: review of toxicity and treatment. *Clinical Cardiology*, 11(10), 716–718.

Gorji, N., Moeini, R., & Memariani, Z. (2018). Almond, hazelnut and walnut, three nuts for neuroprotection in Alzheimer's disease: a neuropharmacological review of their bioactive constituents. *Pharmacological Research*, 129, 115–127.

Greenwood, D. (1992). The quinine connection. *Journal of Antimicrobial Chemotherapy*, 30(4), 417–427.

Gurung, P., & De, P. (2017). Spectrum of biological properties of *Cinchona* alkaloids: a brief review. *Journal of Pharmacognosy and Phytochemistry*, 6(4), 162–166.

Hoffman, A., & Shahidi, F. (2009). Paclitaxel and other taxanes in hazelnut. *Journal of Functional Foods*, 1(1), 33–37.

Humboldt, F.V. (1808). Uber die Chinawälder in Südamerika, 2. *Der Gesellschaft Naturforschender Freunde zu Berlin Magazin für die neuesten Entdeckungen in der gesamten Naturkunde*, 1, 104–120.

Jäger, H. (2004). *Cinchona pubescens. Enzyklopädie der Holzgewächse: Handbuch und Atlas der Dendrologie*, Weinheim, Wiley-VCH, pp. 1–14.

Jakopic, J., Petkovsek, M.M., Likozar, A., Solar, A., Stampar, F., & Veberic, R. (2011). HPLC–MS identification of phenols in hazelnut (*Corylus avellana* L.) kernels. *Food Chemistry*, 124(3), 1100–1106.

Júnior, W.S.F., Cruz, M.P., dos Santos, L.L., & Medeiros, M.F.T. (2012). Use and importance of quina (*Cinchona* spp.) and ipeca (*Carapichea ipecacuanha* (Brot.) L. Andersson): plants for medicinal use from the 16th century to the present. *Journal of Herbal Medicine*, 2(4), 103–112.

Kacprzak, K.M. (2013). Chemistry and biology of Cinchona alkaloids. *Natural Products*, 1, 605–641.

Karres, I., Kremer, J.P., Dietl, I., Steckholzer, U., Jochum, M., & Ertel, W. (1998). Chloroquine inhibits proinflammatory cytokine release into human whole blood. *American Journal of Physiology-Regulatory, Integrative and Comparative Physiology*, 274(4), R1058–R1064.

Kaufman, T.S., & Rúveda, E.A. (2005). The quest for quinine: those who won the battles and those who won the war. *Angewandte Chemie International Edition*, 44(6), 854–885.

Keene, A.T., Anderson, L.A., & Phillipson, J.D. (1983). Investigation of *Cinchona* leaf alkaloids by high-performance liquid chromatography. *Journal of Chromatography A*, 260, 123–128.

Köksal, A.İ., Artik, N., Şimşek, A., & Güneş, N. (2006). Nutrient composition of hazelnut (*Corylus avellana* L.) varieties cultivated in Turkey. *Food Chemistry*, 99(3), 509–515.

Kubiak-Martens, L. (1999). The plant food component of the diet at the late Mesolithic (Ertebolle) settlement at Tybrind Vig, Denmark. *Vegetation History and Archaeobotany*, 8(1), 117–127.

Kull, O., & Niinemets, Ü. (1993). Variations in leaf morphometry and nitrogen concentration in Betula pendula Roth., *Corylus avellana* L. and Lonicera xylosteum L. *Tree Physiology*, 12(3), 311–318.

Kurian, S., and Sankar, M. (2007). *Medicinal Plants*, Vol.2. New Delhi, New India Publishing, pp. 374.

Lee, M.R. (2002). Plants against malaria. Part 1: Cinchona or the Peruvian bark. *The Journal of the Royal College of Physicians of Edinburgh*, 32(3), 189–196.

Leung, A.Y. (1980). *Encyclopedia of Common Natural Ingredients Used in Food, Drugs, and Cosmetics*, 3rd ed.Hoboken, NJ, Wiley, pp. 848.

Li, P. C., & Cheng, S. X. (1979). "Betulaceae," in *Flora Republicae Popularis Sinicae*, Vol. 21, eds. K.-Z. Kuang and P.-C. Li. Henderson, NV, Science Press, pp. 44–137.

Li, Y., & Tian, J. (2016). Evaluation of local anesthetic and antipyretic activities of *Cinchona* alkaloids in some animal models. *Tropical Journal of Pharmaceutical Research*, 15(8), 1663–1666.

Liu, J., Cao, R., Xu, M., Wang, X., Zhang, H., Hu, H., Li, Y., Hu, Z., Zhong, W.,& Wang, M. (2020). Hydroxychloroquine, a less toxic derivative of chloroquine, is effective in inhibiting SARS-CoV-2 infection *in vitro*. *Cell Discovery*, 6(1), 1–4.

Maldonado, C., Barnes, C.J., Cornett, C., Holmfred, E., Hansen, S.H., Persson, C., Antonelli, A., & Rønsted, N. (2017). Phylogeny predicts the quantity of antimalarial alkaloids within the iconic yellow *Cinchona* bark (Rubiaceae: *Cinchona calisaya*). *Frontiers in Plant Science*, 8, 391.

Marzocchi, S., Pasini, F., Verardo, V., Ciemniewska-Żytkiewicz, H., Caboni, M.F., & Romani, S. (2017). Effects of different roasting conditions on physical-chemical properties of Polish hazelnuts (*Corylus avellana* L. var. *Kataloński*). *LWT*, 77, 440–448.

Masullo, M., Cerulli, A., Mari, A., de Souza Santos, C.C., Pizza, C., & Piacente, S. (2017). LC-MS profiling highlights hazelnut (Nocciola di Giffoni PGI) shells as a byproduct rich in antioxidant phenolics. *Food Research International*, 101, 180–187.

Masullo, M., Cerulli, A., Olas, B., Pizza, C., & Piacente, S. (2015). Giffonins A–I, antioxidant cyclized diarylheptanoids from the leaves of the hazelnut tree (*Corylus avellana*), source of the Italian PGI Product "Nocciola di Giffoni". *Journal of Natural Products*, 78(1), 17–25.

Masullo, M., Mari, A., Cerulli, A., Bottone, A., Kontek, B., Olas, B., Pizza, C., & Piacente, S. (2016). Quali-quantitative analysis of the phenolic fraction of the flowers of *Corylus avellana*, source of the Italian PGI product "Nocciola di Giffoni": isolation of antioxidant diarylheptanoids. *Phytochemistry*, 130, 273–281.

Mattonai, M., Licursi, D., Antonetti, C., Galletti, A.M.R., & Ribechini, E. (2017). Py-GC/MS and HPLC-DAD characterization of hazelnut shell and cuticle: insights into possible re-evaluation of waste biomass. *Journal of Analytical and Applied Pyrolysis*, 127, 321–328.

Mehlenbacher, S.A., & Smith, D.C. (2006). Self-compatible seedlings of the cutleaf hazelnut. *HortScience*, 41(2), 482–483.

Mollica, A., Zengin, G., Stefanucci, A., Ferrante, C., Menghini, L., Orlando, G., Brunetti, L., Locatelli, M., Dimmito, M., Novellino, E., & Onaolapo, O.J. (2018). Nutraceutical potential of *Corylus avellana* daily supplements for obesity and related dysmetabolism. *Journal of Functional Foods*, 47, 562–574.

Monagas, M., Garrido, I., Lebron-Aguilar, R., Gómez-Cordovés, M.C., Rybarczyk, A., Amarowicz, R., & Bartolome, B. (2009). Comparative flavan-3-ol profile and antioxidant capacity of roasted peanut, hazelnut, and almond skins. *Journal of Agricultural and Food Chemistry*, 57(22), 10590–10599.

Nair, K.P. (2010). *The Agronomy and Economy of Important Tree Crops of the Developing World*, 1st ed. Amsterdam, Netherlands, Elsevier, pp. 368.

Nair, K.P. (2021). *Tree Crops: Harvesting Cash from the World's Important Cash Crops*. New York: Springer, pp. 129–149.

Nakai, J. (2018). Food and agriculture organization of the United Nations and the sustainable development goals. *Sustainable Development*, 22.

Nikolaieva, N., Brindza, J., Garkava, K., & Ostrovsky, R. (2014). Pollen features of hazelnut (*Corylus avellana* L.) from different habitats. *Modern Phytomorphology*, 6, 53–58.

Noujaim, S.F., Stuckey, J.A., Ponce-Balbuena, D., Ferrer-Villada, T., López-Izquierdo, A., Pandit, S.V., Sánchez-Chapula, J., & Jalife, J. (2011). Structural bases for the different anti-fibrillatory effects of chloroquine and quinidine. *Cardiovascular Research*, 89(4), 862–869.

Oliveira, I., Sousa, A., Morais, J.S., Ferreira, I.C., Bento, A., Estevinho, L., & Pereira, J.A. (2008). Chemical composition, and antioxidant and antimicrobial activities of three hazelnut (*Corylus avellana* L.) cultivars. *Food and Chemical Toxicology*, 46(5), 1801–1807.

Oliveira, I., Sousa, A., Valentão, P., Andrade, P.B., Ferreira, I.C., Ferreres, F., Bento, A., Seabra, R., Estevinho, L., & Pereira, J.A. (2007). Hazel (*Corylus avellana* L.) leaves as source of antimicrobial and antioxidative compounds. *Food Chemistry*, 105(3), 1018–1025.

Parhizgar, A.R., & Tahghighi, A. (2017). Introducing new antimalarial analogues of chloroquine and amodiaquine: a narrative review. *Iranian Journal of Medical Sciences*, 42(2), 115.

Peev, C.I., Vlase, L., Antal, D.S., Dehelean, C.A., & Szabadai, Z. (2007). Determination of some polyphenolic compounds in buds of Alnus and *Corylus* species by HPLC. *Chemistry of Natural Compounds*, 43(3), 259–262.

Pelvan, E., Olgun, E.Ö., Karadağ, A., & Alasalvar, C. (2018). Phenolic profiles and antioxidant activity of Turkish Tombul hazelnut samples (natural, roasted, and roasted hazelnut skin). *Food Chemistry*, 244, 102–108.

Permin, H., Norn, S., Kruse, E., & Kruse, P.R. (2016). On the history of Cinchona bark in the treatment of Malaria. *Dansk medicinhistorisk arbog*, 44, 9–30.

Pfeifer, A. (2019). The Silviculture of Trees Used in British Forestry. *IrIsh Forestry*, 76(1&2), 105–106.

Piccinelli, A.L., Pagano, I., Esposito, T., Mencherini, T., Porta, A., Petrone, A.M., Gazzerro, P., Picerno, P., Sansone, F., Rastrelli, L., & Aquino, R.P. (2016). HRMS profile of a hazelnut skin proanthocyanidin-rich fraction with antioxidant and anti-*Candida albicans* activities. *Journal of Agricultural and Food Chemistry*, 64(3), 585–595.

Popenoe, W. (1949). *Cinchona* cultivation in guatemala: a brief historical review up to 1943. *Economic Botany*, 3(2), 150–157.

Prosperini, S., Ghirardello, D., Scursatone, B., Gerbi, V., & Zeppa, G. (2009). Identification of soluble phenolic acids in hazelnut (*Corylus avellana* L.) kernel. *Acta Horticulturae*, (845), 677–680.

Riethmüller, E., Alberti, Á., Tóth, G., Béni, S., Ortolano, F., & Kéry, Á. (2013). Characterisation of Diarylheptanoid-and Flavonoid-type Phenolics in *Corylus avellana* L. Leaves and Bark by HPLC/DAD–ESI/MS. *Phytochemical Analysis*, 24(5), 493–503.

Roersch van der Hoogte, A., & Pieters, T. (2015). Science, industry and the colonial state: a shift from a German-to a Dutch-controlled cinchona and quinine cartel (1880–1920). *History and Technology*, 31(1), 2–36.

Savarino, A., Boelaert, J.R., Cassone, A., Majori, G., & Cauda, R. (2003). Effects of chloroquine on viral infections: an old drug against today's diseases. *Lancet Infectious Diseases*, 3(11), 722–727.

Shahidi, F., Alasalvar, C., & Liyana-Pathirana, C.M. (2007). Antioxidant phytochemicals in hazelnut kernel (*Corylus avellana* L.) and hazelnut byproducts. *Journal of Agricultural and Food Chemistry*, 55(4), 1212–1220.

Solar, A., Veberič, R., Bacchetta, L., Botta, R., Drogoudi, P., Metzidakis, I., Rovira, M., Sarraquigne, J., & Silva, A.P. (2009). Phenolic characterization of some hazelnut cultivars from different European germplasm collections. *Acta horticulturae*, 845(96), 613–618.

Tinner, W., Hubschmid, P., Wehrli, M., Ammann, B., & Conedera, M. (1999). Long-term forest fire ecology and dynamics in southern Switzerland. *Journal of Ecology*, 87(2), 273–289.

Torello Marinoni, D., Valentini, N., Portis, E., Acquadro, A., Beltramo, C., Mehlenbacher, S.A., Mockler, T., Rowley, E., & Botta, R. (2018). High density SNP mapping and QTL analysis for time of leaf budburst in *Corylus avellana* L. *PLoS One*, 13(4), e0195408.

Tracy, J.W. (1996). Drugs used in the chemotherapy of protozoal infections. In *Goodman and Gilman's the Pharmacological Basis of Therapeutics*, pp. 965–985.

Tunon, H., Olavsdotter, C., & Bohlin, L. (1995). Evaluation of anti-inflammatory activity of some Swedish medicinal plants. Inhibition of prostaglandin biosynthesis and PAF-induced exocytosis. *Journal of Ethnopharmacology*, 48(2), 61–76.

Ustaoglu, B., & Karaca, M. (2014). The effects of climate change on spatiotemporal changes of hazelnut *Corylus avellana* cultivation areas in the Black Sea Region Turkey. *Applied Ecology and Environmental Research (AEER)*, 12(2), 309–324.

Vander Wall, S.B. (2001). The evolutionary ecology of nut dispersal. *Botanical Review*, 67(1), 74–117.

Vincent, M.J., Bergeron, E., Benjannet, S., Erickson, B.R., Rollin, P.E., Ksiazek, T.G., Seidah, N., & Nichol, S.T. (2005). Chloroquine is a potent inhibitor of SARS coronavirus infection and spread. *Virology Journal*, 2(1), 1–10.

Wang, M., Cao, R., Zhang, L., Yang, X., Liu, J., Xu, M.,Shi, Z., Hu, Z., Zhong, W., & Xiao, G. (2020). Remdesivir and chloroquine effectively inhibit the recently emerged novel coronavirus (2019-nCoV) *in vitro*. *Cell Research*, 30(3), 269–271.

Wang, Y., Guo, B., Zhang, F., Yao, H., Miao, Z., & Tang, K. (2007). Molecular cloning and functional analysis of the gene encoding 3-hydroxy-3-methylglutaryl coenzyme A reductase from hazel (*Corylus avellana* L. Gasaway). *BMB Reports*, 40(6), 861–869.

Wangchuk, P. (2004). *Bioactive Alkaloids from Medicinal Plants of Bhutan*. M.Sc. thesis. Department of Chemistry, University of Wollongong, Australia.

Wangchuk, P., Keller, P.A., Pyne, S.G., Taweechotipatr M, Tonsomboon A, Rattanajak R, et al. (2011). Evaluation of an ethnopharmacologically selected Bhutanese medicinal plants for their major classes of phytochemicals and biological activities. *Journal of Ethnopharmacology*, 137, 730–742.
Whitcher, I.N., & Wen, J. (2001). Phylogeny and biogeography of *Corylus* (Betulaceae): inferences from ITS sequences. *Systematic Botany*, 26(2), 283–298.
Xie, D.Y., Sharma, S.B., Wright, E., Wang, Z.Y., & Dixon, R.A. (2006). Metabolic engineering of proanthocyanidins through co-expression of anthocyanidin reductase and the PAP1 MYB transcription factor. *Plant Journal: For Cell and Molecular Biology*, 45(6), 895–907.
Yang, Z., Zhao, T.T., Ma, Q.H., Liang, L.S., & Wang, G.X. (2018). Resolving the speciation patterns and evolutionary history of the intercontinental disjunct genus *Corylus* (Betulaceae) using genome-wide SNPs. *Frontiers in Plant Science*, 9, 1386.
Yang, Z.Y., Huang, Y., Ganesh, L., Leung, K., Kong, W.P., Schwartz, O., Subbarao, K., & Nabel, G.J. (2004). PH-dependent entry of severe acute respiratory syndrome coronavirus is mediated by the spike glycoprotein and enhanced by dendritic cell transfer through DC-SIGN. *Journal of Virology*, 78(11), 5642–5650.
Yuan, B., Lu, M., Eskridge, K.M., Isom, L.D., & Hanna, M.A. (2018). Extraction, identification, and quantification of antioxidant phenolics from hazelnut (*Corylus avellana* L.) shells. *Food Chemistry*, 244, 7–15.
Zeppa, D.G.S.P.G., & Gerbi, V. (2010). Phenolic acid profile and antioxidant capacity of hazelnut (*Corylus avellana* L.) kernels in different solvent systems. *Journal of Food and Nutrition Research*, 49(4), 195–205.

20 *Crataegus laevigata* (Midland Hawthorn) and *Emblica officinalis* (Indian Gooseberry)

Bouabida Hayette and Dris Djemaa

CONTENTS

20.1 Introduction 395
20.2 Description 396
20.3 Traditional Knowledge 397
20.4 Chemical Derivatives (Bioactive Compounds – Phytochemistry) 397
20.5 Potential Benefits, Applications, and Uses 399
20.5.1 Antioxidant Effects 399
20.5.2 Anti-Inflammatory Effects 400
20.5.3 Cardiac and Vascular Effects 401
20.5.4 Antimicrobial 402
20.5.5 Immunomodulatory Effects 403
20.5.6 Hepatoprotective Effect 403
20.6 Conclusion 404
References 404

20.1 INTRODUCTION

For a long time, the use of herbal remedies in therapy was instinctive because there was no information on the causes of diseases or on the mode of action of plants in the body, everything was related to the experience (Kelly, 2009). Awareness of the use of medicinal plants is the result of many years of experience in disease control, and man has researched medicines in different parts of the plant seeds, fruits, leaves, and flowers (Petrovska, 2012). As a result, plants are a major source of food, medicine, and more, (Chandra et al., 2017) and have been used in natural healing for over 5,000 years (Xin et al., 2019). Almost 500 thousand plant species exist on earth (Cowan, 1999). Almost 20% of the total number of known plant species exists in Brazil with the most diverse flora (55 thousand species) (Carvalho et al., 2007). In developing regions like Africa, medicinal plants play an important role in the treatment of diseases (Elansary et al., 2018). The worldwide use of some herbal medicines continues to grow for improving health. New research proves that the health benefits of plants are due to their bioactive compounds such as polyphenols, terpenoids, phenolic acids, alkaloids, flavonoids, glycosides, saponins, and lignans (Sellami et al., 2018). Herbal remedies can affect the pharmacokinetics of some drugs (Moreira et al., 2014). Therefore, the consumption of some medicinal plants causes undesirable effects (hypersensitivity, organic toxicities) (Ekor, 2014). Nevertheless, there have been notable advances in the exploration of scientific methods to analyze the active components in botanicals (Avigan et al., 2016). Plants provide numerous bio-macromolecules such as carbohydrates, lipids, and nucleic acids (Zhang et al., 2020) and active metabolites of terpenoids, alkaloids, phenolic compounds (Ksouri et al., 2007).

The phenolic compounds are the subject of scientific research because of their biological and pharmacological properties: anti-microbial (Anarado et al., 2020), anti-cancer, anti-diabetic,

DOI: 10.1201/9781003205067-20

antibacterial, antifungal, hepatoprotective, anti-inflammatory (Chekuri et al., 2020), Anti-viral, anti-asthmatic, anti-arthritis, anti-venom, anti-obesity, in depression and anxiety (Rathore et al., 2020). In addition to its biological antioxidant (Kumar et al., 2021) and reducing properties (Ahmed and Mustafa, 2020). Indeed, they play an essential role in the fight against free radicals in the human body, cellular oxidative stress, which can lead to various diseases (Jamshidi-Kia et al., 2020).

Crataegus laevigata (Midland hawthorn) is a shrub (common hawthorn) belonging to the Rosaceae family (Moustafa et al., 2019) which grows in North Africa, Europe, and Central Asia, North America (Caliskan, 2015). *Emblica officinalis* is a tree (Family – Euphorbiaceae) commonly known as "Amla" and "Indian gooseberry" (Sampath Kumar et al., 2012) and native to Asia, China, the Mascarene Islands, and Malaysia (Hasan et al., 2016). Scientific evidence has shown that both plants have antioxidant activities due to the presence of different bioactive compounds such as phenolic components, flavonoids, vitamins, and different minerals. These compounds are said to have numerous pharmacological effects, in particular anti-inflammatory, hepatoprotective, cardioprotective, antimicrobial, immunomodulatory etc. Therefore, the purpose of this chapter is to provide a review of the medicinal properties of *E. officinalis* and *Crataegus laevigata* in relation to its mechanisms of action and associated plant components.

20.2 DESCRIPTION

The genus *Crataegus* known as hawthorns (Figure 20.1), is derived from the Greek word "kratos" meaning hardwood (Moustafa et al., 2019), *Crataegus* relates to the Crataegeae tribe and subfamily Maloideae of Rosaceae (28 genera and 940 species) (Evans and Campbell, 2002) and a family Rosaceae comprises about 280 species with economic importance (Hyam and Pankhurst, 1995). *Crataegus laevigata* is a thorny shade-tolerant shrub with a height of 12 m (Thomas et al., 2021). The hawthorn grows on mountains, native to the Mediterranean region, especially in North Africa, Europe, and Central Asia, but also in many parts of North America (Caliskan, 2015). The flowers have two shapes rarely varying from one to five with an equal number of pyrenes in the fruit (Thomas et al., 2021). *Crataegus* fruits are an important food source for birds and small mammals and are found in trees even in winter (González-Varo et al., 2021).

Emblica officinalis (Indian gooseberry) is commonly known "Amla" (Murugesan et al., 2021) belongs to the Euphorbiaceae family (Balasubramanian et al., 2014). It is a deciduous tree (Khosla and Sharma, 2012), with a height of 8–18 meters (Figure 20.2). It has a twisted stem, spreading

FIGURE 20.1 *Crataegus laevigata.*

FIGURE 20.2 *Emblica officinalis.*

branches and greenish-yellow flowers, average fruit weight (60–70 g) with a pale-yellow color and its feathery leaves (Sampath Kumar et al., 2012). In India, Amla trees grow in tropical forests ascending (Rai et al., 2012; Thilaga et al., 2013). It is native to tropical south East Asia, central and southern India and Pakistan, Bangladesh, Sri Lanka, southern China, Mascarenes Islands, and Malaysia (Hasan et al., 2016).

20.3 TRADITIONAL KNOWLEDGE

Recently, numerous scientific studies have been carried out to validate the traditional uses of *C. laevigata* and *E. officinalis.* These studies have shown that the two herbs help treat and prevent a wide range of disorders. Traditional uses of *C. laevigata* and *E. officinalis* are shown in Tables 20.1 and 20.2.

20.4 CHEMICAL DERIVATIVES (BIOACTIVE COMPOUNDS – PHYTOCHEMISTRY)

The flower buds of *Crataegus laevigata* are richer in flavonoids, in particular in flavonols, flavones and phenolic acids (derivatives of hydroxycinnamic acid) and the immature fruits in proanthocyanidins as well as in phenolic acids (hydroxycinnamic acid), compared to the fruit. Ripe gave the highest anthocyanin content (Rodriguesa et al., 2012). A similar study shows that flowers exhibited high levels of tocopherols and ascorbic acid and fatty acids while overripe fruits showed high levels of carbohydrates, sugars, and total saturated fatty acids. Unripe fruits had the highest polyunsaturated fatty acids contents (Barros et al., 2011). The aqueous extract of *Crataegus oxyacantha* is rich in flavonoids: 0.5–0.8% (2.5% in flowers): vitexin and its glycosides, rutoside, quercetin, kaempferol, and other compounds leucoanthocyanidins, (-) (+) epicatechin, (+) catechin, phenolic acids: caffeic, chlorogenic. triterpene acids, amines, adenine, adenosine, guanine, uric acid, essential oils, ß-sitosterol (Konieczynski, 2013). The gas chromatogram of *Crataegus* extract shows the following phenolic compounds identified: 2, 3 dihydro, 5-dihydroxy-6-methyl-4H-pyran-4-one; 1,2-dihydroxybenzene; 2-furanemethanol; 3, 7, 11, 15-tetrametyl-2-hexadene-1-ol. GC-MS data showed the extract to be rich in phenolic compounds, consistent with the results of applying the ABTS test to the extracts (Silva et al., 2000). The ethanolic extract of hawthorn berries represents phenolic compounds 3.54% (gallic acid), the total content of flavonoid aglycones is 0.18%. 0.14% hyperoside is the main component of flavonol, and the content of procyanidins was 0.44% (Tadic et al., 2008).

TABLE 20.1
Traditional Uses of *Crataegus laevigata*

Region	Used Part	Treatment	References
North America	Fruit	Heart problems (hypertension), angina, arrhythmia, and congestive heart failure	Edwards et al., 2012
Native American tribes	Fruit	To treat gastrointestinal ailments and heart problems, and consumed the fruit as food	Edwards et al., 2012
Meskwaki, Blackfoot, Ojibwa, Potawatomi, Okanagon, Okanagan-Colville, Iroquois, and Cherokee	Decoctions of the shoots, roots, and bark	Gastrointestinal ailments	Edwards et al., 2012
Cherokee and Thompson	Decoctions of the bark	Heart problems	Edwards et al., 2012
China	Fruit	Digestive problems, Hyperlipidemia, poor circulation, and dyspnea	Rigelsky & Sweet, 2002
Europe	Leaves and flowers	Cardiotonic, diuretic and antiatherosclerotic agent	Chang et al., 2005
	Fruit	Food (juices, jams, teas)	Thomas et al., 2021
Algiers	Bark and roots	Give a yellow or brown dye and tanning	Thomas et al., 2021

TABLE 20.2
Traditional Uses of *Emblica officinalis*

Region	Used Part	Treatment	References
Ayurvedic (Indian)	Fruits	Anemia, leprosy, cough, asthma, hemorrhages, liver and digestive problems	Tsarong, 1994
Tibetan (South Asia)	Fruits	Reduce body temperature in fever, as an anti-inflammatory and diuretic	Tsarong, 1994
Undefined	Fruit	Digestion problems, strengthens the heart, improves vision, hair and body shine, diuretic, laxative, antipyretic, aphrodisiac, tonic.	Kumar et al., 2012
Undefined	Fresh fruit juice	Tonic, diuretic and anti-bilious remedy. burning sensation, on thirst, dyspepsia, and other digestive disorders	Kumar et al., 2012
Undefined	Dried fruit powder	Hyperacidity, ulcers, and blood impurities	Kumar et al., 2012
India	Fruit	Antinociceptive, Cardioprotective, Cytoprotective, hepatoprotective hypotensive potential, immunomodulatory,	Jain et al., 2016
India	Fruit, juice, leaves, roots, branches	Anticancer activity	Jain et al., 2016
India	Fruit, juice	Lipid-lowering, Antiatherosclerotic	Jain et al., 2016
India	Root, juice (leaves)	Dental problems	Jain et al., 2016
India	Decoction (fruit), root, infusion of leaves, bark	Diarrhea	Jain et al., 2016

Recently, several scientific studies have reported many bioactive compounds of *E. officinalis*. Four commercial extracts of *E. officinalis* consist of total phenol, total flavonoid, and total tannin, (poly) phenolics (ellagic, gallic, and corilagin acids), and ascorbic acid the major component of the fruits of this plant (Poltanov et al., 2009). A comparative study between two varieties of Amla (wild and cultivated) shows that the total content of phenols was 32.32g/100g of gallic acid equivalent for the fruits of the wild variety whereas the cultivated variety (Chakiya) had 24.50g/100 g of gallic acid equivalent (Mishra et al., 2009). A phytochemical screening indicates the presence of alkaloids, saponin, phenols, tannins, flavonoids, and carbohydrates (Mehvish and Barkat, 2018). Similar results in the work of Sowmya and Nanjammanni (2017) showed the presence of carbohydrates, phenol, carboxylic acid, tannin, and flavonoids in the aqueous, alcoholic extract of the pulp of *Emblica officinalis*. Recently, Priyanka and Stanley (2021) indicate the presence of phytochemicals such as anthraquinone, tannins, saponins, flavonoids, glycosides, terpenoids, carbohydrates, and amino acids. The methanolic extract of *Emblica officinalis* fruit contains gallic acid (2.10%), monic acid (4.90%), ellagic acid (2.10%), quercetin (28%) and rutin (3.89%), and glucagonalin (1.46%) by high performance liquid chromatography studies (Middha et al., 2015). It is important to explore and identify the phytochemicals present in specific parts of *Crataegus laevigata* and *Emblica officinalis* and to understand their therapeutic role as well as their pharmacological and physiological mode of action.

Overall, Tables 20.3 and 20.4 summarize some of the chemical components of the parts of the *Crataegus laevigata* and *Emblica officinalis* respectively.

20.5 POTENTIAL BENEFITS, APPLICATIONS, AND USES

20.5.1 Antioxidant Effects

The antioxidant activity of *Crataegus laevigata* and *Emblica officinalis* caused mainly by the high content of polyphenols (Alirezalu et al., 2020; Chakraborty et al., 2020), and can protect cells against oxidative stress caused by free radicals (Pawlowska et al., 2019).

TABLE 20.3
The Chemical Components of the Parts of the *E. officinalis* plant

Plant Part	Phytoconstituents	Analysis	References
Fruits Amla	Alcaloïdes, Saponines, Glycosides, Protéines, Phénols et Phytostérols.	GC-MS	Lovey and Kumar, 2019
Amla extract	tannins, alkaloids, polyphenols, gallic acid, ellagic acid, emblicanin A and B, phyllembeine, quercetin, ascorbic acids, vitamins and minerals Different	Phytochemical	Khurana et al., 2019
Amla extract	Ascorbic acid, caffeic acid, ellagic acid and ferulicacid.	High performance liquid chromatography with UV-visible detector (HPLC-UV-vis)	Kanatt et al., 2018
Tree (root, bark, leaf, flower, fruit, and seed)	secondary metabolites (emblicanin-A and emblicanin-B) tannins, gallic acid, pyrogallol, and pectin, vitamin-C,	Phytochemical	Gantait et al., 2021
Amla extract	hexagalloyl derivative (emblifatmin (E4)), trans-cinnamic acid (E1), methyl gallate (E2), gallic acid (E3), and quercetin-3-O-α-arabinofuranoside (E5),	Phytochemical (MS and NMR spectroscopy)	Abdel Bar et al., 2021

TABLE 20.4
The Chemical Components of the Parts of the *C. laevigata* Plant

Plant Part	Phytoconstituents	Analysis	References
Methanolic leaves extract	chlorogenic acid, rutin, isoquercetin, hesperidin, and catechin	High performance liquid chromatography (HPLC)	Ziouche et al., 2020
Methanolic fruit extract	chlorogenic acid, rutin, isoquercetin, hesperidin, catechin, caffeic acid and quercetin		
Leaf extracts	Vitexin, Hyperoside Naringenin Chlorogenic acid	HPLC-DAD	Lund et al., 2020
Ethanol leaves extracts	polyphenol (473.4 mg GAE g−1) and flavonoids (80.9 mg CE g−1)	HPLC	Belabdelli et al., 2021
Methanolic leaves extract	Total phenolic content (110.41±1.47 mg GAE g extract−1) Flavonoid content (29.94±1.85mg QE g extract−1)	HPLC-PDA-ESI/MS^n	Coimbra et al., 2020

The antioxidant group's procyanidin (procyanidin b2) and hyperoside are the most abundant compounds in the roots of *Crataegus laevigata* (O-LE-21) (Magnus et al., 2020). The total antioxidant capacity in the polyphenolic extract of stressed (drought, cold) *Crataegus laevigata* was improved with increased levels of (-)-epicatechin and hyperoside (Kirakosyan et al., 2003). The results obtained by Karar and Kuhnert (2015) indicate that the *Crataegus laevigata* species is an important source of phenolic components, in particular simple phenolic acids, chlorogenic acids, proanthocyanidins, flavonoids, and flavonoid glycosides in different parts of the plant (leaves, fruits, and drops of herbs). The DPPH radical scavenging capacity of the *Crataegus laevigata* extract is EC50 of 52.04g/ml (the total phenol content of the extract) (Tadic et al., 2008). The DPPH test showed a considerable antioxidant potential of the ethanolic extract of the leaves (IC50=22.50g/ml) (Belabdelli et al., 2021). The antioxidant activity of the methanolic extracts of leaves and fruits was evaluated by the method of radical reduction of DPPH. The leaves of *Crataegus* have a high antioxidant activity compared to the fruit extract (IC50 are respectively 24.52µg/ml and 323.87µg / ml), and the IC50 of ascorbic acid = 2.74 µg/ml (leaves) is lower than that of fruits (Ziouche et al., 2020).

The antioxidant potential of *Emblica officinalis* fruit is 17 times greater than that of pomegranate and Amla extracts have strong antioxidant activity compared to synthetic antioxidants such as butylated hydroxyanisole, quercetin (Hussain et al., 2021). A study shows that the fruit extracts of *Emblica officinalis* inhibited the production of free radicals induced by Cr and restored the activity of the antioxidant (Sai Ram et al., 2002). Rats were treated with arsenic and lead showed increased signs of oxidative stress such as lipid peroxidation, H_2O_2, OH *. However, an oral dose of *E. officinalis* increased antioxidant enzymes and restored altered levels of LPO induced by arsenic and lead (Fazal et al., 2021). Middha et al. (2015) work showed the highest antioxidant activities using DPPH (17.33–89.00%), ABTS (23.03–94.16%), oxide scavenging activity nitric (12.94 to 70.16%), LPO (56.54%), and phosphomolybdenum dosage (142 ± 6.09 g / ml).

Verma, Arun, and Rajkumar (2021) indicate that Amla fruit extract is rich in phenols and therefore can be used in place of chemical nuggets of butylated hydroxytoluene to improve the antioxidant capacity of goat cuts. Free radical scavenging activity present in *Emblica* extract, DPPH assay (EC50 = 15.67 ± 1.41g / ml) and ABTS (EC50 = 18.84 ± 1.02 g / ml) (Gunti et al., 2019).

20.5.2 Anti-Inflammatory Effects

Inflammation and oxidative stress can cause acute and chronic disorders (urticaria, eczema, atopic dermatitis) (Al-Waili, 2003), 70% ethanol extracts of *C. laevigata* is a potential alternative agent for inflammatory treatment. In a recent study, these extracts have protective effects against lipopolysaccharide-induced damage in human keratinocytes by suppressing, mRNA levels of pro-inflammatory chemokines and interleukins in cells stimulated by lipopolysaccharide (Nguyen et al., 2021). The study of Kim et al (2011) shows that the aqueous extract of hawthorn fruit suppresses the expression of inflammatory cytokines (interleukin (IL) -1β, tumor necrosis factor (TNF) -α and IL-6) on mouse macrophages this which explain its beneficial effect in the regulation of inflammatory reactions. A model of carrageenin-induced rat paw edema is reduced after oral administration of *C. laevigata* extract with a dose-response anti-inflammatory effect, the dose (200 mg/kg) showed 72.4% of its activity compared to indomethacin which produces a 50% reduction in rat paw edema (Tadic et al., 2008). Elango and Devaraj (2010) reported that hawthorn extract has an immunomodulatory effect and attenuates pro-inflammatory immune responses, pretreatment of male rats after ischemia, and reperfusion for 15 days with hawthorn (100 mg/kg) resulted in reduced levels of pro-inflammatory cytokines. Asmawi et al. (1993) work shows anti-inflammatory activity in the aqueous fraction of methanol extract from leaves of the *Emblica officinalis* plant by inhibition of edema induced by carrageenan and dextran in rats and a second series of *in vitro* experience shows that Amla extract inhibited the migration of human polymorphonuclear leukocytes at relatively low concentrations. It did not inhibit the synthesis of leukotriene B4 or platelet activating factor in human polymorphonuclear leukocytes or the synthesis of thromboxane B2 in human platelets during coagulation. Among the advantages of the hydroalcoholic extract of *Emblica officinalis*, it has an anti-inflammatory effect with powerful anti-ulcer activity, an experiment induced acute inflammation in mice by intramuscular injection of carrageenan, histamine, serotonin, and prostaglandin E2 and chronic inflammation has been caused by lumbar granulomas. Intraperitoneal administration of the extract (300, 500, and 700 mg/kg) significantly inhibited rat paw edema against all phlogistic factors and also reduced granuloma formation and the dose of 700 mg/kg shows maximum anti-inflammatory activity (Golechha et al., 2014). Methanolic fruit extract of *Emblica officinalis* reduces carrageenin-induced rat paw edema in a dose-dependent manner, the edema reduction was 72.71% after 4 h compared to drug standard (diclofenac: 10 mg/kg) was 61.57%, in addition to treatment with the extract significantly reduced the level of these cytokines compared to group induced by inflammation (Middha et al., 2015).

20.5.3 Cardiac and Vascular Effects

Coronary heart disease is a multifactorial disease and causes morbidity and mortality in humans (Kabiri et al., 2010). Research has shown that eating a diet rich in saturated fat is a major factor in the development of atherosclerosis and coronary heart disease (Jakulj et al., 2007). Many herbal remedies can reduce the oxidation of low-density lipoproteins (LDL). Kanyonga et al. (2011) research investigated the effect of *C. laevigata* in lowering blood lipid levels in animals. *C. laevigata* is mainly rich in flavonoids (vitexin and its derivatives) and other phenolic derivatives, which show promising benefits in the treatment of cardiovascular disorders (Lund et al., 2020). For example, treatment with an ethanolic extract of *C. laevigata* in patients with mild heart failure significantly improved the exercise capacity of the patients (Degenring et al., 2003). A study is also reported that extracts of *C. laevigata*, which could be beneficial in cardiac regeneration after a myocardial infarction (Halver et al., 2019). *Crataegus laevigata* is characterized by the presence of flavonoids such as flavonol and flavanone is mainly attributed to the relaxation and dilation of arteries, coronary arteries, and cardioactive, hypotensive, due to its effectiveness. The German Commission also approved the use hawthorn as a heart remedy. *Crataegus* extract lowers blood pressure and is a powerful antioxidant that delays atherosclerosis, which may also contribute to a healthy vascular system (Silva et al., 2000). Long et al. (2006) showed the comparison between hawthorn extract and cardioactive drugs by a cardiomyocyte test. Hawthorn has an effect different from those recorded with cardioactive drugs (epinephrine; milrinone; ouabain), similar to that recorded for propranolol, and hawthorn extract improves the rhythm of cardiomyocytes, while propanol induces arrhythmia. The work of Patil et al. (2019) shows that ethanolic extracts of *Emblica officinalis* can have cardioprotective actions in mice subjected to a diet high in fat (30% fat) by functionally and histologically modulating cardiac function. Amlamax TM is an extract (35% of gallo-ellagi-tannins, hydrolyzed tannins) applied to human volunteers at two doses of 500 mg and 1000 mg per day for 6 months. Blood tests at 3 and 6 months after treatment with the extract showed a decrease in total and LDL cholesterol and an improvement in the beneficial effects of HDL cholesterol (Antony et al., 2008).

20.5.4 Antimicrobial

The extract of *C. laevigata* showed antimicrobial activity in particular against Gram+ bacteria (*Micrococcus flavus*, *Bacillus subtilis*, and *Listeria monocytogenes*) without effect on *Candida albicans* (Tadic et al., 2008). The methanolic extract of ripe *Crataegus* fruit from the region of Serbia showed antimicrobial activity against *Escherichia coli*, *Pseudomonas aeruginosa*, and *Salmonella abony* (Kostić et al., 2012). The results obtained by Belkhir et al. (2013) shows that polyphenolic extracts strong antibacterial activity against Gram positive bacteria (*S. aureus* and S. *faecalis*) while the activity is low against *Salmonella typhimurium* (Gram negative) compared to ampicillin and oxytetracycline used as positive controls. The antibacterial activity of the *Crataegus* extract was tested on some bacterial strains (*Enterococcus faecalis*, *Bacillus cereus*, *Escherichia coli*, *Staphylococcus aureus*, *Pseudomonas aeruginosa*, and *Listeria monocytogenes*) after 48 h the results show that the extract inhibited only *L. monocytogenes* and *S. aureus* (Nunes et al., 2017). *E. officinalis* extract exhibits antimicrobial activity against Gram-positive bacteria (*S. aureus*, *E. faecalis*, and *L. monocytogenes*) compared to Gram-negative bacteria (*E. coli*) (Gunti et al., 2019). The methanolic extract of the fruits and leaves of *Emblica officinalis* against three respiratory pathogens (*Staphylococcus aureus*, *Klebsiella pneumoniae*, *Streptococcus pyogenes*). The results showed that *Staphylococcus aureus* is more sensitive compared to other bacteria (Javale and Sabnis, 2010). The extract of the fruits of *E. officinalis* shows maximal inhibition against Gram positive bacteria (*Staphylococcus aureus*) against Gram negative bacteria (*Escherichia coli* and *Salmonella typhi*) (Singh et al., 2019). Lovey and Kumar (2019) applied five different solvents (methanol, ethanol, distilled water, chloroform and petroleum ether) for the preparation of Amla fruit extracts effective against *C. albicans* all methanolic extracts of *Emblica officinalis* had effective antimicrobial activity against *Propionibacterium acne*, *Staphylococcus aureus*, *Pseudomonas aeruginosa*, *Escherichia coli*, and *Candida albicans*. The alkaloids extracted from the methanolic extract of the fresh ripe fruit of *Emblica officinalis* exhibited antimicrobial activity against certain bacteria (Gram negative and Gram positive) (Rahman et al., 2004) (Figure 20.3).

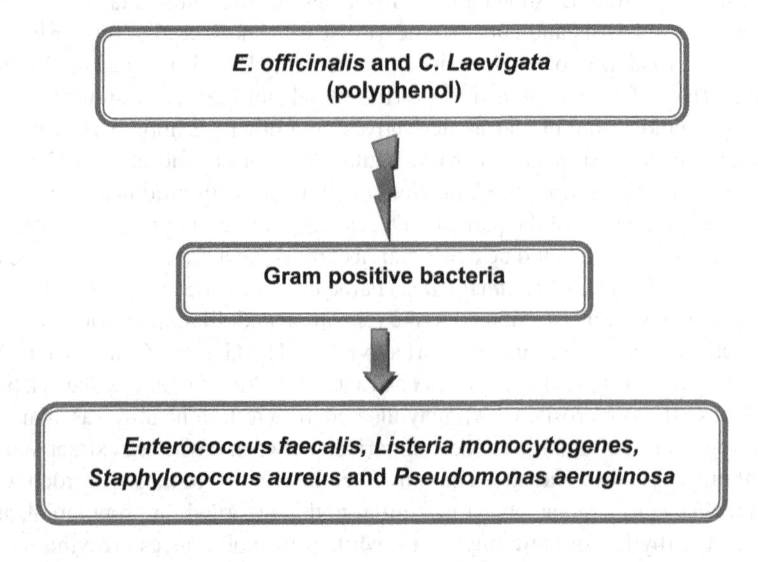

FIGURE 20.3 Common antibacterial activity of two plants *Crataegus laevigata* and *Emblica officinalis*.

20.5.5 Immunomodulatory Effects

Recently, scientific studies have proven that extracts from *Crataegus laevigata* and *Emblica officinalis* plants can be used as an immune regulator due to their effect on lymphocytes and the immune response. The study by Lis et al., (2020) reported the effect of hawthorn phenolic extract on lymphocyte populations in lymphoid organs of unimmunized mice and on humoral immunity. Response of mice immunized with sheep red blood cells. The phenolic extract with the following doses (50, 100, 200mg/kg) was administered orally and then transmitted to sheep erythrocytes, the extract increased the percentage of thymus + CD8+ and CD4+ cells, the absolute number of T and B splenocytes and antibodies. Elango and Devaraj (2010) reported that pretreatment of male Sprague Dawley rats with hawthorn extract applied at a dose of 100mg/kg for 15 days suppressed 40% of cytotoxic T cells therefore decreased cell death by apoptosis.

The aim of Sai Ram et al. (2002) study was to determine the immunomodulatory properties of Amla using chromium (VI) as an immunosuppressive agent. Treatment with Amla inhibited apoptosis and DNA fragmentation induced by Cr, therefore Amla relieved the immunosuppressive effects induced by Cr on lymphocyte proliferation. The immunomodulatory effect of *Emblica officinalis* (Amla) powders, four-day-old chicks were on a diet had amla powder (0.5%) produced maximum lymphoid organ weight (Tomar et al., 2018). The richness of *Emblica officinalis* in vitamin C and especially in ascorbic acid mediates the immunomodulatory effects. Consequently it stimulates the immune system by the improvement of the activity of the natural killer cells and the cellular toxicity dependent on the antibodies (Sowmya and Nanjammanni, 2017). The incorporation of *Emblica officinalis* in the diet of lactating cows reduces the incidence of the disease in particular mastitis by improving the non-specific immunity of perinatal cows (Pandey et al., 2020).

20.5.6 Hepatoprotective Effect

Results based on *in vivo* studies have shown that butanol extract from *Crataegus* leaves has an effect doxorubicin-induced hepatotoxicity in female rats, but oral administration of the extract for 10 days improved glutathione level and glutathione peroxidase activity and oxidative stress parameters (MDA) (Mecheri et al., 2019). Histological observation shows that bark and leaf extract reduces liver damage induced by high cholesterol diet in male Wistar rats therefore hawthorn extracts may improve liver damage caused by high cholesterol (Rezaei-Golmisheh et al., 2015). Chahardahcharic and Setorki (2018) results confirm that treatment with *Crataegus* extract at doses (100, 200, and 400 mg/kg) improved pancreatic damage, reduced blood sugar and TCA as well as increased MDA in male Wistar rats. Another study found that standardized *Crataegus* extract has an effect on carbon tetrachloride (CCl 4)-induced liver damage in rats. *Crataegus* extract with the following doses (10, 20, or 40 mg/kg) and silymarin as a control (25 mg/kg) were administered once daily orally at the same time as CCl 4, Crataegus extract reduces serum alanine aminotransferase levels, Serum aspartate aminotransferase levels and alkaline phosphatase (ALP) compared to the control group (Abdel Salam et al., 2012). The results of Sultana et al., (2005) suggest that *E. officinalis* inhibits hepatic toxicity in Wistar rats, a pretreatment with the fruit extract of *E. officinalis* at doses of 100 and 200 mg/kg reduced levels of enzymes (GP_X, GSH, MDA) and serum peroxides (SGPT, SGOT, LDH) in an animal model induced by necrogenic CCl4 (1ml/kg body weight). *In vivo* studies of Damodara Reddy et al. (2010) indicate that administration of *Emblica officinalis* fruit extract (250 mg/kg body weight) to alcoholic rats returns plasma enzymes to near normal levels, as well as the level of antioxidants enzymatic and nonenzymatic and results in decreased protein carbonyl lipid peroxidation. *Emblica officinalis* extract neutralizes the toxic effects of lead nitrate and aluminum sulfate in a biological model (albino rats) and therefore has a hepatoprotective role by reducing glutathione depletion and inhibiting cytochrome activation P450 (Das et al., 2017).

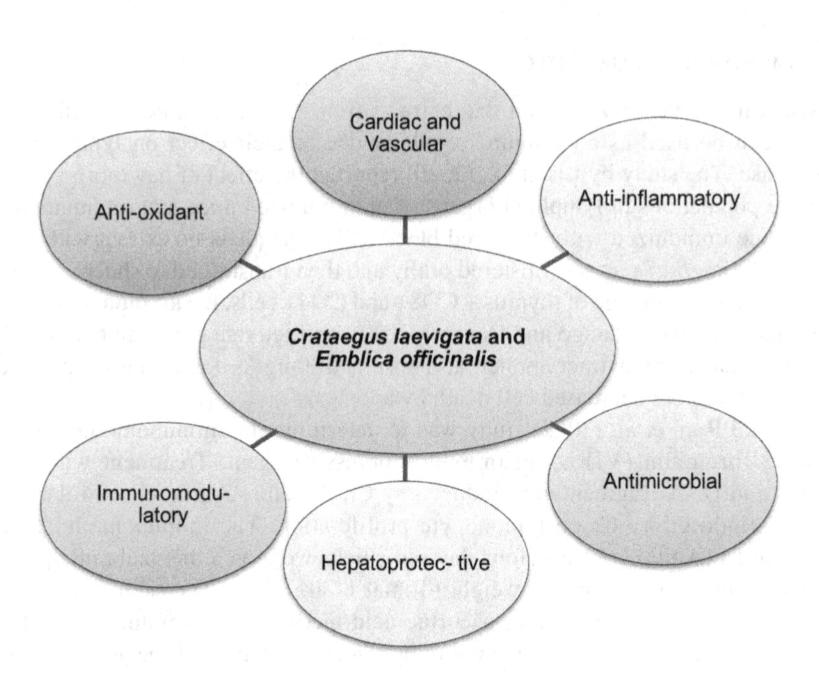

FIGURE 20.4 Common pharmacological effects of two plants *Crataegus laevigata* and *Emblica officinalis.*

20.6 CONCLUSION

Recently, medical research has turned to herbal medicine because of its fewer side effects compared to synthetic drugs. *Crataegus laevigata* and *Emblica officinalis* are two species recognized in traditional medicine and are common edible plants. The trials showed no significant side effects for them. Both plants have also shown various pharmacological effects (Figure 20.4). Scientific evidence has shown that the pharmacological activities of *Crataegus laevigata* and *Emblica officinalis* are due to the presence of many naturally occurring bioactive compounds such as total phenol, total flavonoid, and total tannin. This review showed that *Crataegus laevigata* and *Emblica officinalis* have pharmacological effects such as the prevention of cardiovascular and hepatic diseases and the ability to reduce the inflammatory effect as well as their antioxidant and antibacterial activity. They can be considered as a regulator of the immune system. Therefore, both plants exert multiple beneficial effects for health and many other pharmacological activities, *Crataegus laevigata* and *Emblica officinalis* can be recommended on the search for bioactive constituents, which are most responsible for its pharmacological effects.

REFERENCES

Abdel Bar, F.M., M.M. Abu Habib, and F.A. Badria. 2021. A new hexagalloyl compound from *Emblica offcinalis*Gaertn.: antioxidant, cytotoxicity, and silver ion reducing activities. *Chemical Papers* 75(12):6509–6518.

Abdel Salam O.M.E., A.A. Sleem, and N. Shafee. 2012. Effect of *Crataegus* extract on carbon tetrachloride-induced hepatic damage. *Comparative Clinical Pathology* 21:1719–1726.

Ahmed, R.H., E.M. Damra. 2020. Green synthesis of silver nanoparticles mediated by traditionally used medicinal plants in Sudan. *International Nano Letters* 10:1–14.

Alirezalu, A., N. Ahmadi, P. Salehi, A. Sonboli, K. Alirezalu, A.M. Khaneghah, F.J. Barba, P.E.S. Munekata, and J.M. Lorenzo. 2020. Physicochemical characterization, antioxidant activity, and phenolic compounds of Hawthorn (*Crataegus* spp.) fruits species for potential use in food applications. *Foods* 9:436.

Al-Waili, N.S. 2003. Topical application of natural honey, beeswax and olive oil mixture for atopic dermatitis or psoriasis: partially controlled, single-blinded study. *Complementary Therapies in Medicine* 11:226–234.

Anarado, C.E., C.J.O. Anarado, N.L. Umedum, and Q.M. Ogbodo. 2020. Comparative phytochemical and antimicrobial studies of leaf, stem, root of *Spathodea companulata. Asian Journal of Applied Chemistry Research* 6(1):10–20.

Antony, B., M. Benny, and T.N.B. Kaimal. 2008. A pilot clinical study to evaluate the effect of *Emblica officinalis* extract (amlamaxtm) on markers of systemic inflammation and dyslipidemia. *Indian Journal of Clinical Biochemistry* 23(4):378–381.

Asmawi, M.Z., H. Kankaanranta, E. Moilanen, and H. Vapaatalo. 1993. Anti-inflammatory activities of Emblica officinalis Gaertn leaf extracts. *Journal of Pharmacy and Pharmacology* 45(6):581–584.

Avigan, M.I., R.P. Mozersky, and L.B. Seeff. 2016. Scientific and regulatory perspectives in herbal and dietary supplement associated hepatotoxicity in the United States. *International Journal of Molecular Sciences* 17(3):331.

Balasubramanian, S., D. Ganesh, P. Panchal, M. Teimouri and V.V.S. Surya Narayana. 2014. GC-MS analysis of phytocomponents in the methanolic extract of *Emblica officinalis* Gaertn (Indian Gooseberry). *Journal of Chemical and Pharmaceutical Research* 6(6):843–845.

Barros, L., A.M. Carvalho, and I.C.F.R. Ferreira. 2011. Comparing the composition and bioactivity of 2Crataegus monogyna flowers and fruits used in folk medicine. *Phytochemical, Analysis* 22(2):181–188.

Belabdelli, F., N. Bekhti, A. Piras, F.M. Benhafsa, M. Ilham, S. Adil, L. Anes. 2021. Chemical composition, antioxidant and antibacterial activity of *Crataegus monogyna* leaves extracts. *Natural Product Research* 1–6. https://doi.org/10.1080/14786419.2021.1958215

Belkhir, M., O. Rebai, K. Dhaouadi, F. Congiu, C.I.G. Tuberoso, M. Amri, and S. Fattouch. 2013. Comparative Analysis of Tunisian Wild *Crataegus azarolus* (Yellow Azarole) and *Crataegus monogyna* (Red Azarole) Leaf, Fruit, and Traditionally Derived Syrup: Phenolic Profiles and Antioxidant and Antimicrobial Activities of the Aqueous-Acetone Extracts. *Journal of Agricultural Food Chemistry* 61(40):9594–9601.

Caliskan, O. 2015. Chapter 55 *Mediterranean Hawthorn Fruit (Crataegus) Species and Potential Usage. Novel Nutraceuticals and Edible Plants Used in the Mediterranean Region.*

Carvalho, A.C.B., D.S.G. Nunes, T.G. Baratelli, M.S.A.Q. Shuqair, and E. Machado Netto. 2007. Aspectos da legislação no controle dos medicamentos fitoterápicos. *T&C Amazônia* 5(11):26–32.

Chahardahcharic, S.V., and M. Setorki. 2018. The effect of hydroalcoholic extract of *Crataegus monogyna* on hyperglycemia, oxidative stress and pancreatic tissue damage in streptozotocin-induced diabetic rats. *Journal of Herbmed Pharmacolgy* 7(4):294–299.

Chakraborty, S., S. Ghag, P.P. Bhalerao, J.S. Gokhale. 2020. The potential of pulsed light treatment to produce enzymatically stable Indian gooseberry (*Emblica officinalis* Gaertn.) juice with maximal retention in total phenolics and vitamin C. *Journal of Food Processing and Preservation* 44(12):14932.

Chandra, H., P. Bishnoi, A. Yadav, B. Patni, A.P. Mishra, and A.N. Nautiyal. 2017. Antimicrobial resistance and the alternative resources with special emphasis on plant-based antimicrobials: A review. *Plants* 6(2):16.

Chang, W-T., J. Dao, and Z-H. Shao. 2005. Hawthorn: Potential roles in cardiovascular disease. *The American Journal of Chinese Medicine* 33(1):1–10.

Chekuri, S., L. Lingfa, S. Panjala, K.C. Sai Bindu, and R.R. Anupalli. 2020. *Acalypha indica* L.: an important medicinal plant: A brief review of its pharmacological properties and restorative potential. *European Journal of Medicinal Plants* 31(11):1–10.

Coimbra, A.T., Â.F.S. Luís, M.T. Batista, S.M.P. Ferreira, and A.P.C. Duarte. 2020. Phytochemical characterization, bioactivities evaluation and synergistic effect of Arbutus unedo and *Crataegus monogyna* extracts with amphotericin B. *Current Microbiology* 77:2143–2154.

Cowan, M.M. 1999. Plant products as antimicrobial agents. *Clinical Microbiology Reviews* 12(4):564–82.

Damodara Reddy, V., P. Padmavathi, S. Gopi, M. Paramahamsa, and N. Varadacharyulu. 2010. Protective effect of *Emblica officinalis* against alcohol-induced hepatic injury byameliorating oxidative stress in rats. *Indian Journal of Clinical Biochemistry* 25:419–424.

Das, S.K., A. Das, B. Das, P. Panda, G.C. Bhuyan, and B.B. Khuntia. 2017. Important uses of Amalaki (*Emblica officinalis*) in Indian system of Medicine with Pharmacological Evidence. *Journal of Pharmacology and Pharmacodynamics* 9(4):202–206.

Degenring, F.H., A. Suter, M. Weber, and R. Saller. 2003. A randomised double blind placebo controlled clinical trial of a standardised extract of fresh Crataegus berries (Crataegisan (R)) in the treatment of patients with congestive heart failure NYHA II. *Phytomedicine* 10:363–369.

Edwards, J.E., P.N. Brown, N. Talent, T.A. Dickinson, P.R. Shipley. 2012. A review of the chemistry of the genus *Crataegus*. *Phytochemistry* 79:5–26.

Ekor M. 2014. The growing use of herbal medicines: issues relating to adverse reactions and challenges in monitoring safety. *Frontiers in Pharmacology* 4:177.

Elango, C. and S.N. Devaraj. 2010. Immunomodulatory effect of hawthorn extract in an experimental stroke model. *Journal of Neuroinflammation* 7:1–13.

Elansary, H.O., A. Szopa, P.B. Kubica, H. Ekiert, H.M. Ali, M.S. Elshikh, E.M. Abdel-Salam, M. El-Esawi, and D.O. El-Ansary. 2018. Bioactivities of Traditional Medicinal Plants in Alexandria. *Evidence-Based Complementary and Alternative Medicine* p.13

Evans, R.C, and C.S. Campbell. 2002. The origin of the apple subfamily (Maloideae; Rosaceae) is clarified by DNA sequence data from duplicated GBSSI genes. *American Journal of Botany* 89(9):1478–84.

Fazal, M., V.P. Veeraraghavan, B. Tahreen, S. Jayaraman, and R. Gayathri. 2021. Antioxidant effects of Emblica officinalis and Zingiber officinalis on arsenic and lead induced toxicity on Albino rats. *Bioinformation* 17(2):295–305.

Gantait, S., M. Mahanta, S. Bera, and S.K. Verma. 2021. Advances in biotechnology of Emblica offcinalis Gaertn. syn. Phyllanthus emblica L.: a nutraceuticalsrich fruit tree with multifaceted ethnomedicinal uses. *Biotechnology* 11:62.

Golechha, M., V. Sarangal, S. Ojha, J. Bhatia, and D.S. Arya. 2014. Anti-Inflammatory Effect of *Emblica officinalis* in Rodent Models of Acute and Chronic Inflammation: Involvement of Possible Mechanisms. *International Journal of Inflammation* 6 p.

González-Varo, J.P, A. Onrubia, N. Pérez-Méndez, R. Tarifa, and J.C. Illera. 2021. Fruit abundance and trait matching determine diet type and body condition across frugivorous bird populations. *Nordic Society Oikos* 1–13. https://doi.org/10.1111/oik.08106.

Gunti, L., R.S. Dass, and N.K. Kalagatur. 2019. Phytofabrication of Selenium Nanoparticles FromEmblica officinalis Fruit Extract and Exploring Its Biopotential Applications: Antioxidant, Antimicrobial, and Biocompatibility. *Frontiers in Microbiology* 10:1–17.

Halver, J., K. Wenzel, J. Sendker, C. Carrillo García, C.A. Erdelmeier, E. Willems, M. Mark Mercola, N. Symma, S. Könemann, E. Koch, A. Hensel, and D. Schade. 2019. *Crataegus* Extract WS® 1442 stimulates cardiomyogenesis and angiogenesis from stem cells: A possible new pharmacology for hawthorn?. *Frontiers in Pharmacology* 10:1–18.

Hasan, Md.R., Md.N. Islam, and Md.R. Islam. 2016. Phytochemistry, pharmacological activities and traditional uses of *Emblica officinalis*: A review. *International Current Pharmaceutical Journal* 5(2):14–21.

Hussain, S.Z., B. Naseer, T. Qadri, T. Fatima, T.A. Bhat. Chapter 15 Anola (*Emblica officinalis*): Morphology, Taxonomy, Composition and Health Benefits. Fruits Grown in Highland Regions of the Himalayas. *Nutritional and Health Benefits* 193–206.

Hyam, R, and R. Pankhurst. 1995. Plants and their Names, *Concise Dictionary* 76(4, 1):441.

Jain, P.K., D. Das, N. Pandey, P. Jain. 2016. Traditional Indian herb *Emblica officinalis* and its medicinal importance. *Innovare Journal Ayruvedic Sciences* 4(4):1–15.

Jakulj, F., K. Zernicke, S.L. Bacon, L.E. Van Wielingen, B.L. Key, S.G. West, and T.S. Campbell. 2007. A high-fat meal increases cardiovascular reactivity to psychological stress in healthy young adults. *The Journal of Nutrition* 137(4):935–939.

Jamshidi-Kia, F.J. P. Wibowo, M. Elachouri, R. Masumi, A. Salehifard-Jouneghani, Z Abolhassanzadeh and Z. Lorigooini. 2020. Battle between plants as antioxidants with free radicals in human body. *Journal of Herbmed Pharmacology* 9(3):191–199.

Javale, P., and S. Sabnis. 2010. Antimicrobial properties and phytochemical analysis of *Emblica officinalis. Asian Journal of Experimental Biological sciences* 91–95.

Kabiri, N., S. Asgary, H. Madani, and P. Mahzouni. 2010. Effects of *Amaranthuscaudatus* L. Extract and lovastatin on atherosclerosis in hypercholesterolemic rabbits. *Journal of Medicinal Plants Research* 4(5):354–361.

Kanatt S.R., A. Siddiqui, and S.P. Chawla. 2018. Antioxidant/antimicrobial potential of *Emblica officinalis* gaertn and its application as a natural additive for shelf life extension of minced chicken meat. *Biointerface Research in Applied Chemistry* 8(4):3344–3350.

Kanyonga, M., M. Faouzi, A. Zellou, et al. 2011. Effect of methanolic extract of *Crataegus oxyacantha* on blood haemostasis in rats. *Journal of Chemical and Pharmaceutical Research* 3(3):713–717.

Karar, M.G.E. and N. Kuhnert. 2015. UPLC-ESI-Q-TOF-MS/MS Characterization of Phenolics from *Crataegus monogyna* and *Crataegus laevigata* (Hawthorn) Leaves, Fruits and their Herbal Derived Drops (*Crataegutt Tropfen*). *Journal of Chemical Biology & Therapeutics* 1(1):1–23.

Kelly, K. 2009. History of medicine. New York: Facts on file 29–50.

Ksouri, R., W. Megdiche, A. Debez, H. Falleh, C. Grignon, and C. Abdelly. 2007. Salinity effects on polyphenol content and antioxidant activities in leaves of the halophyte Cakile maritime. *Plant Physiology and Biochemistry* 45(3–4):244–249.

Khosla, S., and S. Sharma. 2012. *Emblica officinalis* in farmokogenetik özellikleri üzerine kısa bir tanımlama. *Spatula DD* 2(3):187–193.

Khurana, S.K., R. Tiwari, K. Sharun, M.I. Yatoo, M.B. Gugjoo, and K. Dhama. 2019. *Emblica officinalis* (Amla) with a Particular Focus on Its Antimicrobial Potentials: A Review. *Journal of Pure and Applied Microbiology* 13(4):1–18.

Kim, S.J., J.Y. Um, S.H. Hong, and J.Y. Lee. 2011. Anti-inflammatory activity of hyperoside through the suppression of nuclear factor-κB activation in mouse peritoneal macrophages. *The American Journal of Chinese Medicine* 39(1):171–181.

Kirakosyan, A., E. Seymour, P.B. Kaufman, S. Warber, S. Bolling, and S.C. Chang. 2003. Antioxidant Capacity of Polyphenolic Extracts from Leaves of *Crataegus laevigata* and *Crataegus monogyna* (Hawthorn) Subjected to Drought and Cold Stress. *Journal of Agriculture Food Chemistry* 51(14):3973–3976.

Konieczynski, P. 2013. Principal component analysis in interpretation of the results of HPLC-ELC, HPLC-DAD and essential elemental contents obtained for medicinal plant extracts. *Central European Journal of Chemistry* 11(4):519–526.

Kostić, D.A., J.M. Velicković, S.S. Mitić, M.N. Mitić, and S.S. Randelović. 2012. Phenolic Content, and Antioxidant and Antimicrobial Activities of *Crataegus Oxyacantha* L (Rosaceae) Fruit Extract from Southeast Serbia. *Tropical Journal of Pharmaceutical Research* 11(1):117–124.

Kumar, K.P.S., D. Bhowmik, A. Dutta, A. Pd. Yadav, S. Paswan, S. Srivastava, and L. Deb. 2012. Recent Trends in Potential Traditional Indian Herbs *Emblica officinalis* and Its Medicinal Importance. *Journal of Pharmacognosy and Phytochemistry* 1(1): 24–32.

Kumar, M., S. Prakash, Radha, N. Kumari, A. Pundir, S. Punia, V. Saurabh, P. Choudhary, S. Changan, S. Dhumal, P.C. Pradhan, O. Alajil, S. Singh, N. Sharma, T. Ilakiya, S. Singh, and M. Mekhemar. 2021. Beneficial Role of Antioxidant Secondary Metabolites from Medicinal Plants in Maintaining Oral Health, *Antioxidants* 10:1061.

Lis, M., M. Szczypka, A. Suszko-Pawłowska, A. Sokół-Łętowska, A. Kucharska, and B. Obmińska-Mrukowicz. 2020. Hawthorn (*Crataegus monogyna*) Phenolic Extract Modulates Lymphocyte Subsets and Humoral Immune Response in Mice. *Journal of Medicinal Plant and Natural Product Research* 86(02):160–168.

Long, S.R., R.A. Carey, K.M. Crofoot, P.J. Proteau, T.M. Filtz. 2006. Effect of hawthorn (*Crataegus oxyacantha*) crude extract and chromatographic fractions on multiple activities in a cultured cardiomyocyte assay. *Phytomedicine* 13:643–650.

Lovey, S., and P.R. Kumar. 2019. Evaluation of Antimicrobial Activity of *Emblica officinalis* against Skin Associated Microbial Strains. *Current Trends in Biotechnology and Pharmacy* 12(4):355–366.

Lund J.A., P.N. Brown, P.R. Shipley. 2020. Quantification of North American and European Crataegus flavonoids by nuclear magnetic resonance spectrometry. *Fitoterapia* 143:104537.

Magnus S., F. Gazdik, N.A. Anjum, E. Kadlecova, Z. Lackova, N. Cernei, M. Brtnicky, J. Kynicky, B. Klejdus, T. Necas and O. Zitka. 2020. Assessment of antioxidants in selected plant rootstocks. *Antioxidants* 9(3):209.

Mecheri, A., W. Benabderrahmane, A. Amrani, N. Boubekri, F. Benayache, S. Benayache, D. Zama. 2019. Hepatoprotective Effects of Algerian Crataegus *oxyacantha* Leaves. *Recent Patents on Food, Nutrition & Agriculture* 10(1):70–75.

Mehvish S., and M.Q. Barkat. 2018. Phytochemical and antioxidant screening of *Amomum subulatum, Elettaria cardamomum, Emblica officinalis, Rosa damascene, Santalum album* and *Valeriana officinalis* and their effect on stomach, liver and heart. *Matrix Science Pharma (MSP)* 2(2):21–26.

Middha, S.K., A.K. Goyal, P. Lokesh, V. Yardi, L. Mojamdar, D.S. Keni, D. Babu, and T. Usha. 2015. Toxicological evaluation of emblica officinalis fruit extract and its anti-inflammatory and free radical scavenging properties. *Pharmacognosy Magazine* 11(3):427–433.

Mishra, P., V. Srivastava, D. Verma, O.P. Chauhan, and G.K. Rai. 2009. Physico-chemical properties of Chakiya variety of Amla (Emblica officinalis) and effect of different dehydration methods on quality of powder. *African Journal of Food Science* 3(10):303–306.

Moreira, D.D.L., S.S. Teixeira, M.H.D. Monteiro, A.C.A. De-Oliveira, and F.J. Paumgartten. 2014. Traditionaluse and safety of herbal medicines. *Rev Bras Farmacogn* 24(2):248–257.

Moustafa, A.A., M.S. Zaghloul, S.R. Mansour, and M. Alotaibi. 2019. Conservation Strategy for protecting Crataegus x sinaica against climate change and anthropologic activities in South Sinai Mountains, Egypt. *The Egyptian Society For Environmental Sciences* 18 (1):1–6.

Murugesan, S., S. Kottekad, I. Crasta, S. Sreevathsan, D. Usharani, M.K. Perumal, and S.N. Mudliar. 2021. Targeting COVID-19 (SARS-CoV-2) main protease through active phytocompounds of ayurvedic medicinal plants – Emblica officinalis (Amla), Phyllanthus niruri Linn. (Bhumi Amla) and Tinospora cordifolia (Giloy) – A molecular docking and simulation study. *Computers in Biology and Medicine* 136:104683.

Nguyen, Q.T.N., M. Fang, M. Zhang, N.Q. Do, M. Kim, S.D. Zheng, E. Hwang, and T.H. Yi. 2021. Crataegus laevigata suppresses LPS-induced oxidative stress during inflammatory response in human keratinocytes by regulating the MAPKs/AP-1, NFκB, and NFAT signaling pathways. *Molecules* 26:869.

Nunes, R., P. Pasko, M. Tyszka-Czochara, A. Szewczyk, M. Szlosarczyk, and I.S. Carvalho. 2017. Antibacterial, antioxidant and anti-proliferative properties and zinc content of five south Portugal herbs. *Pharmaceutical Biology* 55(1):114–123.

Pandey, A., S.V. Singh, N.K. Singh, R. Kant, J.P. Singh, R.K. Gupta, and D. Niyogi. 2020. Effect of feeding Embilica officinalis (Amla) on milk quality in cattle affected with subclinical mastitis. *Journal of Pharmacognosy and Phytochemistry* 9(2):1259–1264.

Patil, B.S., P.S. Kanthe, C.R. Reddy, and K.K. Das. 2019. Emblica officinalis (Amla) ameliorates high-fat diet induced alteration of cardiovascular pathophysiology. *Cardiovascular & Hematological Agents in Medicinal Chemistry* 17:52–63.

Pawlowska, E., J. Szczepanska, A. Koskela, K. Kaarniranta, and J. Blasiak. 2019. Dietary polyphenols in age-related macular degeneration: protection against oxidative stress and beyond. *Oxidative Medicine and Cellular Longevity* 5(6):1–13.

Petrovska, B.B. 2012. Historical review of medicinal plants' usage. *Pharmacognosy Review* 6(11):1–5.

Poltanov, E.A., A.N. Shikov, H.J.D. Dorman, O.N. Pozharitskaya, V.G. Makarov, V.P. Tikhonov, and R. Hiltunen. 2009. Chemical and antioxidant evaluation of Indian gooseberry (Emblica officinalis gaertn., syn. phyllanthus emblica L.) supplements. *Phytotherapy Research* 23(9):1309–1315.

Priyanka, R.C., and S.A. Stanley. 2021. Evaluation of phytochemical activity of Aloe Barbadensis, Emblica Officinalis, Sesamum Indicum and Cocos Nucifera. *International Journal of Progressive Research in Science and Engineering* 2(5):49–52.

Rahman, S., M.M. Akbor, A. Howlader, and A. Jabbar. 2004. Antimicrobial and cytotoxic activity of the alkaloids of Amlaki (Emblica officinalis). *Pakistan Journal of Biological Sciences* 12(16):1152–1155.

Rai, N., Tiwari, L., Sharma, R.K., and Verma, A.K. 2012. Pharmaco-botanical Profile on *Emblica officinalis* Gaertn. – A Pharmacopoeial Herbal Drug. *STM Journals* 1(1): 29–41.

Ram, M.S., D. Neetu, B. Yogesh, B. Anju, P. Dipti, T. Pauline, S.K. Sharma, S.K.S. Sarada, G. Ilavazhagan, D. Kumar, and W. Selvamurthy. 2002. Cyto-protective and immunomodulating properties of Amla (Emblica officinalis) on lymphocytes: an in-vitro study. *Journal of Ethnopharmacology* 81:5–10.

Rathore, S., M. Mukim, P. Sharma, S. Devi, J.C. Nagar, and M. Khalid. 2020. Curcumin: A Review for Health Benefits. *International Journal of Research and Review* 7(1):271–290.

Rezaei-Golmisheh, A., H. Malekinejad, S. Asri-Rezaei, A.A. Farshid, and P. Akbari. 2015. Hawthorn ethanolic extracts with triterpenoids and flavonoids exert hepatoprotective effects and suppress the hypercholesterolemia-induced oxidative stress in rats. *Iranian Journal of Basic Medical Sciences* 18(7):691–699.

Rigelsky, J.M. and B.V. Sweet. 2002. Hawthorn: pharmacology and therapeutic uses. *American Journal of Health-System Pharmacy* 59(5):417–422.

Rodriguesa, S., R.C. Calhelhaa, J.C.M. Barreiraa, M. Dueñas, A.M. Carvalho, R.M.V. Abreu, C. Santos-Buelga, and I.C.F.R. Ferreira. 2012. Crataegus monogyna buds and fruits phenolic extracts: growth inhibitory activity on human tumour cell lines and chemical characterization by HPLC-DAD-ESI/MS. *Food Research International* 49(1):516–523.

Sampath Kumar, K.P., D. Bhowmik, A. Dutta, A. Yadav, S. Paswan, S. Srivastava, L. Deb. 2012. Recent Trends in Potential Traditional Indian Herbs Emblica officinalis and Its Medicinal Importance. *Journal of Pharmacognosy and Phytochemistry* 1(1):24–32.

Sellami, M., O. Slimeni, A. Pokrywka, G. Kuvačić, L.D. Hayes, M. Milic, and J. Padulo. 2018. Herbal medicine for sports: a review. *Journal of the International Society of Sports Nutrition* 15:14.

Silva, A.P.D., R. Rocha, C.M.L. Silva, L. Mira, M.F. Duarte, and M.H. Florêncio. 2000. Antioxidants in medicinal plant extracts. A research study of the antioxidant capacity of Crataegus, Hamamelisand Hydrastis. *Phytotherapy Research* 14(8):612–616.

Singh, S.N., A.S. Moses, and A.D.M. David. 2019. Antimicrobial activity of Emblica officinalis extracts against selected bacterial. *International Journal of Basic and Applied Research* 9(1):1730–1735.

Sowmya, M.N., and Nanjammanni, N. 2017. A phyto-chemical analysis of seedless amalaki fruit (Emblica Officinalis) churna. *International Journal of Pharmaceutical Science Invention* 6(3):09–12.

Sultana, S., S. Ahmad, N. Khan, T. Jahangir. 2005. Effect of Emblica officinalis (Gaertn)on CCl4 induced hepatic toxicity and DNA synthesis in Wistar rats. *Indian Journal of Experimental Biology* 43(5):430–436.

Tadic, V.M., S. Dobric, G.M. Markovic, S.M. Đorevic, I.A. Arsic, N.R. Menkovic, and T. Stevic. 2008. Anti-inflammatory, gastroprotective, free-radical-scavenging, and antimicrobial activities of hawthorn berries ethanol extract. *Journal of Agricultural and Food Chemistry* 56(17):7700–7709.

Thilaga, S., Largia, M.J.V., Parameswari, A., Nair, R.R., and Ganesh, D. 2013. High frequency somatic embryogenesis from leaf tissue of *Emblica officinalis* Gaertn. - A high valued tree for non-timber forest products. *Australian Journal of Crop Science* 7(10): 1480–1487.

Thomas, P.A., T. Leski, N.L. Porta, M. Dering, and G. Iszkuło. 2021. Biological Flora of the British Isles: Crataegus laevigata. *Journal of Ecology* 109:572–596.

Tomar, R.S., R.P.S. Baghel, S. Nayak, A. Khare, and P. Sharma. 2018. Immunomodulatory effect of Withania somnifera, Boerhaavia diffusa and Emblica officinalis in broilers. *Journal of Pharmacognosy and Phytochemistry* 7(3):3303–3306.

Tsarong, T.J. 1994. *Tibetan Medicinal Plants*, 1st ed. India: Tibetan Medical publications.

Verma Arun, K., Rajkumar V. 2021. Antioxidant effect of Amla (Emblica Officinalis) fruit and curry (Murraya Koenijii) leaf extracts on quality of goat meat nuggets. *Indian Journal of Small Ruminants* 27(1):105–112.

Xin T., Y. Zhang, X. Pu, R. Gao, Z. Xu, and J. Song. 2019. *Trends in Herbgenomics. Science China Life Sciences* 62(3):288–308.

Zhang W.J., S. Wang, C.Z. Kang, C.G. Lv, L. Zhou, L.Q. Huang, and L.P. Guo. 2020. Pharmacodynamic material basis of traditional Chinese medicine based on biomacromolecules: a review. *Plant Methods* 16:26.

Ziouche, N. L. Derradji, and Y. Hadef. 2020. Determination of polyphenolic components by high performance liquid chromatography (HPLC) and evaluation of the antioxidant activity of leaves and fruits of Crataegus mongyna Jacq. *GSC Biological and Pharmaceutical Sciences* 13(1):251–256.

21 *Eucalyptus spp.* (Eucalypts) and *Ficus religiosa* (Sacred Fig)

Surendra Pratap Singh, Bhoomika Yadav, and Kumar Anupam

CONTENTS

21.1 Introduction 412
21.2 *Eucalyptus spp.* 412
21.2.1 Morphological Description 412
21.2.2 Distribution 413
21.2.3 Chemical Composition and Derivatives 413
21.2.4 Potential Benefits of *Eucalyptus* 416
21.2.4.1 For Hair 417
21.2.4.2 For Skin 417
21.2.4.3 For Diabetes 418
21.2.4.4 For Fever 418
21.2.4.5 For Teeth 418
21.2.4.6 Relief from Stomach Worms 418
21.2.4.7 Hair Lice Removal 418
21.2.4.8 Relief of Muscle Pain 418
21.2.4.9 Use in Pneumonia 418
21.2.4.10 Relief from Kidney Stones 418
21.2.4.11 Plant Protection 418
21.2.5 Disadvantages of Eucalyptus Oil 418
21.3 *Ficus religiosa* 418
21.3.1 Morphological Description 418
21.3.2 Distribution 419
21.3.3 Common Names of *F. religiosa* in Different Languages 419
21.3.4 Phytochemistry of *F. religiosa* 419
21.3.4.1 Constituents of the Leaves 420
21.3.4.2 Constituents of the Bark 421
21.3.4.3 Constituents of the Fruits and Seeds 421
21.3.5 Potential Benefits of *F. religiosa* 421
21.3.5.1 Beneficial for the Lungs 422
21.3.5.2 Use as an Immunity Booster 422
21.3.5.3 Beneficial for the Liver 422
21.3.5.4 Relief from Phlegm 422
21.3.5.5 Dental Disease Treatment 422
21.3.5.6 Jaundice Treatment 422
21.3.5.7 Controlling Diabetes 422
21.3.5.8 Treat Urinary Disease 422
21.3.5.9 Importance in Infertility Problems 422
21.3.5.10 Benefits in Cracked Heel Problem 422
21.3.5.11 Benefits in Skin Disease Treatment 423
21.3.5.12 Curing Boils 423

DOI: 10.1201/9781003205067-21

21.3.5.13 Benefit in Blood Disorders ... 423
21.3.5.14 Uses in Fighting Fever ... 423
21.3.5.15 Uses in Typhoid ... 423
21.3.5.16 Beneficial in Impotency ... 423
21.3.5.17 Benefits in Snakebite ... 423
21.4 Conclusion and Future Prospects ... 423
References ... 424

21.1 INTRODUCTION

Traditionally, foresters have always believed their forests to be a benign influence on the environment. *Eucalyptus* spp. and *Ficus religiosa* are beneficial in having high economic value, both for timber, pulp, latex, and oil production (Prasad et al., 2006). The selection and cloning of medicinal plantation such *Eucalyptus* spp. and *F. religiosa* has resulted in very substantial increases in latex and oil yields for pharmaceutical use. *Eucalyptus spp.* and *F. religiosa* (Sacred fig) both are enriched with pharmaceutical value. Today both are being seen as the biggest pillar in saving human life to deal with epidemics like COVID-19. Scientists from many countries are researching to know how effective herbal and other medicines can be effectively used in the treatment of epidemic diseases. The Government of India has made people aware of many decoctions and many homeopathic medicines through Ayush. The rapid growth and increase in woody biomass make it useful as a resource for the timber and pulp industries and its secondary metabolites are now being recognized as potential natural and renewable resources for health care products. Many pharmacological studies have therefore been made and, where bioactivity has been demonstrated, it has often been found to be associated with the non-volatile, rather than the volatile, constituents of the plants (Singh and Dwivedi, 1987). Phloroglucinol derivatives in particular (e.g., robustadials, euglobals, macrocarpals, and sideroxylonals) have been found to exhibit strong biological activities. These include antioxidant, antibacterial, anti-inflammatory, anti-tumor, anti-malarial, and HIV-RTase inhibitory promoting activities. From ancient times, the leaves and bark of various species of *Eucalyptus* and *F. religiosa* have been used as traditional ayurvedic medicines for the treatment of such ailments as snake bites, colds, fever, diarrhea, toothache etc. (Duh et al., 2012; Devanesan et al., 2018). The plant kingdom represents the various resources of drugs and medicines. Definitely, medicinal plants have played a major role in healthcare since ancient times. These represent an important resource of new antimicrobial drugs and medicines to cure and protect from multidrug resistant microorganisms. The antimicrobial agents may be hidden in medicinal plant sap, latex, essential oils and extracts. Therefore, the present review summing up the previous research data regarding antimicrobial activity, chemical structure, and other important effects of *Eucalyptus* spp. and *F. religiosa*.

21.2 *EUCALYPTUS SPP.*

21.2.1 Morphological Description

Eucalyptus belongs to the Myrtaceae family. It comprises 140 genera and about 3800 and 5650 species (Kantvilas, 1996). It is called Nilgiri also in the Hindi language. Eucalyptus is notably a fast-growing tree of good form for modern harvesting practices. *Eucalyptus* is intended ultimately for use as pole timber or a source of fiber (Gomide, 2006). The leaves and bark are the by-products of eucalyptus plantations after felling of trees. The use of leaves and bark as biomass resources are considered an important research topic. The leaves are used for extraction of essential oil (Kesharwani et al., 2018). This tree is very tall and thin. It can take the form of a mature tree, a low height shrub, or a very large tree. There are many species of this tree which can be mainly divided into 4 size categories. If we talk about size, then the height of the smallest species of this tree is 10 meters. The medium-sized species reach a height of about 11–30 meters. Large-sized trees reach a

FIGURE 21.1 *Eucalyptus* spp. (A) Tree, (B) green leaf, (C) stem-bark, and (D) immature fruits.

height of 31–60 meters. Its longest species reach more than 60 meters in height (Sillett et al., 2015). This is an evergreen tree. Its leaves are pointed, with knots on the surface. The oil is stored from these lumps. The oil glands are secreted deeply in the leaves, well below the epidermal cuticle and other cells which together form the surface layers of the foliage (Becerra et al., 2018). Its flowers are hairy. The stamens of this flower are covered with an outer covering which is like a pair. Its bark varies according to the species and age of the tree (Oke et al., 2021). Its bark is fibrous in small pieces which can be pulled as shown in Figure 21.1.

21.2.2 Distribution

The *Eucalyptus spp.* is native to Australia and Tunisian. There are also a small number native to Indonesia, Philippines, and New Guinea. It is grown in subtropical and tropical regions of the world. The eucalyptus various species are also cultivated in India and Southern Europe (Paine et al., 2011).

21.2.3 Chemical Composition and Derivatives

The medicinal and aromatic properties of *Eucalyptus* are normally associated with steam volatile components. Consequently, the number of volatile compounds reported from *Eucalyptus* far exceeds the number of non-volatile ones (Abiri et al., 2021). The leaves are the most frequently investigated part of the plant but interesting constituents have also been isolated from the bark and the wood. The characteristics of the different *Eucalyptus* species wood are summarized in Table 21.1, as the chemical composition from their respective holocellulose, lignin, extractive, and ash isolations. The total amount of polysaccharides means holocellulose for *Eucalyptus* species are ranging between 55.4% and 70.1%. The lignin content for the different species of *Eucalyptus* showed a variation in absolute values of 21.6% to 30.8 %. The insoluble lignin is known as Klason

TABLE 21.1
The Chemical Composition of *Eucalyptus* Species as Oven-Dried Basis

Species	Holo-cellulose (%)	Total lignin (%)	Klason lignin (%)	Soluble lignin (%)	Total extractives (%)	Dichloro-methane (%)	Ethanol (%)	Ash (%)
E. botryoides	64.7	27.1	24.3	2.8	9.7	0.4	5.8	0.6
E. camaldulensis	55.4	24	21.8	2.3	18.9	1.1	14.8	0.8
E. globulus	70.1	24.8	21.1	3.7	6.1	0.4	2.3	0.6
E. grandis	66.8	27.8	25.1	2.7	6.7	0.7	4.2	0.4
E. maculata	64.8	21.6	18.5	3.1	10	1.1	6.5	2.2
E. ovata	68.3	26	22.2	3.8	6.4	0.9	2.8	0.6
E. propinqua	62.7	29.9	27.7	2.2	8.8	0.6	6.7	0.4
E. resinífera	61	30.8	28.1	2.8	8.2	0.9	5.2	0.6
E. rudis	59	26.8	23.4	3.5	14.5	1.2	10.7	0.7
E. saligna	64.2	26.6	23.7	2.9	10.1	0.9	7.2	0.8
E. sideroxylon	61.1	26.6	23.3	3.2	13.5	0.7	10.8	0.4
E. viminalis	68.1	26.8	23.2	3.6	6.1	0.8	2.9	0.6

TABLE 21.2
Acetates and Monosaccharides Percentage in the Various *Eucalyptus* Species

Species	Acetates	Arabinose	Galactose	Glucose	Xylose
E. botryoides	7.3	1.1	1.1	68.2	22.3
E. camaldulensis	8.3	1.2	1.8	62.3	26.4
E. globulus	8.1	1.1	1.1	64.3	25.4
E. grandis	7.1	1.1	1.1	68	22.7
E. maculata	8.5	1.1	1.1	63.3	26.1
E. ovata	6.1	1	1	63.6	28.2
E. propinqua	6.6	1.1	2.1	68.6	21.6
E. resinífera	8	1.2	2.4	63.5	24.9
E. rudis	7.7	1.1	2.2	65.5	23.5
E. saligna	7.8	1.1	2.3	65	23.8
E. sideroxylon	7.9	1.1	2.2	61	27.7
E. viminalis	8.5	1.1	1.1	63.1	26.1

lignin (Neiva et al., 2015). The total extractives are generally aliphatic compounds. The Eucalyptus wood contained on average extractives 9.1% and ranging between 6.1 and 18.9%. The extractive soluble in dichloromethane (DCM) is very trace (0.4%), while ethanol soluble extractives reached an average of 6.9%. Ash contents are very low and almost similar for all species ranging from 0.4 to 0.8% (Gomide, 2006).

The acetates and carbohydrates composition for the various *Eucalyptus* spp. are shown in Table 21.2. All *Eucalyptus* spp. have low amounts of galactose and arabinose. The major constituents are xylose (related only to hemicelluloses) and glucose (related mostly to cellulose). The acetates are by-products of the hydrolysis of the acetyl groups and the depolymerization of the hemicelluloses.

The chemical compositions of essential oils extracted from leaves and fruit of *E. camaldulensis* are shown in Table 21.3 (Dogan et al., 2017). The minimum yields of leaf and fruit essential oils are 1.2% (v/w) and 1.0% (v/w) respectively. There are six to seven compounds identified in essential oil; these are representing 93.8% and 99.0%, respectively, of the total oils. The key constituents of essential oil in the leaf are Spathulenol (3.2%), Borneol L (5.5%), Limonene (5.5%), α-terpineol (10.7%), α-pinene (12.7%), Eucalyptol (1,8-cineole) (14.1%), and p-cymene (42.1%) (Ghalem and Mohamed, 2008) and in the fruit are α-pinene (9.0%), α-terpinol (15.1%), p-cymene (30.0%) and eucalyptol (1,8 cineole) (34.5%) (Nikbakht et al., 2015).

TABLE 21.3
The Chemical Composition of Essential Oil Extracted from Leaf and Fruit of *E. camaldulensis*

Compound	Leaf (%)	Fruit (%)
α-pinene	12.7	9
α-terpineol	10.7	15.1
Borneol L	5.5	5.3
Eucalyptol (1,8-cineole)	14.1	34.5
γ-terpinene	–	5.1
Limonene	5.5	–
p-cymene	42.1	30
Spathulenol	3.2	–
Total	**93.8**	**99**

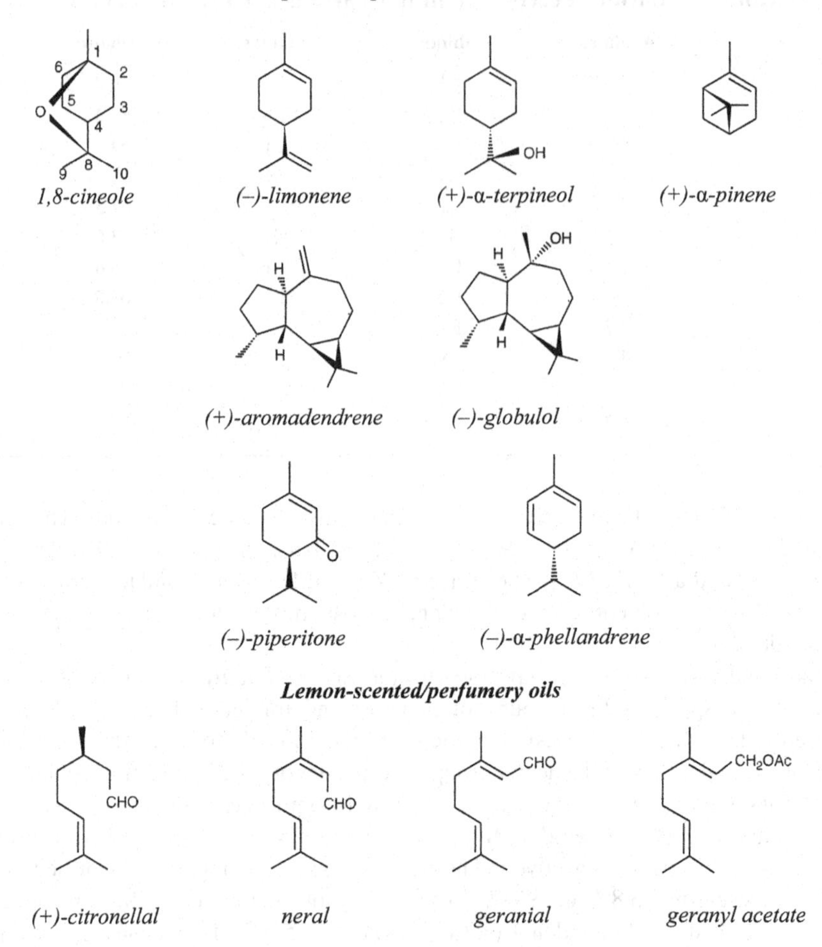

FIGURE 21.2 Chemical structure of some common constituents of *Eucalyptus* oil.

The large and diverse range of compounds found in eucalyptus oils is the monoterpene ether 1,8-cineole (Figure 21.2) (Dhakad et al., 2018). It is used for medicinal, flavor, and fragrance purposes and has significant biological activity (e.g., mosquito repellency). The most common constituents co-occurring with 1,8-cineole are limonene and α-terpineol, both of which can be derived from the menth-1-en-8-yl cation, the same biogenetic precursor from which cineole is derived. The other biogenetic pathways then contribute to other monoterpenes, for example, sesquiterpenes such as aromadendrene, α -pinene, globulol and aromatic constituents (Al-Snafi, 2017a).

The chemistry of the volatile oils is only one facet of Eucalyptus secondary metabolite chemistry. Eucalyptus also contains triterpenes, flavonoids, long chain ketones, acylphloroglucinol derivatives, glycosides, and adducts combining more than one of these chemical entities. Some examples are shown in Figure 21.3.

21.2.4 Potential Benefits of *Eucalyptus*

The essential oil extracted from fresh leaves is used in the treatment of various diseases. The essential oil is extracted by the distillation process, which is a colorless and tasteless liquid.

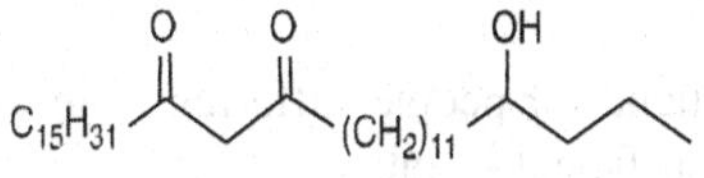

4-hydroxytritriacontane-16,18-dione

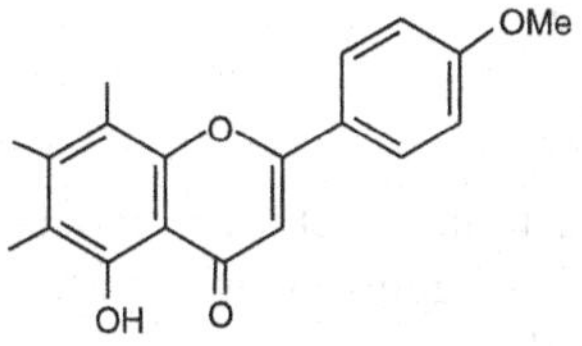

eucalyptin

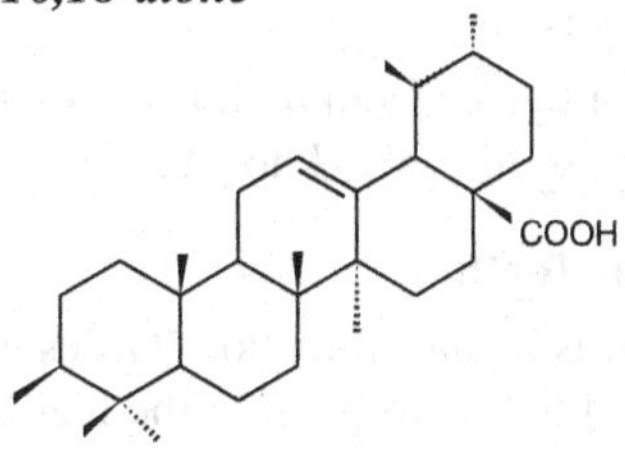

ursolic acid

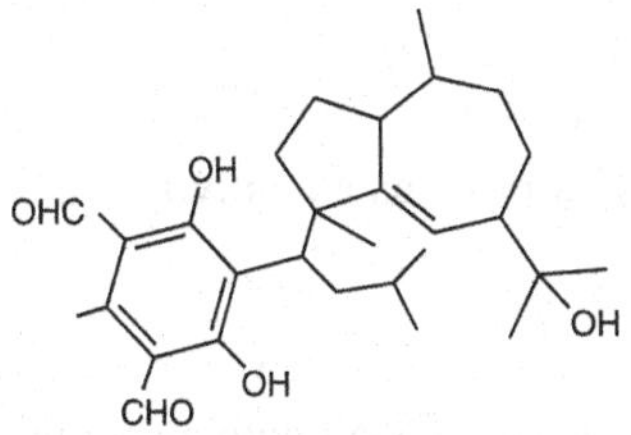

macrocarpal D

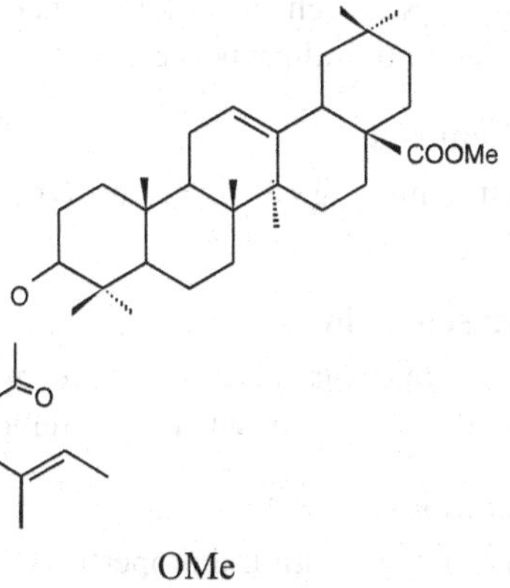

cis-p-methoxycinnamoyloxyoleanolic acid methyl ester

FIGURE 21.3 Chemical structures of some nonvolatile constituents of *Eucalyptus.*

However, it is soluble in alcohol. *Eucalyptus* oil can be used as an antiseptic and a stimulant. It is also beneficial in increasing the heart rate and controlling uncontrolled heart rate. It is also believed that eucalyptus oil has no expiry date. In addition, it is also used to treat malaria. Eucalyptus oil can also be used if there is a complaint of pain in the neck or other parts of the body. *Eucalyptus* leaves are used for the treatment of various diseases and ailments, such as infection, barley, upset stomach, phlegm, and allergies (Silva et al., 2003), and infections of the respiratory tract, whooping cough, asthma, tuberculosis, osteoarthritis, joint pain (arthritis), acne, wound, ulcer, burns, shingles, loss of appetite, cancer (Döll-Boscardin et al., 2012), dental plaque, etc. (Mishra et al., 2010).

21.2.4.1 For Hair

Using eucalyptus oil in the hair gives shine and thick hair, but eucalyptus oil should not be used in excess. It reduces the shine of the hair if applied to the hair for one hour. The hair should be washed thoroughly with shampoo and water.

21.2.4.2 For Skin

If the skin is affected by infection, then eucalyptus oil is applied to that area. It is also used for skin infections like shingles, acne, and chickenpox.

21.2.4.3 For Diabetes

Consuming eucalyptus is very beneficial for people suffering from diabetes. It controls the blood sugar in the body. But use it only in sufficient quantity.

21.2.4.4 For Fever

If a person is suffering from high fever, then eucalyptus oil is used to reduce the temperature of the fever. Eucalyptus oil is also known as fever oil.

21.2.4.5 For Teeth

Eucalyptus oil is an anti-infection (Gilles et al. 2010) and antibacterial (Bachir and Benali, 2012). If there is any kind of problem in the teeth, like inflammation in the gums, worms, eucalyptus oil helps in removing the problem like toothache etc. (Gilles et al. 2010).

21.2.4.6 Relief from Stomach Worms

If worms are found in the stomach of children, then eucalyptus oil helps in removing stomach worms. Eucalyptus oil has antiviral properties.

21.2.4.7 Hair Lice Removal

Eucalyptus oil is used in removing hair lice. Eucalyptus oil kills hair lice as an antiviral and frees the scalp from lice.

21.2.4.8 Relief of Muscle Pain

Eucalyptus oil acts like a sanjeevani herb to remove muscle pain. Massage well with eucalyptus oil on the place of pain. By doing this, muscle pain is removed (Jun et al. 2013).

21.2.4.9 Use in Pneumonia

Eucalyptus oil has antiseptic and antiviral properties (Elaissi et al. 2012). Massaging the chest using this oil helps to clean the lungs, reduce inflammation (Lu et al. 2004), and reduce the symptoms of tuberculosis (Burrow, Eccles, and Jones, 1983).

21.2.4.10 Relief from Kidney Stones

Kidney stones are very painful. Due to its pain, the person becomes very weak. In such a situation, applying eucalyptus oil on the painful area will give relief from pain (Kovar et al. 1987).

21.2.4.11 Plant Protection

It would be of great benefit to be able to employ eucalyptus oil as a natural fungicide, one which was biodegradable and able to control some of the important plant pathogens. The potential use of eucalyptus oils in agriculture has been investigated by Singh and Dwivedi (1987) in attempts to control Sclerotium rolfsii, the causative organism of foot-rot of barley.

21.2.5 Disadvantages of Eucalyptus Oil

Eucalyptus oil has many benefits, but eucalyptus also has some disadvantages. People who are allergic to eucalyptus oil should stop using it immediately. Excessive consumption of eucalyptus oil can be harmful to health. It contains toxic compounds (Dhakad et al., 2018).

21.3 *FICUS RELIGIOSA*

21.3.1 Morphological Description

F. religiosa (Sanskrit: Ashwath; English: Sacred Fig,) is a large, semi-evergreen or dry season-deciduous and multi-branched tree up to 3-meter trunk diameter and up to 30 m high, of the species

FIGURE 21.4 *Ficus religiosa* (A) Tree, (B) stem-bark, (C) dark green leaf, and (D) ripe and matured fruits.

of sycamore or banyan. The bark of the old tree is cracked and white-brown in color. The roots of *F. religiosa* tree are covered with stems inside the ground and spread far and wide.

The leaves are cordate in shape with a distinctive extended drip tip; these are 10–17 cm long and 8–12 cm broad, with a 6–10 cm petiole. Its new leaves are soft, smooth, and light red in color. The fruits are smooth, spherical, and small figs 1–1.5 cm in diameter as shown in Figure 21.4. It is green in raw state and purple in ripe state. *F. religiosa* has a very long lifespan, ranging on average between 900 and 1500 years. *F. religiosa* tree is planted around the temples, due to its religious importance in Hinduism and Buddhism.

21.3.2 Distribution

F. religiosa is native to India, Pakistan, Nepal, Bangladedh, Sri Lanka, Myanmar, Thailand Indonesia, Malaysia, China, and India, including North East region, Eastern Himalaya, and Andaman Nicobar Islands. It is also found in Venezuela, Florida, and Iran (Galil, 1984).

21.3.3 Common Names of *F. religiosa* in Different Languages

The botanical name of Sacred fig is *Ficus religiosa Linn.* (*Ficus religiosa*) Syn-Ficus caudata Stokes; *Ficus peepul Griff.* and it belongs to the Moraceae (Moraceae) family. *F. religiosa* is also known by other names within the country or abroad. *F. religiosa* names in different languages are listed in Table 21.4 (Khare, 2008).

21.3.4 Phytochemistry of *F. religiosa*

According to Ayurveda, each and every part of the *F. religiosa* such as the leaf, bark, shoot, seeds, and fruit have numerous medicinal benefits (Bhavyasree and Xavier, 2021). The phytochemical research carried out on *F. religiosa* had led to the isolation of amino acids, few other classes of secondary metabolites phytosterols, phenolic components, volatile components, aliphatic alcohols, furanocoumarins, and hydrocarbons, from its different parts. The amino acids and phenolic components (flavonoids and tannins) are present in about all the parts of *F. religiosa*. The polyphenolic substances are present only in its root (Ravindran et al., 2020).

TABLE 21.4
***Ficus religiosa* Names in Different Indian Local Languages**

Local language	Common name
Hindi	Sacred fig vriksh. Bodhi tree
English	The tree of intelligence, sacred fig
Sanskrit	Pippal, Kunjarashan, Ashwattha, Bodhi tree, Chaldal, Bodhidrum, Gajashan
Oriya	Jori, Pipplo, Usto pippolo
Urdu	Pipal
Assamese	Anhot
Konkani	Pimpoll
Kannada	Arali
Gujarati	Pipro
Tamil	Arsumaram, Arasu
Telugu	Ravichettu, Ashvatthamu
Nepali	Pipal
Punjabi	Sacredfig
Bengali	Asvatwha
Marathi	Pimple
Malayalam	Arachu, Arasu, Arayal
Manipuri	Sana Khongnang
Persian	Darakhte laranza

21.3.4.1 Constituents of the Leaves

The leaves of *F. religiosa* contain leucine, methionine, serine, tryptophan, tannic acid, isoleucine, threonine, glycine, asparatic acid, and arginine (Behari et al., 1984). The phytosterols (2.8%), like campesterol, sitosterol, stigmasterol, and triterpene alcohols (28.5%), like amyrin and lupeol, have been isolated from the nonsaponifiable fraction of light petroleum leaf extract of *F. religiosa* (Figure 21.5).

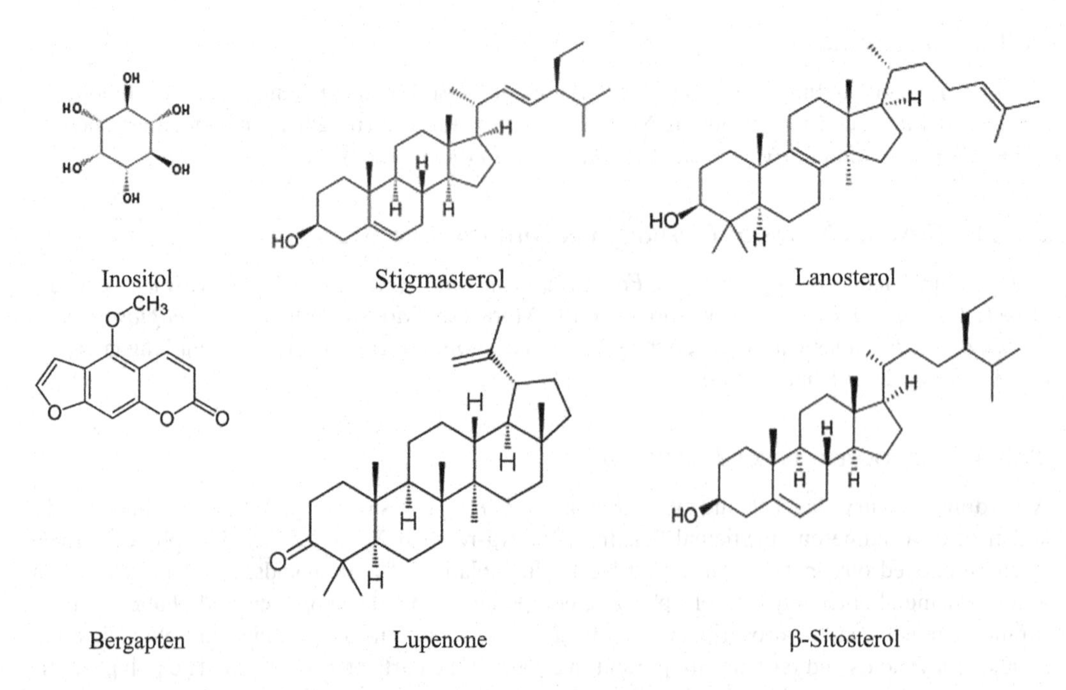

FIGURE 21.5 Chemical structures of the compounds from *Ficus religiosa*.

Along with triterpene and phytosterols 7.1%, of long-chain hydrocarbons (n-nonacosane and n-hentriacontane) and 7.9% of aliphatic alcohols (nhexacosanol and n-octacosanol) have also been isolated. A variety of carbohydrates and proteins are found in the leaves. So, its leaves are good fodder for animals (Bamikole et al., 2003; Choudhary, 2006).

21.3.4.2 Constituents of the Bark

The bark of *F. religiosa* comprises of bergaptol and bergapten. The phytosterols like (Choudhary, 2006), stigmasterol, sitosterol, and its glucoside (sitosteryl-d-glucoside) and lanosterol have been isolated from alcoholic and petroleum ether extracts of the bark (Thomas et al., 2000). The two substituted furanocoumarins, 4-hydroxy-7H-furo [3,2- g] chromen-7-one (Bergaptol) and 4-methoxy-7H-furo [3,2- g] chromen-7-one (Bergapten) are isolated from the benzene extract of the bark (Swami and Bisht, 1996). The isolated furanocoumarins are shown excellent in vitro antimicrobial activity. The carbocyclic polyol "Inositol" is isolated from the alcoholic bark extracts. Vitamin K1, methyl oleonate, n-octacosanol, and lupen-3-one isolated from the petroleum ether extracts of the bark. *F. religiosa* bark comprises around 8.7% total tannin content on average. The neutral detergent fiber (NDF), acid detergent fiber (ADF), acid detergent lignin (ADL), phenolic components, and saponins are found in the inner bark of *F. religiosa* (Mali and Borges, 2003).

21.3.4.3 Constituents of the Fruits and Seeds

The *F. religiosa* fruits are a good source of fiber. The immature fruits contain a considerable amount of ADL, ADF, and NDF (Ali and Qadry, 1987). Serotonin (5-hydroxytryptamine) is found in the fruits. The ripe fruits comprise 51.7 ± 0.9 g/100 g on oven dry basis (four independent determinations) of the total fiber content, that include cellulose (40.3%), hemicellulose (19.0%), lignin (34.9%), and pectin (5.8%). The fruit of *F. religiosa* consists of asparagines and tyrosine. Tyrosine and asparagin are the most abundant amino acids of the fruit pulp of *F. religiosa*. Its fruit pulp also contains aspartic acid, alanine, glycine, threonine, norvaline, and norleucine – these amino acids are found in free form. The protein hydrolysate of the fruit is rich in isoleucine, cystiene, phenylalanine, and serine. The phenolic components like flavonoids constitute a major class of metabolites of the *F. religiosa*. Its fruits contain a considerable number of flavonoids and other phenolic components. Flavonols, like kaempferol, quercetin, and myricetin, are present up to 160.8, 256.3, and 694 mg/kg of the fruit powder, respectively. The condensed tannins are reported in the immature fruits. It is found that the yield of phenolic and flavonoid components from the *F. religiosa* fruits depends on the isolation techniques and extraction solvents. The highest yield is obtained, when extracted with 80% hydro-methanol as compared to absolute ethanol or methanol. In addition, extraction by shaking technique gives more yields of phenolic and flavonoids components as compared to refluxing technique. Lot of volatile components are identified in the *F. religiosa* fruit, such as (Z)-3-hexenol, 1-hexanol, simple aliphatics (undecane, tridecane, tetradecane, and (Z)-3- hexenyl acetate), acyclic monoterpenes (perillene, cyclic, cadinene, (Z)-ocimene, and (E)-ocimene). The main function of these volatile components has been suggested to attract the pollinating wasps. The *F. religiosa* seeds contain alanine, valine, and threonine, which are enriched with secondary metabolites used as antimicrobial and larvicide agents.

21.3.5 Potential Benefits of *F. religiosa*

F. religiosa has medicinal values and is used for treatment of different ailments of animals and humans. It shows a significant role in the poor and deprived people's lives, with few medicinal facilities. It has various pharmacological activities such as antibacterial and antioxidant (Singh, Singh, and Goel, 2011), hypoglycemic, wound-healing activities, hypolipidemic, anticonvulsant, anti-ulcer, immunomodulatory, and antihelminthics have been studied (Devanesan et al., 2018).

21.3.5.1 Beneficial for the Lungs

The *F. religiosa* is beneficial for swelling and constriction in the passages of the lungs, wheezing in the throat, shortness of breath, chest tightness, and cough. *F. religiosa* leaf extract has such special properties, which can show effective effect on bronchospasm. Its use gives relief. Along with this, *F. religiosa* leaves are also effective in increasing oxygen levels (Cagno et al., 2015).

21.3.5.2 Use as an Immunity Booster

F. religiosa leaf increases immunity. Amidst the increasing infection of COVID-19, every person should strengthen the immunity in advance so that the infection cannot dominate (Venkateshwarlu et al., 2014).

21.3.5.3 Beneficial for the Liver

Consuming too much alcohol has a bad effect on the liver. In such a situation, *F. religiosa* leaves can be consumed to keep the liver healthy. Using its extract can protect the liver from damage (Al-Snafi, 2017b).

21.3.5.4 Relief from Phlegm

F. religiosa leaves may be a great option for cough (Ballabh and Chaurasia, 2007). Therapeutic elements are found in *F. religiosa* leaf. Using which can get relief in phlegm. Using *F. religiosa* leaves in the form of juice can get rid of the problem of phlegm. Use of *F. religiosa* leaf is beneficial in eye disease. The white latex that comes out from *F. religiosa* leaves is applied to the eye, and it gives relief from pain in the eyes(Cagno et al., 2015).

21.3.5.5 Dental Disease Treatment

Mix equal quantity of bark of *F. religiosa* and banyan tree and cook it in water. By rinsing it, the diseases of the teeth are cured. Teeth are strengthened by daily brushing with a fresh twig of *F. religiosa*. This kills the bacteria and also reduces the inflammation of the gums. Apart from this, *F. religiosa* twig is used as a plaque-reducing agent and antiseptic (kills the germs that cause bad breath) (Sharma et al., 2016).

21.3.5.6 Jaundice Treatment

Grind 3–4 new *F. religiosa* leaves with sugar candy in 250 ml water and sieve them. Give this syrup to the patient twice. Use it for 3–5 days. It is a panacea for jaundice disease (Tiwari et al., 2019).

21.3.5.7 Controlling Diabetes

Consuming 40 ml decoction of the bark of *F. religiosa* tree is beneficial in diabetes (Singh et al., 2020).

21.3.5.8 Treat Urinary Disease

Taking decoction of *F. religiosa* bark is beneficial in the problem of frequent urination (Rathee et al., 2015).

21.3.5.9 Importance in Infertility Problems

F. religiosa is beneficial in the problem of infertility. After the end of menstruation, consume 1–2 grams powder of dried fruit of *F. religiosa* with raw milk. Infertility may be cured by 14 days of dosing (Suriyakalaa et al., 2021).

21.3.5.10 Benefits in Cracked Heel Problem

Many people complain that the heels of their feet are torn. In such a situation, apply fresh white latex secreted from *F. religiosa* leaves. Apply the juice of *F. religiosa* leaves or white latex on cracked hands and feet. It brings relief (Parate et al.,2015).

21.3.5.11 Benefits in Skin Disease Treatment

Itching and skin diseases spreading on the skin end by eating soft *F. religiosa* leaves. Drinking 40 ml decoction of it gives the same benefit (Chandrasekar et al., 2010; Das, 2020).

21.3.5.12 Curing Boils

The benefits of *F. religiosa* leaves are also available for boils and pimples. Grind the bark of *F. religiosa* in water, apply it on the boils, and tie with a wet bandage. Boils and swelling are cured (Roy et al., 2009).

- Grind the powder of *F. religiosa* bark and mix ghee in it. Applying it to a wound caused by a burn or injury stops bleeding and it is beneficial to heal the wound immediately.
- Spraying the powder of *F. religiosa* bark on the wounds caused by fire burns, the wound dries up immediately.
- Rub the bark of *F. religiosa* in rose water and apply it on old and unhealing wounds.
- Apply a paste of medicine on the wound and cover it with soft leaves of *F. religiosa.* It dries the wound.
- Sprinkle fine powder of freshly fallen *F. religiosa* leaves on the wound, it cures the wound immediately.

21.3.5.13 Benefit in Blood Disorders

F. religiosa bark is beneficial in blood disorders like varicose. Mixing 5 grams of honey in 40 ml decoction of bark should be given in the morning and evening. Licking 1–2 grams powder of *F. religiosa* seed with honey in the morning and evening purifies the blood (Shukla et al., 2012).

21.3.5.14 Uses in Fighting Fever

Consuming 10–20 ml decoction of *F. religiosa* leaves is beneficial in fever (Singh and Jaiswal, 2014).

21.3.5.15 Uses in Typhoid

Consuming 1–2 grams of powder of bark of *F. religiosa* tree with honey in the morning and evening is beneficial in typhoid (Parate et al., 2015).

21.3.5.16 Beneficial in Impotency

Consuming a half spoon of powder of *F. religiosa* fruit with milk thrice a day ends impotence and benefits in semen disorders and weakness (Yadav and Srivastava, 2013).

21.3.5.17 Benefits in Snakebite

In case of snakebite, drink the juice of *F. religiosa* leaves 3–4 times in the amount of 2 tablespoons until the doctor is available. Keep giving *F. religiosa* leaves in the mouth for chewing. The effect of the poison will be less (Ghimire and Bastakoti, 2009).

21.4 CONCLUSION AND FUTURE PROSPECTS

From literature review revealed *Eucalyptus* and *F. religiosa* to be important traditional medicinal plants used for the ethnomedical treatment of diabetes, asthma, inflammatory disorders, epilepsy, diarrhea, sexual disorders, and gastric infections. There is rising interest in the health sciences, sector industries, and academia in aromatic and medicinal plants. The medicinal traditions and trends of ancient civilizations, such as those of India and China, have large armamentaria of plants in their pharmacopeias, which are used all over Southeast Asia. A similar situation exists in South America and Africa. Therefore, a high population of the world depends on aromatic and medicinal plants for their drugs and medicines. The pharmacological and clinical studies carried out on the fresh plant materials, crude extracts, and isolated components of *Eucalyptus* and *F. religiosa* provide practical support for its many traditional and beneficial uses. The current studies have been focused on evaluating the wound-healing, anticonvulsant, antibacterial, antidiabetic, anti-amnesic, and anti-inflammatory activities. Some of its traditional uses have extensively explored in this chapter. Thus, *Eucalyptus spp.* and *F. religiosa* are also green, rapid, nontoxic, eco-friendly, and cost-effective antimicrobial agents. The outcome of the present review will provide realistic support for the modern and futuristic potential clinical use of *Eucalyptus Spp.* and *F. religiosa.*

REFERENCES

Abiri, R., N. Atabaki, R. Sanusi, S. Malik, R. Abiri, P. Safa, N.A.A. Shukor & Abdul-Hamid, H. (2021). New insights into the biological properties of eucalyptus-derived essential oil: A promising green anti-cancer drug. *Food Reviews International*, 1–36.

Al-Snafi, A.E. (2017a). The pharmacological & therapeutic importance of Eucalyptus species grown in Iraq. *IOSR Journal of Pharmacym*, 7(3): 72–91.

Al-Snafi, A.E. (2017b). Pharmacology of Ficus religiosa: A review. *IOSR Journal of Pharmacy*, 7(3): 49–60.

Ali, M. & Qadry, J. (1987). Amino-acid-composition of fruits and seeds of medicinal-plants. *Journal of the Indian Chemical Society*, 64(4): 230–231.

Bachir, R.G. & Benali, M. (2012). Antibacterial activity of the essential oils from the leaves of Eucalyptus globulus against Escherichia coli & Staphylococcus aureus. *Asian Pacific Journal of Tropical Biomedicine*, 2(9): 739–742.

Ballabh, B. & Chaurasia, O. (2007). Traditional medicinal plants of cold desert Ladakh: Used in treatment of cold, cough & fever. *Journal of Ethnopharmacology*, 112(2): 341–349.

Bamikole, M., Babayemi, O., Arigbede, O & Ikhatua, U. (2003). Nutritive value of Ficus religiosa in West African dwarf goats. *Animal Feed Science & Technology*, 105(1–4): 71–79.

Becerra, P.I., Catford, J.A., Inderjit, L., McLeod, M., Andonian, K., Aschehoug, E.T., Montesinos, D. & Callaway, R.M. (2018). Inhibitory effects of Eucalyptus globulus on understorey plant growth & species richness are greater in non-native regions. *Global Ecology & Biogeography*, 27(1): 68–76.

Behari, M., Rani, K., Matsumoto, T., & Shimizu, N. (1984). Isolation of active-principles from the leaves of Ficus religiosa. *Current Agriculture*, 8(1–2): 73–76.

Bhavyasree, P. & Xavier, T. (2021). A critical green biosynthesis of novel CuO/C porous nanocomposite via the aqueous leaf extract of Ficus religiosa & their antimicrobial, antioxidant, & adsorption properties. *Chemical Engineering Journal Advances*, 8: 100152.

Burrow, A., Eccles R. & Jones A. (1983). The effects of camphor, eucalyptus and menthol vapour on nasal resistance to airflow & nasal sensation. *Acta Oto-laryngologica*, 96(1–2): 157–161.

Cagno, V., Civra, A., Kumar, R., Pradhan, S., Donalisio, M., Sinha, B.N., Ghosh M. & Lembo D. (2015). Ficus religiosa L. bark extracts inhibit human rhinovirus and respiratory syncytial virus infection in vitro. *Journal of Ethnopharmacology*, 176: 252–257.

Chandrasekar, S., Bhanumathy, M., Pawar A. & Somasundaram T. (2010). Phytopharmacology of Ficus religiosa. *Pharmacognosy Reviews*, 4(8): 195.

Choudhary, G. (2006). Evaluation of ethanolic extract of Ficus religiosa bark on incision & excision wounds in rats.*Planta Indica*, 2(3): 17–19.

Das, K. (2020). Application of Indian medicinal herbs for skin problems following safety measures against COVID-19. *Iranian Journal of Dermatology*, 23: 24–37.

Devanesan, E.B., A. Anand, V., Kumar, P.S., Vinayagamoorthy P. & Basavaraju P. (2018). Phytochemistry & Pharmacology of Ficus religiosa. *Systematic Reviews in Pharmacy*, 9(1): 45–48.

Dhakad, A.K., Pandey, V.V., Beg, S., Rawat, J.M. & Singh,A (2018). Biological, medicinal and toxicological significance of Eucalyptus leaf essential oil: A review. *Journal of the Science of Food and Agriculture*, 98(3): 833–848.

Dogan, G., Kara, N., Bagci E. & Gur (2017). Chemical composition and biological activities of leaf and fruit essential oils from Eucalyptus camaldulensis. *Zeitschrift für Naturforschung C*, 72(11–12): 483–489.

Döll-Boscardin, P.M., Sartoratto, A., Sales Maia, B.H.L.d.N., Padilha de Paula, J., Nakashima, T., Farago, P.V. & Kanunfre, C.C. (2012). In vitro cytotoxic potential of essential oils of Eucalyptus benthamii & its related terpenes on tumor cell lines. *Evidence-Based Complementary & Alternative Medicine*, 2012: 1–8.

Duh, P.-D., Chen, Z.-T., Lee, S.-W., Lin, T.-P.,Wang, Y.-T., Yen, W.-J., Kuo, L.-F. & Chu, H.-L.(2012). Antiproliferative activity & apoptosis induction of Eucalyptus Citriodora resin & its major bioactive compound in melanoma B16F10 cells. *Journal of Agricultural & Food Chemistry*, 60(32): 7866–7872.

Elaissi, A., Z. Rouis, N.A.B. Salem, S. Mabrouk, Y. ben Salem, K.B.H. Salah, M. Aouni, F. Farhat, R. Chemli & F. Harzallah-Skhiri (2012). Chemical composition of 8 eucalyptus species' essential oils & the evaluation of their antibacterial, antifungal & antiviral activities. *BMC Complementary & Alternative Medicine*, 12(1): 1–15.

Galil, J. (1984). Ficus religiosa L.-the tree-splitter. *Botanical Journal of the Linnean Society*, 88(3): 185–203.

Ghalem, B.R. & Mohamed B. (2008). Antibacterial activity of leaf essential oils of Eucalyptus globulus & Eucalyptus camaldulensis. *African Journal of Pharmacy & Pharmacology*, 2(10): 211–215.

Ghimire, K. & Bastakoti R.R. (2009). Ethnomedicinal knowledge & healthcare practices among the Tharus of Nawalparasi district in central Nepal. *Forest Ecology & Management*, 257(10): 2066–2072.

Gilles, M., Zhao, J., An, M. & Agboola, S. (2010). Chemical composition & antimicrobial properties of essential oils of three Australian Eucalyptus species. *Food Chemistry*, 119(2): 731–737.

Gomide, J.L. (2006). Eucalyptus wood characteristics Brazilian pulping industry. In TAPPI Pulping Conference, Atlanta, GA.

Jun, Y.S., Kang, P., Min, S.S., Lee, J.-M., Kim H.-K. & Seol G.H. (2013). Effect of eucalyptus oil inhalation on pain & inflammatory responses after total knee replacement:A randomized clinical trial. *Evidence-Based Complementary and Alternative Medicine*, 2013.

Kantvilas, G. (1996). The discovery of Tasmanian eucalypts: An historical sketch. *Tasforests*, 8: 1–14.

Kesharwani, V., Gupta, S., Kushwaha, N., Kesharwani, R. & Patel, D.K. (2018). A review on therapeutics application of eucalyptus oil. *International Journal of Herbal Medicine*, 6(6): 110–115.

Khare, C.P. (2008). *Indian Medicinal Plants: An Illustrated Dictionary.* Berlin/Heidelberg, New York: Springer–Verlag, 269–270.

Kovar, K., Gropper, B., Friess, D. & Ammon, H. (1987). Blood levels of 1, 8-Cineole & locomotor activity of mice after inhalation & oral administration of rosemary Oil1. *Planta Medica*, 53(4): 315–318.

Lu, X., Tang, F., Wang, Y., Zhao, T., & Bian, R. (2004). Effect of Eucalyptus globulus oil on lipopolysaccharide-induced chronic bronchitis and mucin hypersecretion in rats. *China Journal of Chinese Materia Medica*, 29(2): 168–171.

Mali, S. & Borges, R.M. (2003). Phenolics, fibre, alkaloids, saponins, & cyanogenic glycosides in a seasonal cloud forest in India. *Biochemical Systematics & Ecology*, 31(11): 1221–1246.

Mishra, A.K., Sahu, N., Mishra, A., Ghosh, A.K., Jha, S. & Chattopadhyay, P. (2010). Phytochemical screening & antioxidant activity of essential oil of Eucalyptus leaf. *Pharmacognosy Journal*, 2(16): 25–28.

Neiva, D., L. Fernandes, S. Araújo, A. Lourenço, J. Gominho, Simões, R. & Pereira, H. (2015). Chemical composition & kraft pulping potential of 12 eucalypt species. *Industrial Crops & Products*, 66: 89–95.

Nikbakht, M., Rahimi-Nasrabadi, M., Ahmadi, F., Gomi, H., Abbaszadeh, S. & Batooli, H. (2015). The chemical composition & in vitro antifungal activities of essential oils of five Eucalyptus species. *Journal of Essential Oil Bearing Plants*, 18(3): 666–677.

Oke, R.A., Adetola, O.O., Owoeye, Y.T., Akemien, N.N., Adaaja, B.O., Bakpolor, V.R. &. Murtala, M.O (2021). Socio-economic and medicinal importance of eucalyptus trees: A critical review. *Global Prosperity*, 1(2): 17–22.

Paine, T.D., Steinbauer, M.J. & Lawson, S.A. (2011). Native & exotic pests of Eucalyptus: A worldwide perspective. *Annual Review of Entomology*, 56: 181–201.

Parate, S., Misar, K. & Chavan, D. (2015)Formulation, development and evaluation of foot cream with Ficus religiosa. *International Journal of Researches in Biosciences, Agriculture and Technology*, 2015: 232–294.

Prasad, P., Subhaktha, P., Narayana, A. & Rao, M.M. (2006). Medico-historical study of "aśvattha"(sacred fig tree). *Bulletin of the Indian Institute of History of Medicine*, 36(1): 1–20.

Rathee, D., Rathee, S., Rathee, P., Deep, A., Anandjiwala, S & Rathee, D. (2015). HPTLC densitometric quantification of stigmasterol and lupeol from Ficus religiosa. *Arabian Journal of Chemistry*, 8(3): 366–371.

Ravindran, D., Bharathi SR, S. & Priyadharshini G, S. (2020). Characterization of surface-modified natural cellulosic fiber extracted from the root of Ficus religiosa tree. *International Journal of Biological Macromolecules*, 156: 997–1006.

Roy, K., Shivakumar, H. & Sarkar, S. (2009). Wound healing potential of leaf extracts of Ficus religiosa on Wistar albino strain rats. *International Journal of PharmTech Research*, 1: 506–508.

Sharma, H., Yunus, G., Mohapatra, A.K., Kulshrestha, R., Agrawal, R. & Kalra, M. (2016). Antimicrobial efficacy of three medicinal plants Glycyrrhiza glabra, Ficus religiosa, & Plantago major on inhibiting primary plaque colonizers & periodontal pathogens: An in vitro study. *Indian Journal of Dental Research*, 27(2): 200.

Shukla, S., Rai, P.K., Chatterji, S., Rai, N.K., Rai, A. & Watal, G. (2012). LIBS based screening of glycemic elements of Ficus religiosa. *Food Biophysics*, 7(1): 43–49.

Sillett, S.C., Van Pelt, R., Kramer, R.D., Carroll, A.L. & Koch, G.W. (2015). Biomass & growth potential of Eucalyptus regnans up to 100 m tall. *Forest Ecology & Management*, 348: 78–91.

Silva, J., Abebe, W., Sousa, S., Duarte, V., Machado, M. & Matos, F. (2003). Analgesic & anti-inflammatory effects of essential oils of Eucalyptus. *Journal of Ethnopharmacology*, 89(2–3): 277–283.

Singh, D., Singh, B. & Goel, R.K. (2011). Hydroethanolic leaf extract of Ficus religiosa lacks anticonvulsant activity in acute electro & chemo convulsion mice models. *Journal of Pharmaceutical Negative Results*, 2(2).

Singh, R. & Dwivedi, R. (1987). Effect of oils on Sclerotium rolfsii causing foot-rot of barley. *Indian Phytopathology*, 40(2): 531–533.

Singh, S. & Jaiswal, S. (2014). Therapeutic properties of Ficus religiosa. *International Journal of Engineering Research and General Science*, 2(5): 149–158.

Singh, T.G., Sharma, R., Kaur, A., Dhiman, S. & Singh, R. (2020). Evaluation of renoprotective potential of Ficus religiosa in attenuation of diabetic nephropathy in rats. *Obesity Medicine*, 19: 100268.

Suriyakalaa, U., Ramachandran, R., Doulathunnisa, J.A., Aseervatham, S.B., Sankarganesh, D., Kamalakkannan, S., Kadalmani, B., Angayarkanni, J., Akbarsha, M.A. & Achiraman, S. (2021). Upregulation of Cyp19a1 & PPAR-γ in ovarian steroidogenic pathway by Ficus religiosa: A potential cure for polycystic ovary syndrome. *Journal of Ethnopharmacology*, 267: 113540.

Swami, K. & Bisht, N. (1996). Constituents of Ficus religiosa & Ficus infectoria and their biological activity. *Journal of the Indian Chemical Society*, 73(11): 631–636.

Thomas, J., Joy, P., Mathew, S., Skaria, B., Duethi, P. & Joseph, T. (2000). *Agronomic Practices for Aromatic & Medicinal Plants*. Calicut: Directorate of Arecanut & Spices Development, 124p.

Tiwari, P., Ansari, V.A., Mahmood, T. & Ahsan, F. (2019). A review on taxonomical classification, phytochemical constituents & therapeutic potential of ficus religiosa (Peepal). *Research Journal of Pharmacy & Technology*, 12(11): 5614–5620.

Venkateshwarlu, G., Eslavath, R., Santhosh, A., Suma, G., Ramakka, D. & Sankirthi, S. (2014). Ficus religiosa-an important medicinal plant: A review of its folklore medicine & traditional uses. *Asian Journal of Research in Pharmaceutical Science*, 4(1): 26–27.

Yadav, Y.C. & Srivastava D. (2013). Nephroprotective & curative effects of Ficus religiosa latex extract against cisplatin-induced acute renal failure. *Pharmaceutical Biology*, 51(11): 1480–1485.

22 *Garcinia indica* (Kokum) and *Ilex aquifolium* (European Holly)

Dicson Sheeja Malar, Mani Iyer Prasanth, Tewin Tencomnao, James Michael Brimson, and Anchalee Prasansuklab

CONTENTS

22.1 Introduction 427
22.2 Botanical Description of the Plants 428
22.3 Traditional Knowledge of the Plants 428
 22.3.1 *Garcinia indica* 428
 22.3.2 *Ilex aquifolium* 429
22.4 Phytochemistry of the Plants 429
 22.4.1 *Garcinia indica* 429
 22.4.2 *I. aquifolium* 429
22.5 Pharmacological Properties of *G. indica* and Its Major Compound Garcinol 430
 22.5.1 Anticancer Activity 430
 22.5.2 Neuroprotective Activity 432
 22.5.3 Antidiabetic and Anti-Obesity Activity 433
 22.5.4 Hepatoprotective Activity 433
 22.5.5 Cardioprotective activity 433
 22.5.6 Anti-Inflammatory Activity 434
 22.5.7 Antimicrobial and Antiviral Activity 434
22.6 Medicinal Properties of *I. aquifolium* and Its Major Compound Ursolic Acid 434
 22.6.1 Anticancer Activity 435
 22.6.2 Neuroprotective Activity 436
 22.6.3 Anti-Inflammatory Activity 436
 22.6.4 Hepatoprotective Activity 437
 22.6.5 Antimicrobial, Antiviral Activity 437
22.7 Conclusion 437
Acknowledgment 438
References 438

22.1 INTRODUCTION

The use of medicinal plants by humans for disease treatment is instinctive from ancient times. The earliest record of the use of medicinal plants dates back to 5000 years from Sumerian and Mesopotamian inscriptions. Ever since several documentations about the plant-based medicines and herbal formulations have been found and reported in Egyptian *Ebers papyrus*, Chinese *Wu Shi Er Bing Fang*, *Shennong Herbal*, *Tang Herbal*, and Indian *Charaka* and *Sushruta Samhitas* (Cragg and Newman, 2005; Spainhour, 2005; Cragg and Newman, 2013). The beginning of the 19th century paved way for the modern medicines with the isolation and identification of morphine. With

DOI: 10.1201/9781003205067-22

time and advancement in technologies, several pharmacologically active compounds were isolated and their application in modern medicine emerged (Yuan et al., 2016; Jones, 2011). Although traditional herbal medicinal practices were replaced by modern synthetic drugs, there is a resurgence in herbal medicine in the recent past due to their easy accessibility, affordability, safety, and efficacy. While the practice of herbal medicine is often undervalued, more than 50% of the pharmaceuticals prescribed in modern medicine were identified from plants (Brahmachari, 2012). The current chapter provides information on the traditional uses, pharmacological properties of the medicinal plants *Garcinia indica, Ilex aquifolium*, and their bioactive compounds.

22.2 BOTANICAL DESCRIPTION OF THE PLANTS

Garcinia indica (Indian mangosteen; Kokum), a native plant of India, is a slow-growing tree with horizontal, drooping branches and grows up to 15 m height, and it belongs to the Guttiferae family. The leaves or oval or elliptical with a length of 5–10 cm and dark green in color. The hermaphroditic or gynodioecious flowers are fleshy and dark pink, which may be solitary or clusters containing 4 sepals and petals. The fruits are round or oval, which occurs as green when raw and turns red or purple when fully ripe (Figure 22.1A). The fruits with whitish pulp are sour to taste containing 3–8 seeds, which accounts for one-quarter of the fruit weight (Janick and Paull, 2008; Nayak et al., 2010a; Baliga et al., 2011; Khare, 2011; Ananthakrishnan and Rameshkumar, 2016).

Ilex aquifolium (Holly) is an evergreen plant belonging to Aquifoliaceae family and is a native of Europe and North America. The dense pyramidal tree grows to a height of 10–12 m with coriaceous and glabrous spiny green leaves of 10 cm long. The plant is normally dioecious with white hypogynous flowers of 6 mm diameter. The fruits are of small size (7–12 mm) and sub-spherical in shape containing yellowish pulp with four seeds (Figure 22.1B) (Peterken and Lloyd, 1967; Obeso, 1998; Hue et al., 2016).

22.3 TRADITIONAL KNOWLEDGE OF THE PLANTS

22.3.1 *Garcinia indica*

The dried fruit rind of *G. indica* is used as coloring, flavoring agent in food preparations, and substitute in wine-making (Jayaprakasha and Sakariah, 2002; Baliga et al., 2011; Kaur et al., 2012). The fruits are also used in the preparation of juice and soup among Konkani people of India (Menezes, 2001). The seeds of the plant are used to obtain kokum butter, an ingredient used in chocolate and cosmetics. Kokum butter is also used as a substitute for ghee among the Indian population (Baliga et al., 2011). Traditionally, the plant is used in the treatment of diarrhea, inflammatory,

FIGURE 22.1 (A) *Garcinia indica* (B) *Ilex aquifolium*.

rheumatic pains, and dermatitis. Fruits are used as cardioprotectant, anti-diabetic, antihelmintic and to treat gastric problems, constipation, and piles. Kokum butter is also used as moisturizer and skin protectant. Leaves were used inflammation of skin, hemorrhoids, dyspepsia, and hyperplasia (Ananthakrishnan and Rameshkumar, 2016; Jeyarani and Reddy, 1999; Padhye et al., 2009).

22.3.2 *ILEX AQUIFOLIUM*

The extracts of *I. aquifolium* have been used in the treatment of inflammatory conditions including rheumatoid arthritis and bronchitis. In folk medicine, the leaves of the plant were used as anti-pyretic, astringent, and diuretic. Leaves and flowers were used in the treatment of cancer, jaundice, and malaria (Nahar et al. 2005; Palu et al. 2019; Müller et al., 1998).

22.4 PHYTOCHEMISTRY OF THE PLANTS

22.4.1 *GARCINIA INDICA*

Garcinol is the major phytocompound present in *G. indica*, which is a polyisoprenylated benzophenone derivative (Yamaguchi et al. 2000). The major organic acid in leaves and rinds has been found to be (–)-hydroxycitric acid present to the extent of 4.1–4.6 and 10.3–12.7%, along with lactone and citric acid in minor quantities (Jayaprakasha and Sakariah, 2002). A sensitive liquid chromatography/electrospray ionization tandem mass spectrometrical (LC/ESI–MS/MS) method and a high-performance liquid chromatography method identified various polyisoprenylated benzophenones including xanthochymol, isoxanthochymol, and camboginol in the extracts of the fruit rinds, stem bark, seed pericarps, and leaves of *G. indica* (Chattopadhyay and Kumar, 2006; Chattopadhyay and Kumar, 2007; Kumar and Chattopadhyay, 2007; Kumar et al., 2009). Euxanthone (1,7-dihydroxy xanthone), volkensiflavone and morelloflavone have been reported from the bark of the plant (Cotterill and Scheinmann, 1977). In addition, Kaur et al., (2012) has identified 14-deoxyisogarcinol and acylphloroglucinol from the fruits of *G. indica* (Kaur et al., 2012). The anthocyanin pigments cyanidin-3-glucoside and cyanidin-3-sambubioside were reported in the fruit rind of the plant (Nayak et al., 2010b). Kokum butter is rich in fatty acids including stearic acid, oleic acids, and glycerides oleodistearin and stearodiolein (Krishnamurthy et al., 1982; Lipp and Adam, 1998). Oil obtained from seed is rich in palmitic acid, and linoleic acid (Hosamani et al., 2009). Some of the major phytoconstituents of the plant have been illustrated in Figure 22.2.

22.4.2 *I. AQUIFOLIUM*

The major classification of compounds identified in *I. aquifolium* were triterpenes including, ursolic acid, α-amyrin, β-amyrin, uvaol, erythrodiol, oleanolic acid, oleanoaldehyde, ursolaldehyde, baurenol, 27-p-Coumaroxyursolic acid, and Ilex lactone from the leaves and fruits of the plant (Palu et al., 2019; Thomas and Budzikiewicz, 1980; Van Genderen and Jaarsma, 1990; Catalano et al., 1978). From the seeds, 2,4-dihydroxyphenylacetic acid and 2,4-dihydroxyphenylacetic acid methyl ester, the phenylacetic acid derivatives were isolated (Nahar et al., 2005). The phenolic acids, vanillic acid, p-hydroxybenzoic acid, chlorogenic acid, and caffeic acid were isolated from the fruits (Thomas and Budzikiewicz, 1980; Ishikura, 1975; Bate-Smith, 1962). The antioxidant rich flavonoids quercetin, kaempferol, and rutin were isolated from the fruit of *I. aquifolium* (Ishikura, 1975; Schindler and Herb, 1955). The cyanogenic glucoside, 2 beta-D-glucopyranosyloxy-p-hydroxy-6,7-dihydromandelonitrile (Willems, 1989), and the anthocyanins, including pelargonidin 3-bioside (Robinson and Robinson, 1931) cyanidin-3-xylosylglucoside (Ishikura, 1971), pelargonidin-3-glucoside (Santamour, 1973) were isolated from fruit. Bohinc (1967) has reported the presence of the alkaloid theobromine from the leaves of the plant. The flowers are reported to be rich in several sterols including sitosterol, stigmasterol, 24-ethylcholesterol, campesterol, 24-ethylidenecholesterol

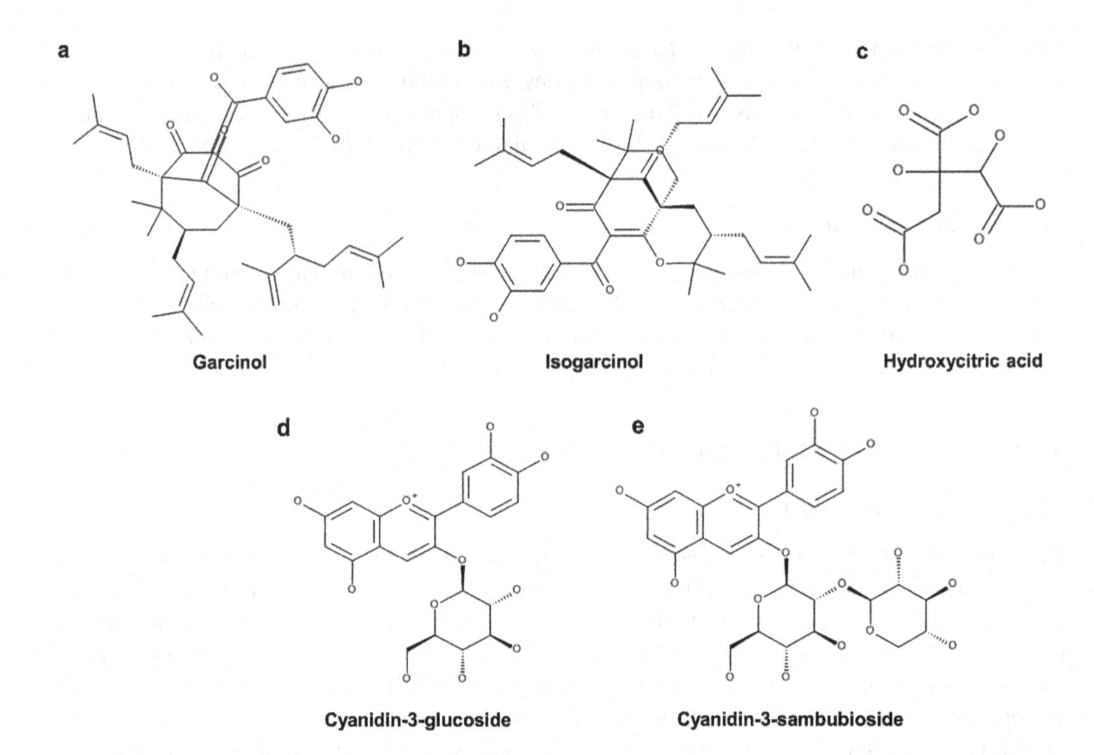

FIGURE 22.2 The major phytoconstituents of *G. indica* (a) garcinol (b) isogarcinol (c) hydroxycitric acid (d) cyanidin-3-glucoside (e) cyanidin-3-sambubioside.

(Knights and Smith, 1977). Apart from the major polyphenols, various fatty acids such as linolenic acid, linoleic acid, behenic acid, stearic acid, lignoceric acid, oleic acid, and gadoleic acid were reported in the leaves of *I. aquifolium* (Catalano et al., 1978; Crombie, 1958). The major active compounds of *I. aquifolium* are represented in Figure 22.3.

22.5 PHARMACOLOGICAL PROPERTIES OF *G. INDICA* AND ITS MAJOR COMPOUND GARCINOL

22.5.1 Anticancer Activity

Methanol extract of *G. indica* showed significant cytotoxic activity against colon cancer, which was attributed to the presence of polyisoprenylated benzophenones (Kumar and Chattopadhyay, 2007). Further, the major polyisoprenylated benzophenone derivative of *G. indica*, garcinol exerted anticancer activity against several cancers by modulating and interfering with various cellular pathways. Garcinol administration in experimental animals significantly lowered proliferating cell nuclear antigen index, colonic aberrant crypt foci, while elevating the activities of detoxifying enzymes of glutathione S-transferase, quinone reductase in the liver, and modulating ERK, PI3K, and Wnt signaling pathways thereby exhibiting chemopreventive potential (Tanaka et al., 2000; Tsai et al., 2014). Garcinol inhibited cell invasion, decreased tyrosine phosphorylation, and inhibited activation of the Src, MAPK/ERK, and PI3K/Akt signaling pathways in human colorectal cancer cell line, HT-29. It also modified the level of Bcl-2, Bax, and caspase-3 along with decreasing MMP-7 aiding in anticancer effect (Liao et al., 2005b; Hong et al., 2007).

Garcinol acts as a potent inhibitor of histone acetyltransferases, as it inhibited chromatin transcription and induced apoptosis in HeLa cells (Balasubramanyam et al., 2004). In addition, garcinol mediated downregulation of NF-kB signaling pathway aided in caspase-mediated apoptosis

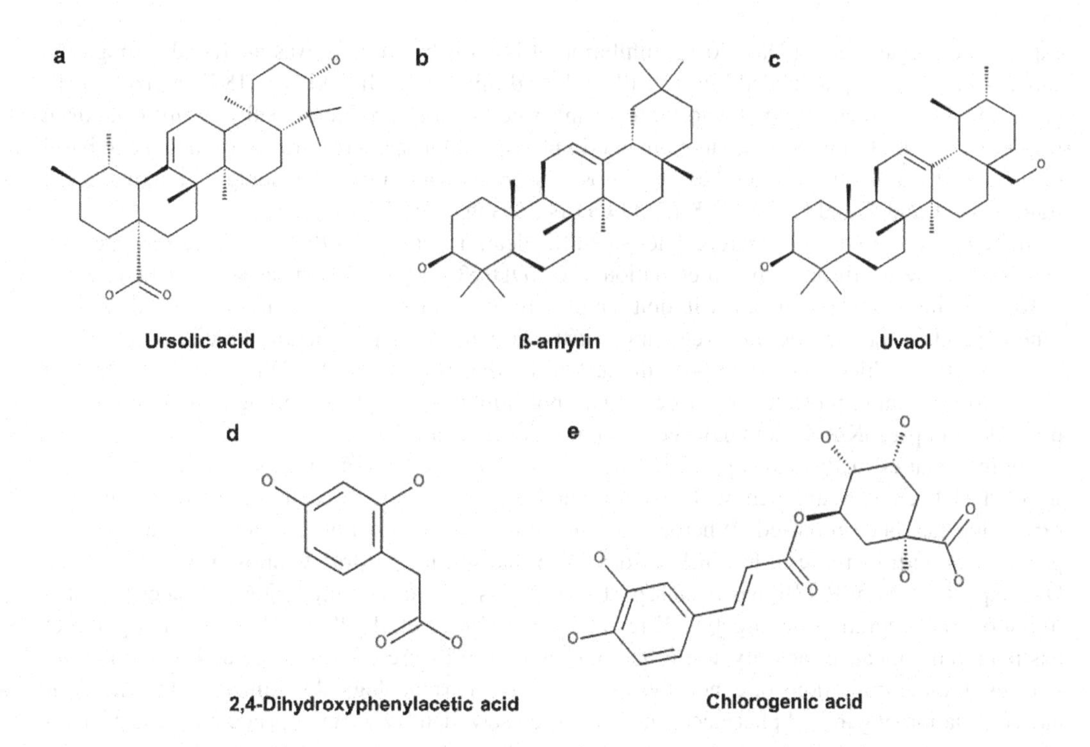

FIGURE 22.3 The major phytoconstituents of *I. aquifolium* (a) ursolic acid (b) β-amyrin (c) uvaol (d) 2,4-dihydroxyphenylacetic acid (e) chlorogenic acid.

indicating its chemoprotective potential (Ahmad et al., 2010). Garcinol was reported to target signal transducer and activator of transcription-3 (STAT-3) signaling pathway both under *in vitro* and *in vivo* conditions to inhibit the tumor growth and invasion of breast cancer (Ahmad et al., 2012a). Further, it induced mesenchymal-to-epithelial transition (MET) in aggressive triple-negative MDA-MB-231 and BT-549 breast cancer cells, mediated by upregulation of epithelial marker E-cadherin and downregulation of mesenchymal markers vimentin, ZEB-1, and ZEB-2. This was mechanistically linked with the deregulation of miR-200s, let-7s, NF-kB, and Wnt signaling pathways, as confirmed in mice models as well (Ahmad et al., 2012b).

Garcinol exhibited morphological changes and inhibited the proliferation of human non-small cell lung carcinoma (NSCLC) cells. Garcinol induced G1 cell cycle arrest was mediated through the upregulation of CDK inhibitors p21Waf1/Cip1 and p27KIP1. Moreover, cyclin-dependent kinase 2 (CDK2), cyclin-dependent kinase 4 (CDK4), cyclin D1, and cyclin D3 were decreased, whereas cyclin E and cyclin-dependent kinase 6 (CDK6) were increased along with inhibition of ERK and p38-MAPK (Yu et al., 2014). In A549 cells, garcinol enriched DNA damage-inducible transcript 3 (DDIT3), altered DDIT3-CCAAT-enhancer-binding proteins beta (C/EBPβ) interaction resulting in the attenuation of the prognostic cancer cell marker Aldehyde Dehydrogenase 1 Family Member A1 (ALDH1A1) expression (Wang et al., 2017a). In addition, garcinol significantly diminished the ability of the NSCLC cells to form spheres and form colonies, by impairing phosphorylation of LRP6, a co-receptor of Wnt and STAT3, downregulating β-catenin, Dvl2, Axin2, and cyclin D1 expressions, suggesting its ability to regulate the Wnt/β-catenin signaling pathway (Huang et al., 2018). Garcinol induced the sensitivity of A549 cells toward TRAIL and induced apoptosis mediated through upregulation of DR5 and downregulation c-FLIP (Kim et al., 2018).

Garcinol treatment significantly inhibited the growth and proliferation and colony formation of oral squamous cell carcinoma cells with a concomitant induction of apoptosis and cell cycle arrest. It exerts anti-proliferative, pro-apoptotic, cell-cycle regulatory, and anti-angiogenic effects by reducing the expression of STAT-3, c-Src, JAK1, and JAK2, NK-κB, and COX-2 besides inhibiting VEGF expression (Aggarwal and Das, 2016). Inhibition of NK-κB by garcinol was mediated through the suppression of TGF-β activated kinase 1 (TAK1) and inhibitor of IkB kinase (IKK) activation (Li et al., 2013a). Further, garcinol also targets cancer cell energy producing pathway mitochondrial respiration by inhibiting ATP production, maximal respiration, spare respiration capacity, and basal respiration. Garcinol treatment reflexively boosted glycolysis apart from the upregulation of glucose transporter 1 and 4, and HIF-1α, AKT, and PTEN (Zhang et al., 2019a).

In hepatocarcinoma cells, garcinol activated the death receptor and ROS and ER stress induced apoptotic pathways through the upregulation of GADD153, caspase-3, and caspase-9 (Cheng et al., 2010). Garcinol could significantly inhibit the growth of gallbladder carcinoma cells in a dose- and time-dependent manner, by downregulation of the activity of matrix metalloproteinase 2 (MMP-2) and MMP-9, which could have been modulated by the attenuation of STAT-3 and Akt (Duan et al., 2018). In human prostate cancer cells, garcinol inhibited autophagy through the activation of p-mTOR and p-PI3K/AKT and induced apoptosis (Wang et al., 2015).

Bioinformatics analysis has predicted the role of STAT3, STAT5 in glioblastoma conditions, which has been substantiated with experimental studies, where several fold-increase in their expression has been reported. Whereas, garcinol treatment showed attenuation of stem cell-like phenotypes with increase in hsa-miR-181d/STAT3 and hsa-miR-181d/5A ratios (Liu et al., 2019). Overexpression of miR-181d has been reported to be associated with inhibition of cancer cell proliferation and migration by targeting K-ras and Bcl-2 (Wang et al., 2012b). Even though garcinol has potential anticancer activity, use of garcinol is still in its pre-clinical stage and this is mainly attributed to the limitations of conclusive evaluation of pharmacological parameters. This necessitates evaluation of garcinol pharmacokinetics to precisely identify an appropriate dose and route of administration, tolerability, and potency under physiological conditions along with characterization of a therapeutic index (Aggarwal et al., 2020).

22.5.2 Neuroprotective Activity

Neurodegenerative diseases including Alzheimer's disease, Parkinson's disease, and Huntington disease involve the abnormal generation of ROS, accumulation of respective proteins impeding with normal cellular process and resulting in neuronal degeneration. *G. indica* and its metabolites due to their high antioxidant potential exert significant neuroprotective efficacy. For instance, *G. indica* exhibited preventive effect against 6-hydroxydopamine-induced parkinsonism in Wistar rats by improving the biochemical and behavioral changes (Antala et al. 2012). Likewise, *G. indica* fruit rind extract administration in Swiss albino mice showed anxiolytic and anti-depressant effect by modulating monoaminergic pathway (Dhamija et al., 2017).

Apart from the extract, the active compound garcinol showed neuroprotective effect by inducing neurite outgrowth and enhancing the expression of neuronal proteins, microtubule-associated protein 2 (MAP-2), and glial fibrillary acidic protein (GFAP). Furthermore, the neuronal marker, high-molecular-weight subunit of neurofilaments (NFH), was highly expressed after garcinol treatment which was mediated by extracellular signal-regulated kinase (ERK) pathway (Liao et al., 2005a; Weng et al., 2011). Infusion of garcinol into lateral amygdala (LA) alleviates fear and anxiety conditions in post-traumatic stress disorder in Sprague-Dawley rats (Maddox et al., 2013). In pentylenetetrazole-induced epileptic mice model, garcinol-reduced seizure scores, enhanced cognition and memory by acting through BDNF/TrkB pathway and upregulating $GABA_A$, GAD65 expression (Hao et al., 2016).

In an ischemia-reperfusion rat model, garcinol decreased cerebral I/R-induced inflammatory cytokines (IL-1β, IL-6, tumor necrosis factor-α (TNF-α), oxidative stress through the suppression of toll-like receptor (TLR) 4 and nuclear NF-κB (p65) expression (Kang et al., 2020). Garcinol may be used to sustain abstinence from drug abuse, especially cocaine. Systemic injection of garcinol significantly impaired the reconsolidation of the cocaine-associated cue memory. Further testing revealed that garcinol had no effect on drug-induced cocaine-seeking, but was capable of blocking the initial conditioned reinforcing properties of the cue and prevents the acquisition of a new response. Additionally, the effects of garcinol are specific to reactivated memories only, temporally constrained, cue-specific, long-lasting, and persist following extended cocaine access (Monsey et al., 2017).

22.5.3 Antidiabetic and Anti-Obesity Activity

Traditionally, fruits of *G. indica* was reported to exhibit antidiabetic and anti-obesity effect. The folklore knowledge was further supported by several experimental analysis. Aqueous extract of *G. indica* fruits was observed to significantly decrease both the fasting and postprandial blood glucose in type 2 diabetic rats, along with restoring the erythrocyte GSH (Kirana and Srinivasan, 2010). Polyherbal formulation containing *G. indica* in streptozotocin-induced diabetic rats, exhibited antidiabetic and antihyperlipidemic properties. Blood glucose level showed significant reduction after 28 days of treatment with formulation at 200 and 400 mg/kg, apart from significant decrease in the activities of gluconeogenic enzymes thereby reversing most of the blood and tissue changes toward the normal level (Subhasree et al., 2015). *G. indica* extract alone or in combination with *Coleus forskohlii* was able to decrease body weight and adipocyte size by promoting fatty acid β-oxidation and modulating gut microbiota in high fat diet induced obese mice thereby inducing anti-obesity effects (Tung et al., 2021).

Garcinol could reduce high fat diet fed mice body weight gain and relative visceral adipose tissue fat weight, the plasma levels of glutamate pyruvate transaminase, total cholesterol, and triacylglycerol. The 16S rRNA gene sequence data indicated that garcinol reversed high fat induced gut dysbiosis by modulating the Firmicutes-to-Bacteroidetes ratios and controlled inflammation by increasing the intestinal commensal bacteria, *Akkermansia*, indicating its role as a novel gut microbiota modulator to induce anti-obesity effects (Lee et al., 2019).

22.5.4 Hepatoprotective Activity

G. indica was able to incorporate hepatoprotection in rats treated with ethanol by modulating the antioxidant enzyme status, attenuating the level of aspartate aminotransferase (AST), alanine aminotransferase (ALT), and alkaline phosphatase (ALP). Similarly, treatment with garcinol significantly decreased serum ALT and AST levels in lipopolysaccharide (LPS)-induced hepatic injury in D-galactosamine (D-Gal)-sensitized mice and dimethylnitrosamine (DMN)-induced liver fibrosis in rats. Further, garcinol significantly reduced the acetylation level of NF-κB, without altering the levels of TNF-α or IL-6 in plasma or liver tissue and attenuated LPS/D-Gal-induced hepatic apoptosis thereby protecting liver from injury (Jing et al., 2014; Hung et al., 2014). Also, garcinol in combination with curcuminoids were able to prevent the progression of liver steatosis to inflammation and fibrosis against non-alcoholic steatohepatitis in mice by decreasing the mRNA levels of MCP1 and CRP and both mRNA and protein levels of TNF-α, NF-kB (Majeed et al., 2020).

22.5.5 Cardioprotective activity

Polyherbal formulation SJT-HT-03, containing *G. indica*, exhibited antihypertensive activity by producing direct depressant effect on heart, inhibition of angiotensin converting enzyme, aldosterone antagonistic as well as diuretic effect and thereby act on multiple targets to achieve optimal effect (Ghelani et al., 2014). Oral administration of garcinol enriched fraction in high fat diet fed mice reduces hyperlipideamia by reducing the level of low-density cholesterol, triglycerides and total cholesterol, apart from reducing inflammatory response and oxidative stress, together reducing the risk of arteriosclerosis (Barve, 2019).

Garcinol treatment increased the heart rate and improved the maximum rate of increase in left-ventricle pressure, maximum rate of decrease in pressure, left ventricle ejection fraction, and systolic pressure in rats with induced heart failure. Garcinol treatment reversed body, liver, and heart weight changes, resulting in returns to near-normal levels. Additionally, the number of broken fibers, extent of inflammatory cell infiltration, and rate of apoptosis remained within normal ranges. Garcinol reduced the cross-sectional areas of cardiomyocytes, and reduced interstitial fibrosis to a normal level. The mRNA and protein levels of cleaved caspase-3, caspase-3, and Bax were reduced, whereas those of Bcl-2 were increased, following garcinol treatment (Li et al., 2020b).

22.5.6 Anti-Inflammatory Activity

Garcinol enriched fraction shows anti-arthritic activity in experimental animals, as it reduced paw swelling and arthritis index thereby exhibiting anti-inflammatory potential. It also improved the motility and stair climbing ability of experimental animals thus reducing hyperalgesia (Warriar et al., 2019). Garcinol can inhibit the receptor activator of NF-κB ligand (RANKL)-induced osteoclastogenesis, osteoclastogenesis-related gene expression, the f-actin ring, and resorption pit formation. In addition, garcinol abrogated RANKL-induced osteoclastogenesis by attenuating the degradation of the MAPK, NF-κB, and PI3K-AKT signaling pathway as well as downstream factors c-jun, c-fos, and NFATC1. *In vivo*, suppression of osteoclastogenesis by garcinol was evidenced by marked inhibition of lipopolysaccharide-induced bone resorption. Garcinol could potentially inhibit the RANKL-induced osteoclastogenesis by suppressing the MAPK, NF-κB, and PI3K-AKT signaling pathways and thus has potential as a novel therapeutic option for osteolytic bone diseases (Jia et al., 2019).

Garcinol inhibited the expression of inducible nitric oxide synthase (iNOS) and cyclooxygenase-2 (COX-2) in lipopolysaccharide (LPS)-activated macrophages *in vitro* by lowering the level of ROS thereby aiding the activation of eukaryotic transcription factor nuclear factor-kappa B (NF-kB) mediated by p38 mitogen-activated kinase (MAPK) (Liao et al., 2004). The intrathecal administration of garcinol in neuropathic pain rat model suppressed microglial activation as well as the expression of interleukin (IL)-1β, IL-6, inducible nitric oxide synthase (iNOS)/nitric oxide (NO), and cyclooxygenase-2 (COX-2)/prostaglandin E2 (PGE2) in the spinal cord of rats. It also reduced the nuclear translocation of NF-κB by decreasing acetyl-p65 protein expression. Similar effects were also observed in LPS-challenged microglia indicating the protective effect against neuropathic pain that is associated with the inhibition of neuroinflammation in microglia (Wang et al., 2017b).

22.5.7 Antimicrobial and Antiviral Activity

G. indica mediated biogenic silver nanoparticles alone and in combination with Tetracyclin showed antibacterial activity, along with good antioxidant activities (Sangaonkar and Pawar, 2018). The polyphenols isolated from the ethyl acetate soluble of methanol extract of stem bark of *G. indica* showed antibacterial activity against *Staphylococcus aureus* and *Salmonella typhi*, which was confirmed by paper disc method (Lakshmi et al., 2011). Docking of garcinol into the active site of the HIV-1-RNase H enzyme showed higher inhibition than the known inhibitor RDS1759 and retained full potency against the RNase H of a drug-resistant HIV-1 reverse transcriptase form. Docking calculations confirmed these findings and suggested this moiety to be involved in the chelation of metal ions of the active site. On the basis of its HIV-1 reverse transcriptase-associated RNase H inhibitory activity, garcinol is worth being further explored concerning its potential as a cost-effective treatment for HIV patients (Corona et al., 2021).

22.6 MEDICINAL PROPERTIES OF *I. AQUIFOLIUM* AND ITS MAJOR COMPOUND URSOLIC ACID

The evaluation of pharmacological properties of *I. aquifolium* is very limited. However, one of the major compounds ursolic acid (UA) has been extensively studied for its varied activities.

22.6.1 Anticancer Activity

The ethyl acetate extract of *I. aquifolium* leaves exhibits anticancer activity against colon carcinoma cells and glioblastoma cells (Frédérich et al., 2009). UA, the bioactive component of *I. aquifolium* exhibits potential anticancer effects. It inhibits angiogenesis, invasion, and migration of NSCLC cells by attenuating programmed death ligand-1 (PD-L1) expression, that mediates metastasis through the EGFR/JAK2/STAT3 pathway (Kang et al., 2021). Invasion of lung cancer cells was inhibited by the diminution of epithelial–mesenchymal transition by downregulating the expression of astrocyte-elevated gene-1 (AEG-1) by targeting NFκB pathway (Liu et al., 2013). Further, UA attenuated the expression of DNMT1 and EZH2, which are involved in tumor progression through SAP/JNK pathway (Wu et al., 2015). In NSCLC cells with EGFR T790M mutations, UA suppressed the expression of the proto-oncogene CT45A2 gene, which is involved in tumorigenesis through β-catenin/TCF4 signaling pathway (Yang et al., 2019). Moreover, UA increased the sensitivity of chemoresistant NSCLC cells by inhibiting miR-149-5p/MyD88 signaling (Chen et al., 2020a). UA specifically binds and inhibits activity of vaccinia-related kinase 1 (VRK1) kinase domain and induces DNA damage in lung cancer cells (Kim et al., 2015).

Connectivity map approach indicates that UA exhibit inhibitory effect against breast cancer cells by modulating some of the key pathways including PLK1, IKK/NF-κB, and RAF/ERK pathways (Guo et al., 2020). In addition, downregulation of Nrf-2, PI3K/AKT, JNK pathway were also reported (Zhang et al., 2020; Luo et al., 2017; Yeh et al., 2010). UA sensitized the triple negative breast cancer cells to doxorubicin and enhances apoptosis by targeting ZEB1-AS1/miR-186-5p/ABCC1 axis (Lu et al., 2021). ATP-binding cassette transporter C (ABCC1) are transporters that mediates the efflux of drugs from cancer cells, preventing their accumulation and causing chemoresistance. UA inhibits the growth of breast cancer stem-like cells by attenuating sFRP4 mediated Wnt/β-catenin pathway and suppressing miRNA-499a-5p involved in invasion and metastasis (Mandal et al., 2021).

Network pharmacological approach suggests that UA acts against colon cancer cells by targeting 113 proteins including MAPK3, VEGFA, TNF, and STAT-3 indicating its multitargeted potential (Zhao et al., 2021). Further UA is also reported to target MMP9/CDH1, Akt/ERK, COX-2/PGE2, p300/NF-κB/CREB2 simultaneously to inhibit cell proliferation (Wang et al., 2013). Studies also report that UA inhibits the metastasis and invasiveness of colon cancer cells, which could be due to the attenuation of epithelial-mesenchymal transition, TGF-β1/Smad/FAK signaling pathways, and upregulation of miR-200a/b/c (Zheng et al., 2021; Prasad et al., 2012; Zhang et al., 2019b; Wang et al., 2019a).

In prostate and gastric cancer cells, UA inhibits the mitochondrial translocation of cofilin-1 by targeting ROCK/PTEN signaling pathway leading to apoptosis (Gai et al. 2016; Mu et al., 2018; Li et al., 2014; Tang et al., 2014). UA shows apoptotic effect in prostate cancer cells through the inhibition of Wnt/β-catenin signaling (Park et al., 2013), PI3K/Akt/mTOR (Meng et al., 2015), while augmenting JNK pathway (Zhang et al., 2010a). Activation of JNK could induce the phosphorylation of Bcl-2 leading to its degradation, thereby reverting the resistance of cancer cells to apoptosis (Zhang et al., 2009; Zhang et al., 2010b). Invasion and metastasis of prostate cancer cells were found to be inhibited by UA through the modulation of pro-inflammatory pathways and inhibition of CXCR4/CXCL12 signaling axis in mouse models (Shanmugam et al., 2011; Shanmugam et al., 2012).

In a dose-dependent manner, UA inhibits the invasive property of renal carcinoma and gastric cancer cells through the activation of NLRP3 inflammasome (Chen et al., 2020b; Chen et al., 2020c). UA activates Hippo pathway (Mst1/2, WW45, LATS1/2, Yap, and Mobl) via Rassf1 to inhibit invasion, migration and induce apoptosis in gastric cancer cells (Kim et al., 2019b). In addition, through the downregulation of Axl/NF-κB pathway, UA inhibits the epithelial-mesenchymal transition in inhibiting the migration of gastric cancer cells (Li et al., 2019). UA induces the generation of ROS thereby downregulating the survival-dependent genes and metastasis genes to cause apoptosis in intestinal cancer cells (Rawat and Nayak, 2021).

UA acts as an activator of sterol regulatory element-binding protein 2 (SREBP2) to lower cholesterol in hepatocellular carcinoma cells and attenuates growth signaling pathways including ERK1/2, AKT, MEK to promote cell cycle arrest and apoptosis (Kim et al., 2019a). Further, UA blocks PI3/AKT, STAT-3 pathway to downregulate survivin leading to apoptosis (Tang et al., 2009; Liu et al., 2017). Moreover, dose-dependent AMPKα-mediated inhibition of DNMT1, whose overexpression results in hypermethylation and oncogenic activation is also observed upon treatment with UA (Yie et al., 2015). Exposure of UA to glioblastoma cells caused necrosis through the increased production of ROS, impairment of ATP level, and mitochondrial permeability transition pore opening (Lu et al., 2014). Further, proliferation of glioblastoma cells was inhibited by attenuating TGF-β1/miR-21/PDCD4 pathway and inducing JNK-dependent lysosomal associated cell death (Wang et al., 2012a; Conway et al., 2021). In addition, UA has been found to increase the chemosensitivity of cancer cells toward drugs in various resistant cancer cells (Lin et al., 2020; Li et al., 2020c; Chen et al., 2020a) indicating it can also be supplemented along with already existing anti-cancer drugs to improve their activity.

22.6.2 Neuroprotective Activity

UA can bind to human dopamine receptor and enhance the expression of *dop-1*, *dop-3* expression in knock-out *C. elegans* model to offer neuroprotection (Naß et al., 2021a). In addition, UA with its strong antioxidant property interacts with Skn-1 and peroxiredoxin-2, which are part of the detoxification mechanism and exerts anti-depressive effect in *Caenorhabditis elegans* (Naß et al., 2021b). Against Aβ, D-Galactose-induced AD mice model, UA improved memory, cognition, and induced neurogenesis besides augmenting the expression of neurogenesis markers, GAP43, and suppressing oxidative stress (Mirza et al., 2021; Liang et al., 2016; Lu et al., 2007). In rotenone, MPTP-induced Parkinsonism animal models, UA protected the tyrosine hydroxylase positive neurons from degeneration, restored motor functions by alleviating oxidative stress, inflammation, α-Synuclein expression and promoting mitochondrial biogenesis, phosphorylation of AKT, ERK (Peshattiwar et al., 2020; Zahra et al., 2020; Rai et al. 2016). During post traumatic brain injury and subarachnoid hemorrhage, UA alleviated cerebral edema, reduced blood brain barrier permeability, and inhibited neuronal apoptosis by attenuating oxidative stress through Nrf-2 dependent mechanism and activating AKT signaling pathway (Ding et al., 2017; Zhang et al., 2014). Upon ischemic stroke, UA improved neurological deficits, increased infarct volume, and increased the expression of PPARγ. Further, activation of Nrf-2 pathway, decreased inflammatory response, an increase in TIMP1 along with downregulation of MMP2, 9 and MAPK (p38, ERK1/2, JNK) pathway was also observed (Wang et al., 2016b; Li et al., 2013b). Under high fat diet-induced cognitive impairments, UA exhibited neuroprotection through the attenuation of ER stress and blocking of inflammatory signaling besides negatively regulating SOCS3, PTP1B, STAT-3, IRS-1 to enhance insulin, mTOR signaling in the mice hippocampus (Lu et al., 2011).

22.6.3 Anti-Inflammatory Activity

Ethanol extract of leaves of *I. aquifolium* inhibited leukotriene biosynthesis and lipid peroxidation in bovine Polymorphonuclear neutrophil leukocyte indicating the anti-inflammatory properties (Müller et al., 1998). UA attenuated allergic inflammation by inhibiting FcεRI-mediated degranulation of mast cells (Dhakal et al., 2021). Upon autoimmune thyroiditis stimulation in Nthy-ori 3-1 cells by inhibiting MALAT1/miR-206/PTGS1 network and NF-κB signaling pathway (Mou et al., 2021). Metastasis-associated lung adenocarcinoma transcript 1 (MALAT1) regulates inflammatory cytokine production and the aberrant expression of which is associated with inflammatory diseases (Li et al., 2020a; Masoumi et al., 2019). UA also protected chondrocytes from TNF-α injury by suppressing MMP13, IL-1β, IL-6, and PTGS2 along with the inhibition of NF-κB/NLRP3 inflammasome pathway (Wang et al., 2020). UA alleviates LPS-induced macrophage inflammation and atherogenesis in mice by suppressing cytokine production and enhancing autophagy (Leng et al., 2016). Similarly, UA exerted anti-inflammatory potential against *Mycobacterium tuberculosis* and concanavalin-A treatment in RAW264.7 cells through the inhibition of TNF-α, IL-1β, and IL-6 (Zerin et al., 2016).

22.6.4 Hepatoprotective Activity

UA has been reported to protect liver from ethanol-mediated damage in experimental animals. Pioneering studies indicate that UA could inhibit LPS/TLR4 signaling pathway, modulate gut-liver microbiome and induce NQO-1, GSTA2 upon alcohol toxicity in liver (Yan et al., 2021a, Yan et al., 2021b). Further hepatocellular apoptosis upon alcohol toxicity was reverted by UA by binding with CASP3 and irreversibly inhibiting CASP3 activity in mice (Ma et al., 2021).

22.6.5 Antimicrobial, Antiviral Activity

The leaf extracts of *I. aquifolium* were reported to exhibit antibacterial activity against *Staphylococcus aureus*, *Pseudomonas aeruginosa*, *Enterobacter aerogenes*, *Proteus vulgaris*, *Salmonella typhimurium*, and anti-fungal activity against *Candida albicans* (Erdemoglu et al., 2009; do Nascimento et al., 2014). *I. aquifolium* extract also showed inhibition against *Mycobacterium tuberculosis* which might be due to the potential inhibition of the survival protein, the heat shock protein16.3 by the active compound UA (Erdemoglu et al., 2009; Jee et al., 2018). UA inhibits ATP production by interfering with the regulatory enzymes of glycolysis pathway, peptidoglycan synthesis pathways and exhibits antimicrobial activity against *S. mutans* UA159 (Park et al., 2018; Park et al., 2015). In methicillin-resistant *Staphylococcus aureus* UA treatment induced membrane disruption, interfered with protein translation, metabolic pathways of glycolysis, and elicited oxidative stress to exhibit antibacterial effect (Wang et al., 2016a). UA showed antibiofilm activity by affecting the virulence factors of *L. monocytogenes* and inhibited listeriolysin O activity (Kurek et al., 2014). Further, UA inhibits HIV-1 protease activity to attenuate the retroviral process of HIV (Quéré et al., 1996).

22.7 CONCLUSION

Despite the pharmacological actions of medicinal plants and their derived compounds, there always exists a certain level of toxicity beyond the optimal dose, which emphasizes the need for performing toxicological analysis. In this regard, subchronic toxicity studies with standardized 40% Garcinol in experimental rodents showed no abnormal clinical signs/behavioral changes, reproductive and developmental parameters, indicating the NOAEL dose to be 100 mg/kg/day (Majeed et al., 2018). Berries of *I. aquifolium* are reported to show adverse effects including nausea, vomiting, drowsiness, and allergy, which may be due to the presence of saponins, which would negatively interact with cellular membranes (Nelson et al., 2007; Evens and Stellpflug, 2012). Repeated dose of up to 1000 mg/kg b.w UA for 90 days did not exhibit any toxicological changes in Wistar rats indicating NOAEL to be greater than 1000 mg/kg/day (Geerlofs, et al. 2020). Pharmacokinetic studies suggest that garcinol and UA have poor bioavailability and fast metabolic clearance restricting their application in clinical conditions (Wang et al., 2019b; Sun et al., 2020). Systematic pharmacokinetics studies of plant extract/compounds are yet to be done to establish an appropriate route of administration and its effective concentration range under physiological conditions. These studies are essential in translating the preclinical findings of *G. indica* and *I. aquifolium* from cell line models and animal species to humans, thereby facilitating dose selection, the characterization of the therapeutic index, identification of a metabolic pathway, and the determination of potency and tolerability.

ACKNOWLEDGMENT

DSM wishes to thank the Second Century Fund (C2F) for Postdoctoral Fellowship, Chulalongkorn University, Thailand, for the support.

REFERENCES

Aggarwal, S., and S.N. Das. 2016. Garcinol inhibits tumour cell proliferation, angiogenesis, cell cycle progression and induces apoptosis via NF-κB inhibition in oral cancer. *Tumour Biology* 37(6):7175–84.

Aggarwal, V., H.S. Tuli, J. Kaur, D. Aggarwal, G. Parashar, N. Chaturvedi Parashar, S. Kulkarni, G. Kaur, K. Sak, M. Kumar, and K.S. Ahn. 2020. Garcinol exhibits anti-neoplastic effects by targeting diverse oncogenic factors in tumor cells. *Biomedicines* 8(5):103.

Ahmad, A., S.H. Sarkar, A. Aboukameel, S. Ali, B. Biersack, S. Seibt, Y. Li, B. Bao, D. Kong, S. Banerjee, R. Schobert, S.B. Padhye, and F.H. Sarkar. 2012a. Anticancer action of garcinol *in vitro* and *in vivo* is in part mediated through inhibition of STAT-3 signaling. *Carcinogenesis* 33(12):2450–6.

Ahmad, A., S.H. Sarkar, B. Bitar, S. Ali, A. Aboukameel, S. Sethi, Y. Li, B. Bao, D. Kong, S. Banerjee, S.B. Padhye, and F.H. Sarkar. 2012b. Garcinol regulates EMT and Wnt signaling pathways *in vitro* and *in vivo*, leading to anticancer activity against breast cancer cells. *Molecular Cancer Therapeutics* 11(10):2193–201.

Ahmad, A., Z. Wang, R. Ali, M.Y. Maitah, D. Kong, S. Banerjee, S. Padhye, and F.H. Sarkar. 2010. Apoptosis-inducing effect of garcinol is mediated by NF-kappaB signaling in breast cancer cells. *Journal of Cellular Biochemistry* 109(6):1134–41.

Ananthakrishnan, R., and K.B. Rameshkumar. 2016. Phytochemicals and bioactivities of *Garcinia indica* (Thouars) Choisy: A review. In *Diversity of Garcinia Species in the Western Ghats: Phytochemical Perspective*, pp.142–50.

Antala, B.V., M.S. Patel, S.V. Bhuva, S. Gupta, S. Rabadiya, and M. Lahkar. 2012. Protective effect of methanolic extract of *Garcinia indica* fruits in 6-OHDA rat model of Parkinson's disease. *Indian Journal of Pharmacology* 44(6):683–7.

Balasubramanyam, K., M. Altaf, R.A. Varier, V. Swaminathan, A. Ravindran, P.P. Sadhale, and T.K. Kundu. 2004. Polyisoprenylated benzophenone, garcinol, a natural histone acetyltransferase inhibitor, represses chromatin transcription and alters global gene expression. *Journal of Biological Chemistry* 279(32):33716–26.

Baliga, M.S., H.P. Bhat, Pai, R.J, Boloor, R, Palatty, P.L. 2011. The chemistry and medicinal uses of the underutilized Indian fruit tree *Garcinia indica* Choisy (kokum): A review. *Food Research International* 44(7):1790–9.

Barve, K. 2019. Garcinol enriched fraction from the fruit rind of *Garcinia indica* ameliorates atherosclerotic risk factor in diet induced hyperlipidemic C57BL/6 mice. *Journal of Traditional and Complementary Medicine* 11(2):95–102.

Bate-Smith, E.C. 1962. The phenolic constituents of plants and their taxonomic significance I. Dicotyledons. *Journal of the Linnean Society (Botany)* 58(371):95–173.

Bohinc, P. 1967. Determination of theobromine in *Ilex aquifolium. Farmaceutski Vestnik* 18:9–20.

Brahmachari, G. 2012. Natural products in drug discovery: Impacts and opportunities: An assessment. In *Bioactive Natural Products: Opportunities and Challenges in Medicinal Chemistry*, 1st ed. Singapore: World Scientific Publishers, pp.1–114.

Catalano, S., A. Marsili, I. Morelli, L. Pistelli, and V. Scartoni. 1978. *Constituents of the leaves of Ilex aquifolium* L. *Planta Medica* 33:416–7.

Chattopadhyay, S.K., and S. Kumar. 2006. Identification and quantification of two biologically active polyisoprenylated benzophenones xanthochymol and isoxanthochymol in Garcinia species using liquid chromatography-tandem mass spectrometry. *Journal of Chromatography B Analytical Technologies in the Biomedical and Life Sciences* 844(1):67–83.

Chattopadhyay, S.K., and S. Kumar. 2007. Liquid chromatography-tandem mass spectrometry method for identification and quantification of two biologically active polyisoprenylated benzophenones, isoxanthochymol and camboginol, in Garcinia species. *Biomedical Chromatography* 21(11):1159–65.

Chen, Q., J. Luo, C. Wu, H. Lu, S. Cai, C. Bao, D. Liu, and J. Kong. 2020a. The miRNA-149-5p/MyD88 axis is responsible for ursolic acid-mediated attenuation of the stemness and chemoresistance of non-small cell lung cancer cells. *Environmental Toxicology* 35(5):561–9.

Chen, Y.M., B.X. Tang, W.Y. Chen, and M.S. Zhao. 2020b. Ursolic acid inhibits the invasiveness of A498 cells via NLRP3 inflammasome activation. *Oncology Letters* 20(5):170.

Chen, Z., Q. Liu, Z. Zhu, F. Xiang, M. Zhang, R. Wu, and X. Kang. 2020c. Ursolic acid protects against proliferation and inflammatory response in LPS-treated gastric tumour model and cells by inhibiting NLRP3 inflammasome activation. *Cancer Management and Research* 12:8413–24.

Cheng, A.C., M.L. Tsai, C.M. Liu, M.F. Lee, K. Nagabhushanam, C.T. Ho, and M.H. Pan. 2010. Garcinol inhibits cell growth in hepatocellular carcinoma Hep3B cells through induction of ROS-dependent apoptosis. *Food and Function* 1(3):301–7.

Conway, G.E., D. Zizyte, J.R.M. Mondala, Z. He, L. Lynam, M. Lecourt, C. Barcia, O. Howe, and J.F. Curtin. 2021. Ursolic acid inhibits collective cell migration and promotes JNK-dependent lysosomal associated cell death in glioblastoma multiforme cells. *Pharmaceuticals* 14(2):91.

Corona, A., S. Sebastian, D. Schallar, R. Schobert, A. Volkamer, B. Biersack, and E. Tramontano. 2021. Garcinol from *Garcinia indica* inhibits HIV-1 reverse transcriptase-associated ribonuclease H. *Archiv der Pharmazie*:e2100123.

Cotterill, P.J., and F. Scheinmann. 1977. Phenolic compounds from the heartwood of *Garcinia indica. Phytochemistry* 16:148–9.

Cragg, G.M., and D.J. Newman. 2005. Biodiversity: A continuing source of novel drug leads. *Pure and Applied Chemistry* 77:7–24.

Cragg, G.M., and D.J. Newman. 2013. Natural products: A continue source of novel drugs. *Biochimica et Biophysica Acta* 1830(6):3670–95.

Crombie, W.M. 1958. Fatty acids in chloroplasts and leaves. *Journal of Experimental Botany* 9:254–61.

Dhakal, H., M.J. Kim, S. Lee, Y.A. Choi, N. Kim, T.K. Kwon, D. Khang, and S.H. Kim. 2021. Ursolic acid inhibits FcεRI-mediated mast cell activation and allergic inflammation. *International Immunopharmacology* 99:107994.

Dhamija, I., M. Parle, and S. Kumar. 2017. Antidepressant and anxiolytic effects of *Garcinia indica* fruit rind via monoaminergic pathway. *3 Biotech* 7(2):131.

Ding, H., H. Wang, L. Zhu, and W. Wei. 2017. Ursolic acid ameliorates early brain injury after experimental traumatic brain injury in mice by activating the Nrf2 pathway. *Neurochemical Research* 42(2):337–46.

do Nascimento, P.G., T.L. Lemos, A.M. Bizerra, A.M. Arriaga, D.A. Ferreira, G.M. Santiago, R. Braz-Filho, and J.G. Costa. 2014. Antibacterial and antioxidant activities of ursolic acid and derivatives. *Molecules* 19(1):1317–27.

Duan, Y.T., X.A. Yang, L.Y. Fang, J.H. Wang, and Q. Liu. 2018. Anti-proliferative and anti-invasive effects of garcinol from *Garcinia indica* on gallbladder carcinoma cells. *Die Pharmazie* 73(7):413–7.

Erdemoglu, N., G. Iscan, B. Sener, and P. Palittapongarnpim. 2009. Antibacterial, antifungal, and antimycobacterial activity of *Ilex aquifolium* leaves. *Pharmaceutical Biology* 47(8):697–700.

Evens, Z.N., S.J. Stellpflug. 2012. Holiday plants with toxic misconceptions. *The Western Journal of Emergency Medicine* 13(6):538–42.

Frédérich, M., A. Marcowycz, E. Cieckiewicz, V. Mégalizzi, L. Angenot, and R. Kiss. 2009. *In vitro* anticancer potential of tree extracts from the Walloon Region forest. *Planta Medica* 75(15):1634–7.

Gai, W.T., D.P. Yu, X.S. Wang, P.T. Wang. 2016. Anti-cancer effect of ursolic acid activates apoptosis through ROCK/PTEN mediated mitochondrial translocation of cofilin-1 in prostate cancer. *Oncology Letters* 12(4):2880–5.

Geerlofs, L., Z. He, S. Xiao, and Z.C. Xiao. 2020. Repeated dose (90 days) oral toxicity study of ursolic acid in Han-Wistar rats. *Toxicological Reports* 7:610–23.

Ghelani, H.S., B.M. Patel, R.H. Gokani, and M.A. Rachchh. 2014. Evaluation of polyherbal formulation (SJT-HT-03) for antihypertensive activity in albino rats. *Ayu* 35(4):452–7.

Guo, W., B. Xu, X. Wang, B. Zheng, J. Du, and S. Liu. 2020. The analysis of the anti-tumor mechanism of ursolic acid using connectively map approach in breast cancer cells line MCF-7. *Cancer Management and Research* 12:3469–76.

Hao, F., L.H. Jia, X.W. Li, Y.R. Zhang, and X.W. Liu. 2016. Garcinol upregulates GABAA and GAD65 expression, modulates BDNF-TrkB pathway to reduce seizures in pentylenetetrazole (PTZ)-induced epilepsy. *Medical Science Monitor* 22:4415–25.

Hong, J., S.J. Kwon, S. Sang, J. Ju, J.N. Zhou, C.T. Ho, M.T. Huang, and C.S. Yang. 2007. Effects of garcinol and its derivatives on intestinal cell growth: Inhibitory effects and autoxidation-dependent growth-stimulatory effects. *Free Radical Biology and Medicine* 42(8):1211–21.

Hosamani, K.M., V.B. Hiremath, and R.S. Keri. 2009. *Renewable energy sources from Michelia champaca* and *Garcinia indica* seed oils: A rich source of oil. *Biomass Bioenergy* 33:267–70.

Huang, W.C., K.T. Kuo, B.O. Adebayo, C.H. Wang, Y.J. Chen, K. Jin, T.H. Tsai, and C.T. Yeh. 2018. Garcinol inhibits cancer stem cell-like phenotype via suppression of the Wnt/β-catenin/STAT3 axis signalling pathway in human non-small cell lung carcinomas. *Journal of Nutritional Biochemistry* 54:140–50.

Hue, G.N., Caudullo, G., de Rigo, D., 2016. *Ilex aquifolium* in Europe: Distribution, habitat, usage and threats. In *European Atlas of Forest Tree Species*. Luxembourg: Publications Office of the European Union, pp.e011fbc.

Hung, W.L., M.L.Tsai, P.P. Sun, C.Y. Tsai, C.C. Yang, C.T. Ho, A.C. Cheng, and M.H. Pan. 2014. Protective effects of garcinol on dimethylnitrosamine-induced liver fibrosis in rats. *Food and Function* 5(11):2883–91.

Ishikura, N. 1971. Paper chromatographic analysis of anthocyanins in the red epicarp of *Ilex aquifolium*. *Botanical Magazine Tokyo* 84:113–7.

Ishikura, N. 1975. Distribution of anthocyanins in Aquifoliaceae and Celastraceae II. *Phytochemistry* 14:743–5.

Janick J, Paull RE. 2008. *The Encyclopedia of Fruit and Nuts. CABI publishing Series*. Wallingford, UK: CABI Publishers, pp.262.

Jayaprakasha, G.K., and Sakariah, K.K. 2002. Determination of organic acids in leaves and rinds of Garcinia indica (Desr.) by LC. *Journal of Pharmaceutical and Biomedical Analysis* 28:379–84.

Jee, B., S. Kumar, R. Yadav, Y. Singh, A. Kumar, and N. Sharma. 2018. Ursolic acid and carvacrol may be potential inhibitors of dormancy protein small heat shock protein16.3 of Mycobacterium tuberculosis. *Journal of BiomolecularStructure and Dynamics* 36(13):3434–43.

Jeyarani, T., and Reddy, Y.S. 1999. *Heat-resistant cocoa butter extenders from mahua (Madhuca latifolia)* and kokum (*Garcinia indica*) fats. *Journal of the American Oil Chemists' Society* 76(12):1431–6.

Jia, Y., J. Jiang, X. Lu, T. Zhang, K. Zhao, W. Han, W. Yang, and Y. Qian. 2019. Garcinol suppresses RANKL-induced osteoclastogenesis and its underlying mechanism. *Journal of Cellular Physiology* 234(5):7498–509.

Jing, Y., Q. Ai, L. Lin, J. Dai, M. Jia, D. Zhou, Q. Che, J. Wan, R. Jiang, and L. Zhang. 2014. Protective effects of garcinol in mice with lipopolysaccharide/D-galactosamine-induced apoptotic liver injury. *International Immunopharmacology* 19(2):373–80.

Jones, A.W. 2011. Early drug discovery and the rise of pharmaceutical chemistry. *Drug Testing and Analysis* 3(6):337–44.

Kang, D.Y., N. Sp, J.M. Lee, and K.J. Jang. 2021. Antitumor effects of ursolic acid through mediating the inhibition of STAT3/PD-L1 signaling in non-small cell lung cancer cells. *Biomedicines* 9(3):297.

Kang, Y., Y. Sun, T. Li, and Z. Ren. 2020. Garcinol protects against cerebral ischemia-reperfusion injury *in vivo* and *in vitro* by inhibiting inflammation and oxidative stress. *Molecular and Cellular Probes* 54:101672.

Kaur, R., S.K. Chattopadhyay., S. Tandon and S. Sharma. 2012. Large scale extraction of the fruits of *Garcinia indica* for the isolation of new and known polyisoprenylated benzophenone derivatives. *Industrial Crops and Products* 37:420–6.

Khare CP. 2011. *Indian Herbal Remedies: Rational Western Therapy, Ayurvedic and Other Traditional Usage, Botany*. New York, USA: Springer, pp.229–230.

Kim, G.H., S.Y. Kan, H. Kang, S. Lee, H.M. Ko, J.H. Kim, and J.H. Lim. 2019a. Ursolic acid suppresses cholesterol biosynthesis and exerts anti-cancer effects in hepatocellular carcinoma cells. *International Journal of Molecular Sciences* 20(19):4767.

Kim, S., S.U. Seo, K.J. Min, S.M. Woo, J.O. Nam, P. Kubatka, S. Kim, J.W. Park, T.K. Kwon. 2018. Garcinol enhances TRAIL-induced apoptotic cell death through up-regulation of DR5 and down-regulation of c-FLIP expression. *Molecules* 23(7):1614.

Kim, S.H., H. Jin, R.Y. Meng, D.Y. Kim, Y.C. Liu, O.H. Chai, B.H. Park, and S.M. Kim. 2019b. Activating hippo pathway via rassf1 by ursolic acid suppresses the tumorigenesis of gastric cancer. *International Journal of Molecular Sciences* 20(19):4709.

Kim, S.H., H.G. Ryu, J. Lee, J. Shin, A. Harikishore, H.Y. Jung, Y.S. Kim, H.N. Lyu, E. Oh, N.I. Baek, K.Y. Choi, H.S. Yoon, and K.T. Kim. 2015. Ursolic acid exerts anti-cancer activity by suppressing vaccinia-related kinase 1-mediated damage repair in lung cancer cells. *Scientific Reports* 5:14570.

Kirana, H., and B. Srinivasan. 2010. Aqueous extract of *Garcinia indica* choisy restores glutathione in type 2 diabetic rats. *Journal of Young Pharmacists* 2(3):265–8.

Knights, B.A., and A.R. Smith. 1977. Sterols and triterpenes of *Ilex aquifolium*., *Phytochemistry* 16:139–40.

Krishnamurthy, N., Y.S. Lewis, and B. Ravindranath. 1982. Chemical constitution of Kokum fruit rind. *Journal of Food Science and Technology* 19:97–100.

Kumar, S., and S.K. Chattopadhyay. 2007. High-performance liquid chromatography and LC-ESI-MS method for the identification and quantification of two biologically active polyisoprenylated benzophenones xanthochymol and isoxanthochymol in different parts of *Garcinia indica*. *Biomedical Chromatography* 21(2):139–63.

Kumar, S., S. Sharma, and S.K. Chattopadhyay. 2009. High-performance liquid chromatography and LC-ESI-MS method for identification and quantification of two isomeric polyisoprenylated benzophenones isoxanthochymol and camboginol in different extracts of Garcinia species. *Biomedical Chromatography* 23(8):888–907.

Kurek, A., K. Markowska, A.M. Grudniak, W. Janiszowska, and K.I. Wolska. 2014. The effect of oleanolic and ursolic acids on the hemolytic properties and biofilm formation of Listeria monocytogenes. *Polish Journal of Microbiology* 63(1):21–5.

Lakshmi, C., K.A. Kumar, T.J. Dennis, and T.S. Kumar. 2011. Antibacterial activity of polyphenols of *Garcinia indica. Indian Journal of Pharmaceutical Sciences* 73(4):470–3.

Lee, P.S., C.Y. Teng, N. Kalyanam, C.T. Ho, and M.H. Pan. 2019. Garcinol reduces obesity in high-fat-diet-fed mice by modulating gut microbiota composition. *Molecular Nutrition & Food Research* 63(2):e1800390.

Leng, S., S. Iwanowycz, F. Saaoud, J. Wang, Y. Wang, I. Sergin, B. Razani, and D. Fan. 2016. Ursolic acid enhances macrophage autophagy and attenuates atherogenesis. *Journal of Lipid Research* 57(6):1006–16.

Li, F., M.K. Shanmugam, L. Chen, S. Chatterjee, J. Basha, A.P. Kumar, T.K. Kundu, and G. Sethi. 2013a. Garcinol, a polyisoprenylated benzophenone modulates multiple proinflammatory signaling cascades leading to the suppression of growth and survival of head and neck carcinoma. *Cancer Prevention Research* 6(8):843–54.

Li, J., C. Dai, and L. Shen. 2019. Ursolic acid inhibits epithelial-mesenchymal transition through the Axl/NF-κB pathway in gastric cancer cells. *Evidence Based Complementary and Alternative Medicine* 2019:2474805.

Li, J., M. Wang, L. Song, X. Wang, W. Lai, and S. Jiang. 2020a. LncRNA MALAT1 regulates inflammatory cytokine production in lipopolysaccharide-stimulated human gingival fibroblasts through sponging miR-20a and activating TLR4 pathway. *Journal of Periodontal Research* 55(2):182–90.

Li, L., X. Zhang, L. Cui, L. Wang, H. Liu, H. Ji, and Y. Du. 2013b. Ursolic acid promotes the neuroprotection by activating Nrf2 pathway after cerebral ischemia in mice. *Brain Research* 1497:32–9.

Li, M., X. Li, and L. Yang. 2020b. Cardioprotective effects of garcinol following myocardial infarction in rats with isoproterenol-induced heart failure. *AMB Express* 10(1):137.

Li, R., X. Wang, X.H. Zhang, H.H. Chen, and Y.D. Liu. 2014. Ursolic acid promotes apoptosis of SGC-7901 gastric cancer cells through ROCK/PTEN mediated mitochondrial translocation of cofilin-1. *Asian Pacific Journal of Cancer Prevention* 15(22):9593–7.

Li, W., L. Luo, W. Shi, Y. Yin, and S. Gao. 2020c. Ursolic acid reduces Adriamycin resistance of human ovarian cancer cells through promoting the HuR translocation from cytoplasm to nucleus. *Environmental Toxicology* 36(2):267–75.

Liang, W., X. Zhao, J. Feng, F. Song, and Y. Pan. 2016. Ursolic acid attenuates beta-amyloid-induced memory impairment in mice. *Arquivos de Neuro-Psiquiatria* 74(6):482–8.

Liao, C.H., C.T. Ho, and J.K. Lin. 2005a. Effects of garcinol on free radical generation and NO production in embryonic rat cortical neurons and astrocytes. *Biochemical and Biophysical Research Communications* 329(4):1306–14.

Liao, C.H., S. Sang, C.T. Ho, and J.K. 2005b. Garcinol modulates tyrosine phosphorylation of FAK and subsequently induces apoptosis through down-regulation of Src, ERK, and Akt survival signaling in human colon cancer cells. *Journal of Cellular Biochemistry* 96(1):155–69.

Liao, C.H., S. Sang, Y.C. Liang, C.T. Ho, and J.K. Lin. 2004. Suppression of inducible nitric oxide synthase and cyclooxygenase-2 in downregulating nuclear factor-kappa B pathway by Garcinol. *Molecular Carcinogenesis* 41(3):140–9.

Lin, J.H., S.Y. Chen, C.C. Lu, J.A. Lin, and G.C. Yen. 2020. Ursolic acid promotes apoptosis, autophagy, and chemosensitivity in gemcitabine-resistant human pancreatic cancer cells. *Phytotherapy Research* 34(8):2053–66.

Lipp, M., and E. Adam. 1998. Review of cocoa butter and alternative fats for use in chocolate-Part A. Compositional data. *Food Chemistry* 62(1):73–97.

Liu, H.W., P.M. Lee, O.A. Bamodu, Y.K. Su, I.H. Fong, C.T. Yeh, M.H. Chien, I.H. Kan, and C.M. Lin. 2019. Enhanced I-miR-181d/p-STAT3 aIHsa-miR-181d/p-STAT5A ratios mediate the anticancer effect of garcinol in *STAT3/5A*: Addicted glioblastoma. *Cancers* 11(12):1888.

Liu, K., L. Guo, L. Miao, W. Bao, J. Yang, X. Li, T. Xi, W. Zhao. 2013. Ursolic acid inhibits epithelial-mesenchymal transition by suppressing the expression of astrocyte-elevated gene-1 in human nonsmall cell lung cancer A549 cells. *Anticancer Drugs* 24(5):494–503.

Liu, T., H. Ma, W. Shi, J. Duan, Y. Wang, C. Zhang, C. Li, J. Lin, S. Li, J. Lv, and L. Lin. 2017. Inhibition of STAT3 signaling pathway by ursolic acid suppresses growth of hepatocellular carcinoma. *International Journal of Oncology* 51(2):555–62.

Lu, C.C., B.R. Huang, P.J. Liao, and G.C. Yen. 2014. Ursolic acid triggers nonprogrammed death (necrosis) in human glioblastoma multiforme DBTRG-05MG cells through MPT pore opening and ATP decline. *Molecular Nutrition & Food Research* 58(11):2146–56.

Lu, J., D.M. Wu, Y.L. Zheng, B. Hu, W. Cheng, Z.F. Zhang, and Q. Shan. 2011. Ursolic acid improves high fat diet-induced cognitive impairments by blocking endoplasmic reticulum stress and IκB kinase β/nuclear factor-κB-mediated inflammatory pathways in mice. *Brain Behavior and Immunity* 25(8):1658–67.

Lu, J., Y.L. Zheng, D.M. Wu, L. Luo, D.X. Sun, and Q. Shan. 2007. Ursolic acid ameliorates cognition deficits and attenuates oxidative damage in the brain of senescent mice induced by D-galactose. *Biochemical Pharmacology* 74(7):1078–90.

Lu, Q., W. Chen, Y. Ji, Y. Liu, and X. Xue. 2021. Ursolic acid enhances cytotoxicity of doxorubicin-resistant triple-negative breast cancer cells via ZEB1-AS1/miR-186-5p/ABCC1 Axis. *Cancer Biotherapy and Radiopharmaceuticals.* doi: 10.1089/cbr.2020.4147. (Ahead of Print).

Luo, J., Y.L. Hu, H. and Wang. 2017. Ursolic acid inhibits breast cancer growth by inhibiting proliferation, inducing autophagy and apoptosis, and suppressing inflammatory responses via the PI3K/AKT and NF-κB signaling pathways *in vitro. Experimental and Therapeutic Medicine* 14(4):3623–31.

Ma, X.Y., M. Zhang, G. Fang, C.J. Cheng, M.K. Wang, Y.M. Han, X.T. Hou, E.W. Hao, Y.Y. Hou, G. Bai. 2021. Ursolic acid reduces hepatocellular apoptosis and alleviates alcohol-induced liver injury via irreversible inhibition of CASP3 in vivo. *Acta Pharmacologica Sinica* 42(7):1101–10.

Maddox, S.A., C.S. Watts, V. Doyère, and G.E. Schafe. 2013. A naturally-occurring histone acetyltransferase inhibitor derived from *Garcinia indica* impairs newly acquired and reactivated fear memories. *PLoS One* 8(1):e54463.

Majeed, M., S. Bani, B. Bhat, A. Pandey, L. Mundkur, and P. Neupane. 2018. Safety profile of 40% Garcinol from *Garcinia indica* in experimental rodents. *Toxicological Reports* 19(5):750–8.

Majeed, M., S. Majeed, K. Nagabhushanam, L. Lawrence, and L. Mundkur. 2020. Novel Combinatorial Regimen of Garcinol and Curcuminoids for Non-alcoholic Steatohepatitis (NASH) in Mice. *Scientific Reports* 10(1):7440.

Mandal, S., N. Gamit, L. Varier, A. Dharmarajan, and S. Warrier. 2021. Inhibition of breast cancer stem-like cells by a triterpenoid, ursolic acid, via activation of Wnt antagonist, sFRP4 and suppression of miRNA-499a-5p. *Life Science* 265:118854.

Masoumi, F., S. Ghorbani, F. Talebi, W.G. Branton, S. Rajaei, C. Power, and F. Noorbakhsh. 2019. Malat1 long noncoding RNA regulates inflammation and leukocyte differentiation in experimental autoimmune encephalomyelitis. *Journal of Neuroimmunology* 328:50–9.

Menezes, M.T. 2001. *The Essential Goa Cookbook*. New Delhi, India: Penguin Books Ltd, pp.1–229.

Meng, Y., Z.M. Lin, N. Ge, D.L. Zhang, J. Huang, and F. Kong. 2015. Ursolic acid induces apoptosis of prostate cancer cells via the PI3K/Akt/mTOR pathway. *American Journal of Chinese Medicine* 43(7):1471–86.

Mirza, F.J., S. Amber, D. Hassan, T. Ahmed, and S. Zahid. 2021. Rosmarinic acid and ursolic acid alleviate deficits in cognition, synaptic regulation and adult hippocampal neurogenesis in an Aβ1-42-induced mouse model of Alz'eimer's disease. *Phytomedicine* 83:153490.

Monsey, M.S., H. Sanchez, and J.R. Taylor. 2017. The naturally occurring compound *Garcinia indica* selectively impairs the reconsolidation of a cocaine-associated memory. *Neuropsychopharmacology* 42(3):587–97.

Mou, L., L. Liao, Y. Zhang, D. Ming, and J. Jiang. 2021. Ursolic acid ameliorates Nthy-ori 3–1 cells injury induced by IL-1β through limiting MALAT1/miR-206/PTGS1 ceRNA network and NF-κB signaling pathway. *Psychopharmacology* 238(4):1141–56.

Mu, D., G. Zhou, J. Li, B. Su, and H. Guo. 2018. Ursolic acid activates the apoptosis of prostate cancer via ROCK/PTEN mediated mitochondrial translocation of cofilin-1. *Oncology Letters* 15(3):3202–6.

Müller, K., K. Ziereis, and D.H. Paper. 1998. *Ilex aquifolium*: Protection against enzymatic and non-enzymatic lipid peroxidation. *Planta Medica* 64(6):536–40.

Nahar, L., W.R. Russell, M. Middleton, M. Shoeb, and S.D. Sarker. 2005. Antioxidant phenylacetic acid derivatives from the seeds of *Ilex aquifolium. Acta Pharmaceutica* 55:187–93.

Naß, J., and T. Efferth. 2021. Ursolic acid ameliorates stress and reactive oxygen species in *C. elegans* knockout mutants by the dopamine Dop1 and Dop3 receptors. *Phytomedicine* 81:153439.

Naß, J., S. Abdelfatah, and T. Efferth. 2021. The triterpenoid ursolic acid ameliorates stress in *Caenorhabditis elegans* by affecting the depression-associated genes *skn-1* and *prdx2*. *Phytomedicine* 88:153598.

Nayak, C.A., N.K. Rastogi, and K.S.M.S Raghavarao. 2010a. Bioactive constituents present in *Garcinia indica* Choisy and its potential food applications: A review. *International Journal of Food Properties* 13:441–53.

Nayak, C.A., P. Srinivas, and N.K. Rastogi. 2010b. Characterisation of anthocyanins from *Garcinia indica* Choisy. *Food Chemistry* 118:719–24.

Nelson, L.S., R.D. Shih, M.J. Balick. 2007. *Handbook of Poisonous and Injurious Plants*, 2nd ed. New York, USA: Springer, pp.187–188.

Obeso, J.R. 1998. Patterns of variation in *Ilex aquifolium* fruit traits related to fruit consumption by birds and seed predation by rodents. *Ecoscience* 5(4):463–9.

Padhye, S., Ahmad, A., Oswal, N. and Sarkar, F.H. 2009. Emerging role of garcinol, the antioxidant chalcone from *Garcinia indica* Choisy and its synthetic analogs. *Journal of Hematology & Oncology* 2:1–13.

Palu, D., A. Bighelli, J. Casanova, M. Paoli. 2019. Identification and quantitation of ursolic and oleanolic acids in *Ilex aquifolium* L. leaf extracts using 13C and 1H-NMR spectroscopy. *Molecules* 24(23):4413.

Park, J.H., H.Y. Kwon, E.J. Sohn, K.A. Kim, B. Kim, S.J. Jeong, J.H. Song, J.S. Koo, and S.H. Kim. 2013. Inhibition of Wnt/β-catenin signaling mediates ursolic acid-induced apoptosis in PC-3 prostate cancer cells. *Pharmacological Reports* 65(5):1366–74.

Park, S.N., S.J. Ahn, and J.K. Kook. 2015. Oleanolic acid and ursolic acid inhibit peptidoglycan biosynthesis in *Streptococcus mutans* UA159. *Brazilian Journal of Microbiology* 46(2):613–7.

Park, S.N., Y.K. Lim, M.H. Choi, E. Cho, I.S. Bang, J.M. Kim, S.J. Ahn, and J.K. Kook. 2018. Antimicrobial Mechanism of Oleanolic and Ursolic Acids on *Streptococcus mutans* UA159. *Current Microbiology* 75(1):11–19.

Peshattiwar, V., S. Muke, A. Kaikini, S. Bagle, V. Dighe, and S. Sathaye S. 2020. Mechanistic evaluation of Ursolic acid against rotenone induced Par'inson's disease- emphasizing the role of mitochondrial biogenesis. *Brain Research Bulletin* 160:150–61.

Peterken, G.F, and P.S. Lloyd. 1967. Biological flora of British Isles: *Ilex aquifolium* L. *Journal of Ecology* 55(3):841–58.

Prasad, S., V.R. Yadav, B. Sung, S. Reuter, R. Kannappan, A. Deorukhkar, P. Diagaradjane, C. Wei, V. Baladandayuthapani, S. Krishnan, S. Guha, and B.B. Aggarwal. 2012. Ursolic acid inhibits growth and metastasis of human colorectal cancer in an orthotopic nude mouse model by targeting multiple cell signaling pathways: Chemosensitization with capecitabine. *Clinical Cancer Research* 18(18):4942–53.

Quéré, L., T. Wenger, and H.J. Schramm. 1996. Triterpenes as potential dimerization inhibitors of HIV-1 protease. *Biochemical and Biophysical Research Communications* 227(2):484–8.

Rai, S.N., S.K. Yadav, D. Singh, and S.P. Singh. 2016. Ursolic acid attenuates oxidative stress in nigrostriatal tissue and improves neurobehavioral activity in MPTP-induced Parkinsonian mouse model. *Journal of Chemical Neuroanatomy* 71:41–9.

Rawat, L., and V. Nayak. 2021. Ursolic acid disturbs ROS homeostasis and regulates survival-associated gene expression to induce apoptosis in intestinal cancer cells. *Toxicological Research* 10(3):369–75.

Robinson, G.M., and R. Robinson. 1931. Survey of anthocyanins I. *Biochemical Journal* 25:1687–705.

Sangaonkar, G.M., and K.D. Pawar. 2018. *Garcinia indica* mediated biogenic synthesis of silver nanoparticles with antibacterial and antioxidant activities. *Colloids and Surfaces B Biointerfaces* 164:210–7.

Santamour, F.S. 1973. Anthocyanins of holly fruits. *Phytochemistry* 12:611–5.

Schindler, H., and M. Herb. 1955. Chemistry of *Nex aquifolium* I. Isolation of ursolic acid and rutin from the leaves. *Archiv der Pharmazie* 288:372–77.

Shanmugam, M.K., K.A. Manu, T.H. Ong, L. Ramachandran, R. Surana, P. Bist, L.H. Lim, A.P. Kumar, K.M. Hui, and G. Sethi. 2011. Inhibition of CXCR4/CXCL12 signaling axis by ursolic acid leads to suppression of metastasis in transgenic adenocarcinoma of mouse prostate model. *International Journal of Cancer* 129(7):1552–63.

Shanmugam, M.K., T.H. Ong, A.P. Kumar, C.K. Lun, P.C. Ho, P.T. Wong, K.M. Hui, and G. Sethi. 2012. Ursolic acid inhibits the initiation, progression of prostate cancer and prolongs the survival of TRAMP mice by modulating pro-inflammatory pathways. *PLoS One* 7(3):e32476.

Spainhour, C.B. 2005. *Natural Products, Drug Discovery Handbook*. New York, USA: Wiley, pp.11–72.

Subhasree, N., A. Kamella, I. Kaliappan, A. Agrawal, G.P. and G.P. Dubey. 2015. Antidiabetic and antihyperlipidemic activities of a novel polyherbal formulation in high fat diet/streptozotocin induced diabetic rat model. *Indian Journal of Pharmacology* 47(5):509–13.

Sun, Q., M. He, M. Zhang, S. Zeng, L. Chen, L. Zhou, and H. Xu. 2020. Ursolic acid: A systematic review of its pharmacology, toxicity and rethink on its pharmacokinetics based on PK-PD model. *Fitoterapia* 147:104735.

Tanaka, T., H. Kohno, R. Shimada, S. Kagami, F.Yamaguchi, S. Kataoka, T. Ariga, A. Murakami, K. Koshimizu, and H. Ohigashi. 2000. Prevention of colonic aberrant crypt foci by dietary feeding of garcinol in male F344 rats. *Carcinogenesis* 21(6):1183–9.

Tang, C., Y.H. Lu, J.H. Xie, F. Wang, J.N. Zou, J.S. Yang, Y.Y. Xing, and T. Xi. 2009. Downregulation of survivin and activation of caspase-3 through the PI3K/Akt pathway in ursolic acid-induced HepG2 cell apoptosis. *Anticancer Drugs* 20(4):249–58.

Tang, Q., Q. Ji, Y. Tang, T. Chen, G. Pan, S. Hu, Y. Bao, W. Peng, P. Yin. 2014. Mitochondrial translocation of cofilin-1 promotes apoptosis of gastric cancer BGC-823 cells induced by ursolic acid. *Tumour Biology* 35(3):2451–9.

Thomas, H., and H. Budzikiewicz. 1980. Components of the Celastrales species, part 6. Constituents of the fruits of *Zlex aquifolium* L. *Zeitschrift fiir Pflanzenphysiologie* 99:271–6.

Tsai, M.L., Y.S. Chiou, L.Y. Chiou, C.T. Ho, and M.H. Pan. 2014. Garcinol suppresses inflammation-associated colon carcinogenesis in mice. *Molecular Nutrition & Food Research* 58(9):1820–9.

Tung, Y.C., Y.A. Shih, K. Nagabhushanam, C.T. Ho, A.C. Cheng, and M.H. Pan. 2021. *Coleus forskohlii* and *Garcinia indica* extracts attenuated lipid accumulation by regulating energy metabolism and modulating gut microbiota in obese mice. *Food Research International* 142:110143.

Van Genderen, H.H., and J. Jaarsma. 1990. Triterpenes and alkanes in developing variegated and albino leaves of *Ilex aquifolium* L. (Aquifoliaceae). *Plant Science* 72:165–72.

Wang, C., Y. Gao, Z. Zhang, C. Chen, Q. Chi, K. Xu, and L. Yang. 2020. Ursolic acid protects chondrocytes, exhibits anti-inflammatory properties via regulation of the NF-κB/NLRP3 inflammasome pathway and ameliorates osteoarthritis. *Biomedicine & Pharmacotherapy* 130:110568.

Wang, C.M., Y.L. Jhan, S.J. Tsai, and C.H. Chou. 2016a. The pleiotropic antibacterial mechanisms of ursolic acid against methicillin-resistant *Staphylococcus aureus* (MRSA). *Molecules* 21(7):884.

Wang, J., L. Liu, H. Qiu, X. Zhang, W. Guo, W. Chen, Y. Tian, L. Fu, D. Shi, J. Cheng, W. Huang, and W. Deng. 2013. Ursolic acid simultaneously targets multiple signaling pathways to suppress proliferation and induce apoptosis in colon cancer cells. *PLoS One* 8(5):e63872.

Wang, J., L. Wang, C.T. Ho, K. Zhang, Q. Liu, and H. Zhao. 2017a. Garcinol from *Garcinia indica* downregulates cancer stem-like cell biomarker ALDH1A1 in nonsmall cell lung cancer A549 cells through DDIT3 activation. *Journal of Agricultural Food Chemistry* 65(18):3675–83.

Wang, J., Y. Li, X. Wang, and C. Jiang. 2012a. Ursolic acid inhibits proliferation and induces apoptosis in human glioblastoma cell lines U251 by suppressing TGF-β1/miR-21/PDCD4 pathway. *Basic and Clinical Pharmacology and Toxicology* 111(2):106–12.

Wang, L., M. Wang, H. Guo, and H. Zhao. 2019. Emerging role of garcinol in targeting cancer stem cells of non-small cell lung cancer. *Current Pharmacology Report* 5:14–19.

Wang, X., T. Wang, F. Yi, C. Duan, Q. Wang, N. He, L. Zhu, Q. Li, and W. Deng. 2019. Ursolic acid inhibits tumor growth via epithelial-to-mesenchymal transition in colorectal cancer cells. *Biological and Pharmaceutical Bulletin* 42(5):685–91.

Wang, X.F., Z.M. Shi, X.R. Wang, L. Cao, Y.Y. Wang, J.X. Zhang, Y. Yin, H. Luo, C.S. Kang, et al. 2012. MiR-181d acts as a tumor suppressor in glioma by targeting K-ras and Bcl-2. *Journal of Cancer Research and Clinical Oncology* 138:573–84.

Wang, Y., M.L. Tsai, L.Y. Chiou, C.T. Ho, and M.H. Pan. 2015. Antitumor activity of garcinol in human prostate cancer cells and xenograft mice. *Journal of Agricultural Food Chemistry* 63(41):9047–52.

Wang, Y., Z. He, and S. Deng. 2016b. Ursolic acid reduces the metalloprotease/anti-metalloprotease imbalance in cerebral ischemia and reperfusion injury. *Drug Design, Development and Therapy* 10:1663–74.

Wang, Y.W., X. Zhang, C.L. Chen, Q.Z. Liu, J.W. Xu, Q.Q. Qian, W.Y. Li, and Y.N. Qian. 2017b. Protective effects of Garcinol against neuropathic–pain - Evidence from in vivo and in vitro studies. *Neuroscience Letters* 647:85–90.

Warriar, P., K. Barve, and B. Prabhakar. 2019. Anti-arthritic effect of garcinol enriched fraction against adjuvant induced arthritis. *Recent Patents on Inflammation & Allergy Drug Discovery* 13(1):49–56.

Weng, M.S., C.H. Liao, S.Y. Yu, and J.K. Lin. 2011. Garcinol promotes neurogenesis in rat cortical progenitor cells through the duration of extracellular signal-regulated kinase signaling. *Journal of Agricultural Food Chemistry* 59(3):1031–40.

Willems, M. 1989. Quantitative determination and distribution of a cyanogenic glucoside in *Ilex aquifolium*. *Planta Medica* 2:195.

Wu, J., S. Zhao, Q. Tang, F. Zheng, Y. Chen, L. Yang, X. Yang, L. Li, W. Wu, and S.S. Hann. 2015. Activation of SAPK/JNK mediated the inhibition and reciprocal interaction of DNA methyltransferase 1 and EZH2 by ursolic acid in human lung cancer cells. *Journal of Experimental & Clinical Cancer Research* 34(1):99.

Yamaguchi, F., T. Ariga, Y. Yoshimura, and H. Nakazawa. 2000. Antioxidative and anti-glycation activity of garcinol from *Garcinia indica* fruit rind. *Journal of Agricultural and Food Chemistry* 48(2):180–5.

Yan, X., X. Liu, Y. Wang, X. Ren, J. Ma, R. Song, X. Wang, Y. Dong, Q. Fan, J. Wei, A. Yu, H. Sui, and G. She. 2021a. Multi-omics integration reveals the hepatoprotective mechanisms of ursolic acid intake against chronic alcohol consumption. *European Journal of Nutrition* doi: 10.1007/s00394-021-02632-x. Epub ahead of print.

Yan, X., X. Ren, X. Liu, Y. Wang, J. Ma, R. Song, X. Wang, Y. Dong, Q. Fan, J. Wei, A. Yu, and G. She. 2021. Dietary ursolic acid prevents alcohol-induced liver injury via gut-liver axis homeostasis modulation: The key role of microbiome manipulation. *Journal of Agricultural Food Chemistry* 69(25):7074–83.

Yang, K., Y. Chen., J. Zhou, L. Ma, Y. Shan, X. Cheng, Y. Wang, Z. Zhang, X. Ji, L. Chen, H. Dai, B. Zhu, C. Li, Z. Tao, X. Hu, and W. Yin. 2019. Ursolic acid promotes apoptosis and mediates transcriptional suppression of CT45A2 gene expression in non-small-cell lung carcinoma harbouring EGFR T790M mutations. *British Journal of Pharmacology* 176(24):4609–24.

Yeh, C.T., C.H. Wu, and G.C. Yen. 2010. Ursolic acid, a naturally occurring triterpenoid, suppresses migration and invasion of human breast cancer cells by modulating c-Jun N-terminal kinase, Akt and mammalian target of rapamycin signaling. *Molecular Nutrition & Food Research* 54(9):1285–95.

Yie, Y., S. Zhao, Q. Tang, F. Zheng, J. Wu, L. Yang, S. Deng, and S.S. Hann. 2015. Ursolic acid inhibited growth of hepatocellular carcinoma HepG2 cells through AMPKα-mediated reduction of DNA methyltransferase 1. *Molecular and Cellular Biochemistry* 402(1–2):63–74.

Yu, S.Y., C.H. Liao, M.H. Chien, T.Y. Tsai, J.K. Lin, and M.S. Weng. 2014. Induction of p21(Waf1/Cip1) by garcinol via downregulation of p38-MAPK signaling in p53-independent H1299 lung cancer. *Journal of Agricultural Food Chemistry* 62(9):2085–95.

Yuan, H., Q. Ma, L. Ye, and G. Piao. 2016 The traditional medicine and modern medicine from natural products. *Molecules* 21(5):559.

Zahra, W., S.N. Rai, H. Birla, S.S. Singh, A.S. Rathore, H. Dilnashin, R. Singh, C. Keswani, R.K. Singh, and S.P. Singh. 2020. Neuroprotection of rotenone-induced parkinsonism by ursolic acid in PD mouse model. *CNS & Neurologiical Disorders: Drug Targets* 19(7):527–40.

Zerin, T., M. Lee, W.S. Jang, K.W. Nam, and H.Y. Song. 2016. Anti-inflammatory potential of ursolic acid in *Mycobacterium tuberculosis*-sensitized and concanavalin A: Stimulated cells. *Molecular Medicine Reports* 13(3):2736–44.

Zhang, G., J. Fu, Y. Su, X. Zhang. 2019a. Opposite effects of garcinol on tumor energy metabolism in oral squamous cell carcinoma cells. *Nutrition and Cancer* 71(8):1403–11.

Zhang, L., Q.Y. Cai, L. Liu, J. Peng, Y.Q. Chen, T.J. Sferra, and J.M. Lin. 2019b. Ursolic acid suppresses the invasive potential of colorectal cancer cells by regulating the TGF-β1/ZEB1/miR-200c signaling pathway. *Oncology Letters* 18(3):3274–82.

Zhang, T., J. Su, K. Wang, T. Zhu, and X. Li. 2014. Ursolic acid reduces oxidative stress to alleviate early brain injury following experimental subarachnoid hemorrhage. *Neuroscience Letters* 579:12–17.

Zhang, X., T. Li, E.S. Gong, and R.H. Liu. 2020. Antiproliferative activity of ursolic acid in MDA-MB-231 human breast cancer cells through Nrf2 pathway regulation. *Journal of Agricultural and Food Chemistry* 68(28):7404–15.

Zhang, Y., C. Kong, Y. Zeng, L. Wang, Z. Li, H. Wang, C. Xu, and Y. Sun. 2010a. Ursolic acid induces PC-3 cell apoptosis via activation of JNK and inhibition of Akt pathways in vitro. *Molecular Carcinogenesis* 49(4):374–85.

Zhang, Y.X., C.Z. Kong, H.Q. Wang, L.H. Wang, C.L. Xu, and Y.H. Sun. 2009. Phosphorylation of Bcl-2 and activation of caspase-3 via the c-Jun N-terminal kinase pathway in ursolic acid-induced DU145 cells apoptosis. *Biochimie* 91(9):1173–9.

Zhang, Y.X., C.Z. Kong, L.H. Wang, J.Y. Li, X.K. Liu, B. Xu, C.L. Xu, and Y.H. Sun. 2010b. Ursolic acid overcomes Bcl-2-mediated resistance to apoptosis in prostate cancer cells involving activation of JNK-induced Bcl-2 phosphorylation and degradation. *Journal of Cellular Biochemistry* 109(4):764–73.

Zhao, J., P. Leng, W. Xu, J.L. Sun, B.B. Ni, G.W. Liu. 2021. Investigating the multitarget pharmacological mechanism of ursolic acid acting on colon cancer: A network pharmacology approach. *Evidence Based Complementary and Alternative Medicine* 2021:9980949.

Zheng, J.L., S.S. Wang, K.P. Shen, L. Chen, X. Peng, J.F. Chen, H.M. An, and B. Hu. 2021. Ursolic acid induces apoptosis and anoikis in colorectal carcinoma RKO cells. *BMC Complementary Medicine and Therapies* 21(1):52.

23 *Alnus glutinosa* (Alder) and *Moringa oleifera* (Drumstick Tree)

Devashree N. Patil, Swati T. Gurme, Pankaj S. Mundada, and Jyoti. P. Jadhav

CONTENTS

23.1 Introduction 447
23.2 Description 448
23.3 Traditional Knowledge 449
23.4 Chemical Derivatives 450
23.5 Potential Benefits, Applications and Use 451
23.5.1 *A. glutinosa* 451
23.5.1.1 Antibacterial Activity 451
23.5.1.2 Antioxidant Activity 452
23.5.1.3 Anticancer Activity 452
23.5.1.4 Chemoprotective Agent 452
23.5.1.5 Anti-inflammatory Activity 452
23.5.1.6 Nitrogen Fixation 453
23.5.1.7 Insecticidal Activity 453
23.5.1.8 Dyeing Property 453
23.5.2 *M. oleifera* 453
23.5.2.1 Moringa in Water Treatment 453
23.5.2.2 Antidiabetic Activity 453
23.5.2.3 Antimicrobial Activity 453
23.5.2.4 Anti-Obesity Activity 454
23.5.2.5 Anticancer Activity 454
23.5.2.6 Anti-Inflammatory Activity 454
23.5.2.7 Anti-Asthmatic Property 454
23.5.2.8 Neuroprotective Property 455
23.5.2.9 Hepatoprotective Property 455
23.6 Conclusion 455
Acknowledgment 456
References 456

23.1 INTRODUCTION

For a long time, medicinal plants have played an essential role in the evolution of human culture. Medicinal plants since ages are at the forefront of nearly all societies of civilizations as a source of medicine and are considered rich sources of traditional medicines, deriving many modern drugs from them (Dar et al., 2016). Plant-based ingredients have a wide variety of therapeutic properties. About 60–80% of people worldwide depend on herbal medicine for their preliminary health care

DOI: 10.1201/9781003205067-23

needs, especially in countries with high medical care costs (Wanjohi et al., 2020). Numerous plants are possessing massive pharmacological activities to cure several ailments. The reason behind the actions is the several phytochemical constituents, especially secondary metabolites, to treat the enormous number of diseases. Polyphenols, including flavonoids, phenols, tannins, saponins, and terpenes, are the exclusive entities possessing several medicinal properties.

The Alnus genus is a tree and shrub, including 35 species worldwide, growing wild in the northern hemisphere's temperate zones. *Alnus glutinosa (L.)* Gaertn., also known as "black alder" or "European alder", is a native of Europe, ranging from mid-Scandinavia to northern Morocco and Algeria (Altınyay et al., 2016). In North Africa, Asia Minor, and Western Siberia, this species grows in moist forests, pastures, and streams (Mushkina et al., 2013). The *A. glutinosa* (Alder) is a hardy and fast-growing deciduous and short-lived tree native to much of Europe. The tree is considered a valuable forest species because of its various silviculture and wood industry uses. Alnus includes species such as *A. japonica, A. hirsuta, A. incana*, and *A. rubra, A. maritima, A. acuminata, A. rubra*, and *A. alnobetula*. Diarylheptanoids, triterpenes, tannins, and flavonoids are the most common secondary metabolites found in this group. They have a wide range of biological activities, including antiviral, hepatoprotective, cytotoxic, antioxidative, anti-Helicobacter pylori, and prostaglandin E2 inhibition (Novakovic et al., 2013).

In Asia, Africa, and Madagascar, the genus of Moringa includes 13 species that have been broadly cultivated for their multiple uses (Rani et al., 2018). It is also known as the "drumstick" or "horseradish". The species included in this genus are *M. arborea* Verdc, *M. Borziana* Mattei, *M. concanensis, M. Drouhardii* Jum. *M. hildebrandtii, M. longituba, M. oleifera, M. Ovalifolia* Dinter and Berger, *M. peregrina, M. pygmaea, M. rivae, M. ruspoliana*, and *M. stenopetala*. It is a common vegetable that is sometimes referred to as the "miracle tree" because of its remarkable therapeutic abilities for a variety of disorders, including some long-lasting ailments (AbdullRazis et al., 2014). *M. oleifera* is a rapid-growing deciduous tree from the Moringaceae family with a trunk diameter of 45 cm, and a height of 10–12 m. *M. oleifera* is a versatile herbal plant with nutritional and health benefits that is used as a portion of human food and a medicinal alternative all over the world (AbdullRazis et al., 2014). This tree is regarded as the most valuable tree on the planet because every component, including the leaves, stem, bark, root, pod, flower, seeds, gum, and seed oil, is beneficial. The plant is commonly used in African and Asian folkloric medicine to treat ulcers, wounds, inflammation, heart disease, cell proliferation inhibition, stroke, obesity, anemia, and liver damage (Aja et al., 2014). When Moringa is eight months old, it grows cream-colored flowers, and the flowering season lasts from January to March. The pods are 30–50 cm long and have a triangular cross-section, with a sticky, black-winged seed inside. The fruit ripens from April to June (Agbogidi and Ilondu, 2012). Various parts contain a profile of essential minerals, amino acids, proteins, quercetin, zeatin, vitamins, β -carotene, β -sitosterol, caffeoylquinic acid, and kaempferol are all found in abundance in the Moringa plant (Anwar et al., 2007). Many nations, including India, Pakistan, Philippines, Hawaii, and many parts of Africa, use the leaves, berries, flowers, and immature pods of this tree as a highly nutritious vegetable (Basit et al., 2015).

23.2 DESCRIPTION

A. glutinosa from the Betulaceae family usually overgrows in wet locations. Gummy young twigs and leaves of black Alder and obovate to rounded, glossy dark green leaves with doubly toothed margins and blunt to occasionally notched apices distinguish this small to a medium-sized deciduous tree. Alder typically lives about 60 years old with a maximum of 160 years depending on the region and grows between 10 and 25 meters tall, with exceptional individuals growing to be 35–40 meters tall (Houstan Durrant et al., 2016). The bark starts brown and lustrous, but it becomes darker, rougher, and fissured as it ages. Leaves rotund or broadly ovate to ellipsoid, 4–9 cm long, 3–7 cm wide, basally rounded; petiole 1–2 cm long; stipules obtuse, soon deciduous; Fruits are rounded with winged seeds. The latin name glutinosa refers to the resinous gum that coats the buds and young leaves, making them slightly sticky (Mushkina et al., 2013). It was used in flood management, riverbank stabilization, and the functioning of river ecosystems; also, alder needs a lot of light and moisture to regenerate naturally, which is usually only possible on disturbed sites (Claessens et al., 2010). Unlike most hygrophilous tree species, black alder can thrive in drier habitats, which acts as a pioneer species. Because of its symbiosis with the actinomycete *Frankia* sp., which is responsible for nitrogen fixation from the atmosphere, it is a very efficient tree species in soil nitrate enrichment (Deptula et al., 2020). Because of the presence of naphthoquinones, the bark is also used for dyeing purposes (Wangkheirakpam and Laitonjam, 2016). It is also the dominant species in several European priority habitats, making it both a conservation goal and a significant forestry species (Rodríguez-González et al., 2014).

M. oleifera is known as "mother's best friend" because it is used to increase a woman's milk production and is often advised to treat anemia. (Anwar et al., 2007). *M. oleifera* has been used in traditional food for 5000 years and is now cultivated for industrial and medicinal purposes all over the world. It can even tolerate harsh drought conditions. *M. oleifera* is also known as the horseradish flower, ben oil tree, drumstick, miracle tree, sohanjna, marango, moonga, kelor, nebeday, and mother's best friend (Chhikara et al., 2021). The fruits are three-sided, linear, pendulous pods with nine longitudinal ridges, fragrant, bisexual flowers surrounded by five unequally veined yellowish-white petals typically 20–50 cm long but can be up to 1 m long and 2.0–2.5 cm wide; it contains nearly 20 globular seeds (Chukwuebuka, 2015). *M. oleifera* flowers contain stimulants, aphrodisiacs, abortifacients, and cholagogues. Piles, fevers, sore throats, bronchitis, scurvy, and catarrh are all treated with the leaves as a poultice. Laxative, abortifacient, vesicant, carminative, and anti-inflammatory properties are all present in the root of *M. oleifera*. The stem bark is used to treat eye diseases because it has Rubefacient and vesicant properties (Basit et al., 2015).

M. oleifera seeds are the best natural coagulants, with antimicrobial and antioxidant properties, and are often used as a cooking oil due to their high nutritional value (Yameogo et al., 2011). Leaves of Moringa contain calcium, magnesium, potassium, sodium, iron, zinc, and phosphorous (Yameogo et al., 2011). Dry leaves contain around 23.6% carbohydrates as an energy source, 35.0% crude fiber (a type of fiber that helps in bowel movement), 10.0% moisture (which makes up around 80% of blood and is a vital medium for oxygen transport), 10.0% ash, and 30.29% crude protein (a measure of mineral content in food and is needed by the body), (Chukwuebuka, 2015). *M. oleifera* bark is boiled in water and dissolved in alcohol and the resulting extract is used to treat stomach disorders (assisting digestion, easing stomach pain, and easing ulcers), poor vision, joint pain, diabetes, anemia, hypertension, toothache, hemorrhoids, and uterine disorders (Gandji et al., 2018). Seeds of *M. oleifera* can also be consumed. Moringa seeds contain a lot of oil, which can be extracted by pressing them known as ben oil, has nourishing assets and features similar to olive oil (Khor et al., 2018).

23.3 TRADITIONAL KNOWLEDGE

Many ethnic groups have used medicinal plants in their traditional information systems since ancient times. Herbal drugs, also known as ethnomedicines, are now widely used. In today's world, it's the source of a lot of necessary medicines. Plants have been used to treat and rebuild people since olden times and continue to do so today and treat diseases in many parts of the world. Alnus plants have been linked to several positive pharmacological effects like anticancer, antiviral, antibacterial infections, especially against the throat and mouth (Dahija et al., 2016). Plants were also used to treat rheumatism, dental abscess, hemorrhoids, skin diseases, inflammation, and wound-healing (Dinić et al., 2015) (Altınyay et al., 2016).

A. glutinosa is a tree that grows in the United States (L.) Gaertn. has a long medical history and focuses on extensive research into their vital functions (Ren et al., 2017). The bark, leaves, and cones are used as folk herbal medicine to treat various ailments such as swelling, inflammation, and rheumatism have also been treated with a decoction of *A. glutinosa* barks. It's also been used as an astringent, bitter, emetic, and hemostatic to treat sore throats and pharyngitis (Middleton et al., 2005). Due to its strong chemo-protective, antioxidant, and antimicrobial properties, it is often taken as a dietary supplement to help prevent the onset of various chronic dermatological conditions. It has also become a study hotspot (Ren et al., 2017). In the Republic of Belarus, the leaves of this plant are used as an antioxidant source (Mushkina et al., 2013). The decoctions of the leaves are used locally for gargling in the treatment of sore throats, pharyngitis, and internally in the treatment of intestinal bleeding due to their astringent and hemostatic properties (Mushkina et al., 2013). The bark of the black alder has recently become available as a food source for the action of various skin conditions (Dinić et al., 2015).

Moringa spp. is included in the Varunadi group of herbs, which are beneficial to the urinary system, according to Sushruta and Vagbhata. Moringa spp. comes in three types, according to Ayurveda: Shyama (black), Shveta (white), and Rakta (red) (Mohanty et al., 2021). *M. oleifera* was introduced to many parts of the world, including Algeria, where it is grown in some Saharan areas (Bachar and tamanrasset). Its edible leaves and seeds (fresh, powdered, or cooked) are high in essential sulfur amino acids, protein, and minerals. Since ancient times, the plant's many resources—leaves, bulbs, seeds, pods, bark, and roots have been used in cooking and herbal medicine to treat various ailments. Traditionally, it was used to treat wounds and illnesses such as colds and diabetes. *M. oleifera* species are well-known for their antioxidant, anti-inflammatory, anticancer, and anti-hyperglycemic properties. The bulk of their biological activity is due to their high content of flavonoids, glucosides, and glucosinolates. Other popular uses involve treating skin infections, anxiety, asthma, wounds, fever, diarrhea, and sore throats and boosting one's resilience and relieving pain and stress when serving in the military (Rani et al., 2018). *M. oleifera* roots are used in Ayurvedic medicine to treat epilepsy, hysteria, heart disorders, colic, and flatulence, as well as anxiety, leaves in the treatment of catarrh, bronchitis, sore throat, headaches, constipation, and seeds in the treatment of hypertension, neuralgia indicating). All the above mentioned diseases correspond to vatta, and kapha (Table 23.1).

23.4 CHEMICAL DERIVATIVES

Diarylheptanoids, natural products with a 1,7-diphenylheptane structural skeleton and appear in linear and cyclic forms, are the group's most prominent constituents (Ren et al., 2017). *A. glutinosa* bark includes several diarylheptanoids;5-O-β-D-glucopyranosyl-heptan-3-one, hirsutenone, hirsutanonol-5-O-β-D-glucopyranoside, platyphyllonol-5-O-β-D-xylopyranoside, oregonin, platyphylloside, rubranoside A, rubranoside B, hirsutanonol, aceroside VII, alnuside A and B, 1,7-bis-(3,4-dihydoxyphenyl)-5- hydroxy-heptane-3-O-β-D-xylopyranoside, (5S)-1-(4-hydroxyphe

TABLE 23.1
General Use of *Alnus glutinosa* and *Moringa oleifera*

Plant	Part	Therapeutic Use
A. glutinosa	Leaves	Antioxidants, anticancer
	Bark	Antioxidants, anti-inflammatory
	Seeds	Antibacterial
M. oleifera	Leaves	Antidiabetic, anti-obesity, hepatoprotective, anti-Parkinson's, anti-inflammatory, anticancer, antimicrobial
	Roots	Antimicrobial, hepatoprotective, anti-inflammatory
	Stem	Antifungal, urine infection
	flowers	Anti-inflammatory, anticancer, antimicrobial
	Seeds	Antidiabetic, anti-obesity, anti-Alzheimer's, anti-asthmatic, anti-inflammatory, antimicrobial
	pods	Antidiabetic, chemoprotective

nyl)-7-(3,4-dihydroxyphenyl) and (5S)-1,7-bis-(3,4-dihydroxyphenyl)-5- O-β-D-[6-(3,4-dimeth oxycinnamoylglucopyranosyl)]- heptan-3-one (Novaković et al., 2013). The seeds of *A. glutinosa* contain dirylheptanoid derivatives, hirsutanonol, oregonin, and genkwanin, a flavone (O'Rourke et al., 2005). Ellagitannin, glutinoin, pedunculagin, and praecoxin D were detected from the cones of *A. glutinosa* (Ivanov et al., 2012). Brassinolide and castasterone were detected from pollens of *A. glutinosa* and confirmed using mass spectroscopy (Plattner et al., 1986). Hirsutanonol, glutinic acid, rhododendron, {3-(4-hydroxyphenyl)-l-methylpropyl-β-D-glucopyranoside}, genkwanin, and oregoninbioactives were also detected from *A. glutinosa* (Sati et al., 2011). Platyphylloside, platyphyllonol-5-O-b-D-xylopyranoside, hirsutenone, andalnuside B were noticed and confirmed using nuclear magnetic resonance, high resolution electrospray ionization mass spectrometry, UV, infrared, and circular dichroism (Novaković et al., 2013). Lupenon, Taraxerone, β-Sitosterol, Lupeol, Simiarenol, Betulin L upenylacetate, and Betulinic acid were detected using bark extracts GC-MS (Felföldi-Gáva et al., 2012). Uvaol, betulinicaldehid, betulinic acid, ursolic acid, betulin, β-amyrin were spotted in bark extract using Liquid chromatography-mass spectroscopy analysis (Felföldi-Gáva et al., 2012). Tannic acid (16–20%), triterpenes (lupenon, glutinon, taraxerol), and sterols (β-sitosterol) are the constituents present in the bark of *A. glutinosa* (Mushkina et al., 2013). Foliar buds of *A. glutinosa* contain ferulic acid, quercetin, p-coumaric acid, hyperoside from hydrolyzed and nonhydrolyzed samples (Peev et al., 2007).

In case of *M. oleifera* methanolic leaf extract, it showed the presence of L-(+)-ascorbic acid-2,6-dihexadecanoate, 14–methyl-8-hexadecenal, phytol, 9-octadecenoic acid, octadecamethylcyclononasiloxane, 1, 2-benzene dicarboxylic acid, 4-hydroxyl-4-methyl-2-pentanone, 3-ethyl-2, 4-dimethylpentane,3, 4-epoxyethanone, N-(-1-methylethyllidene)-benzene ethanamine, 4, 8, 12, 16 tetramethylheptadecan-4-olide, 3-5-bis (1, 1-dimethylethyl)-phenol, 1-hexadecanol, 3, 7, 11, 15-tetramethyl-2 hexadecene-1-ol, hexadecanoic acid and 1, 2, 3-propanetriyl ester-9 octadecenoic acid. Whereas, methanolic seed extract contains L-(+) – ascorbic acid-2, 6-dihexadecanoate, methyl ester-hexadecanoicacid, oleic acid, 9-octadecenoic acid, and 9-octadecenamide (Aja et al., 2014). Pods of *M. oleifera* showed the presence of Nitriles, an isothiocyanate, thicarbamates, methyl-phydroxybenzoate, nitrites, β-sitosterol Isothiocyanate, O-[20-hydroxy-30-(200- heptenyloxy)]-propylundecanoate, O-ethyl-4-[(α-L-rhamnosyloxy)-benzyl] carbamate, thiocarbamates, O-(1heptenyloxy) propyl undecanoate, O-ethyl-4-(alpha-L-rhamnosyloxy) benzyl carbamate, methyl p-hydroxybenzoate, beta-sitosterol (Mohanty et al., 2021). *M. oleifera* leaves have been reported to be a rich source of β-carotene, protein, vitamin C, calcium, and potassium with promising source of natural antioxidants (Anwar et al., 2007). 4(α-L-rhamnosyloxy)-benzyl isothiocyanate, niazimicin, 3-*O*-(6′-*O*-oleoyl-β-D-glucopyranosyl)- β-sitosterol, β-sitosterol-3-*O*- β-D-glucopyranoside, niazirin, glycerol-1-(9-octadecanoate), β -sitosterol bioactives has been isolated from moringa seeds (Anwar et al., 2007). Moringine, quercetin, kaempferol, 4-(α-L-rhamnopyranosyloxy)-benzyl glucosinolate, 4(α-L-Rhamnosyloxy) benzyl isothiocyanate, 4-(4'-O-acetyl-α-L-rhamnosyloxy) benzyl isothiocyanate, 4-(α-L-rhamnosyloxy) phenyl acetonitrile, gallic acid, ellagic acid, R = H → β-sitosterol, Pterygospermin have been isolated from *M. oleifera* (Chhikara et al., 2021). Even, roots of *M. oleifera* noted the presence of moringine, moringinine, spirachin, 4-(α-L-rhamnopyranosyloxy)- benzylglucosinolate, and benzylglucosinolate Moringine, moringinine, spirachin, 1,3-dibenzyl urea, alpha-phellandrene, p-cymene, Deoxy-niazimicine, 4-(alpha-L-rhamnopyranosyloxy) benzylglucosinolate bioactives (Mohanty et al., 2021).

23.5 POTENTIAL BENEFITS, APPLICATIONS AND USE

23.5.1 *A. GLUTINOSA*

23.5.1.1 Antibacterial Activity

Methanolic extract of *A. glutinosa* seeds displayed antibacterial activity against *Staphylococcus aureus* NCTC 10788, *Pseudomonas aeruginosa* NCTC 6750, *Escherichia coli* NCIMB 8110,

Lactobacillus plantarum NCIMB 6376, *Escherichia coli* NCIMB 4174, *Klebsiella aerogenes* NCTC 9528, *Citrobacter freundii* NCTC 9750, and *Staphylococcus aureus* NCTC 11940 (methicillin-resistant) (Middleton et al., 2005).

23.5.1.2 Antioxidant Activity

Leaf extract possesses the highest antioxidant activity in comparison with bark extract. The antioxidant function of the extract is related to the presence of phenolics and flavonoids (Dahija et al., 2014). The bioactives isolated from *A. glutinosa* cones including ellagitannin, glutinoin, pedunculagin, and praecoxin D executed radical-scavenging activity (Ivanov et al., 2012). Even leaf extract of *A. glutinosa* extracted in solvent water:ethanol at ratio 7:3 and 3:7 showed radical-scavenging activity by 1,1-diphenyl-2 picrylhydrazyl (DPPH) radical scavenging method ranged from 49.21% to 49.42%. The radical-scavenging activity in the leaf extract is due to the phenolics present ranging from 17.82% and 18.96% (Mushkina et al., 2013). The whole bark of *A. glutinosa* revealed 276.55±28.57 (mg g^{-1}) radical-scavenging activity with 21.57±4.08 and 29.00±5.33 of total flavonoids and phenolic content (mg g^{-1}) equivalent to quercetin and gallic acid from the continental zone (Skrypnik et al., 2019). Whereas radical-scavenging activity from the coastal region was 60.96±6.12 and flavonoid and phenolic content was 9.68±1.84 and 12.18±1.53 (Skrypnik et al., 2019). Thus, the activity of alder bark was higher from the continental zone than the coastal region, therefore permitting to elevate the production of extractive biomolecules.

23.5.1.3 Anticancer Activity

5(S)-1,7-di(4-hydroxyphenyl)-3-heptanone-5-O-β-D-glucopyranoside and 5(S)- 1,7-di(4-hydroxyphenyl)-5-O-β-D-[6-(E-p-coumaroylglucopyranosyl)] heptane-3-one neutralizes reactive oxygen species. Even non-small cell lung carcinoma development was inhibited with a minimum inhibitory concentration of less than 30 μM (Dinić et al., 2014). Thus, these compounds can act as standard cell protectors during chemotherapy without significantly reducing the chemotherapeutic's effectiveness. Platyphlloside, platyphyllonol-5-O-b-D-xylopyranoside, alnuside B, and hirsutenone all had substantially better anticancer efficacy than diarylheptanoid curcumin, which was used as a positive control against lung carcinoma cells (Novaković et al., 2014). The PC-3 and HeLa cell lines were found to be cytotoxic to the leaf extract. HT-29 human colon carcinoma cells were cytotoxic to the hirsutanone compound isolated from leaves (León-Gonzalez et al., 2014). *A. glutinosa* leaf extract had antiproliferative activity against the C6 and Hela cell lines, with IC50 values of 45.7ug/ml and 53.8ug/ml, respectively (Sahin Yaglıoglu et al., 2016).

23.5.1.4 Chemoprotective Agent

Platyphylloside - 5(*S*)-1,7-di(4-hydroxyphenyl)-3- heptanone-5-O-*β*-D-glucopyranoside and 5(*S*)-1,7-di(4-hydroxyphenyl)-5-O-*β*-D-[6-(*E-p*-coumaroylglucopyranosyl)] heptane-3 one compounds protected noncancerous human keratinocytes (HaCaT) against doxorubicin-induced DNA damage in cells (Dinić et al., 2015). As a result, during chemotherapy, it acts as a chemoprotective agent for noncancerous dividing cells.

23.5.1.5 Anti-inflammatory Activity

Activated inflammatory or immune cells mediate inflammation is a multistep operation. Macrophages are important players in this process because they facilitate immunopathological changes such as the overproduction of proinflammatory mediators and cytokines caused by activated tumor necrosis factor and cyclooxygenase 2 (Kalinkevich et al., 2014). The effect of *A. glutinosa* stem bark on TNF-α levels suggests that it is linked to a reduction in inflammatory processes through cytokine deprivation in Hela cells (HL-60) and may be used in ethnopharmacology (Acero and Muñoz-Mingarro, 2012). Even *A. glutinosa* cones inhibited TNF- α and PGE2 by 96 ± 0.6 and 25 ± 7.1% at concentration 1mg/ml (Kalinkevich et al., 2014).

23.5.1.6 Nitrogen Fixation

In terms of biological soil regeneration, *A. glutinosa* has shown promising results because of an aerenchyma, that allows oxygen to diffuse to the roots via diffusion and stem photosynthesis, as its roots may penetrate anaerobic soils (Warlo et al., 2019). It basically lives in symbiosis with nitrogen-fixing actinomycetes *Frankia* sp. As a result of its nitrogen-fixing powers, it can help to increase soil fertility.

23.5.1.7 Insecticidal Activity

Secondary metabolites present in plants have an antibacterial and fungistatic effect on insects, functioning as repellents, antifeedants, poisons, or deterrents to feeding or oviposition. *A. glutinosa* aqueous extract has executed insufficient insecticidal properties against *Phyllotreta* sp. and *Plutellaxylostella* larvae and pupae (Jankowska and Wojciechowicz-Żytko, 2016).

23.5.1.8 Dyeing Property

Natural hair dyes contain few chemical components, and many people prefer them to synthetic hair colors. Tannin content present in *A. glutinosa* executes hair dyeing ability (Onal and Demir, 2009). As a result, this tree is crucial in the creation of cosmetics.

23.5.2 *M. oleifera*

23.5.2.1 Moringa in Water Treatment

Moringa seeds have long been used to clarify drinking water and reduce the health risks associated with high turbidity in many rural areas. Through extensive testing, *M. oleifera* seeds are an important water-clarifying agent in various colloidal suspensions (Arora et al., 2013). Its pods have been used as a coagulant for water treatment and as a cheap and efficient sorbent for organics removal. *M. oleifera* biomass can be utilized in a cost-effective and efficient method in the dairy industry effluent treatment. Removal efficiencies of up to 98% for both color and turbidity were achieved using 0.2 g Moringa extract and 0.2 L of 1.0 g/L sorbate solution (Vieira et al., 2010). In treating wastewater containing water-soluble reactive red2, *M. oleifera* seed protein-montmorillonite composite showed promise as an adsorbent (Mi et al., 2019).

23.5.2.2 Antidiabetic Activity

Both Type 1 and Type 2 diabetes can be treated with *M. oleifera*. Polyphenols, including rutin, quercetin-3-glycoside, kaempferol, and glycosides, help to lower blood sugar levels. *M. oleifera* alkaloids including, terpenoids, flavonoids, glycosides, and carotenoids have been found to have antidiabetic assets (Chhikara et al., 2021). Methanolic extract of *M. oleifera* pods reduced serum glucose and nitric oxide levels while increasing antioxidant levels in pancreatic tissue by restoring histoarchitectural impairment to the islet cells (Gupta et al., 2012). The activity of α-amylase and α-glucosidase was significantly decreased by aqueous extract *M. oleifera* leaf extract, with increased antioxidant capacity, glucose tolerance, and glucose uptake rate in yeast cells and also, in Streptozotocin-induced diabetes, it protects the pancreas from ROS-mediated damage by increasing cellular antioxidant defences and lowering hyperglycemia (Khan et al., 2017). Even *M. oleifera* seed powder was effective against streptozotocin-induced diabetic rats at low doses by restoring the level of lipid peroxide, IL-6, immunoglobulins, urine parameters, and liver enzymes. (Al-Malki and El Rabey, 2015).

23.5.2.3 Antimicrobial Activity

Moringa has been validated against several bacteria and proved to be efficient against several microorganisms. Different parts of *M. oleifera* executed antimicrobial activity against *Staphylococcus aureus, Scenedesmus obliquus, Pseudomonas aeruginosa, Proteus mirabilis, Mycobacterium phlei,*

Bacillus subtilis, Klebsiella pneumoniae, Salmonella typhi A, and *Escherichia coli*. The activity is due to the chemical constituents, such as pterygospermin, 4-(4'-O-acetyl-a-L-rhamnopyranosyloxy) benzylisothiocyanate, 4-(a-L-rhamnopyranosyloxy) benzylisothiocyanate, niazimicin, benzyl isothiocyanate, 4-(a-L-rhamnopyranosyloxy) benzyl, spirochin, and anthonine present in Moringa to inhibit the microbes (Fozia Farooq, 2012).

23.5.2.4 Anti-Obesity Activity

Obesity is a global health issue that is regarded as a significant cause of illness and mortality. Moringa has been proved to contain exclusive polyphenols and may respond to alleviate the variations related to obesity. In obese rats, *M. oleifera* ethanolic extract reduced leptin and resistin mRNA expression while increasing adiponectin gene expression in female Wistar rats. This change in gene expression was accompanied by weight loss and improvements in the atherogenic index, coronary artery index, and glucose level (Metwally et al., 2017). Also, in Sprague-Dawely male rats, *M. oleifera* seed oil extract significantly improved hematological and metabolic disturbances and lowered leptin and resistin levels, reducing lipid peroxidation and inflammatory cytokines (Kilany et al., 2020). Ethanolic leaf extract of *M. oleifera* was fed with a dose of 300mg/kg to male Wistar rats, which reduces body weight gain and improved glucose, lipid fractions, and metabolic hormone levels, showing that it has anti-obesity properties and can reduce tissue insulin resistance. Moringa treatment also reduced oxidative stress and increased antioxidants in the liver, and improved liver function enzymes, demonstrating that the plant has antioxidant potential (Othman et al., 2019). Improving redox balance pathways is a good therapeutic choice in cases of obesity and its harmful consequences.

23.5.2.5 Anticancer Activity

Studies are focused on the anticancer activity of different *M. oleifera* parts extract due to high nutritional benefits from Moringa. At doses of 22.61 g/mL and 6.25 g/mL, Prostate cancer cells treated with *M. oleifera* flower extract displayed a 50% inhibition in PC3 cells after 24 and 48 hours. *M. oleifera* flower extract induced the accumulation of G1 phase cell cycle arrest and apoptosis using annexin V staining (Ju et al., 2018). Ethanolic extract of *M. oleifera* leaves and bark was effective against MDA-MB-231 and HCT-8 cancer cell lines, reducing cell survivability. The number of apoptotic cells in MDA-MB-231 cell lines treated with leaves increased from 27% to 46%, and in bark treated cell lines increased from 27% to 29%, while control cells showed only 5% late apoptotic cells. In HCT-8 cell lines, there was a similar rise in late apoptotic cells (Al-Asmari et al., 2015). Seed extract had no anticancer activity against the cancer cell lines MDA-MB-231 and HCT-8.

23.5.2.6 Anti-Inflammatory Activity

Inflammation is a host defense mechanism that defends against bacteria, stress, and tissue damage, contributing to the progression of many chronic diseases. *M. oleifera* seed extract reduced carrageenan-induced rat paw edema by 33% at 500 mg/kg dose, which was equal to aspirin's reduction by 27% at 300 mg/kg; with this even *in-vitro*, studies dramatically reduced nitric oxide generation, and gene expression of LPS-inducible nitric oxide synthase, interleukins 1β and 6 (Jaja-Chimedza et al., 2017). Also, *M. oleifera* leaf extract at a concentration 100 μg/ml also lowered nitric oxide production than standard aspirin on RAW 264.7 macrophage cells (Xu et al., 2019). Even ethanolic flower extract possessed anti-inflammatory activity, and it was proved by *in-vitro* protein denaturation assay (Alhakmani et al., 2013).

23.5.2.7 Anti-Asthmatic Property

During a trial to assess the efficacy and safety of seed kernels for the management of asthmatic patients, *M. oleifera* seed kernels showed promising action in treating bronchial asthma. The activity is probably due to the moringine alkaloid, which relaxes bronchioles. Studies have found a significant reduction in the intensity of asthma symptoms and improvements in respiratory function (Fozia Farooq, 2012).

23.5.2.8 Neuroprotective Property

Alzheimer's disease is the most common neurodegenerative disease, characterized by the build-up of beta-amyloid plaques and neurofibrillary tangles in the brain, resulting in memory loss. Ethanolic *M. oleifera* extract ameliorates scopolamine-induced memory and learning impairments by activating the Akt, ERK1/2, and CREB signaling pathways and improving the cholinergic neurotransmission system and neurogenesis (Zhou et al., 2018). Even leaf extract possessed the anti-Alzheimer property by improving spatial memory and neurodegeneration in the dentate gyrus of the hippocampus and lowering malondialdehyde levels and AChE activity while increasing superoxide dismutase and catalase activity (Sutalangka et al., 2013). Also, pretreatment with isothiocyanate derived from *M. oleifera* seed extract for one week not only altered the signaling pathways for inflammation, apoptosis, and oxidative stress in a sub-acute Parkinson's disease animal model by 1-methyl-4-phenyl-1,2,3,6-tetrahydropyridine but also altered the signaling pathways associated with PD (Kou et al., 2018).

23.5.2.9 Hepatoprotective Property

Liver illness, caused by drugs, toxic substances, or excessive alcohol intake, is one of the most complex health disorders to overcome. The liver plays a significant role in xenobiotics and the detoxification of several drugs. *M. oleifera* leaves effectively prevented rat hepatotoxicity caused by carbon tetrachloride by restoring serum marker enzymes (Singh et al., 2014). The noted activity may be due to the polyphenols present in the leaf extract having radical-scavenging property. Similarly, pretreatment with *M. oleifera* leaf extract for 28 days will protect rats from cadmium-induced hepatotoxicity by suppressing increased biochemical marker enzymes (Toppo et al., 2015) (Figure 23.1).

23.6 CONCLUSION

Herbal medicine is now well established for therapeutic purposes and is universally recognized as safe and effective. This chapter outlined research advances in the use of *A. glutinosa* and *M. oleifera* extracts in various bioscience areas depicting the usefulness of this plant. Thus, *A. glutinosa* and *M. oleifera* can help with a range of health issues, including illnesses and malnutrition, while also serving as a valuable natural resource for the population and industry. The availability of accurate

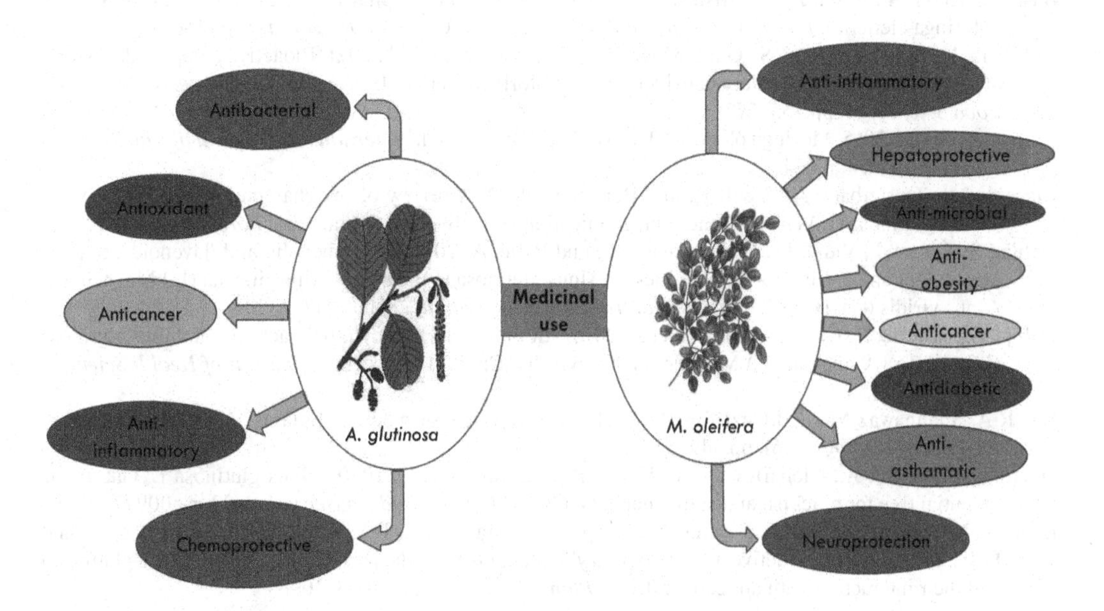

FIGURE 23.1 Medicinal use of *Alnus glutinosa* and *Moringa oleifera*.

analytical instruments with significant developments in metabolic engineering will purpose in the growth of life-saving, beneficial pharmaceuticals from both *A. glutinosa* and *M. oleifera*.

ACKNOWLEDGMENT

DST-INSPIRE's meritorious fellowship to Ms. Patil D. N. from DST, New Delhi, India is gratefully acknowledged.

REFERENCES

Acero, N., and Muñoz-Mingarro, D. 2012. Effect on tumor necrosis factor-α production and antioxidant ability of black alder, as factors related to its anti-inflammatory properties. *Journal of Medicinal Food*, 15, pp.542–548.

Agbogidi O.M., and Ilondu E.M. (2012) Moringa oleifera lam: Its potentials as a food security and rural medicinal item. *Journal of Bio Innovation*, 6, pp.156–167.

Aja, P.M., Nwachukwu, N., Ibiam, U.A., Igwenyi, I.O., Offor, C.E., and Orji, U.O. 2014. Chemical constituents of Moringa oleifera leaves and seeds from Abakaliki, Nigeria. *American Journal of Phytomedicine and Clinical Therapeutics*, 2, pp.310–321.

Al-Asmari, A.K., Albalawi, S.M., Athar, M.T., Khan, A.Q., Al-Shahrani, H., and Islam, M. 2015. Moringa oleifera as an anticancer agent against breast and colorectal cancer cell lines. *PLoS One*, 10, pp.1–14.

Al-Malki, A., and El Rabey, H.A. 2015. The antidiabetic effect of low doses of Moringa oleifera lam. Seeds on Streptozotocin induced diabetes and diabetic nephropathy in male rats. *BioMed Research International*, 2014, pp.1–14.

Alhakmani, F., Kumar, S., Khan, S.A. 2013. Estimation of total phenolic content, in-vitro antioxidant and anti-inflammatory activity of flowers of Moringa oleifera. *Asian Pacific Journal of Tropical Biomedicine*, 3, pp.623–627.

Altınyay, C., Süntar I., Altun L., Keleş H., and Akkol E.K. Phytochemical and biological studies on *Alnus glutinosa* subsp. *glutinosa*, *A. orientalis* var. *orientalis* and *A. orientalis* var. *pubescens* leaves. *Journal of Ethnopharmacology*, 192, pp.148–160.

Anwar, F., Latif, S., Ashraf M., and Gilani, A.H., 2007. *Moringa oleifera*: A food plant with multiple medicinal uses. *Phytotherapy Research*, 21, pp.17–25.

Arora, D.S., Onsare, J.G., and Kaur, H. 2013. Bioprospecting of Moringa (Moringaceae): Microbiological Perspective. 1, pp.193–215.

Basit, A., Rizvi, A., Alam, J., and Mishra, A. 2015. Phytochemical and pharmacological overview of Sahajan (Moringa oleifera). *International Journal of Pharma And Chemical Research*, 1, pp.156–164.

Chhikara, N., Kaur, A., Mann, S., Garg, M.K., Sofi, S.A., and Panghal, A. 2021. Bioactive compounds, associated health benefits and safety considerations of Moringa oleifera L.: An updated review. *Nutrition and Food Science*, 51, pp.255–277.

Chukwuebuka, E. 2015. Moringa oleifera "The Mother's Best Friend." *International Journal of Nutrition and Food Sciences*, 4, p.624.

Claessens H., Oosterbaan A., Savill P., and Rondeux J.2010. A review of the characteristics of black alder (Alnus glutinosa (L.) Gaertn.) and their implications for silvicultural practices. *Forestry*, 83, p.2.

Dahija, S., Čakar, J., Vidic, D., Maksimović, M., and Parić, A. 2014. Total phenolic and flavonoid contents, antioxidant and antimicrobial activities of Alnus glutinosa (L.) Gaertn., Alnus incana (L.) Moench and Alnus viridis (Chaix) DC. extracts. *Natural Product Research*, 28, pp.2317–2320.

Dahija, S., Haverić, S., Čakar, J., and Parić, A. 2016. Antimicrobial and cytotoxic activity of Alnus glutinosa (L.) Gaertn., A. incana (L.) Moench, and A. viridis (Chaix) DC. extracts. *Journal of Health Sciences*, 6, pp.100–104.

Dar R.A, Shahnawaz M., and Qazi PH. 2016. General overview of medicinal plants: A review. *Journal of Phytopharmacology*, 6(6), pp.349–351.

Deptuła M., Piernik A., Nienartowicz A., Hulisz P., and Kaminski D. 2020. Alnus glutinosa L. Gaertn. as potential tree for brackish and saline habitats. *Global Ecology and Conservation*, 22, p.e00977.

Dinić, J., Novaković, M., Podolski-Renić, A., Stojković, S., Mandić, B., Tešević, V., Vajs, V., Isaković, A., and Pešić, M. 2014. Antioxidative activity of diarylheptanoids from the bark of black alder (Alnus glutinosa) and their interaction with anticancer drugs. *Planta Medica*, 80, pp.1088–1096.

Dinić, J., Randelović, T., Stanković, T., Dragoj, M., Isaković, A., Novaković, M., and Pešić, M. 2015. Chemoprotective and regenerative effects of diarylheptanoids from the bark of black alder (Alnus glutinosa) in human normal keratinocytes. *Fitoterapia*, 105, pp.169–176.

Farooq, F. 2012. Medicinal properties of Moringa oleifera: An overview of promising healer. *Journal of Medicinal Plant Research*, 6, pp.4368–4374.

Felföldi-Gáva, A., Szarka, S., Simándi, B., Blazics, B., Simon, B., and Kéry, Á. 2012. Supercritical fluid extraction of Alnus glutinosa (L.) Gaertn. *Journal of Supercritical Fluids*, 61, pp.55–61.

Gandji, K., Chadare, F.J., Idohou, R., Salako, V.K., and Assogbadjo, A.E. 2018. Status and utilisation of Moringa oleifera Lam: A review. *African Crop Science Journal*, 26, pp.137–156.

Gupta, R., Mathur, M., Bajaj, V.K., Katariya, P., Yadav, S., Kamal, R., and Gupta, R.S. 2012. Evaluation of antidiabetic and antioxidant activity of Moringa oleifera in experimental diabetes. *Journal of Diabetes*, 4, pp.164–171.

Houston Durrant, T., de Rigo, D., and Caudullo, G. 2016. Alnus glutinosa in Europe: Distribution, habitat, usage and threats. *European Atlas of Forest Tree Species*, 1, 64–65.

Ivanov, S.A., Nomura, K., Malfanov, I.L., and Ptitsyn, L.R. 2012. Glutinoin, a novel antioxidative ellagitannin from Alnus glutinosa cones with glutinoic acid dilactone moiety. *Natural Product Research*, 26, pp.1806–1816.

Jaja-Chimedza, A., Graf, B.L., Simmler, C., Kim, Y., Kuhn, P., Pauli, G.F., and Raskin, I. 2017. Biochemical characterization and anti-inflammatory properties of an isothiocyanate-enriched moringa (Moringa oleifera) seed extract. *PLoS One*, 12, pp.1–21.

Jankowska, B., and Wojciechowicz-Żytko, E. 2016. Efficacy of aqueous extracts of black alder (Alnus glutinosa GAERTN.) and black elderberry (Sambucus nigra L.) in reducing the occurrence of Phyllotreta spp., some lepidopteran pests and diamondback moth parasitoids on white cabbage. *Polish Journal of Entomology*, 85, pp.377–388.

Ju, J., Gothai, S., Hasanpourghadi, M., Nasser, A.A., Aziz Ibrahim, I.A., Shahzad, N., et al. 2018. Anticancer potential of Moringa oleifera flower extract in human prostate cancer PC-3 cells via induction of apoptosis and downregulation of AKT pathway. *Pharmacognosy Magazine*, 14, pp.477–81.

Kalinkevich, K., Karandashov, V.E., and Ptitsyn, L.R. 2014. In vitro study of the anti-inflammatory activity of some medicinal and edible plants growing in Russia. *Russian Journal of Bioorganic Chemistry*, 40, pp.752–761.

Khan, W., Parveen, R., Chester, K., Parveen, S., and Ahmad, S. 2017. Hypoglycemic potential of aqueous extract of moringa oleifera leaf and in vivo GC-MS metabolomics. *Frontiers in Pharmacology*, 8, p.577.

Khor, K.Z., Lim, V., Moses E.J., and Samad N.A., 2018. The in vitro and in vivo anticancer properties of moringa oleifera. *Evidence-Based Complementary and Alternative Medicine*, 2018, pp.1–14.

Kilany, O.E., Abdelrazek, H.M.A., Aldayel, T.S., Abdo, S., and Mahmoud, M.M.A. 2020. Anti-obesity potential of Moringa olifera seed extract and lycopene on high fat diet induced obesity in male Sprauge Dawely rats. *Saudi Journal of Biological Sciences*, 27, pp.2733–2746.

Kou, X., Li, B., Olayanju, J.B., Drake, J.M., and Chen, N. 2018. Nutraceutical or pharmacological potential of Moringa oleifera Lam. *Nutrients*, 10(3), p.343.

León-Gonzalez, A.J., Acero, N., Munoz-Mingarro, D., López-Lázaroa, M., Martín-Cordero, C. 2014. Cytotoxic activity of hirsutanone, a diarylheptanoid isolated from Alnus glutinosa leaves. *Phytomedicine*, 21, pp.866–870.

Metwally, F.M., Rashad, H.M., Ahmed, H.H., Mahmoud, A.A., Abdol Raouf, E.R., and Abdalla, A.M. 2017. Molecular mechanisms of the anti-obesity potential effect of Moringa oleifera in the experimental model. *Asian Pacific Journal of Tropical Biomedicine*, 7, pp.214–221.

Mi, X., Shang, Z., Du, C., Li, G., Su, T., Chang, X., Li, R., Zheng, Z., and Tie, J. 2019. Adsorption of an anionic azo dye using moringa oleifera seed protein montmorillonite composite. *Journal of Chemistry*, 2019, pp.1–8.

Middleton, P., Stewart, F., Al-Qahtani, S., Egan, P., Rourke, C., Abdulrahman, A., Byres, M., et al. 2005. Antioxidant, antibacterial activities and general toxicity of Alnus glutinosa, Fraxinus excelsior and Papaver rhoeas. *Iranian Journal of Pharmaceutical Research*, 2, pp.81–86.

Mohanty, M., Mohanty, S., Bhuyan, S.K., and Bhuyan, R. 2021. Phytoperspective of Moringa oleifera for oral health care: An innovative ethnomedicinal approach. *Phytherapy Research*, 35, pp.1345–1357.

Mushkina, O.V., Gurina, N.S., Konopleva, M.M., Bylka, W., and Matlawska, I. 2013. Activity and total phenolic content of alnus glutinosa and alnus incana leaves. *Acta Scientiarum Polonorum, Hortorum Cultus*, 12, pp.3–11.

Novakovic, D., Feligioni, M., Scaccianoce, S., Caruso, A., Piccinin, S., Schepisi, C., Errico, F., Mercuri, N.B., Nicoletti, F., and Nisticò, R. 2013. Profile of gantenerumab and its potential in the treatment of Alzheimer's disease. *Drug Design, Development and Therapy*, 7, pp.1359–1364.

Novaković, M., Pešić, M., Trifunović, S., Vučković, I., Todorović, N., Podolski-Renić, A., Dinić, J., Stojković, S., Tešević, V., Vajs, V., and Milosavljević, S. 2014. Diarylheptanoids from the bark of black alder inhibit the growth of sensitive and multi-drug resistant non-small cell lung carcinoma cells. *Phytochemistry*, 97, pp.46–54.

Novaković, M., Stanković, M., Vučković, I., Todorović, N., Trifunović, S., Tešević, V., Vajs, V., and Milosavljević, S. 2013. Diarylheptanoids from alnus glutinosa bark and their chemoprotective effect on human lymphocytes DNA. *Planta Medica*, 79, pp.499–505.

O'Rourke, C., Byres, M., Delazar, A., Kumarasamy, Y., Nahar, L., Stewart, F., and Sarker, S.D. 2005. Hirsutanonol, oregonin and genkwanin from the seeds of Alnus glutinosa (Betulaceae). *Biochemical Systematics and Ecology*, 33, pp.749–752.

Onal, A., and Demir, B. 2009. Preparation of dyeing prescription and investigation of natural hair dyeing properties of walnut (Juglans regia L.), logwood (Alnus glutinosa L.), alkanet (Alkanna tinctoria L.) madder red (Rubai tinctorum L.) and wouw (Reseda luteola L.) extracts. *Asian Journal of Chemistry*, 21, pp.449–1452.

Othman, A.I., Amer, M.A., Basos, A.S., and El-Missiry, M.A. 2019. Moringa oleifera leaf extract ameliorated high-fat diet-induced obesity, oxidative stress and disrupted metabolic hormones. *Clinical Phytoscience*, 5(1), pp.1–10.

Peev, C.I., Vlase, L., Antal, D.S., Dehelean, C.A., and Szabadai, Z. 2007. Determination of some polyphenolic compounds in buds of Alnus and Corylus species by HPLC. *Chemistry of Natural Compounds*, 43, pp.259–262.

Plattner, R., Taylo, S.L., and Grove, M.D. 1986. Detection of brassinolide and castasterone in Alnus glutinosa (European Alder) pollen by mass spectrometry/mass spectrometry. *Journal of Natural Products*, 49, pp.540–545.

Rani, N.Z.A., Husain, K., and Kumolosasi, E. 2018. Moringa genus: A review of phytochemistry and pharmacology. *Frontiers in Pharmacology*, 9, pp.1–26.

Razis, A.F.A., Ibrahim, M.D., and Kntayya S.B. 2014. Health benefits of *Moringa oleifera*. *Asian Pacific Journal of Cancer Prevention*, 15(20), pp.8571–8576.

Ren, X., He, T., Chang, Y., Zhao, Y., Chen, X., Bai, S., Wang, L., Shen, M., and She, G. 2017. The genus Alnus, a comprehensive outline of its chemical constituents and biological activities. *Molecules*, 22(8), p.1383.

Rodríguez-González, P.M., Campelo, F., Albuquerque, A., Rivaes, R., Ferreira, T., and Pereira, J.S. 2014. Sensitivity of black alder (Alnus glutinosa [L.] Gaertn.) growth to hydrological changes in wetland forests at the rear edge of the species distribution. *Plant Ecology*, 215, pp.233–245.

Sahin Yaglıoglu, A., Eser, F., Tekin, S., and Onal, A. 2016. Antiproliferative activities of several plant extracts from Turkey on rat brain tumor and human cervix carcinoma cell lines. *Frontiers in Life Science*, 9, pp.69–74.

Sati, S.C., Sati, N., and Sati, O.P. 2011. Bioactive constituents and medicinal importance of genus Alnus. *Pharmacognosy Reviews*, 5(10), pp.174–183.

Singh, D., Arya, P.V., Aggarwal, V.P., and Gupta, R.S. 2014. Evaluation of antioxidant and hepatoprotective activities of Moringa oleifera lam. leaves in carbon tetrachloride-intoxicated rats. *Antioxidants*, 3, pp.569–591.

Skrypnik, L., Grigorev, N., Michailov, D., Antipina, M., Danilova, M., and Pungin, A. 2019. Comparative study on radical scavenging activity and phenolic compounds content in water bark extracts of alder (Alnus glutinosa (L.) Gaertn.), oak (Quercus robur L.) and pine (Pinus sylvestris L.). *European Journal of Wood and Wood Products*, 77, pp.879–890.

Sutalangka, C., Wattanathorn, J., Muchimapura, S., and Thukham-Mee, W. 2013. Moringa oleifera mitigates memory impairment and neurodegeneration in animal model of age-related dementia. *Oxidative Medicine and Cellular Longevity*, 2013, pp.1–9.

Toppo, R., Roy, B.K., Gora, R.H., Baxla, S.L., and Kumar, P. 2015. Hepatoprotective activity of Moringa oleifera against cadmium toxicity: In rats. *Veterinary World*, 8, pp.537–540.

Vieira, A.M.S., Vieira, M.F., Silva, G.F., Araújo, Á.A., Fagundes-Klen, M.R., and Veit, M.T., Bergamasco, R. 2010. Use of Moringa oleifera seed as a natural adsorbent for wastewater treatment. *Water, Air, & Soil Pollution*, 206, pp.273–281.

Wangkheirakpam, S.D., and Laitonjam, W.S. 2016. Studies on the uses of some plants for medicinal and dyeing properties. *International Journal of Chemical Studies*, 5(1), pp.93–102.

Wanjohi, B.K., Sudoi, V., Njenga, E.W., and Kipkore, W.K. 2020. An ethnobotanical study of traditional knowledge and uses of medicinal wild plants among the marakwet community in Kenya. *Evidence-based Complement. Alternative Medicine*, 2020, pp.1–8.
Warlo, H., von Wilpert, K., Lang, F., and Schack-Kirchner, H. 2019. Black alder (Alnus glutinosa (L.) Gaertn.) on compacted skid trails: A trade-off between greenhouse gas fluxes and soil structure recovery? *Forests*, 10(9), p.726.
Xu, Y.B., Chen, G.L., and Guo, M.Q. 2019. Antioxidant and anti-inflammatory activities of the crude extracts of Moringa oleifera from Kenya and their correlations with flavonoids. *Antioxidants*, 8, p.296.
Yameogo, C.W., Bengaly, M.D., Savadogo, A., Nikiema, P.A., and Traore, S.A. 2011. Determination of chemical composition and nutritional values of Moringa oleifera leaves. *Pakistan Journal of Nutrition*, 10, pp.264–268.
Zhou, J., Yang, W.S., Suo, D.Q., Li, Y., Peng, L., Xu, L.X., Zeng, K.Y., Ren, T., Wang, Y., Zhou, Y., Zhao, Y., Yang, L.C., and Jin, X. 2018. Moringa oleifera seed extract alleviates scopolamine-induced learning and memory impairment in mice. *Frontiers in Pharmacology*, 9, pp.1–11.

24 *Madhuca longifolia* (Mahuwa) and *Santalum album* (Indian Sandalwood)

Surendra Pratap Singh, Bhoomika Yadav, and Kumar Anupam

CONTENTS

24.1 Introduction 461
24.2 Description....................462
24.2.1 Traditional Knowledge 462
24.2.2 Geographical Distribution 462
24.2.3 Morphological Description....................463
24.2.4 Phytochemistry: Bioactive Compounds....................463
24.2.5 Therapeutic Benefits/Traditional Applications....................464
24.3 *Santalum album*....................465
24.3.1 Traditional Knowledge 465
24.3.2 Geographical Distribution 466
24.3.3 Morphological Description....................466
24.3.4 Phytochemistry: Bioactive Compounds....................466
24.3.5 Therapeutic Benefits/Traditional Applications....................468
24.4 Conclusion 468
References....................468

24.1 INTRODUCTION

The term "medicinal plants" includes various types of plants in herbal medicine. The word "herb" has been derived from the Latin word "herba". Earlier the term herb was only applied to non-woody plants, including those that come from trees and shrubs but nowadays, herb refers to any part of the plant like fruit, seed, bark, flower, leaf, stigma, or root as well as a non-woody plant. These medicinal plants are also used as food, flavonoid, medicine, or perfume and also in certain spiritual activities. Treatment with medicinal plants is considered very safe, as there are no, or minimal, side effects. These remedies are in sync with nature, which is the biggest advantage. The golden fact is that use of herbal treatment is independent of age group or gender.

Medicinal plants like Mahua (*Madhuca indica*) and sandalwood (*Santalum album*) cure several common ailments. These are used as home remedies for treatments of ailments in various parts of the country. *M. indica* is a member of the Sapotaceae family and is found in most subtropical regions of the indo-Pakistan subcontinent. Mahua trees are grown in the state of Gujarat, Uttar Pradesh, Bihar, Jharkhand, Chhattisgarh, Madhya Pradesh, Maharashtra, Orissa, Andhra Pradesh, West Bengal, Deccan, Chota Nagpur, Siwaliks, and Karnataka (Suryawanshi and Mokat, 2021). It is commonly known as butter tree, oil-nut, madhūka, madkam mahwa, mahuwa, mahua, mohulo, or Iluppai or vippa chettu. It is a fast-growing shady, large, deciduous tree that grows to approximately 20–25 meters in height and possesses semi-evergreen or evergreen foliage. One of the fruits of *M. indica* is the herbal medications, which are among the most significant and frequently used medicines in the Ayurvedic medicine systems in the world. *M. indica* is a prominent tree in tropical

DOI: 10.1201/9781003205067-24

mixed deciduous forests in India. The leaves are used as a poultice to relieve eczema. Mahua is valued for its oil-bearing seeds and flowers, which are utilized for alcoholic beverage production. Its seeds are of economic importance as they are enriched with edible fats. The bark is a good medicine for itch, swelling, snakebite poisoning, and fractures. The distilled flower juice is considered a tonic – both nutritional and cooling. It is also used in the treatment of acute and chronic tonsillitis, pharyngitis, helminthes, and bronchitis ailments. Earlier phytochemical studies on mahua included characterization of glycosides, flavonoids, saponins, sapogenins, triterpenoids, and steroids.

Santalum album (family: Santalaceae) is a tiny evergreen tree that grows to a height of 4 m in Australia, but it may grow to a height of 20 m in India, with a diameter of up to 2.4 m with slender, drooping branchlets (Pullaiah and Swamy, 2021). The bark of elder trees is tight, dark brown, reddish, dark grey, or practically black, smooth in young trees, rough with deep vertical fractures in older trees, and red on the inside. Leaves are thin, generally opposite, oblong or ovate elliptic in shape, 3–8 × 3–5 cm, glabrous and brilliant green above, glaucous and somewhat lighter underneath; tip rounded or pointy; stem grooved, 5–15 cm long; venation clearly reticulate. Flowers in axillary or terminal paniculate cymes are purplish-brown, tiny, straw-colored, reddish, green, or violet, approximately 4–6 mm long, up to 6 in small terminal or axillary clusters, odorless. Fruit is a globose, fleshy drupe that is crimson, purple, or black when ripe, approximately 1 cm in diameter, with a hard-ribbed endocarp and a scar on top, nearly stalkless, smooth, and single-seeded. Flowers bloom in India from March to April and ripen in the winter; flowers appear in Australia from December to January, as well as June to August, and mature fruit is accessible from June to September. Birds disperse seeds, which feed on the outside fleshy pericarp, allowing the species to spread quickly. When the tree is 5 years old, it begins to produce viable seeds. The generic name is taken from the Greek word "santalon", which means "sandalwood," and the species name is derived from the Latin word "albus", which means "white", in reference to the bark. Overall, the objective of this chapter is to highlight the current understanding of *M. indica* and *S. album*, their chemical constituents, benefits, and applications.

24.2 DESCRIPTION

24.2.1 Traditional Knowledge

M. indica has been given a prominent place in Ayurveda. *M. indica* or oil-nut tree is the sacred tree of various temples in South India, including Tirukkodimaada Senkundrur at Tiruchengode, Iluppaipattu Neelakandeswarar Temple, and Mumbai Mahaleswarar. Wine distilled from *M. indica* flowers finds mention in several Buddhist and Hindu literature works. *M. indica* is one of the resourceful forest tree species that provide fuel, fodder, and food. Flowers and fruits of mahua are edible and enriched with high nutritious values. *M. indica* flower is having an important role in the life of the poor and tribal peoples in many parts of India. It is consumed in a raw, cooked, or fried form. It is also used in the preparation of distilled liquors. The freshly prepared distilled liquor has a strong and smoky odor, which disappears after aging. It is also used as a universal panacea in the Indian tribal and poor people's medicine. The paste of the mahua tree bark is used to cure bone fractures. The raw fruit is nonedible, obtained from the mahua tree at age of 4 to 7 years, and contains one to two kidney-shaped kernels. The oil is extracted from kernels and used for several different purposes. It is used for biodiesel production also (Ghadge and Raheman, 2006). The wood of the mahua tree is utilized in the furniture utility like window and door making.

24.2.2 Geographical Distribution

Nutritionally as well as economically, it is an important tree, growing throughout the subtropical region of the indo-Pakistan subcontinent. *M. indica* trees are mainly grown in the state of Gujarat, Uttar Pradesh, Bihar, Jharkhand, Chhattisgarh, Madhya Pradesh, Maharashtra, Orissa, Andhra

FIGURE 24.1 *M. indica* (A) Tree, (B) green leaf, (C) flower and (D) seeds.

Pradesh, West Bengal, Deccan, Chota Nagpur, Siwaliks, and Karnataka. Apart from many dense forests of the country, it is also found in rural areas.

24.2.3 Morphological Description

M. indica is a medium-sized to large-growing deciduous shady tree that grows about 16–20 meters tall (Figure 24.1). The plant is found mostly growing under dry subtropical and tropical climatic conditions. The plant grows well on rocky, gravelly, saline, and sodic soils, even in pockets of soil between crevices of barren rock. The plant has a deep, strong taproot and short, stout trunk, 80 cm in diameter. The crown is rounded with multiple branches. The bark possesses yellowish grey to dark brown color, vertically cracked and wrinkled, exfoliating in thin scales, and has a milky substance inside. Leaves are thick, leathery having 10–30 cm length, lanceolate, narrowed at both ends, glabrous, distinctly nerved, and clustered at the end of the branches. It exudes a milky sap when broken. Young leaves are pinkish and wooly underneath. Flowers are small and fleshy, dull or pale white in color. Corollas possess tubular, fleshy, pale-yellow aromatic, and caduceus. Flowering generally occurs March to April. Among its innumerable uses, the flower is used to make the famous drink "mahua wine" in central India (Patel and Naik, 2010).The flowers are followed by fleshy berries that are 2–6 cm long, oval shape having 1–4 seeds that are brown to black-colored (Azam, Waris, and Nahar, 2005). They are greenish when young and adopt a pinkish yellow color when ripe. The air-dried flower of *M. indica* is used in wine production. Besides wine production, it has innumerable benefits from increasing physical strength to removing epilepsy. *M. indica* is a wild tree enriched in carbohydrates, fat, vitamin-C, protein, iron, calcium, phosphorus, etc. Apart from many dense forests of the country, they are also found in rural areas. Apart from alcohol production from the flower, the leaves, seeds, and bark of this tree are very useful in many diseases.

24.2.4 Phytochemistry: Bioactive Compounds

The most prevalent components of *M. indica* fruit, fixed oils, phosphatides, essential oils, tannins, minerals, vitamins, amino acids, fatty acids, glycosides, and other phytochemicals (Figure 24.2),

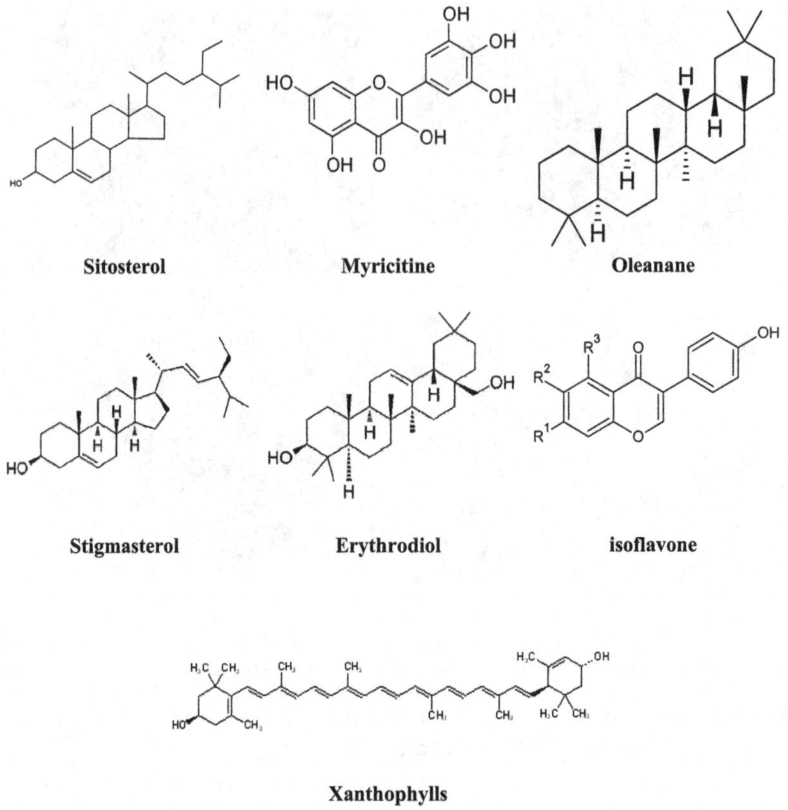

Sitosterol **Myricitine** **Oleanane**

Stigmasterol **Erythrodiol** **isoflavone**

Xanthophylls

FIGURE 24.2 Bioactive Compounds of *M. indica*.

have all been isolated from this plant(Khatoon and Reddy, 2005). The leaves of Mahua tree contain saponin, an alkaloid, and glycoside. Saprogenic and other basic acids are found in the seeds. Various phytochemical studies on Mahua include characterization of sapogenin, triterpenoids, steroids, saponin, flavonoids, and glycosides. In view of the aides and attributed medicinal properties, new components including madhucic acid (penta cyclic triterpenoids), madhushazone, four new oleanane-type triterpene glycosides, and madhucosides A and B. The fresh flower contains 2- acetyl-l- 1- pyrroline, the aroma molecule. They also contain polysaccharide which on hydrolysis gives D-galactose, D-glucose, L-arabininose, Lrhamose, D-xylose, and D-glucuronic acid. The oil content of dried mahua seeds is about 50% weight (Goud, Patwardhan, and Pradhan, 2006). The oil is characterized by saturated fatty acids such as palmitic (17.8 wt%) and stearic (14.0% wt) acids and a comparatively low percentage of free fatty acid (FFA) content of around 20%wt (Puhan, Vedaraman, Rambrahamam, and Nagarajan, 2005). The remaining fatty acids are mostly spread among unsaturated components such as oleic (46.3% wt) and linoleic (17.9% wt) acids. The relatively high percentage of saturated fatty acids (35.8% wt) found in *M. indica* oil results in relatively poor low-temperature properties of the parallel methyl esters, as evidenced by PP value of 6 °C (Singh and Singh, 1991).

24.2.5 Therapeutic Benefits/Traditional Applications

In India, the majority of the plant components of *M. indica* are utilized in traditional medicine (Gupta, Chaudhary, and Sharma, 2012). Carbohydrates, fats and proteins, calcium, phosphorus, iron, carotene, and vitamin C are found in mahua. There are many dietary benefits of mahua, due to its many nutrients. Vegetable oil is also extracted from the mahua seeds. This oil is used for cooking purposes in rural areas. It is also applied on the body as skin moisturizer. Its barks are used in dental and tonsillitis problems. Oil extracted from the seeds of the Mahua plant is applied over the area affected with skin diseases and body pain. Nasal administration of the fresh juice of the flowers of Mahua is used in diseases of vitiated pitta dosha, like headache, burning sensation of the eyes, etc. Dried flowers of Mahua are boiled in milk and administered in a dose of 40–50 ml to treat weakness of the nerves and diseases of the neuromuscular system. Decoction prepared from the bark of the tree is given in dose of 30–40 ml to treat irritable bowel syndrome and diarrhea. Fresh juice of the flowers is given in a dose of 20–25 ml to treat hypertension, hiccups, and dry cough. Externally, the seed oil massage is very effective to alleviate pain. In skin diseases, the juice of flowers is rubbed for foliation. It is also beneficial as a nasya (nasal drops) in diseases of the head due to pitta, like sinusitis. The mahua have several pharmacological potencies and it is being used from the tradition (Awasthi, Bhatnagar, and Mitra, 1975). The crude methanolic extract of mahua is used for analgesic, antipyretic, and anti-inflammatory purposes (Shekhawat and Vijayvergia, 2010).

24.3 *SANTALUM ALBUM*

24.3.1 Traditional Knowledge

Sandalwood (Figure 24.3) is mentioned throughout Indian mythology, folklore, and scripture. It is referenced in Indian literature dating back to 200 BC MilindaPahna, Patanjali Mahabhasaya (100 BC), Dhamma Pada, Anguttara, and Vinaya Pitaka (400–300 BC). In his *Arthasastra*, Kautilya describes a type of sandalwood (200 BC). The epics Ramayana and Mahabharata both mention sandalwood. Some of the allusions to sandalwood in ancient Indian literature seem unlikely to refer to *Pterocarpus santalinus*, often known as Red Sandalwood. *S. album* has been grown in India for the last 25 centuries, according to a large body of evidence. Sandalwood oil is highly valued throughout the world, with India serving as the primary exporter. Sandalwood, also known as "Bai tan xiang" in China and "Byaku-dan" in Japan and has been mentioned in ancient Sanskrit and Chinese scriptures and manuscripts. It's also utilized in a variety of initiation ceremonies to prepare the mind of the student for consecration. It burns on sacred flames in Zoroastrian temples to relieve the woes of all humanity. Because of its wide range of qualities, it is employed by Jews, Buddhists, Hindus, and nearly every other faith system.

FIGURE 24.3 *Santalum album* (A) Tree, (B) green leaf and immature fruit, (C) bark, and (D) seeds.

24.3.2 Geographical Distribution

S. album is native to the Indian peninsula's tropical belt, as well as eastern Indonesia and northern Australia. The species is found mostly in India's dry tropical areas and the Indonesian islands of Timor and Sumba. It is indigenous to the southern Indian highlands, with the main sandal tracts including most of Karnataka and surrounding districts of Maharashtra, Tamil Nadu, and Andhra Pradesh. In these areas, the species is primarily found in dry deciduous and scrub woods. A typical monsoon vine thicket growing on clean sand is the plant type. It's been seen on sand dunes along the shore, just above the typical high tide line and adjacent to mangroves. It may also be found on the beach's low lateritic cliffs. The tree is an obligatory hemiparasite on *Cassia siamea*, *Pongamia glabra*, and *Lantana acuminata*, among other hosts. India, China, Sri Lanka, Indonesia, Malaysia, the Philippines, and Northern Australia have all planted it. Sandalwood grows naturally in a number of tropical locations and tolerates a wide range of site conditions. However, it grows more vigorously in some than others. *S. album* grows best in well-drained loamy soil, particularly on sun-exposed hill slopes. It needs a minimum of 20–25 inches of annual rainfall. It is neither cold nor waterlogged tolerant, although it is drought-tolerant and a light demander in sapling and later stages. Drought and fire destroy the trees over time.

24.3.3 Morphological Description

S. album is an evergreen tree with slender drooping branchlets that grows up to 20 meters tall and a diameter of up to 2.4 meters (Figure 24.3). The tree's bark is dense, dark brown, reddish, dark grey, or almost black in color, smooth in young trees, rough with deep vertical fractures in older trees, and red on the interior (Sundharamoorthy, Govindarajan, Chinnapillai, and Raju, 2018). Heartwood is yellowish to dark brown and highly fragrant, whereas sapwood is white and odorless (González, Suaza-Gaviria, and Pabón-Mora, 2021). Leaves are thin, generally opposite, oblong or ovate elliptic in shape, 3–8 × 3–5 cm, glabrous and brilliant green above, glaucous and somewhat lighter underneath; tip rounded or pointy; stem grooved, 5–15 cm long; venation clearly reticulate. Flowers in auxiliary or terminal paniculate cymes are purplish-brown, tiny, straw-color, reddish, green, or violet, approximately 4–6 mm long, up to 6 in small terminal or axillary clusters, odorless. Fruit is a globose, fleshy drupe that is crimson, purple, or black when ripe, approximately 1 cm in diameter, with a hard ribbed endocarp and a scar on top, virtually stalkless, and single-seeded (Benencia F, 1999). In India, flower panicles bloom from March to April, while fruits mature over the winter. Flowers bloom in Australia from December to January, as well as from June to August, while fruits ripen from June to September. Seed dispersion allows the species to expand quickly. When the tree reaches the age of five years, it begins to produce viable seeds. Circumference of trees older than 30 years can range from 18 to 38 inches.

24.3.4 Phytochemistry: Bioactive Compounds

Sandalwood, the most expensive wood and essential oil, has had its chemical components well studied (Figure 24.4). Only after 30 years of development under natural conditions can sandalwood oil collect in the heartwood. The age of the tree, the color of the heartwood, organ development, individual tree, placement within the tree, environmental signals, and genetic variables all have an impact on the quantity and composition of any essential oil (Kumar et al., 2019). Indian Sandalwood (*S. album*) and Australian Sandalwood (*S. spicatum*) are the two most common commercial species of sandalwood. *In vitro* grown callus cultures provided a renewable resource of biomass which is easily culturable and the conditions are very much standardized under laboratory conditions (Harbaugh, 2007). Additionally, the physicochemical and phytochemical characterization of sandalwood callus extracts has been reported by Misra and Dey (2012).

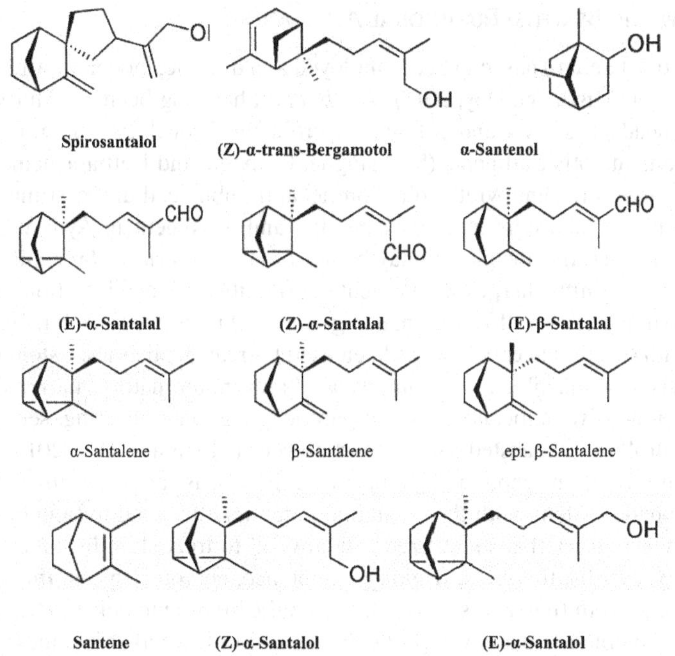

Spirosantalol (Z)-α-trans-Bergamotol α-Santenol

(E)-α-Santalal (Z)-α-Santalal (E)-β-Santalal

α-Santalene β-Santalene epi- β-Santalene

Santene (Z)-α-Santalol (E)-α-Santalol

FIGURE 24.4 Chemical structures of important constituents of sandalwood.

The steam distillate of the heartwood is traditionally marketed as marketable sandalwood essential oil, which is an age-old practice (Sindhu, Upma, and Arora, 2010). The essential oil yield from an old developed tree varies between 2.5 and 6%, depending on the tree's age, heartwood color, individual tree under investigation, position inside the tree, and growing environment (Kucharska, Frydrych, Wesolowski, Szymanska, and Kilanowicz, 2021). Sesquiterpene alcohols such as -α and -βsantalols ($C_{15}H_{24}O$), bergamotols, and several of their stereoisomers are major constituents of commercially available sandalwood oil, whereas minor constituents include lanceol, nuciferol, bisabolol, and sesquiterpene hydrocarbons such as –α and -βsantalenes ($C_{15}H_{24}$), bergamotenes, α, β, and γ-curcumenes, β-bisabolene, and phenylpropanoids. A-santalol is usually more prevalent than β-santalol. Sesquiterpene alcohols, cis-α-santalol and cis-β-santalol, α-transbergamotol, and epi-cis-β-santalol were identified as significant essential oil components by Verghese and colleagues (Gramaje, León, Pérez-Sierra, Burgess, and Armengol, 2014).Trans-β-santalol and cis-lanceol are minor components, as are the hydrocarbons santene (C_9H_{14}), α-santalene, β-santalene, α-bergamotene, epi-β-santalene, α-curcumene, β-curcumene,γ-curcumene,β-bisabolene, α-bisabolol, and heterocyclics. Other substances found in sandalwood oil include alcohols santenol ($C_9H_{16}O$) and teresantalol ($C_{10}H_{16}O$); aldehydes nor-tricycloekasantalal ($C_{11}H_{16}O$) and isovaleraldehyde; ketones l-santenone ($C_9H_{14}O$), and santalone ($C_{11}H_{16}O$); acids teresantalic acid ($C_{10}H_{14}O$) occurring partially free and partially esterified and α- and β-santalic acids ($C_{15}H_{22}O_2$). Sesquiterpenes, led by the two principal sesquiterpene alcohols, α-santalol and β-santalol, as well as E, E-farnesol, and α-bisabolol, are the main components of Australian Sandalwood oil. A total of 53 chemicals were discovered in the oil, accounting for 99.9% of the total, including 30 sesquiterpenols (78.5%), 9 sesquiterpenes (7.8%), a terpenoic acid (0.4%), and 5 sesquiterpenoid isomers (4.4 percent). α- and β-santalol were the most common components in the essential oils, accounting for 19.6% and 16.0 percent, respectively (Kucharska et al., 2021). The content of α-santalol was lower than the suggested range of 41–55%; nevertheless, the level of β-santalol was near to the 18% specification. Bisabolenols A, B, C, and D and their isomers were found in significant concentrations in the oil, accounting for 25.0% of the total.

24.3.5 Therapeutic Benefits/Traditional Applications

White sandalwood (Chandan) has long been employed as a diuretic, moderate stimulant, and demulcent in the Ayurveda (Misra and Dey, 2013). *S. album* oil has long been used for mouth and throat irritation, burns, headaches, liver and gallbladder problems, bronchitis, common colds, fever, urinary tract infection, and other ailments (Subasinghe, Gamage, and Hettiarachchi, 2013). The oil is used as a cooling, antiseptic, antipyretic, diaphoretic, antiscabietic, diuretic, stimulant, expectorant, cicatrisant, carminative, aphrodisiac, antispasmodic, and astringent in Ayurveda. As it includes antifungal and antibacterial principles, it may be used to treat psoriasis, bronchitis, sunstroke, palpitations, urethritis, vaginitis, herpes zoster, acute dermatitis, urinary infection, dysuria, boosting activity, gastric (Ahmed et al., 2013), mucin, and gonorrheal recovery (Dikshit, 1984). *S. album* oil, in combination with other plant extracts, has been used to treat elephantiasis, stomach ailments, and lymphatic filariasis. Pharmacological testing revealed anti-inflammatory, antiremorogenic, antimitotic, antiviral, hypotensive, anticancerous, antipyretic, ganglionic blocking, sedative, and insecticide effects in hydrolyzed exhausted sandalwood powder (Misra and Dey, 2013). Sandalwood oil was traditionally used to treat venous and lymphatic stasis such as varicose veins and enlarged lymph nodes in the lymphatic system, with the medicinal potential attributed to santalols, which have an anti-inflammatory action. Herbalists utilized sandalwood to treat skin disorders, acne, dysentery, gonorrhea, anxiety, cystitis, tiredness, frigidity, impotence, nervous tension, immune-booster, dermatitis, stomachache, vomiting, and stress in traditional Chinese medicine (TCM). Sandalwood is used in Chinese medicine to treat any sort of chest discomfort, whether it comes from the lungs or the heart (Gramaje et al., 2014). Angina discomfort is relieved by the oil's regulating and dispersing effect. Sandalwood is also mentioned in *De Materia Medica* by Discorides. In addition, the German Commission E monograph recommends 1/4 teaspoon (1–1.5 g) of sandalwood oil for the supportive treatment of urinary tract infections, as well as for aches, fevers, and heart strengthening.

24.4 CONCLUSION

Researchers sought to compile botanical, pharmacological and phytochemical information on *M. indica* and *S. album* therapeutic plants utilized in Indian medicine, in the current review. Due to its varied pharmacological characteristics, *M. indica* is one of the most often utilized Ayurvedic medicinal herbs to treat a variety of human diseases. Its anti-inflammatory and antioxidative capabilities, however, are its most remarkable characteristics. Various pharmacological properties of *M. indica* have been researched scientifically, indicating the plant's therapeutic relevance. For more than a century, biochemists and chemists interested in the chemistry, structure, synthesis, and biological origins of sandalwood trees. The delightfully diverse collections of chemicals have been challenged by the study of sesquiterpenoids in sandalwood. Recently, an increase in research efforts to validate traditional health care applications using experimental techniques *in vivo* or *in vitro*, in-depth mechanistic investigations and pharmacological for essential oil components, as well as potential clinical trials, has gained attraction. In the near future, *M. indica* and *S. album* might be used as varied topics for complementary and alternative medicine research and development. The numerous possible treatments may be identified from the leaves, fruits, roots, and barks of these plants based on its traditional knowledge, implying that they serve as natural sources for future pharmaceutical ingredient development. As a result, rigorous scientific research should be conducted to investigate the therapeutic potential of these important Ayurvedic and other herbal-based medications.

REFERENCES

Ahmed, N., Khan, M.S.A., Jais, A.M.M., Mohtarrudin, N., Ranjbar, M., Amjad, M.S., ... Chincholi, A. 2013. Anti-ulcer activity of sandalwood (*Santalum album* L.) stem hydro-alcoholic extract in three gastric-ulceration models of wistar rats. *Boletín Latinoamericano y del Caribe de Plantas Medicinales y Aromáticas*, 12(1), pp.81–91.

Awasthi, Y., Bhatnagar, S., and Mitra, C. 1975. Chemurgy of sapotaceous plants: Madhuca species of India. *Economic Botany*, 29, pp.380–389.

Azam, M.M., Waris, A., and Nahar, N. 2005. Prospects and potential of fatty acid methyl esters of some non-traditional seed oils for use as biodiesel in India. *Biomass and Bioenergy*, 29(4), pp.293–302.

Benencia F, and Courreges, M. 1999. Antiviral activity of sandalwood oil against Herpes Simplex viruses 1 and 2. *Phytomedicine*, 6(2), pp.119–123.

Dikshit, A. 1984. Antifungal action of some essential oils against animal pathogens. *Fitoterapia*, 55, pp.171–176.

Ghadge, S.V., and Raheman, H. 2006. Process optimization for biodiesel production from mahua (*Madhuca indica*) oil using response surface methodology. *Bioresource Technology*, 97(3), pp.379–384.

González, F., Suaza-Gaviria, V., and Pabón-Mora, N. 2021. Floral development and morphology of the mistletoe Antidaphne viscoidea: A case of extreme flower reduction in the sandalwood family (Santalaceae). *Australian Journal of Botany*, 69(3), pp.152–161.

Goud, V.V., Patwardhan, A.V., and Pradhan, N.C. 2006. Studies on the epoxidation of mahua oil (Madhumica indica) by hydrogen peroxide. *Bioresource Technology*, 97(12), pp.1365–1371.

Gramaje, D., León, M., Pérez-Sierra, A., Burgess, T., and Armengol, J. 2014. New Phaeoacremonium species isolated from sandalwood trees in Western Australia. *IMA Fungus*, 5(1), pp.67–77.

Gupta, A., Chaudhary, R., and Sharma, S. 2012. Potential applications of mahua (Madhuca indica) biomass. *Waste and Biomass Valorization*, 3(2), pp.175–189.

Harbaugh, D.T. 2007. A taxonomic revision of Australian northern sandalwood (Santalum lanceolatum, Santalaceae). *Australian Systematic Botany*, 20(5), pp.409–416.

Khatoon, S., and Reddy, S.R.Y. 2005. Plastic fats with zero trans fatty acids by interesterification of mango, mahua and palm oils. *European Journal of Lipid Science and Technology*, 107(11), pp.786–791.

Kucharska, M., Frydrych, B., Wesolowski, W., Szymanska, J.A., and Kilanowicz, A. 2021. A comparison of the composition of selected commercial sandalwood oils with the international standard. *Molecules*, 26(8), p.2249.

Kumar, G.R., Chandrashekar, B., Rao, M.S., Ravindra, M., Chandrashekar, K., and Soundararajan, V. 2019. Pharmaceutical importance, physico-chemical analysis and utilisation of Indian sandalwood (Santalum album Linn.) seed oil. *Journal of Pharmacognosy and Phytochemistry*, 8(1), pp.2587–2592.

Misra, B.B., and Dey, S. 2012. Phytochemical analyses and evaluation of antioxidant efficacy of in vitro callus extract of east Indian sandalwood tree (Santalum album L.). *Journal of Pharmacognosy and Phytochemistry*, 1(3), pp.7–16.

Misra, B.B., and Dey, S. 2013. Biological activities of East Indian sandalwood tree, Santalum album. *PeerJ PrePrints*. 1, pp.1–30. https://doi.org/10.7287/peerj.preprints.96v1

Patel, M., and Naik, S. 2010. Flowers of Madhuca indica J. F. Gmel.: Present status and future perspectives. *Indian Journal of Natural Products and Resources,* 1(4), pp.438–443.

Puhan, S., Vedaraman, N., Rambrahamam, B., and Nagarajan, G. 2005. Mahua (Madhuca indica) seed oil: A source of renewable energy in India. *Journal of Scientific and Industrial Research,* 64, pp. 890–896.

Pullaiah, T., and Swamy, M.K. 2021. Sandalwood: The green gold. In *Sandalwood: Silviculture, Conservation and Applications* (pp. 1–8). New York: Springer.

Shekhawat, N., and Vijayvergia, R. 2010. Investigation of anti-inflammatory, analgesic and antipyretic properties of Madhuca indica GMEL. *European Journal of Inflammation*, 8(3), pp.165–171.

Sindhu, R.K., Upma, K.A., and Arora, S. 2010. Santalum album linn: A review on morphology, phytochemistry and pharmacological aspects. *International Journal of PharmTech Research*, 2(1), pp.914–919.

Singh, A., and Singh, I. 1991. Chemical evaluation of mahua (Madhuca indica) seed. *Food Chemistry*, 40(2), pp.221–228.

Subasinghe, U., Gamage, M., and Hettiarachchi, D. 2013. Essential oil content and composition of Indian sandalwood (Santalum album) in Sri Lanka. *Journal of Forestry Research*, 24(1), pp.127–130.

Sundharamoorthy, S., Govindarajan, N., Chinnapillai, A., and Raju, I. 2018. Macro-microscopic atlas on heartwood of Santalum album L.(Sandalwood). *Pharmacognosy Journal*, 10(4), 730–733.

Suryawanshi, Y.C., and Mokat, D.N. 2021. Morphophysiological seed variability in Mahua trees from Western Ghats and its impact on tribal life. *Proceedings of the National Academy of Sciences, India Section B: Biological Sciences*, 91(1), pp. 227–239.

Index

Note: Locators in *italics* represent figures and **bold** indicate tables in the text.

AA metabolites, *see* Arachidonic acid metabolites
AAMI, *see* Age-associated memory impairment
ACC deaminase activity, *see* 1-Aminocyclopropane-1-carboxylate deaminase activity
ACCX, *see* 25-Acetylcimigenol xylopyranoside
ACE, *see* Angiotensin-converting enzyme
Acetogenins, 346
25-Acetylcimigenol xylopyranoside (ACCX), 66
Achyranthes aspera L., 12, *12*
Acokanthera schimperi, 14, *14*
Actaea racemosa, see Cimicifuga racemosa
AD, *see* Alzheimer's disease
Adhatoda schimperiana, 13
Adhatoda vasica, 303
  anthelmintic effect, 327
  antibacterial effect, 325
  anticestodal effect, 328
  antidiabetic effect, 327–328
  antifungal effect, 327
  anti-inflammatory effect, 326
  antimicrobial effect, 325–326
  antioxidant effect, 325
  antituberculosis effect, 328
  antitussive effect, 326
  antiviral effect, 326
  chemical derivatives, 310–313
  description, 305
  ethnopharmacological uses of, **308–309**
  hepatoprotective effect, 326
  hepato-protective effect, 328
  immunomodulatory effect, 329
  important phytoconstituents of, **318–319**
  phytocompounds present in, *320–321*
  radio-modulatory effect, 328–329
  reproductive organs, effect on, 329
  thrombolytic effect, 327
  traditional knowledge, 306–307
  uterine effect, 327
Adipocyte hypertrophy, 188
Advanced glycated end products (AGEs), 65
Aedes aegypti, 62, 293, 345
AEG-1, *see* Astrocyte-elevated gene-1
Aegle marmelos, *340–342*, 346, **348**
  anticancer agents, 351
  antimalarial, antidiabetic activities, 352
  antimicrobial activity, 350
  antioxidant activity, 346–347
  antipyretic, anti-inflammatory and analgesic activities, 354
  chemical derivatives, their source, and function from, **345**
  COVID-19 perspective of, 354–355
  hepatoprotective and cardioprotective activities, 353
  phytochemicals associated with, 344–345
  traditional knowledge, 342
    commercial, 344
    medicinal, 343–344
    nutritional, 342–343
Aeglemarmelosine isolate, 345
African soapberry, *see Phytolacca dodecandra*
Age-associated memory impairment (AAMI), 38
AGEs, *see* Advanced glycated end products
Ajmalicine, 323
Alanine aminotransferase (ALT), 433
Alanine transferase (ALT), 252
Albizia schimperiana Oliv., 16, *16*
Alder, *see Alnus glutinosa*
Alexandrian Senna, 271, 276
Alkaline phosphatase (ALP), 353, 433
Alkaloids, 21
Alnus glutinosa
  antibacterial activity, 451–452
  anticancer activity, 452
  anti-inflammatory activity, 452
  antioxidant activity, 452
  chemical derivatives, 450–451
  chemoprotective agent, 452
  description, 448–449
  dyeing property, 453
  general use of, **450**
  insecticidal activity, 453
  medicinal use of, *455*
  nitrogen fixation, 453
  traditional knowledge, 449–450
ALP, *see* Alkaline phosphatase
ALT, *see* Alanine aminotransferase; Alanine transferase
Alzheimer's disease (AD), 185, 324
  Nardostachys jatamansi in, 171–172
  Ocimum tenuiflorum in, 172
Amarogentin, 294
American Ginseng (AG), *see Panax quinquefolium*
American ginseng berry extract (AGBE), 187
Amino (5-(4-methoxyphenyl)-2-methyl- 2-(thiophen-2-yl)-2,3-dihydrofuran-3-yl) methanol (AMTM), 255
Amino acids, 296
1-Aminocyclopropane-1-carboxylate (ACC) deaminase activity, 255
AMP-activated protein kinase (AMPK), 68
Amyloid precursor protein (APP), 227
Andrographis paniculata, 28
  analgesic activity, 32
  antidiabetic activity, 32
  antifertility activities, 31
  anti-HIV and cytotoxic activity, 31
  antimicrobial activity, 31
  anti-neurodegenerative activities, 31
  antioxidant activity, 32
  bioactive compounds, 29–30
  clinical trials of therapeutic agents, 31–32
  distribution and common names, 28
  morphology, 28, *29*
  pharmacological activities, 30–32

phytoconstituents in, *29*
Angiotensin-converting enzyme (ACE), 243–244
Annona squamosal, 346, **349**
anticancer agents, 351–352
antimalarial, antidiabetic activities, 352–353
antimicrobial activity, 350–351
antioxidant activity, 349–350
antipyretic, anti-inflammatory and analgesic activities, 354
chemical derivatives, their source and function from, **347**
COVID-19 perspective of, 354–355
hepatoprotective and cardioprotective activities, 353–354
phytochemicals associated with, 345–346
traditional knowledge, 342
commercial, 344
medicinal, 343–344
nutritional, 342–343
Anopheles stephensi, 62
APP, *see* Amyloid precursor protein
Arabinoxylan, 243
Arachidonic acid (AA) metabolites, 326
Arogyavardhinigutika, 209
Ashwagandha, *see Withania somnifera*
Ashwini Kumars, 50
Asparagin, 421
Asparagus racemosus, 95–96
antenatal tonic, 105
anti-aflatoxigenic activity, 105
antibacterial activity, 103
anticancer property, 105–106
antidepressant activity, 104–105
antidiarrheal activity, 103
antihepatotoxic activity, 104
antileshmanial activity, 103
antilithiatic effects, 104
antioxidant effects, 104
anti-plasmodial activity, 103
antiprotozoal activity, 103
antisecretory and antiulcer activity, 103
antitussive effect, 103
antiulcer, 105
aphrodisiac activity, 105
cardiovascular effects, 104
description, 96–97
galactogogue effect, 103
gastrointestinal effects, 103–104
immunomodulatory activity, 104
memory enhancement and protection against amnesia, 105
neurodegenerative disorders, 104
phytochemicals, 100–101
potential benefits, applications, and pharmacological activities, 102–106
traditional knowledge, 98–99
uterus, effect on, 104
Aspartate aminotransferase (AST), 433
AST, *see* Aspartate aminotransferase
Astrocyte-elevated gene-1 (AEG-1), 435
Atorvastatin, 170
Avellana nux sylvestris, 379
Avellana's wild nut, 379
Azadirachta indica, 365, **372**
agronomic applications, 371–372
chemical derivatives, 369–370
description, 366–367
food, 372
major phytoconstituents in, *369*
medical applications, 370–371
morphology of, *367*
traditional knowledge, 368–369

Bacillus cereus, 383
Bacillus subtilis, 253, 270, 325, 351
Bacopa monnieri, 32, 37, 172
analgesic activities, 38
anticancer activity, 37
antidepressant and antianxiety effects, 36
antidiabetic activity, 37
antiepileptic effects, 36
antimicrobial activity, 37
anti-ulcerative activity, 36–37
bioactive compounds, 34
cardiovascular activities, 37–38
cognitive activities, 38
distribution and common names, 34
morphology, 33, *33*
pharmacological activities, 34–38
phytoconstituents in, *35*
sedative and tranquilizing properties, 36
Bacosides, 36
Bael fruit, *see Aegle marmelos*
Balys, 116
Basalis magnocellularis, 38
BCRP, *see* Breast cancer resistant protein
Berberine, 157
Berberis aristata, 365
chemical derivatives, 369–370
description, 366–367
major phytoconstituents in, *370*
morphology of, *368*
potential benefits, **373**
potential benefits, applications and uses, 372
traditional knowledge, 368–369
Betulinic acid, 266
BHA, *see* Butylated hydroxyanisole
BHT, *see* Butylated hydroxytoluene
Bifidobacterium adolescentis, 279
Bioactive compounds, 21
Biological amines, 310
Biomphalaria pfeifferi, 206
Bisphosphonates, 66
Black Cohosh, *see Cimicifuga racemosa*
Brassinolide, 451
Breast cancer resistant protein (BCRP), 123
Bruguda syndrome, 244
Butylated hydroxyanisole (BHA), 124
Butylated hydroxytoluene (BHT), 124

Caenorhabditis elegans, 436
Caffeic acid, 67, 381
Caffeine, 172
Caitha, 49
Calabar Bean (CB), *see Physostigma venenosum*
Calabarine, 184
Campesterol, 429
Cancer, 186, 317

Candida albicans, 325, 437
Carbenoxolone, 136
Carboxymethylated fruit gum, 345
Carcinoma, 317
Carissa spinarum L., 14–15, *15*
Carotenoids, 21
Cassia angustifolia, 277
Cassia siamea, 466
Castasterone, 451
CAT, *see* Catalase
Catalase (CAT), 57, 353
Catechin, 382
Cathachunine, 319
Catharanthus roseus, *see Vinca rosea*
CCRAS, *see* Central Council for Research in Ayurvedic Science
Central Council for Research in Ayurvedic Science (CCRAS), 293
Cerebral ischemia
    Nardostachys jatamansi in, 171
    Ocimum tenuiflorum in, 171
Chagas disease, 368
Chemical constituents and uses, 3–12
Chemical derivatives of herbs, shrubs, and trees, 18–21
Chirantin, *see* Amarogentin
Chirata, *see Swertia chirata*
Chirayita, *see Swertia chirata*
Chlorogenic acid, 168
Chlorophytum borivilianum, *47*
    analgesic effect, 61
    anthelmintic effect, 61
    anticancer effect, 60
    antidiabetic effect, 60
    antidyslipidemic effect, 62
    antimicrobial effect, 61
    antioxidant effect, 54–58
    anti-stress effect, 61
    anti-ulcerative effect, 60–61
    anxiolytic effect, 63
    aphrodisiac effect, 62–63
    chemical compounds isolated from, *52*
    chemical derivatives, 51–52
    hepatoprotective effect, 63
    immunomodulatory effect, 58–60
    larvicidal effect, 62
    plant description, 47–49
    toxicity, 63
    traditional knowledge, 49–50
    vernacular names of, **49**
Chloroquine (CQ), 388–389
Cholinergic dysfunction, 191
Chromium, 254
Chyawanprash, 50
Cimicifuga racemosa, *48*
    anti-allergic effect, 64
    anticancer effect, 67–68
    antidiabetic effect, 68–69
    antiestrogenic activity, 64
    antihyperglycemic effect, 65–66
    anti-inflammatory effect, 69
    antimicrobial effect, 66
    anti-osteoporosis effect, 66–67
    antioxidant effect, 67
    anxiolytic effect, 69
    chemical compounds isolated from, *57*
    chemical derivatives, 52–54
    estrogenic activity (EA), 64–65
    GABA receptor modulating effect, 69–70
    menopause effect, 70–71
    neuroprotective effect, 71
    plant description, 49
    serotonin receptor effect, 71–72
    traditional knowledge, 50–51
Cimicifuga racemosa extracts (CRE), 66
Cimicifugic acid A, 67
Cimiracemate B, 67
Cinchona calisaya, 385–386
Cinchona ledgeriana, 386
Cinchona officinalis, 386
    botanical description, 384
    chemical constituents of, 387–388
    COVID-19 treatment, 388–389
    cultivation, 386
    etymology and common names, 384
    future perspectives, 389
    history, 384
    medicinal uses, 386–387
    taxonomy, 384–386
    toxicology of *Cinchona*, 388
Cinchona pubescens, *385*, 386
Cinchona tree, *see Corylus officinalis*
Cinchonidine, 387
Cinchonine, 387
Cinchonism, 388
Cirsilineol, 168
Cis-diamminedichloroplatinum, 187
Cisplatin, *see* Cis-diamminedichloroplatinum
Clostridium perfringens, 279
Clostridium sphenoides, 279
Cocaine, 87
Coca plant, *see Erythroxylum coca*
Coffee, 88
Cognitive enhancers, *see* Nootropic drugs
Coleus forskohlii, 433
Common hazel, *see Cinchona avellana*
Convolvulus pluricaulis, *84*
    antiulcer and anti-catatonic properties, 88
    beauty, enhancing, 88
    body cholesterol, reducing, 88
    description of the plant, 84–85
    hypertension, reducing, 87–88
    memory, enhancing, 87
    phytochemistry, 86–87
    reproductive system, improving, 88
    thyroid gland, effects on, 88
    traditional knowledge, 85–86
Corylus avellana, 378
    biological activities, 383–384
    botanical aspects, 378
    chemical composition of, 381
        diarylheptanoids, 382
        flavonoids, 381–382
        lignans, 382–383
        phenolics, 381
        tannins and proanthocyanidins, 382
        taxanes, 383
        volatile compounds, 383
    future perspectives, 389

habitat and ecology, 380
hazelnuts' uses in traditional medicine, 380–381
importance and usage, 380
taxonomy, 379–380
Costunolide, 268
Costus oil, 266
Costus roots, 265
bioactive phytoconstituents of, 265–266
COX-2, *see* Cyclooxygenase-2
CPK, *see* Creatine phosphokinase
CQ, *see* Chloroquine
Crataegus laevigata, 396
anti-inflammatory effects, 400–401
antimicrobial, 402
antioxidant effects, 399–400
cardiac and vascular effects, 401–402
chemical derivatives, 397–399
description, 396–397
hepatoprotective effect, 403
immunomodulatory effects, 403
traditional knowledge, 397
traditional uses of, **398**
Crataegus oxyacantha, 397
CRE, *see Cimicifuga racemosa* extracts
Creatine phosphokinase (CPK), 170
Croton macrostachyus Del., 15, *16*
Crude drugs, 2, 3
Culex quinquefasciatus, 62, 254, 345
Custard apple, *see Annona squamosal*
Cyanidin-3-glucoside, 429
Cyanidin-3-sambubioside, 429
β-Cyclocostunolide, 266
Cyclooxygenase-2 (COX-2), 434
Cynaropicrin, 267, 269

DACD, *see* Diabetes-related cognitive decline
Dalton's lymphoma, 347
Daru haldhi and Chitra, *see Berberis aristata*
DCM, *see* Dichloromethane
Decreased glutathione reductase (GSH), 353
DEN, *see* Diethylnitrosamine
Diabetes, 68
Diabetes-related cognitive decline (DACD), 327
Diabetic encephalopathy, 327
Diarylheptanoids, 382
Dichloromethane (DCM), 415
Diethylnitrosamine (DEN), 354
Dihydro costunolide, 266
Dihydropiperlonguminine, 227
2,4-Dihydroxyphenylacetic acid, 429
7,12-Dimethylbenz (a) anthracene (DMBA), 123
Dioscorea villosa, 96
description, 97–98
phytochemicals, 101–102
potential benefits, applications, and pharmacological activities, 106
traditional knowledge, 99
Dioscorine, 106
9,10-Dioxoanthracene, 271
1,1-Diphenyl-2-picrylhydrazyl (DPPH), 142, 452
Divya aushad, 49
DMBA, *see* 7,12-Dimethylbenz (a) anthracene
Dodonaea angustifolia L.F., 15, *15*
DPPH, *see* 1,1-Diphenyl-2-picrylhydrazyl
Drumstick tree, *see Moringa oleifera*

Embelia ribes
application in Ayurveda, 117
application in traditional uses, 117–118
bioactive molecule in, *116*
description of, 114
formulations of, 116
fresh and dry fruits of, *115*
pharmacological activity of, **117**
pharmacological uses, 118
anthelmintic, 118–119
antibacterial, 118
antidiabetic, 119
antifertility, 119–120
antitumor, 120
hepatoprotective, 119
phytochemical constituents of, 116
toxicological effect of, 120
Embelin, 120
Emblica officinalis, 397
anti-inflammatory effects, 400–401
antimicrobial, 402
antioxidant effects, 399–400
cardiac and vascular effects, 401–402
chemical derivatives, 397–399
description, 396–397
hepatoprotective effect, 403
immunomodulatory effects, 403
traditional knowledge, 397
traditional uses of, **398**
Enterobacter aerogenes, 437
Enterococcus faecalis, 325
Epicatechin, 382
Epigallocatechin, 382
7-Epiloganin, 244
ER, *see* Estrogen receptors
ERK pathway, *see* Extracellular signal-regulated kinase pathway
Erythroxylum coca, 85, 88
altitude illness, relieving, 89
description of the plant, 85
environmental stress, relieving, 89
fast-acting antidepressant, 89
hunger, alleviating, 89
phytochemistry, 87
traditional knowledge, 86
treating gastrointestinal disorders, oral sores, and toothaches, 88
Escherichia coli, 270, 325, 383
Eseroline, 192
(*L*)-Eseroline, 184
Essential oils, 204
Estrogen receptors (ER), 64
Ethnomedicine, 2–3
24-Ethylcholesterol, 429
24-Ethylidenecholesterol, 429
Eubacterium spp., 279
Eucalypts, *see Eucalyptus* spp.
Eucalyptus camaldulensis, 415, **415**
Eucalyptus oil, 417
common constituents of, *416*

nonvolatile constituents of, *417*
Eucalyptus spp., *413*
acetates and monosaccharides percentage in, **415**
chemical composition and derivatives, 413–416, **414**
for diabetes, 418
distribution, 413
eucalyptus oil, disadvantages of, 418
for fever, 418
for hair, 417
hair lice removal, 418
kidney stones, relief from, 418
morphological description, 412–413
muscle pain, relief of, 418
plant protection, 418
pneumonia, use in, 418
for skin, 417
stomach worms, relief from, 418
for teeth, 418
Eugenol, 168
European Holly, *see Ilex aquifolium*
Extracellular signal-regulated kinase (ERK) pathway, 192, 432

False Black Pepper, *see Embelia ribes*
Ferulic acid, 67, 381
Ficus religiosa, 418, *419*
blood disorders, benefit in, 423
chemical structures of the compounds from, *420*
common names of, 419, **420**
cracked heel problem, benefits in, 422
curing boils, 423
dental disease treatment, 422
diabetes, controlling, 422
distribution, 419
in fighting fever, 423
future prospects, 423–424
impotency, beneficial in, 423
infertility problems, importance in, 422
jaundice treatment, 422
for liver, 422
for lungs, 422
morphological description, 418–419
phlegm, relief from, 422
phytochemistry of, 419
bark, constituents of, 421
fruits and seeds, constituents of, 421
leaves, constituents of, 420–421
skin disease treatment, benefits in, 423
snakebite, benefits in, 423
typhoid, uses in, 423
urinary disease, treating, 422
Flavonoids, 18, 21, 270, 381–382
Fluoxetine, 193
Folium Sennae Ethanolic (FSE), 281
Fructo-oligosaccharides, 60
FSE, *see* Folium Sennae Ethanolic
Fukinolic acid, 67
Fusobacterium nucleatum, 190

GABA, *see* Gamma amino butyric acid
Gamma amino butyric acid (GABA), 36
Gamma-hydroxybutyrate (GHB), 193
Garcinia indica (Kokum), *428*
anticancer activity, 430–432
antidiabetic and anti-obesity activity, 433
anti-inflammatory activity, 434
antimicrobial and antiviral activity, 434
botanical description of the plants, 428
cardioprotective activity, 433–434
hepatoprotective activity, 433
major phytoconstituents of, *430*
neuroprotective activity, 432–433
phytochemistry of the plants, 429
traditional knowledge of the plants, 428–429
Garcinol, 430, 432
Gastrointestinal nematodes, 118
Gendarussa schimperiana, 13
GFAP, *see* Glial fibrillary acidic protein
GH, *see* Glutathione
GHB, *see* Gamma-hydroxybutyrate
Giloy, *see Tinospora cordifolia*
Ginkgo biloba, 172
Ginsenosides, 181
Glial fibrillary acidic protein (GFAP), 432
Glucose transporter 4 (GLUT4) expression, 65, 187
Glutathione (GH), 57
Glutathione peroxidase (GPx) activity, 187
Glycosides, 18
Glycyrrhetinic acid, 135
Glycyrrhiza glabra, 133, *134*
antibacterial activity, 136
anticarcinogenic and antimutagenic activity, 137
anticoagulant activity, 137
antidiabetic, 137–138
antifungal activity, 136
antioxidant and anti-inflammatory, 136
antitussive and expectorant activity, 136
antiulcer activity, 136
antiviral, 136
clinical studies, 138, **138**
ethnobotanical description
chemical constituents, 135–136
ethnobotanical uses, 135
macroscopic, 134
microscopic, 134–135
scientific studies, 136–138
hepatoprotective activity, 137
immunomodulatory activity, 136
traditional actions and uses of, **135**
Glycyrrhizin, 133
Goldenseal, *see Hydrastis canadensis*
GPx activity, *see* Glutathione peroxidase activity
GSH, *see* Decreased glutathione reductase
Gumarin, 121
Gurmar, *see Gymnema sylvestre*
Gurmar herbal toothpaste, 144
Gymnemagenin, 141
Gymnemagenol, 143
Gymnemasaponin III, 141
Gymnema sylvestre, 138, *139*
bioactive components in, *121*
chemical constituents of, 120–121
clinical studies, 144
clinical studies of, **144**
description of, 115
ethnobotanical description
anti-arthritic activity, 142
antibiotic and antimicrobial activity, 142

anticancer and cytotoxic activity, 143
antidiabetic property, 141
antihyperlipidemic activity, 143
anti-inflammatory activity, 142
antioxidant activity, 142
chemical constituents, 140–141
dental caries, treatment of, 143–144
ethnobotanical uses, 140
hepatoprotective activity, 143
immunostimulatory activity, 143
macroscopic, 139
microscopic, 139
pharmacological activities, 141–144
wound-healing activity, 143
morphological image of, *115*
pharmacological actions
antiarthritic activity, 124
anticancer activity, 123
anti-inflammatory activity, 125
antimicrobial activity, 123–124
antioxidant activity, 124
diabetes mellitus, application in, 122–123
hepatoprotective activity, 125
immunomodulating effect of *G. sylvestre*, 124–125
lipid-lowering activity, 123
toxicological effect of, 125
traditional actions and uses of, **140**
traditional use, 122
Gymnemic acid, 120
Gymnemic acid A, 140

Haemonchus contortus, 118
HAGs, *see* Hydroxyanthracene glycosides
Hawthorns, 396
HCQ, *see* Hydroxychloroquine
Heart-leaved moonseed, *see Tinospora cordifolia*
Henna, *see Lawsonia inermis*
Herbal medicine, 22
Herbs
chemical derivatives of, 18–21
morphological descriptions of, 12–13
traditional knowledge of, 17–18
uses and bioactive constituents of, **4–11**, 22
Hippocampus, 71
Holy Basil, *see Ocimum tenuiflorum*
Hormone replacement therapy (HRT), 70
HRT, *see* Hormone replacement therapy
Human umbilical vein endothelial cells (HUVECs), 268
HUVECs, *see* Human umbilical vein endothelial cells
Hydrastis canadensis
acne and psoriasis, 158
antibacterial effect, 154, 157
cardiovascular effects, 158
chemical derivatives of
chlamydia and herpes, 157
compounds found in, *155–157*
description, 152
diabetes, 157
different parts of, *153*
immune modulation, 158
oral health, 158
traditional knowledge, 152–153
upper respiratory tract infection and colds, 157
2-Hydroxy-1,4-napthoquinone, 154
Hydroxyanthracene glycosides (HAGs), 278
Hydroxychloroquine (HCQ), 388–389
Hypercholesterolemia, 253

IAA, *see* Indole acetic acid
IBD, *see* Irritable bowel diseases
ICAM-1, *see* Intercellular adhesion molecule-1
Ilex aquifolium, *428*
botanical description of the plants, 428
major phytoconstituents of, *431*
medicinal properties of, 434
anticancer activity, 435–436
anti-inflammatory activity, 436–437
antimicrobial, antiviral activity, 437
hepatoprotective activity, 437
neuroprotective activity, 436
phytochemistry of the plants, 429–430
traditional knowledge of the plants, 429
Indian barberry, *see Berberis aristata*
Indian ginseng, 50
Indian gooseberry, *see Emblica officinalis*
Indian sandalwood, *see Santalum album*
Indian snakeroot, *see Rauvolfia serpentine*
Indigenous knowledge, 2
Indole acetic acid (IAA), 255
Indole terpene alkaloids, 324
Inducible nitric oxide synthase (iNOS), 267, 434
Influenza viruses, 326
iNOS, *see* Inducible nitric oxide synthase
Intercellular adhesion molecule-1 (ICAM-1), 269
Inula racemosa, 264
Ionizing radiation (IR), 187
IR, *see* Ionizing radiation
Irritable bowel diseases (IBD), 245
Isabgol, *see Plantago ovate*
Isoflavones-8-methoxy-5, 6, 4-trihydroxy isoflavone-7 0-beta-D-glucopyranoside, 100
Isoprenoids, 20
Isothymonin, 168
Isothymusin, 168

Justicia schimperiana (Hochst. Ex Nees) T., 13, *14*

Kaempferol 3-rhamnoside, 382
Kalanchoe laciniata (L.) DC, 12, *12*
Kamla, 116
Katukadyaghrita, 209
Klason lignin, 413–415
Klebsiella pneumoniae, 270, 325
Kokum, *see Garcinia indica*
Kust, *see Saussurea costus*
Kutki, *see Picrorhiza kurroa*

LA, *see* Lateral amygdala
Lactobacillus brevis, 279
Lantana acuminata, 466
Lateral amygdala (LA), 432
Lawsonia inermis
abortifacient activity, 159
anti-aging properties, 158–159
antibacterial activity, 159
antifungal activity, 159
antioxidant activity, 159
chemical derivatives of, 154

compounds found in, *155–156*
description, 152
different parts of, *154*
hepatoprotective activity, 159
hypoglycemicactivity, 159
immunomodulatory effect, 159
memory enhancement, 158
traditional knowledge, 153
wound healing, 158
Lazy-bowel syndrome, 280
LH, *see* Luteinizing hormone
Licorice, *see Glycyrrhiza glabra*
Lignans, 382–383
Lipopolysaccharide (LPS), 66, 434
Liquid chromatography/electrospray ionization tandem mass spectrometrical (LC/ESI–MS/MS) method, 429
Lochnera rosea, 304
Long pepper, *see Piper longum*
Lorazepam, 36
LPS, *see* Lipopolysaccharide
Lued-Ngam, 166
Lupeol, 167
Luteinizing hormone (LH), 71

Madagascar periwinkle, *see Vinca rosea*
Madhuca indica, *463*
bioactive compounds of, *464*
geographical distribution, 462–463
morphological description, 463
phytochemistry, 463–464
therapeutic benefits/traditional applications, 464
traditional knowledge, 462
Mahatiktakaghrita, 209
Mahuwa, *see Madhuca indica*
Malabar nut, *see Adhatoda vasica*
Malassezia furfur, 253
MAP-2, *see* Microtubule-associated protein 2
Mast cells, 64
Matrix metalloproteinase-2 (MMP-2), 249
Matrix metalloproteinase-9 (MMP-9), 249
Maximal electroshock (MES), 270
Medhya rasayan drugs, 28
Medicinal plants, 1
general importance of, 2
global level, significance at, 2–3
Menorrhagia, 306
Mentat, 103
Menth-1-en-8-yl cation, 416
MES, *see* Maximal electroshock
Mesenchymal-to-epithelial transition (MET), 431
MET, *see* Mesenchymal-to-epithelial transition
Methanolic fraction (MF), 327
Methyl caffeate, 67
Methylglyoxal (MG), 65
MF, *see* Methanolic fraction
MG, *see* Methylglyoxal
MIC, *see* Minimum inhibitory concentration
Microtubule-associated protein 2 (MAP-2), 432
Midland hawthorn, *see Crataegus laevigata*
Minimum inhibitory concentration (MIC), 61
Mitochondrial permeability transition (MPT), 267
Mitomycin C (MMC), 187
Mitotic spindle poisons, 317
MMC, *see* Mitomycin C
MMP-2, *see* Matrix metalloproteinase-2
MMP-9, *see* Matrix metalloproteinase-9
MOA, *see* Monoamine oxidase-A
Moko lactone, 266
Monoamine oxidase-A (MOA), 135, 372
Mootravaah, 116
Moringa oleifera
anti-asthmatic property, 454
anticancer activity, 454
antidiabetic activity, 453
anti-inflammatory activity, 454
antimicrobial activity, 453–454
anti-obesity activity, 454
chemical derivatives, 450–451
description, 448–449
general use of, **450**
hepatoprotective property, 455
medicinal use of, *455*
neuroprotective property, 455
traditional knowledge, 449–450
in water treatment, 453
Moringa treatment, 454
Morphological descriptions
of herbs, 12–13
of shrubs, 13–15
of trees, 15–17
MPT, *see* Mitochondrial permeability transition
Musli, *see Chlorophytum borivilianum*
Myricetin 3-rhamnoside, 381

Nardostachys jatamansi, *165*
in Alzheimer's disease, 171–172
in cerebral ischemia, 171
chemical derivatives, 166–168
description, 164
hepatoprotective activity and cardioprotective activity, 170
nootropic activity, 173
in Parkinson's disease, 171
potential benefits, applications, and uses, 170
structures of compounds found in, *167*
traditional knowledge, 164–166
Neem, *see Azadirachta indica*
Neurofilaments (NFH), 432
Neurological severity score (NSS), 71
Neuronal proteins, 432
NFH, *see* Neurofilaments
Nine Kots, 165
Nitric oxide, 267
Non-heterocyclic alkaloids, 310
Non-small cell lung carcinoma (NSCLC) cells, 431
Nootropic drugs
NSCLC cells, *see* Non-small cell lung carcinoma cells
NSS, *see* Neurological severity score

Ocimum tenuiflorum
in Alzheimer's disease, 172
in cerebral ischemia, 171
chemical derivatives, 166–168
description, 164
hepatoprotective activity and cardioprotective activity, 170
nootropic activity, 173

potential benefits, applications, and uses, 170
structures of compounds found in, *169*
traditional knowledge, 164–166
Ocotillol-type aglycone, 183
Oligospirostanoside, 100
Oncovin, 317
Opiates, 70
Osteoporosis, 66
Ovariectomized (OVX), 71
OVX, *see* Ovariectomized
Oxalis corniculata L., 12–13, *13*

Pachan, 116
Pakshaghaat, 116
Palmatine, 227
Palmitic acid, 223
Panax ginseng, 181
Panax notoginseng, 181
Panax quinquefolium, 181
antiaging properties, 189–190
anticancer activity, 186–187
antidiabetic activity, 187–188, *189*
antimicrobial activity, 190–191
bioactive components of, 182–183
bioactive constituents of, *184*
cardioprotective activity, 186
description, 181
multiple sclerosis (MS) prevention, 190
neuronal protection, 185–186
obesity, prevention of, 188–189, *189*
traditional knowledge, 182
Pancreatic duodenal homeobox-1 (PDX-1), 65
Para-nitrophenyl phosphate (pNPP), 322
Parathyroid hormone (PTH), 66
Parkinson's disease, 171, 227
PCa, *see* Prostate cancer
p-coumaric acid, 381
PDMs, *see* Plant-derived molecules
PDX-1, *see* Pancreatic duodenal homeobox-1
Pentylenetetrazol (PTZ), 270
Peptic ulcer disease, 251
Peptostreptococcus intermedius, 279
Periwinkle, *see Vinca rosea*
Peroxisomal receptor antagonists (PPARs), 137
p-F11, *see* Pseudoginsenoside F11
Phenolic acids, 21
Phenolics, 21, 205, 381
Phenols, 244
Phenserine, 184, 192
Phenylpropanoids, 18, *19*
Pheretima posthuma, 118
Phloretin 2′-O-glucoside, 382
Phorbol myristate acetate (PMA), 269
Phrasa-Kaprao, 166
Phrasa-Mawang, 166
Physostigma venenosum, 181
antidote agent, 193–194
beneficial effects of, *194*
bioactive components of, 183–185
bioactive constituents of, *185*
description, 181–182
glaucoma treatment, 194
neuronal protection, 191–193
traditional knowledge, 182
Physostigmine, 183–184, 191, 193, 194
Physostigmine-triggered carbamylation, 194
Phytochemicals, 3, 18
Phytolacca dodecandra, *204*
anthelminthic property, 206
antimalarial property, 207
antimicrobial property, 206–207
antiviral property, 207
bioactive compounds and phytochemistry, 205–206
botanical aspects and habitat, 204–205
general description, 204
hepatoprotective property, 207
molluscicidal property, 206
traditional knowledge, 205
Phytonutrients, 180
Picrorhiza kurroa, *208*
anticarcinogenic and antineoplastic activity, 210
antidiabetic activity, 210
anti-inflammatory and antiallergic activities, 210
antioxidant activity, 210
bioactive compounds and phytochemistry, 209
botanical aspects and habitat, 208
general description, 207–208
immunostimulatory activity, 210
nephroprotective activity, 210
secondary metabolites, potential production of, 211
traditional knowledge, 208–209
Picrotoxin, 270
Piper longum, *219*
activities of, **227**
anti-amebic and antihelminthic activity, 225
anti-apoptotic and antioxidant activity, 226
antiasthmatic and analgesic activity, 226–227
anticancer and antitumor activity, 226
antidepressant activity, 226
antidiabetic activity, 227
antiepileptic and therapeutic activity for Alzheimer's disease, 227
anti-inflammatory and anti-arthritic activity, 226
antimicrobial activity, 225
anti-obesity and hypocholesterolemic activity, 225–226
antiparkinsonian activity, 227
antiplatelet activity, 225
anti-snake venom activity, 225
antiulcer activity, 225
chemical derivatives, 222–223
coronary vasodilation and cardioprotective activity, 225
description, 219
hepatoprotective activity, 226
immunomodulatory and activity against COVID-19, 226
insecticidal activity, 225
major phytochemicals present in, *223*
melanin-inhibiting and antifertility activity, 227
radioprotective activity, 226
traditional knowledge, 220–221
Piperlonguminine, 227
Pipli, *see Piper longum*
Pistillate flowers, 98
Pitta shodhaka, 277
PKC, *see* Protein kinase C
Plantago ovate, *237*

antibacterial activity, 246
anticancer activity, 247
anticorrosive activity, 248
antidiabetic activity, 249
anti-inflammatory activity, 249
antileishmanial effect, 247–248
anti-nematode activity, 249
anti-obesity activity, 247
antioxidant activity, 249
anti-ulcer activity, 248–249
bioedible films, 249
chemical derivatives, 241–243
coagulation activity, 247
description and distribution, 237–239
food industry, uses in, 250
gastrointestinal functions, 245–246
hepatoprotective activity, 249
hypolipidemic activity, 247
immunomodulatory actions, 246
against industrial pollution, 250
lead biosorbent, activity as, 248
mineral and vitamin composition of, **243**
natural super disintegrant, 248
against Parkinson's and Alzheimer's diseases, 250
phytochemical composition of, **241**
phytochemicals and carbohydrates of, *242*
side effects
absorption of drugs, 251
absorption of minerals, 251
bloating, 251
traditional knowledge, 240
used for making natural eye drops, 248
wound-healing activity, 250
Plant-derived molecules (PDMs), 244
Plasmodicum falciparum, 352
PMA, *see* Phorbol myristate acetate
pNPP, *see* Para-nitrophenyl phosphate
POF, *see* Premature ovarian failure
Polycythemia, 89
Polypeptide, 141
Pongamia glabra, 466
Porphyromonas endodontalis, 190
Porphyromonas gingivalis, 190
PPARs, *see* Peroxisomal receptor antagonists
PPD, *see* Protopanaxadiol
PPT, *see* Protopanaxatriol
Preadipocyte hyperplasia, 188
Premature ovarian failure (POF), 189
Prevotella intermedia, 190
Proanthocyanidins, 382
Propionibacterium acnes, 253
Prostate cancer (PCa), 67
Protein kinase C (PKC), 192
Proteus vulgaris, 325, 437
Proto-alkaloids, 310
Protopanaxadiol (PPD), 183
Protopanaxatriol (PPT), 183
Protoxylems, 97
Pseudoginsenoside F11 (p-F11), 183
Pseudomelanosis coli, 280
Pseudomonas aeruginosa, 118, 270, 323, 325, 383, 437
Psyllium, 238, 240, 247
Pterocarpus santalinus, 465
PTH, *see* Parathyroid hormone
PTZ, *see* Pentylenetetrazol
Purgation therapy, 277

Quercetin 3-rhamnoside, 381
Quinidine, 387
Quinine, 386, 387
Quinine hydrochloride, 387
Quinine sulphate, 387

Raktashodhal, 116
RANKL, *see* Receptor activators of nuclear factor-kappa-B
Rasayana, 221
Rauvolfia serpentine, 239
anti-Alzheimer's activity, 252
antibacterial activity and herbal gels, 252–253
anticancer and antitumor activity, 253
anticorrosive activity, 254
antidiabetic activity, 252
antidiarrheal activity, 251
anti–larvicidal activity, 254
anti-oxidative activity and anti-heavy metal toxicity, 254
anti-SARS activity, 254
anti-ulcer activity, 251
associated microbes, 254–255
chemical derivatives, 243–244
chemical structures of alkaloids in, *245*
description and distribution, 239–240
hypolipidemic activity, 253
nanoparticle formations, 254
phytochemical composition of, **244**
side effects, 255
traditional knowledge, 240–241
vitamin and mineral composition of, **246**
Reactive oxygen species (ROS), 64, 67, 186, 267
Receptor activators of nuclear factor-kappa-B (RANKL), 66
Red Sandalwood, 465
Reserpine, 243, 253–255
Reverse-phase high-pressure anion exchange (RP-HPAE) chromatography, 51
Rheumatism, 306
Rho-kinase 2 (ROCK-II) enzyme, 63
Rivastigmine, 185, 192
ROCK-II enzyme, *see* Rho-kinase 2 enzyme
Rootstock, 96
ROS, *see* Reactive oxygen species
Rosmarinic acid, 168
RP-HPAE chromatography, *see* Reverse-phase high-pressure anion exchange chromatography
Rumex nepalensis Spreng, 13, *13*

Sacred fig, *see Ficus religiosa*
Safed musli, *see Chlorophytum borivilianum*
Salmonella typhi, 434
Salmonella typhimurium, 323, 437
Salvage pathway, 68
SAM, *see* Senescence-accelerated mice
Sandalwood, *see Santalum album*
Santalum album, 465
geographical distribution, 466
morphological description, 466
phytochemistry, 466–467
therapeutic benefits/traditional applications, 468

traditional knowledge, 465
Saponin, 51, 244
Sarvajvaraharalauha, 209
Saussurea costus, 262, *264*
antibacterial and antifungal activity, 270
anticancer and antitumor, 267–268
antiepileptic or anticonvulsant activity, 270
anti-inflammatory, 266–267
anti-ulcerogenic and cholagogic activity, 268
botanical description, 264–265
cardioprotective activity, 270
hepatoprotective, 268
immunomodulatory activity, 268–269
as insect and pest repellent, 270–271
perfumery, 271
pharmacological properties of, *267*
phytochemistry, 265
costus oil, 266
costus roots, bioactive phytoconstituents of, 265–266
respiratory diseases and asthma, 269
taxonomic hierarchy, 264
traditional knowledge, 265
for treating blood-related disorders, 269
for treating skin ailments, 269
SCE, *see* Sodium chloride extract
Selective estrogen receptor modulator (SERM), 65
Selective serotonin reuptake inhibitors (SSRIs), 71
Senescence-accelerated mice (SAM), 186
Senna, *see Senna alexandrina*
Senna alexandrina, 271, *277*
antibacterial activity, 280
anticancer, 281
antifungal activity, 280–281
antihelminthic activity, 281
anti-obesity and antidiabetic activities, 281
antioxidant activity, 281
botanical description, 276
chemical derivatives, 278
pharmacological properties of, *279*
senna as a laxative drug, 279–280
traditional knowledge, 276–277
Senna leaf (sanamakki), 277
SERM, *see* Selective estrogen receptor modulator
Serotonin, 36
Serum glutamate oxaloacetate transaminase (SGOT), 353
Serum glutamate pyruvate transaminase (SGPT), 353
Sesquiterpene alcohols, 467
Sesquiterpene lactones, 265
SGOT, *see* Serum glutamate oxaloacetate transaminase
SGPT, *see* Serum glutamate pyruvate transaminase
Shankhpushpi, *see Convolvulus pluricaulis*
Shatavari, *see Asparagus racemosus*
Shigella flexneri, 118
Shrubs
chemical derivatives of, 18–21
morphological descriptions of, 13–15
parts of the plants, uses, and bioactive constituents of, 22
traditional knowledge of, 17–18
Signal transducer and activator of transcription-3 (STAT-3) signaling pathway, 431
Sinapic acid, 381
Sitosterol, 429
β-Sitosterol, 167
Sivadari, 85
Smart drugs, *see* Nootropic drugs
SOD, *see* Superoxide dismutase
Sodium chloride extract (SCE), 247
Spikenard, *see Nardostachys jatamansi*
SREBP2, *see* Sterol regulatory element-binding protein 2
SSRIs, *see* Selective serotonin reuptake inhibitors
Staphylococcus aureus, 252, 270, 323, 325, 351, 383, 434, 437
Staphylococcus epidermidis, 253, 325
STAT-3 signaling pathway, *see* Signal transducer and activator of transcription-3 signaling pathway
Sterol regulatory element-binding protein 2 (SREBP2), 436
Stigmasterol, 51, 429
Streptococcus pneumonia, 253
Streptococcus pyogenes, 118
Streptozotocin (STZ), 65
STZ, *see* Streptozotocin
Sugar apple, *see Annona squamosal*
Sugar Destroyer, *see Gymnema sylvestre*
Superoxide dismutase (SOD), 57, 229, 353
Swertia chirata, *292*
botanical features and habitat, 292
chemical constituents, 293–294
functional in mild to moderate cases of COVID-19, 293
importance and uses, 292–293
major phytoconstituents in, *294*
pharmacological activity of, 294
traditional knowledge of, 293
Synergetic effect, 21
Syzygium guineense (Willd.) DC., 16–17, *17*

T2DM, *see* Type 2 diabetes mellitus
Tannic acid, *see* Tannins
Tannins, 21, 382
Taxanes, 383
TBARs, *see* Thiobarbituric acid reactive substances
TBI, *see* Traumatic brain injury
Terpenes, 20, *20*
Terpenoids, 20, 21, 205
Tetrahydrofurobenzofuran, 192
Tetrahydropiperic acid, 223
Thin liquid chromatography (TLC), 327
Thiobarbituric acid reactive substances (TBARs), 270, 353
Thrips crawfordi, 96
Tiktakaghrita, 209
Tinnevelly Senna (*Cassia angustifolia* Vahl), 271, 276
Tinospora cordifolia, *220*
activities of, **230**
anti-arthritic and anti-osteoporotic activity, 229
anticancer and antitumor activity, 228
antidiabetic and hypolipidemic activity, 229
anti-HIV and wound-healing activity, 229
anti-inflammatory and anti-stress activity, 228
antimicrobial activity, 228
antioxidant and antitoxic activity, 229
antiparkinsonian and memory-enhancing activity, 229
anti-ulcer activity, 228
chemical derivatives, 223–224
description, 219
hepatoprotective and anti-amebic activity, 228–229

immunomodulatory and activity against COVID-19, 228
major phytochemicals present in, *224*
traditional knowledge, 221–222
against urinary calculi and uremia, 229
TLC, *see* Thin liquid chromatography
TNF-α, *see* Tumor necrosis factor-alpha
Traditional herbal medicines, 21
Traditional knowledge of herbs, shrubs, and trees, 17–18
Traditional medicinal knowledge, 1–2
Traditional plant-derived medicines, 21
Traumatic brain injury (TBI), 71
Trees
chemical derivatives of, 18–21
defined, 15
morphological descriptions of, 15–17
traditional knowledge of, 17–18
uses and bioactive constituents of, 22
Tree turmeric, *see Berberis aristata*
Tridoshaghna, 221, 222
Tumor necrosis factor-alpha (TNF-α), 66
Type 2 diabetes mellitus (T2DM), 187
Tyrosine, 421

UA, *see* Ursolic acid
Udrashool, 116
Upper respiratory tract (URT), congestion in, 136
Ursolic acid (UA), 167, 168, 319, 322, 434

Vardhaman pippali, 221
Vasaka, *see Adhatoda vasica*
Vascular endothelial growth factor (VEGF), 268
Vasicine, 329
VEGF, *see* Vascular endothelial growth factor
Vernonia anthelmintica, 118
Vinblastin, 317
Vinblastine, 304, 319
Vinblastine sulfate, 319
Vincamine, 324
Vinca rosea, 302–303, *302*, *304*
anticancer effect, 317, 319, 321
antidiabetic effect, 321–322
antidiarrheal effect, 322
antihelminthic effect, 322
anti-HIV effect, 323
antihypertensive effect, 323
antimicrobial effect, 323
antimycobacterium tuberculosis effect, 324
antioxidant effect, 316–317
antiplatelet aggregation effect, 324
anti-ulcer effect, 324
bioactive compounds present in, *314–316*
chemical derivatives, 310
cytotoxic effect, 321
description, 303–305
hypolipidemic effect, 322
neuroprotective effect, 324
phytochemical analysis of, **311–313**
safety aspects, 325
traditional knowledge, 305–306
traditional uses of, **307**
vernecular names of, **305**
wound-healing effect, 324
Vincristine, 303, 304, 319
Vindogentianine, 319
Vindolicine, 319, 322
Vindolidine, 319
Vindoline, 304, 319, 324
Vindolinine, 319
Vinoceptine, 324
Virechna, 277
Vitamins, 21
Volatile compounds, 383

White ginseng, 181
Wild yams, *see Dioscorea villosa*
Withania somnifera
botanical description, 295–296
chemical constituents present in, 296
habitat and cultivation, 296
major phytoconstituents in, *297*
morphology of, *295*
potential benefits, applications, and uses of, 297
traditional knowledge, 296

Yangambin, 227

Printed in the United States
by Baker & Taylor Publisher Services

Printed in the United States
by Baker & Taylor Publisher Services